D0204310

EX LIBRIS
McGraw Hill
NAME

CALCULUS

An Integrated Approach

Churchill-Brown Series

Complex Variables and Applications
Fourier Series and Boundary Value Problems
Operations Mathematics

International Series in Pure and Applied Mathematics

Ahlfors: *Complex Analysis*
Bender and Orszag: *Advanced Mathematical Methods for Scientists and Engineers*
Boas: *Invitation to Complex Analysis*
Buck: *Advanced Calculus*
Colton: *Partial Differential Equations*
Conte and deBoor: *Elementary Numerical Analysis: An Algorithmic Approach*
Edelstein and Keshet: *Mathematical Models in Biology*
Hill: *Experiments in Computational Matrix Algebra*
Lewin and Lewin: *An Introduction to Mathematical Analysis*
Morash: *Bridge to Abstract Mathematics*
Parzynski and Zipse: *Introduction to Mathematical Analysis*
Pinter: *A Book of Abstract Algebra*
Ralston and Rabinowitz: *A First Course in Numerical Analysis*
Ritger and Rose: *Differential Equations with Applications*
Rudin: *Principles of Mathematical Analysis*
Small and Hosack: *Calculus: An Integrated Approach*
Vanden Eynden: *Elementary Number Theory*
Simmons: *Differential Equations with Applications and Historical Notes*
Walker: *Introduction to Abstract Algebra*

McGraw-Hill Series in Higher Mathematics

Rudin: *Functional Analysis*
Rudin: *Real and Complex Analysis*

CALCULUS
An Integrated Approach

Donald B. Small
Colby College

John M. Hosack
University of the South Pacific

McGraw-Hill Publishing Company
New York St. Louis San Francisco Auckland Bogotá Caracas Hamburg
Lisbon London Madrid Mexico Milan Montreal New Delhi Oklahoma City
Paris San Juan São Paulo Singapore Sydney Tokyo Toronto

CALCULUS: An Integrated Approach

2 3 4 5 6 7 8 9 0 DOC DOC 9 4 3 2 1 0

ISBN 0-07-058264-5

The editor was Robert Weinstein;
the cover designer was John Hite;
the cover photograph was taken by H. Westover / F. Stop Pictures;
the production supervisor was Denise L. Puryear.
R.R. Donnelley & Sons Company was printer and binder.

Library of Congress Catalog Card Number: 89-063765

To

Margaret Small

for her love, support, and unlimited patience

Contents

PREFACE

To the Instructor

This text presents an *integrated* approach to calculus by integrating the development of major concepts with respect to both single and multivariable functions. The emphasis is on developing a conceptual and unified approach. A heavy reliance on approximation and error bound analysis provides a unifying theme throughout the text. Chapters one through four treat the Differential Calculus and Chapters five through nine the Integral Calculus. It is expected that this material represents two semester's worth of material (i.e., 45 to 55 class periods per semester).

The intended audience for this text are well prepared students. In particular, students who have successfully completed a full year course in calculus in high school. The Mathematical Association of America (MAA) recommends that colleges design calculus courses for these students that:

1. acknowledge and build on the high school (calculus) experiences of the students,
2. provide necessary review opportunities to ensure an acceptable level of understanding of Calculus I topics,
3. are *clearly different* from high school calculus courses (in order that students do not feel that they are essentially just repeating their high school work),
4. result in an equivalent of a one semester advanced placement.

This text satisfactorily addresses each of these four recommendations. In particular, integrating the development of major concepts with respect to single and multivariable functions makes the text (and course) *clearly different* from high school courses.

The authors' primary concerns are with the development of the student's ability to learn, analyze, and solve problems; to learn how to learn mathematics; and to become involved in developing mathematics. These concerns are reflected in the following aspects that are emphasized throughout the text.

- The Basic Approximation Process (approximate–improve approximation–generate sequence of approximations–take limit). This process underlies the development of every major concept in the Calculus.

- The "natural" role of sequences in the study of convergence. The limit of a sequence is the basic limit idea. The limit of a function is obtained by first composing the function with an appropriate sequence and then evaluating the limit of the resulting composition sequence.

- Development of an inquisitive approach on the part of the reader. In addition to the homework exercises, there are three categories of questions included in the text that are designed to test understanding and provoke thought.

  - Questions that are posed and answered in the body of the text.
  - Questions for which no answers are given. These are often used to call the reader's attention to missing steps in a calculation.
  - Questions that are set off by themselves in the text for which detailed answers are given in the Appendix.

- The structure for the development of a mathematical concept. There are four major stages in the development process:

  - Motivational stage. Particular instances "rooted" in student experiences are used to establish a "desire" and a "need" to develop the particular concept in question.
  - Definition stage. This is the concluding step in the Basic Approximation Process.
  - Algebraic stage. Determining how the concept "behaves" with respect to the standard arithmetic and functional operations.
  - Application stage. Employing the concept in particular applications as well as considering the concept as part of a mathematical theory.

- Development of conceptual understanding rather than a compilation of facts and techniques.

- Non-routine exercises that require the student to approach material in different ways: True-False questions, Give an example or show why no example can exist, Prove or Disprove the given statement, Analysis by graphing, and Projects. In order to actively involve students in developing mathematics, the proofs of several important theorems have been converted into exercises (with ample hints). "Project" exercises are open-ended exercises that lend themselves to small group efforts. The authors suggest that these exercises culminate in written reports.

- The text contains the usual routine exercises, although we are not in the competition for the text with the largest number of exercises. The routine exercises consist of drill exercises on important manipulations and word problems in which the student must convert a verbal description into a mathematical problem. These problems provide the student with useful practice in interpretation as well as suggesting the wide range of possible applications.

The importance of questioning, conjecturing, and working numerical examples cannot be over emphasized. (Mathematics is not a spectator sport!) Surely an inquisitive approach and a willingness to "struggle" with a problem are marks of an educated person. Students are encouraged to work in small groups as well as individually, and to utilize both the computer and hand calculators for their computations.

Additional features that characterize this work are:

- Numerical methods (e.g., the bisection algorithm, numerical integration) are used in the constructive development of concepts rather than appended in optional sections.

- Exercises involving the use of several levels of computational resources. It is generally recognized that numerical methods are best carried out with the aid of a calculator or computer. Computer Algebra Systems (such as Maple, MACSYMA, muMATH, Mathematica, SMP, Derive, etc.) are becoming available and will become equally important for carrying out algorithmic processes involving symbolic manipulation such as differentiation. There are exercises throughout the text whose basic aim is to increase student involvement with mathematics through exploratory exercises involving a calculator, graphing program, or computer. Students gain little by passively watching a computer print out an answer; they can gain a great deal by actively using the computer to explore, conjecture, and verify.

- Careful development of the function concept (source, domain, range, target) and consistent use of the notation $f : \mathbb{R}^n \to \mathbb{R}^m$. Helping students learn how to analyze functions is a major goal of this text.

- Special emphasis is placed on interpreting graphs. It is expected that students will learn to use graphs to "guide their analysis" rather than following the traditional approach of using analysis to produce a graph. For example, extremal values are introduced by analyzing the "high" and "low" points of graphs. Other examples include studying the relationship between a function and its derivative by sketching the graph of one from the graph of the other. (What does the location of an extreme value of the derivative function imply about the graph of the function?)

- A conscious effort to guide students in understanding the logic involved in the statement of a theorem and its proof. An initial step in learning how to prove theorems is practicing making up examples. (Most exercise sections contains several such questions.) A "next" step is to understand the meaning of implication. Another step is to be able to identify where and how the hypothesis and conclusion are used in the proof. (Several exercise sections have questions that lead the student through this process.) It is expected that students will study the proofs in the text as "worked examples" and will become cognizant of direct and indirect proofs. It is also expected that students will be able to complete a proof given an outline with extensive hints. Several homework exercises are of this nature.

The length of the first chapter (Functions) requires an explanation. Since a major objective of the text is to help guide students in learning how to analyze functions, it is essential to establish a firm basis for this analysis at the beginning of the course. The authors have found that the time spent on this chapter pays major dividends as the course unfolds.

The lengths of sections is determined by the concept being discussed and not by the amount of material that can be presented in one class. Thus several sections (may) require more than one class period. This feature of the text provides flexibility for the instructor to "tailor" the material to an individual class by determining the "class breaks" rather than relying on the authors doing it. This is particularly important in the first chapter, considering the intended audience. The chapter contains many examples; thus the instructor can vary the pace depending upon the backgrounds of the students.

Acknowledgments

We are grateful for family and friends who have been a constant source of support and encouragement through the six year development period of this text. We especially wish to recognize and thank Ken Lane, our colleague during the initial stage of development, for his enthusiasm, inspiration, and numerous contributions. Several classes of Colby students have been closely involved with the development of this text. We value and appreciate their contributions. In particular, we wish to thank Colby student Mehmet Darmar for his careful checking of all of the exercises.

Special thanks are extended to the following reviewers for their many valuable suggestions and criticisms: John Eidswick, California State University at San Bernadino; John Wilson, Centre College; Melvyn Jeter, Illinois Wesleyan University; H. E. Lacey, Texas A&M University; Nancy Baxter, Dickinson College; H. G. Mushenheim, University of Dayton; D. R. LaTorre, Clemson University; Stephen L. Davis, Davidson College; Doug Child, Rollins College; Theodore G. Faticoni, Fordham University; George Mitchell, Indiana University of Pennsylvania; Bonnie Gold, Wabash College, and Hillel Gershenson, University of Minnesota. We want to especially express our appreciation to George Schultz and his students at St. Petersburg Jr. College, and John T. Hardy and his students at the University of Houston for class testing our text.

To the Student

This text is primarily concerned with the study and analysis of functions. Questions such as the following will motivate and guide our work.

- What is a function, f?
- What are the source and target, domain and range of f?
- What does it mean for a function to converge to a limit?
- Is f continuous?
- Is f differentiable? If so, how do we differentiate it?
- Is f integrable? If it is, how do we integrate it?
- For what values of x is $f(x) = 0$?
- How can $f(x)$ be computed for a given x?
- How can a function be approximated, if not computed exactly?
- How can a function be expressed as a series?

The emphasis will be on the process of developing concepts and their extensions to higher dimensions rather than on techniques. Reasoning rather than computation is expected. Sequences and the Basic Approximation Process will be our most important tools. A student successfully completing a course based on this text will be well equipped to pursue additional study in mathematics either independently or in formal courses. More importantly, the student should be able to effectively apply the Basic Approximation Process in problem solving and in developing concepts in other areas of mathematics as well as in related fields: physics, engineering, chemistry, biology, and economics.

An inquisitive attitude is fostered through numerous questions and exercises interspersed throughout the text. Such an attitude is a key for developing and applying mathematics. Effective problem solving approaches consist of asking and answering questions (e.g., What is the question asking?, What do I know?, What do I need to know?). In a broader and more fundamental sense, in the views of the authors, an inquisitive attitude is a necessary part of being an educated person.

A strong emphasis on making up examples are basic ingredients in the drive for conceptual understanding, for problem solving, and for establishing a basis for logical arguments.

Students will find the major concepts (functions, convergence, continuity, differentiation, integration, series representation) developed in this text appearing and reappearing in future studies. For example, convergence and approximation are the "backbones" of analysis courses (e.g., differential equations, complex variables, real analysis). Differentiation and integration are the concepts needed to study change, the source of most mathematical applications. Power series (e.g. geometric series) are used in practically every undergraduate mathematics course.

Clear instructions and models of how to develop and extend concepts along with a firm founding in the basic calculus concepts is the legacy this text leaves to the student.

Notation and Numbering Used in the Text

Theorems, definitions, examples, questions, and figures are numbered consecutively within sections. For example, definition 1.2.3 will be in chapter 1 section 2 immediately before example 1.2.4. The end of examples, proofs, and questions are indicated by $\triangle$, $\square$, $\diamondsuit$respectively. Theorems and definitions are set in italics to emphasize their importance. Standard functions, such as sin and log are set in standard type, while other functions, such as f, are set in italics.

Questions should be worked by the student as they are encountered as a check on understanding. Completely worked answers to the questions appear in the Appendix.

Exercises appear at the end of each section. They are often in odd/even pairs, with the answer to the odd numbered exercises given in the Appendix. Exercises involving the use of a calculator or computer are marked with a "microcomputer" icon in the margin.

Don Small
John Hosack

Post Script

We are, in fact, giving a dishonest picture of Mathematics if we do not allow the student to participate in finding the right problem or theorem. It has often been said that, once a mathematician knows what he is trying to prove, his job is half over. This may be an over simplification; nevertheless, the normal state of mathematical activity is one that involves (a) a situation that is crying out for understanding, and (b) a search for the right way to look at it. Unfortunately, we often exclude this intuitive discovery aspect of mathematics from our teaching. A carefully organized course in mathematics is sometimes too much like a hiking trip in the mountains that never leaves the well-worn trails. The tour manages to visit a steady sequence of the "high spots" of the natural scenery. It carefully avoids all false starts, dead-ends, and impossible barriers, and arrives by five o'clock every evening at a well-stocked cabin. The order of difficulty is carefully controlled, and it is obviously a most pleasant way to proceed. However, the hiker misses the excitement of risking an enforced camping out, of helping locate a trail, and of making his way cross-country with only intuition and a compass as a guide. "Cross-country" mathematics is a necessary ingredient of a good education.

Henry Pollak

Chapter 1

FUNCTIONS

1.1 Introduction

A driving force in the development of calculus was the need to solve the age-old problems of how to analyze rate of change and how to measure length, area, and volume. These two categories of problems motivated the development of the two main branches of calculus, differentiation (rate of change) and integration (measure). The realization in the seventeenth century that these problems are, in general, inverses of one another is truly one of the great intellectual discoveries of all time. This result is formalized in the Fundamental Theorem of Calculus, which connects the two major concepts of the calculus, integration and differentiation.

The central role of calculus, and hence mathematics, in both liberal arts and technical education over the past two hundred years is, in large part, a reflection of the fact that rate of change and measurement are basic to many activities. The ability to analyze and quantify rates of change and to approximate measurements within a stated degree of accuracy are important in every field of human endeavor.

Rates of change involve dependency relationships: effects are dependent on the causes, outputs are dependent on the inputs, experimental results are compared to control results. In mathematics, dependency relations are called functions. Thus we think of a function as relating cause(s) to effect(s), input(s) to output(s), or control results to experimental results. The major goal of this text is to aid the reader in learning how to analyze functions.

Functions modeling change can be grouped into four general categories:

1. *Polynomial functions.* These are functions that are generated from real numbers and one or more variables through (repeated) applications of addition, subtraction, and multiplication operations. For example,

$$f(x, y) = 3x^4y^2 - 6xy^5 + 2x^3 + 7y - 13$$

is a polynomial in x and y.

2. *Periodic functions.* Periodic change is seen in the ocean tides or felt when taking one's pulse count or heard in the sound of a musical instrument. Numerous examples of periodic change are found in the patterns and seasons of nature, e.g., seasonal temperature graphs. The hands on a clock demonstrate periodic change, as does a swinging pendulum. Certain election laws are designed to guarantee periodic change. Probably the most familiar periodic functions are the sine and cosine functions from which are derived the other four *trigonometric functions.*

The Sine Function

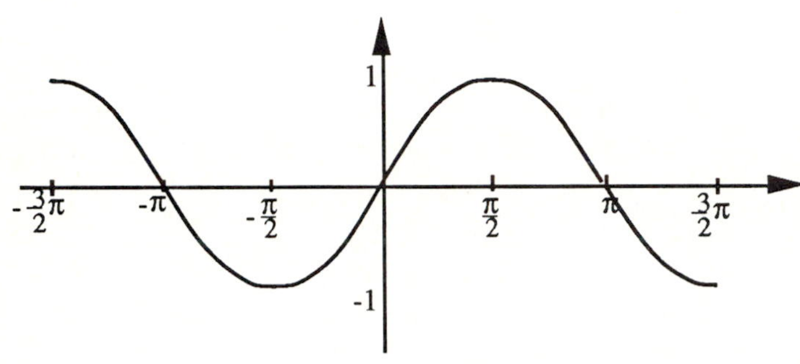

Figure 1.1.1

3. *Exponential functions.* Exponential change is characterized by the rate of change being proportional to the quantity present, such as the growth of a savings account through the accumulation of interest, the decay of radium 14, the melting of an ice cube, or population growth. The mathematical model for exponential change is an *exponential function* such as that defined by the relation

$$y = 2^x$$

where the change in y as x increases by 1 is equal to the value of y before the change:

$$2^{x+1} - 2^x = 2^x$$

The following figure illustrates the effect of changing the base from 2 to $\frac{1}{2}$.

Graph of $y = 2^x$ and $y = (1/2)^x = 2^{-x}$

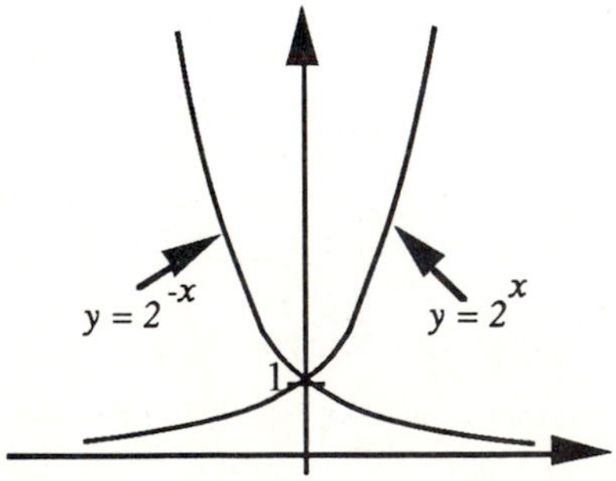

Figure 1.1.2

We have plotted $y = 2^x$ for integer values of x (which are easy to compute, e.g., $2^{-3} = 1/8$) and then "filled in" the graph. What allows us to fill in the graph, e.g., compute $2^{\sqrt{2}}$? We will consider the problem of defining exponential functions in Chapters 1 and 2, although a complete solution will is not given until Chapter 7.

4. *Discrete functions* (in contrast to continuous functions). The term *impulse* is often used to describe discrete change. The beep of a radar monitor is an example of a discrete change as compared to the continuous change exhibited by the change from daylight to darkness. Since the changes in a discrete system are "isolated," they can be numbered and hence ordered (e.g., first beep, second beep, etc.). For example, the "arrival numbers" issued at a pizza counter orders the arrival of customers (discrete change). Such an ordering (of events) is called a *sequence.* A sequence may be finite as in the case of pizza customers or infinite as in the set of positive integers. Another example of an infinite sequence is the following set of approximations to $\sqrt{2}$:

$$\{1.4, 1.41, 1.414, 1.4142, 1.41421, 1.414213, \ldots\}$$

The nth term is the approximation to $\sqrt{2}$ accurate to n decimal places. *Sequences* are the mathematical functions associated with modeling discrete changes. Sequences are often used to "sample" change. Most surveys and discrete data collection are of this nature, as is a piano teacher's checking a student's preparation for a recital.

The graph of a sequence is a set of isolated points rather than a continuous curve as was the case for the sine function. The function given by the relation $y = 2^n$, where n is an integer, is an example of a sequence.

The reader should pause at this point and make up several examples of discrete changes that can be modeled (described) by sequences.

The primary use of sequences in this text will be to approximate the behavior of (other) functions. For example, the sequence of hourly measurements of the volume of water flowing over a dam provides an approximation to the continually changing volume of water flowing over the dam.

We speak of the constant functions, the identity function [i.e. $f(x) = x$], the sine and cosine functions, the exponential functions, and sequences as constituting a set of "building blocks" for the generation of functions. All of the functions in this text can be obtained from these building blocks by utilizing the operations of addition, subtraction, multiplication, division, composition, raising to powers, taking inverses, differentiation, integration, and taking limits. It is therefore very important that the reader pay particular attention to understanding the geometric and analytic properties of our building blocks and the various ways of combining them to produce other functions.

In this chapter, we shall concentrate on developing the function concept, looking at basic properties, special classes, and common operations. We will be particularly interested in the geometrical analysis of functions. Chapaters 2, 3, and 4 are primarily devoted to the analysis of rates of change, while Chapters

5 through 8 are concerned with measuring (length, area, volume). Chapter 9 (Series) deals with the approximation of functions by polynomials.

Mathematics is sometimes described as a language. Thus it is important that we begin our study with a section devoted to defining the terminology and notation that will be used throughout the text.

1.2 Sets of Real Numbers

We will use the terminology of sets throughout the text and will start with a brief review of the needed terminology. A *set* is a collection of elements.

1. Set description. There are two common ways of denoting a set:
(a) Explicit listing of the elements enclosed by curly braces. For example, $S = \{a, b, c\}$ or $P = \{Sally, Sam, Suzy, Sim, Sue\}$. The order of the listing is not important. Thus $S = \{a, b, c\} = \{b, a, c\} = \{c, b, a\}$.
(b) Stating a condition that an element must satisfy in order to belong to the set in question. For example, $K = \{x : 1 < x < 7, x \text{ a positive integer}\}$ is the set $\{2, 3, 4, 5, 6\}$. The form is $\{x : x \text{ satisfies a condition}\}$. The colon ":" is read "such that." We will call this the set builder notation.

2. Set membership. We write $b \in S$ to indicate that b is an element of S. For example, if $S = \{a, b, c\}$, then $b \in S$.

3. Subset (set inclusion). We write $A \subseteq B$ to indicate that set A is contained in set B. Formally, $A \subseteq B$ provided that every element of A is an element of B. For example, $\{a, b\} \subseteq \{a, b, c\}$, but {a,b} is not contained in {b,c,d,e}.

Most of the functions we discuss will involve the real numbers. Several sets of real numbers occur often enough to be given special names and symbols. The set of *natural numbers* $\{1, 2, 3, \ldots\}$ is denoted by $\mathbb{N}$. The natural numbers are the foundation for mathematics. Notice that zero is not considered a "natural number." Many civilizations with substantial mathematical sophistication, such as the Greek, did not have the concept of zero. The first known use of the number zero was in ninth century Hindu mathematics. The nonnegative integers are obtained by adding zero to the above set, $\mathbb{N}_0 = 0, 1, 2, 3, \ldots$. The numbers can next be extended to the set of *integers* 0, 1, -1, 2, $-2, \ldots$ which we will denote by $\mathbf{Z}$, to the set of *rational numbers* (ratios of integers) which we will denote by $\mathbf{Q}$, and then to the set of *real numbers* which we will denote by $\mathbb{R}$. If we wish to divide by a real number, we must use the nonzero real numbers $\mathbb{R}^*$.

The most frequently used sets of real numbers will be intervals. An *interval* in $\mathbb{R}$ is a continuum of numbers, that is, a range of numbers without gaps. A *closed interval* contains its endpoints; an *open interval* contains neither of its endpoints. An interval that contains some, but not all, of its endpoints is neither open nor closed. The *interior* of an interval is all points in the interval, except any endpoints.

Note that since the interior of an interval does not contain the endpoints, it is always an open interval. If the interval is $[a, a]$, then its interior is the empty set, which can be viewed as a trivial interval. In the following example, the reader should pay particular attention to the usage of brackets, parentheses, and braces.

Example 1.2.1 (a) The interval $[1,4] = \{r : 1 \leq r \leq 4\}$ is a closed interval containing both of its endpoints 1 and 4.

Figure 1.2.2

Its interior is the open interval (1,4).

(b) The interval $[1, 4) = \{r : 1 \leq r < 4\}$ contains one endpoint, 1, but not the other endpoint, 4.

Figure 1.2.3

Thus it is neither a closed interval nor an open interval, but its interior (1,4) is open.

(c) The interval $[1, \infty) = \{r : 1 \leq r\}$ contains its only endpoint 1, so it is a closed interval.

Figure 1.2.4

(d) Finally, $(-\infty, \infty) = \mathbb{R}$ has no endpoints and by convention is considered both an open interval and a closed interval.

Figure 1.2.5

Note that it is a special case, the only interval with no endpoints. It is equal to its interior.

Note the different uses of "{,}" and "(,)"— $\{1, 2\}$ is a set of two elements; (1,2) is the interval of all real numbers greater than 1 and less than 2. △

Many of the results of calculus depend on whether or not a set is bounded. Intuitively, the interval $[1, \infty)$ is not bounded, since the values in the interval go off to infinity. The intervals [1,4] and [1,4) are bounded. The precise definition follows.

Definition 1.2.6 *Let S be a set of real numbers. The number r is said to be an* upper bound *for the set S if $s \leq r$ for every number s in S. Similarly, r is a* lower bound *for the set S if $r \leq s$ for every number s in S. A set S is* bounded *if it has both an upper bound and a lower bound. A number $r > 0$ is a* bound *for S if $-r \leq s \leq r$ for every s in S.*

Example 1.2.7 (a) Let $S = \{-1, 0, 1, 2\}$. Since -1 is the smallest element in S, any number ≤ -1 is a lower bound for S (e.g., -2 is a lower bound). Furthermore, -1 is the greatest (largest) lower bound. The least (smallest) upper bound is 2. Also the number 2 is a bound for S, as is any larger number.

(b) The closed interval $[-1, 4]$ is bounded, with lower and upper bounds -1 and 4 contained in the interval. Any number less than -1 is also a lower bound; any number greater than 4 is also an upper bound. The number 4 is a bound for the interval.

(c) The interval $[-5, 4)$ is bounded with lower and upper bounds -5 and 4. The upper bound and endpoint 4 is not contained in the interval, so the interval is not closed. Although 4 is an upper bound, it is not a bound for the interval. (Why?)

(d) The interval $[1, \infty)$ is closed since it contains its only endpoint, 1, and is unbounded.

(d) The open interval $(-\infty, 1)$ is unbounded.

(e) The open interval $(-\infty, \infty) = \mathbb{R}$ is unbounded and has neither upper nor lower bounds. △

Note that if r is an upper bound of a set and $r' > r$, then r' is also an upper bound for the set. Similarly, if r is a lower bound for a set and $r'' < r$, then r'' is also a lower bound for the set.

Definition 1.2.8 *Given a set of real numbers S, r is the* greatest lower bound *(abbreviated glb) for S if r is a lower bound and r is greater than any other lower bound.*

We need two conditions to hold for r to be the glb for a set:
(1) $r \leq s$ for every number s in the set S, and
(2) if t is any lower bound for S ($t \leq s$ for every number s in the set S), then $t \leq r$.

Similarly, r is a least upper bound *(lub) for the set S if r is an upper bound and r is less than any other* upper bound for S.

Question 1.2.9 Write out the two conditions for least upper bound which are analogous to the two conditions for greatest lower bound. ◇

One might expect that if a number r is the greatest lower bound of a set S, then it is in the set. This need not be true as is seen when we consider an interval such as $(-1, 2]$, which has no minimum element in it. The greatest lower bound for $(-1, 2]$ is -1, but -1 is not in $(-1, 2]$.

Example 1.2.10 (a) Let $S = \{1/n : n = 1, 2, 3, \ldots\} = \{1, 1/2, 1/3, \ldots\}$. The lub of S is 1. Clearly 0 is a lower bound of S. To see that 0 is the glb of S, we note that if $r > 0$, then there is an n such that $r > 1/n > 0$. Thus no positive number can be a lower bound of S.
(b) Let $S = \mathbb{R}$. Although S has no upper or lower bounds, it is sometimes said that the lub of S is infinity, ∞, and the glb of S is negative infinity, $-\infty$. △

Higher Dimensional Spaces

The set of real numbers that we have been discussing is usually identified with the one dimensional real number line. An origin, 0, is chosen with the negative numbers to the left of 0 and the positive numbers to the right of 0.

To identify higher dimensional spaces with sets of numbers we need to introduce the concept of the Cartesian product (named after the French philosopher and mathematician René Descartes, 1596–1650).

Definition 1.2.11 (Cartesian Product) *If A and B are two sets, then the* Cartesian product *of A and B, written $A \times B$, is the set of all ordered pairs with the first element from A and the second element from B. That is,*

$$A \times B = \{(a, b) : a \in A \text{ and } b \in B\}$$

If $p = (a, b)$ and $q = (c, d)$ are two elements of $A \times B$, then $p = q$ if and only if the corresponding elements are equal, i.e., $a = c$ and $b = d$.

Notation: We will write $A \times A$ as A^2.

Example 1.2.12
(a) Let $A = \{a, b\}$ and $B = \{d, e\}$. Then $A \times B = \{(a, d), (a, e), (b, d), (b, e)\}$.
(b) Let $A = \{1, 2\}$. Then $A^2 = A \times A = \{(1, 1), (1, 2), (2, 1), (2, 2)\}$. △

Question 1.2.13
What is the Cartesian product of $A = \{x, y\}$ and $B = \{z\}$? ◇

The concept of Cartesian product can be extended to more than two sets: $A \times B \times C = \{(a, b, c) : a \in A, b \in B, \text{ and } c \in C\}$.

Example 1.2.14 (a) Let $A = \{a, b\}$, $B = \{c, d\}$, and $C = \{e, f\}$. Then

$$\begin{aligned} A \times B \times C &= (A \times B) \times C = \{(a, c), (a, d), (b, c), (b, d)\} \times \{e, f\} \\ &= \{(a, c, e), (a, c, f), (a, d, e), (a, d, f), (b, c, e), (b, c, f), \\ &\quad (b, d, e), (b, d, f)\} \end{aligned}$$

Notice that we identify ((a,c),e) with (a,c,e).

(b) Let $A = \{1,2\}$ and $B = \{a,b\}$. Then

$$\begin{aligned} A \times A \times B &= (A \times A) \times B = \{(1,1),(1,2),(2,1),(2,2)\} \times \{a,b\} \\ &= \{(1,1,a),(1,1,b),(1,2,a),(1,2,b),(2,1,a), \\ &\quad (2,1,b),(2,2,a),(2,2,b)\} \end{aligned}$$

Notice that we identify ((1,1),a) with (1,1,a). △

We will often use the Cartesian product of a set with itself. For example, $A^3 = A \times A \times A$ is all triples of elements of A, i.e., $A^3 = \{(a_1, a_2, a_3) : a_i \in A \text{ for all } i = 1,2,3\}$. Especially important is the Cartesian product of the real numbers with itself. $\mathbb{R}^2 = \mathbb{R} \times \mathbb{R} = \{(a,b) : a \text{ and } b \in \mathbb{R}\}$. $\mathbb{R}^2$ is identified with the two-dimensional plane (or *Cartesian plane*) in the following way originated by Descartes. An origin is chosen and two perpendicular axes are selected, one horizontal and the other vertical. We will use the term *xy-plane* when referring to the plane with a rectangular coordinate system, where the horizontal axis is labeled x and the vertical axis is labeled y. A point in the plane is identified with the pair of real numbers (a,b) which represent the coordinates of the point with respect to the two axes. The first value, a, is called the *x-coordinate* (or *x-component*) of the point; b is the *y-coordinate* (or *y-component*) of the point.

Cartesian Coordinate System

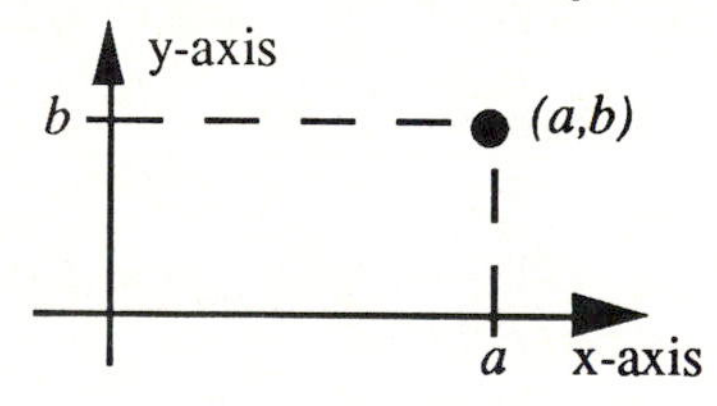

Figure 1.2.15

We can extend the concept of intervals from the real line to $\mathbb{R}^n$.

Definition 1.2.16 *An* n-dimensional rectangle *in* $\mathbb{R}^n$ *is the Cartesian product of* n *one-dimensional intervals in* $\mathbb{R}$. *A* closed rectangle *is one that is the Cartesian product of closed one-dimensional intervals, and an* open rectangle *is one that is the Cartesian product of open one-dimensional intervals.*

Example 1.2.17 (a) $A = [1,2] \times [1,3] = \{(x,y) : 1 \leq x \leq 2 \text{ and } 1 \leq y \leq 3\}$ is a closed two-dimensional rectangle in $\mathbb{R}^2$.

(b) $B = [a,b] \times (c,d] = \{(x,y) : a \leq x \leq b \text{ and } c < y \leq d\}$ is a rectangle in $\mathbb{R}^2$ which is neither open nor closed. Why?

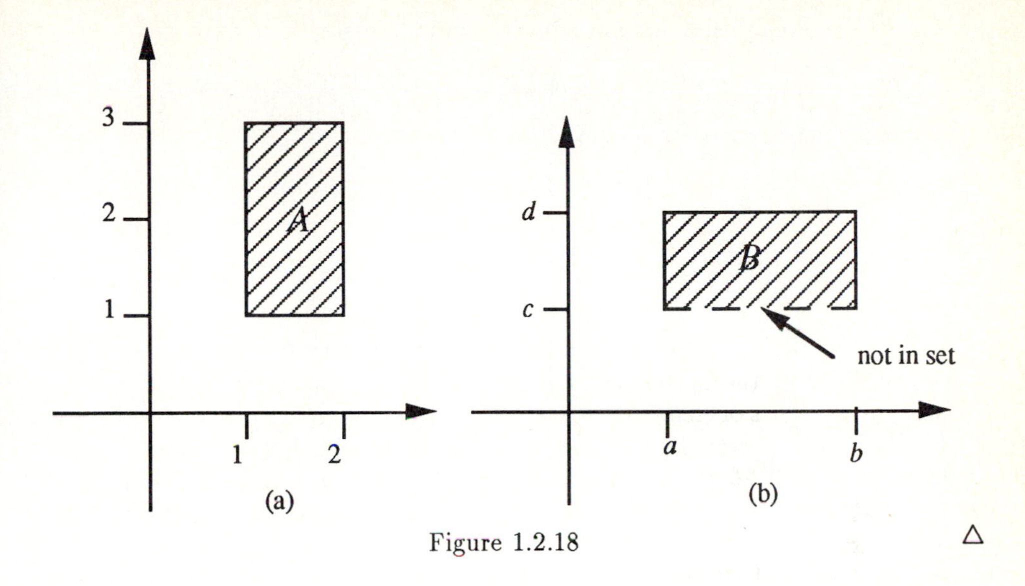

Figure 1.2.18

$\triangle$

Notice that for a rectangle to be closed, all its edges must be in the rectangle.

Notation: In our diagrams we use a solid line to indicate that the edge is contained within the figure and a dashed line to indicate that the edge is not contained within the figure.

Example 1.2.19 The first quadrant $\{(x, y) : x \geq 0 \text{ and } y \geq 0\} = [0, \infty) \times [0, \infty)$ is also a rectangle in the plane, as is the plane itself, $\mathbb{R}^2$, since

$$\mathbb{R}^2 = \mathbb{R} \times \mathbb{R} = (-\infty, \infty) \times (-\infty, \infty)$$

$\triangle$

Just as the plane can be identified with $\mathbb{R}^2$ by establishing an $x-y$ coordinate system, three-dimensional space can be identified with $\mathbb{R}^3 = \{(x, y, z) : x, y, z \in \mathbb{R}\}$ by establishing a rectangular coordinate system with three mutually perpendicular axes. Such systems are referred to as "right-handed" or "left-handed." A system is right-handed provided that when the fingers of your right hand are curled from the positive x-axis toward the positive y-axis, your thumb points in the direction of the positive z-axis. A system that is not right-handed is left-handed. Right-handed systems are the most common and will be used in this text. The designation of a screw as being "right-threaded" or "left-threaded" is based on the same criterion.

Cartesian Coordinate System in 3 Space

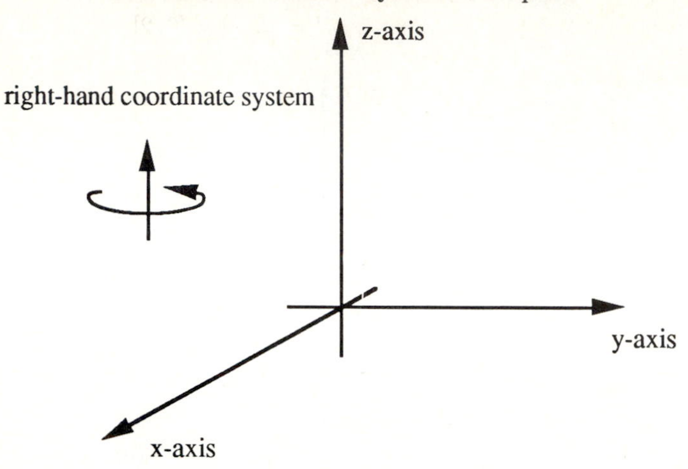

Figure 1.2.20

Traditionally, the z-axis is vertical, the y-axis is horizontal, and the x-axis is coming out of the page.

In general, we identify n dimensional space with $\mathbb{R}^n$, and write the coordinates of a point x in $\mathbb{R}^n$ as $x = (x_1, x_2, \ldots, x_n)$. When we are dealing with n-dimensional space, $n > 1$, we will often use subscripts rather than giving individual names $(x, y, z, w, \ldots)$ to each coordinate.

Note that x is used in two ways: as the name of a coordinate in $\mathbb{R}^2$ or $\mathbb{R}^3$, or as the name of a point in $\mathbb{R}^n$.

Example 1.2.21 We will locate the point p in $\mathbb{R}^3$ given by $p = (1, 2, 3)$.

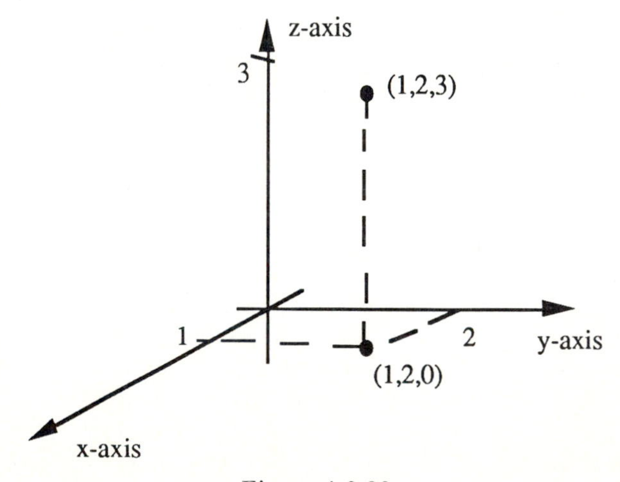

Figure 1.2.22 △

Example 1.2.23 A rectangle in $\mathbb{R}^3$ is the Cartesian product of three one-dimensional intervals in $\mathbb{R}$. For example, $S = [0,1] \times [0,2] \times [0,3]$ is a closed rectangle in $\mathbb{R}^3$.

Rectangle in Three Dimensions

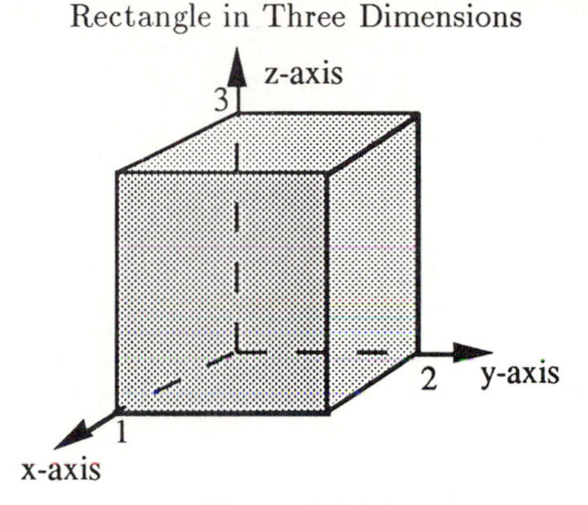

Figure 1.2.24

△

Notice that the term "rectangle" is used even though S has the shape of a box. We could use the term "box" since it describes three-dimensional sets in $\mathbb{R}^3$ better than the term "rectangle." But then when we consider a four-dimensional rectangle such as $S = [0,1] \times [1,2] \times [0,2] \times [1,3]$ in $\mathbb{R}^4$, the term "box" doesn't fit well.

Distances Between Points in n-dimensional Space

Introducing a concept of "distance" is often one of the first steps in quantifying a subject. Distance is not only an important concept in the physical sciences, but also in the biological and social sciences.

Definition 1.2.25 *If r is a real number, then $|r|$, read as "the absolute value (or magnitude) of r," is the* distance *from r to the origin, 0. If a and b are two real numbers, then $|a - b| = |b - a|$ is the* distance *between a and b.*

Example 1.2.26 The distance between -1 and 3 is

$$|(-1) - 3| = |3 - (-1)| = 4$$

△

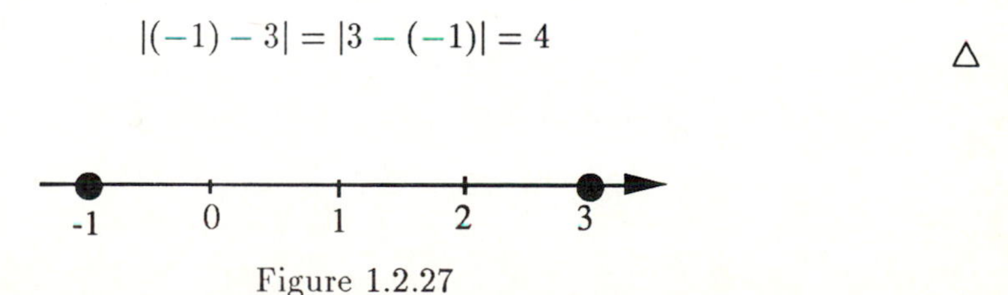

Figure 1.2.27

When a and b are real numbers, the definition of $|a-b|$ is sometimes expressed as:

$$|a-b| = \begin{cases} a-b & \text{if } a-b \geq 0 \\ -(a-b) & \text{if } a-b < 0 \end{cases}$$

A combination of the absolute value and inequality notations provides a convenient way to describe an interval. For example,

$$|x-2| < 5$$

represents the set of all real numbers whose distance from 2 is less than 5. This is just the interval $(-3,7) = \{x : -3 < x < 7\}$.

Theorem 1.2.28 *The basic properties of absolute value are:*
(a) $|ab| = |a||b|$
(b) $|a+b| \leq |a| + |b|$ (the triangle inequality).

Proof: To show that $|a+b| \leq |a|+|b|$ it is sufficient to show that $a+b \leq |a|+|b|$ and $-(a+b) \leq |a|+|b|$ [since either $|a+b| = a+b$ or $|a+b| = -(a+b)$]. Since $a \leq |a|$ and $b \leq |b|$, we can add these inequalities to obtain $a+b \leq |a|+|b|$. (Why can we add inequalities?) Similarly $-a \leq |a|$ and $-b \leq |b|$ give $-(a+b) \leq |a|+|b|$. These two inequalities give $|a+b| \leq |a|+|b|$. □

In $\mathbb{R}^2$, the two-dimensional plane, the distance between two points $x = (x_1, x_2)$ and $y = (y_1, y_2)$ is obtained by setting up a right triangle with legs parallel to the coordinate axes and then applying the Pythagorean theorem.

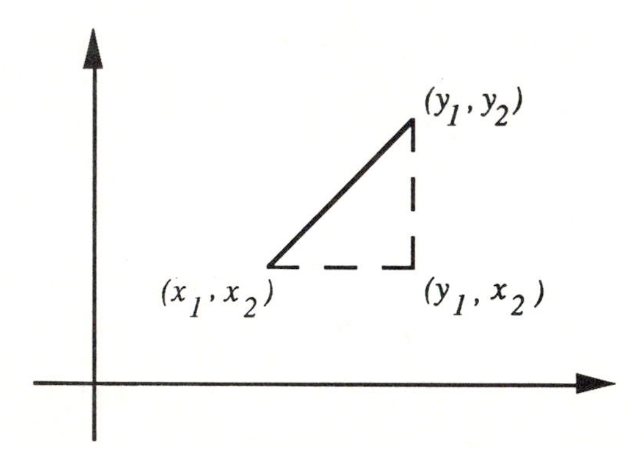

Figure 1.2.29

The distance between $x = (x_1, x_2)$ and $y = (y_1, y_2)$ is denoted by

$$|x-y| = [(x_1-y_1)^2 + (x_2-y_2)^2]^{1/2}$$

which is also equal to $|y-x|$.

Example 1.2.30 (a) We will find the distance between (1,2) and (−1,3). The distance is $[(1-(-1))^2+(2-3)^2]^{1/2} = (4+1)^{1/2} = \sqrt{5}$

(b) The distance from (3,4) to (0,0) is

△

$$|(3,4)-(0,0)| = [(3-0)^2+(4-0)^2]^{1/2} = 5$$

In three dimensions the distance from a point $x = (x_1, x_2, x_3)$ to $y = (y_1, y_2, y_3)$ is the length of the diagonal of the three-dimensional rectangle:

Distance in Three Dimensions

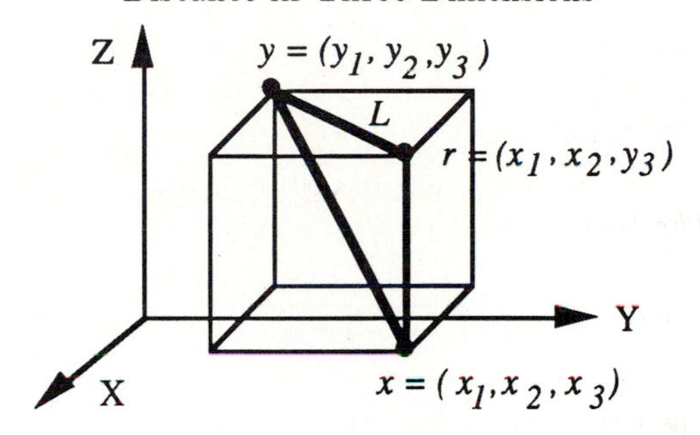

Figure 1.2.31

The distance is determined by two applications of the Pythagorean theorem. The first application determines the length of the diagonal in the $z = y_3$ plane from y to r:

$$L = [(x_1-y_1)^2+(x_2-y_2)^2]^{1/2}.$$

The second application applied to the right triangle yrx gives the distance from x to y:

$$\begin{aligned} |x-y| &= |(x_1,x_2,x_3)-(y_1,y_2,y_3)| \\ &= [L^2+(x_3-y_3)^2]^{1/2} \\ &= [(x_1-y_1)^2+(x_2-y_2)^2+(x_3-y_3)^2]^{1/2} \end{aligned}$$

In n dimensions the Pythagorean Theorem is applied $n-1$ times to obtain the *distance* from a point $x = (x_1, \ldots, x_n)$ to a point $y = (y_1, \ldots, y_n)$ as

$$|x-y| = [(x_1-y_1)^2+(x_2-y_2)^2+\ldots+(x_n-y_n)^2]^{1/2}$$

Example 1.2.32 We will find the distance between $x = (1,2,3,-4)$ and $(-1,0,-2,3)$ in $\mathbb{R}^4$. The distance is

$$\begin{aligned} [(-1-1)^2+(0-2)^2+(-2-3)^2+(3-(-4))^2]^{1/2} &= [4+4+25+49]^{1/2} \\ &= \sqrt{82} \end{aligned}$$

△

Notice that the distance from x to the origin is a special case of the distance between x and y, where y is the origin.

Example 1.2.33 The distance from the point (1,2,3) in $\mathbb{R}^3$ to the origin is

$$|(1,2,3)| = |(1,2,3) - (0,0,0)| = [1^2 + 2^2 + 3^2]^{1/2} = \sqrt{14}$$ △

Question 1.2.34 What is the distance from the point (1,2,3,4) in $\mathbb{R}^4$ to the origin (0,0,0,0)? ◇

The distance from a point to the origin is less than the sum of the distances along the coordinate axes from the origin to the point. In $\mathbb{R}^2$, $|(x,y)| \leq |x| + |y|$ by the Pythagorean theorem.

Distance from a Point to the Origin

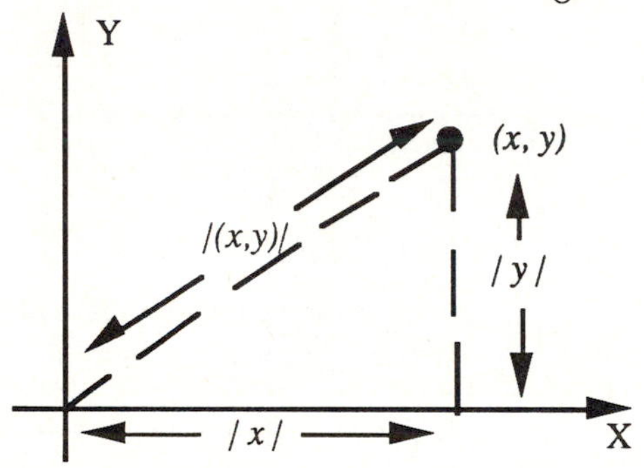

Figure 1.2.35

This extends to $\mathbb{R}^n$ in the following result.

Theorem 1.2.36 (The Triangle Inequality (in $\mathbb{R}^n$))
If $x = (x_1, \ldots, x_n) \in \mathbb{R}^n$ then $|x| \leq |x_1| + \cdots + |x_n|$.

Example 1.2.37 We will find a point in the plane that is two units from (1,3). The point must be on the circle about (1,3) of radius 2. If the point sought has coordinates (x, y), then the square of the distance between (x, y) and (1,3) must equal 4, i.e. $(x-1)^2 + (y-3)^2 = 4$. Any solution to this equation will do. There are infinitely many solutions, so we can pick an easy one by moving two units from (1,3) parallel to the x-axis to the point with coordinates (1+2,3) = (3,3). The point with coordinates (3,3) is two units from the point with coordinates (1,3). △

A Digression on Proofs

We pause in our discussion of functions to describe the structure of a mathematical theorem and the logic involved in proving a theorem.

Every mathematical theorem can be written as

1. A conditional statement: "If (statement A), then (statement B)" (sometimes

written "statement A $\Longrightarrow$ statement B")
2. Or a biconditional statement: "(Statement A) if and only if (Statement B)" [sometimes written "statement A iff statement B" or A $\Longleftrightarrow$ B)].

In a conditional statement, the first statement (statement A) is called the *hypothesis* and the second statement (statement B) is called the *conclusion.*

The truth value of a conditional statement is defined in terms of the truth values of the hypothesis and conclusion. Since a mathematical statement is either true or false, there are four cases to consider as shown in the following truth table:

hypothesis	conclusion	conditional statement
True	True	True
True	False	False
False	True	True
False	False	True

Thus the conditional statement, "If A, then B" is true *except* when A is true and B false.

To prove a mathematical theorem is to show that the conditional statement (or statements in the case of the biconditional form) is true in all cases. Such a conditional statement is called an implication.Therefore, to show that a conditional statement is an implication requires showing that the "A true, B false" case can never exist.

To repeat the above for emphasis, a proof of a mathematical theorem consists of showing that the conclusion can never be false whenever the hypothesis is true.

A biconditional statement is two conditional statements connected with "and." Thus "A if and only if B" has the same meaning as "If A, then B," and "if B, then A."

Two basic proof techniques will be frequently used in this text:
1. *Direct proof.* The hypothesis is assumed to be true and the conclusion is then shown to be true.
2. *Indirect proof or proof by contradiction.* The conclusion is assumed to be false, and then it is shown that this leads to a contradiction of the hypothesis. (The contradiction establishes the fact that the case of "hypothesis true and conclusion false" cannot exist.)

The reader should view the theorems and proofs in this text as "worked examples." In studying a theorem and its proof, one should identify the hypothesis, the conclusion, and the proof technique, and understand how the proof argument shows that the "hypothesis true - conclusion false" case cannot exist.

Bounded Sets in $\mathbb{R}^n$

We can now extend the concept of a bounded set to higher dimensions. Again, the intuitive idea is that a set is bounded if it does not extend off toward infinity. Since the distance of a point $x \in \mathbb{R}^n$ from the origin is $|x|$, a set in $\mathbb{R}^n$ is bounded if the distances of its points from the origin do not go to infinity. The origin is a convenient point to use in measuring distances, but any point would do.

Definition 1.2.38 *Let A be a subset of* $\mathbb{R}^n$. *A is* bounded *if there is a real number B such that $|a| \leq B$ for all a in A.*

Notice that a subset A of $\mathbb{R}^n$ is bounded iff $\{|a| : a \in A\}$ is a bounded subset of $\mathbb{R}$.

Theorem 1.2.39 *A set in $\mathbb{R}^n$ is bounded if and only if each set of components is bounded.*

Proof: This theorem is a biconditional statement. Thus we have two tasks to perform: show that "if a set in $\mathbb{R}^n$ is bounded, then each set of components is bounded" and show that "if each set of components is bounded, then the set is bounded."

First we assume the hypothesis "A is a bounded set in $\mathbb{R}^n$." We want to conclude that each set of components is bounded. By the remark below the definition, if $A \subseteq \mathbb{R}^n$ is bounded, then $\{|a| : a \in A\}$ is bounded by some number B, so the absolute values of the kth components $\{|a_k| : a \in A\}$ are also bounded by B since $|a_k| \leq |a|$.

Now we prove that if each set of components is bounded, then the set is bounded. Here we assume that for each k, the set of kth components is bounded by some number B_k. Then by Theorem 1.2.36, $|a| \leq |a_1| + \cdots + |a_n|$, so $\{|a| : a \in A\}$ is bounded by $B = B_1 + \cdots + B_n$. Thus the set A is bounded. □

Example 1.2.40 $A = [a, b] \times [c, d] \times [r, s] = \{(x, y, z) : a \leq x \leq b, c \leq y \leq d, \text{ and } r \leq z \leq s\}$ is a closed bounded rectangle in $\mathbb{R}^3$. △

Example 1.2.41 Some unbounded subsets of the plane are

(a) The plane $\mathbb{R}^2$ itself

(b) The first quadrant $\{(x, y) : x \geq 0 \text{ and } y \geq 0\}$

(c) The positive x-axis, $\{(x, 0) : x > 0\}$ △

Section 1.2 Exercises

1. Sketch and label the vertices of the rectangles in $\mathbb{R}^2$ defined by

   (a) $S = (1, 3) \times (-1, 1)$

   (b) $S = [1, 2] \times (-1, 0]$

2. Sketch and label the vertices of the rectangle in $\mathbb{R}^3$ defined by $S = [1,2] \times (1,3) \times [2,3]$

3. Express the following rectangles in terms of Cartesian products.

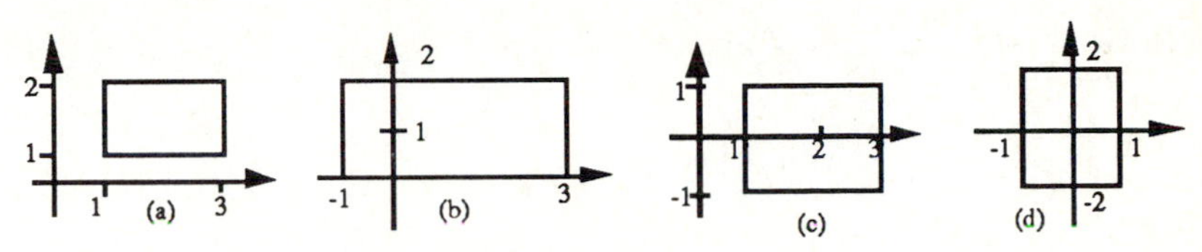

Figure 1.2.42

4. Using "set builder" notation, describe the set of points in the marked regions of the following figures:

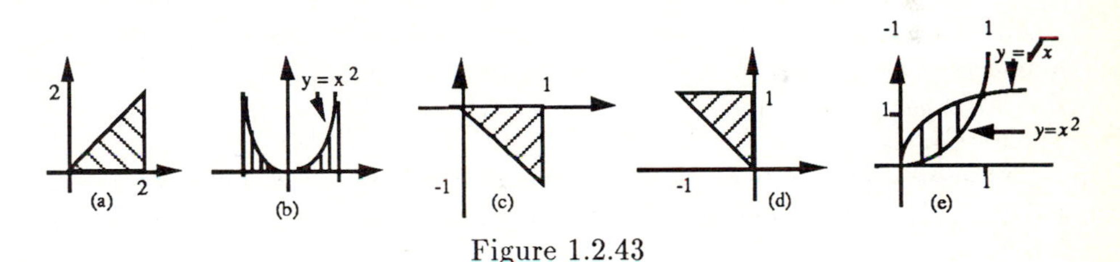

Figure 1.2.43

5. Mark each of the following statements true (T) or false (F).

   (a) $\{a, b, 2, 3\} = \{a, b, 3, 2\}$.

   (b) $[1,2] \times (0,1) = \{(x,y) : 1 \leq x \leq 2, 0 \leq y \leq 1\}$.

   (c) Let $A = \{a, b\}$, then $A^3 = A \times A \times A$ has eight elements.

   (d) If A is a closed rectangle in $\mathbb{R}^2$, then A is a bounded set.

   (e) 4 is a bound for $\{-5, 2, 3, -2\}$.

6. Find the greatest lower bound and least upper bound of each of the following sets.

   (a) $[-3, 0)$ (b) (99,1000] (c) $(-12, 10)$

   (d) [2,210] (e) $\{1, .5, .25, .125, ...\}$ (f) $\{0, 1, 2, 3, 4, ...\}$

   (g) $\{1, -1, 2, 4, 3, 5\}$ (h) $\{4, 2, 6, 3, 8, 5\}$

7. List all elements in the following Cartesian products.

(a) $\{1,2\} \times \{3,6\}$

(b) $\{a,b\} \times \{2,-1\}$

(c) $\{a,b\} \times \{a,b\}$

(d) $\{1,3,7\} \times \{2,6\}$

(e) $\{2,6\} \times \{1,3,7\}$

(f) $\{3,1\} \times \{1,3\}$

(g) $\{2,4\} \times \{1,3\} \times \{a,b\}$

(h) $\{1,3\} \times \{2,4\} \times \{a,b\}$

(i) $\{a\} \times \{b,c\} \times \{c,d,e\} \times \{1,2\}$

(j) $\{2,1\} \times \{3,a,2\} \times \{s\} \times \{d\} \times \{1,3\}$

8. Show that 1 is the lub of S in part (a) of Example 1.2.10.

9. Let R be the closed rectangle in the plane with vertices $(-1,3)$, $(-1,-1)$, (2,3), and $(2,-1)$.

   (a) Describe R as a Cartesian product.
   (b) Describe the interior of R as a Cartesian product.
   (c) Prove or disprove that R is a bounded set.

10. Determine whether or not rectangle $A = [1,3] \times [-2,2]$ is contained in the rectangle $B = [-1,4] \times [0,3]$. Explain your reasoning.

11. The purpose of this exercise is to lead the reader to conjecture a formula for the number of elements in a cross product.

    (a) List and count the number of elements in
        (i) $\{a,b,c\} \times \{1\}$
        (ii) $\{a,b,c\} \times \{1,2\}$
        (iii) $\{a,b\} \times \{x,y\} \times \{1\}$
        (iv) $\{a,b\} \times \{x,y\} \times \{1,2\}$
    (b) Given that A has k elements, B has m elements, C has n elements, and D has p elements, conjecture the number of elements in
        (i) $A \times B \times C$ (ii) $A \times B \times C \times D$

12. Find the distance between the following pairs of points in $\mathbb{R}^n$.

    (a) (1,2,3) and $(-1,-2,-1)$

    (b) (1,1,1) and $(2,-1,3)$

    (c) (2,0) and $(-1,4)$

    (d) $(-1,-1)$ and $(-3,-4)$

    (e) (0,0,0,0) and (1,1,1,1)

    (f) (1,0,2,3) and (2,1,3,1)

13. Give examples of each of the following or state why no such example can be found. (Hint: Example 1.2.10 may help with some of these.)

(a) A set with no least upper bound.
(b) A set that has a least upper bound that is not in the set itself.
(c) A set whose least upper bound and greatest lower bound are the same.
(d) A rectangle in $\mathbb{R}^5$.
(e) A set of real numbers whose greatest lower bound is 1.
(f) A set of real numbers, not including 0, whose least upper bound is 0.
(g) A point in $\mathbb{R}^4$ that is distance 5 from $(1, 2, 5, -1)$.

14. Show that there is no point in $\mathbb{R}^2$ whose distance from (0,0) is 1 and whose distance from (3,3) is also 1.

15. For pairs of real numbers a and b, explain why each of the following relations is true or give an example showing why the relation is false:

(a) If $|a| \leq |b|$, then $a^2 \leq b^2$.
(b) If $|a| \leq |b|$, then $a^3 \leq b^3$.
(c) If $a^2 \leq b^2$, then $a \leq b$.
(d) If $a^3 \leq b^3$, then $|a| \leq |b|$.
(e) $a^2 - 2ab + b^2 \geq 0$.
(f) If $a \geq b$, then $a = (a + b + |a - b|)/2$.
(g) $||a| - 1| \geq |a - 1|$.
(h) If $0 < a < b$, then $a^2 < b^2$.
(i) If $ab \geq 0$, then $b \geq 0$.
(j) If $0 < a < 1$ and $0 < b < 1$, then $0 < a + b < 1$.
(k) If $|a| > |b|$ and $ab > 0$, then $a > b$.
(l) If $ab > 0$ and $a + b < 0$, then $a < 0$.
(m) If $1/a > 1/b$, then $a > b$.

16. This exercise illustrates a standard way of proving that two sets are equal.

Two sets are equal provided they have the same elements. Thus to show that sets A and B are equal, show that $A \subseteq B$ and $B \subseteq A$. I.e. Show that if $x \in A$ then $x \in B$ and if $x \in B$ then $x \in A$. Show that $A = \{x : x \text{ is a prime number } < 20\}$ is equal to $B = \{2, 3, 5, 7, 11, 13, 17, 19\}$.

1.3 The Function Concept

"Cause–effect" and "input–output" are popular phrases used in describing certain types of dependence relations between two sets. In mathematics, the term "function" is used to denote a dependence relation satisfying a uniqueness condition. In this section, we discuss functions, notation, and ways of describing functions.

Consider the following:

1. The amount of sales tax on an item depends on the cost of the item.
2. The cost of mailing a UPS package depends on both the weight of the package and the distance it is to be sent.
3. The position of a shell fired from a cannon with a fixed initial velocity and a fixed angle of projection depends on the time elapsed since firing.
4. In a telephone listing, the telephone number is related to the name.

Relations 1, 2, and 3 have the property that each associates a *single* "output" value with the given "input" value. Such a relation is called a "functional relationship," or just a "function." Relation 4 is not a functional relation since two subscribers with the same name will have different telephone numbers. We view a function as a machine that turns "input" into "output."

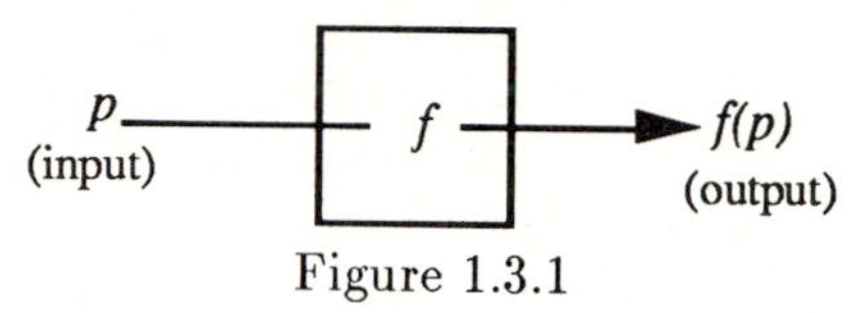

Figure 1.3.1

Note that the input and output elements of a function need not be real numbers. In 1 above, input and output values are real numbers. In 2, the input is an ordered pair (weight, distance), and the output is a single number, cost. In 3, the input is a single number, time, and the output is an ordered triple that locates the position of the shell in three-dimensional space.

Example 1.3.2 Consider a square with sides of length s. We will express the perimeter of the square in terms of the length of a diagonal.

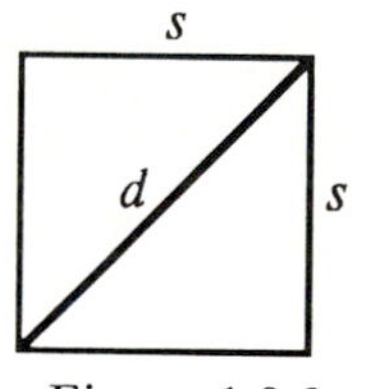

Figure 1.3.3

The perimeter is $4s$, and by the Pythagorean theorem, $s^2 + s^2 = d^2$,. Thus $2s^2 = d^2$ or $s = d/\sqrt{2}$. Thus the perimeter is $4s = 4d/\sqrt{2} = 2\sqrt{2}d$. The relation $PERIM(d) = 2\sqrt{2}d$ where $d \in (0, \infty)$, expresses the perimeter of the square in terms of its diagonal. △

The set of inputs is called the *domain* of the function, and the set of outputs is called the *range* of the function. For convenience we will formally define the concept of a function in terms of two sets called the *source* and *target* which contain the domain and range, respectively.

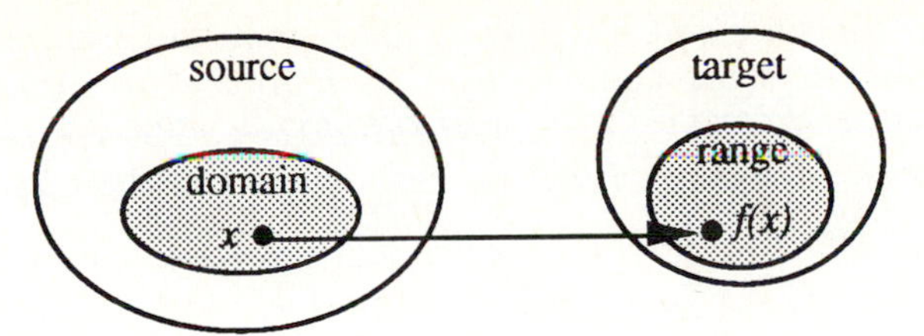

Figure 1.3.4

Definition 1.3.5 (Function) *A* function *from a set A into a set B is a relation that assigns to each element in some subset of A exactly one element in B. We call set A the* source *of the function and set B is called the* target. *The subset of A to which the relation assigns elements in B is called the* domain *of the function. The set of elements in B which are assigned to elements in the domain is called the* range *of the function.*

The *uniqueness condition* that distinguishes a function from other types of relations between sets is the requirement that an element in the domain is assigned to *exactly one* element in the range.

The symbolism

$$f : A \to B$$

means that f is a function from source A into target B. Notice that f denotes the relation. For any element p in the domain, $f(p)$ denotes the unique element in B corresponding to p and is called the *image* of p under f.

Functions are usually denoted (named) by one or more letters. In mathematics, single letters are usually used, f, g, and h being the most popular. In computer science it is normal to use several letters to allow for a meaningful name. This seems like a reasonable practice, and we will often follow it.

In the previous example, the relation $PERIM(d) = 2\sqrt{2}d$, is a function with domain and range both equal to $(0, \infty)$.

Given a function $f : A \to B$, we denote

1. domain of f by "Domain(f)."

2. range of f by "Range(f)" or "$f(A)$" [$f(A)$ is also called the image of A under the function f.]

Note that Domain(f) is a subset of the source of f and $f(A)$ is contained in the target of f; see Figure 1.3.4.

Functions are also called *mappings.* When the word mapping is used, functions are often displayed using a diagram, such as the preceeding one.

It is important to notice that if p is a point in the domain of a function f, then there is *exactly* one corresponding element $f(p)$ in the range. It is also important to be able to distinguish the four sets associated with a function: source, domain, range, and target. These sets are explicitly identified in the following examples.

Example 1.3.6 We will define a function CHAR from the natural numbers $\mathbb{N}$ to the lower case letters of the alphabet by $CHAR(n) = c$ if c is the nth letter in the alphabet. Thus $CHAR(1) = a, \ldots, CHAR(26) = z$. The source of CHAR is the natural numbers, Domain($CHAR$) is $\{1, \ldots, 26\}$, and the target and range of $CHAR$ are both $\{a, \ldots, z\}$. △

Example 1.3.7 Let $PROJX : \mathbb{R}^2 \to \mathbb{R}$ be defined by $PROJX(x, y) = x$. $PROJX$ is the *projection* from $\mathbb{R}^2$ onto $\mathbb{R}$. The source and domain of $PROJX$ are both $\mathbb{R}^2$, the range and target of $PROJX$ are both $\mathbb{R}$. △

Example 1.3.8 A table giving the daily low and high temperatures defines a function $f : \mathbb{R}^3 \to \mathbb{R}^2$. The source is $\mathbb{R}^3$, the domain is the set of ordered triples (day, month, year), and the relation assigns to each ordered triple in the domain an ordered pair of temperatures. The range is the set of ordered pairs (low temperature, high temperature), and the target is $\mathbb{R}^2$. Thus f(2,5,53) = (40,73) means that on May 2, 1953 the low temperature was 40 and the high temperature was 73. △

Example 1.3.9 (Natural Domain) Suppose we want to define a function $h : \mathbb{R} \to \mathbb{R}$ such that $h(x) = |1/x|$. The function h cannot be defined at 0, since 1/0 is not defined. The *natural domain* of h, the largest subset of the source on which h can be defined, is

$$\mathbb{R}^* = \{x \in \mathbb{R} : x \neq 0\}$$

so let the domain of h be $\mathbb{R}^*$. △

If we wish, we could restrict our definition of the domain of h to the subset $\{x : x > 0\}$ of $\mathbb{R}^*$. This would give a different function $g : \mathbb{R} \to \mathbb{R}$, which has the same range as h, but is a different function since the domains are different.

Definition 1.3.10 (Equality of Functions) *Two functions $f : A \to B$ and $g : A \to B$ are the same, $f = g$, if they are defined on* exactly *the same points (i.e., have the same domain) and $f(p) = g(p)$ for every one of these points.*

Example 1.3.11 In the Example 1.3.9, $h \neq g$. △

Question 1.3.12 In order to understand the "roles" of the sets associated with a function, the reader should pause to make up several examples such as the following:

- Two different functions defined by the same formula.

- Two different functions having the same domains and same ranges.
- A function with domain $\neq$ source and range = target.

It is necessary to be careful when evaluating a function. Consider the function $f : \mathbb{R} \to \mathbb{R}$ with domain $\mathbb{R}$ given by $f(x) = 3 + 2x$. If we wish to evaluate f at an expression, e.g., $1 + x$, we must remember to substitute the expression for the variable on both sides of the defining formula. The value of $f(1 + x)$ is obtained by replacing x by $1+x$ on both sides of the equation, giving $f(1+x) = 3 + 2(1 + x) = 5 + 2x$. If we wish to evaluate $f(f(1 + x))$, we can do it in two ways. We could begin by evaluating the innermost application of the function and proceed outward:

$$f(f(1+x)) = f(3 + 2(1 + x)) = f(5 + 2x) = 3 + 2(5 + 2x) = 13 + 4x$$

Or we could work inward:

$$\begin{aligned} f(f(1+x)) &= 3 + 2f(1 + x) = 3 + 2(3 + 2(1 + x)) = 3 + 6 + 4(1 + x) \\ &= 13 + 4x \end{aligned}$$

Both approaches will give the same answer if carried out correctly.

Graphs of Functions

The graph of a function gives a picture of the function.

Definition 1.3.13 (Graph of a Function) *If $f : A \to B$, then the* graph *of f, denoted $Graph(f)$, is the subset of the Cartesian product $A \times B$ given by*

$$Graph(f) = \{(x, f(x)) : x \in Domain(f)\}.$$

If $A = B = \mathbb{R}$, then $A \times B = \mathbb{R}^2$ is the plane and the graph of $f : A \to R$ is a subset of the plane.

Example 1.3.14 Let $h : \mathbb{R} \to \mathbb{R}$ be defined by $h(x) = |1/x|$ for all nonzero real x. When x is near 0, $h(x)$ is very far from 0, and when x is far from 0, $h(x)$ is near 0. The graph of h consists of the two curves:

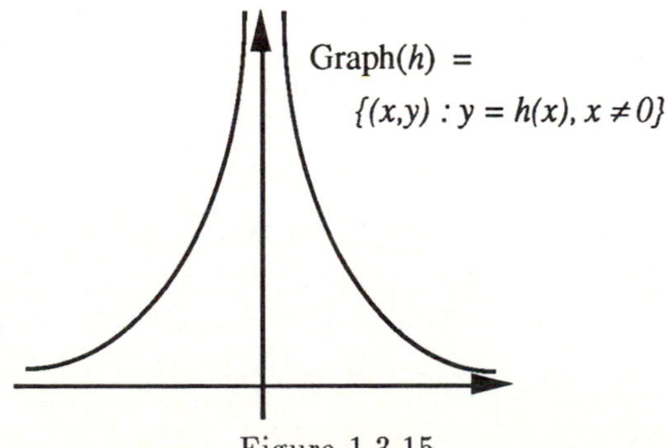

Figure 1.3.15

Note that the vertical projection of Graph(h) onto the x-axis is the domain of h and the horizontal projection of Graph(h) onto the y-axis is the range of h. $\triangle$

Suppose G, a subset of the plane, is the graph of a function $g : A \to B$. If $(x, y) \in G$ and $(x, z) \in G$ then y and z must *both* be equal to $g(x)$. Since only one element in B corresponds to x, we must have $y = z$. This says that there cannot be two different points (x, y) and (x, z) on the graph; i.e., no vertical line cuts the graph at more than one point. (This fact is sometimes called the "Vertical Line Test.") The following subset of the plane cannot be the graph of any function f with $y = f(x)$, since the vertical line through the points P_1 and P_2 intersects the curve at two points.

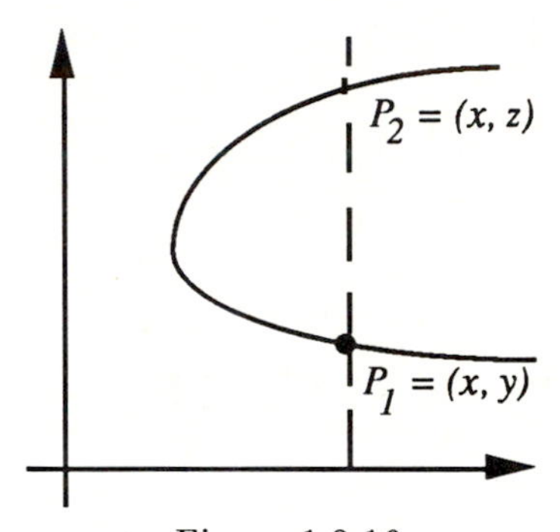

Figure 1.3.16

If G is a subset of $A \times B$ with the property "if $(x, y) \in G$ and $(x, z) \in G$, then $y = z$," then G is the graph of a function from A to B.

Example 1.3.17 (a) Consider three sets, $A = \{1, 2, 3\}$, $B = \{x, y, z\}$, and $S = \{(1, x), (2, y), (2, x)\}$. S is a subset of the Cartesian product $A \times B$. It is not the graph of any function $f : A \to B$ since two distinct values, x and y, correspond to 2.

(b) Let $A = \{1, 2, 3\}$, $B = \{x, y, z\}$, and $S = \{(1, x), (2, y)\}$. S is a subset of the Cartesian product $A \times B$. It is the graph of a function $g : A \to B$. The domain of g is $\{1, 2\}$, and the range of g is $\{x, y\}$, $g(1) = x$ and $g(2) = y$. Note that the graph of g consists of just two points. $\triangle$

The next two examples illustrate how a graph can define a function.

Example 1.3.18 (A Function Defined by Its Graph) Let G be the part of the unit circle lying in the first quadrant of the plane. Thus $G = \{(x, y) \in \mathbb{R}^2 : x \geq 0, y \geq 0, \text{ and } x^2 + y^2 = 1\}$. How can we define a function g whose graph is G?

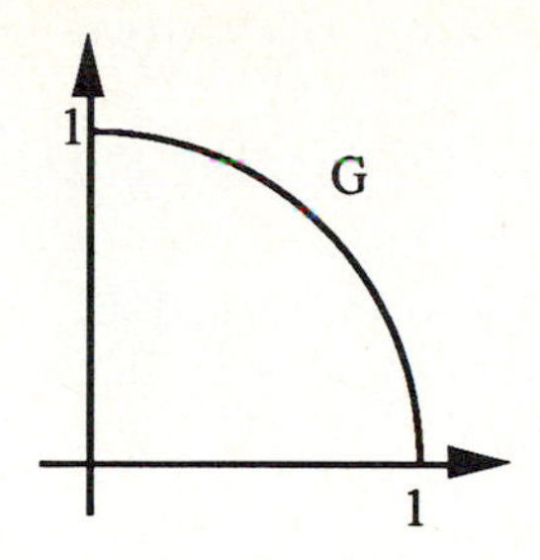

Figure 1.3.19

For every x in [0,1] there is a unique value y such that (x, y) is in G. Solving $x^2 + y^2 = 1$, $x \geq 0, y \geq 0$, for y in terms of x we obtain $y = (1 - x^2)^{1/2}$. Define $g : \mathbb{R} \to \mathbb{R}$ by setting Domain(g) equal to [0,1] and $g(x) = (1 - x^2)^{1/2}$ for x in [0,1]. Now we have Graph(g) equal to G. △

Example 1.3.20 (Paraboloid) Let G be the set of points in $\mathbb{R}^3$ satisfying $x^2 + y^2 = z$.

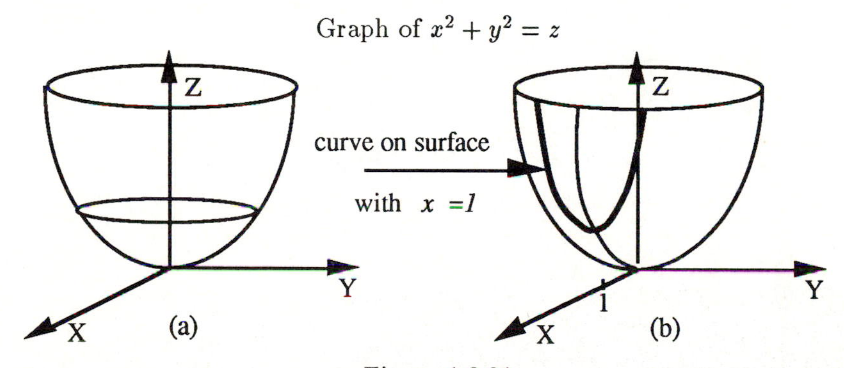

Figure 1.3.21

This figure is called a *paraboloid* since if we hold x or y constant, we get a parabola. For example, if we set $y = 0$, we get $x^2 = z$, and if we set $x = 1$, we get $1 + y^2 = z$ (in Figure 1.3.21 (b), the curve on the paraboloid where $x = 1$ is highlighted). G is the graph of the function $f : \mathbb{R}^2 \to \mathbb{R}$ defined by $f(x, y) = x^2 + y^2$ for all $(x, y) \in \mathbb{R}^2$:

$$\begin{aligned} Graph(f) &= \{(p, f(p)) : p \in \mathbb{R}^2\} \\ &= \{(x, y, f(x, y)) : x, y \in \mathbb{R}\} \\ &= \{(x, y, z) : z = f(x, y) = x^2 + y^2\} \end{aligned}$$ △

We emphasize that the graph of a function $f : A \to B$ is a *subset of* $A \times B$. The point we are making is that the graph of a function is not always a single curve. In the last few examples, we have seen graphs of functions consisting of a single curve, two curves, two points, and a curved surface in three space.

Some Examples of the Uses of Functions

Example 1.3.22 (A Function Defined by a Table) In this example the domain will be a finite set.

Consider a person who in the years 1960, 1961, 1962, and 1963 owes 100.00, 120.50, 130.40, and 120.50 dollars in taxes to the municipality in the respective years. We can display this information in a table:

Tax(T)	100.00	120.50	130.40	120.50
Year(Y)	1960	1961	1962	1963

We use this table to define a function $TAX : Y \to T$. The following diagram displays this function as a mapping. On the left-hand side is the set that is the domain of the function TAX. On the right-hand side is the range of the function TAX. In this example the range and the target coincide and the domain and the source coincide.

TAX as a Mapping

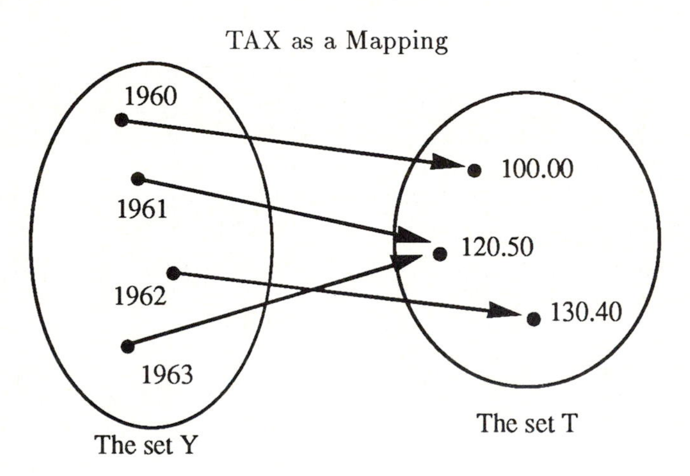

Figure 1.3.23

The TAX function is an example of a *many-to-one* function. That is, a function in which two or more elements in the domain correspond to the same element in the range.

A final way to picture a function is to use the graph of a function. The graph of TAX is the set of points (1960,100.00), (1961,120.50), (1962,130.40), (1963,120.50). Since the domain and range of TAX are subsets of the real numbers, the plane can be used to show the graph. The domain is shown along the horizontal coordinate axis, and the range is shown along the vertical axis. The ordered pairs of the graph of the function are the points on the plane with the corresponding rectangular coordinate values.

The Graph of TAX

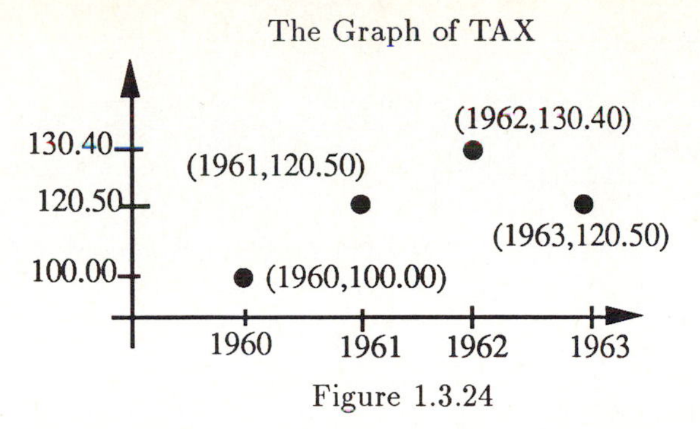

Figure 1.3.24

Notice that we do not connect the point (1960,100.00) and the point (1961,120.50). If we did, it would incorrectly imply that the function TAX is defined for real numbers, such as 1960.5, which lie between 1960 and 1961. △

In many publications (especially newspapers and magazines), functions are usually given by graphs or tables. In mathematical and scientific publications, functions are often defined by formulas. In the "word problems" popular in algebra texts, functions are usually given implicitly, such as in Example 1.3.2. We will use all these methods in describing functions.

Sometimes a function is partially specified, without explicitly giving the domain, or even the source and target. In these cases the source and target are determined from the context, and the domain is the largest possible subset of the source for which the function relation can be defined, the natural domain. Occasionally some assumptions have to be made as to the meaning of the specifications.

Example 1.3.25 (Implicit Domains of Functions) (a) Let $f : \mathbb{R} \to \mathbb{R}$ be given by $f(x) = x + 1$. The expression $x + 1$ makes sense for all real numbers, so the natural domain is $\mathbb{R}$.

(b) Let f with source [0, 1] be given by $f(x) = 1/x$. Here the target is not specified, but since the reciprocal $1/x$ of a real number x in (0, 1] is a real number, the target must be $\mathbb{R}$. The domain will be (0, 1], the natural domain.

(c) Define a function by $f(x, y) = x + y$. Here we are adding elements, so x and y are probably real numbers. The specification is ambiguous however, since the specifier may have had integers in mind. Usually integers are denoted by letters i, j,..., n, and real numbers by letters later in the alphabet such as $r, s, \ldots$. So we will assume that we are working with pairs of real numbers (x, y) rather than restricting ourselves to pairs of integers. The expression $x + y$ takes a pair of real numbers, (x, y) and gives a real number $x + y$. Thus we assume that the source is $\mathbb{R}^2$, the target is $\mathbb{R}$, and the domain is the natural domain $\mathbb{R}^2$. △

Section 1.3 Exercises

1. (a) Graph the function given by the table

$f(x)$	0	-1	2	4	2
x	1	2	3	4	5

   (b) Show f as a mapping of one value to the corresponding value.

   (c) What is the domain of f?

   (d) What is the range of f?

2. (a) Graph the function given by the table

$f(x)$	-3	1	3	0	3
x	1	2	3	4	5

   (b) Show f as a mapping of one value to the corresponding value.

   (c) What is the domain of f?

   (d) What is the range of f?

3. For each of the following subsets of the Cartesian product of $A = \{0, 1, 3, 4\}$ and $B = \{x, y, c, d\}$, is the set the graph of a function?

   (a) $\{(0, x), (1, x), (3, c), (4, d)\}$

   (b) $\{(0, c), (3, d), (1, c), (4, y)\}$

   (c) $\{(1, d), (3, c), (0, x), (1, y)\}$

   (d) $\{(1, x), (3, x), (0, c), (3, d)\}$

4. For each of the following, find the appropriate formula for the function, as well as specifying the domain.

   (a) The area A of a circle as a function of its circumference C.

   (b) The area A of a square as a function of its perimeter P.

   (c) The radius r of a sphere as a function of its volume V. (Recall that the volume of a sphere of radius r is $\frac{4}{3}\pi r^3$.)

   (d) The volume V of a sphere as a function of its surface area S.

   (e) The surface area S of a sphere as a function of its volume V.

   (f) The length d of the diagonal of a square as a function of its area A.

   (g) The length d of the internal diagonal of a cube as a function of the volume V of the cube.

   (h) The area of a rectangle as a function of its perimeter P and its height h.

   (i) The volume of a closed box as a function of its surface area S (including sides, top, and bottom), its height h, and its width w.

5. In each of the following, make sure to specify the domain.

   (a) A square is inscribed in a circle. Express the area S of the square as a function of the area A of the circle.

   (b) A square is circumscribed about a circle. Express the area S of the square as a function of the area A of the circle.

   (c) A right circular cylinder is inscribed in a cone whose height is 7 and base radius is 4. Express the volume V of the cylinder as a function of its radius r.

   (d) A rectangle is inscribed in a circle of radius 3. Express the area A of the rectangle as a function of the length x of the side of the rectangle.

   (e) The centers of the sides of a rectangle of height h and width w are joined to form a "diamond". Express the area of the diamond as a function of the height and the width of the rectangle.

6. Consider the function $f : \mathbb{R} \to \mathbb{R}$ with domain $\mathbb{R}$ given by $f(x) = 5 - 3x$. Evaluate f at the specified points.

   (a) $f(0)$ (b) $f(-1)$ (c) $f(2)$ (d) $f(3x)$

   (e) $f(x-1)$ (f) $f(2t)$ (g) $f(y)$ (h) $f(x^3)$

   (i) $f(y^2)$ (j) $f(-x)$ (k) $f(x+y)$ (l) $f(f(x))$

7. Evaluate the function specified by the rule $f(x) = x^3$, where Domain$(f) = \mathbb{R}$..

   (a) $f(1)$ (b) $f(3)$ (c) $f(2x)$ (d) $f(3t)$

   (e) $f(2y)$ (f) $f(x+2)$ (g) $f(1-x)$ (h) $f(x^3)$

   (i) $f(y^4)$ (j) $f(-x)$ (k) $f(f(x))$ (l) $f(1-f(x))$

8. Evaluate for the function specified by the rule $f(x,y) = x + 2y$, where Domain$(f) = \mathbb{R}^2$.

   (a) $f(-1,1)$ (b) $f(3,0)$ (c) $f(x,x)$

   (d) $f(3,t)$ (e) $f(2y-1,y)$ (f) $f(x^3,2)$

   (g) $f(y^3,x^2)$ (h) $f(1+x,1-y)$ (i) $f(2,f(1,3))$

   (j) $f(f(1,1),f(2,2))$ (k) $f(f(0,0),0)$ (l) $f(1+f(0,1),f(1,0))$

9. Evaluate the following for the function $f : \mathbb{R}^2 \to \mathbb{R}^2$ defined by $f(x,y) = (x^2, 2xy)$ on all of $\mathbb{R}^2$.

(a) $f(1,0)$ (b) $f(2,1)$ (c) $f(3,x)$ (d) $f(t,t)$

(e) $f(y,x)$ (f) $f(x+1,y-2)$ (g) $f(x,xy)$ (h) $f(x^2,y)$

(i) $f(0,y^{10})$ (j) $f(f(1,2))$ (k) $f(f(t,1))$ (l) $f(f(f(1,1)))$

10. Evaluate and simplify the expression $(f(x+h)-f(x))/h$ for f defined on all of $\mathbb{R}$.

(a) $f(x)=3x$ (b) $f(x)=x^2$ (c) $f(x)=x^3$

(d) $f(x)=(x-1)^2$ (e) $f(x)=8$ (f) $f(x)=2x-5x^2$

11. State the source, target, and natural domain of the following.

(a) $f(x)=(x-1)^{-1}$ (b) $f(x)=(x-2)^{-2}$

(c) $f(x)=(x,x+1)$ (d) $f(x)=(x,x-2,x+5)$

(e) $f(x,y)=\frac{1}{xy}$ (f) $f(x)=\frac{1}{(x-1)(x-2)}$

12. (a) A cow transportation ship can carry 400 cows. The ship will not sail if there are less than 200 cows. When there are exactly 200 cows, the cost per cow is \$550. If there are more than 200 cows, then the cost per cow is reduced by \$2 for every cow over 200. Express the amount of money collected by the transportation company as a function of the number of cows.

(b) Cows obtain food from a company that delivers weekly a kilogram of cow food for \$0.10. If the order exceeds 200 kilograms, then the price is reduced by \$0.01 per kilogram on that part of the order exceeding 200 kilograms. Each cow consumes 2 kilograms of food and 1 kilogram of water each day. What is the weekly cost of cow food delivery as a function of the number of cows? (Assume that cow food cannot be stockpiled.)

(c) A cow needs a new stall. The cost of the front is \$10 per linear foot; the cost of the sides and rear is \$7 per linear foot. Express the cost of a stall as a function of the dimensions of the stall (include the domain of the cost function).

(d) Cows are transported to market by a trucking company that charges \$1 per head for small cows and \$3 per head for large cows. A truck can hold at most 30 small cows or 15 large cows. Construct a function which will give the amount received by the company as a function of the number of small and large cows in a truck. (Include the domain of the function and the assumptions you make in constructing your function.)

13. (a) Sketch the graph of a function $f : \mathbb{R} \to \mathbb{R}$ that has the property that $f(x) > x$.
    (b) Make up a word problem whose solution is the graph you created for (a).

14. For each of the following, give an example or show that no example can exist.

    (a) Two different functions, f and g, that have the same domain.
    (b) Two different functions, f and g, that have the same range.
    (c) Two different functions, f and g, that have the same domain and the same range.
    (d) Two different functions, f and g, such that the range of f is a proper subset of the domain of g.
    (e) A function with a bounded interval for its domain and an unbounded interval for its range.
    (f) A function with an unbounded interval for its domain and a bounded interval for its range.

15. This exercise will explore the behavior of a family of functions. The object is for you to use a function graphing program to examine several examples and then to draw conclusions from these examples about the family of functions. If you are using an automated program for the graphing, it is easy for you to examine as many examples as you need in order to formulate your conclusions.

    (a) Consider functions defined on $[-1, 1]$ of the form $f(x) = x^n$ where n is a positive odd integer. Graph these for $n = 1$, 3, and 5. What does the graph of $f(x) = x^{21}$ look like? Check your conjecture with your graphing program. How does the graph of $f(x) = x^n$ vary with n, a positive odd integer? Does your conclusion change if the interval is enlarged?
    (b) Consider functions defined on $[-1, 1]$ of the form $f(x) = x^n$ where n is a positive even integer. Graph these for $n = 2$, 4, and 6. What does the graph of $f(x) = x^{20}$ look like? Check your conjecture with your graphing program. How does the graph of $f(x) = x^n$ vary with n, a positive even integer? Does your conclusion change if the interval is enlarged?

1.4 Invertibility and Boundedness of Functions

Function is a fundamental concept in mathematics. As mentioned earlier, our major goal is to learn how to analyze functions. In this section, we study the basic functional properties of invertibility and boundedness.

Example 1.4.1 (Celsius to Fahrenheit Temperature) A common "car game," played while driving past a bank sign with a print-out that cycles through the temperature in the Celsius scale, time of day, and temperature in the Fahrenheit scale, is to convert the temperature from Celsius to Fahrenheit before the sign does.

What is the relationship between the two temperature scales? What is the function that converts (maps) temperature readings in the Celsius scale to the Fahrenheit scale?

On the Fahrenheit(F) scale water freezes at 32 degrees and boils at 212 degrees, while on the Celsius(C) scale water freezes at 0 degrees and boils at 100 degrees. Assuming a constant rate of change from C to F, we note that a change of 100 degrees C produces a change of 180 degrees F. Or, for each 1 degree change in C there is a $\frac{9}{5}$ degree change in F. That is, the rate of change from C to F is $\frac{9}{5}$. Knowing that F is 32 when C is 0 allows us to define the conversion function mapping C to F, $CTOF : \mathbb{R} \to \mathbb{R}$ defined by $CTOF(C) = \frac{9}{5}C + 32$.

Now that we have a way to convert from Celsius temperatures to Fahrenheit temperatures, it is natural to ask if it is possible to reverse the process. That is, given any Fahrenheit temperature F, can we find a Celsius temperature C such that $CTOF(C) = F$, i.e., is F the image under $CTOF$ of some value C? Clearly the answer is yes, just solve

$$\frac{9}{5}C + 32 = F$$

for C. This yields the inverse relation

$$C = \frac{5}{9}(F - 32)$$

The important question is now "Does the inverse relation define a function?" That is, for a given temperature value F, is there exactly one temperature value C such that $CTOF(C) = F$? △

Although the inverse relation above is functional, it is not always true that the inverse relation of a function is functional. In the TAX Example 1.3.22 of the previous section, TAX(1961) = 120.50 and TAX(1963) = 120.50. Hence the inverse relation associates two outputs, namely 1961 and 1963, with the input 120.50 and therefore is not a functional relation. The crucial condition that guarantees that the inverse relation will be functional is given by the following definition.

Definition 1.4.2 (One-to-One Function) *A function $f : A \to B$ is* one-to-one, *if for every element in the range of f there is only one corresponding element in the domain of f. Equivalently, two distinct elements of the domain cannot correspond to the same element in the range.*

As a mapping, if f is one-to-one then distinct points x and y in the domain are mapped to distinct points u and v in the range.

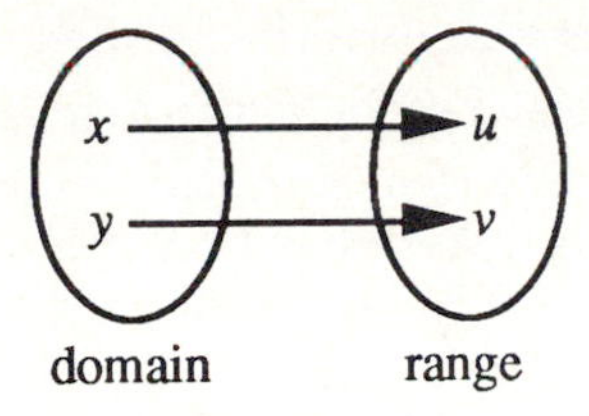

Figure 1.4.3

An equivalent form of the definition of one-to-one is: a function f is one-to-one if whenever $f(x) = f(y)$ for two elements x and y in the domain of f, then $x = y$.

This formulation of the one-to-one concept is usually the easiest to apply when checking to see if a given function is one-to-one. Part (b) of the following example illustrates the procedure.

Example 1.4.4 (One-to-One Functions) (a) The function $CHAR$ of Example 1.3.6 is one-to-one.

(b) Consider $f : \mathbb{R} \to \mathbb{R}$ defined by $f(x) = 2^x$ with domain $\mathbb{N}$. From the previous paragraph, f is one-to-one provided $f(x) = f(y)$ implies that $x = y$. Now

$$\begin{aligned} f(x) = f(y) &\Longrightarrow 2^x = 2^y \Longrightarrow \frac{2^x}{2^y} = 1 \text{ (dividing both sides by } 2^y\text{)} \\ &\Longrightarrow 2^{x-y} = 1 \Longrightarrow x - y = 0 \Longrightarrow x = y \end{aligned}$$

Thus f is a one-to-one function. △

Clearly the base 2 in the above argument could be replaced by any positive number $b \neq 1$ without changing the result. Thus $f(x) = b^x$, $b > 0$ and $b \neq 1$, is a one-to-one function. (What happens when $b = 1$?)

If f is *not* one-to-one, then there is a z in the range of f such that $z = f(x)$ and $z = f(y)$ for some x and y in the domain.

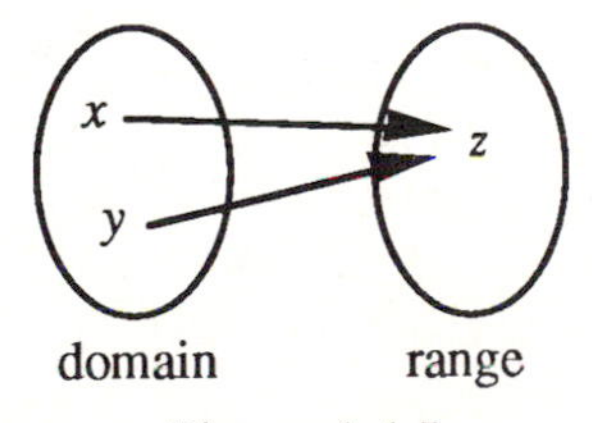

Figure 1.4.5

Example 1.4.6 (Functions not One-to-One) (a) The projection function of Example 1.3.7 above is not one-to-one since $PROJX(1, y) = 1$ for all real numbers y.

(b) Consider $f : \mathbb{R} \to \mathbb{R}$ defined by $f(x) = x^2$. This function is not one-to-one, since, e.g., $f(-1) = f(1) = 1$ △

If a function $f : A \to B$ is one-to-one, then given an element b in the range of f there is only one corresponding element in the domain of f. This allows us to define a new function f^{-1}, the inverse function of a one-to-one function, which undoes the action of f.

Definition 1.4.7 (Inverse Function) *If $f : A \to B$ is one-to-one, then the* inverse function *to f, usually denoted f^{-1}, is a function $f^{-1} : B \to A$ defined for b in Range(f) by setting $f^{-1}(b)$ equal to the unique a in A such that $f(a) = b$. Thus $f^{-1}(b) = a$ if and only if $f(a) = b$.*

If $f : \mathbb{R} \to \mathbb{R}$ is one-to-one and we write $f(x) = y$, then $f^{-1}(y) = x$. Thus we can try to find the inverse function by solving for x in terms of y. [It may be necessary to restrict the domain of a function in order to define an inverse function. For example, $f(x) = \sin(x)$ $-\pi/2 \leq x \leq \pi/2$ has an inverse: $f^{-1}(x) = \sin^{-1}(x) = \arcsin(x)$.]

Example 1.4.8 (Inverse Functions) (a) The function $CHAR$ of Example 1.3.6 is one-to-one, so it has an inverse function $CHAR^{-1}$, which might be called ORDER. So,

$$ORDER(a) = CHAR^{-1}(a) = 1, \ldots, ORDER(z) = 26$$

Now Domain($ORDER$) = Range($CHAR$) = $\{a, \ldots, z\}$ and Range($ORDER$) = Domain($CHAR$) = $\{1, \ldots, 26\}$.

(b) Let $f : \mathbb{R} \to \mathbb{R}$ be defined by $f(x) = 2x + 6$. Let the domain of f be the natural domain $\mathbb{R}$, the largest subset of the source $\mathbb{R}$ on which f can be defined. Then, solving $f(x) = y$, or $2x + 6 = y$, for x in terms of y, we obtain $x = (y - 6)/2 = y/2 - 3$. Thus $f^{-1}(y) = x = y/2 - 3$. Since the symbol y is arbitrary, we can replace y by x in $f^{-1}(y) = y/2 - 3$ and write

$$f^{-1}(x) = \frac{x}{2} - 3 \text{ for all real numbers x}$$ △

If we have the graph of a one-to-one function, we can use it to find the graph of the inverse function. If the point (x, y), $y = f(x)$, is on the graph of f, then (y, x), $x = f^{-1}(y)$, is on the graph of f^{-1}. For example, if $(1, a)$ is on the graph of f, then $(a, 1)$ is on the graph of f^{-1}. What we need to do is to interchange the roles of x and y. This is done geometrically by reflecting the graph of f in the line $y = x$.

Example 1.4.9 Consider a simple example, $y = f(x) = 2x$. The function f is one-to-one on $\mathbb{R}$. The inverse function is $f^{-1}(x) = x/2$. Note that the graph of f^{-1} is the reflection in the line $y = x$ of the graph of f. △

Graphs of $y = x/2$ and $y = 2x$

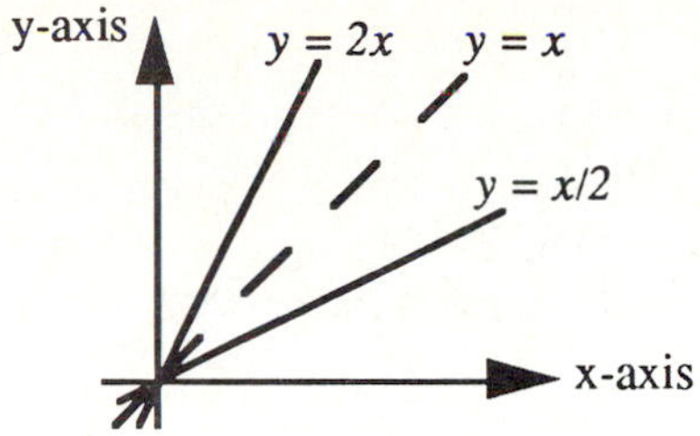

Figure 1.4.10

Question 1.4.11 Let $A = \{1, 2, 3, 4\}$ and $B = \{a, b, c, d\}$. For each of the following subsets of $A \times B$, is the subset the graph of a function, and if so, is the function one-to-one? If it is one-to-one, what is the graph of the inverse function?

(a) $\{(1, c), (2, d), (3, b), (4, a)\}$

(b) $\{(1, a), (2, a), (3, c), (4, d)\}$

(c) $\{(1, a), (2, a), (1, c), (3, b), (4, d)\}$ ◇

Example 1.4.12 (A Mapping into the Plane) Let $f(t) = (t + 1, t^2)$ for all real numbers t. The natural domain of f is all of $\mathbb{R}$. The target of f is $\mathbb{R}^2$, since f assigns to a real number t a pair of real numbers $(x, y) = (t + 1, t^2)$ in $\mathbb{R}^2$.

To show that f is one-to-one, we need to show that if $f(t) = f(s)$, then $t = s$. Suppose $f(t) = f(s)$, then $(t + 1, t^2) = (s + 1, s^2)$, so $t + 1 = s + 1$ and $t = s$. Thus f is one-to-one. △

Example 1.4.13 (A Function of Two Variables) Here is an example of a function with domain in $\mathbb{R}^2$ and range in $\mathbb{R}$.

Let $f(x, y) = 1 - x/2 - y/3$ for all real numbers x and y. Then f maps a pair of real numbers (x, y) to a real number $1 - x/2 - y/3$. Thus $f : \mathbb{R}^2 \to \mathbb{R}$ and the natural domain of f is $\mathbb{R}^2$.

To check if f is one-to-one, we set $f(x, y) = f(u, v)$. Thus

$$1 - \frac{x}{2} - \frac{y}{3} = 1 - \frac{u}{2} - \frac{v}{3}$$

It is clear that f is not one-to-one, since $f(2, 3) = f(4, 0)$ and $(2, 3) \neq (4, 0)$.

The graph of f is all points $((x, y), z)$ such that $f(x, y) = z$. Again we identify $((x, y), z)$ with (x, y, z), so that the graph of f is in $\mathbb{R}^3$. The graph, all (x, y, z) with $z = 1 - x/2 - y/3$, is a plane in $\mathbb{R}^3$. A plane is determined by three points, so we can pick points on the coordinate axes to determine the plane. If we set $x = y = 0$, then $z = 1$ so (0,0,1) is one point. If $x = z = 0$, then $y = 3$ so $(0, 3, 0)$ is a second point. If $y = z = 0$, then $x = 2$ so $(2, 0, 0)$ is a third point. △

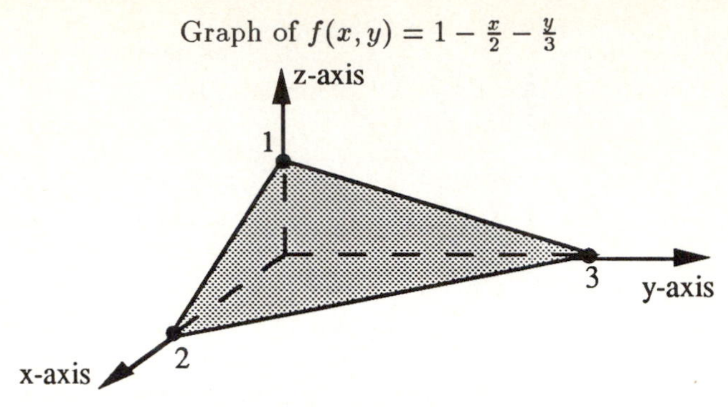

Figure 1.4.14

Example 1.4.15 (There Are as Many Even Integers as Integers) A one-to-one function is sometimes called a "matching" function since each element of the domain is matched with a unique element of the range and conversely each element of the range is matched with a unique element of the domain. Matching provides an alternative to counting as a method of comparing the relative sizes of two sets. To compare the size of the freshmen class to the sophomore class, one could match (pair) a freshmen with a sophomore, then match another freshmen with another sophomore, then match another freshmen with another sophomore, etc. The class having unmatched members at the end of the process would be the larger class. If there are no unmatched members left, then both classes are the same size.

What happens when both sets to be compared are infinite? Note that we cannot count the elements in an infinite set, so we must use matching: two sets are of the same size if there is a matching between them.

The function $f : \mathbb{N} \to \mathbb{N}$ defined by $f(n) = 2n$ is clearly one-to-one and its range is the even integers. Thus to each integer (n) , there corresponds a unique even integer $(2n)$ and hence f^{-1} exists. Therefore, Domain(f^{-1}) = Range(f) the even integers, and Range(f^{-1}) = Domain(f) = $\mathbb{N}$. Hence every even integer maps onto an integer [e.g., $f^{-1}(-26) = -13$]. Thus there are as many even integers as integers! △

Functions, such as the one in the previous example, with the property that the range equals the target are called *onto* functions.

Definition 1.4.16 (Onto Functions) *A function is* onto *if its range is equal to its target.*

Example 1.4.17 (Onto Functions) (a) The TAX function in the previous section is an onto function (although not one-to-one).

(b) The $PROJX$ function of Example 1.3.7 is onto.

(c) The function $f : \mathbb{R}^2 \to \mathbb{R}$ defined by $f(x,y) = x + y + 1$ was shown in an earlier example not to be one-to-one. However, it is onto, since any real number can be expressed in the form $x + y + 1$.

(d) $f : \mathbb{R} \to \mathbb{R}^2$ defined by $f(t) = (t+1, t^2)$ was shown to be one-to-one, but is clearly not onto since $(0, -3)$ is not in Range(f). $\triangle$

Often situations involving sizes of infinite sets, such as the Example 1.4.15 above, are counterintuitive (How can a subset that is not the whole set be the same size as the whole set?). The difficulty lies in the fact that our intuition is primarily based on (relatively small) finite sets. In comparing very large or infinite sets, we need to *rely on mathematics*, in particular, the function properties of one-to-one and onto. The following statement is true for all sets: Two sets A and B have the same number of elements if and only if there exists a function $f : A \to B$ with Domain(f) $= A$ that is both one-to-one and onto.

We now pause in our development of function properties to list a few questions. The reader is reminded that a primary goal of a calculus course, as of any course, is to develop an inquisitive approach on the part of the student.

Given a function f with source A and target B, there are several questions about f that naturally arise.

1. What is the range of f, $f(A)$? Does $f(A) = B$, i.e., is the function onto? If f is onto, then the equation $f(a) = b$ has at least one solution for any b in B.

2. Is f one-to-one? If whenever $f(x) = f(y)$ we must have $x = y$, then f is one-to-one. If f is one-to-one, then the equation $f(x) = b$ has at most one solution for any b in B. If f is both one-to-one and onto, then the equation $f(x) = b$ has exactly one solution for any b in B.

3. If the range of f is a subset of $\mathbb{R}^n$, then we can ask if f is bounded. By definition, a function is bounded if its range is bounded (see below).

4. If the function has as its range a subset of the real numbers, then we may be interested in the maximum value in the range of the function, and the value or values in the domain that are sent to the maximum value in the range. We can ask similar questions about the minimum value in the range. In the TAX example it is natural to ask what the maximum tax was and when it was paid. Looking at the values in the range we see that the maximum value is 130.40 and that this value occurred in the year 1962. We also see that the minimum value of 100.00 occurred in the year 1960.

5. Another question is where a particular value is taken on. That is, given a function f and a value b, we want to find one (or all) x in the domain of f such that $f(x) = b$. In the *TAX* Example 1.3.22, when is the value 120.50 taken on? From the definition, we see that 120.50 is taken on at the years 1961 and 1963. Frequently we wish to find when a function vanishes, $\{x : f(x) = 0\}$. If $f(x) = 0$, then x is called a *root* or *zero* of the function f.

If we ask when the function $f(x) = x^2 - 1$ is equal to zero, then we need to find the roots of $x^2 - 1 = 0$, or equivalently, solve the equation $x^2 = 1$. The roots are $x = 1$ and $x = -1$.

If the function f is one-to-one, and thus has an inverse f^{-1}, then finding x such that $f(x) = b$ is equivalent to evaluating the inverse function $f^{-1}(b) = x$. Notice that when f is one-to-one, there can be exactly one x such that $f(x) = b$

for b in the range of f. This situation occurs with the function CTOF of Example 1.4.1: given a Fahrenheit temperature F, if we want to find a corresponding Celsius temperature C so that $CTOF(C) = F$, then C is given by $C = CTOF^{-1}(F) = FTOC(F) = 5(F - 32)/9$.

The above questions introduced a new property of functions, boundedness. Recall from Section 1.2 that a set A is bounded if there is a real number B such that $|a| \leq B$ for all a in A. We shall say a function is bounded if its range is bounded.

Definition 1.4.18 (Bounded Function) *A function $f : A \to \mathbb{R}^n$ is* bounded *if the range of f, $f(A)$, is a bounded subset of $\mathbb{R}^n$.*

Example 1.4.19 (Bounded and Unbounded Functions) (a) Let $f : \mathbb{R} \to \mathbb{R}$ be defined by $f(x) = 1$ for all real numbers x. The domain of f is $\mathbb{R}$, which is not bounded; the graph of f is $\{(x, 1) : x \in \mathbb{R}\}$, which is not bounded (see the graph below); however, the range of f is $\{1\}$, which is a bounded set. Thus f is a bounded function.

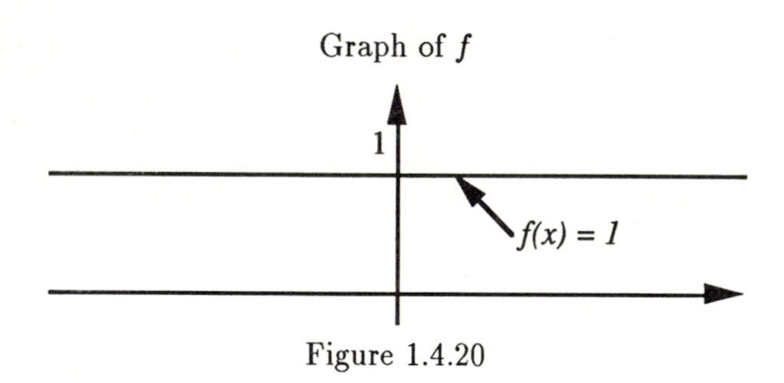

Figure 1.4.20

(b) Let $g : [0, 1] \to \mathbb{R}$ be defined by $g(x) = 1/x$. Here the domain is all real numbers in $[0, 1]$ except 0. Even though the domain of g is bounded, the graph of g (see below) clearly indicates that Range(g) and thus g is unbounded. To show analytically that g is unbounded, it suffices to show that given any (large) positive number M, there exists an x_0 in Domain(g) = (0,1] such that $g(x_0) > M$. To do this, we set $g(x) = 1/x > M$ and solve for x. That is, $x < 1/M$. Thus let $x_0 = \frac{1}{2}M$. Since $g(x_0) = 2M > M$, g is an unbounded function.

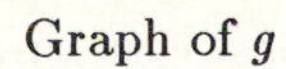

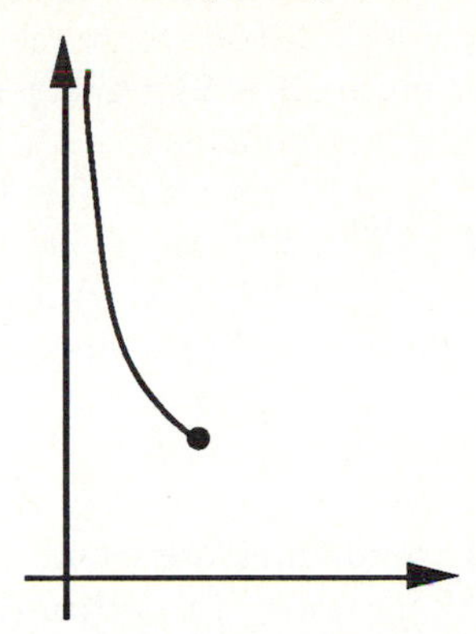

Figure 1.4.21

(c) Let $h : \mathbb{R} \to \mathbb{R}^2$ be defined by $h(t) = (1, \sin(t))$ for all real numbers t. Since the distance from $(1, \sin(t))$ to the origin (0, 0) is

$$|(1, \sin(t))| = [1 + \sin^2(t)]^{1/2} \leq [1 + 1]^{1/2} = \sqrt{2}$$

for all t, the range of h is bounded (it is inside the disk of radius $\sqrt{2}$). Thus h is a bounded function.

Note from these examples that the function may be bounded and the graph may be unbounded (as in Example 1.4.19 a above). However, if the graph is a bounded set, then the range, and hence the function, must be bounded. △

Question 1.4.22 Can a function be both increasing and bounded? How do you tell if a function is increasing? ◇

Example 1.4.23 We will use the concepts in this section to analyze some functions.

(a) Let $A = \{x, y\}$ and $B = \{1, 2, 3\}$. Let $S = \{(x, 2), (y, 1)\}$, a subset of the Cartesian product $A \times B$. S is the graph of the function $f : A \to B$ given by $f(x) = 2$ and $f(y) = 1$. The function f is one-to-one and bounded. It is not onto its target B since 3 is not in the range.

(b) Let $f(x) = 1/(1 - x)$. The source and target are $\mathbb{R}$. The function f can be defined for all real numbers x except $x = 1$, since when $x = 1$ the denominator $1 - x = 0$. The natural domain for f is $\mathbb{R} - \{1\}$, all real numbers except 1. Is f one-to-one? Suppose

$$f(x) = \frac{1}{1 - x} = \frac{1}{1 - y} = f(y)$$

Taking the reciprocal of both sides, we have $1 - x = 1 - y$, or $x = y$. Thus f is one-to-one. Is f onto the target $\mathbb{R}$? That is, for any real number b can we find an x such that $b = f(x) = 1/(1-x)$? Solving for x in terms of b we have $x = 1 - 1/b$. This makes sense if $b \neq 0$. Thus for any b except 0 we can find an x such that $f(x) = b$. Thus the range of f is $\mathbb{R}^*$, and f is not onto. This function is not bounded, since its range is not bounded.

(c) Let $f(x, y) = (x + y, x, y)$. The source of f is $\mathbb{R}^2$ and the target of f is $\mathbb{R}^3$. The natural domain of f is all of $\mathbb{R}^2$: it can be defined for all points (x, y) in $\mathbb{R}^2$. Is f onto? Its range is all points (u, v, w) in $\mathbb{R}^3$ such that $u = x + y$, $v = x$, and $w = y$. By eliminating x and y from these three equations we see that every point (u, v, w) in the range satisfies $u = v + w$. This set of points is a plane in $\mathbb{R}^3$. Since a plane is two-dimensional and $\mathbb{R}^3$ is three-dimensional, f cannot be onto. Is f one-to-one? Suppose $f(x, y) = (x + y, y, x) = (r + s, s, r) = f(r, s)$. Then $y = s$ since the second coordinates are equal and $x = r$ since the third coordinates are equal. Thus, if $f(x, y) = f(r, s)$ then $(x, y) = (r, s)$. So f is one-to-one. This function is not bounded, since its range, a plane, is not a bounded subset of $\mathbb{R}^3$. △

Section 1.4 Exercises

1. (a) Graph the function given by the table

$f(x)$	9	8	7	9	2
x	1	2	3	4	5

(b) Show f as a mapping.

(c) What is the domain of f?

(d) What is the range of f?

(e) Is f one-to-one?

2. (a) Graph the function given by the table

$f(x)$	-1	0	1	2	3
x	1	2	3	4	5

(b) Show f as a mapping.

(c) What is the domain of f?

(d) What is the range of f?

(e) Is f one-to-one?

3. For each of the following subsets of the Cartesian product of $A = \{1, 2, 3, 4\}$ and $B = \{a, b, c, d\}$, is the set the graph of a function, and if so is the function one-to-one or onto? If it is one-to-one, what is the graph of the inverse function?

(a) $\{(1, a), (2, b), (3, c), (4, d)\}$

(b) $\{(2,a),(3,c),(1,a),(4,b)\}$

(c) $\{(1,d),(2,c),(3,b),(4,a)\}$

(d) $\{(1,a),(1,b),(2,c),(3,d)\}$

4. Analyze the function $f(x,y) = 3x + 4y$. Give the source, target, natural domain, and range. Is it one-to-one or onto?

5. Analyze the function $f(x) = \dfrac{1}{(x-2)^2}$. Give the source, target, natural domain, and range. Is it one-to-one or onto?

6. Analyze the function $f(x) = (x^2, x+1)$. Give the source, target, natural domain, and range. Is it one-to-one or onto?

7. (a) A cow is walking along a mountain path. Her altitude at time t is $t^2 - 2t + 2$ for t between 0 and 1. Will the cow be at the same altitude at two different times?

   (b) A cow is walking along a mountain path. Her altitude at time t is $t^3 - 6t^2 + 11t + 2$ for t between 0 and 4. Will the cow be at the same altitude at two different times?

   (c) A disabled cow is being dropped cow food from a balloon. The position the food lands is given by $p(x,y) = (x+1, x-y^3+1)$ as a function of the aircraft's position. If the cow is at (x_0, y_0), where should the balloon be to drop the food?

   (d) A cow is attempting to swat a bug on her broad back with her tail. The point hit is given by $p(x,y) = (x+2, y^2+2)$ in terms of the tail's release point, (x,y). Is there a safe place for the bug?

8. Analyze the function $f(x,y) = \dfrac{1}{(x-1)(y+1)}$. Give the source, target, natural domain, and range. Is f one-to-one or onto?

9. For each of the following, give an example or explain why no example exists:

   (a) An unbounded function with domain [1,2]

   (b) An unbounded function with range $\mathbb{R}$

   (c) An unbounded function with graph $\{(x,y) : x^2 + y^2 = 1, y > 0\}$

   (d) A bounded function with domain $\{x : x < 0\}$

   (e) A bounded function with range [0,1]

   (f) A bounded function with graph $\{(x,y) : x = y\}$

   (g) A bounded function with domain $\mathbb{R}^2$

   (h) An unbounded function with domain $[-1,1] \times [-1,1]$

10. Prove or disprove that the exponential function $f : \mathbb{R}^2 \to \mathbb{R}$ defined by $f(x, y) = 2^{x+y}$, where x and y are integers, is a one-to-one function.

11. Prove or disprove that the exponential function $f(x) = 2^x$, where x is an integer, is a bounded function.

12. Prove or disprove that a function given by a table is always bounded.

13. State precisely the most general statement you can concerning the boundedness of polynomials.

14. Prove or disprove that there are as many odd integers as there are integers (see Example 1.4.15).

15. Project: Show that there is a one-to-one correspondence between the set of rational numbers and the set of integers. Some references are Howard Eves, "Lecture Thirty-four: Beyond the Finite," in *Great Moments in Mathematics (after 1650),* MAA, 1981, and E. Kamke, *Theory of Sets,* Dover, 1950.

16. Project: Show that there is no one-to-one correspondence between the set of rational numbers and the set of all real numbers. Thus, the infinity of real numbers is "larger" than the infinity of rational (or integer) numbers. See the above references.

1.5 Special Classes of Functions

In this section we will look at some important special classes of functions corresponding to the polynomials, trigonometric, exponential, and discrete classes of functions introduced in Section 1.1.

Sequences

In the introduction, sequence functions were introduced as models for sampling change. (Note that zero change is represented by a constant function.) A characteristic of a sampling function, as well as sequences in general, is the natural ordering of elements in the domain. We can speak of the "first" sample, "second" sample, "third" sample, etc.. Any activity that can be described in terms of ordered "steps" can be modeled by a sequence function. For example, a sequence of moves in chess, scheduling a day's events, constructing a team's schedule of games, preparing a meal, getting dressed, observing the growth of a bean plant, recording temperatures, or outlining a paper can all be mathematically modeled with sequence functions.

Definition 1.5.1 (Sequence) *A* sequence *is a function whose domain is a set of integers.*

A sequence is finite if its domain is a finite set and is infinite if its domain is an infinite set.

Example 1.5.2 (Finite and Infinite Sequences) (a) A student's summary grade report can be modeled with a finite sequence (finite domain, the student's semester):

$$r : \{1,2,3,4,5,6,7,8\} \to \mathbb{R}^4$$

where

$$r_n = (\text{Sem. Credits, Sem. GPA, Total Credits, Total GPA}).$$

Since the domain is finite, this is a finite sequence.

(b) The function $s : \mathbb{N} \to \mathbb{N}$ defined by $s(n) = n^2$ for all $n \in \mathbb{N}$ is an infinite sequence. It is not a bounded function, since its range $\{n^2 : n \in \mathbb{N}\}$ is not a bounded subset of $\mathbb{R}$. △

When dealing with a sequence s, the notation s_n is often used in place of $s(n)$. Thus our example above is defined by $s_n = s(n) = n^2$ for positive integers n. Often when talking about a sequence we will refer to $\{s_n\} = \text{Range}(s)$ rather than s. When we refer to $\{s_n\}$, the domain of s (the possible values for n) may not be explicitly stated. In these cases the domain will be the natural numbers $\mathbb{N}$ or occasionally the natural numbers with zero added, $\mathbb{N}_0 = \{0, 1, 2, ...\}$.

Example 1.5.3 (Fibonacci Model for Growth) We model a rabbit population in which rabbits 1 year or older are classed as "mature" and those less than 1 year old are classed as "young". Assume that all pairs of rabbits consist of one male and one female. Suppose every pair of mature rabbits has one pair of offspring every year and that young rabbits have no offspring. The offspring of one year are not counted until the next year, and they become mature (and have their own offspring) the year following that. We will assume that rabbits live indefinitely. Let y_n denote the number of pairs of young rabbits in year n, let m_n denote the number of pairs of mature rabbits in year n, and let r_n denote the total number of pairs of rabbits in year n. We want to determine the sequence $r_1, r_2, r_3, \ldots$.

We establish some relationships between r_n, m_n, and y_n. Our first relation is "the whole is equal to the sum of the parts"

$$r_n = m_n + y_n$$

Now, any rabbit, young or mature, alive (and counted) in year $n - 1$ will be mature in year n. This gives

$$m_n = r_{n-1}$$

Also, a young rabbit in year n must be the offspring of a pair of rabbits that were mature in year $n - 1$. Thus

$$y_n = m_{n-1} = r_{n-2}$$

Substituting for m_n and y_n in the first equation gives

$$r_n = r_{n-1} + r_{n-2}$$

This relation defines a *recursive* sequence. In a recursive sequence, a term is defined by an expression containing earlier terms in the sequence. To prevent infinite regression, some initial terms must be given.

Suppose in this example that there is one young pair of rabbits in year 1, so that

$$y_1 = 1 \quad m_1 = 0 \quad r_1 = 1$$

and

$$y_2 = 0 \quad m_2 = 1 \quad r_2 = 1$$

Using the recurrence relation

$$r_n = r_{n-1} + r_{n-2}$$

we compute $r_3 = r_2 + r_1 = 1 + 1 = 2$, $r_4 = 2 + 1 = 3$, $r_5 = 5$, and so on.

The recurrence relation

$$r_n = r_{n-1} + r_{n-2}$$

is called the *Fibonacci relation* and the resulting sequence $\{r_n\}$ the *Fibonacci sequence* (named after the Italian mathematician Leonardo of Pisa, called "Fibonacci," 1170-1250). The Fibonacci sequence occurs in many settings in nature and mathematics. An example is the spiral pattern of leaves around a blossom. On a oak or apple stem, there are 5 (r_5) leaves for every 2 (r_3) spiral turns; on pear stems, 8 (r_6) leaves for every 3 (r_4) turns; and on willow stems, 13 (r_7) leaves for every 5 (r_5) spiral turns. There is a mathematics journal, *The Fibonacci Quarterly*, devoted to Fibonacci and related recurrence relations. △

The technique of recursive definition is often useful. In a recursive definition, two cases must be specified: the base case and the inductive step. The base case tells us how to start the sequence. The inductive step tells us how to get one term from the previous terms. Another example of this is the *factorial* function.

Example 1.5.4 (Factorial Defined Recursively)
We define $FACT : \mathbb{N}_0 \to \mathbb{N}$ by

$$\begin{aligned} FACT(0) &= 1 \quad \text{(the base case)} \\ FACT(n) &= n\,FACT(n-1) \quad \text{for } n > 0 \text{ (the inductive step).} \end{aligned}$$

Thus $FACT(3) = 3 \cdot FACT(2) = 3 \cdot 2 \cdot FACT(1) = 3 \cdot 2 \cdot 1 \cdot FACT(0) = 3 \cdot 2 \cdot 1 \cdot 1 = 6$. $FACT(n)$, usually written $n!$, is the product of the first n natural numbers. △

Population growth models provide numerous examples of sequences in $\mathbb{R}^n$. In these examples, a population (ants, bears, maple trees, etc.) is partitioned into stages (based on age, size, function, etc.) and assumptions are made concerning birth and death rates for each stage. The population in the different stages is then predicted over a set of time periods. The sequence is the function that maps a time period into the ordered tuple of population stages.

Example 1.5.5 (Raccoon Age Distribution) Consider modeling raccoons with three age groups:

$$\begin{aligned} a &= \text{young raccoons, up to 2 years old} \\ b &= \text{midlife raccoons, between 2 and 4 years old} \\ c &= \text{old raccoons, 4 to 6 years old} \end{aligned}$$

Suppose that young raccoons do not have any offspring, each midlife raccoon gives birth to 4 young, and each old raccoon gives birth to 1 young (our model is only considering female raccoons). Suppose also that 40% of the young raccoons survive through their first 2 years to become midlife raccoons, 60% of the midlife raccoons survive to become old raccoons, and that no old raccoons survive past age 6. Finally, we assume that we initially have 100 young, 50 midlife, and 30 old raccoons.

The sequence model showing the population of each stage as well as the total population for each time period of 2 years is $s : \mathbb{N}_0 \to \mathbb{R}^4$. The following table illustrates a few values of this sequence function:

Period	Young	Midlife	Old	Total
0	100	50	30	180
1	230	40	30	300
2	190	92	24	306
3	392	76	55	523
4	359	157	46	563
5	674	144	94	916
6	669	269	86	1024
7	1162	266	161	1589
8	1232	465	160	1857
9	2021	493	279	2793
10	2250	808	295	3353
...	...	...	...	...
19	33873	9553	4602	48028
20	42815	13549	5732	62096

Note the cycling behavior in the young and midlife raccoons in the early periods and also the larger changes in the populations between the even and odd periods for young raccoons and between the odd and even periods for both the midlife and old raccoons. These observations lead to several questions concerning long-term trends: Will the cycling continue? Will the uneven growth patterns persist?

What can be said about long-term distribution of the total population between the three stages? What can be said about growth rate? If we continued the above table through several more periods, we would observe long-term population distribution converging to 70% young, 21% midlife, and 9% old. We would also see the growth rate converging to 33.4%.

Sequences and convergence questions concerning long-term behavior of sequences will play a prominent role throughout this text. △

Example 1.5.6 (Chaotic Behavior) This example is a simple model which can exhibit chaotic behavior. Let P_n be the number of insects alive in year n; P_0 is the initial population. The number of insects alive in year $n+1$ is dependent upon the number alive in year n: insects alive in one year reproduce larva which survive the winter, but the adults die during the winter. A simple model is

$$P_{n+1} = aP_n - bP_n^2$$

Here $a > 0$ represents the birthrate: the higher the birthrate, the more insects there will be in the next generation. The $b > 0$ coefficient represents the effects of crowding, which is proportional to the square of the population size. By an appropriate choice of units we can set $a = b = \lambda$ and our model becomes

$$P_{n+1} = \lambda P_n(1 - P_n)$$

[The function $f(x) = \lambda x(1-x)$ is called the *quadratic map* and will appear in several examples.] In these units $0 < P_n < 1$. The behavior of P_n as λ increases from 0 to 4 is quite surprising. When $\lambda < 1$, the population dies out; setting $\lambda = 0.5$ and $P_0 = 0.8$ we have

n	P_n
0	0.8
1	0.08
2	0.0368
3	0.0177
4	0.0087

For $3 < \lambda < 3.4$, P_n oscillates between two values. For $3.4 < \lambda < 3.57$, P_n oscillates between 4, then 8, then 16, ...values. When $\lambda > 3.57$, the behavior of P_n is chaotic: the-long term behavior of P_n is extremely sensitive to the initial value P_0. For $0 < \lambda < 3$, the long term behavior is independent of the initial value. The reader will be asked to verify these statements in the exercises. △

Example 1.5.7 (The Bisection Algorithm) The bisection algorithm is a method of generating a sequence of approximations to a number with a desired property. Suppose that we know that a number with the desired property lies in an interval $[a, b]$. We will generate a sequence of approximations $\{s_n\}$ lying in shrinking intervals $[a_n, b_n]$. The initial interval is $[a_1, b_1] = [a, b]$. Let $m = (a + b)/2$ be the midpoint of the interval $[a, b]$. Then $s_1 = m$ is the initial approximation to a number with the desired property. If we can determine which half-interval, $[a, m]$ or $[m, b]$, contains a number with the desired property, then we can reduce the uncertainty about the location of a number with the desired property by half, from being in an interval of length $b - a$ to being in an interval of length $(b - a)/2$. The bisection algorithm continues this process of bisecting the interval, obtaining a sequence $s_1, s_2, \ldots$ of approximations in shrinking intervals $[a_n, b_n]$.

We will demonstrate this process by computing an approximation to $\sqrt{2}$ which is accurate within 0.01 ($|error_n| = |\sqrt{2} - s_n|$).

Step(n)	a_n	b_n	$b_n - a_n$	s_n	$\lvert error_n \rvert$
1	1.000	2.000	1.000	1.500	0.086
2	1.000	1.500	0.500	1.250	0.164
3	1.250	1.500	0.250	1.375	0.039
4	1.375	1.500	0.125	1.438	0.024
5	1.375	1.438	0.062	1.406	0.008
6	1.406	1.438	0.031	1.421	0.007
7	1.406	1.421	0.016	1.412	0.002
8	1.412	1.421	0.008	1.412	0.002

Thus $s_8 = 1.412$ is an approximation to $\sqrt{2}$, which is accurate to within 0.01. Notice in the above table that the size of the error can increase from step to step, as between steps 1 and 2. However, the size of the error is never more than one-half the width of the interval $b_n - a_n$, since s_n is in the center of the interval $[a_n, b_n]$, which also contains the value being approximated. △

Sequences of approximations will be a major topic in Chapter 2.

Step Functions

We will now consider some functions defined on intervals. There is an important class of functions, called *step functions,* where the domain is an interval but the range is not. They will be used in the chapter on integration. Step functions are constant on intervals and then "step" up or down at the ends of the interval.

Example 1.5.8 An example of a step function is found in the postage rate tables. If we wish to mail a 10-kg package, then we might find a table of the following form:

Postage Rates

Cost	1.25	1.50	1.75
Distance	0 to 200 km	201 to 300 km	301 to 400 km

Here we have a function whose value is constant over the interval from 0 to 200 km. Immediately above 200 km the value "steps up" to 1.50.

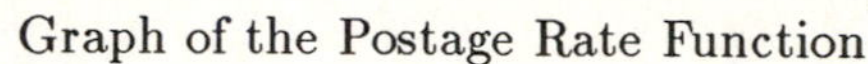

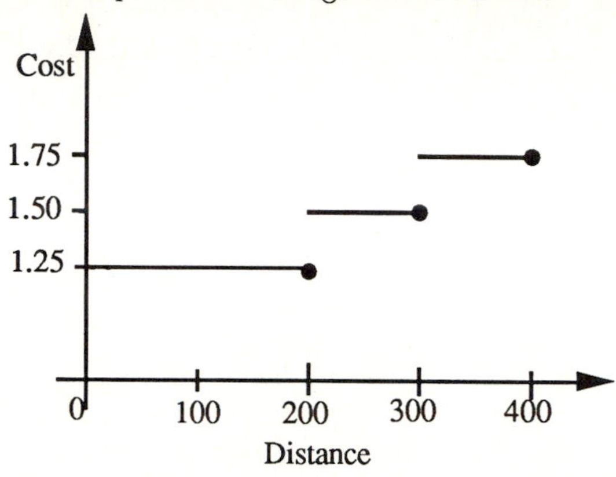

Figure 1.5.9

In this graph the symbol " " shows the value of the function at a step point. When the distance is 200, the value of the function is 1.25 not 1.50. △

Example 1.5.10 (The *FLOOR* Function) Another example of a step function is the *greatest integer (FLOOR) function.* If x is a real number, then define $[x]$ to be the greatest integer less than or equal to x. In computer science $[x]$ is called the floor of x. For example, $[1.2] = 1$, $[1.99] = 1$, $[2.00] = 2$, $[-1.69] = -2$. If we define $FLOOR : \mathbb{R} \to \mathbb{N}$ by $FLOOR(x) = [x]$ then we have a step function which has the constant value n on the interval $[n, n+1)$.

Graph of the Function FLOOR

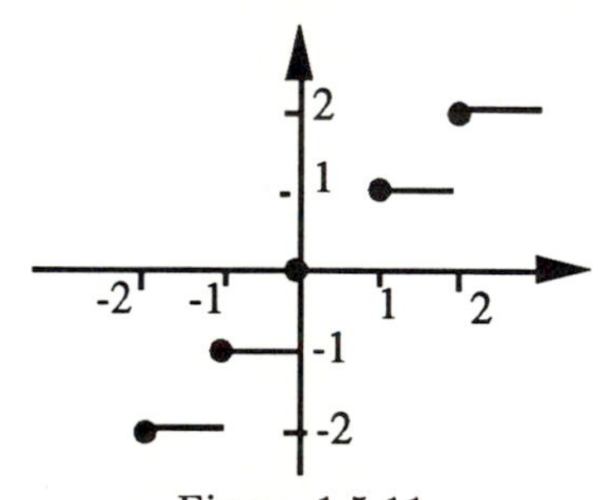

Figure 1.5.11 △

The functions in each of the last two examples were "increasing" in the sense that, as the domain values increased, the corresponding range values increased (or remained the same). Such functions are called *monotonic.* Geometrically, a monotonic function is a function from the reals into the reals whose graph does not change direction.

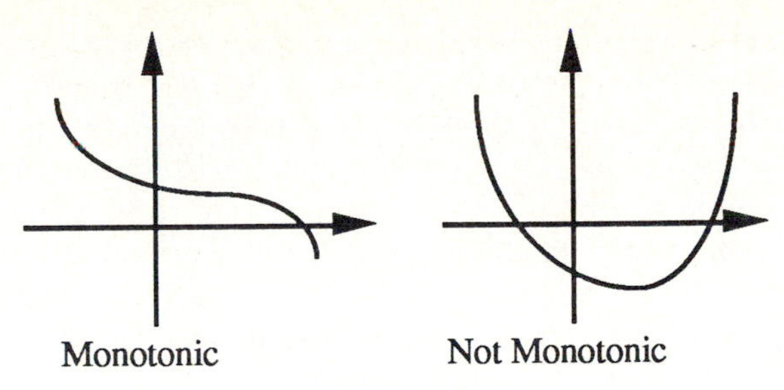

Figure 1.5.12

Definition 1.5.13 (Monotone Functions) *Let A and B be subsets of the real numbers. A function $f : A \to B$ is* strictly increasing (strictly monotonically increasing) *if $x < y$ implies $f(x) < f(y)$.*

If $x < y$ implies $f(x) \le f(y)$, then f is called increasing (monotonically increasing).

If $x < y$ implies $f(x) > f(y)$, then f is strictly decreasing (strictly monotonically decreasing).

If $x < y$ implies $f(x) \ge f(y)$, then f is decreasing *(monotonically decreasing).*

A function is called monotonic *if it is either monotonically increasing or monotonically decreasing.*

The graph of a monotonically increasing function "rises to the right" and the graph of a monotonically decreasing function "falls to the right".

Example 1.5.14 Let $f : \mathbb{R} \to \mathbb{R}$ be defined by $f(x) = x^2$. By looking at the graph of f (a parabola), it is obvious that f is not monotonic. However, if the domain of f were restricted to the negative real numbers, then f would be strictly (monotonically) decreasing. If the domain were restricted to the positive real numbers, then f would be strictly (monotonically) increasing.

$\triangle$

Example 1.5.15 (Sales Tax) The sales tax function is a monotonically increasing function. However, since it is a step function, it is not strictly increasing. Not all step functions are increasing or decreasing. For instance, the unemployment function is usually approximated by a step function (i.e., assume that unemployment is constant during each time period) which is neither increasing nor decreasing. $\triangle$

Theorem 1.5.16 *If $f : \mathbb{R} \to \mathbb{R}$ is strictly increasing, then f is one-to-one.*

Proof: Note that the form of this theorem is a conditional implication (see Section 1.2). In order to prove the theorem we must show that if the hypothesis holds, then the conclusion must hold. We start by assuming that we have a strictly increasing function f. Then what we need to show is that if $f(x) = f(y)$ for some x and y in Domain(f), then $x = y$. That is, if two points x and y are mapped to the same point, then x and y must actually be identical. Since x and y are real numbers, one of the relations $x = y$, $x < y$, or $x > y$ must hold. If $x < y$, then by the definition of "strictly increasing," $f(x) < f(y)$. This is impossible, since we have assumed that x and y map to the same point, $f(x) = f(y)$. Similarly, if $y < x$, then $f(y) < f(x)$, which is also impossible. Thus the only possibility is $x = y$, which is what we wanted to show. □

Question 1.5.17 (a) State an analogous theorem for strictly decreasing functions to the one just proved.

(b) Prove your theorem given in part (a). ◇

Example 1.5.18 (a) Let $f(x) = x^2$ on Domain(f) = $(-\infty, 0]$. f is strictly decreasing and has inverse function $f^{-1}(x) = -\sqrt{x}$, Domain(f^{-1}) = $[0, \infty)$ = Range(f).

(b) Let $g(x) = x^2$ on Domain(g) = $[0, \infty)$. g is strictly increasing and has inverse function $g^{-1}(x) = \sqrt{x}$, Domain(g^{-1}) = $[0, \infty)$ = Range(g).

(c) Let $CUBE(x) = x^3$, Domain($CUBE$) = $\mathbb{R}$. $CUBE$ is strictly increasing on all of $\mathbb{R}$ and has as an inverse the cube root function $CUBERT : \mathbb{R} \to \mathbb{R}$ given by $CUBERT(x) = x^{1/3}$, Domain($CUBERT$) = $\mathbb{R}$. △

Example 1.5.19 (Absolute Value Function) An important function whose graph is similar to that of $f(x) = x^2$, with two monotonic pieces, is the absolute value function. If x is a real number, then we define the absolute value of x by

$$|x| = \begin{cases} x & \text{if } x \geq 0 \\ -x & \text{if } x < 0 \end{cases}$$

The absolute value function $ABS : \mathbb{R} \to \mathbb{R}$ is defined by $ABS(x) = |x|$, Domain(ABS) = $\mathbb{R}$.

Graph of Absolute Value

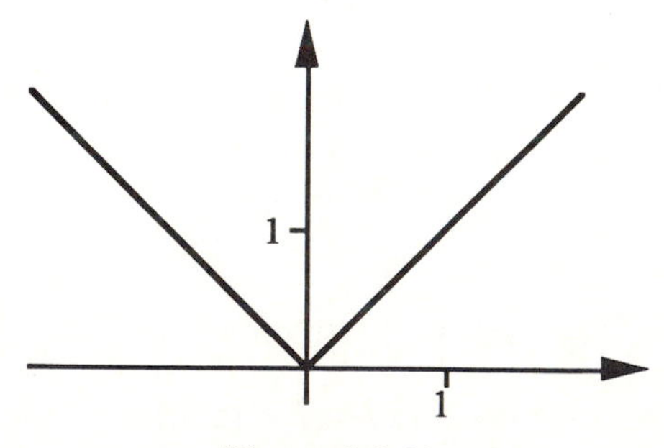

Figure 1.5.20

Recall from Section 1.2 that $ABS(a-b) = |a-b|$ is defined to be the distance between two real numbers a and b.Thus,

$$|a-b| = \begin{cases} a-b, & \text{if } a-b \geq 0 \\ -(a-b) & \text{if } a-b < 0 \end{cases}$$ △

Trigonometric Functions

This is one of the four major classes of functions discussed in the introduction. The following concept of periodicity gives trigonometric functions much of their importance.

Definition 1.5.21 (Periodic Function) *A function $f : \mathbb{R} \to B$ is* periodic *of period p if $f(x+p) = f(x)$ for all x in the domain of f, and p is the smallest positive number for which this is true.*

If $f(x+p) = f(x)$ for all x, then f repeats itself over intervals of length p. In the definition we must have $p > 0$ since $f(x+0) = f(x)$ is always true.

It is very helpful to discover periodicity in a function, since then we only need to analyze the function over one period rather than over the whole domain.

The most important examples of periodic functions are the trigonometric functions. Recall that if θ is a real number, considered as an angle in radians, then $\cos(\theta) = x$ and $\sin(\theta) = y$ where (x, y) is the point on the unit circle in the $x-y$ plane at angle θ counterclockwise from the x-axis. (If you don't recall the trigonometric functions, then see the Appendix for a brief review.)

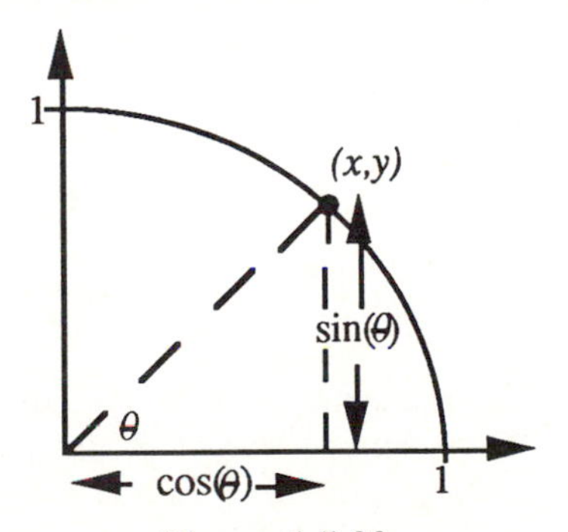

Figure 1.5.22

As θ varies from 0 to 2π radians ($\pi = 3.14159265\ldots$), $\sin(\theta)$ increases from 0 to 1 (when $\theta = \pi/2$), decreases to -1, and then increases again to 0.

Graph of Sine

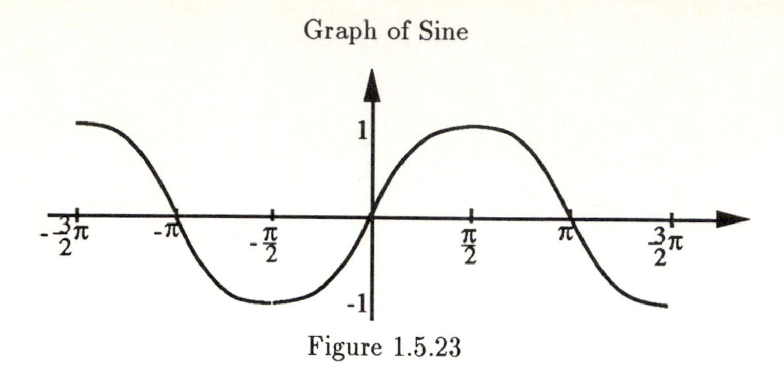

Figure 1.5.23

For the sine function, $\sin(x + 4\pi) = \sin(x)$, so sine repeats itself over an interval of length 4π. However, 4π is not the smallest value possible. By observing the graph of sine, we see that $\sin(x + 2\pi) = \sin(x)$ for all x. Sine repeats itself over an interval of length 2π and does not repeat itself over any smaller interval. Thus sine has period 2π. The maximum value of sine is 1 and occurs at $\pi/2$ within the closed interval $[0, 2\pi]$. Since sine is periodic with period 2π, the maximum value of 1 must be taken on at $\pi/2 + n2\pi$ for any integer n. Similarly, the minimum value of sine is -1 and is taken on at $3\pi/2 + n2\pi$ for any integer n.

The graph of cosine is identical to the graph of sine, shifted left by $\pi/2$ radians. Thus cosine is periodic of period 2π and has range $[-1, 1]$.

Graph of Cosine

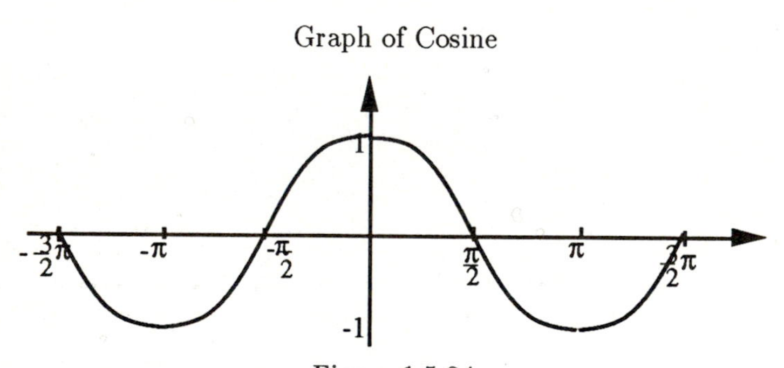

Figure 1.5.24

The other trigonometric functions are defined in terms of the sine and cosine. The tangent function $\tan : \mathbb{R} \to \mathbb{R}$ is defined by $\tan(x) = \sin(x)/\cos(x)$. The natural domain of $\tan(x)$ is all real numbers x such that $\cos(x) \neq 0$. Recall that $\cos(x) = 0$ when $x = \pi/2 + n\pi$ for any integer n. If $x > \pi/2$ and x is close to $\pi/2$, then $\cos(x)$ is a negative number near 0 and $\sin(x) > 0$, so $\tan(x)$ is very negative. If $x < \pi/2$ and x is close to $\pi/2$, then $\cos(x)$ is a positive number near 0 and $\sin(x) > 0$, so $\tan(x)$ is very positive. The graph of tangent has period π.

Graph of Tangent

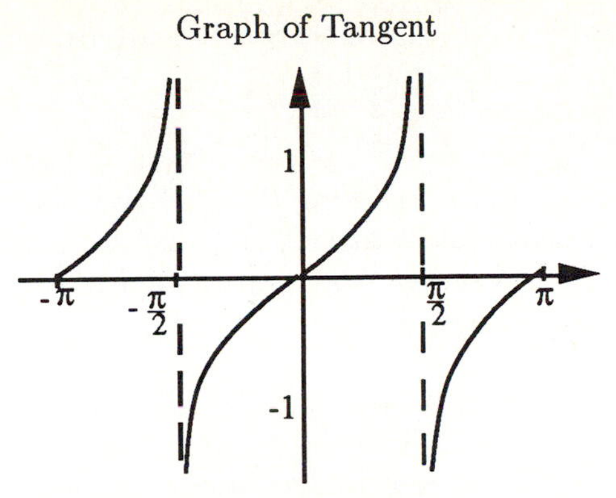

Figure 1.5.25

In the closed interval $[-\pi/2, \pi/2]$, sine is strictly increasing. Thus by Theorem 1.5.16 sine restricted to this interval is one-to-one and so has an inverse function $\sin^{-1}$, called the *arcsin*. From the definition, $\sin(x) = y$ if and only if $\sin^{-1}(y) = x$. For example, $\sin(0) = 0$ implies $\arcsin(0) = 0$, $\sin(\pi/4) = 1/\sqrt{2}$ implies $\arcsin(1/\sqrt{2}) = \pi/4$, and $\sin(-\pi/2) = -1$ implies $\arcsin(-1) = -\pi/2$. The sine function is strictly monotonic on other intervals, which also could be used to determine an inverse function, but the common choice is $[-\pi/2, \pi/2]$.

Now let's determine the graph of arcsin. As in Example 1.4.9, we need to reflect the graph of sine in the line $y = x$. Domain(arcsin) = Range(sin) = $[-1, 1]$ and Range(arcsin) = $[-\pi/2, \pi/2]$. The graph of sine, restricted to $[-\pi/2, \pi/2]$, begins at $(-\pi/2, -1)$ and increases monotonically to $(\pi/2, 1)$. The graph of arcsin will begin at $(-1, -\pi/2)$ and increases monotonically to $(1, \pi/2)$.

Graph of Sine and Arcsine

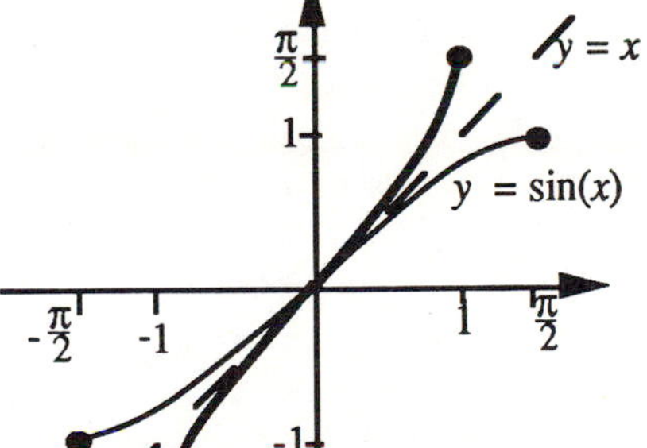

Figure 1.5.26

Inverse functions for the other trigonometric functions can be obtained in a similar manner.

Question 1.5.27 (a) How can we restrict the domain of cosine to obtain a one-to-one function?

(b) If we call the inverse of the restricted function arccos, what are arccos(+1), arccos($1/\sqrt{2}$), arccos(0), and arccos(−1)? ◊

The most important relation involving the sine and cosine functions is the Pythagorean theorem, which says that $\sin^2(t) + \cos^2(t) = 1$ [see the discussion of relationship (1) in the Appendix]. Note: $\sin^2(t) = (\sin(t))^2$, which is different from $\sin(t^2)$ and from $\sin(\sin(t))$.

Example 1.5.28 Consider the function $f : \mathbb{R} \to \mathbb{R}$ given by $f(x) = \cos(4x)$, Domain$(f) = \mathbb{R}$. The function f is closely related to the cosine function. As x varies between 0 and $\pi/2$, $4x$ varies between 0 and 2π. Since the period of cosine is 2π, f has a period of $\pi/2$. Thus the period of $f(x) = \cos(4x)$ is one-fourth the period of cosine. △

Polynomial Functions

The polynomial category of functions was one of four function categories described in the introduction (see Section 1.1). Polynomials are important because we can easily compute their values, since they are defined in terms of the operations of addition and multiplication. In fact, polynomials are essentially the only large class of functions whose values we can easily compute. Computations with other types of functions, such as the trigonometric functions, are usually reduced to computing with polynomials that approximate the desired function. We will see how this is done later.

Polynomials are finite sums of terms where each term consists of a coefficient and one or more variables raised to nonnegative integer powers. The sum of the powers of the variables in a term is called the degree of the term. The *degree of a polynomial* is defined to be the largest degree of all the terms.

Polynomial functions are of the form $p : \mathbb{R}^n \to \mathbb{R}$ for $n \geq 1$ where the natural domain is all of $\mathbb{R}^n$. For example, $p(x, y, z) = 3x^2y^7z^3 - 13x^4 + 4z$ is a polynomial function of degree 12 mapping $\mathbb{R}^3$ into $\mathbb{R}$.

Zero degree polynomials are *constant functions*. These are the functions that model zero change. First-degree polynomials model constant (nonzero) change. In one variable, first-degree polynomials, $p(x) = mx + b$, are particularly nice. They are unbounded, strictly monotonic, one-to-one, and onto functions. Their graphs are straight lines. The rate of change of a first-degree polynomial is called the *slope* of the line. The slope of a straight line in the $x - y$ plane is the change in the y-coordinate divided by the change in the x-coordinate. The Δ (delta) symbol is usually used to denote change. If we take the points (x_1, y_1) and (x_2, y_2), then the slope is

$$\frac{\Delta y}{\Delta x} = \frac{y_2 - y_1}{x_2 - x_1} = \frac{mx_2 + b - mx_1 - b}{x_2 - x_1} = m$$

A vertical line (x = constant) is said not to have a slope or is said to have an infinite slope.

Since the slope of the graph of a first-degree polynomial is just the tangent of the angle formed by the straight line and the positive x-axis, the slope is sometimes referred to as the "direction" of the curve (graph). Because each (nonvertical) straight line has a unique slope, two points (e.g., the x and y intercepts) or a point and slope will uniquely determine a straight line.

Example 1.5.29 (a) We will find the equation of the straight line passing through the points (2,3) and (7,1). Let (x, y) be an arbitrary point on the desired line. Since the line has a unique slope, we have

$$\text{slope} = \frac{y-3}{x-2} = \frac{1-3}{7-2} = -\frac{2}{5}$$

Thus

$$y - 3 = -\frac{2}{5}(x-2)$$

or

$$y = -\frac{2}{5}x + \frac{19}{5}$$

(b) We will find the equation of the line with slope 4 that passes through the point (3,5). Again, since the slope of the desired line is unique, we have

$$\text{slope} = \frac{y-5}{x-3} = 4 \quad \text{or} \quad y = 4x - 7$$

△

The correct terminology for a general first-degree polynomial, $p(x) = mx + b$, is *affine* function. Since the graph of an affine function is a straight line, the term "linear" is sometimes applied to affine functions. But strictly speaking, a linear function must map zero to zero, which implies that the constant term must be zero and the graph must pass through the origin.

A concept related to that of linear function is the *linear combination* of functions. If f_1, f_2, ..., f_n are functions, then a linear combination of these functions is a function g of the form

$$g = c_1 f_1 + \cdots + c_n f_n$$

where c_1, ..., c_n are constants. A polynomial is a linear combination of power functions. For example,

$$g(x) = 3x^4 + 5x^2$$

is a linear combination of $f_1(x) = x^4$ and $f_2(x) = x^2$ with constants 3 and 5. We will see that the integral and derivative of a linear combination of functions can be computed if we know the derivative or integral of the individual functions.

We may also speak of affine functions of several variables. The graph of a first-degree polynomial (affine function) in two variables is a plane (see Figure 1.4 for a diagram).

Higher Degree Polynomials

Quadratic polynomials are those of the second degree: the exponents of the variables in any term add to 2 or less. If p is a polynomial of one variable, then it is quadratic and is of the form $p(x) = ax^2 + bx + c$ for some real constants a, b, and c, $a \neq 0$.

In two variables a polynomial of the second degree has the form

$$p(x, y) = ax^2 + bxy + cy^2 + dx + ey + f$$

where a, b, c, d, e, and f are fixed real numbers. The terms ax^2, bxy, and cy^2 are all of second degree, since the sum of the exponents of the variables in each of these terms is two.

A polynomial of degree n in one variable x has the form

$$p(x) = a_n x^n + a_{n-1}x^{n-1} + \cdots + a_1 x + a_0$$

where a_0, a_1, ..., a_{n-1}, and $a_n \neq 0$ are fixed real numbers. We can express this sum using the *summation notation.* This is a way of writing summations in a compact form, where the subscript is allowed to vary. In our example,

$$p(x) = \sum_{k=0}^{n} a_k x^k = a_0 + a_1 x + a_2 x^2 + \cdots + a_{n-1}x^{n-1} + a_n x^n$$

Here we let k take on all the integer values between 0 and n. These $n + 1$ values of k replace the "k" in the general term $a_k x^k$, giving $n + 1$ terms $a_0 x^0$, $a_1 x$, $a_2 x^2, \ldots,$ $a_n x^n$. We then add the $n + 1$ terms together.

Rational Functions

Definition 1.5.30 (Rational Functions) *A* rational function *is a ratio of polynomials.*

The primary difficulty with rational functions is that the denominator will vanish at the roots of the polynomial in the denominator.

Example 1.5.31 (a) Consider $f(x) = (x^2 + 4x + 3)/(x^2 + x - 2) = (x^2 + 4x + 3)/((x - 1)(x + 2))$. The polynomial in the denominator vanishes at $x = 1$ and $x = -2$, so the natural domain of f is $\mathbb{R} - \{1, -2\}$. We will be able to sketch the graph f after we have studied continuity in the next chapter.

(b) The function $f(x, y) = (x^2+y^2)/(x^2y+x^2+y+1) = (x^2+y^2)/((x^2+1)(y+1))$ is a rational function of two variables. The polynomial in the denominator vanishes only when $y = -1$, thus the natural domain is all of $\mathbb{R}^2$ except the line $y = -1$. △

Example 1.5.32 (Supply-Demand Curves) In economics, the point of intersection of the *supply* and *demand* curves is called the *equilibrium point*, the point where supply and demand are equal. We shall show that when the supply and demand curves are straight lines, the coordinates of the equilibrium point are given by a function of four variables. The situation is pictured below.

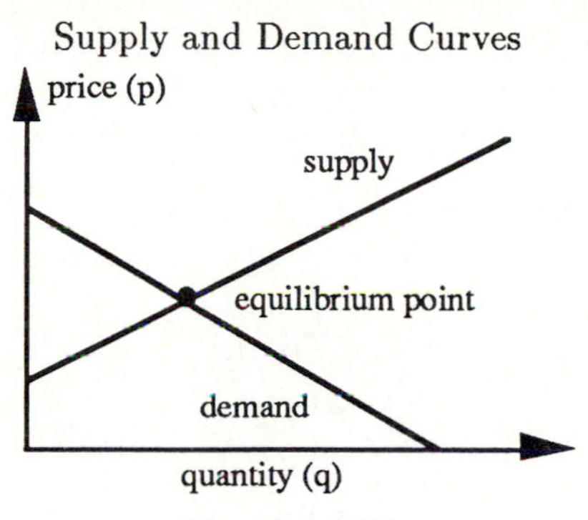

Figure 1.5.33

We will let p denote price and q denote quantity. Then we have

$$\text{supply curve:} \quad p_s = m_s q + c_s$$

$$\text{demand curve:} \quad p_d = m_d q + c_d$$

where m_s and m_d are the slopes of the supply and demand curves respectively. We now solve for the q coordinate of the equilibrium point. At this point $p_s = p_d$ and thus the right-hand sides of the above equations must be equal. We have

$$\begin{aligned} p_s = p_d \quad &\Rightarrow \quad m_s q + c_s = m_d q + c_d \\ &\Rightarrow \quad q = \frac{c_d - c_s}{m_s - m_d} \end{aligned}$$

The value of the corresponding p coordinate of the equilibrium point is found by substituting the expression for q into the equation for either the supply or demand curve. The result (using the supply curve) is $p = m_s \frac{c_d - c_s}{m_s - m_d} + c_s$. The equilibrium point function

$$EP : \mathbb{R}^4 \to \mathbb{R}^2$$

is defined by

$$EP(m_s, m_d, c_s, c_d) = (\frac{c_d - c_s}{m_s - m_d}, \frac{m_s(c_d - c_s)}{m_s - m_d} + c_s)$$

The interested economics student can use this EP function to discuss the (economic) effects of changing one or more of the input variables. △

Exponential and Logarithm Functions

Although exponential functions may be defined for several variables, we will be primarily concerned with exponential functions of a single variable. That is, functions of the form $f : \mathbb{R} \to \mathbb{R}$ defined by $f(x) = b^x$, where $b > 0$ and Domain$(f) \subseteq \mathbb{R}$. The number b is called the *base* of the exponential function.

We will first consider exponential functions defined on the integers. We saw in Example 1.4.4 that $f(x) = 2^x$ with domain $\mathbb{N}$ was strictly increasing. If the base b of the exponential function is greater than 1, then $f(x) = b^x$ with domain $\mathbb{N}$ is strictly increasing. If $b < 1$, then $f(x) = b^x$ is strictly decreasing.

Graphs of $f(x) = 2^x$ and $g(x) = (\frac{1}{2})^x = 2^{-x}$

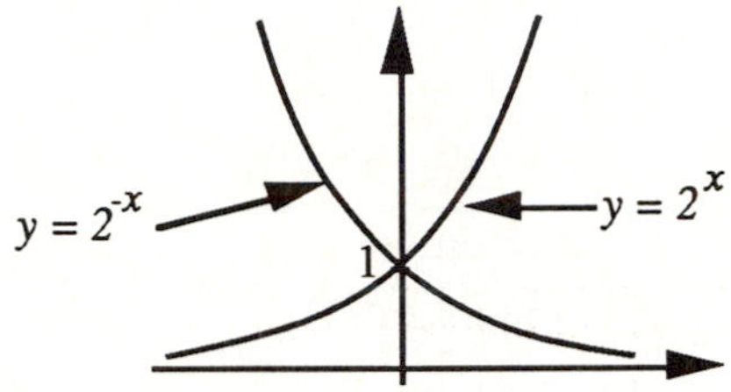

$y = 2^{-x}$ $y = 2^x$ 1

Figure 1.5.34

Since the exponential function is one-to-one, it has an inverse function, which is called the *logarithm function* and written log. Recall that the graph of an inverse function can be obtained by reflecting ("flipping") the graph of the original function in the line $y = x$.

Graph of Exponential and Log

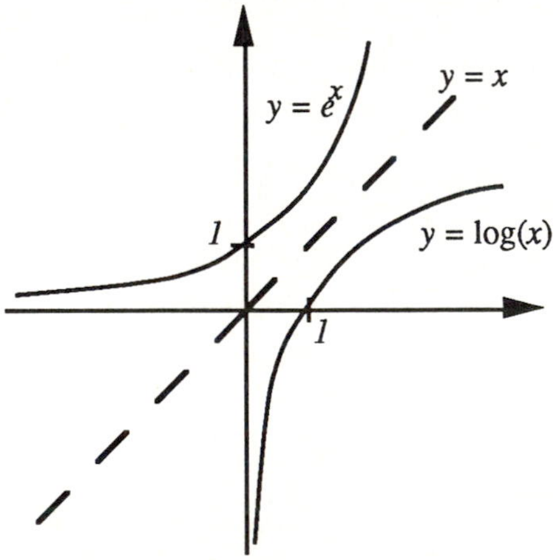

$y = e^x$ $y = x$ 1 $y = \log(x)$ 1

Figure 1.5.35

Note that the logarithm function is unbounded both below and above, whereas the exponential function was only unbounded above.

In elementary algebra, 10 is usually chosen as the base for an exponential function, while in computer science, 2 is usually the base. However, as we shall see in Chapter 7, the irrational number e ($e = 2.7182\ldots$) is the usual base in mathematics. Thus, $f(x) = e^x$ is referred to as *the* exponential function (or natural exponential function), when the base is not explicitly given. Similarly, the inverse function, $\log_e(x)$, defined by

$$e^x = y \quad \text{if and only if} \quad \log_e(y) = x$$

is usually denoted by $\log(x)$ or $\ln(x)$ and is called the (natural) logarithm function.

The exponential expression and (common, base 10) logarithm expression are related by

$$10^x = y \quad \text{if and only if} \quad x = \log_{10}(y)$$

For bases other than 10, this can be extended to

$$b^x = y \quad \text{if and only if} \quad x = \log_b(y) \quad (b > 0 \quad \text{and} \quad b \neq 1)$$

Let $b > 1$. Initially we can define b^x for x an integer. How can we extend the definition to rational numbers (filling in the gaps in the graph of b^x)? If r is a rational number, then $r = m/n$ where m and n are integers and $n > 0$. For the law of exponents to hold, we need

$$b^r = b^{m/n} = (b^m)^{1/n}$$

Thus we can define rational exponentials if we can determine roots, $x^{1/n}$. We have seen a way to approximate any root using the bisection algorithm in Example 1.5.7. Thus we can extend the domain of $f(x) = b^x$ from the integers $\mathbb{N}$ to the rational numbers $\mathbb{Q}$. The next step is to extend the domain of the exponential function to all of $\mathbb{R}$. We will see an elegant way to define the exponential function in Chapter 7, but let us consider how the problem might be approached using the idea of approximations.

As a concrete example, suppose we wish to approximate $2^{\sqrt{2}}$. We can approximate $\sqrt{2}$ as in Example 1.5.7 to within 0.01 as $1.412 = 1412/1000$. We now can approximate $2^{\sqrt{2}}$ as $2^{1412/1000}$.

Section 1.5 Exercises

1. Give the 10th term of the Fibonacci sequence.

2. Consider the recursively defined sequence $DEQ : \mathbb{N} \to \mathbb{R}$, given by $DEQ(1) = 2$, $DEQ(2) = -1$, and $DEQ(n) = DEQ(n-1) + 2DEQ(n-2)$ for $n = 3, 4, 5, \ldots$. Determine the first five terms of this sequence.

3. A function related to $FLOOR$ is the *least integer (ceiling) function.* If x is a real number, then define $CEIL(x)$ to be the smallest integer greater than or equal to x. For example, $CEIL(2.3) = 3$, $CEIL(-2.3) = -2$, $CEIL(1) = 1$. What are $CEIL(10.1)$, $CEIL(-2.9)$, $CEIL(-2.1)$, and $CEIL(9.8)$?

4. For each of the following, give an example or state why no example can be found:

   (a) A bounded sequence
   (b) An unbounded sequence
   (c) A sequence that has no upper bound and no lower bound
   (d) A finite sequence
   (e) A periodic sequence

5. Sketch the graphs of the following functions. Consider the domain to be the interval $[-3, 4]$.

   (a) The absolute value function
   (b) The greatest integer or $FLOOR$ function
   (c) The ceiling function (see Exercise 3 above)
   (d) The sine function
   (e) The cosine function

6. Sketch the graphs of the following functions.

   (a) The arcsine function over its natural domain
   (b) The arccosine function over its natural domain
   (c) The arctangent function over its natural domain
   (d) $f : [-2, 7] \to \mathbb{R}$ where $f(x) = \sin(x) + \cos(x)$
   (e) $m : [-3, 4] \to \mathbb{R}$ where $m(x) = FLOOR(x) + CEIL(x)$

7. Mark each of the following statements true (T) or false (F).

   (a) The greatest integer function is a step function.
   (b) $FLOOR(x + 1) = CEIL(x)$.
   (c) The ceiling function is a one-to-one function.
   (d) No nonconstant, even- degree polynomial is monotonic.
   (e) Every polynomial is a bounded function over its natural domain.
   (f) Every odd-degree polynomial is a one-to-one function.
   (g) The range of a polynomial of odd degree is all of $\mathbb{R}$.
   (h) $\sin(x + y) = \sin(x) + \sin(y)$
   (i) An even degree polynomial can be bounded below, but not above.
   (j) $\sec^2(x) = \tan^2(x) + 1$

8. For any of the following functions that are periodic, give the period.

   (a) $f(x) = \sin(3x)$ (b) $g(x) = 3x + \sin(x)$

   (c) $f(x) = (\sin(x))(\cos(x))$ (d) $g(x) = \sin^2(x)$

   (e) $h(x) = \sin^2(x) + \cos^2(x)$ (f) $g(x) = 3\cos(5x)$

9. Find an affine function whose graph passes through

   (a) (1,1) and (4,0).

   (b) $(-1, 0)$ and (2,3).

10. Sketch the graphs of the affine functions:

    (a) $f(x, y) = 2x + 3y + 6$

    (b) $f(x, y) = 3x - 4y + 12$

11. A first-degree polynomial, $p(x) = ax + b$, changes direction zero times; a second-degree polynomial, $p(x) = ax^2 + bx + c$, changes direction once.

    (a) How many times can a polynomial of degree 3 change directions?

    (b) How many times can a polynomial of degree 4 change directions?

    (c) How many times can a polynomial of degree n change directions?

12. What is the natural domain of the rational function

    (a) $f(x) = (3x^3 + 9x^2 + 2x + 1)/(x^2 + 4x + 3)$

    (b) $f(x) = (x^3 + 3x^2 + 5)/(x^2 + 6x + 9)$

    (c) $f(x, y) = (xy^3 + 3xy + 5)/(xy + y - 2x - 2)$

    (d) $f(x, y) = (x^2 + xy^2 + 4x + 1)/(xy - 3y + x - 3)$

13. For each of the following, give an example or explain why no example exists.

    (a) A linear function of three variables.

    (b) An affine function of two variables that is also a linear function of two variables.

    (c) A polynomial in one variable that is of degree 4 and has no second-degree term.

    (d) An affine function of one variable that has no inverse.

    (e) A third-degree polynomial in one variable that is one-to-one.

    (f) A rational function whose domain is all of $\mathbb{R}$.

    (g) An affine function whose inverse is not an affine function.

14. For each of the following, give an example or state why no example can be found.

    (a) A polynomial of degree n whose graph does not touch the x-axis. Does it make a difference if n is even or odd?

    (b) A polynomial of degree n whose graph crosses the x-axis exactly once. Does it make a difference if n is even or odd?

    (c) A polynomial of degree n whose graph crosses the x-axis exactly k times ($k \leq n$). Does it make a difference if n is even or odd?

    (d) A polynomial of degree n whose graph touches the x-axis exactly once, but does not cross it. Does it make any difference if n is even or odd?

    (e) A polynomial of degree n whose graph touches but does not cross the x-axis exactly k times. What restriction needs to be placed on k?

15. For what value of x is

    (a) $\log(x) = 1$ (b) $\log(x) = 0$

16. Simplify the expressions:

    (a) $\log(e^{x+2})$ (b) $e^{\log(x^2+2)}$

17. Assuming the "standard" laws of exponents, prove that

    (a) $\log(ab) = \log(a) + \log(b)$

    (b) $\log(a/b) = \log(a) - \log(b)$

    (c) $\log(a^n) = n\log(a)$ [Hint: Let $u = \log(a)$ and $v = \log(b)$, then $e^u = a$ and $e^v = b$.]

18. Is $f : \mathbb{R} \to \mathbb{R}$ defined by $f(x) = 1/x$ a monotonic function? Why?

19. Mark each of the following statements true (T) or false (F).

    (a) The functions $f : \mathbb{R} \to \mathbb{R}$ defined by $f(x) = x^2$ and $g : \mathbb{R} \to \mathbb{R}$ defined by $g(x) = \sqrt{x}$ are inverses of one another.

    (b) The functions $f : \mathbb{R} \to \mathbb{R}$ defined by $f(x) = x^3$ and $g : \mathbb{R} \to \mathbb{R}$ defined by $g(x) = x^{1/3}$ are inverses of one another.

    (c) If $\sin(a) = b$, then $\arcsin(b) = a$.

    (d) If $\arcsin(a) = b$, then $\sin(b) = a$.

    (e) The inverse of the sine function is a bounded function.

20. (a) A cow pond has a boat ramp which is 10 feet along the ramp and extends to a height of 3 feet above the water. If the edge of the water is at $x = 0$, what function will model a side view of the area where the ramp enters the water?

(b) A cow barn roof has a saw-tooth shape with three teeth: $\triangle$. Each tooth has a base of 10 meters and a height of 3 meters. Construct a function to model the roofline. (Include the domain of the function.)

21. A number x_0 is called a *fixed point* of a function f if $f(x_0) = x_0$. Thus 0 is a fixed point of $f(x) = x^2$ and 3 is a fixed point of $f(x) = x^2 - 6$. Finding a fixed point of f is equivalent to finding a solution to the equation $f(x) = x$.

    (a) What does it mean graphically to solve the equation $f(x) = x$? (Hint: Consider the graphs of the two examples above.)

    Find the fixed points for the following functions.

    (b) $f : \mathbb{R} \to \mathbb{R}$ defined by $f(x) = |x|/x$

    (c) $f : \mathbb{R} \to \mathbb{R}$ defined by $f(x) = x^2 - 2$

    (d) $f : \mathbb{R}^2 \to \mathbb{R}^2$ defined by $f(x, y) = (\sin(x), FLOOR(y))$

    (e) $f(x) = \cos(x)$. (Approximate a fixed point to an estimated seven digits of accuracy; see Exercise 22. We will discuss the theory of approximation in Chapter 2.)

22. Consider the function $f(x) = \tan(x) - 1$.

    (a) Graph f to see that it must have a fixed point between 0 and $\pi/2$.

    (b) Observe that solving $f(x) = x$ is equivalent to solving $y = y(x) = f(x) - x = 0$. Since $y(0) = -1$ and $y(1.4) \approx 3.40$, it seems reasonable that the graph $y = y(x)$ must cross the x-axis somewhere between 0 and 1.4. (In Chapter 2 will study the theoretical basis for such an assumption.) We can use the bisection algorithm to estimate where the crossing point, and thus our fixed point, lies. Starting with $a_1 = 0$ and $b_1 = 1.4$ as the initial interval, successively divide the interval in half, restricting your attention to the half containing the crossing point (zero of y). Thus $s_1 = (0 + 1.4)/2 = 0.7$ and $y(0.7) \approx -0.86$. Thus the zero of y must lie in the interval between 0.7 and 1.4. Continue this process (with the aid of your calculator or computer) until the interval containing the zero of y has a length of less than 0.002. Then the midpoint of this interval must be within 0.001 of the fixed point.

23. What is the relation between the graphs of $f(x) = 2^x$, $g(x) = e^x$, and $h(x) = 3^x$?

24. Extend Figure 1.5.35 by adding to the same set of axes the graphs of $y = x^3$, $y = x^{1/3}$, $y = x^5$, and $y = x^{1/5}$.

25. Superimpose the graphs of $y = x^n$ for several values of n, both positive and negative.

(a) What effect does increasing the size of n have on the graph of $y = x^n$?

(b) What effect does decreasing the size of n have on the graph of $y = x^n$?

26. Superimpose the graphs of $y = 2^x$, $y = 3^x$, $y = 4^x$, $y = e^x$, and $y = x$ on the same set of axes. What effect does changing the size of n have on the graph of $y = e^x$?

27. Superimpose the graphs of $y = e^x$, $y = e^{1/x}$, $y = e^{-x}$ on the same set of axes. Explain the behavior of the graph of $y = e^{1/x}$.

28. Consider the statement: (*) "A periodic function is a bounded function."

    (a) Write statement (*) as a conditional statement.

    (b) Show that the conditional statement of part (a) is not an implication by showing that the "hypothesis true – conclusion false" case can exist. Hint: Consider the tangent function.

    (c) As a result of parts (a) and (b), explain why the statement (*) is not a mathematical theorem.

29. Consider the statement: (*) "The graph of a polynomial of degree n can touch the x-axis in at most n places".

    (a) Write statement (*) as a conditional statement.

    (b) Give an indirect proof of the conditional statement in part (a). Hint: Assume that f is a polynomial of degree n and that the graph of f touches the x-axis at points $\{p_1, p_2, \ldots, p_{n+1}\}$. Now argue that the degree of f must be at least the degree of $g(x)$ where $g(x) = (x - p_1)(x - p_2)\cdots(x - p_{n+1})$.

    (c) As a result of parts (a) and (b), explain why statement (*) is a mathematical theorem.

30. Project: Define a "comfort" function of two or more variables for stairways. Let one of the variables represent the slope of the stairway. Let the range of your function be the set {uncomfortable, comfortable, dangerous}. (Your definition should be based on your analysis of at least 10 different stairways.)

31. This exercise examines the model of chaotic growth, Example 1.5.6. These exercises can best be done with a graphing program or with a program which allows repetition.

    (a) Choose several values of λ between 0 and 1. Verify that P_n dies out as n increases.

    (b) Choose $\lambda = 2$ and several other values of λ between 1 and 3. How does P_n behave as n increases? How is the behavior dependent upon the initial population P_0 or the value of λ? (See Exercise 32 below.)

(c) Choose several values of λ between 3 and 3.57. How does P_n behave as n increases? The behavior should be periodic. How is the behavior dependent upon the initial population P_0 or the value of λ?

(d) Choose several values of λ greater than 3.57. How does P_n behave as n increases? How is the behavior dependent upon the initial population P_0 or the value of λ?

32. The function involved in Example 1.5.6 and the above exercise is the quadratic map $f(x) = \lambda x(1-x)$, $x \in [0,1]$. Some of its behavior can be explained by examining its fixed points (see an earlier exercise in this section for an introduction to fixed points).

    (a) Graph the quadratic map. Note that it is a parabola with the axis along $x = 0.5$. Any fixed point must lie on the intersection of $y = x$ with $y = f(x)$. What fixed point does f always have?

    (b) Find the other fixed point of f in [0,1] as a function of λ. How is this fixed point related to the behavior investigated in Exercise 31(b)?

1.6 Operations on Functions

Transforming the Graph — Shifting and Stretching

There are four transformations we will consider: vertical shifts, horizontal shifts, vertical stretches, and horizontal stretches. We will use the function $y = f(x) = x^2$ to introduce the concepts. The initial parabola is shown on the left-hand side of Figure 1.6.1.

1.

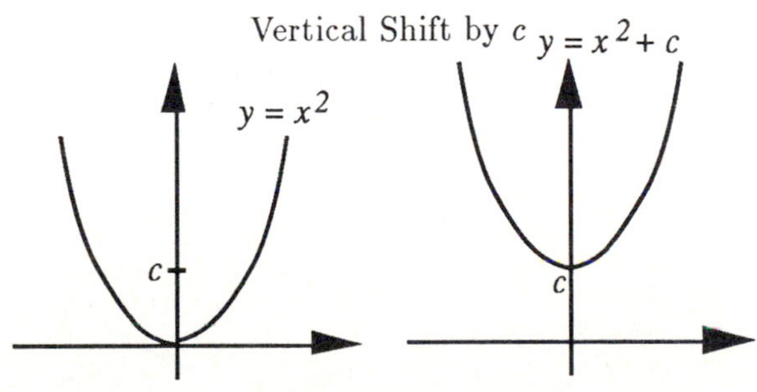

Figure 1.6.1

To shift the graph of the function vertically by c units, we simply add c to the function, obtaining $y = x^2 + c$ whose graph is on the right-hand side of Figure 1.6.1. Note that if c is negative, the graph will be shifted down.

2.

Horizontal Shift by c

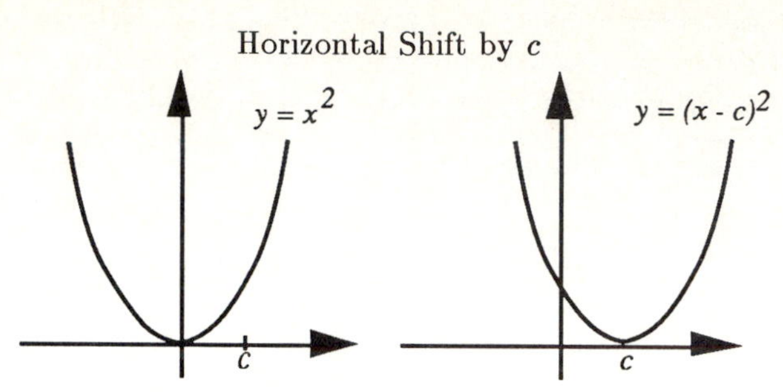

Figure 1.6.2

To shift the graph of a function horizontally, the natural thing to try is to replace the variable with the variable plus a constant. If we want to shift $y = x^2$ one unit to the right, should we add $+1$, obtaining $y = (x+1)^2$? No, we want the root to be at $x = 1$, but $y = (x+1)^2$ has a root at $x = -1$. Thus we should add -1 to shift the graph one unit to the right. In general, to shift to the right by c units, we subtract c from the variable x: $y = (x-c)^2$. The graph is shown in Figure 1.6.2.

3.

Vertical Stretch by Factor c

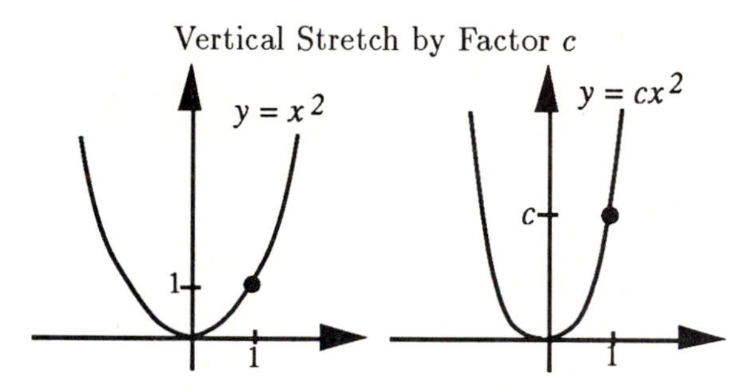

Figure 1.6.3

To change the y value by a factor c, we multiply the function by c. Our example $y = f(x) = x^2$ becomes $y = cf(x) = cx^2$.

4.

Horizontal Stretch by Factor c

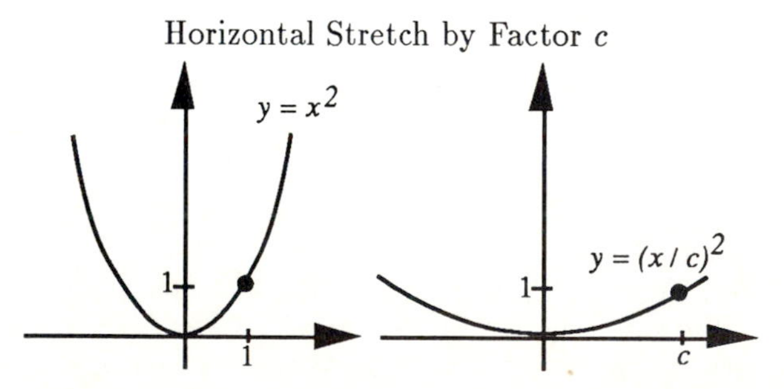

Figure 1.6.4

The horizontal stretch transformation is a bit more difficult to determine. Initially, when $y = x^2$, $y = 1$ when $x = 1$. To have $y = 1$ when $x = 3$, then we want to stretch the graph horizontally by a factor of 3. If we try $y = (3x)^2$, then when $x = 3$, $y = 81$: we used the wrong factor! If we try $y = (x/3)$, then $y = 1$ when $x = 3$. In general, to stretch the graph horizontally by a factor c, we divide the variable by c. Note that if $c < 0$, then the graph will be stretched and reflected in the y-axis.

Let's combine transformations. Naturally, the order in which the operations are done effects the result. For example, if we shift left by 2 units and then stretch horizontally by a factor of 3, we have

$$y = x^2 \qquad y = (x+2)^2 \qquad y = (\frac{x}{3} + 2)^2$$

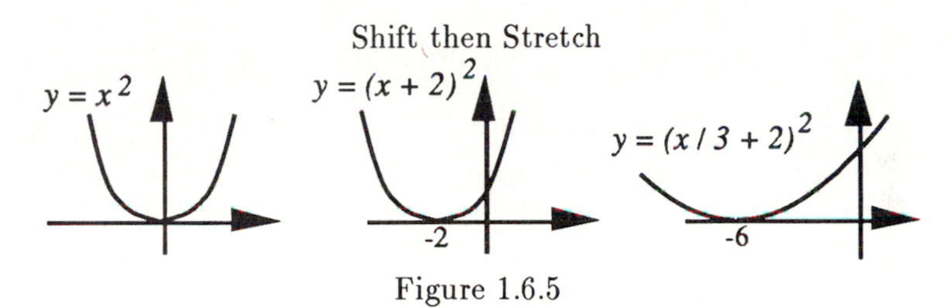

Figure 1.6.5

If we stretch horizontally by a factor of 3 then shift left by 2 units, we have:

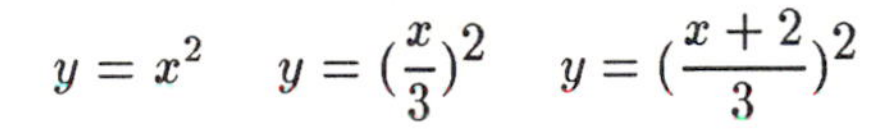

$$y = x^2 \qquad y = (\frac{x}{3})^2 \qquad y = (\frac{x+2}{3})^2$$

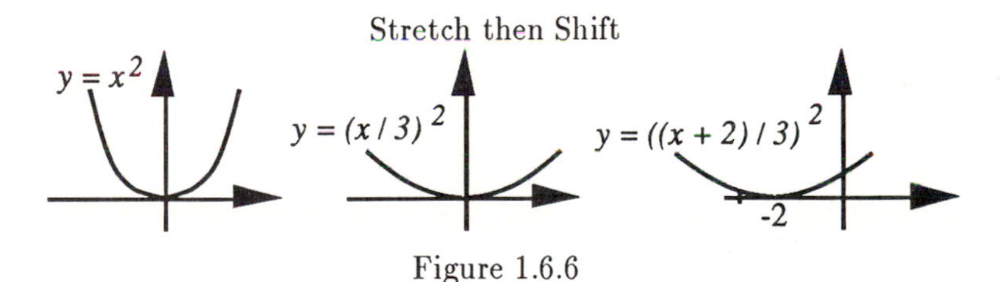

Figure 1.6.6

Example 1.6.7 We will construct a periodic function whose range is $[-3, 5]$ and whose period is 7. We will start with the basic periodic function, sine. The period of sine is 2π and the period of f defined by $f(x) = \sin(x/c)$ is $2\pi c$. (Why?) We want $2\pi c = 7$, or $c = 7/(2\pi)$. So $f(x) = \sin(2\pi x/7)$ has period 7. The range of f is the range of sine, $[-1, 1]$, which has a width of 2 units. To get a range of $[-3, 5]$ we need a width of 8 units. We can stretch the graph of f vertically by a factor of 4. Let $g(x) = 4f(x) = 4\sin(2\pi x/7)$. The function g has period 7 and range $[-4, 4]$. We want the range to be $[-3, 5]$, so we can shift g vertically 1 unit. Define $h(x) = g(x) + 1 = 4\sin(2\pi x/7) + 1$. Now h has period 7 and range $[-3, 5]$. $\triangle$

Example 1.6.8 The trigonometric functions are important in many physical applications, where the basic sine function is used to describe a wave. (For example, radio waves, water waves, etc.) If we set

$$f(x) = A\sin[B(x - C)]$$

then A is called the *amplitude,* B is called the *frequency,* and C is called the *phase (shift)* of the wave. △

Example 1.6.9 We will sketch the graph of $f(x) = 3\sin(2x+1) - 5$. We first rewrite the equation to place it in the form $f(x) = A\sin[B(x - C)]$, i.e., $f(x) = 3\sin(2[x - (-1/2)]) - 5$. The graph of f is the graph of sine that has been: stretched horizontally by a factor of $\frac{1}{2}$ (since $B = 2$), giving a period of π; then shifted left by $\frac{1}{2}$ units (since $C = -\frac{1}{2}$); then stretched vertically by a factor of 3 (since $A = 3$) giving a range interval length of 6; and finally shifted vertically down by 5 units (because of the -5).

The range of f will be $[-3-5, 3-5] = [-8, -2]$. Since $\sin(x)$ is at the midpoint of its range when $x = 0$, f is at the midpoint of its range when $2(x - \frac{-1}{2}) = 2x + 1 = 0$, i.e., when $x = -\frac{1}{2}$.

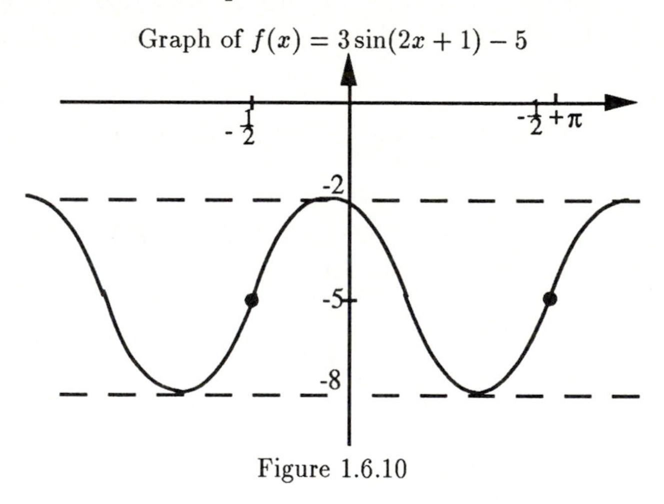

Figure 1.6.10 △

Question 1.6.11 Hours of Sunlight. Determine a function that maps a day of the year into the (approximate) maximum number of hours of sunlight that Bangor, Maine, can experience on that day. Sketch the graph of your function. The following values for the number of hours of sunlight have been rounded off to full days.

Date	Day	Hrs. of Sunlight	
Mar. 21	80	12	(spring equinox)
June 22	174	15	(summer solstice)
Sept. 21	264	12	(fall equinox)
Dec. 22	356	9	(winter solstice)
Mar. 21	445	12	(spring equinox)

Hint: Consider transforming a sine curve to fit the data for the time period from spring equinox to the following spring equinox, [80, 445], and then shifting the graph 80 units to the left. ◇

Composition of Functions

The previous transformations on functions were related to changes in the graph of the function. We will now consider ways of combining functions to get new functions. One way is to *compose* functions that have compatible range and domain.

Definition 1.6.12 (Composition of Functions) *If we have two functions $f : A \to B$ and $g : B \to C$, then we can define a new function, the* composition *$g \circ f$ of f and g, where $g \circ f : A \to C$ is defined by $g \circ f(a) = g(f(a))$ if $f(a)$ is in Domain(g). The domain of the composition is Domain$(g \circ f) = \{a : a \in$ Domain(f) and $f(a) \in$ Domain$(g)\}$. Note that we must fit together the range of f and the domain of g.*

Example 1.6.13 Consider the two functions f and g with $f : A \to B$ and $g : B \to C$.

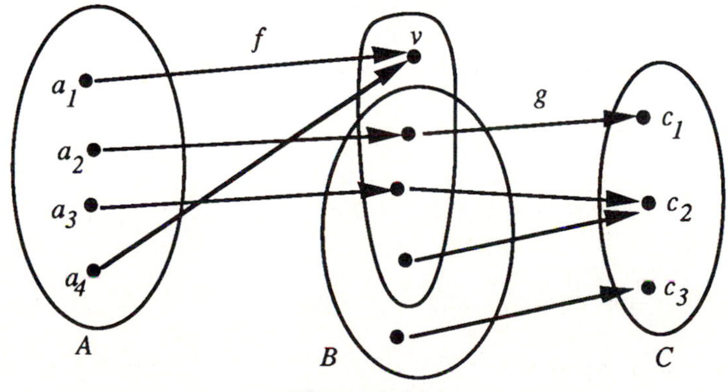

Figure 1.6.14

The composite function $g \circ f : A \to C$ has domain $\{a_2, a_3\}$.

$$g \circ f : A \to C$$

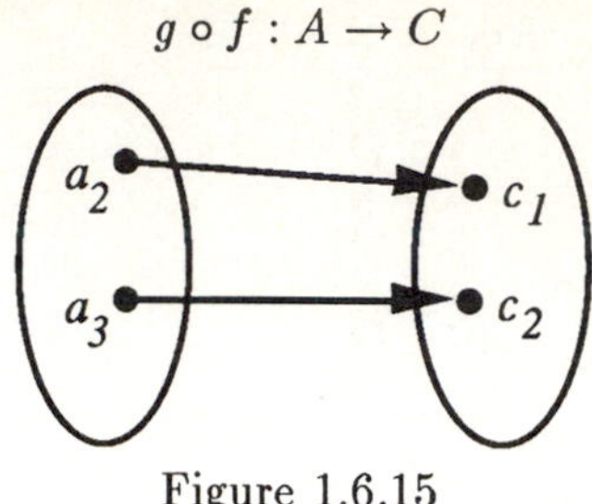

Figure 1.6.15

△

Note that when we write $g \circ f$, the function g is applied to the result of first applying f, that is, $g \circ f(x) = g(f(x))$. The order of application is the opposite of the order in which they are written.

Example 1.6.16 Let $f : \mathbb{R} \to \mathbb{R}$ be defined by $f(x) = x^2 + x$ and $g : \mathbb{R} \to \mathbb{R}$ be defined by $g(x) = x^3$. Then

$$\begin{aligned} g \circ f(x) &= g(f(x)) = g(x^2 + x) = (x^2 + x)^3 \\ &= (x^2 + x)(x^4 + 2x^3 + x^2) \\ &= x^6 + 3x^5 + 3x^4 + x^3 \end{aligned}$$

If we compose the functions in the other order, we obtain an entirely different function:

$$f \circ g(x) = f(g(x)) = f(x^3) = (x^3)^2 + (x^3) = x^6 + x^3$$

△

Example 1.6.17 We will find the composition of the sequence s, $s_n = 1/n$, where n is a positive integer, and the function $SQ : \mathbb{R} \to \mathbb{R}$ defined by $SQ(x) = x^2$ for all real numbers x. Let $f = SQ \circ s$. Since SQ is defined for all real numbers and the range of s is a subset of real numbers, there is no difficulty fitting together the range of s and the domain of SQ, Domain(f) = Domain(s) = $\mathbb{N}$. For a positive integer n,

$$f(n) = SQ \circ s(n) = SQ(s_n) = SQ(\frac{1}{n}) = \frac{1}{n^2}$$

△

Example 1.6.18 (The Spiral Function) Define $SPIRAL : \mathbb{R} \to \mathbb{R}^3$ by $SPIRAL(t) = (3\cos(t), 3\sin(t), t)$. If we label the coordinate axes x, y, and z,then the range of $SPIRAL$ is a spiral of radius 3 in $\mathbb{R}^3$ space whose axis is the z-axis. The range of $SPIRAL$ looks like a coiled spring.

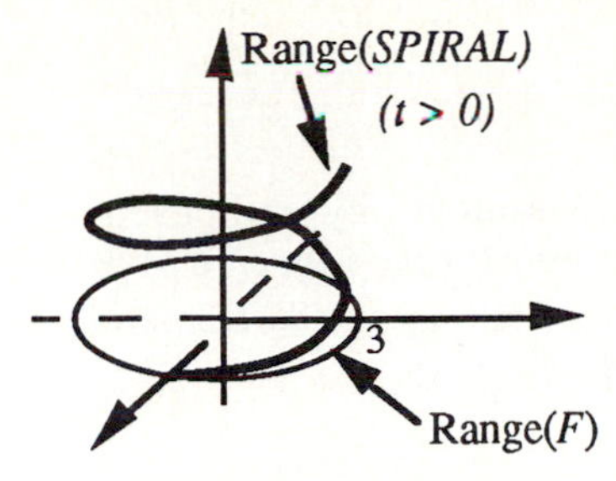

Figure 1.6.19

A composition may make this clearer. Let $PROJXY : \mathbb{R}^3 \to \mathbb{R}^2$ be defined by $PROJXY(x, y, z) = (x, y)$. $PROJXY$ is the projection from $\mathbb{R}^3$ onto $\mathbb{R}^2$, where we project onto the $x - y$ plane (see Example 1.3.7 in Section 1.3). If we compose these two functions,

$$F = PROJXY \circ SPIRAL : \mathbb{R} \to \mathbb{R}^2$$

then $F(t) = (3\cos(t), 3\sin(t))$. If we set $F(t) = (x(t), y(t)) = (3\cos(t), 3\sin(t))$, then for all real t,

$$x(t)^2 + y(t)^2 = 9\cos^2(t) + 9\sin^2(t) = 9(\cos^2(t) + \sin^2(t)) = 9$$

The range of F is the circle of radius 3 in the $x-y$ plane. The range of $SPIRAL$ (for $t > 0$) is thus above the circle about the origin of radius 3. △

Example 1.6.20 (Compositions of Functions Defined by Tables) We will determine the composite function $f \circ g$ and list the elements in its domain where functions f and $g : \mathbb{R} \to \mathbb{R}$ are defined by the following tables.

x	$g(x)$
3	7
7	2
2	8
5	3
4	10

t	$f(t)$
−2	1
3	2
5	3
7	0
10	−4
8	−5

Applying the definition of composition yields

x	$f(g(x))$
3	0
2	−5
5	2
4	−4

$$\text{Domain}(f \circ g) = \{3, 2, 5, 4\}$$

If we extend the above example by composing $\log_2$ with $f \circ g$, we have $\text{Domain}(\log_2 \circ f \circ g) = \{5\}$ and $\text{Range}(\log_2 \circ f \circ g) = \{1\}$. Composing the exponential function $h(x) = 2^x$ with $f \circ g$ yields $\text{Domain}(h \circ f \circ g) = \{3, 2, 5, 4\}$ and $\text{Range}(h \circ f \circ g) = \{1, 1/32, 4, 1/16\}$. △

Example 1.6.21 (Composition of Functions Defined by Graphs) Consider the functions f and g given by

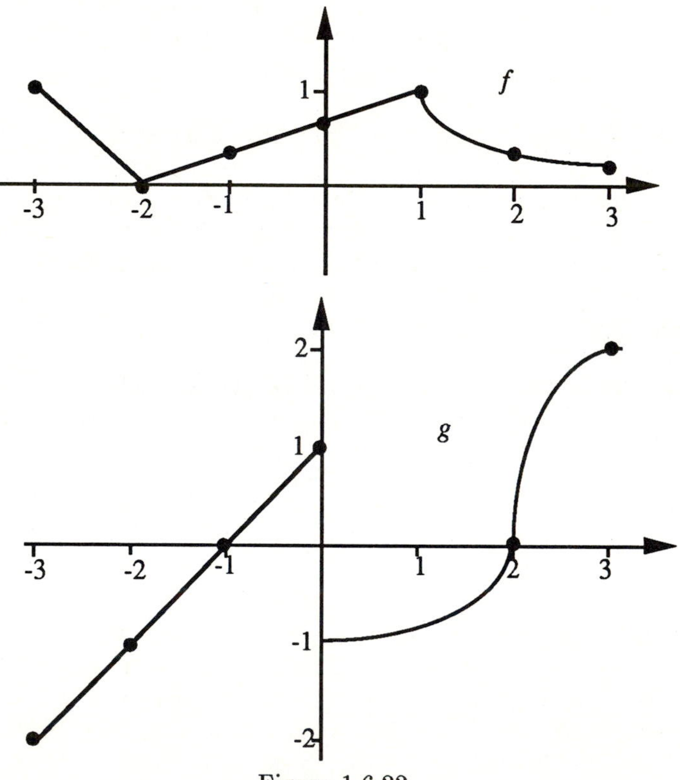

Figure 1.6.22

We can sketch the graph of $f \circ g$ on $[-3, 3]$ by plotting $f \circ g$ at the points $-3, -2, \ldots, 2, 3$. This gives

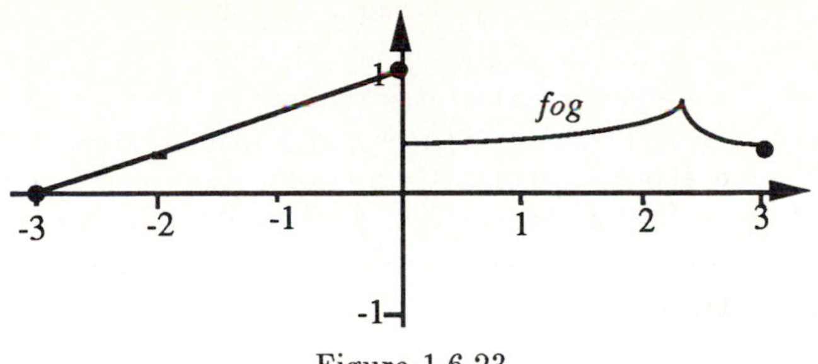

Figure 1.6.23 △

Warning: If a function is composed with itself, $f \circ f$, then the notation f^2 is sometimes used. Thus $f^2(x) = f(f(x))$. This notation is also used to denote the square of the function, as when we write $\cos^2(x) = \cos(x)\cos(x)$. These two notations are inconsistent; however, the meaning can usually be determined from the context.

When a function is composed with itself, we say that the function is *iterated*. A computer or calculator with a function key affords an interesting way to generate recursive sequences through iteration. The following example illustrates the process with the square root function.

Example 1.6.24 (Iterating the Square Root Function) Let x_0 be a positive number. Let $f(x) = \sqrt{x}$. We will generate the sequence $f(x_0)$, $f(f(x_0))$, $f(f(f(x_0)))$, etc. The sequence can be defined recursively by letting x_0 be the initial value (the base case) and letting $x_n = f(x_{n-1})$ (the inductive step). With a calculator (or computer) enter x_0 and then press the square root key repeatedly. The screen displays form a recursive sequence. If $x_0 = 2$, then the first few terms of the recursive sequence are

$$2, 1.414213562, 1.189207115, 1.090507733, 1.044273782, 1.021897149, 1.010889286, 1.005429901, 1.002711275, 1.00135472, 1.000677131, 1.000338508, \ldots$$

In this example, 2 is the base case and applying the square root function is the inductive step. △

The elements in the recursive sequence of the previous example seem to be converging (i.e., getting close to) to one. Is this result dependent or independent on the choice of 2 for the base case? The reader should experiment redoing the previous example using other values for the base case and then form a conjecture concerning the long term behavior of these sequences.

The iteration process for generating a recursive sequence has an interesting graphical interpretation. We will illustrate it for the previous example. Draw the graphs of the square root function, $y = f(x) = \sqrt{x}$, and the identity function, $y = i(x) = x$ on the same set of axes. Starting with the base case, x_0, move vertically to $(x_0, f(x_0))$ on the square root function curve and then move horizontally to $(f(x_0), f(x_0))$ on the identity function curve. This point represents the first iteration. Now repeat this two step process, i.e., move vertically (from $(f(x_0), f(x_0)))$ to $(f(x_0), f(f(x_0)))$ on the square root function curve and then move horizontally to $(f(f(x_0)), f(f(x_0)))$ on the identity function curve. This point represents the second iteration. The continuation of this process yields what is called a "cobweb" graph.

Cobweb Graph for the Square Root Function

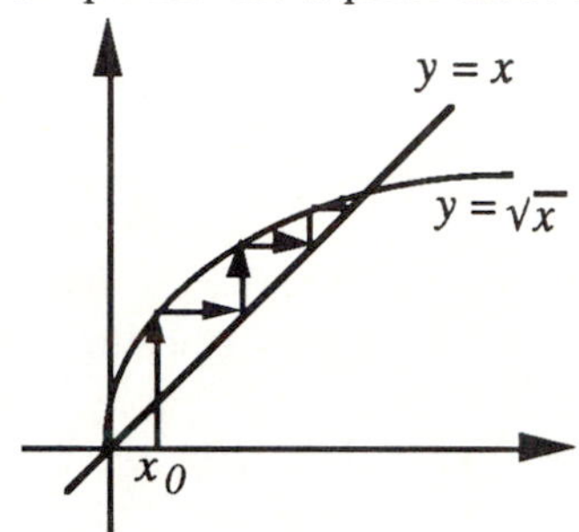

Figure 1.6.25

The reader should give a geometric argument based on the above cobweb graph that the sequence in the previous example does converge to one.

There are further opportunities for work with recursive sequences formed by iteration in the exercises.

Arithmetic Operations on Functions

If we have two functions $f : A \to \mathbb{R}$ and $g : A \to \mathbb{R}$ with the same domain and the reals as target, then we have a final way of combining them by using the *arithmetic operations* of addition, subtraction, multiplication, and division. Whenever we have a set of elements that we can combine using operations like addition and multiplication, then we say that we have an *algebra.* We use the algebra of real numbers as an example of how to define an algebra of real valued functions.

Definition 1.6.26 (Algebra of Functions) *Assume $f : A \to \mathbb{R}$ and $g : A \to \mathbb{R}$, with common domain D contained in A.*

Define $(f+g) : A \to \mathbb{R}$ by $(f+g)(a) = f(a) + g(a)$ for every a in D.

Define $(f-g) : A \to \mathbb{R}$ by $(f-g)(a) = f(a) - g(a)$ for every a in D.

Define $(fg) : A \to \mathbb{R}$ by $(fg)(a) = f(a)g(a)$ for every a in D.

If $g(a) \neq 0$ for all a in D, then we define $(f/g) : A \to \mathbb{R}$ by $(f/g)(a) = f(a)/g(a)$ for every a in D.

Example 1.6.27 Consider the function $SQ : \mathbb{R} \to \mathbb{R}$ defined by $SQ(r) = r^2$ for all real r. The function cosine is also defined on the reals, so we can construct the functions SQ + cos, $SQ -$ cos, and $SQ \cdot \cos : \mathbb{R} \to \mathbb{R}$. For example,

$$\begin{aligned} (SQ \cdot \cos)(0) &= SQ(0)\cos(0) = 0 \cdot 1 = 0 \\ (SQ + \cos)(\frac{\pi}{4}) &= SQ(\frac{\pi}{4}) + \cos(\frac{\pi}{4}) \\ &= \frac{\pi^2}{16} + (\frac{1}{\sqrt{2}}) \end{aligned}$$

Since $SQ(0) = 0$, we cannot divide cosine by SQ. We can, however, restrict our domain to the nonzero real numbers, $\mathbb{R}^* = \{x \in \mathbb{R} : x \neq 0\}$. Define $RATIO : \mathbb{R}^* \to \mathbb{R}$ by $RATIO(x) = \cos(x)/SQ(x)$ for x in $\mathbb{R}^*$. △

Section 1.6 Exercises

1. (a) Give an example of a periodic function of period 5.
   (b) Give an example of a periodic function of period 1.
   (c) Describe (the graph of) the function $f(x) = \sin(x + 1)$. What is the range?
   (d) Describe (the graph of) the function $f(x) = 3\sin(x)$. What is the range?
   (e) Describe (the graph of) the function $f(x) = 4\sin(x) + 2$. What is the range?
   (f) Describe (the graph of) the function $f(x) = 3\sin(x) + 2$. What is the range?

2. Given that $f(x) = \sqrt{x - 2}$, $g(x) = 3x + 1$, and $h(x) = 4 - x^2$, evaluate each of the following.

   (a) $(f \circ g)(x)$ (b) $(g \circ f)(x)$ (c) $(f \circ h)(x)$

   (d) $(h \circ f)(x)$ (e) $(g \circ g)(x)$ (f) $(h \circ f \circ g)(x)$

3. Given that $f(x) = 2x + 1$ for x in $(-3, 4]$, $g(x) = 3 - x^2$ for x in $[-2, 3)$, and $h(x) = 7 - 3x$ for x in [1,5], evaluate the following. State "undefined" if the composition is not defined.

   (a) $(f \circ g)(1)$ (b) $(f \circ h)(1)$ (c) $(g \circ f)(1)$

   (d) $(h \circ f)(2)$ (e) $(f \circ f)(2)$ (f) $(h \circ h)(1)$

   (g) $(f \circ (g \circ h))(2)$ (h) $(f \circ (h \circ g))(2)$ (i) $(h \circ (h \circ g))(-1)$

   (j) $(h \circ (g \circ f))(-1)$ (k) $(g \circ (g \circ g))(1)$ (l) $(h \circ (h \circ h))(2)$

4. Let the functions f and g be defined by the following graphs.

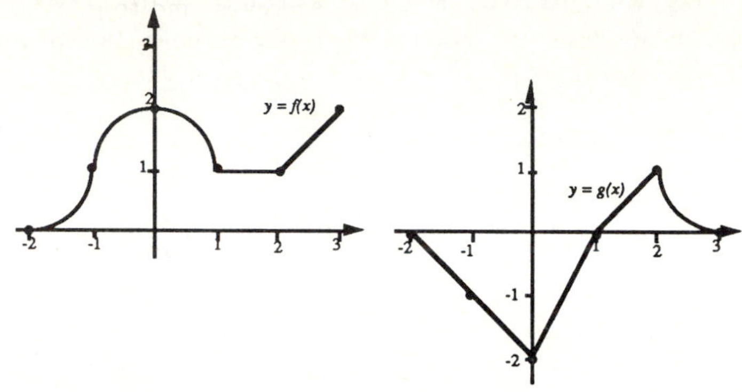

Figure 1.6.28

Evaluate

(a) $f(g(-1))$ (b) $f(g(1))$ (c) $f(g(2))$

(d) $g(f(1))$ (e) $g(f(3))$

5. Using the functions f and g defined in Exercise 4, sketch the graph of $f+g$.

6. Using the functions of Exercise 4 above:

   (a) Sketch the graph of $f \circ g$.

   (b) Sketch the graph of $g \circ f$.

7. Given the graphs of the functions f and g on $[-3, 3]$:

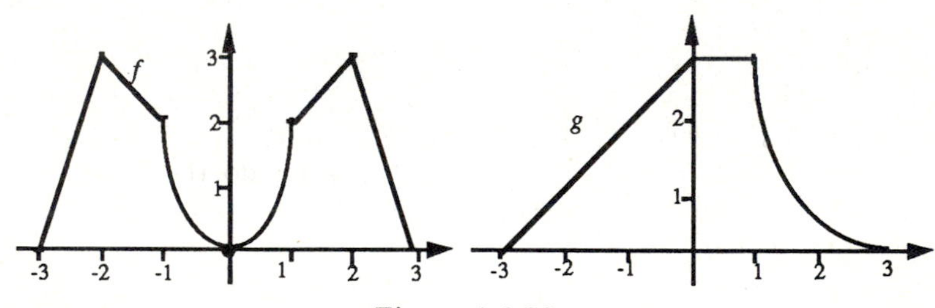

Figure 1.6.29

   (a) Sketch the graph of $f \circ g$.

   (b) Sketch the graph of $g \circ f$.

8. (a) Give an example of two affine functions whose product is an affine function.

   (b) Give an example of two polynomials whose ratio is a rational function with natural domain $\mathbb{R} - \{1, 2\}$.

Exercises 9 through 12 are designed to reinforce the discussion on transformations of graphs by leading you to discover the properties, and to prepare you for the next group of exercises. You may wish to use a computer to help with the graphing, and to reread the discussion *after* doing the exercises.

9. Reflections of graphs.

   (a) Sketch the graphs of $f(x) = x^2$, $f(x) = -x^2$, $f(x) = x^3$, and $f(x) = -x^3$ on the interval $[-2, 2]$. How does the graph of $y = -f(x)$ depend geometrically upon the graph of $y = f(x)$? Can you describe how to get from one to the other by a reflection? Test your conclusions on $f(x) = \sin(x)$ and $f(x) = \log_{10}(x)$.

   (b) Sketch the graphs of $f(x) = x^2$, $f(x) = (-x)^2$, $f(x) = x^3$, and $f(x) = (-x)^3$ on the interval $[-2, 2]$. How does the graph of $y = f(-x)$ depend geometrically upon the graph of $y = f(x)$? Can you describe how to get from one to the other by a reflection? Test your conclusions on $f(x) = \sin(x)$ and $f(x) = (x+1)^2$.

10. Horizontal and vertical translations.

    (a) Sketch the graphs of $f(x) = x^2$, $f(x) = (x-1)^2$, $f(x) = (x-2)^2$, $f(x) = (x+2)^2$, and $f(x) = (x+1)^3$ on the interval $[-3, 3]$. How does the graph of $y = f(x+a)$ depend upon the graph of $y = f(x)$ and a? Test your conjecture with $f(x) = x^{1/2}$ and several positive and negative values of a.

    (b) Sketch the graphs of $f(x) = x^2$, $f(x) = x^2 + 1$, $f(x) = x^2 - 2$, and $f(x) = x^3 + 1$ on the interval $[-3, 3]$. How does the graph of $y = f(x) + a$ depend upon the graph of $y = f(x)$ and a? Test your conjecture with $f(x) = x^{1/2}$ and several positive and negative values of a.

    (c) Combine your results from parts (a) and (b) to make a statement on how the graph of $y = f(x+a) + b$ depends upon the graph of $y = f(x)$, a and b. (Again, you should do this *without* rereading the section on transformation of graphs.) Test your conclusion with $f(x) = \log_{10}(x)$, $a = 1$, $b = 2$.

11. Scaling (stretching) graphs.

    (a) Sketch the graphs of $f(x) = x^2$, $f(x) = 2x^2$, $f(x) = (x-1)^2$, $f(x) = 2(x-1)^2$, $f(x) = \log_{10}(x)$ and $f(x) = 2\log_{10}(x)$ on $[-2, 2]$. How does the graph of $y = af(x)$ depend upon the graph of $y = f(x)$ and a. How does the sign of a effect the result? Test your answer to this question on $f(x) = (x-1)^{1/2}$, $a = -2$. and $f(x) = (x+1)^2$, $a = \frac{1}{2}$.

    (b) Sketch the graphs of $f(x) = x^2$, $f(x) = (2x)^2$, $f(x) = (\frac{1}{2}x)^2$, $f(x) = (x-1)^2$, $f(x) = (2x-1)^2$, $f(x) = \log_{10}(x)$ and $f(x) = \log_{10}(2x)$ on $[-2, 2]$. How does the graph of $y = f(ax)$ depend upon the graph of

$y = f(x)$ and $a > 0$? How does the sign and magnitude of a affect the result? Test your answer to this question on $f(x) = (x-1)^{1/2}$, $a = -2$.

12. Based on your results in Exercises 3 through 5, how do the graphs of $y = f(x) + a$, $y = f(x+a)$, $y = af(x)$, and $y = f(ax)$ depend upon the graph of $y = f(x)$ and a. How does the sign of a affect the result?

In Exercises 13 through 16, sketch the graph of the function (consider the function as a transformation of a standard function). (You should *not* use a graphing program for these.)

13. $f(x) = 2(x+9)^2$.

14. $f(x) = 3(4x+8)^3$.

15. $f(x) = 3\cos(4x-5)+2$.

16. $f(x) = 5\cos(2x+5)-1$.

17. Using the function f defined in Exercise 4, sketch the graph of $y = f(x-2)$.

18. Using the function g defined in Exercise 4, sketch the graph of $y = 3g(2x)$.

In Exercises 19 through 22, write a transformed function to obtain the desired results.

19. Transform $f(x) = x^2$ so that the minimum is at (1, 2).

20. Transform $f(x) = x^4$ so that the minimum is at $(3, -4)$.

21. Transform $f(x) = \sin(x)$ so that the period is two and the range is $[2, 4]$.

22. Transform $f(x) = \cos(x)$ so that the period is 4 and the range is $[-2, 3]$.

23. (a) A cow pond has waves whose tops are 9 feet apart and whose tops are 2 feet above the troughs. Find a function to model the surface of this storm-wracked pond.

    (b) A cow's back is left by a wasp, which flies outward, spiraling away from the cow higher and higher (tracing out a cylindrical path). The wasp disappears from sight in 10 seconds, when it is 30 feet (horizontally) from the cow and at an altitude of 20 feet. Find a function to model the position of the wasp at time t.

(c) A cow's water basin is 2 feet wide and 6 feet long. The vertical cross-sections perpendicular to the length are parabolas; the vertical cross-sections parallel to the length are trapezoids; the horizontal cross-sections are rectangles. The large end has a depth of 2 feet and the small end has a depth of 1 foot. Construct a function which will give the depth as a function of the xy-coordinates measured from a corner at the small end.

(d) A cow has an infestation of lice. Initially (in the morning of the first day) there was but one louse, but the number tripled by the end of each day, until at the end of the sixth day an ointment was applied. This treatment reduced the number of lice linearly until by the end of the tenth day all the lice were gone. Construct a function which gives the number of lice on the cow at the end of a day.

24. Consider the statement: (*) "The product of two affine functions is an affine function".

    (a) Write statement (*) as a conditional statement.

    (b) Show that the conditional statement in part (a) is not an implication by showing that the "hypothesis true—conclusion false" case can exist. Hint: Find two affine functions whose product is not an affine function.

    (c) As a result of parts (a) and (b), explain why statement (*) is not a mathematical theorem.

25. Project: Let $f(x) = ax^2 + b$ be a polynomial of degree 2. Under what conditions (if any) can there be a polynomial g of degree 3 such that $f \circ g = g \circ f$?

Exercises 26 through 28 involve iteration.

26. Given a function $f : \mathbb{R} \to \mathbb{R}$ a sequence can be obtained by *iteration*. Start with a real number x_0 in the domain of f. Define $x_1 = f(x_0)$, $x_2 = f(x_1)$, and, in general, $x_n = f(x_{n-1})$.

    (a) Find the first four iterates (x_1 through x_4) of $f(x) = 2x$ with $x_0 = 1$, expressed as decimal integers.

    (b) Find the first four iterates of $f(x) = 2^x$ with $x_0 = 1$, expressed as decimal integers.

27. Let $f(x) = -x+1$. Is the sequence of iterations $f(x), f(f(x)), f(f(f(x))), \ldots$ periodic? Explain.

28. Carefully draw the graphs of $f(x) = x^2 - 2$ and $y = x$ on the same set of axes. Using a cobweb graph, describe the behavior of the iterates of $f(x_0)$ for

    (a) $x_0 = -2.5$ (b) $x_0 = 0$ (c) $x_0 = 2.5$

29. Using a computer or a calculator with a function key and a repeat key, determine the behavior of the iterates of the following functions. Choose various initial values x_0, compute iterates as described in Exercise 26 and then graphically display the results with cobweb graphs. Do you find any consistent behavior?

    (a) $f(x) = \sin(x)$

    (b) $f(x) = \cos(x)$

    (c) $f(x) = 2^x$

30. This exercise involves the quadratic map introduced in Example 1.5.6. The values P_n are the iterates of the quadratic map, $f(x) = \lambda x(1-x)$.

    (a) Draw the cobweb graph for f when $\lambda = 0.5$. Does it display the behavior discussed in Example 1.5.6?

    (b) Draw the cobweb graph for f when $\lambda = 2$. Does it display the behavior discussed in Exercises 31(b) and 32 of Section 1.5?

Chapter 2

LIMITS AND CONTINUITY

2.1 The Need for Approximations

Approximation is a central theme in calculus. It gives a solution to the problem of computing difficult to define quantities: find an easily computed quantity which is sufficiently close to the desired quantity.

Both in the development of concepts and in numerous applications, a crucial step often involves approximating a given expression to within a stated accuracy. As you know, a function f is a rule that assigns a definite point $f(x)$ in the range of f to each point x in the domain of f. To find the value of $f(x)$ exactly, we must know x exactly. This point seems trivial until we realize that in most situations we have only approximations for x available! This reality often arises as a result of imperfections in measuring devices and other data-gathering mechanisms. More importantly, as we shall see, the necessity of approximation is an artifact of the number system and cannot be avoided. The following historical digression will provide some insights into the problem.

During the 6th century B.C. the Greek scholar Pythagoras founded a school for the study of philosophy, mathematics, and natural science at Crotona, in southern Italy. Central to the Pythagorean philosophy was the notion that whole numbers are responsible for various qualities of man and matter. These scholars knew a great deal about the properties of the integers and ascribed mystical properties to numbers they found interesting. (The famous Pythagorean Theorem, with which you are no doubt familiar, was known by the Babylonians some 1000 years earlier. However, the first general proof of this Theorem was probably given by Pythagoras or one of his followers.)

A natural generalization of the integers is $\mathbf{Q}$, the collection of rational numbers. A rational number is any real number that can be expressed as a ratio of integers. The rational numbers are important because they arise so naturally in the world around us. To the Pythagoreans, rationals were legitimate because of their relationship with the integers: rational numbers are related to integers through the simple operation of division. For our purposes, rational numbers are convenient because arithmetic operations with rationals can be done exactly.

The discovery of numbers that were *not* rational surprised and disturbed the Pythagoreans. Such numbers are called *irrational* numbers. A good example arises from the Pythagorean Theorem. Consider an isosceles right triangle whose sides have length 1 unit. In the language of Chapter 1 the hypotenuse should have length $\sqrt{2}$.

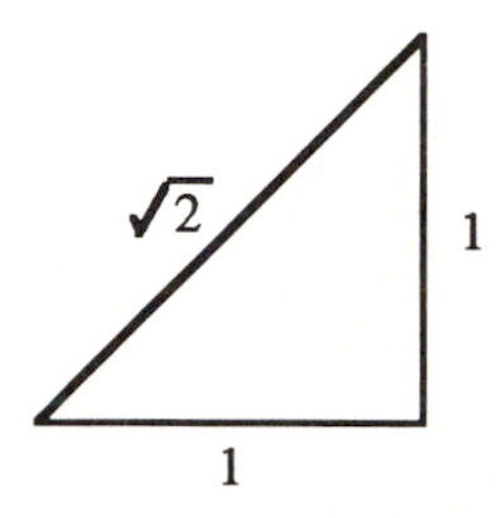

Figure 2.1.1

We shall illustrate the method of *proof by contradiction* in establishing that $\sqrt{2}$ is an irrational number. When using this method one assumes that the negation of the conclusion is true and then shows that this assumption leads to a logical contradiction.

Theorem 2.1.2 *$\sqrt{2}$ is an irrational number.*

Proof (by contradiction): Assume that $\sqrt{2}$ is a rational number. Thus $\sqrt{2}$ can be expressed as a reduced fraction, that is $\sqrt{2} = a/b$. By reduced fraction, we mean that a and b have no factors in common. Therefore

$$a = b\sqrt{2} \quad \text{or} \quad a^2 = 2b^2$$

Since a^2 is 2 times an integer (namely b^2) , we see that a^2, and hence a, must be even. (Why?) Letting $a = 2p$ we get,

$$4p^2 = 2b^2 \quad \text{or} \quad 2p^2 = b^2$$

We must conclude that b^2, and hence b, is even. However, a and b cannot both be even; they have no common factors! Thus we have obtained a logical contradiction and hence are forced to conclude that the assumed rationality of $\sqrt{2}$ is impossible. Therefore $\sqrt{2}$ is an irrational number. □

Using a similar argument, we could show that any integer that is not a perfect square has an irrational square root. Although the proof is considerably more difficult, it can be shown that π is irrational.

Question 2.1.3 Prove that $\sqrt{3}$ is irrational.

Hint: If a prime number (like 3) divides a product, then it must divide one of the factors of the product. ◇

Since irrational numbers do not have finite decimal representations, most computations involving irrational numbers must involve approximations. It is important to realize that in using approximations one needs to be conscious of how good the approximations are and thus how reliable is the final result. As we shall see, we can find arbitrarily good approximations for $\sqrt{6}$, π, and other rational and irrational real numbers. A major focus of this chapter is the question:

Under what conditions do "good approximations" for x yield "good approximations" for $f(x)$?

Exactly what does "good approximation" mean?

Section 2.1 Exercises

1. Describe three situations in "everyday" life in which you use approximations.
2. Describe a situation in "everyday life" which involves a sequence of approximations (to some mental or physical construct).
3. Consider the theorem: If b^2 is an odd number, then b is an odd number.
   (a) Identify the hypothesis and the conclusion.
   (b) Explain what it means to say that b^2 is an odd number.
   (c) Give an indirect proof (i.e., proof by contradiction) of the theorem.
4. Consider the statement: (∗) a^2 is an even integer if and only if a is an even integer.
   (a) Rewrite statement (∗) as two conditional statements.
   (b) Show that each of the conditional statements in part (a) is an implication (i.e., show that the "hypothesis true–conclusion false" case can not exist).
   (c) Is statement (∗) a mathematical theorem? Why?
5. Show that $\sqrt{5}$ is an irrational number.
6. Project: Show that π is irrational. References: M.M. Schiffer and L. Bowden, "The Role of Mathematics in Science," MAA, 1984, pp 59–69. D. Castellanos, "The Ubiquitous Π (Part 1)," *Mathematics Magazine,* 61 (2) April 1988, pp 67–98.
7. Project: Investigate the relationship between the arithmetic and geometric means and their applications. Reference: M.M. Schiffer and L. Bowden, "The Role of Mathematics in Science," MAA, 1984, pp 59–69.

2.2 Approximations and Convergence

Sequences are important in the study of approximations as indicated by their role in the *Basic Approximation Process.* This process, which we will apply repeatedly throughout the course, consists of the following steps:

1. Approximate the unknown quantity with a known property.
2. Determine a way to obtain a better approximation.
3. Generate a convergent *sequence* of approximations such that each is a better approximation than the preceding ones.
4. Define the desired property to be the common limit of all possible sequences in step 3. When there is no common limit, the desired property is said not to exist.

In this section we will make the idea of the limit of a sequence precise.

In Chapter 1 we defined a sequence as a special type of function (one whose domain consists of a subset of the integers). Recall that sequences may be described by a formula, for example

$$s : \mathbb{N} \to \mathbb{R},\ s(n) = s_n = \frac{1}{n}$$

or by a recursive relation, such as the factorial function and the Fibonacci sequence. In this chapter, we shall also see sequences described algorithmically. Although the source of a sequence is the set of integers, the target may be $\mathbb{R}$ or $\mathbb{R}^m$, $m > 1$. For example, the raccoon age distribution, Example 1.5.5, involved a sequence $s : \mathbb{N}_0 \to \mathbb{R}^4$.

We will use a recursive sequence defined by iterating a function and its associated cobweb graph to introduce the major questions that are addressed in this section.

Example 2.2.1 Consider the function $f : \mathbb{R} \to \mathbb{R}$ defined by $f(x) = -x^3$. Let u_n, v_n, w_n be the three sequences obtained by iterating f on the base cases $1.1, -1, 0.8$. Thus $u_0 = 1.1$ and $u_n = f(u_{n-1})$ for $n \geq 1$; $v_0 = -1$ and $v_n = f(v_{n-1})$ for $n \geq 1$; and $w_0 = 0.8$ and $w_n = f(w_{n-1})$ for $n \geq 1$, respectively. Computing the first few elements of these sequences yields:

$$\{u_n\} = \{1.1, -1.331, 2.357948, -13.109994, 2253.240236, \ldots\}$$

$$\{v_n\} = \{-1, 1, -1, 1, -1, 1, -1, ...\}$$

$$\{w_n\} = \{0.8, -0.512, 0.13421773, -0.00241785, 0.00000001, \ldots\}$$

The "cobweb" graphs of the three sequences are shown in Figure 2.2.2 (also see Figure 1.6.25).

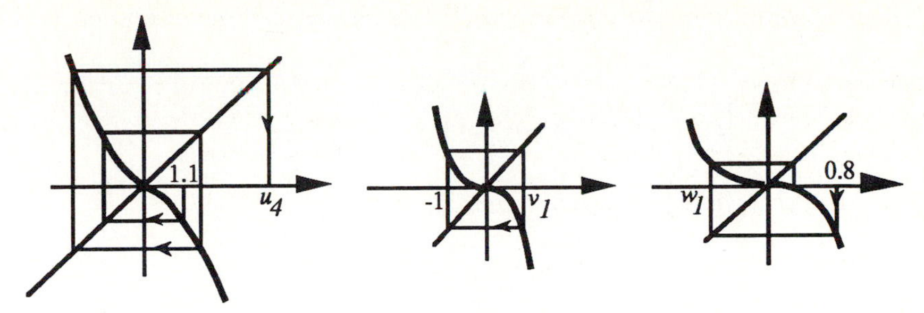

Figure 2.2.2

$\{u_n\}$ is an example of an *unbounded, divergent* sequence. Note that the elements of $\{u_n\}$ become very large in both the positive and negative sense and thus the elements will not eventually cluster about any number.

$\{v_n\}$ is an example of a *bounded, divergent* sequence. $\{v_n\}$ is also an example of a *periodic* sequence. Note that the elements of a periodic sequence will never eventually cluster about a *single* number.

$\{w_n\}$ is an example of a *convergent* sequence. Even though the elements alternate in sign, they seem to be closely *approximating* the number zero. Thus, given an approximation w_n, a better approximation is obtained by applying f to w_n to obtain $w_{n+1} = f(w_n)$. △

Question 2.2.3 Generate three sequences by iterating the function $g : \mathbb{R} \to \mathbb{R}$ defined by $g(x) = x^3$ on the base cases: 1.3, -1, 0.9. Using a computer or calculator, compute the first few values of each of the sequences. Sketch the cobweb graphs to illustrate the behavior of the sequences and then intuitively analyze the results (as was done in Example 2.2.1). ◇

Questions concerning convergence, the role of boundedness in relation to convergence, and the use of sequences to obtain approximations are the major concerns of this section.

We now consider the problem of using a sequence to determine an approximation of a given number to any desired degree of accuracy.

Example 2.2.4 Consider the sequence $s : \mathbb{N} \to \mathbb{R}$, $s_n = 0.333\ldots3$ (n threes), for $n = 1, 2, 3, \ldots$ as a *sequence of approximations* to $\frac{1}{3}$. Following the Basic Approximation Process, we have an initial approximation, $s_1 = 0.3$, obtain a better approximation, $s_2 = 0.33$, and then a sequence of approximations, $\{s_n\}$. The "limit" of this sequence should be the desired quantity, $\frac{1}{3}$. Before giving a formal definition of limit, in this example and in the following we will explore the idea of limit by looking at the error between the desired quantity and the approximation.

The error associated with approximation s_1 is the distance between s_1 and $\frac{1}{3}$, or

$$\left|s_1 - \frac{1}{3}\right| = \left|0.3 - \frac{1}{3}\right| = \left|\frac{3}{10} - \frac{1}{3}\right| = \frac{1}{30}$$

The error associated with approximation s_2 is

$$\left|s_2 - \frac{1}{3}\right| = \left|0.33 - \frac{1}{3}\right| = \frac{1}{3 \cdot 10^2}$$

and in general the error associated with approximation s_n is

$$\left|s_n - \frac{1}{3}\right| = \frac{1}{3 \cdot 10^n}$$

Thus given any accuracy ACC, we can find an integer n such that

$$\left|s_n - \frac{1}{3}\right| < ACC$$

For example, if $ACC = 0.0007$, then

$$|s_n - \frac{1}{3}| = \frac{1}{3 \cdot 10^n} < 0.0007$$

whenever

$$\frac{1}{10^n} < 0.0021 = \frac{21}{10^4} = \frac{2.1}{10^3}$$

or $n > 4$. △

The notion that the sequence $\{s_n\}$ approximates a number L with accuracy ACC can be phrased geometrically as follows:

Definition 2.2.5 (Approximation with Given Accuracy) *A sequence* $\{a_n\}$ approximates L with accuracy ACC *if there is an integer* N, *such that if* $n > N$ *then* $|s_n - L| < ACC$.

The statement can be rephrased that eventually (after $n > N$) all terms are within a distance ACC of L.

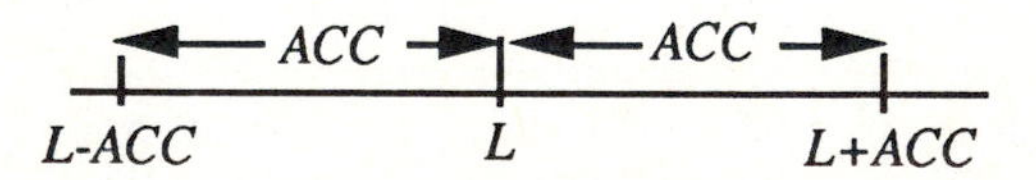

Figure 2.2.6

Note that the expression $|s_n - L| < ACC$ has exactly the same meaning as $L - ACC < s_n < L + ACC$.

Example 2.2.7 Consider the sequence $\{s_n\}$ defined by

$$s_n = 1 + \left(\frac{-1}{2}\right)^n \qquad for n = 1, 2, 3, \ldots$$

Intuitively, it appears that the larger the value chosen for n, the closer the term of the sequence is to 1. Recall from Chapter 1 that the distance between two numbers was described using the absolute value function. Thus our intuitive observation can be rigorously expressed by writing

$$|s_n - 1| = \left|1 + \left(\frac{-1}{2}\right)^n - 1\right| = \left|\left(\frac{-1}{2}\right)^n\right| = \frac{1}{2^n}$$

Since $1/2^n$ is a decreasing function of n, if we choose a specific value for n, say $n = 10$, then

$$|s_n - 1| < \frac{1}{2^{10}} \qquad \text{for all n} = 11,\ 12,\ 13,\ \ldots$$

Thus all terms of the sequence with $n > 10$ lie in the open interval $(1 - 1/2^{10}, 1 + 1/2^{10})$. This is exactly what is required to show that $\{s_n\}$ approximates 1 with accuracy $1/2^{10}$.

Similarly, this sequence approximates 1 with accuracy $1/2^{1000}$ since all terms of the sequence with $n > 1000$ lie in the open interval $(1 - 1/2^{1000}, 1 + 1/2^{1000})$. △

In the plane a disk replaces the interval. Thus $\{s_n\}$ approximates L with accuracy ACC if there is a disk D of radius ACC such that the terms s_n eventually lie within D. In the figure below, the terms s_n approximate L through disks of smaller and smaller radii ACC.

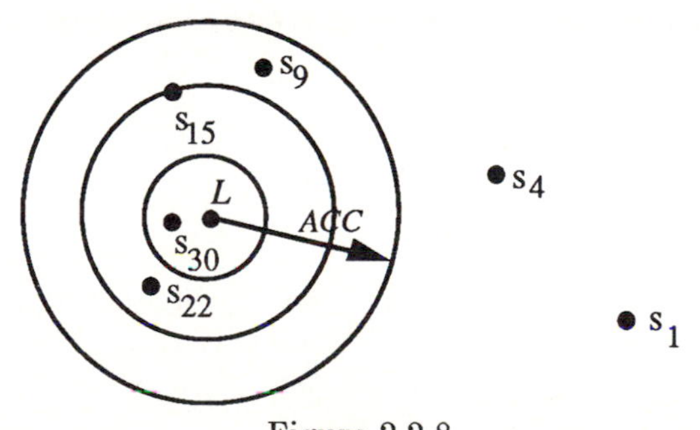

Figure 2.2.8

Example 2.2.9 We will form a sequence of approximations to $\left(\frac{1}{3}, \frac{4}{7}\right)$ and determine a term of the sequence that approximates $\left(\frac{1}{3}, \frac{4}{7}\right)$ with an accuracy of 0.00001 (i.e., an approximation whose error is at most 0.00001).

Let $s_n : \mathbb{N} \to \mathbb{R}^2$ be defined by $s_n = (a_n, b_n)$ where a_n is the approximation to $\frac{1}{3}$ obtained by truncating the decimal expansion of $\frac{1}{3}$ after the nth decimal place. Let b_n be the analogous approximation to $\frac{4}{7}$. The error associated with the nth approximation is the distance $|s_n - (\frac{1}{3}, \frac{4}{7})|$ (see Section 1.2). We shall use the triangle inequality (Theorem 1.2.36) to obtain the error bound. Thus:

$$\left|s_n - \left(\frac{1}{3}, \frac{4}{7}\right)\right| = \left|(a_n, b_n) - \left(\frac{1}{3}, \frac{4}{7}\right)\right| = \left|\left(a_n - \frac{1}{3}, b_n - \frac{4}{7}\right)\right| \le \left|a_n - \frac{1}{3}\right| + \left|b_n - \frac{4}{7}\right|$$

Since $\dfrac{1}{3} = 0.33333\ldots$ and $\dfrac{4}{7} = 0.57142857\ldots$, we have

$s_1 = (0.3,\ 0.5)$

and $$\left|s_1 - \left(\frac{1}{3}, \frac{4}{7}\right)\right| \le \left|0.3 - \frac{1}{3}\right| + \left|0.5 - \frac{4}{7}\right| \approx 0.104762 = \text{first error bound}$$

$s_2 = (0.33,\ 0.57)$

and $$\left|s_2 - \left(\frac{1}{3}, \frac{4}{7}\right)\right| \le \left|0.33 - \frac{1}{3}\right| + \left|0.57 - \frac{4}{7}\right| \approx 0.004762 = \text{second error bound}$$

$s_3 = (0.333,\ 0.571)$

and $$\left|s_3 - \left(\frac{1}{3}, \frac{4}{7}\right)\right| \le \left|0.333 - \frac{1}{3}\right| + \left|0.571 - \frac{4}{7}\right| \approx 0.000762 = \text{third error bound}$$

$s_4 = (0.3333,\ 0.5714)$

and $$\left|s_4 - \left(\frac{1}{3}, \frac{4}{7}\right)\right| \le \left|0.3333 - \frac{1}{3}\right| + \left|0.5714 - \frac{4}{7}\right| \approx 0.000062 = \text{fourth error bound}$$

$s_5 = (0.33333,\ 0.57142)$

and $$\left|s_5 - \left(\frac{1}{3}, \frac{4}{7}\right)\right| \le \left|0.33333 - \frac{1}{3}\right| + \left|0.57142 - \frac{4}{7}\right| \approx 0.000012 = \text{fifth error bound}$$

$s_6 = (0.333333,\ 0.571428)$

and $$\left|s_6 - \left(\frac{1}{3}, \frac{4}{7}\right)\right| \le \left|0.333333 - \frac{1}{3}\right| + \left|0.571428 - \frac{4}{7}\right|$$

$$\approx \quad 0.0000001 = \text{sixth error bound}$$

Since the distance in $\mathbb{R}^2$ between s_6 and $(\frac{1}{3}, \frac{4}{7})$ is less than 0.00001, we have found the desired approximation. Since subsequent terms in the sequence will be closer to $(\frac{1}{3}, \frac{4}{7})$, they will also approximate $(\frac{1}{3}, \frac{4}{7})$ to within 0.00001. △

When we say that a sequence $\{a_n\}$ approximates a number L with accuracy ACC, we mean that if we examine terms of the sequence, we eventually get to the point where all the remaining terms will differ from L by less than ACC. We say that a sequence *converges* to a number L if the sequence can be used to approximate L to any desired *degree of accuracy.* Alternatively, the sequence $\{a_n\}$ converges to the number L if for any desired number of digits, L can be approximated by $\{a_n\}$. This description of convergence, while possessing intuitive clarity, lacks the precision necessary for mathematical use. We can, however, give these statements precise and formal meanings.

Definition 2.2.10 (Convergence) *A sequence s_n* converges *to the point L in $\mathbb{R}^m$ if for any $ACC > 0$ we can find an N such that*

$$|s_n - L| < ACC \qquad \textit{for all } n > N$$

N usually depends on the value of ACC.

Recall from Section 1.2 that if s_n and L are in $\mathbb{R}^m$, $m > 1$, then

$$|s_n - L| = [(s_{n,1} - L_1)^2 + \cdots + (s_{n,m} - L_m)^2]^{1/2}$$

where L is an m-tuple $(L_1, \ldots, L_m)$ and s_n is an m-tuple $(s_{n,1}, \ldots, s_{n,m})$.

Note the double subscripts in the above expression are necessary because s_n being in $\mathbb{R}^m$, has m components. Thus in order to show that the components vary with n we need *two* subscripts: the first showing the n and the second indicating the component of s_n. Thus $s_{2,3}$ means the third component of s_2.

The "sequence $\{s_n\}$ converges to the point L" is formally denoted by the expression:

$$\lim_{n\to\infty} s_n = L$$

and informally by the shorthand notation

$$s_n \to L$$

The point L is called the *limit of the sequence.*

The reader should note that the three expressions

1. $\{s_n\}$ converges to the point L,

2. $\lim_{n\to\infty} s_n = L$,

3. $\{s_n\}$ approximates the point L for all $ACC > 0$

are all equivalent. These expressions will be used interchangeably throughout the text.

A sequence is said to be *convergent* if it converges to a limit and *divergent* if it does not converge to a limit. Let us consider some examples.

Example 2.2.11 (A Constant Sequence) Consider the constant sequence $s_n = c$ for all n. Then $|s_n - c| = 0$ for all n, and so $\lim_{n\to\infty} s_n = c$. △

Example 2.2.12 (a) The sequence defined by $z_n = 1/n$ converges to 0. To show this using the definition, we let ACC denote any accuracy (which must be a positive number) and then find an N such that

$$|z_n - 0| = \left|\frac{1}{n}\right| = \frac{1}{n} < ACC \qquad \text{for all } n > N$$

Solving the inequality $1/n < ACC$ for n, we find that $n > 1/ACC$ must hold. (Note that we need $ACC > 0$ for this to make sense.) We now set N equal to the minimum value of n for which this is true: $N = 1/ACC$.

Now $|z_n - 0| = 1/n < ACC$ for all $n > N$, and thus $z_n \to 0$

(b) Similarly, if c is any constant, $\lim_{n\to\infty} c/n = 0$. △

Example 2.2.13 The sequence defined by $b_n = (2+n)/n$ converges to 1. To show this using the definition, given any accuracy ACC we must find an N such that if $n > N$,

$$|b_n - 1| = \left|\frac{2+n}{n} - 1\right| = \left|\frac{2}{n}\right| = \frac{2}{n} < ACC$$

Solving the inequality $2/n < ACC$ for n tells us that we must have $n > 2/ACC$. Analogously to the example above, we set $N = 2/ACC$. Then for $n > N$, $|b_n - 1| < ACC$, and so $b_n \to 1$. △

Example 2.2.14 (Convergence in $\mathbb{R}^2$) Let us now illustrate the definition of convergence in the case of a sequence converging to a point in $\mathbb{R}^m$. For example, let $s : \mathbb{N} \to \mathbb{R}^2$ be defined by

$$s_n = \left(\frac{3}{n}, \frac{2n^2 + 127n}{n^2}\right)$$

We shall show that $\{s_n\}$ converges to $L = (0, 2)$.

Let ACC be an arbitrary positive number. Then

$$|s_n - L| = \left|\left(\frac{3}{n} - 0, \frac{2n^2 + 127n}{n^2} - 2\right)\right|$$

Now, using Theorem 1.2.36 (the triangle inequality in several variables) we get

$$\begin{aligned} |s_n - L| &\leq \left|\frac{3}{n} - 0\right| + \left|\frac{2n^2 + 127n}{n^2} - 2\right| \\ &\leq \left|\frac{3}{n}\right| + \left|\frac{2n^2 + 127n - 2n^2}{n^2}\right| \\ &\leq \left|\frac{3}{n}\right| + \left|\frac{127}{n}\right| \end{aligned}$$

Now

$$\left|\frac{3}{n}\right| < \frac{ACC}{2} \quad \text{if } n > \frac{6}{ACC} \quad \text{and} \quad \left|\frac{127}{n}\right| < \frac{ACC}{2} \quad \text{if } n > \frac{254}{ACC}$$

Thus if we choose N to be any integer satisfying

$$N \geq \max\left\{\frac{6}{ACC}, \frac{254}{ACC}\right\}$$

we have

$$|s_n - L| < \frac{ACC}{2} + \frac{ACC}{2} = ACC \quad \text{for } n > N$$

Note that each of the two component sequences of $\{s_n\}$ converges to the corresponding component in L. That is,

$$\left\{\frac{3}{n}\right\} \to 0, \quad \left\{\frac{2n^2 + 127n}{n^2}\right\} \to 2$$

△

This example illustrates the fact that if a sequence converges componentwise, then it converges. Conversely, if a sequence converges, then each component sequence must converge. (Why?) For emphasis, we state this result as a theorem.

Theorem 2.2.15 *A sequence in $\mathbb{R}^m$ converges if and only if it converges component-wise.*

Example 2.2.16 The sequence $s : \mathbb{N} \to \mathbb{R}^3$, defined by

$$s_n = \left(1 + \left(\frac{-1}{2}\right)^n, \frac{1}{n}, \frac{2+n}{n}\right)$$

converges to (1, 0, 1) since each of the component sequences converges (see the previous examples). △

Example 2.2.17 The sequence defined by $d_n = (1, 1/n^2, (-1)^n)$ diverges because the third component sequence does not converge (even though the first two component sequences do converge). Why does the third component sequence diverge? (See Example 2.2.22 for the answer, but you should try finding the reason yourself.) △

Theorem 2.2.18 *If a sequence in* $\mathbb{R}^m$ *has a limit, then the limit is unique.*

Proof: We first prove this theorem for a sequence in $\mathbb{R}$ using the method of contradiction (i.e., assume that two different limits exist and then derive a contradiction).

Assume that there exists a sequence s_n that converges to both L and M, $L \neq M$. Choose small intervals I_1 and I_2 centered at L and M that do not intersect.

Figure 2.2.19

If the sequence converges to L, then after a certain point in the sequence all the terms will lie in the interval I_1 so that none of the terms will lie in I_2. Consequently, the sequence cannot also converge to M. Thus we have a logical contradiction of our assumption, and therefore the number to which a sequence converges (if it converges at all) is unique.

Now consider a sequence in $\mathbb{R}^m$, for $m > 1$. By Theorem 2.2.15 the sequence converges if and only if the component sequences converge. Since the component sequences are in $\mathbb{R}$, their limits must be unique. Thus the limit of a sequence in $\mathbb{R}^m$ must also be unique. □

Since sequences are functions, the definitions related to boundedness (glb, lub) and monotonicity (increasing, decreasing, strictly increasing, strictly decreasing) in Chapter 1 apply to sequences. For sequences in $\mathbb{R}^m, m > 1$, these notions are interpreted *componentwise.*

For example, $(a, b) < (c, d)$ if and only if $a < c$ and $b < d$.

Example 2.2.20 The sequence $\{(n, (n-1)/n, 2)\}$ is increasing, but not strictly increasing since the third component is not strictly increasing. The sequence $\{(\sin(n), n, n^2)\}$ is not increasing because the first component is not an increasing function of n. △

Example 2.2.21 The unbounded sequence $\{t_n\}$ defined by $t_n = n$, $n = 1, 2, 3, \ldots$, is divergent since it does not get close to any number L, i.e., no number L exists, such that given any accuracy ACC, there is an N with the property that $|t_n - L| < ACC$ for all $n > N$. △

A little reflection should convince us that any unbounded sequence is divergent. However, the converse of this statement, which says that a divergent sequence is unbounded, is not true. Consider the following example.

Example 2.2.22 The sequence defined by $d_n = (-1)^n$ is divergent. Its values switch back and forth between -1 and $+1$, not staying close to either one. Note, however, that $\{d_n\}$ is bounded.

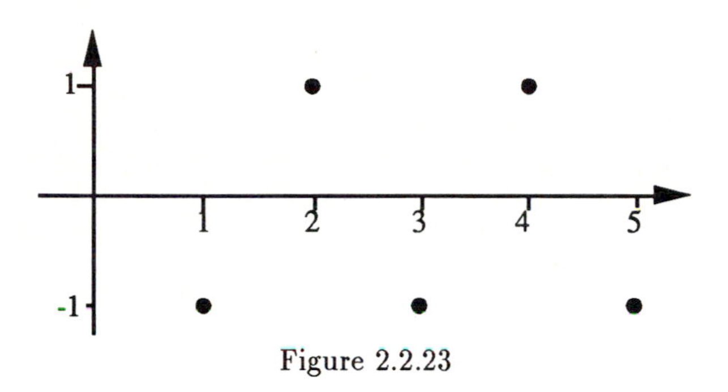

Figure 2.2.23

△

The relationship between convergence and boundedness is given in the next two theorems.

Theorem 2.2.24 *If a sequence in* $\mathbb{R}^m$ *is convergent, then the sequence is bounded.*

Proof: By Theorem 1.2.39 a sequence is bounded if and only if each component sequence is bounded. Thus we need only prove the theorem for sequences in $\mathbb{R}$.

Let $s_n \to L$. For $ACC = 1$, there exists an N such that

$$|s_n - L| < 1 \quad \text{for all } n > N$$

or

$$L - 1 < s_n < L + 1 \quad \text{for all } n > N$$

Only some of the values $s_1, s_2, s_3, \ldots, s_N$ may be outside the interval $[L-1, L+1]$. Let the largest of the absolute values of these be A, so that $|s_n| \leq A$ for $n \leq N$, or

$$-A < s_k < A \quad \text{for } k = 1, 2, 3, \ldots, N$$

Thus if we let $B = \max\{A, |L-1|, |L+1|\}$, then s_n is bounded by B for all n.
□

Example 2.2.25 Phred, who is standing one unit away from a wall, steps toward the wall in such a way that each step moves her halfway to the wall. If d_n denotes the distance Phred has moved after n steps, then

$$\begin{aligned} d_1 &= \frac{1}{2} \\ d_2 &= \frac{1}{2} + \frac{1}{2^2} \\ &\vdots \\ d_n &= \frac{1}{2} + \frac{1}{2^2} + \cdots + \frac{1}{2^n} \end{aligned}$$

Clearly d_n is an increasing sequence that is bounded above by 1. (Why?) Physical considerations dictate that d_n converges. The next theorem, moreover, shows that the convergence is a result of d_n being both bounded and increasing. $\triangle$

The proof of this next theorem depends on the following axiom of the real number system. Recall that the least upper bound of a set of real numbers is the smallest upper bound and the greatest lower bound is the largest lower bound.

Completeness Axiom. Every bounded, nonempty set of real numbers has a least upper bound (lub) and a greatest lower bound (glb).

Theorem 2.2.26 *If $\{s_n\}$ is a bounded, monotonic sequence, then $\{s_n\}$ is a convergent sequence.*

Proof: (The proof will only be given for monotonically increasing sequences in $\mathbb{R}$.)

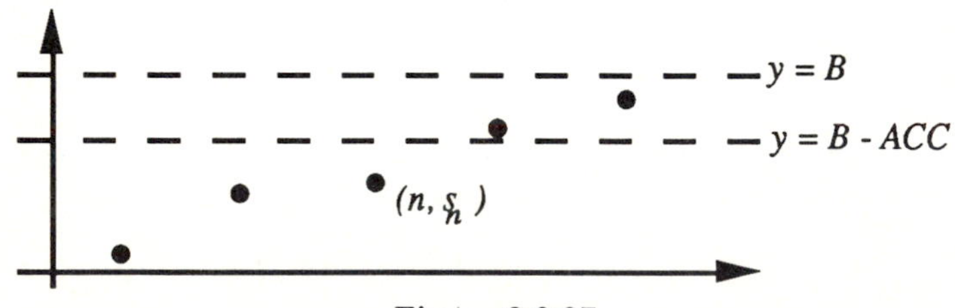

Figure 2.2.27

Let $\{s_n\}$ be a monotonic increasing sequence in $\mathbb{R}$. Let B be the lub for $\{s_n\}$. We will show that, in fact,

$$\lim_{n \to \infty} s_n = B$$

Consider any accuracy, $ACC > 0$. Since B is the lub of $\{s_n\}$, $B - ACC$ is not an upper bound of $\{s_n\}$ and thus there exists some n, say n^*, such that

$$B - ACC < s_{n^*} < B$$

Since $\{s_n\}$ is increasing,

$$B - ACC < s_{n^*} < s_n < B \quad \text{for all } n > n^*$$

That is,

$$|s_n - B| < ACC \quad \text{for all } n > N = n^*$$

Thus $\{s_n\}$ converges to B. □

A similar argument, left as an exercise (Exercise 19), can be given to show that a bounded, monotonically decreasing sequence converges to its greatest lower bound. For emphasis, we restate this theorem in a more explicit form.

Theorem 2.2.28 *If $\{s_n\}$ is increasing and bounded above, then $s_n \rightarrow lub\{s_n\}$. If $\{s_n\}$ is decreasing and bounded below, then $s_n \rightarrow glb\{s_n\}$.*

This last theorem is particularly useful because it allows us to determine convergence (when the sequence is bounded and monotonic) without having to guess the value of the limit. Note that in all the preceding examples of convergent sequences it was possible to guess the value of the limit. In general, this will not be possible.

Example 2.2.29 Does the sequence $x_n = 1 + 1/2! + 1/3! + \cdots + 1/n!$ converge?

Since $\{x_n\}$ is an increasing sequence, Theorem 2.2.26 says that $\{x_n\}$ converges provided that $\{x_n\}$ is bounded. Comparing $\{x_n\}$ to the sequence in Example 2.2.25, we see that $x_n < d_n + 1$ for all n. Thus, since $\{d_n\}$ is bounded, $\{x_n\}$ is bounded (i.e., $x_n < d_n + 1 < 2$). Hence $\{x_n\}$ is a convergent sequence. △

Since the convergence of a sequence is really determined by the behavior of the "tail end" of the sequence, the hypothesis of the theorem can actually be weakened by replacing the monotonic condition with the requirement that the sequence is monotonic for n greater than some K.

Example 2.2.30 (Recursively Defined Sequence) The sequence $\{t_n\}$ defined recursively by

$$t_0 = 1, t_n = t_{n-1} + \left(\frac{1}{10}\right)^n \quad \text{for} \quad 1 \leq n \leq 100, t_n = t_{n-1} - \left(\frac{1}{10}\right)^n \quad \text{for} \quad n > 100$$

is decreasing for $n > 100$. Since $\{t_n\}$ is bounded below by zero (Why?), the sequence is convergent (although we do not know the limit). △

Example 2.2.31 (Sequence Diverging to Negative Infinity) The sequence $b_n = 13n - n^2$ is clearly decreasing for n greater than 13. Since $\{b_n\}$ is unbounded below, the sequence diverges. △

An unbounded, increasing sequence is said to *diverge to positive infinity* and an unbounded, decreasing sequence is said to *diverge to negative infinity.* The sequence in the last example diverges to negative infinity.

In many of the theoretical applications of convergent sequences to be found throughout this text, we will be interested in sequences that converge to 0. The proof of the following fact about such sequences will illustrate the utility of the definition of convergence as well as the structure of an "if and only if" theorem.

Theorem 2.2.32 *Let* $\{s_n\}$ *be a sequence. Then* $\lim_{n\to\infty} s_n = 0$ *if and only if* $\lim_{n\to\infty} |s_n| = 0$.

Proof: This proof holds for $\mathbb{R}^m$ if the "0" in "$\lim_{m\to\infty} s_n = 0$" is considered to be the m-tuple $(0,\ldots,0)$.

As with any theorem involving a biconditional implication (see Section 1.2) of the form "(statement 1) if and only if (statement 2)," we must prove that the truth of either statement implies the truth of the other. First, suppose $\lim_{n\to\infty} s_n = 0$. We must show that $\lim_{n\to\infty} |s_n| = 0$. That is, given any $ACC > 0$, we must find an N, such that for $n > N$,

$$||s_n| - 0| < ACC$$

Let $ACC > 0$ be given. Since $\lim_{n\to\infty} s_n = 0$, we can find an N such that for $n > N$,

$$|s_n - 0| < ACC$$

However, from simple facts about zero and absolute value we have

$$|s_n - 0| = |s_n| = ||s_n|| = ||s_n| - 0|$$

Therefore, if $n > N$, then $||s_n| - 0| < ACC$. In terms of the geometry of the situation, if s_n is within ACC of 0, then $|s_n|$ is within ACC of 0.

Now, to prove the other "half" of the theorem, suppose $\lim_{n\to\infty} |s_n| = 0$ and that $ACC > 0$ has been given. From the convergence of $\{|s_n|\}$, we can find an N, such that for all $n > N$, $||s_n| - 0| < ACC$. Therefore, if $n > N$, $|s_n - 0| < ACC$. □

Note that the limit in the above theorem is zero. The statement is not necessarily true when the limit is not zero. For example,

$$s_n = -2 + \frac{1}{n} \to -2$$

However,

$$|s_n| = \left|-2 + \frac{1}{n}\right| \to 2$$

Section 2.2 Exercises

1. For each of the following sequences, list the first five elements and the tenth element of the sequence.

   (a) $a_n = 1 - 2/n, \quad n = 1, 2, 3, \ldots$
   (b) $b_n = 5n^2/(3n+1), \quad n = 1, 2, 3, \ldots$
   (c) $c_n = (-1)^n/(10n), \quad n = 1, 2, 3, \ldots$
   (d) $d_n = (\cos(n\pi), (n+1)/n), \quad n = 1, 2, 3, \ldots$
   (e) $f_n = ((-1)^n, 2/n, \sin(n\pi)), \quad n = 1, 2, 3, \ldots$

2. For each sequence in Exercise 1, determine if the sequence converges or diverges. If a sequence converges, give an expression for N in terms of ACC (see the definition of convergence). If a sequence diverges, explain why it diverges.

3. Determine which of the following sequences converge. Justify your answers.

   (a) $x_n = n/(n+1) - (n+1)/n, \quad n = 1, 2, 3, \ldots$
   (b) $x_n = (-1)^n - (-1)^{n+1}, \quad n = 1, 2, 3, \ldots$
   (c) $x_n = n^3/(n^2+3n)^2 - n^2/(n+1), \quad n = 1, 2, 3, \ldots$
   (d) $x_n = (n^3, 1/n, \sin(n)/n^4), \quad n = 1, 2, 3, \ldots$
   (d) $x_n = (n/(n+17)^4, 2^n), \quad n = 1, 2, 3, \ldots$

4. Determine which of the following sequences converge. Justify your answers.

   (a) $z_n = (n+2)/n^2 + n/(n-1), \quad n = 1, 2, 3, \ldots$
   (b) $z_n = 3^n, \quad n = 1, 2, 3, \ldots$
   (c) $z_n = (\sqrt{n}, \sin(n)/n), \quad n = 1, 2, 3, \ldots$
   (d) $z_n = (\sqrt{n}/(n+1), 1/(\sin(n)+2), n/(n^2+1)), \quad n = 1, 2, 3, \ldots$
   (e) $z_n = (1/\sqrt{n}, \sqrt{n+3}/(n+2), 1/2^n), \quad n = 1, 2, 3, \ldots$

5. Determine which of the following sequences converge. Justify your answers.

   (a) $a_n = (2/n, n/10^2), \quad n = 1, 2, 3, \ldots$
   (b) $b_n = ((n-1)/n, \tan(n), \cos(n\pi)), \quad n = 1, 2, 3, \ldots$
   (c) $c_n = (n + (-1)^n)/(n+1), \quad n = 3, 4, 5, \ldots$
   (d) $d_n = (n^2/(n+1), 7, 4, 1/n), \quad n = 4, 5, 6, \ldots$
   (e) $e_n = 4^n/(3^n - 1,000), \quad n = 1, 2, 3, \ldots$

6. Find a term of $\{(2^n - 1)/2^n\}$ that approximates 1 with accuracy 0.001.

7. Find a sequence of approximations to $\frac{7}{9}$ and then determine the first term of the sequence that approximates $\frac{7}{9}$ with $ACC = 0.001$.

8. Find a sequence of approximations to $\frac{1}{11}$ and then determine the first term of the sequence that approximates $\frac{1}{11}$ with $ACC = 0.000005$.

9. Find a sequence of approximations to $(\frac{1}{3}, \frac{5}{8})$ and then determine the first term of the sequence that approximates $(\frac{1}{3}, \frac{5}{8})$ with $ACC = 0.000005$.

10. (Mel Henriksen) Joe Zyxx's girlfriend Oxtyl signals to him across a 10-foot-long room to indicate that she will kiss him goodnight. Joe's elation gives way to a feeling of depression when he realizes that, in order to reach Oxtyl, he must first get halfway to her, then he must traverse half that distance again, etc. So he must pass over an "infinite" number of successive points to get his goodnight kiss. Can he do it? (This dilemma was noticed first by the Greek philosopher Zeno, who raised many difficult questions about motion.) After examining a certain sequence, Joe realized that he could get close enough to Oxtyl for all practical purposes. Set up an appropriate sequence and indicate how many "midpoints" Joe must reach to get within 1 inch of Oxtyl.

11. For each of the following, give an example or explain why no example exists.

    (a) A sequence that diverges to positive infinity.
    (b) A bounded sequence that diverges to positive infinity.
    (c) A bounded sequence in $\mathbb{R}^3$.
    (d) An unbounded sequence that converges to 6.
    (e) A divergent, monotonic sequence.
    (f) A recursively defined sequence that converges to (1,2).
    (g) A sequence that converges to both 1 and -1.
    (h) A divergent sequence that has a least upper bound of 5.
    (i) An increasing sequence that is bounded above.
    (j) A sequence that is bounded below, but not above.
    (k) A monotonic sequence in $\mathbb{R}^2$.
    (l) An unbounded sequence in $\mathbb{R}^2$.
    (m) A divergent sequence whose least upper bound is -2.
    (n) A decreasing sequence whose greatest lower bound is -3.

12. Show that $\{s_n\}$, given by $s_n = n/(n+1), n > 0$, is a strictly increasing sequence.

13. Show that $\{t_n\}$ in Example 2.2.30 is bounded below by zero.

14. Consider the sequence:

$$s_1 = \frac{1}{2}, \quad s_n = s_{n-1} + \frac{1}{2}(1 - s_{n-1}), \quad n \geq 2$$

(a) Determine whether you think that the sequence converges or diverges. Explain your thinking.

(b) Describe what you must show in order to establish your determination in part a.

(c) Based on your answers to parts (a) and (b), prove or disprove that the sequence converges.

15. Consider the theorem: Every unbounded sequence in $\mathbb{R}^n$ is divergent.

(a) Rewrite the statement of the theorem as a conditional statement.

(b) What does it mean to say that a sequence in $\mathbb{R}^n$ is unbounded?

(c) What must be true of at least one component of a sequence in $\mathbb{R}^n$ if the sequence is divergent?

(d) Prove the theorem.

16. Consider the theorem: Every bounded monotonically decreasing sequence in $\mathbb{R}$ is convergent.

(a) Rewrite the statement of the theorem as a conditional statement.

(b) Prove the theorem. (Hint: Theorem 2.2.26 may be helpful.)

17. Consider the theorem: A convergent sequence in $\mathbb{R}^3$ is bounded.

(a) Rewrite the statement of the theorem as a conditional statement.

(b) What must be shown to establish that the conditional statement in part a is an implication?

(c) What must be true of each of the component sequences if a sequence in $\mathbb{R}^3$ is bounded?

(d) If a sequence $\{s_n\}$ in $\mathbb{R}^3$ is convergent, what can be said about the behavior of each component sequence?

(e) Prove the theorem.

18. (a) A cow wishes to spiral in on the circumference of a circular cow pond of radius 8, stopping at the four cardinal directions (NWSE) to rest. Find a sequence of resting points which does this. (The cow does not go into the pond!)

(b) A cow pond is cirular of radius 8 and of uniform depth 3. A fish in the pond wishes to start at the top of the pond, one unit from the Northern end, and to spiral in to the center bottom of the pond, stopping to rest at exactly k points on each circuit. Find a sequence of resting points which does this.

19. Prove that if $\{s_n\}$ is a bounded, monotonic sequence in $\mathbb{R}^2$, then $\{s_n\}$ is a convergent sequence.

20. Prove that if a sequence in $\mathbb{R}^m$ converges, then each component sequence must converge (see Theorem 2.2.15).

21. Consider the recursive sequence $\{s_n\}$ obtained by iterating the function $f : \mathbb{R} \to \mathbb{R}$ defined by $f(x) = x^2$ applied to the number x_0.

    (a) Draw two cobweb graphs for $\{s_n\}$, one with $x_0 = -0.9$ and one with $x_0 = 1.1$.

    (b) Using a computer or calculator, iterate eight times for each starting value, $x_0 = -2, -0.6, 0.9, 1.1$.

    (c) Using the results of parts a and b, make a conjecture concerning the convergence or divergence of $\{s_n\}$ with respect to the size of the absolute value of x_0.

22. Let $\{s_n\}$ be the recursive sequence obtained by iterating the function $f : \mathbb{R} \to \mathbb{R}$ defined by $f(x) = x^2$ applied to x_0, $-1 < x_0 < 1$.

    (a) Determine an expression for s_n.

    (b) Show that $\{s_n\}$ is a decreasing sequence by showing that $s_{n+1}/s_n < 1$.

    (c) Use Theorem 2.2.26 to show that s_n is convergent.

    (d) Use a computer or calculator to find the smallest value of n such that s_n approximates its limit with $ACC = 0.000005$.

23. Let $\{s_n\}$ be the recursive sequence obtained by iterating the square root function applied to 2.

    (a) Determine an expression for s_n.

    (b) Show that $\{s_n\}$ is a decreasing sequence by showing that $s_{n+1}/s_n < 1$.

    (c) Use Theorem 2.2.26 to show that $\{s_n\}$ is convergent.

    (d) Use a computer or calculator to find the smallest value of n such that s_n approximates its limit with $ACC = 0.000005$.

24. Consider the sequence $A_n = (2 \cdot 4 \cdot 6 \cdots 2n) \ / \ (1 \cdot 3 \cdot 5 \cdots (2n-1))$.

    (a) Show that $A_n = ((2^n n!)^2)/(2n)!$.

    (b) Does A_n converge? Investigate the behavior of A_n for several values of n.

2.3 The Algebra of Limits of Sequences

Frequently we will have occasion to combine sequences algebraically. For example, the summary grade report for a graduating class could be the "average" of the summary grade reports for each of the students. In Example 1.5.2 a student's summary grade report was modeled by a sequence. Thus to obtain the class average, the grade reports (sequences) are "added" and the resulting sequence "multiplied" by $1/n$, where n is the number of students in the class. As another example, suppose the different books in a bookstore are numbered and x_n denotes the number of books of title n in the inventory on January first. If b_n and s_n are the sequences of books bought and sold during January, then $x_n + b_n - s_n$ is the inventory on February first of books of title n.

In this section, we extend the algebraic operations on finite sequences to limits of infinite sequences.

In the study of calculus, it is typical to follow the motivation and definition of a concept with theorems showing how the "concept behaves" with respect to the standard arithmetic operations of addition and multiplication. The following theorem is such a theorem for the concept of the limit of a sequence. This theorem, and the ones like it found later in the text, greatly simplifies the computation of certain limits in that it allows us to compute the limit of a combination of functions in terms of the limits of the individual functions. Above all, it will help us avoid the direct use of the definition in computing the limits of certain sequences.

Theorem 2.3.1 (Algebra of Limits) *If* $\{a_n\}$ *and* $\{b_n\}$ *are convergent sequences in* $\mathbb{R}$ *with* $\lim_{n\to\infty} a_n = L$ *and* $\lim_{n\to\infty} b_n = M$, *then*

(a) $\lim_{n\to\infty}(a_n + b_n) = L + M$

(b) $\lim_{n\to\infty} a_n b_n = LM$

(c) $\lim_{n\to\infty} ka_n = kL$ *for any constant* k

(d) If $M \neq 0$ *and* b_n *is never* 0, *then* $\lim_{n\to\infty} a_n/b_n = L/M$

Verbally, the first two parts of the theorem are "the limit of a sum (of convergent sequences) is the sum of the limits" and "the limit of a product (of convergent sequences) is the product of the limits."

Proof: We prove part (a) of the theorem to give an idea of how it is done. We shall first give an intuitive geometric argument to indicate our thinking.

Consider an open interval I with center at $L + M$ and radius ACC.

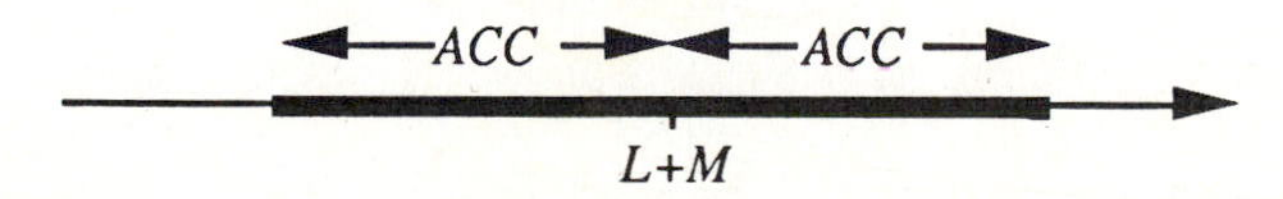

Figure 2.3.2

Let I_1 and I_2 be open intervals half as long as I centered at L and M, respectively.

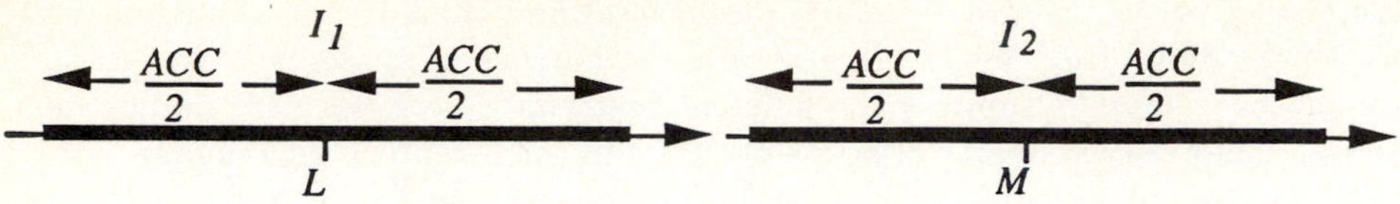

Figure 2.3.3

If for some value of n, a_n lies in I_1 and b_n in I_2, then we have

$$L - \frac{ACC}{2} < a_n < L + \frac{ACC}{2}$$

$$M - \frac{ACC}{2} < b_n < M + \frac{ACC}{2}$$

and adding we get

$$L + M - ACC < a_n + b_n < L + M + ACC$$

which means that $a_n + b_n$ is in I.

Now for a rigorous analytical argument consider an arbitrary $ACC > 0$. Since $\lim_{n \to \infty} a_n = L$, there exists an N_1 such that

$$|a_n - L| < \frac{ACC}{2} \quad \text{for all } n > N_1$$

Similarly, since $\lim_{n \to \infty} b_n = M$, there exists an N_2 such that

$$|b_n - M| < \frac{ACC}{2} \quad \text{for all } n > N_2$$

Let $N = max\{N_1, N_2\}$. Now, applying the Triangle Inequality (see Chapter 1), we have

$$\begin{aligned} |(a_n + b_n) - (L + M)| &= |(a_n - L) + (b_n - M)| \\ &\leq |a_n - L| + |b_n - M| \\ &< \frac{ACC}{2} + \frac{ACC}{2} \quad \text{for all } n > N \\ &< ACC \quad \text{for all } n > N \end{aligned}$$

Thus according to the definition, $\lim_{n\to\infty} a_n + b_n = L + M$. □

It is important to note that only parts (a) and (c) of Theorem 2.3.1 apply to convergent sequences in $\mathbb{R}^m$, $m > 1$, since *neither multiplication nor division is defined in* $\mathbb{R}^m$, $m > 1$.

Example 2.3.4 To see how useful this theorem can be, consider the limit of the sequence defined by

$$s_n = \frac{n^2 + n + 1}{4n^2 + 2n + 2} \quad \text{for } n = 1, 2, 3, \ldots$$

Since n is never 0, divide the numerator and denominator by n^2 and get

$$\lim_{n\to\infty} s_n = \lim_{n\to\infty} \frac{1 + \frac{1}{n} + \frac{1}{n^2}}{4 + \frac{2}{n} + \frac{2}{n^2}}$$

In general, whenever we need to compute $\lim_{n\to\infty} p(n)/q(n)$, where p and q are polynomials, divide both numerator and denominator by the highest power of n.

By part (d) of the theorem we may compute the limits of the numerator and denominator separately and obtain

$$\lim_{n\to\infty} s_n = \frac{\lim_{n\to\infty}(1 + \frac{1}{n} + \frac{1}{n^2})}{\lim_{n\to\infty}(4 + \frac{2}{n} + \frac{2}{n^2})}$$

Using part (a) of the theorem on the numerator and denominator (extended to three terms rather than two terms), Example 2.2.12 on the limit of $\{1/n\}$, and parts (b) and (c) of the theorem (to get the limit of $\{2/n^2\}$ equal to zero), we get

$$\lim_{n\to\infty} s_n = \frac{1 + 0 + 0}{4 + 0 + 0} = \frac{1}{4}$$ △

It is important to understand the processes (not just the mechanics) that were involved in the previous example. We should be able to obtain the following results by "inspection" as well as being able to justify them by referring to appropriate definitions and theorems.

1. $\lim_{n\to\infty}(2n - 4n^3 + 13n^2)/(2n^3 - 77n^2) = -2$ (divide by n^3).

2. $\lim_{n\to\infty}(4 - 2n^3)/(n^2 + 3n^4) = 0$ (divide by n^4).

3. $s_n = (8n^5 + 17)/(n^3 + 4)$ diverges to positive ∞.

4. $t_n = (n - 4n^3)/(n^2 + 13)$ diverges to negative ∞.

The following theorem summarizes these results.

Theorem 2.3.5 *Let $r : \mathbb{N} \to \mathbb{R}$ be a rational sequence in n, $r(n) = \frac{p(n)}{q(n)}$ where p and q are polynomials in n. Then*

(a) $r(n) \to 0$ if and only if the degree $p(n)$ is less than the degree of $q(n)$.

(b) $r(n)$ diverges if and only if the degree of $p(n)$ is greater than the degree of $q(n)$.

(c) $r(n) \to a/b$ if and only if the ratio of the highest powered term of $p(n)$ to the highest powered term of $q(n)$ reduces to a/b.

The following example illustrates a limit problem that cannot be addressed by Theorem 2.3.5 or earlier limit theorems.

Example 2.3.6 We will find the limit (if it exists) of the sequence $s_n = \sin(n)/n$. Since s_n is defined as a quotient, we are tempted to apply part (d) of Theorem 2.3.1. However, the limit of the numerator, $\sin(n)$, does not exist (Why?) and therefore the hypothesis of that theorem is not satisfied. Since the quotient is not a rational function [$\sin(n)$ is not a polynomial], the process illustrated in Example 2.3.4 cannot be applied. Although $\{s_n\}$ is a bounded sequence, it is not monotonic and thus we cannot rely on Theorem 2.2.26 which requires a sequence to be both bounded and monotonic. Furthermore, it is not clear how we could use the definition to determine convergence. What can be done?

Since $\sin(n)$ is bounded by 1, we have

$$\frac{-1}{n} \le s_n = \frac{\sin(n)}{n} \le \frac{1}{n}$$

Furthermore since $(-1/n) \to 0$ and $1/n \to 0$, s_n must also converge to 0. This is intuitively clear, since s_n is "squeezed" between two sequences converging to zero. Thus convergence is established, not by working directly with the sequence $\{\sin(n)/n\}$, but by showing that certain sequences of upper and lower bounds of $\sin(n)/n$ converge to the *same* number. △

The "squeeze" method in the previous example is one of the most useful processes in calculus for finding limits. We formalize this process in the following theorem called the *squeeze theorem.*

Theorem 2.3.7 (Squeeze Theorem) *If $\{a_n\}$, $\{b_n\}$, and $\{c_n\}$ are sequences in $\mathbb{R}^m$ with*

$$\lim_{n\to\infty} a_n = L, \quad \lim_{n\to\infty} b_n = L, \quad \text{and} \quad a_n \le c_n \le b_n$$

for all n greater than some N, then $\lim_{n\to\infty} c_n = L$.

Proof: (The proof will only be given for sequences in $\mathbb{R}$.)

Let $ACC > 0$ be given. We know that we can find an N_1 so that for $n > N_1$, a_n will lie within distance ACC of L. Also, we can find an N_2 so that if $n > N_2$, b_n will also be within ACC of L. Letting N be the larger of N_1 and N_2, then for $n > N$ we have both a_n and b_n within ACC of L. Since c_n is between a_n and b_n, it must also be within ACC of L. □

The reader should realize that the proof or application of this theorem in $\mathbb{R}^m$ for $m > 1$ is more involved than what may be apparent at first glance. The increased difficulty is due to the componentwise order relation in $\mathbb{R}^m$. Note that two points are ordered if and only if each of the pairs of corresponding components is ordered in the same manner. Since in $\mathbb{R}$ there is only one component, every pair of points is ordered. That is, given any two numbers a and b exactly one of the following is true:

$$a < b, \quad a = b, \quad \text{or} \quad a > b$$

This is not the case in $\mathbb{R}^m$ for $m > 1$. For example,

$$(3, -2) < (5, 0) \quad \text{since} \quad 3 < 5 \quad \text{and} \quad -2 < 0$$

but $(3, -2)$ and $(2,0)$ cannot be ordered componentwise since $3 > 2$ and $-2 < 0$.

Section 2.3 Exercises

1. Which of the following sequences converge?

   (a) $a_n = 2/n$ for $n = 1, 2, 3, \ldots$

   (b) $b_n = (n - 1)/n$ for $n = 1, 2, 3, \ldots$

   (c) $c_n = (n + (-1)^n)/(n + 1)$ for $n = 1, 2, 3, \ldots$

   (d) $d_n = (0.8)^n$ for $n = 1, 2, 3, \ldots$

   (e) $e_n = n^2/(n + 1)$ for $n = 1, 2, 3, \ldots$

   (f) $f_n = (n + 1)/n^2$ for $n = 20, 21, 22, \ldots$

   (g) $g_n = 4^n/(3^n + 1000)$ for $n = 1, 2, 3, \ldots$

   (h) $h_n = 4^n/(1{,}000 - 3^n)$ for $n = 1, 2, 3, \ldots$

   (i) $i_n = (2^n + 1{,}000{,}000)/3^n$ for $n = 1, 2, 3, \ldots$

   (j) $j_n = (n + 1)^2/n^2$ for $n = 1, 2, 3, \ldots$

   (k) $k_n = \cos(n\pi)$ for $n = 0, 2, 4, \ldots$

   (l) $l_n = 3^n/n^{20}$ for $n = 1, 2, 3, \ldots$

   (m) $m_n = (3/n, (n + 1)/n)$ for $n = 1, 2, 3, \ldots$

   (n) $N_n = (0.2^n, (n^2 + 1)/n)$ for $n = 1, 2, 3, \ldots$

   (o) $o_n = (\sin(n)/n, 0.9^n)$ for $n = 1, 2, 3, \ldots$

2. Find the limits of the sequences in Exercise 1 that converge.

3. Which of the sequences in Exercise 1 diverge to positive infinity? To negative infinity? Neither?

4. For each of the following give an example or explain why no example exists.

   (a) A sequence converging to $(2, 3, -4)$.

   (b) A bounded sequence that does not converge.

   (c) An unbounded sequence that converges to $\left(\frac{1}{7}, \tan(3)\right)$.

   (d) A monotone sequence that is bounded above and does not converge.

   (e) A strictly increasing sequence that is bounded above and does not converge.

   (f) A divergent sequence $\{s_n\}$ which is absolutely convergent, that is, $\{|s_n|\}$ converges.

   (g) A sequence in $\mathbb{R}^2$ whose first component converges to 13 and whose second component diverges to positive infinity.

   (h) A sequence that converges to both -3 and 1.

5. Determine, by *inspection,* the convergence or divergence of the sequences defined by the following formulas. In the case of convergence, give the value of the limit. In the case of divergence, state whether the divergence is to positive infinity, negative infinity, or neither.

   (a) $a_n = (n - 3n^6)/(4n^5 - 45n^3 + 17n)$

   (b) $b_n = (3n^3 + 4n - 6)/(3n^2 + 14n - 72n^5)$

   (c) $c_n = (n \sin(n) + 16)/(n^2 + 32n)$

   (d) $d_n = (n(n^2 + 2) - 3n^3)/(n(n + 1)(n + 2))$

   (e) $e_n = ((n^2 + 3n + 2) - n^2)/(4n + 5)$

   (f) $f_n = ((3n^2 + 4n)/(2n^3 + 6n), (4n^3 + 2n^2)/(2n^2 + 6n^4))$

   (g) $g_n = ((5n^3 + 3n^4)/(3n^4 + 2n^3), 1/n)$

6. Let $s_n = (bn^p + 36n^3 - 7)/(n^{13} + 4n^2)$. For each of the following, determine values for b and p that will satisfy the given condition. The values are not necessarily unique.

   (a) $\{s_n\}$ diverges to positive infinity.

   (b) $\{s_n\}$ converges to 17.

   (c) $\{s_n\}$ converges to 0.

   (d) $\{s_n\}$ diverges to negative infinity.

7. Make up an example to illustrate an application of the squeeze theorem in $\mathbb{R}^2$.

8. Using the definition of convergence of a sequence, prove part (c) of Theorem 2.3.1 (for sequences in $\mathbb{R}$).

9. Consider the true statement: If $\lim_{n\to\infty} s_n = L$, then $\lim_{n\to\infty} |s_n| = |L|$.

   (a) What must be shown to establish that $\lim_{n\to\infty} |s_n| = |L|$? Hint: refer to the definition of convergence.

   (b) Is $||s_n| - |L|| \leq |s_n - L|$? Hint: Consider the four possible cases for the numerical signs of s_n and L (e.g., s_n positive and L negative).

   (c) Prove the statement.

10. Does the converse of the statement in Exercise 9 hold? Give a proof or a counterexample. (Note that the converse of the conditional statement "If A, then B" is the conditional statement "If B, then A.")

11. Consider part (a) of Theorem 2.3.5.

    (a) Rewrite the biconditional statement as two conditional statements.

    (b) Give an indirect proof (i.e., proof by contradiction) for each of the conditional statements in part (a). Hint: If $p(n)/q(n)$ does not converge to 0, then there is a number $k > 0$ such that $|p(n)/q(n)| > k$ for large values of n. This means that $|p(n)| > k|q(n)|$ for large values of n. Is this last inequality possible if the degree of $p(n)$ is less than the degree of $q(n)$?

12. A cow and his friend (who has a weight problem and is twice as heavy) are on opposite sides of a plank balanced on a crossbar. Our cow is moving outward, at distances $s_n = 6n^3 + 4n + 1$ from the crossbar. The friend is also moving outward, at distances $t_n = 3n^3 + 6n + 1$. In the long run, are they tending toward a position of balance? (In order for the plank to be balanced, the products of the weights and distances from the balance point must be equal.)

2.4 Limits of Functions

Convergence, the "limiting behavior" of a sequence as the elements in the domain become arbitrarily large (i.e., $n \to \infty$), was the central issue of Section 2.3. In this section, we shall extend the convergence concept to functions defined over intervals (in contrast to sequences that are only defined on the integers). Our approach will be to transform questions about convergence of functions into questions about convergence of sequences by composing the function f in question with an appropriate sequence. This will lead us to the definition of the limit of a function, the concept on which most of the calculus rests.

Our approach is to begin with the notion of approximation, as we did with sequences. The motivating question is:

Given the function $f : \mathbb{R} \to \mathbb{R}$, what number (if any) does $f(x)$ approximate when x approximates a?

Example 2.4.1 Let $f : \mathbb{R} \to \mathbb{R}$ be defined by $f(x) = (x^2 + x)/(x+1)$. What number (if any) does $f(x)$ approximate when x approximates 2?

To transform the question into one concerning sequences, we compose the function f with the sequence $x_n = 2 + z_n$ where $z_n \to 0, z_n \neq 0$. This yields the sequence:

$$f_n = f(x_n) = \frac{(2+z_n)^2 + (2+z_n)}{(2+z_n)+1}$$

Now, applying the theorem on the algebra of limits of sequences yields

$$\lim_{n\to\infty} f_n = \lim_{n\to\infty} \frac{(2+z_n)^2 + (2+z_n)}{(2+z_n)+1} = \frac{\lim_{n\to\infty}(4 + 4z_n + z_n^2 + 2 + z_n)}{\lim_{n\to\infty}(2 + z_n + 1)} = 2$$

since $z_n \to 0$. Thus $f(x)$ approximates 2 whenever x approximates 2. △

It is very important that the reader understand the composition process and how it transforms the approximation question for the function f into a convergence question for a sequence. Let us consider a few examples before formally stating a definition.

Example 2.4.2 Let the function $f : \mathbb{R} \to \mathbb{R}$ be defined by $f(x) = \frac{x^2-1}{x-1}$. What number (if any) does $f(x)$ approximate when x approximates 1? Note that f is not defined at $x = 1$.

We compose the function f with the sequence $x_n = 1 + z_n$ where $z_n \to 0$, $z_n \neq 0$. This yields the sequence

$$f_n = \frac{(1+z_n)^2 - 1}{(1+z_n)-1} = 2 + z_n$$

Note that since $z_n \neq 0$, the denominator of $f_n \neq 0$.

Now applying the theorem on the algebra of the limits of sequences yields

$$\lim_{n\to\infty} f_n = \lim_{n\to\infty} (2 + z_n) = 2$$

since $z_n \to 0$. Thus $f(x)$ approximates 2 whenever x approximates 1 even though f is not defined when $x = 1$. △

Example 2.4.3 Let the function $f : \mathbb{R} \to \mathbb{R}$ be defined by $f(x) = 6/(x+1)$. What number does $f(x)$ approximate (if any) when x approximates -1?

We compose the function f with the sequence $x_n = -1 + z_n$ and obtain the sequence

$$f_n = \frac{6}{(-1+z_n)+1} = \frac{6}{z_n}.$$

Since f_n diverges as $z_n \to 0, z_n \neq 0$, f does not approximate any number when x approximates -1.

Note that if $\{z_n\}$ were chosen to be $\{1/n\}$, then f_n would diverge to positive infinity, while if $\{z_n\}$ were chosen to be $\{-1/n\}$, f_n would diverge to negative infinity. Thus it is correct to say that f_n diverges, but it would not be correct to say that f_n diverges to either positive or negative infinity. △

Example 2.4.4 Consider the *FLOOR* function, $f(x) = [x]$, over the open interval $(-1, 3)$. What number (if any) does f approximate when x approximates 2?

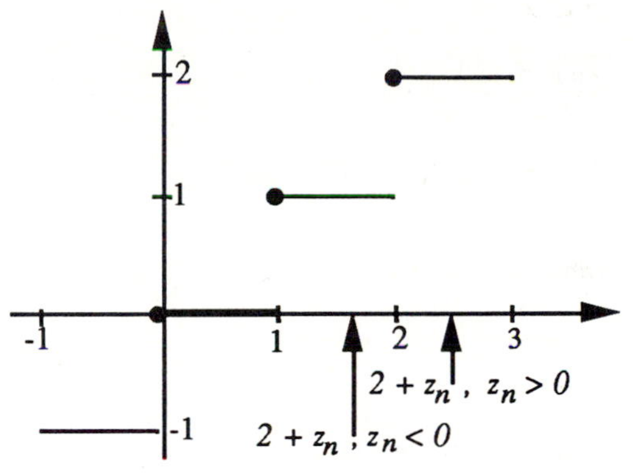

Figure 2.4.5

Let $x_n = 2 + z_n$ where $z_n \to 0, z_n \neq 0$, and consider the composition $f_n = f(x_n)$. If z_n is a positive sequence, then

$$f_n = [2 + z_n] \to 2.$$

However if z_n is a negative sequence, then

$$f_n = [2 + z_n] \to 1$$

Thus the sequence f_n does not converge (Why?) and so f does not approximate any number when x approximates 2. Note, however, that f does approximate the number 2 whenever z_n is *restricted* to be positive and f does approximates the number 1 whenever z_n is *restricted* to be negative. △

The previous examples not only illustrate the process of transforming an approximation problem for functions into a convergence problem for sequences, but also raises two basic questions that underlie our work on limits:

1. Given a function $f : \mathbb{R} \to \mathbb{R}^m, m \geq 1$, under what conditions will $\{f(x_n)\}$ converge to a point L when $\{x_n\}$ converges to a?

2. Given a function $f : \mathbb{R} \to \mathbb{R}^m, m \geq 1$, if we know that $\{f(x_n)\}$ converges to *some* point when $\{x_n\}$ converges to a, how do we determine the *actual* point?

In most instances, the answers to the above questions will be immediate consequences of the work in the last section, particularly the definition for the convergence of a sequence. The limit concept, based on the following definition, is the fundamental concept on which all the calculus rests.

Definition 2.4.6 (Limit of a Function) *Let* $f : \mathbb{R}^k \to \mathbb{R}^m$. *Let* a *be a point in* $\mathbb{R}^k$ *with* $a = \lim_{n\to\infty} x_n$ *for* some *sequence* $\{x_n\}$ *in Domain(f)* , $a \notin \{x_n\}$. *Then, the* limit of f as x approaches a is L, *written*

$$\lim_{x\to a} f(x) = L$$

if for all *sequences* $\{x_n\}$ *in Domain(f) converging to* a, $a \notin \{x_n\}$,

$$\lim_{n\to\infty} f(x_n) = L$$

We make six observations about this definition:

1. The definition does not require that the function be defined at $x = a$.
2. The limit is not required to be $f(a)$, even if f is defined at a.
3. The sequence $\{x_n\}$, $x_n \to a$, has none of its values equal to a.
4. The limit must be L *for all* sequences $\{x_n\}$ in the domain of f converging to a, not just for some.
5. The condition that some sequence $\{x_n\}$ in the domain of f, $x_n \neq a$, converges to a is to avoid having to worry about isolated points. For example, define $f : \mathbb{R} \to \mathbb{R}$ by

$$f(x) = \begin{cases} 1 & \text{for } x = 0 \\ x & \text{for } x \in [1, 2] \end{cases}$$

   where Domain(f) is $\{x : x = 0 \text{ or } 1 \leq x \leq 2\}$. Zero is an isolated point of the domain of f. Suppose isolated points were allowed. Then since no sequence $\{x_n\}$ in the domain of f, $x_n \neq 0$, converges to $a = 0$, the condition "if $\lim_{n\to\infty} x_n = 0$ for $\{x_n\}$ in Domain(f) , $x_n \neq 0$, then $\lim_{n\to\infty} f(x_n) = L$" is trivially valid since the "if condition" never holds. Thus for every L $\lim_{x\to 0} f(x) = L$, clearly an undesirable result.

6. There is an equivalent definition, the *epsilon-delta* definition, which does not mention sequences. To state that $f(x)$ is close to L, we require that the distance between them be less than an arbitrary "small" number, epsilon: $|f(x) - L| < \epsilon$. To state that x is close to (but not equal to) a, we require that the distance be less than a "small" positive number delta but greater than zero: $0 < |x - a| < \delta$. Thus the epsilon-delta definition of $\lim_{x\to a} f(x) = L$ is: for any $\epsilon > 0$ there is a $\delta > 0$ such that if $0 < |x-a| < \delta$, then $|f(x) - L| < \epsilon$.

The next example illustrates these observations.

Example 2.4.7 We will evaluate $\lim_{x\to 1} f(x)$ where $f : \mathbb{R} \to \mathbb{R}$ is defined by

$$f(x) = \begin{cases} x^2 + 4x & \text{for } x \neq 1 \\ 3 & \text{for } x = 1 \end{cases}$$

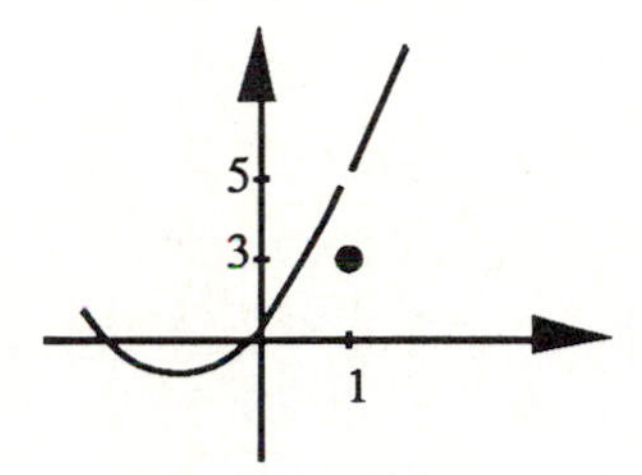

Figure 2.4.8

Since the existence and value of the limit is independent of the value of the function at $x = 1$ (see observation 1 above), we shall only consider values of $x \neq 1$. We compose the function f with the sequence $x_n = 1 + z_n$ where $z_n \to 0$, $z_n \neq 0$, and obtain

$$f_n = f(x_n) = (1 + z_n)^2 + 4(1 + z_n)$$

We now apply the theorem on the algebra of limits of sequences and obtain

$$\lim_{n\to\infty} f(x_n) = 5$$

Thus, $\lim_{x\to 1} f(x) = 5$. Note that the limit $5 \neq f(1) = 3$ (observation 2). Also note that if the definition of limit did not prohibit x_n from being 1, we would have needed to consider the sequence $x_n = 1$. Then, since

$$\lim_{n\to\infty} f(x_n) = \begin{cases} 5 & \text{for } x_n \neq 2 \text{ for all } n \\ 3 & \text{for } x_n = 2 \text{ for all } n \end{cases}$$

which would contradict observation 4. △

We will now establish the convention that whenever the sequence $\{z_n\}$ is used with a limit, it will be understood that $z_n \to 0$ and that $z_n \neq 0$.

Example 2.4.9 Consider the function $h : [0, 4] \to \mathbb{R}$ defined by

$$h(x) = \begin{cases} x^2 - 2 & \text{for } 0 \leq x < 2 \\ 2x - 3 & \text{for } 2 \leq x \leq 4 \end{cases}$$

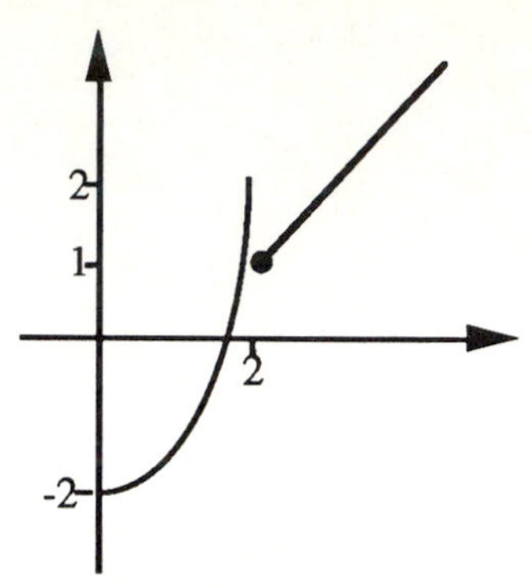

Figure 2.4.10

We will show that the limit of h does not exist as $x \to 2$. At $x = 2$ the defining formula for f changes; thus we consider how the function "behaves" as x converges to 2 by a sequence of values that are greater than 2 and also by a sequence of values that are less than 2.

First, we compose h with the sequence $x_n = 2 + z_n$ where z_n converges to 0 through positive values. We get

$$h(2 + z_n) = 2(2 + z_n) - 3$$

Thus the limit of h as $x \to 2$ through values greater than 2 is

$$\lim_{n \to \infty} h(2 + z_n) = \lim_{n \to \infty} (2(2 + z_n) - 3) = 1$$

Second, we compose h with the sequence $x_n = 2 + t_n$ where t_n converges to 0 through negative values. This yields:

$$h(2 + t_n) = (2 + t_n)^2 - 2$$

Thus the limit of h as $x \to 2$ through values less than 2 is

$$\lim_{n \to \infty} h(2 + t_n) = \lim_{n \to \infty} (2 + t_n)^2 - 2 = 2$$

Since the limits of the two sequences do not agree, the limit of h as x approaches 2 *does not exist.* (Note that Definition 2.4.6 of the limit of a function requires that $\lim_{n \to \infty} f(x_n)$ must be the same value for *all* sequences $\{x_n\}$ converging to 2.) △

This example illustrates a situation where a function fails to have a limit at a prescribed point; however "one-sided" limits exist in the sense that restricting x to approach 2 from only one side does yield a limit. We formalize this type of situation with the following two definitions.

Definition 2.4.11 (Left-Hand Limit) *Let* $f : \mathbb{R} \to \mathbb{R}^m$. *Let* a *be a point in* $\mathbb{R}$ *with* $a = \lim_{n\to\infty} x_n$ *for* some *sequence of points* x_n *in the domain of* f *with* $x_n < a$. *The* limit of f as x approaches a from the left is L *if for* all *sequences* $\{x_n\}$ *converging to* a *with* $x_n < a$, *we have* $\lim_{n\to\infty} f(x_n) = L$. *We write*

$$\lim_{x\to a^-} f(x) = L$$

and call L the left-hand limit of f at a.

Definition 2.4.12 (Right-Hand Limit) *Let* $f : \mathbb{R} \to \mathbb{R}^m$. *Let* a *be a point in* $\mathbb{R}$ *with* $a = \lim_{n\to\infty} x_n$ *for* some *sequence of points* x_n *in the domain of* f *with* $x_n > a$. *The* limit of f as x approaches a from the right is L *if for* all *sequences* $\{x_n\}$ *converging to* a *with* $x_n > a$, *we have* $\lim_{n\to\infty} f(x_n) = L$. *We write*

$$\lim_{x\to a^+} f(x) = L$$

and call L *the* right-hand limit of f at a.

The following theorem provides us with a technique for determining the behavior of a function at points arbitrarily close to the specified point. The technique is to compute the left- and right-hand limits.

Theorem 2.4.13 *The limit of* $f(x)$ *as* x *approaches* a *exists and is equal to* L *if and only if both the left- and right-hand limits exist and are equal to* L.

Example 2.4.14 Let $f : \mathbb{R} \to \mathbb{R}$ be defined by

$$f(x) = \begin{cases} x^2 + 3x & \text{for } x \le 2 \\ -5x + 20 & \text{for } x > 2 \end{cases}$$

We will determine the limit of f at $x = 2$ if it exists.

Consider an arbitrary sequence $\{z_n\}$, $z_n \to 0$ and $z_n > 0$. The left-hand limit is:

$$\lim_{x\to 2^-} f(x) = \lim_{n\to\infty} f(2 - z_n) = \lim_{n\to\infty} [(2 - z_n)^2 + 3(2 - z_n)] = 10$$

The right-hand limit is:

$$\lim_{x\to 2^+} f(x) = \lim_{n\to\infty} f(2 + z_n) = \lim_{n\to\infty} [-5(2 + z_n) + 20] = 10$$

Since both left- and right-hand limits exist and equal 10, $\lim_{x\to 2} f(x) = 10$. $\triangle$

So far all our examples have involved functions whose source and target are both $\mathbb{R}$, although the definition *did not* impose this restriction. After stating a limit theorem for functions $f : \mathbb{R} \to \mathbb{R}^m$ that is analogous to Theorem 2.2.15, we shall consider some examples of functions whose source or target is not $\mathbb{R}$. It is very helpful to keep the following two important facts in mind when applying the limit concept to functions: $\mathbb{R}^k \to \mathbb{R}^m$, where m or $k > 1$.

1. A sequence converges in $\mathbb{R}^m$ if and only if it converges componentwise.

2. A limit of a function exists at a specified point if and only if the same limit value is obtained along *every curve* leading to the specified point. In $\mathbb{R}$ there are only two possible curves to consider, one approaching from the left and one approaching from the right, while in $\mathbb{R}^m$, $m > 1$, there are infinitely many.

We will state the first as a theorem. The proof, which uses Theorem 2.2.15, is left to the exercises (Exercise 13).

Theorem 2.4.15 *If* $f : \mathbb{R} \to \mathbb{R}^m$, *then* $\lim_{x\to a} f(x)$ *is computed componentwise.*

Example 2.4.16 The function $f : \mathbb{R} \to \mathbb{R}^3$ defined by $f(x) = (x, \sin(x), 1/x)$ diverges as $x \to 0$ since the third component diverges. That is, $\lim_{x\to 0} 1/x$ does not exist. △

Example 2.4.17 Let $f : \mathbb{R}^2 \to \mathbb{R}$ be defined by

$$f(x, y) = \frac{xy - 2x}{3 - x} \quad \text{for } x \neq 3$$

We will evaluate $\lim_{(x,y)\to(1,3)} f(x, y)$.

We compose f with the sequence $\{(1 + u_n, 3 + v_n)\} \to (0, 0)$, where $u_n \to 0$, $v_n \to 0$, and $(u_n, v_n) \neq (0, 0)$. The sequence $\{(u_n, v_n)\}$ "plays the role" of $\{z_n\}$ in the previous examples. Now

$$\begin{aligned}
\lim_{(x,y)\to(1,3)} f(x, y) &= \lim_{n\to\infty} f((1 + u_n, 3 + v_n)) \\
&= \lim_{n\to\infty} \frac{(1 + u_n)(3 + v_n) - 2(1 + u_n)}{3 - (1 + u_n)} \\
&= \lim_{n\to\infty} \frac{3 + 3u_n + v_n + u_n v_n - 2 - 2u_n}{2 - u_n} \\
&= \frac{1}{2}
\end{aligned}$$

△

Example 2.4.18 Let $f : \mathbb{R}^3 \to \mathbb{R}^2$ be defined by

$$f(x, y, z) = (\sin(xy), \frac{x}{y - z}) \quad \text{for } y \neq z$$

We will evaluate the limit of f as $(x, y, z) \to (0, 1, 2)$.

We compose f with the sequence $\{(u_n, 1 + v_n, 2 + w_n)\}$ where $\{u_n\} \to 0$, $\{v_n\} \to 0$, $\{w_n\} \to 0$, and $(u_n, v_n, w_n) \neq (0, 0, 0)$. Now

$$\begin{aligned}\lim_{(x,y,z)\to(0,1,2)} f(x, y, z) &= \lim_{n\to\infty} f(u_n,\ 1 + v_n,\ 2 + w_n) \\ &= \lim_{n\to\infty} (\sin(u_n(1 + v_n)),\ \frac{u_n}{(1 + v_n) - (2 + w_n)}) \\ &= (0, 0)\end{aligned}$$

△

We now give two examples to illustrate that a function may behave very differently along different curves leading to the same point.

Example 2.4.19 (Dependence of the Limit on Approach Direction) Consider the behavior of the function $f(x, y) = xy/(x^2 + y^2)$ as $(x, y) \to (0, 0)$. This function cannot be evaluated at $(0, 0)$. We consider the line $x = 0$ (the y-axis) and approach $(0, 0)$ along this line. A sequence $\{s_n\}$ converging to $(0, 0)$, with elements on the y-axis will have the form $(0, y_n)$, with $\lim_{n\to\infty} y_n = 0$. For any such sequence $s_n = (0, y_n)$ we have

$$\lim_{n\to\infty} f_n = \lim_{n\to\infty} f(s_n) = \lim_{n\to\infty} \frac{0y_n}{0 + y_n^2} = 0$$

In this same fashion we could approach $(0, 0)$ along the line $y = 0$ (the x-axis). Setting $s_n = (x_n, 0)$ we have

$$\lim_{n\to\infty} f_n = \lim_{n\to\infty} f(s_n) = \lim_{n\to\infty} \frac{x_n 0}{x_n^2 + 0} = 0$$

We might now be tempted to say that the limit of f as (x, y) approaches $(0, 0)$ is 0. Beware of hasty conclusions! Consider an approach along the line $x = y$. To accomplish this, select any sequence $s_n = (x_n, y_n)$ converging to $(0, 0)$ with $x_n = y_n \neq 0$. Note that for any such sequence

$$\lim_{n\to\infty} f(s_n) = \lim_{n\to\infty} \frac{x_n y_n}{x_n^2 + y_n^2}$$

Since $x_n = y_n$ we get

$$\lim_{n\to\infty} f(s_n) = \lim_{n\to\infty} \frac{x_n x_n}{x_n^2 + x_n^2} = \frac{1}{2}$$

We must conclude that f has *no limit* at $(0, 0)$. △

The issue is more complex than this example has shown. We realize that agreement of limits along two different lines does not guarantee the existence of the limit. Perhaps, however, if the limits along *all straight lines agree, then the limit of the function will exist.* This seems eminently reasonable. Too bad it is false! Consider the next example.

Example 2.4.20 (Dependence of Limit on Approach Curve)
We will find the limit, if it exists, of $f(x,y) = x^2y/(x^4+y^2)$ as $(x,y) \to (0,0)$. Note that the function is not defined at $(0,0)$. Suppose we approach $(0,0)$ along *any* line of the form $y = cx$, where c is a nonzero constant. A sequence $\{s_n\}$ converging to the origin, having elements on the line will have the form $s_n = (x_n, cx_n)$, with $\lim_{n\to\infty} x_n = 0$. Composing f with $\{s_n\}$ we obtain

$$\lim_{n\to\infty} f(s_n) = \lim_{n\to\infty} \frac{x_n^2 cx_n}{x_n^4 + c^2x_n^2}$$

Dividing numerator and denominator by x_n^2 yields

$$\lim_{n\to\infty} f(s_n) = \lim_{n\to\infty} \frac{cx_n}{{x_n}^2 + c^2}$$

The numerator of this expression converges to 0 and the denominator converges to c^2. By Theorem 2.3.1

$$\lim_{n\to\infty} f(s_n) = \frac{0}{c^2} = 0$$

Now, consider an approach along the parabola $y = x^2$. Substituting the appropriate sequence $s_n = (x_n, y_n) = (x_n, x_n^2)$ into f, we get

$$\lim_{n\to\infty} f(s_n) = \lim_{n\to\infty} \frac{x_n^2 x_n^2}{{x_n}^4 + x_n^4} = \frac{1}{2}$$

Thus f has no limit at $(0,0)$. △

The moral of these last two examples is that one needs to consider the behavior of the function along *every* curve leading to the specified point. This, of course, is the natural generalization of Theorem 2.4.13.

As with Theorem 2.3.1, we have an *algebra of limits of functions* that allows us to simplify the computation of limits of functions.

Theorem 2.4.21 (Algebra of Limits of Functions) *Let $f, g : \mathbb{R}^k \to \mathbb{R}$ for $k \geq 1$. If $\lim_{x\to a} f(x) = L$ and $\lim_{x\to a} g(x) = M$, then*

(a) $\lim_{x\to a}(f(x) + g(x)) = L + M$

(b) $\lim_{x\to a} f(x)g(x) = LM$

(c) $\lim_{x\to a} cf(x) = cL$, *for all constants c*

(d) $\lim_{x\to a} f(x)/g(x) = L/M$ *provided $M \neq 0$.*

This theorem is a direct consequence of the analogous theorem for sequences (Theorem 2.3.1) and is used in much the same manner. As was true for Theorem 2.3.1, only parts (a) and (c) can be extended to the situation where the target for the functions f and g is $\mathbb{R}^m, m > 1$.

Theorem 2.4.22 (Squeeze Theorem in Several Dimensions) *If* $f, g, h : \mathbb{R} \to \mathbb{R}^m$ *and if* $f(x) \le g(x) \le h(x)$ *and* $\lim_{x\to a} f(x) = \lim_{x\to a} h(x) = L$, *then* $\lim_{x\to a} g(x) = L$.

Applications of this theorem are illustrated in the next two examples.

Example 2.4.23 We will show that $\lim_{x\to 0} \sin(x) = 0$. (Our argument will be partially a geometric one.)
Let OAB be a sector of a circle of radius one and center 0.

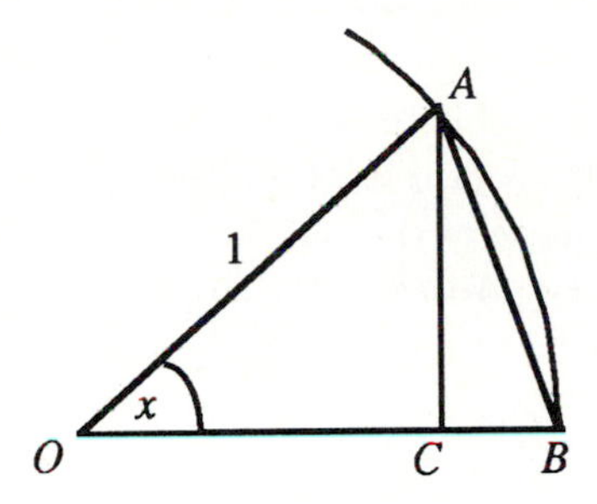

Figure 2.4.24

Recall that the length of an arc of a circle is equal to the length of the radius times the measure of the central angle (in radians) subtended by the arc. Thus the length of the arc AB is $|OA|x = x$. Since $|OA| = 1$, $\sin(x) = |AC|/|OA| = |AC|$. Since $|AC| \le chordAB \le arcAB$, we can apply the Squeeze Theorem with $f(x) = 0$, $g(x) = |AC|$, and $h(x) = arcAB$. Thus

$$0 \le \sin(x) \le x$$

and since

$$\lim_{x\to 0} h(x) = \lim_{x\to 0} x = 0, \quad \lim_{x\to 0} \sin(x) = 0 \qquad \triangle$$

As an immediate consequence of this result and the fundamental trigonometric identity, $\sin^2(x) + \cos^2(x) = 1$, we have

$$\lim_{x\to 0} \cos^2(x) = \lim_{x\to 0}(1 - \sin^2(x)) = 1$$

and thus

$$\lim_{x\to 0} \cos(x) = 1$$

Why is this last limit a positive one rather than a positive or negative one?

Example 2.4.25 We will show that $\lim_{x\to 0} \sin(x)/x = 1$.

We shall give a geometric argument combined with an application of the Squeeze Theorem and the definition of limit. Let OAB be a sector of a circle of radius 1 and center O.

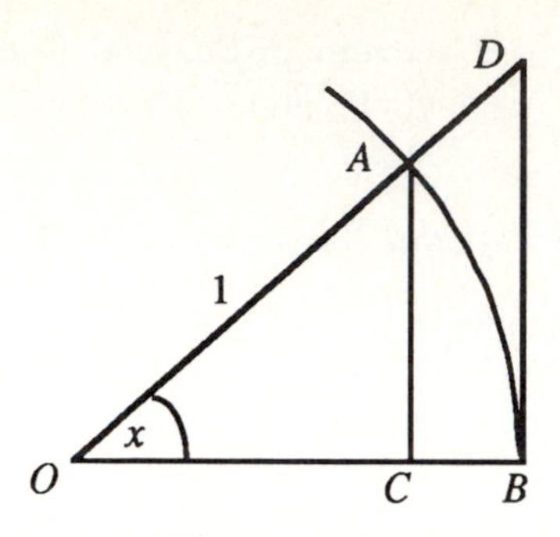

Figure 2.4.26

Recall that the area of a sector of a circle is one-half the measure of the central angle x times the square of the radius of the circle. Thus the area of the sector OAB is $\frac{1}{2}|OB|^2x = \frac{1}{2}x$. Also recall that

$$\sin(x) = \frac{|AC|}{|OA|} = |AC|, \quad \text{since } |OA| = 1$$
$$\cos(x) = \frac{|OC|}{|OA|} = |OC|, \quad \text{since } |OA| = 1$$

and

$$\tan(x) = \frac{|DB|}{|OB|} = |DB|, \quad \text{since } |OB| = 1$$

The fact that the sector OAB contains the triangle OAC and is contained in the triangle ODB yields the following double inequality that in turn will lead to an application of the Squeeze Theorem.

$$area(\Delta OAC) \leq area(sector OAB) \leq area(\Delta ODB)$$

Thus

$$\frac{1}{2}|OC||AC| \leq \frac{1}{2}x \leq \frac{1}{2}|OB||BD|$$

or

$$\frac{1}{2}\cos(x)\sin(x) \leq \frac{1}{2}x \leq \frac{1}{2}\tan(x)$$

Dividing by $\sin(x)$ and multiplying by 2, we obtain

$$\cos(x) \leq \frac{x}{\sin(x)} \leq \frac{1}{\cos(x)}$$

or, taking reciprocals,

$$\frac{1}{\cos(x)} \leq \frac{\sin(x)}{x} \leq \cos(x)$$

Here we need $0 < x < \pi/2$ to make all terms positive and thus maintain the direction of the inequality under the operations.

Now, since $\cos(x) \to 1$ as $x \to 0$, we apply the Squeeze Theorem to obtain

$$\lim_{x \to 0+} \frac{\sin(x)}{x} = 1$$

Clearly, a similar argument with $x < 0$ yields $\lim_{x \to 0-} \sin(x)/x = 1$. Thus by Theorem 2.4.13,

$$\lim_{x \to 0} \frac{\sin(x)}{x} = 1$$

△

Example 2.4.27 As an application of the use of the preceding example, we will evaluate $\lim_{x \to 0} \tan(x)/x$. Note that the function $f(x) = \tan(x)/x$ is not defined at $x = 0$.

Following the approach illustrated in the previous examples, we compose f with an arbitrary sequence $\{z_n\}$ which converges to 0, $z_n \neq 0$, and obtain

$$f(z_n) = \frac{\tan(z_n)}{z_n}$$

Thus

$$\begin{aligned} \lim_{x \to 0} \frac{\tan(x)}{x} &= \lim_{n \to \infty} \frac{\tan(z_n)}{z_n} \\ &= \lim_{n \to \infty} \frac{\sin(z_n)}{z_n \cos(z_n)} \\ &= \lim_{n \to \infty} \frac{\sin(z_n)}{z_n} \frac{1}{\cos(z_n)} \\ &= (1)(1) = 1 \end{aligned}$$

△

Question 2.4.28 Give an intuitive argument that $\lim_{x \to 0} \sin(x)/x = 1$ implies that x is a good approximation to $\sin(x)$ for small values of x. "Check" your reasoning by using a computer or calculator to plot the graphs of x and $\sin(x)$ over the interval $[-0.5, 0.5]$. ◇

The final result in this section involves the limit of a function when one or more of the variables approaches positive or negative infinity. We shall illustrate the situation with examples and leave the construction of a formal definition as an exercise (Exercise 7).

Example 2.4.29 Let $f : \mathbb{R} \to \mathbb{R}$ be defined by $f(x) = x\sin(x)/(x^2+2)$. Does f approximate any number as $x \to \infty$?

In contrast to the sequences $\{z_n\} \to 0$ in previous examples, consider an arbitrary sequence $\{t_n\} \to \infty$. Composing f with $\{t_n\}$ and then taking the limit of the resulting sequence, yields

$$\lim_{x\to\infty} f(x) = \lim_{n\to\infty} f(t_n) = \lim_{n\to\infty} \frac{t_n \sin(t_n)}{{t_n}^2+2}$$

Since

$$0 \le \left|\frac{t_n \sin(t_n)}{t_n^2+2}\right| \le \frac{t_n}{t_n^2+2}$$

the Squeeze Theorem implies that $\lim_{x\to\infty} |f(x)| = 0$ and thus by Theorem 2.2.32 $\lim_{x\to\infty} f(x) = 0$. △

Example 2.4.30 Let $f : \mathbb{R}^2 \to \mathbb{R}$ be defined by $f(x,y) = (xy^2+3y)/(x^2+y^2)$. We will evaluate the limit, if it exists, as $x \to 2$ and $y \to \infty$.

Composing f with the sequence $\{(2+z_n, t_n)\}$ where $z_n \to 0$, $z_n \ne 0$, and $t_n \to \infty$ yields

$$f((2+z_n,t_n)) = \frac{(2+z_n)t_n^2+3t_n}{(2+z_n)^2+t_n^2}$$

Following the method illustrated in Example 2.3.4, we divide numerator and denominator by t_n^2 to obtain

$$\lim_{n\to\infty} f((2+z_n,t_n)) = \lim_{n\to\infty} \frac{2+z_n+3/t_n}{(2+z_n)^2/t_n^2+1} = 2$$

by Theorem 2.3.1. △

Section 2.4 Exercises

1. Compose the given function with the given sequence and list the first five elements of the new sequence.

    (a) $f(x) = 2x-1, \quad s_n = 1/n \quad$ for$n = 1, 2, 3, \ldots$

    (b) $g(x) = x - 1/x, \quad a_n = n+2 \quad$ for $n = 1, 2, 3, \ldots$

    (c) $f(x) = (x-1)/x, \quad b_n = (-1)^n/n \quad$ for $n = 1, 2, 3, \ldots$

    (d) $f(x) = (x-1)/x, \quad c_n = 1/n^2 \quad$ for $n = 1, 2, 3, \ldots$

    (e) $g(x) = \begin{cases} x^2 & \text{for } x < 0 \\ 1+x & \text{for } x > 0 \end{cases} \quad d_n = 1/n \quad$ for $n = 1, 2, 3, \ldots$

(f) $g(x) = \begin{cases} x^2 & \text{for } x < 0 \\ 1+x & \text{for } x > 0 \end{cases}$ $\quad d_n = (-1)^n/n \quad$ for $n = 1, 2, 3, \ldots$

2. For each of the following, compose the given function with an appropriate sequence and then compute the limit (if it exists).

(a) $\lim_{x\to\infty}(x^2-1)/(x^2+1)$ (b) $\lim_{x\to\infty}(x^2-1{,}000)/(x+1{,}000{,}000)$

(c) $\lim_{t\to-\infty}\sin(\pi t)/(1-t)$ (d) $\lim_{x\to\infty}(1/x-1)/(x-1)$

(e) $\lim_{x\to 2}(2x-1)$ (f) $\lim_{x\to 0}(2-5x)$

(g) $\lim_{x\to -2} x^2$ (h) $\lim_{x\to 1} 3/(x+1)$

(i) $\lim_{t\to 0}|t|$. (j) $\lim_{s\to 0} s/|s|$

(k) $\lim_{x\to 3}(2x-6)/(x-3)$ (l) $\lim_{x\to 2}(x^2-4)/(x-2)$

(m) $\lim_{x\to 1}(x^3-1)/(x-1)$ (n) $\lim_{x\to 0}(2x-5x^2)/x$

(o) $\lim_{x\to 0} f(x)$ where $f(x) = \begin{cases} -x^2 & \text{for } x < 0 \\ x^2 & \text{for } x > 0 \end{cases}$

(p) $\lim_{x\to 0} g(x)$ where $g(x) = \begin{cases} x^2 & \text{for } x < 0 \\ 1+x & \text{for } x > 0 \end{cases}$

(q) $\lim_{x\to 2}((2x+1)/(x^2-2), (x^2-4)/(x-2), 2)$

(r) $\lim_{x\to 0}(2-\sqrt{4-x}/x)$

(s) $\lim_{(x,y)\to(0,0)}(x+y)(x+3y)$

(t) $\lim_{x\to 1}(1-x^3)/(1-x)$

(u) $\lim_{(x,y)\to(1,2)}(xy+3x)/(x\sin(y))$

(v) $\lim_{x\to 0}(\sqrt{x+2}-\sqrt{2})/x$

(w) $\lim_{(x,y)\to(3,1)}(x^2y-3y+4)/(xy^2-2xy)$

(x) $\lim_{(x,y)\to(-1,0)}(|x^2-1{,}000y|\sin(xy))/(x^2+y^2)$

(y) $\lim_{x\to -2}(x^2, 1/(x+2), |x|, x+2)$

(z) $\lim_{x\to\infty}(\sin(x)/x^2, \cos(x)/x, \sin^2(x)+\cos^2(x))$

3. Compute the one-sided limits, when they exist.

   (a) $\lim_{x\to 1^-} \frac{1}{x-2}$

   (b) $\lim_{x\to 1^-} FLOOR(x)$

   (c) $\lim_{x\to -3^+} \frac{|x+3|}{x+3}$

   (d) $\lim_{x\to 0^-} x|x-1|$

   (e) $\lim_{(x,y)\to(2^-,3^+)} \left(\frac{x-2}{|x-2|}, \frac{y-3}{|y-3|}\right)$

4. Compute the limit of f as (x, y) approaches (0,0) along the indicated curves.

$$f(x,y) = \frac{x^2 - y^2}{x^2 + y^2}$$

   (a) $y = x$ (b) $y = x/2$ (c) $y = x^2$ (d) $y = 2x$

5. Compute the limit of g as (x, y) approaches (1,1) along the indicated curves.

$$g(x,y) = \frac{x - y^4}{x^3 - y^4}$$

   (a) $x = 1$ (b) $y = 1$ (c) $x = y^4$

6. For each of the following, give an example or explain why no example exists.

   (a) A function with a left-hand limit of 0 at 1 and a right-hand limit of 2 at 1.
   (b) A function that has no limit at 0.
   (c) A function whose limit at 0 is 0 and whose left hand-limit at 0 is 1.
   (d) A function whose limit at 2 is 0.

7. Let $f : \mathbb{R} \to \mathbb{R}$. State a formal definition for the limit of $f(x)$ as $x \to \infty$.

8. Use the Squeeze Theorem to show that

$$\lim_{x\to\infty} \frac{\sin(x)\cos(x)}{x^2 + 1} = 0$$

9. Define $f : \mathbb{R} \to \mathbb{R}$ by the relation $f(x) = \sin(1/x)$.

   (a) Compute $\lim_{n\to\infty} f(s_n)$ for the sequence $s_n = 2/((1 + 4n)\pi)$.
   (b) Compute $\lim_{n\to\infty} f(t_n)$ for the sequence $t_n = 2/((3 + 4n)\pi)$.

(c) Using the results of parts (a) and (b), prove that $\lim_{x\to 0} f(x)$ does not exist.

10. Prove that $\lim_{x\to a} x^4 = a^4$ in two ways. First, using the definition of limit, and second, using the theorem on the algebra of limits, Theorem 2.4.21, and the fact that $\lim_{x\to a} x = a$.

11. A cow barn's roof that projects into a square on the ground with 8-foot sides, is to be described mathematically. The outer edge is 10 feet above the ground. The roof is to gently rise in a parabolic shape as it approaches the centerline, which is 15 feet above the ground. At the very center is a flagpole which rises an additional 10 feet. Construct a function h which gives the height of the roof and flagpole. At what points (a, b) is $\lim_{(x,y)\to(a,b)} h(x,y) \neq h(a,b)$?

12. Rewrite Theorem 2.4.13 as two conditional statements joined with the conjunction "and."

13. Prove Theorem 2.4.15.

14. Give an intuitive argument that $\lim_{n\to\infty} 2^{z_n} = 1$ if $z_n \to 0$ by finding an n_0 such that if $n > n_0$ then $|2^{1/n} - 1| < 0.01$. (The argument is not complete, since the definition of limit requires that the limit be 1 for *all* sequences z_n converging to zero, not just $z_n = 1/n$.)

15. (Mike Henle) As x approaches 0, $\sin(\sin(x))/x^k$ is defined and nonzero for exactly one value of the constant k. What is the value of k and what is the limit?

2.5 Continuity

A function is a relation or a mapping which assigns elements in the range set to elements in the domain set. Functions are often classified according to the properties of the domain that are preserved under the mapping. Since one-to-one functions send distinct elements into distinct elements, one-to-one functions *preserve* distinctness. Increasing functions *preserve* order since if $a < b$, then $f(a) < f(b)$. Having noted that the limit concept is the basic concept upon which all the calculus rests, it is not surprising that the class of functions that preserve the limit [i.e., if x converges to a, then $f(x)$ converges to $f(a)$] is among the most important of all classes of functions in mathematics. Such functions are called *continuous functions.* That is, if a function preserves or "continues the limit" at the point a, it is said to be *continuous at* $x = a$. Geometrically, the graph of a function is "unbroken" at $x = a$ provided f is continuous at $x = a$. The following example illustrates three typical situations with respect to the question of preserving or continuing the limit. You should pay particular attention to the behavior of the function at x = 0, 3, and 5.

Example 2.5.1

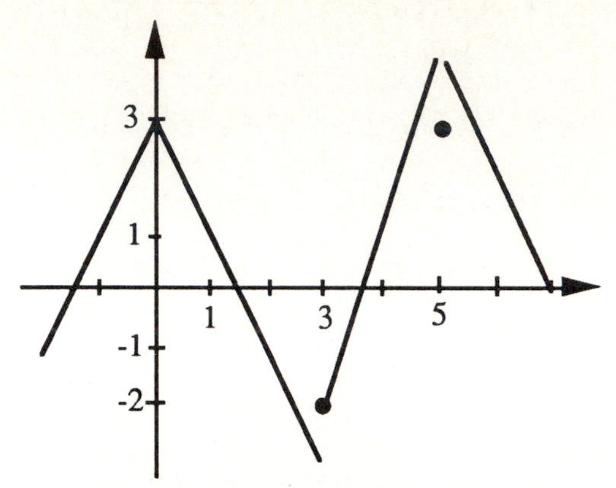

Figure 2.5.2

Consider the function $f : \mathbb{R} \to \mathbb{R}$, given by

$$f(x) = \begin{cases} 2x+3 & \text{for } x < 0 \\ -2x+3 & \text{for } 0 \leq x < 3 \\ 3x-11 & \text{for } 3 \leq x < 5 \\ 3 & \text{for } x = 5 \\ -2x+14 & \text{for } 5 < x \end{cases}$$

Although the formula changes at $x = 0$, there is no break in the graph at this point. In contrast, at $x = 3$ the graph "jumps." Technically, the jump is the result of the fact that the one-sided limits of f do not agree at $x = 3$. The problem at $x = 5$ is not caused by a disagreement between the one-sided limits. The problem here is that $\lim_{x\to 5} f(x) \neq f(5)$. Thus f does not continue the limit and hence is not continuous at $x = 3$ or 5, but is continuous for every other point in the domain of f. △

Definition 2.5.3 (Continuity at a Point) *Let $f : \mathbb{R}^k \to \mathbb{R}^m$ and let p be a point in Domain(f). The function f is* continuous at p *if* $\lim_{x\to p} f(x) = f(p)$.

Usually our functions will be defined on intervals or rectangles. However, note that by our definition of the limit of a function, Definition 2.4.6, p cannot be an isolated point of Domain(f). Thus the function $f : \mathbb{R} \to \mathbb{R}$ with Domain(f) = $\{x : x = 0, \text{ or } 1 \leq x \leq 2\}$ and $f(x) = 1$ for all x in Domain(f) is not continuous at $x = 0$, since 0 is an isolated point, but f is continuous at all other points in its domain.

If f is continuous at each *point* in its domain, then f is said to be a continuous function. Note that since $\lim_{x\to p} x = p$, we could (symbolically) emphasize the property of continuing the limit by writing f is continuous at $x = p$ if and only if

$$\lim_{x\to p} f(x) = f(\lim_{x\to p} x)$$

Since we can only talk about continuity properties of a function at points where the function is defined (i.e., its domain), it does not make any sense to say that a function is continuous or is not continuous at a point where the function is not defined. This rather obvious restriction is illustrated in the next example.

Example 2.5.4 Consider the hyperbola given by $h(x) = 1/x$.

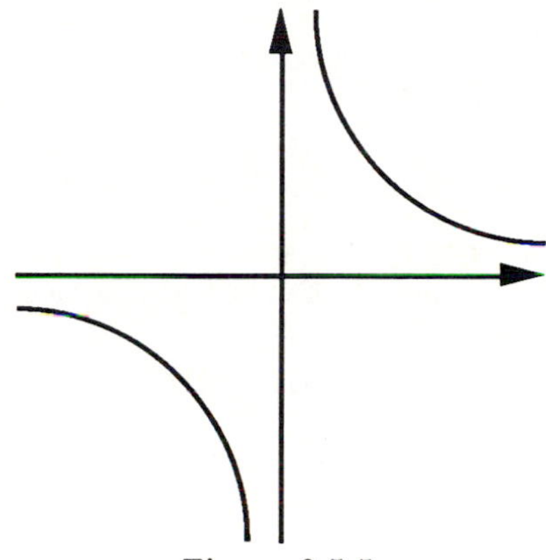

Figure 2.5.5

The function is continuous even though the graph is in two parts. This is a reflection of the fact that the domain of h consists of two disjoint intervals (i.e., 0 is *not* in the domain of h, so the question of the continuity of h at 0 does not arise). △

Basically, a function is continuous at the point p if $f(x)$ closely approximates $f(p)$ when x closely approximates p. Continuous functions are, in some sense, predictable functions.

A consequence of Theorem 2.4.13 provides a practical method for determining the continuity of a function at a point.

Theorem 2.5.6 *A function* $f : \mathbb{R} \to \mathbb{R}^m$ *is continuous at the point* p *if*

(a) $\lim_{x \to p^-} f(x)$ *exists*

(b) $\lim_{x \to p^+} f(x)$ *exists, and*

(c) $\lim_{x \to p^-} f(x) = \lim_{x \to p^+} f(x) = f(p)$

Notes: (a) If the one-sided limits exist at p and do not agree, then the function cannot be continuous at a.

(b) If the domain of f is $[a, b]$, then the left-hand limit does not exist at $x = a$ no matter what f is. Thus f may be continuous at $x = a$ without both left- and right- hand limits existing. As an example, consider $f(x) = 1$ with Domain(f) = [0, 1]. The function is continuous at all points in its domain, but right-hand limits do not exist at 1 and left-hand limits do not exist at 0.

(c) A function $f : \mathbb{R} \to \mathbb{R}^m$ is continuous at a point p if and only if each component function is continuous at p.

Example 2.5.7 Consider the function f of Example 2.5.1. To establish the continuity of f at 0 consider

$$\lim_{x\to 0^-} f(x) = \lim_{x\to 0^-}(2x+3) = 2(0)+3 = 3$$

$$\lim_{x\to 0^+} f(x) = \lim_{x\to 0^+}(-2x+3) = -2(0)+3 = 3$$

and $\quad f(0) = -2(0) + 3 = 3$

Similarly, to establish that f is not continuous at 3 consider

$$\lim_{x\to 3^-} f(x) = \lim_{x\to 3^-}(-2x+3) = -2(3)+3 = -3$$

$$\lim_{x\to 3^+} f(x) = \lim_{x\to 3^+}(3x-11) = 3(3)-11 = -2$$

Since the one-sided limits do not agree at 3, f is not continuous at 3. △

Example 2.5.8 The constant function $f : \mathbb{R} \to \mathbb{R}$, given by $f(x) = c$ is continuous on all of its domain $\mathbb{R}$. △

Example 2.5.9 The identity function $i : \mathbb{R} \to \mathbb{R}$ given by $i(x) = x$ is a continuous function on $\mathbb{R}$. To establish this consider

$$\lim_{x\to p} i(x) = \lim_{x\to p} x = p = i(p)$$

for any value of p. △

We combine the simple facts of the last two examples with the following familiar-looking theorem to establish the continuity of large classes of functions.

Theorem 2.5.10 (Algebra of Continuous Functions) *Let $f, g : \mathbb{R}^k \to \mathbb{R}$ be continuous functions at the point p. Then,*

(a) $f + g$ is continuous at p,

(b) fg is continuous at p.

(c) If $g(p) \neq 0$, then f/g is continuous at p.

Note that $f - g = f + (-g)$, thus subtraction is included.

The importance of this theorem and the two preceding examples cannot be overemphasized for they imply that:

1. All polynomial functions of one or more variables are continuous.

2. All rational functions (quotients of polynomials) are continuous at every point in their domain. (Recall that rational functions are undefined at points where the denominator is zero.)

3. All trigonometric functions are continuous. To see this, suppose we want to show that sine is continuous at p. We must show that if $z_n \to 0$, $z_n \neq 0$, then $\lim_{n\to\infty} \sin(p + z_n) = \sin(p)$.

We know that $\lim_{n\to\infty} \sin(z_n) = 0 = \sin(0)$ and $\lim_{n\to\infty} \cos(z_n) = 1 = \cos(0)$ (by Example 2.4.23 and its consequences), so

$$\begin{aligned} \lim_{n\to\infty} \sin(p + z_n) &= \lim_{n\to\infty} [\sin(p)\cos(z_n) + \sin(z_n)\cos(p)] \\ &= [\sin(p)](1) + (0)[\cos(p)] \\ &= \sin(p) \end{aligned}$$

Similarly cosine is continuous (see the exercises), and thus by the algebra of continuous functions, $\tan(x) = \sin(x)/\cos(x)$ and the other trigonometric functions are continuous.

We state the following theorem now in order to complete (with respect to continuity) the basic classes of functions even though a complete discussion must await Chapter 7.

Theorem 2.5.11 *The exponential and logarithmic functions are continuous.*

Thus the vast majority of functions that we have encountered in all our mathematics courses are continuous functions!

Another way of viewing continuous functions is to note that these are functions *whose limits can be found by substitution.* For example, consider $f(x) = x^3 - 2x + 6$. Since f is a polynomial and hence continuous,

$$\lim_{x\to 2} f(x) = f(2) = 10$$

Example 2.5.12 We will evaluate $\lim_{(x,y)\to(1,2)}(x^3y^4 - 3xy^2 + 2x - 4)$.

Since $p(x, y) = (x^3y^4 - 3xy^2 + 2x - 4)$ is a polynomial in x and y, it is continuous and thus

$$\lim_{(x,y)\to(1,2)} p(x, y) = p(1, 2) = 2$$

Easy! △

Example 2.5.13 Let $f : \mathbb{R}^2 \to \mathbb{R}^3$ be defined by

$$f(x, y) = (x \sin(y), \cos(x), x^2y^3)$$

Is f a continuous function? Let (a, b) be an arbitrary point in the domain of f, $\mathbb{R}^2$. Since each component is a continuous function, the limit of each component can be obtained by substitution. Hence

$$\lim_{(x,y)\to(a,b)} f(x, y) = (a \sin(b), \cos(a), a^2b^3) = f(a, b)$$

Thus, by the definition, f is a continuous function. △

Example 2.5.14 Let $f : \mathbb{R} \to \mathbb{R}$ be defined by

$$f(x) = \begin{cases} x^2 + 3x & \text{for } x \leq 2 \\ -x + b & \text{for } x > 2 \end{cases}$$

For what value of b will f be continuous at $x = 2$?

For f to be continuous at $x = 2$, $\lim_{x\to 2} f(x) = f(2) = 10$. Thus by Theorem 2.4.13,

$$\lim_{x\to 2^-} f(x) = 10 \quad \text{and} \quad \lim_{x\to 2^+} f(x) = 10$$

That is,

$$\lim_{x\to 2^-} (x^2 + 3x) = 10 \quad \text{and} \quad \lim_{x\to 2^+} (-x + b) = 10$$

Since $-x + b$ is a polynomial and thus continuous,

$$\lim_{x\to 2^+} (-x + b) = -2 + b = 10.$$

Thus $b = 12$. △

An important question is: Does the composition of continuous functions yield a continuous function? Let us consider the composition of two functions

$$f : A \to B \quad g : B \to C \quad g \circ f : A \to C$$

Let $h = g \circ f$. Thus $h(x) = g(f(x))$. If f and g are continuous, is h continuous? Paraphrasing, a good approximation of x yields a good approximation of $f(x)$. If g is evaluated at this good approximation of $f(x)$, will this yield a good approximation of $h(x)$? Sounds reasonable! The following theorem guarantees it; the proof is left to the exercises.

Theorem 2.5.15 *The composition of continuous functions is continuous.*

Note that it is *not possible* to compose two real-valued functions of several variables (Why?).

The next result shows the power of this last theorem as well as establishing a very useful result. The proof illustrates a common proof technique in mathematics, that of "breaking up" a result into parts, establishing each part, and then "gluing" the parts back together.

Theorem 2.5.16 *Let $g : \mathbb{R} \to \mathbb{R}$ be a continuous function and suppose p and q are integers. Then $[g(x)]^{p/q}$ is a continuous function.*

Proof: We express $[g(x)]^{p/q}$ as a composite function. If $h(x) = x^p$ and $k(x) = x^{1/q}$, then $h(k(g(x))) = [g(x)^{1/q}]^p = [g(x)]^{p/q}$.

The proof will now be broken down into two parts. Part 1 will establish that $k(x) = x^{1/q}$ is a continuous function. Then using Theorem 2.5.15, Part 2 will show that $k(g(x)) = [g(x)]^{1/q}$ is continuous. A second application of Theorem 2.5.15 will yield that $h(k(g(x))) = [g(x)]^{p/q}$ is continuous, the desired result.

Part 1: We shall establish the result for $q = 2$, show how the argument extends to the case for $q = 3$, and leave the general result to the more adventurous reader. Thus we consider the question: Is $k(x) = x^{1/2}$ a continuous function? That is, is

$$\lim_{x \to p} k(x) = p^{1/2} \quad \text{for } p > 0?$$

Equivalently, is

$$\lim_{x \to p} [k(x) - p^{1/2}] = 0 \quad \text{for } p > 0?$$

Now,

$$\begin{aligned}
\lim_{x \to p} [k(x) - p^{1/2}] &= \lim_{n \to \infty} [k(p + z_n) - p^{1/2}] \\
&= \lim_{n \to \infty} [(p + z_n)^{1/2} - p^{1/2}] \\
&= \lim_{n \to \infty} [(p + z_n)^{1/2} - p^{1/2}] \left[\frac{(p + z_n)^{1/2} + p^{1/2}}{(p + z_n)^{1/2} + p^{1/2}} \right]
\end{aligned}$$

This last step is called "multiplying and dividing by the *conjugate.*" Continuing we get

$$\lim_{x \to p} [k(x) - p^{1/2}] = \lim_{n \to \infty} \frac{p + z_n - p}{(p + z_n)^{1/2} + p^{1/2}} = \lim_{n \to \infty} \frac{z_n}{(p + z_n)^{1/2} + p^{1/2}}$$

Now we can apply the Squeeze Theorem. Since

$$0 \leq \left| \frac{z_n}{(p + z_n)^{1/2} + p^{1/2}} \right| \leq \left| \frac{z_n}{p^{1/2}} \right|$$

and

$$\lim_{n\to\infty}\left|\frac{z_n}{p^{1/2}}\right| = 0$$

then

$$\lim_{n\to\infty}\left|\frac{z_n}{(p+z_n)^{1/2}+p^{1/2}}\right| = 0$$

Thus, by Theorem 2.2.32,

$$\lim_{n\to\infty}\frac{z_n}{(p+z_n)^{1/2}+p^{1/2}} = 0$$

and therefore

$$\lim_{x\to p} k(x) = p^{1/2} = k(p)$$

Hence k is a continuous function.

The only difference in the argument for the case $q = 3$ is in the expression for the conjugate. The conjugate of

$$(p+z_n)^{1/3} - p^{1/3}$$

is

$$(p+z_n)^{2/3} + (p+z_n)^{1/3}p^{1/3} + p^{2/3}$$

Part 2: Since k and g are continuous functions, their composition $k(g(x)) = [g(x)]^{1/q}$ is a continuous function (Theorem 2.5.15). Now, h is a polynomial function and hence is continuous. Thus, applying Theorem 2.5.15 again, $h(k(g(x))) = [g(x)]^{p/q}$ is a continuous function. □

It is often very instructive to consider (both analytically and geometrically) how a function can fail to satisfy a given property. We illustrate this type of analysis by looking at the different ways that a function can fail to be continuous at $x = 2$.

1. A function has a limit at $x = 2$, but is not continuous there since $\lim_{x\to 2} f(x) \neq f(2)$. For example, consider

$$f(x) = \begin{cases} x & \text{for } x \neq 2 \\ 1 & \text{for } x = 2 \end{cases}$$

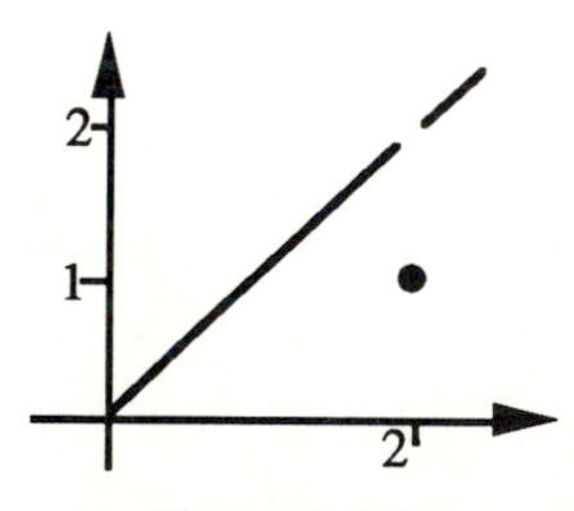

Figure 2.5.17

2. A function is not continuous because the limit does not exist even though f is defined at $x = 2$ and both one sided limits exist as $x \to 2$. Thus f is bounded in a neighborhood of $x = 2$. For example, consider

$$f(x) = \begin{cases} x & \text{for } x \leq 2 \\ 3 & \text{for } x > 2 \end{cases}$$

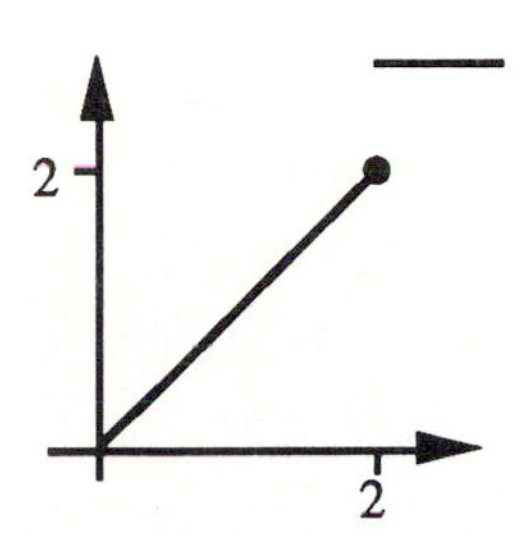

Figure 2.5.18

3. A function f is not continuous because the limit does not exist and f is not bounded in a neighborhood of $x = 2$. For example, consider

$$f(x) = \begin{cases} 1/(x-2)^2 & \text{for } x \neq 2 \\ 1 & \text{for } x = 2 \end{cases}$$

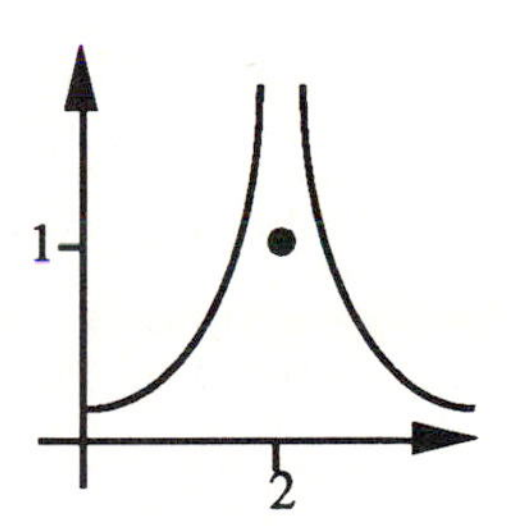

Figure 2.5.19

Geometrically, we see that a function is discontinuous at a point if the function is defined at the point and its graph is "broken" at that point. Furthermore, a graph can be broken by having a single point "misplaced" as in 1 or by having a "finite jump" as in 2 or by having an "infinite jump" as in 3.

Section 2.5 Exercises

1. State where each of the following functions are continuous.

(a) $f(x) = x^2 - 2x + 3$ (b) $g(x) = 1/(|x| + 1)$

(c) $h(x) = x/(x^3 - 1)$ (d) $h(x) = x/(x^2 + 1)$

(e) $f(x, y) = (x - y)/(x^2 + xy)$ (f) $g(x, y) = (xy + 2)/|xy|$

(g) $h(x) = (\sin(x)/x, \cos(x), x^2)$ (h) $f(x, y) = (x/y, \tan(x), x - y)$

(i) $g(x, y, z) = (xz, y/(x - z))$ (j) $h(x, y) = (y/(x + 1), \sin(x), \tan(y))$

2. Let $f(x) = \begin{cases} x^2 & \text{for } x < 1 \\ Ax - 3 & \text{for } x \geq 1 \end{cases}$

   If f is continuous at 1, what is the value of the constant A?

3. Let $g(x) = \begin{cases} A^2x^2 & \text{for } x \leq 2 \\ (1 - A)x & \text{for } x > 2 \end{cases}$

   For what values of A is g continuous at 2?

4. Let $h(x) = \begin{cases} Ax - B & \text{for } x \leq 1 \\ 3x & \text{for } 1 < x < 2 \\ Bx^2 - A & \text{for } 2 \leq x \end{cases}$

   Find values for A and B for which h is continuous at 1 and discontinuous at 2.

5. Let $f(x, y) = \begin{cases} (Ax + 2, 3y, 7) & \text{for } x < 2 \text{ and all } y \\ (4y, 4B, 7) & \text{for } x \geq 2 \text{ and all } y \end{cases}$

   For what values of A and B is f continuous at (2,13)?

6. Give an example or state why no example exists.

   (a) A discontinuous function whose left- and right-hand limits agree as $x \to 2$.

   (b) A polynomial in x and y that is not continuous at (2,3).

   (c) A continuous function that is negative for $x < 3$ and positive for $x \geq 3$.

   (d) A continuous function that is not monotonic.

   (e) A function that is continuous at $x = 2$, but is unbounded in every interval containing 2.

7. Consider the function f: $\mathbb{R} \to \mathbb{R}$ defined by $f(x) = \cos(x)$.

   (a) What must be shown to establish that f is continuous?

   (b) Compute the $\lim_{x \to p} f(x)$ by composing f with an appropriate sequence and then evaluating the limit of the resulting composition.

   (c) Prove that $f(x) = \cos(x)$ is a continuous function.

8. Consider the function $f : \mathbb{R} \to \mathbb{R}$ defined by $f(x) = x/(x^2 + 1)$.

   (a) What must be shown to establish that f is a continuous function?

   (b) Compute the $\lim_{x \to p} f(x)$ by composing f with an appropriate sequence and then evaluating the limit of the resulting composition.

   (c) Prove that f is a continuous function.

9. A cow barn's square roof with 8-foot sides is to be described mathematically. The outer edge is 10 feet above the ground. The roof is to gently rise as it approaches the centerline, which is 15 feet above the ground. At the very center is a flagpole which rises an additional 10 feet. Construct a *continuous* function h which gives the height of the roof.

10. Prove by the definition that $f(x) = (x^2 + 2, x - 3)$ is a continuous function. Hint: If the component functions are continuous, is f continuous? Why?

11. Prove that $f(x) = |x|$ is a continuous function. Hint: Follow the proof development outline [parts (a), (b), and (c) of Exercise 7].

12. Prove that $f(x, y) = (xy, \sin(x), \tan(y))$ is continuous at (0,0).

13. There are several ways in which a function can fail to be continuous. If the one-sided limits of a function agree with one another at a point but *do not* agree with the value of the function at that point, then the function is said to have a *removable singularity* at the point. If the one-sided limits do not agree with each other, then the function is said to have an *essential singularity* at the point in question. Give an example of each of the following.

    (a) A function with a removable singularity at -1.

    (b) A function with an essential singularity at 0.

    (c) A function with an essential singularity at 2 and a removable singularity at 4.

    (d) A function with an infinite number of removable singularities.

14. Prove Theorem 2.5.15. [Hint: Let f and g denote the two continuous functions and let $h(x) = f(g(x))$. Apply the definition of continuity twice, first to $g(x)$ and then to $f(g(x))$].

15. Prove that $f : \mathbb{R} \to \mathbb{R}$ defined by

$$f(x) = \begin{cases} 0 & \text{if } x \text{ is irrational} \\ 1 & \text{if } x \text{ is rational} \end{cases}$$

    is discontinuous at every real number. (Hint: Note that between any two real numbers there is both a rational number and an irrational number. Then use proof by contradiction.)

2.6 Three Important Theorems

In this section we present three important theorems that describe frequently used properties of continuous functions. In addition, we present an application of the "Bisection Algorithm." This algorithm has probably been used (informally) by everyone several times. It is the basis for effectively playing the game of Twenty Questions, carrying out a search (for a lost person or a needle), or implementing a sorting procedure in computer science.

The first theorem illustrates the fundamental difference between a function being continuous at a point and merely having a limit at that point. It is often referred to as the *sign-preserving* property of continuous functions.

Theorem 2.6.1 *Let $f : \mathbb{R} \to \mathbb{R}$ be a continuous function at $x = p$ and let $f(p) \neq 0$. Then there exists an open interval centered at p over which $f(x)$ has the same numerical sign as $f(p)$.*

Proof: (By contradiction) We shall prove the case where $f(p) > 0$. The case where $f(p) < 0$ will be left as an exercise.

Assume that the conclusion is false. That is, in every open interval centered at p, there exists an x in Domain(f) such that $f(x) \leq 0$. Let $\{I_n\}$ be the sequence of open intervals, each one centered at p,

$$I_n = (p - \frac{1}{n}, p + \frac{1}{n})$$

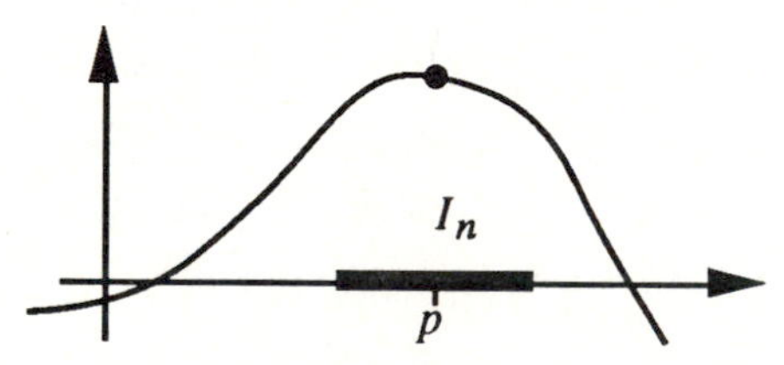

Figure 2.6.2

Let x_n be a point in I_n and in Domain(f) for which $f(x_n) \leq 0$. Since

$$p - \frac{1}{n} < x_n < p + \frac{1}{n}, \quad \text{then } x_n \to p$$

by the Squeeze Theorem. Thus the continuity of f guarantees that $\{f(x_n)\}$ converges to the *positive* number $f(p)$. However $f(x_n)$, being nonpositive, is bounded above by zero. Thus we have a logical contradiction since $\{f(x_n)\}$ being convergent and bounded above by zero must converge to a nonpositive number. □

Example 2.6.3 To show that the continuity of f at p is a necessary condition in the preceding theorem rather than just having the limit exist as $x \to p$, consider the function $f : \mathbb{R} \to \mathbb{R}$, defined by

$$f(x) = \begin{cases} 0 & \text{for } x < 0 \\ 1 & \text{for } x = 0 \\ 0 & \text{for } x > 0 \end{cases}$$

Although f is positive at $x = 0$ and the limit of f exists as $x \to 0$, there is no open interval centered at 0 over which f is positive, since any open interval containing 0 also contains $x < 0$ and $x > 0$ where the function is zero. △

The second important theorem justifies a procedure that you have probably used numerous times in high school in finding the roots (zeros) of an equation. For example, if you were asked to find a root of $x^3 - x - 3 = 0$, you might have said that since $f(x) = x^3 - x - 3$ is negative for $x = 0$ and positive for $x = 2$, there must be a root between 0 and 2. Geometrically, you were saying that since the graph of f passes through the points $(0, -3)$ and (2,3), it must cross the x-axis somewhere between $x = 0$ and $x = 2$. This is certainly true, and the reason is that f, being a polynomial, is a *continuous* function and functions continuous over an interval have the property that their graphs are unbroken. The next theorem, called the *Intermediate Value Theorem,* makes this notion precise.

Theorem 2.6.4 (Intermediate Value Theorem) *Let $f : \mathbb{R} \to \mathbb{R}$ be continuous on $[a, b]$ and let K be any number between $f(a)$ and $f(b)$. Then there is at least one number c in $[a, b]$ for which $f(c) = K$.*

A picture makes the statement of this theorem clear. The following figure shows the graph of a continuous function f with $f(a) > f(b)$. As x moves from a to b, the graph of f must touch the line $y = K$ *at least* once (in the figure, the graph actually touches the line twice).

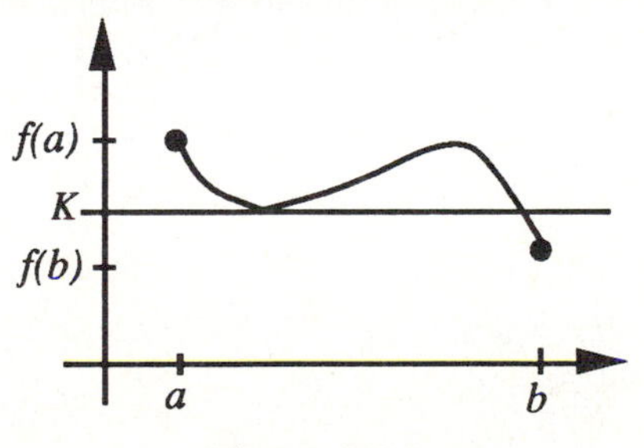

Figure 2.6.5

For the situation pictured, there are two possible choices for the "c" of the theorem.

The theorem asserts that a continuous function takes on all values between any two values it assumes, or, equivalently, if f is continuous on the closed interval $[a, b]$, then the closed interval with $f(a)$ and $f(b)$ as endpoints is in the range of f. (Suppose that $f(a) \leq f(b)$, then the closed interval with $f(a)$ and $f(b)$ as endpoints is $[f(a), f(b)]$.)

Proof (of the Intermediate Value Theorem): Suppose f is continuous on $[a, b]$. Exactly one of the following must be true:

$$\begin{aligned}&\text{(a) } f(a) = f(b)\\&\text{(b) } f(a) < f(b)\\&\text{(c) } f(a) > f(b)\end{aligned}$$

In case (a) the Intermediate Value Theorem is true since there are no values between $f(a)$ and $f(b)$ except $f(a) = f(b)$. We will concentrate our efforts on case (b). The analysis of case (c) is analogous.

Given some value K with $f(a) < K < f(b)$ we are going to ***algorithmically*** construct three sequences, $\{L_n\}$, $\{R_n\}$, and $\{M_n\}$. The method used will be the Bisection Algorithm introduced in Chapter 1. To begin the process we let $L_0 = a$ and $R_0 = b$ (L and R denote left and right, respectively). We let M_0 be the midpoint of $[L_0, R_0]$, $M_0 = (L_0 + R_0)/2$. If $f(M_0) = K$, we have found the desired c, namely $c = M_0$.

If $f(M_0) \neq K$, we reason that the desired value of c must be either to the left of M_0 or to the right of M_0. Therefore, compute L_1 and R_1 as follows:

$$\begin{aligned}&\text{If } f(M_0) < K, \text{ then } L_1 = M_0 \text{ and } R_1 = R_0\\&\text{If } f(M_0) > K, \text{ then } L_1 = L_0 \text{ and } R_1 = M_0\end{aligned}$$

We now compute $M_1 = (L_1 + R_1)/2$. If $f(M_1) = K$, we are done. If not, the process is repeated. That is,

$$\begin{aligned}&\text{If } f(M_1) < K, \text{ then } L_2 = M_1 \text{ and } R_2 = R_1\\&\text{If } f(M_1) > K, \text{ then } L_2 = L_1 \text{ and } R_2 = M_1\end{aligned}$$

We can construct three sequences by continuing in this manner. Of course, if at any point $f(M_n) = K$, $c = M_n$ is the sought after value.

These sequences can be defined formally with recursion:

$$\begin{aligned}L_0 = a,\ R_0 = b,\ M_0 &= (a+b)/2\\\text{If } f(M_n) = K, \text{ then } L_{n+1} = R_{n+1} &= M_n\\\text{If } f(M_n) < K, \text{ then } L_{n+1} = M_n \text{ and } R_{n+1} &= R_n\\\text{If } f(M_n) > K, \text{ then } L_{n+1} = L_n \text{ and } R_{n+1} &= M_n\\M_{n+1} &= (L_n + R_n)/2\end{aligned}$$

At each stage of the algorithm (for each value of n) we bisect the interval and continue to work with the remaining half that must contain a solution.

Several questions must be asked. Do these sequences converge? If they do converge, to what do they converge? How does the algorithm solve the problem of finding c with $f(c) = K$? How does any of this prove the Intermediate Value Theorem?

First, if $f(M_n) = K$ for some value of n, then the bisection process stops and we have proven the theorem by finding the desired $c = M_n$. Thus we may assume that $f(M_n)$ is never K and the intervals will continue to be bisected. At each stage the length of the closed interval $[L_n, R_n]$ is one-half the length of $[L_{n-1}, R_{n-1}]$. Since the length of $[a, b]$ is $|b - a|$, the length of $[L_n, R_n]$ is $(b - a)/2^n$, that is

$$R_n - L_n = \frac{b - a}{2^n}$$

Hence we have

$$\lim_{n \to \infty} (R_n - L_n) = 0$$

Now, $\{R_n\}$ is a decreasing sequence that is bounded below (by a, for example). Thus $\{R_n\}$ must converge. Furthermore, as was just pointed out, since the terms of $\{L_n\}$ become arbitrarily close to those of $\{R_n\}$, $\{L_n\}$ must have the same limit as $\{R_n\}$. Since each term of $\{M_n\}$ is "between" the corresponding terms of $\{L_n\}$ and $\{R_n\}$, by the Squeeze Theorem $\{M_n\}$ converges to the common limit of the other two sequences. Call this common limit c.

Thus far we have not used the continuity of f. But, since f is continuous at c we know that $\lim_{x \to c} f(x) = f(c)$. From the definition of the limit of a function at a point we know that since $\lim_{n \to \infty} L_n = c$ and $\lim_{n \to \infty} R_n = c$,

$$\lim_{n \to \infty} f(L_n) = \lim_{n \to \infty} f(R_n) = f(c)$$

Note also that left endpoints were always chosen so that $f(L_n) < K$ and right endpoints were chosen so that $f(R_n) > K$. This guarantees that

$$\lim_{n \to \infty} f(L_n) \leq K \quad \text{and} \quad \lim_{n \to \infty} f(R_n) \geq K$$

Both of these limits are $f(c)$ so we may write

$$f(c) \leq K \quad \text{and} \quad f(c) \geq K$$

so that $f(c) = K$.

The questions have now been answered. The sequences always converge to some number c with $f(c) = K$. The theorem is proven by the existence of the limit! □

The Bisection Algorithm has served as the theoretical basis for proving the Intermediate Value Theorem. As a practical and useful tool, the Bisection Algorithm is a good numerical technique for finding roots of equations as indicated in Chapter 1.

As an application of the Intermediate Value Theorem, we shall show in the following theorem that every positive real number has a positive nth root. Recall that $b = a^{1/n}$ if and only if $b^n = a$.

Theorem 2.6.6 *Let n be a positive integer and let k be a positive real number. Then there is exactly one positive real number c such that $c^n = k$.*

Proof: Choose a number b greater than one such that $0 < k < b$ and define a function $f : [0, b] \to \mathbb{R}$ by $f(x) = x^n$. Since f is continuous and

$$f(0) = 0 < k < b < b^n = f(b)$$

the Intermediate Value Theorem implies that there is at least one number, say c, in $(0, b)$ for which $f(c) = k$. Since f is strictly increasing, it is a one-to-one function and thus there cannot be a second number d such that $d^n = a$. □

Example 2.6.7 Alpha and a friend go on an overnight camping trip to the top of Mount View, a distance of 6 miles. They leave at 8 o'clock in the morning and arrive on top at noon. The next morning they again leave at 8 o'clock and arrive home at 11. Over lunch, Alpha's friend poses an interesting question. He asks that since we set out at 8 o'clock each morning, was there any time when we were at exactly the same point on the trail both days? Alpha, who had completed a calculus course the previous semester, was quick to say "of course, the Intermediate Value Theorem guarantees it." Here is Alpha's reasoning, check to see if it is correct.

Let $c(t)$ be a climbing function that measures the distance traveled by time t [i.e., $c(4)$ is the distance to the top]. Let $d(t)$ be a descending function that measures the distance left to go at time t [e.g., $d(0) = 6$, $d(3) = 0$]. Now let $f(t) = c(t) - d(t)$. We assume that functions c and d (and hence f) are continuous. Since

$$f(0) = c(0) - d(0) = 0 - 6 < 0 \quad \text{and} \quad f(3) = c(3) - d(3) = c(3) - 0 > 0$$

the Intermediate Value Theorem guarantees that there is a number t_0, $0 \leq k \leq 3$ such that $f(t_0) = 0$. Thus on both mornings at time $t = t_0$ Alpha and his friend were at the same point on the trail. △

Example 2.6.8 We will use the Bisection Algorithm to find a solution to $x^5 - 10 = 0$ with an accuracy of 0.01. We had a similar problem in Example 1.5.7 of Chapter 1. Here we will use the Intermediate Value Theorem to justify the method. Let $f(x) = x^5 - 10$. Since $f(0) = -10$ and $f(2) = 22$, the Intermediate Value Theorem tells us that there is a point p in $[0, 2]$ with $f(p) = 0$, i.e., $p^5 = 10$. The Bisection Algorithm gives us a sequence of points M_1, M_2,...obtained by bisecting the interval. The distance between M_1 and p is less than 1, the half-width of the interval. Each time we bisect the interval the maximum possible error is cut in half. If we do n bisections, the error will be cut in half n times, so the error will be reduced to less than $2/2^n$. If $n = 8$, then after 8 bisections the error will be less than $2/2^8 = 1/128 < 0.01$. Thus we perform the bisection 8 times to obtain an approximate solution to $x^5 - 10 = 0$ with an error of at most 0.01. The values of M are:

$$\begin{aligned}
M_1 &= 1.00\\
M_2 &= 1.50\\
M_3 &= 1.75\\
M_4 &= 1.625\\
M_5 &= 1.5625\\
M_6 &= 1.59375\\
M_7 &= 1.578125\\
M_8 &= 1.5853975
\end{aligned}$$

The value p is within 0.01 of 1.5853975. △

Before stating the third important theorem of this section, consider the following question:

Question 2.6.9 Given a real-valued function of a real variable $f : \mathbb{R} \to \mathbb{R}$ with domain A, is there some number c in A such that $f(c) \geq f(x)$ for all x in A? The value $f(c)$ is a maximum of the range of f. In other words, does f attain a maximum on A?

We might also ask if f attains a minimum. The next several examples show the pathologies that might prevent this.

Example 2.6.10 Let $f : [-1, 2] \to \mathbb{R}$ be defined by

$$f(x) = \begin{cases} x & \text{for } -1 \leq x < 1 \\ -x^2 - 1 & \text{for } \ \ 1 \leq x \leq 2 \end{cases}$$

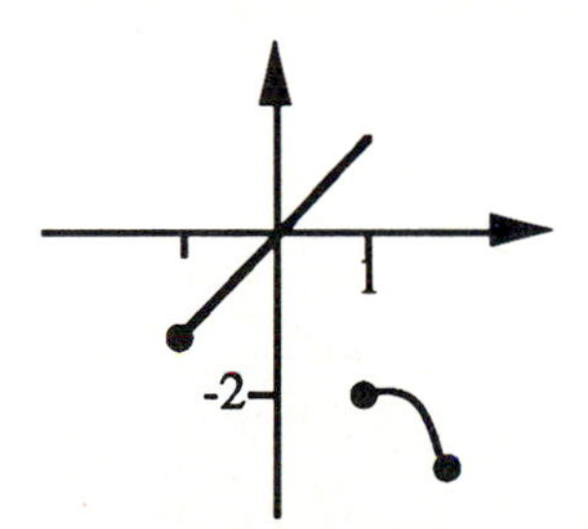

Figure 2.6.11

Although f is bounded above, f has no maximum. The least upper bound of $f(A)$ is 1 but there is no x in A with $f(x) = 1$. Specifically, the range of f is $[-5, 1)$. The function does attain a minimum at $x = 2$. Note that f is not continuous at $x = 1$. △

Example 2.6.12 Let $g(x) = x$ for $0 < x \leq 1$.

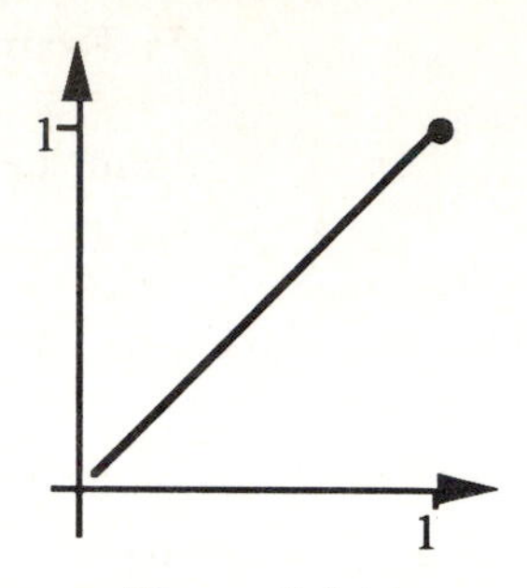

Figure 2.6.13

This function is bounded above and below but does not attain a minimum on its domain. The range of g is (0,1]. △

Example 2.6.14 A function that is not bounded above cannot attain a *maximum.* This can occur with a bounded or unbounded domain.

Define $r : \mathbb{R} \to \mathbb{R}$ with Domain(r) equal to $\mathbb{R}$ by $r(x) = x$.

Define $s : \mathbb{R} \to \mathbb{R}$, with Domain($s$) equal to (0,1], by $s(x) = 1/x$.

The function r has an unbounded domain and is not bounded above or below. It does not attain a maximum or minimum but is continuous. By contrast, s has bounded domain and a range that is not bounded above. It is continuous but does not attain a maximum.

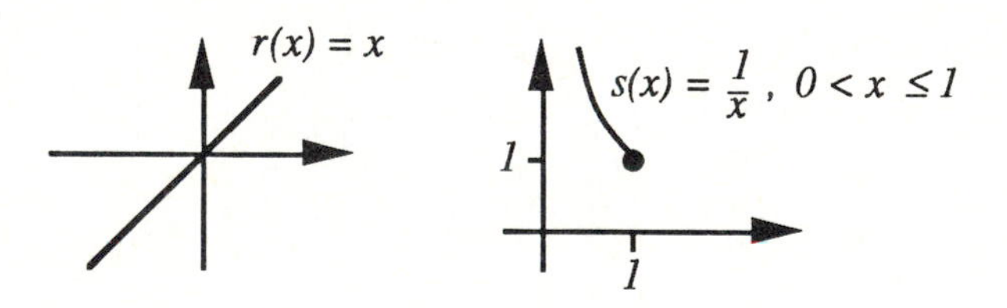

Figure 2.6.15 △

What then, is the secret!? Under what condition is a function *guaranteed* to attain a maximum and a minimum? The conditions required are those that appear in the hypotheses of many of the major theorems of calculus. It is important to notice that the form of the domain of the function is crucial in this theorem.

Theorem 2.6.16 (Extreme Value Theorem) *If $f : \mathbb{R} \to \mathbb{R}$ is continuous on domain(f) $= [a, b]$, then f attains a maximum and a minimum on $[a, b]$.*

Note that minimum and maximum values are called *extreme values* or *extrema.* The theorem tells us that a function which is continuous on a *closed* and *bounded* interval attains extreme values on that interval.

We enumerate the problems with the functions in the previous examples. In Example 2.6.10, f is not continuous at 1. In Example 2.6.12, the domain of g does not contain the left endpoint. In Example 2.6.14, the domain of r is unbounded and the domain of s does not contain the left endpoint.

To prove the Extreme Value Theorem we use a variation of the Bisection Algorithm. Our algorithm will produce a sequence that converges. The limit of the sequence will be a point at which the function attains a maximum. A completely analogous algorithm will yield minima.

Proof (Extreme Value Theorem): Let $f : \mathbb{R} \to \mathbb{R}$ be continuous on the closed interval $[a, b]$. Define sequences $\{a_n\}$, $\{b_n\}$, and $\{c_n\}$ with the following recursions

$$a_1 = a, \quad b_1 = b, \quad c_1 = \frac{a+b}{2}$$

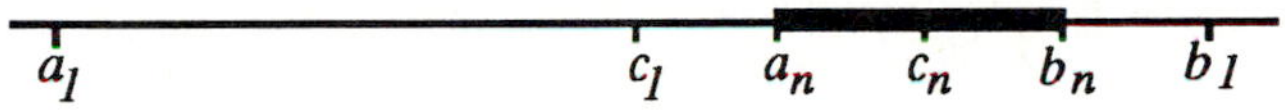

a_1 c_1 a_n c_n b_n b_1

Figure 2.6.17

If there is an x in $[a_n, c_n]$ such that $f(x) \geq f(y)$ for all y in $[c_n, b_n]$, then define

$$a_{n+1} = a_n, \quad b_{n+1} = c_n$$

otherwise define

$$a_{n+1} = c_n, \quad b_{n+1} = b_n$$

In all cases

$$c_{n+1} = \frac{a_{n+1} + b_{n+1}}{2}$$

and at each stage of the algorithm, $[a_n, b_n]$ contains the maximum of f on $[a, b]$.

Basically, at each stage we consider two halves of an interval. If *some* number in the left-half interval yields a larger function value than *every* number in the right-half interval, the right-half interval is discarded and the bisection occurs again. Otherwise, some number in the right-half interval dominates the left-half interval and the left-half is discarded. The process continues and the intervals become progressively smaller.

Note that the algorithm does *not* provide a method for deciding which half interval has a number dominating the other half interval. Although the algorithm will prove the theorem, it is not a good practical method for finding extreme values.

As with the Bisection Algorithm, the sequences $\{a_n\}$, $\{b_n\}$, and $\{c_n\}$ must all converge to the same point. Let their common limit be denoted by c. It remains for us to prove that $f(c) \geq f(x)$ for all x in $[a, b]$.

Choose any point x_1 in [a,b]. If $x_1 = c$, the inequality is certainly true, so assume $x_1 \neq c$. Since x_1 and c are *different* real numbers, there must be some point in the algorithm when x_1 and c are no longer in the same interval. At this point the algorithm chooses the interval containing c, not x_1. Therefore, there must be a number x_2 in the same half interval containing c with $f(x_2) \geq f(x_1)$.

We proceed with x_2 just as we did with x_1. If $x_2 = c$, then $f(c) \geq f(x_1)$ as desired; if not there is some stage where x_2 and c are separated and there is an x_3 such that $f(x_3) \geq f(x_2)$. Continuing the process we generate the sequence

$$f(x_1) \leq f(x_2) \leq f(x_3) \leq f(x_4) \ldots$$

If the process ends at stage n with $x_n = c$, then $f(x_1) \leq f(c)$ as we wished to show. If not, we have an infinite sequence $\{x_n\}$ converging to c. Since f is *continuous,*

$$\lim_{n \to \infty} f(x_n) = f(\lim_{n \to \infty} x_n) = f(c)$$

With $\{f(x_n)\}$ an increasing sequence, all elements of $\{f(x_n)\}$ are less than or equal to the limit $f(c)$. In particular, $f(x_1) \leq f(c)$, as was to be shown. □

In Chapter 3 we will discover a way to find where f attains its maximum that avoids a lot of the difficulty involved in the proof of this theorem.

We now state (without proof) an Intermediate Value Theorem and an extreme value theorem for real-valued functions of several real variables.

Theorem 2.6.18 (The Intermediate Value Theorem in $\mathbb{R}^n$) *Let $f : \mathbb{R}^n \to \mathbb{R}$ be a continuous function with domain, a closed and bounded rectangle in $\mathbb{R}^n$. If*

$$a = (a_1, a_2, \ldots, a_n) \quad \text{and} \quad b = (b_1, b_2, \ldots, b_n)$$

are points in Domain(f), then for every real number K between $f(a)$ and $f(b)$ there must be some point

$$c = (c_1, c_2, \ldots, c_n)$$

in Domain(f) with $f(c) = K$.

Theorem 2.6.19 (The Extreme Value Theorem in $\mathbb{R}^n$)
Let $f : \mathbb{R}^n \to \mathbb{R}$ be a continuous function, with domain a closed and bounded rectangle in $\mathbb{R}^n$. Then f attains both a maximum and a minimum on its domain.

These theorems can be proven using a multivariate generalization of the Bisection Algorithm. However, more general techniques belonging to a subject known as *topology* are usually employed.

Section 2.6 Exercises

In Exercises 1 through 4, use the Intermediate Value Theorem to show that there is a root (zero) of the function in the indicated interval.

1. $f(x) = 3x^4 - 2x - 17$, [0,2]
2. $f(x) = \sin(x)$, [1,5]
3. $f(x) = (x^2 - 6x + 4)/(x - 3)$, [0,1]
4. $f(x) = \tan(x)$, $[-1, 1]$
5. Give an example or state why no example exists.
   (a) A function of two variables defined over the unit square $\{0 \le x \le 1, 0 \le y \le 1\}$ which has a minimum, but no maximum value.
   (b) A function defined over [0,1] that obtains its maximum value at $x = 0$ and $x = 1$.
   (c) A function $f : \mathbb{R} \to \mathbb{R}$ that has a limit at every point, but does not satisfy the "sign preserving" property.
   (d) A continuous function $f : (0, 1) \to \mathbb{R}$ that has neither a maximum nor a minimum value.
   (e) Two functions $f, g : \mathbb{R} \to \mathbb{R}$ such that

$$\max\{f(x) + g(x)\} > \max\{f(x)\} + \max\{g(x)\}$$

6. Explain that there was a time in your life when you were exactly 29 inches tall.
7. Let p be a polynomial with at least two terms. If the coefficients of the highest powered term and the lowest powered term have opposite signs, prove that $p(x) = 0$ for at least one x.
8. Let $f(x) = \tan(x)$. Even though $f(\pi/4) = 1$ and $f(3\pi/4) = -1$, there is no point a in $[\pi/4, 3\pi/4]$ for which $f(a) = 0$. Why does this not contradict the Intermediate Value Theorem?
9. Let $f(x) = (x + 1)/(x^2 - 9)$. Now $f(0) = -1/9$ and $f(4) = 5/7$. However, there is no point a between 0 and 4 for which $f(a) = 0$. Why does this not contradict the intermediate value theorem?
10. If $f : \mathbb{R} \to \mathbb{R}$ is a continuous function with the property that its range is contained in the set of integers, prove that f must be a constant function.
11. Use the Bisection Algorithm to find a solution to the given equation to the given accuracy. Use a computer or calculator to do your calculations.

(a) $x^3 + 7 = 0$ on $[-3, -1]$; accuracy 0.1

(b) $x^5 + x^2 + 1 = 0$ on $[-2, -1]$; accuracy 0.01

(c) $x^4 - 3x - 1 = 1$ on [1,2]; accuracy 0.02

(d) $x^2 - 2/x = 4$ on [1,3]; accuracy 0.001

12. Let $f : [0, 1] \to \mathbb{R}$ be a continuous function bounded below by 0 and above by 1. Prove that there is at least one point a in [0,1] which f maps into itself. That is, $f(a) = a$. Such a point is called a *fixed point* of f. (For definition of "fixed point," see Exercise 21 in Section 1.5.) Hint: Apply the Intermediate Value Theorem to $h(x) = f(x) - x$.

13. If you lost a contact lens in a rectangular room that measured 10 feet by 20 feet, how many "yes-no" questions would you need to ask in order to be able to describe a section of the floor 1 foot square that was guaranteed to contain your lens? Hint: Develop a two dimensional bisection process.

14. Explain the statement: "Thus the continuity of f guarantees that $\{f(x_n)\}$ converges to the *positive* number $f(p)$" in the proof of Theorem 2.6.1.

15. Prove Theorem 2.6.1 for the case where $f(p) < 0$.

16. State and prove a theorem as to the maximum and minimum values attained by the exponential function $f(x) = b^x$, $b > 1$.

17. Using your result in Exercise 16, what can one conclude about the logarithm function $\log_b$?

18. Prove or disprove that $\cos(x) = x$ has at least one solution. Hint: Apply the Intermediate Value Theorem to $f(x) = \cos(x) - x$.

19. Consider a cookie with an even bead of icing around the edge. Is it possible to divide the cookie into two equal parts? Is it possible to divide the cookie into two equal parts in such a way that will also divide the icing equally? Explain your answers.

20. In this exercise we will examine the error analysis of the Bisection Algorithm and how to use it in finding solutions to equations. Suppose that a continuous function $f : \mathbb{R} \to \mathbb{R}$ satisfies $f(a)f(b) < 0$, where $a < b$.

    (a) Why must f have a root in $[a, b]$?

    (b) If we wish to determine the root with an accuracy of at least ACC, how many bisections may be necessary? (Express your answer as a function of a, b, and ACC.)

    (c) How many bisections are necessary to find a root of $x^5 + 4x^3 + 2x - 8$ in $[1, 2]$ with an accuracy of 0.001?

    (d) Find a root of $x^5 + 4x^3 + 2x - 8$ in $[1, 2]$ with an accuracy of 0.001.

2.7 Curve Sketching in the Large

In Chapter 1, our curve sketching consisted primarily of plotting points and then connecting the points with a smooth (continuous) curve. This last operation was an "act of faith" for without continuity we could not justify sketching an unbroken curve. In this section, we shall show that useful curve sketching can be done "in the large" using only intercept points, asymptotes, and properties of continuous functions. The intermediate value property discussed in the last section is the key continuity property.

Definition 2.7.1 *An* asymptote *of a graph of a function is a straight line that the graph approximates arbitrarily closely for large positive or negative values of either x or y.*

To determine a horizontal asymptote, evaluate the limit of $f(x)$ as x approaches positive or negative infinity. If the limit is L, then $y = L$ is a horizontal asymptote. The graph of a function may have no horizontal asymptote. For example, $f(x) = x$ is such a function. A graph may have a single horizontal asymptote, such as in the case of $f(x) = (x+1)/x$. Or, the graph of a function may have two horizontal asymptotes. Consider the graph of $f(x) = \tan^{-1}(x)$.

$f(x) = \tan^{-1}(x)$

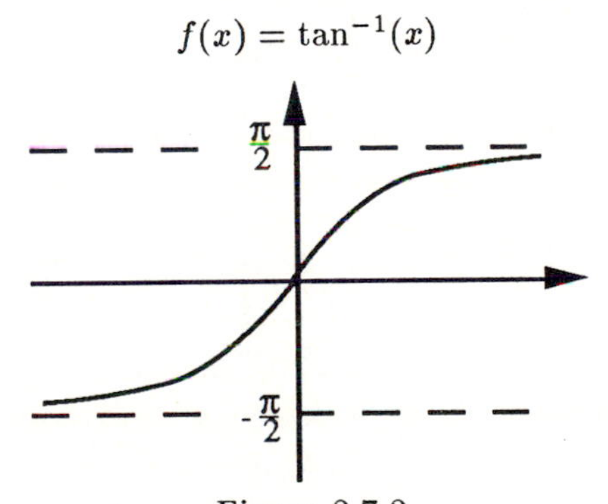

Figure 2.7.2

To determine vertical asymptotes, find the values that make the denominator of $f(x)$ equal to zero. Suppose b is such a number. If $f(x)$ becomes unbounded as $x \to b$ from *either* side, $x = b$ is a vertical asymptote. A graph may have several vertical asymptotes. For example, the graph of

$$f(x) = \frac{2x^3 - 3}{(x-1)(x+3)(x-4)}$$

has three vertical asymptotes ($x = 1$, $x = -3$, and $x = 4$) and one horizontal asymptote ($y = 2$).

The graph of $f(x) = \tan(x)$ has an infinite number of vertical asymptotes. Every vertical line of the form $x = \pi/2 + n\pi$, n an integer, is a vertical asymptote.

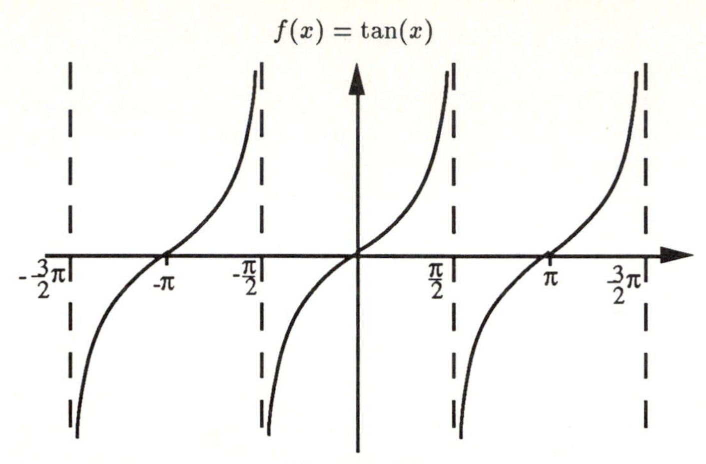

Figure 2.7.3

Note that a graph may cross a horizontal asymptote, but not a vertical asymptote. Why?

The objective in curve sketching is to obtain a useful sketch with a minimum of effort. The accuracy required depends, of course, on the problem being analyzed. In this section we are concerned with sketching the behavior of a function "in the large." For example, we will want to know whether a function diverges to positive infinity or to negative infinity as x approaches a vertical asymptote, whether the curve lies above, below, or intersects a horizontal asymptote, but we will allow considerable leeway in finding the actual point of intersection. We shall not be concerned with "local properties" such as local maxima or minima or concavity (these considerations will be taken up in Chapter 3).

Mastering the art of curve sketching involves making use of the obvious properties that can be obtained by inspection, asking "What has to happen, given the information I know?" and *a lot of experience.* We shall walk through some examples.

Example 2.7.4 We will sketch the graph of

$$f(x) = \frac{(2x^3 - 3)}{(x-1)(x+3)(x-4)}$$

We first check for asymptotes. By inspection (mentally divide by x^3 and then take the limit as $x \to \infty$) , we see that $y = 2$ is a horizontal asymptote and $x = 1$, -3, and 4 are vertical asymptotes. Note that the vertical asymptotes partition the plane into vertically unbounded regions. Since it is always helpful to determine how a function behaves at the endpoints of its domain, we consider how the function behaves as x approaches each of the vertical asymptotes from either side. If the function is not defined at a vertical asymptote, then its behavior is that it either diverges to positive infinity or to negative infinity. Thus we need merely to determine the numerical sign of $f(x)$ as x approaches a vertical asymptote.

For example, consider for $x \to 4^+$, $f(x) = (2x^3 - 3)/((x-1)(x+3)(x-4))$. We have x approaching 4 through values greater than 4. Since the numerator $(2x^3 - 3)$ is positive and each of the factors in the denominator is positive for $x > 4$, $f(x)$ is positive and thus the curve will approach the vertical asymptote from the right, "going (up) to positive infinity."

Next consider for $x \to 4^-$, $f(x) = (2x^3-3)/((x-1)(x+3)(x-4))$. Since only one factor is negative in the defining expression for $f(x)$ when $x < 4$, but close to 4, $f(x)$ is negative and thus the curve will approach the vertical asymptote from the left "going (down) to negative infinity."

Now, consider $x \to 1^+$. Since the numerator is negative, two factors in the denominator are positive, and one is negative when $x > 1$, but close to 1, $f(x)$ is positive (even number of negative factors) and thus the curve will approach the vertical asymptote $x = 1$ from the right "going to positive infinity."

Similar analysis yields that:

The curve approaches $x = 1$ from the left going to negative infinity.
The curve approaches $x = -3$ from the right going to negative infinity.
The curve approaches $x = -3$ from the left going to positive infinity.

We indicate how the curve approaches the vertical asymptotes by drawing small arrows as in the following sketch.

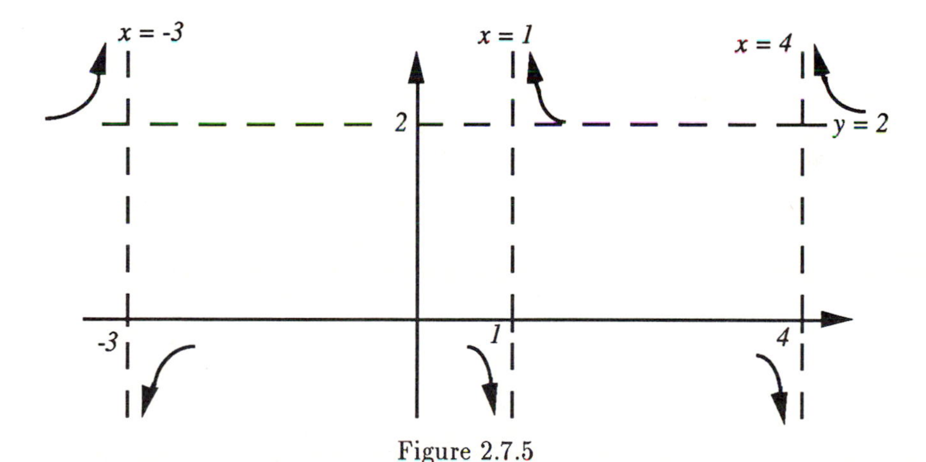

Figure 2.7.5

Next, we find or approximate the intercept points (if it is easy to do). That is, the points where the curve meets an x or y axis or a horizontal asymptote. To find the x intercept, find the roots of the equation $f(x) = 0$. Since

$$f(x) = \frac{2x^3 - 3}{(x-1)(x+3)(x-4)}$$

$f(x) = 0$ implies that the numerator must be zero. That is, $2x^3 - 3 = 0$, or $x^3 = \frac{3}{2}$. A calculator gives (to four decimal places) $x = 1.1447$ for one root. Any approximation by a number between 1 and 1.5 would serve just as well.

To find the y intercept, evaluate $f(0)$. We get $f(0) = -\frac{1}{4}$. To find where the curve intersects the horizontal asymptote (if it does), find the roots of the equation $f(x) - 2 = 0$. Combining the two terms, expanding the multiplication, and simplifying yields

$$f(x) - 2 = \frac{4x^2 + 22x - 27}{(x-1)(x+3)(x-4)} = 0$$

Now setting the numerator equal to zero, we have

$$4x^2 + 22x - 27 = 0$$

The use of the quadratic formula and a hand calculator gives the roots $x = 1.0332, -6.5332$.

Next, we plot the intercept points. Using the information obtained, the fact that f is continuous, and the Intermediate Value Theorem, we are ready to complete our sketch. We fill in the sketch region by region.

For $x > 4$: Since f diverges to positive infinity for x near 4 and the curve does not cross the asymptote $y = 2$ (no intercept points) in this interval, the intermediate value theorem implies that the graph of f must lie above the asymptote. Hence the curve approaches $y = 2$ asymptotically from above as $x \to \infty$.

For $1 < x < 4$: Since we know how the curve approaches the vertical asymptotes at the endpoints 1 and 4 and also where it crosses the horizontal asymptote and the x-axis, we use the continuity of f to sketch a smooth curve joining the two arrows and passing through the two known intersection points.

For $-3 < x < 1$: Since we know how the curve approaches the vertical asymptotes at the endpoints -3 and 1 and also where it crosses the y-axis, we use the fact that f is continuous to sketch a smooth curve joining the two arrows and passing through the y intersection point. Since the curve does not cross the horizontal asymptote, the Intermediate Value Theorem implies that it must lie entirely under the horizontal asymptote.

For $x < -3$: We know how the curve approaches the vertical asymptote at the right endpoint -3 and where it crosses the horizontal asymptote. Since it only crosses the asymptote once, the Intermediate Value Theorem implies that the curve must lie below the horizontal asymptote for all values of x less than the intercept value. Thus the curve approaches the line $y = 2$ asymptotically from below as $x \to -\infty$.

Our sketch now looks like

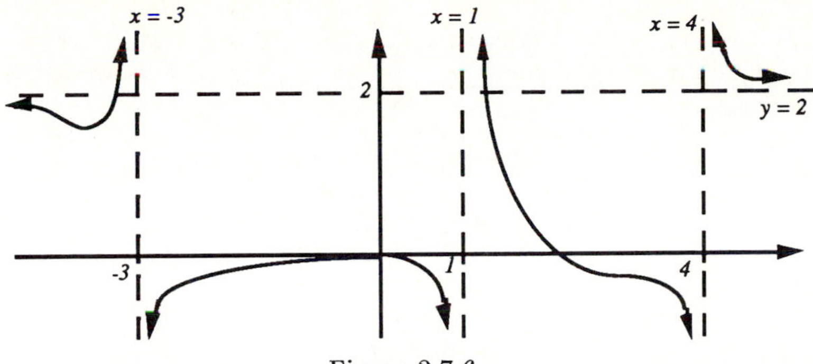

Figure 2.7.6

It is important to realize that our sketch is only reliable for the portions near the asymptotes and intersection points. $\triangle$

This next example illustrates the importance of canceling common factors in the numerator and denominator before checking for asymptotes.

Example 2.7.7 We will sketch the graph of the function defined by

$$f(x) = \begin{cases} 1/x & \text{for } x < 2, x \neq 0 \\ (4 - x^2)/(x^2 - 2x) & \text{for } x > 2 \end{cases}$$

Now checking for asymptotes, we see by inspection for $x < 2$ that $y = 0$ (x-axis) is a horizontal asymptote for the curve as $x \to -\infty$ and $x = 0$ (y-axis) is a vertical asymptote. For $x > 2$, we need to use caution. Now

$$\lim_{x \to \infty} \frac{4 - x^2}{x^2 - 2x} = -1$$

implies that $y = -1$ is a horizontal asymptote, and it appears that $x = 2$ is a vertical asymptote since the denominator is zero for $x = 2$. However, for $x > 2$, the defining expression for $f(x)$ can be simplified by canceling the common factor in the numerator and denominator to yield

$$f(x) = \begin{cases} 1/x & \text{for } x < 2, x \neq 0 \\ -(2 + x)/x & \text{for } x > 2 \end{cases}$$

Thus $x = 2$ is *not* a vertical asymptote since the denominator of the simplified expression is not zero for $x = 2$ and thus f is not unbounded as $x \to 2^+$ (nor is f unbounded as $x \to 2^-$).

We now sketch the coordinate axes and the three asymptotes (the negative x-axis and the y-axis are asymptotes).

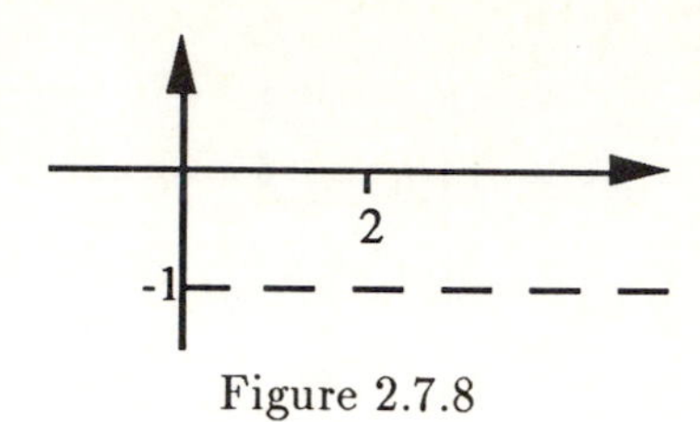

Figure 2.7.8

Note that there are no x or y intercepts. Furthermore, $f(x)$ is negative for x negative and thus the curve must lie below the horizontal asymptote. Also, by inspection, we see that $f(x)$ is positive for $0 < x < 2$ and is negative for $x > 2$. However, it is more important for sketching purposes to determine where the curve lies with respect to the horizontal asymptote (for $x > 2$) than it is to locate the curve with respect to the x-axis. For $x > 2$,

$$f(x) = \frac{2+x}{-x} = -1 - \frac{2}{x}$$

Thus the curve lies below the horizontal asymptote ($y = -1$) for $x > 2$. From these observations, we can draw in arrows on our sketch to indicate how the curve approaches each of the asymptotes.

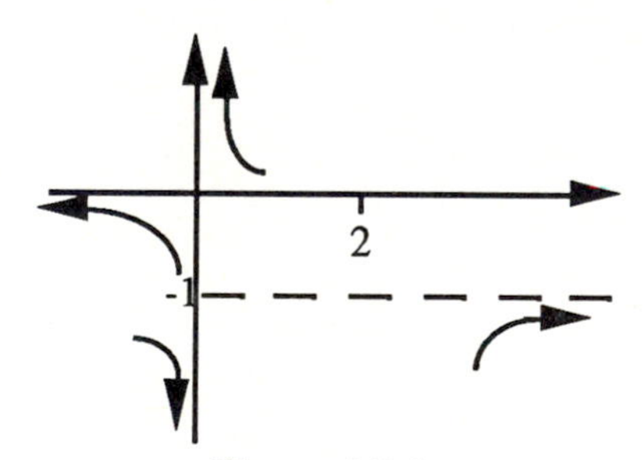

Figure 2.7.9

The only additional information that is needed before we complete the sketch is to determine how the function behaves as $x \to 2$. Since the defining expressions for $f(x)$ are continuous functions, we may compute (mentally) the left- and right-hand limits as $x \to 2$. Thus the curve approaches the point $(2, \frac{1}{2})$ from the left and the point $(2, -2)$ from the right. We now use the continuity property (unbroken curve) to sketch the graph over each of the three intervals of the domain of f.

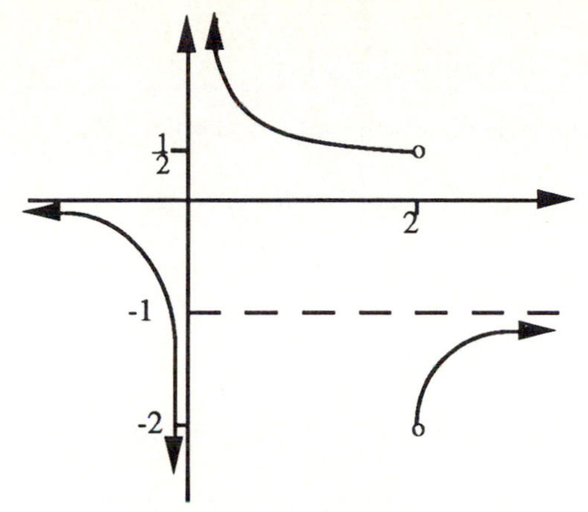

Figure 2.7.10 △

Let us recall a couple of facts about repeated roots that we learned in earlier courses. A root r of a polynomial $p(x)$ is said to be a *root of multiplicity* k if there is a polynomial $q(x)$ such that $p(x) = (x - r)^k q(x)$ and $q(r) \neq 0$. For example, 3 is a root of multiplicity 2 of $p(x) = (x-3)^2(x-7)$. At a root of even multiplicity, the graph of a polynomial touches but does not cross the x-axis. (Consider the parabola, $y = x^2$. Zero is a root of multiplicity two.) On the other hand, at a root of odd multiplicity, the graph does cross the x-axis.

For the final two examples of this section, we shall "reverse" the process and ask for a function whose graph has certain specified properties.

Example 2.7.11 We will construct a function whose graph satisfies the following conditions:

1. $y = 1$ is a horizontal asymptote.
2. $x = 1$ and 2 are vertical asymptotes.
3. The graph crosses the horizontal asymptote at (4,1).

We begin by noting that condition 2 requires that the denominator contain $(x-1)$ and $(x-2)$ as factors and condition 1 requires that the highest powered term of the numerator is the same as the highest powered term of the denominator. Thus,

$$f(x) = \frac{x^2 + bx + c}{(x-1)(x-2)}$$

will satisfy the first two conditions for any constants b and c. Now to guarantee that the graph crosses the horizontal asymptote at (4,1) means that the equation: (function) $-$ (asymptote) $= 0$, is true when $x = 4$. That is, we need to determine constants b and c such that $f(4) - (1) = 0$. We know that $f(4) - 1 = 0$ implies that

$$\frac{16 + 4b + c}{(3)(2)} - 1 = 0$$

or

$$10 + 4b + c = 0.$$

Thus one possibility is to set $b = 0$ and $c = -10$. Hence,

$$f(x) = \frac{x^2 - 10}{(x-1)(x-2)}$$

satisfies the three stated conditions. Can you find other functions that would also satisfy the three conditions? △

Example 2.7.12 We will construct a rational function whose graph has the following properties:

1. $y = -2$ is a horizontal asymptote.
2. $x = -1$ and 2 are vertical asymptotes.
3. $f(x)$ diverges in the same way as x approaches -1 from both the left and right.
4. The graph has an x-intercept at $x = -3$.

Note that properties not listed are not necessarily excluded. For example, the graph may have more than just the one x-intercept that is specified.

We start by observing that the specified vertical asymptotes require that the denominator contain $(x + 1)$ and $(x - 2)$ as factors. Furthermore, since $f(x)$ diverges in the same way as x approaches -1 from either side, $(x + 1)$ must be raised to an even power (Why?), say 2 (i.e., -1 must be a root of even multiplicity). To have a horizontal asymptote at $y = -2$ requires that $\lim_{x\to\infty} f(x) = -2$, and therefore the highest powered term in the numerator must be -2 times the highest powered term in the denominator. Thus conditions 1, 2, and 3 would be satisfied by:

$$\frac{-2x^3}{(x+1)^2(x-2)}$$

Now consider condition 4. This requires that the numerator contain a factor of $(x+3)$. So replace the $-2x^3$ in the numerator by $-2x^2(x+3)$. Note that this preserves the condition on the highest powered term in the numerator. Thus,

$$f(x) = \frac{-2x^2(x+3)}{(x+1)^2(x-2)}$$

satisfies all four conditions. △

The reader is urged to "play around" with sketching rational functions and making up functions whose graphs satisfy given conditions.

Section 2.7 Exercises

The purpose of Exercises 1 through 3 is to understand the "role" of an individual factor in the sketch of a rational function.

1. (a) Sketch the graph of $f(x) = (2(x-1)(x+2))/((x+1)(x-2)^2)$.
   (b) In part (a), replace the $(x+1)$ factor with $(x+1)^2$ and sketch the graph of the resulting function. Compare the graphs in parts (a) and (b). Describe the effect of changing the multiplicity of the $(x+1)$ factor.
   (c) In part (a), replace the $(x+1)$ factor with $(x+1)^3$ and sketch the graph of the resulting function. Compare the graphs in parts (a), (b), and (c). Describe the effect of changing the multiplicity of the $(x+1)$ factor.
   (d) In part (a), replace the $(x+1)$ factor with $(x+1)^4$ and sketch the graph of the resulting function. Compare the graphs in parts (a), (b), (c), and (d). Does the multiplicity of a factor have any graphical implications? If yes, what are they?
2. Repeat Exercise 1 using the $(x-1)$ factor in place of the $(x+1)$ factor.
3. (a) State a conjecture concerning the graph of a rational function in a neighborhood of a vertical asymptote that is given by a factor of even (odd) multiplicity.
   (b) Prove or disprove the conjecture given in part (a).
4. Find the horizontal and vertical asymptotes (if there are any) and sketch the graph. Label the asymptotes and intercept points.
   (a) $f(x) = 3 + 1/x$
   (b) $g(x) = 2/x - 1$
   (c) $h(x) = 2x/(x^2 + x - 6)$
   (d) $f(x) = -x/(x^2 - 2x + 1)$
   (e) $g(x) = (4x^2 + 4x + 1)/(2x^2 - x - 1)$
   (f) $h(x) = (x^2 - 4x + 4)/(2x^2 - 4x)$
   (g) $f(x) = (x-1)(x+2)/(x-3)(x+1)(x-2)$
   (h) $f(x) = x^2(x-1)/(x-1)(x+1)(x+3)$
   (i) $g(x) = (x+2)/\sqrt{1-x}$
   (j) $f(x) = \sqrt{x}/(x-2)$
5. For each of the following give an example or state why no example exists. Sketch the graphs of your examples.

(a) A rational function whose graph has no horizontal asymptotes.

(b) A rational function whose graph has no vertical asymptotes.

(c) A rational function whose graph has an x-intercept at $x = 2$ and a y-intercept at $y = 3$.

(d) A polynomial whose graph has a vertical asymptote at $x = 2$.

(e) A trigonometric function whose graph has a vertical asymptote at $x = \pi/2$.

6. Give an example or state why no example exists. Sketch the graph of your example.

(a) A function whose graph has vertical asymptotes at $x = -1, 3$, horizontal asymptote at $y = 2$, and an x-intercept at $x = 2$.

(b) A function whose graph has vertical asymptotes at $x = 1, 4$, horizontal asymptote at $y = 1$, and the curve approaches positive infinity as x approaches either of the asymptotes from either side.

(c) A function whose graph has horizontal asymptotes at $y = -1, 1$, vertical asymptote at $x = 2$, and only one intercept point which is at $y = 4$.

(d) A function whose graph has a vertical asymptote at $x = 1$ and has a maximum value for some value of $x > 2$.

(e) A function whose graph has a vertical asymptote at $x = 1$ and has a minimum value for some value of $x > 2$.

7. Sketch the graph for each of the following functions. Label each asymptote and intercept.

(a) $f(x) = 3(x-2)^2(x+3)/[(x+1)(x+2)^2(x-7)^3]$

(b) $f(x) = 2(x+3)^5(x-2)(x^2+1)/[(x-1)^2(x+1)^3(x-3)^2]$

(c) $f(x) = -3(x^2+2)^3(x+1)^3(x-3)^2/[(x^2+1)(x-1)^3(x+3)(x-5)^2]$

(d) $f(x) = 5x^2(x+2)^4/[2(x-1)^3(x+1)^2]$

8. For each of the following, use a computer or calculator to approximate the asymptotes and intercepts, and then sketch the graphs. Label the asymptotes and intercepts.

(a) $f(x) = (x^3 + 3x - 3)/(x^3 - 4x + 2)$

(b) $f(x) = (2x^4 + 3x)/(x^4 + x - 4)$

(c) $f(x) = (x^3 - 4x + 2)/(x^4 + x - 4)$

(d) $f(c) = (x^5 - 3x^2 + x - 2)/(x^5 - 30x^4 + 226x^3 - 31x^2 + 255x - 225)$

9. Project: Investigate the behavior of a rational function in the neighborhood of a root (of numerator or denominator) in terms of the multiplicity of the root (even or odd). Hints: Follow steps (a) through (f) for the roots of the numerator and then repeat the sequence of steps for the roots of the denominator.

   (a) Make up an example of a rational function that has a root of multiplicity 2 at $x = 3$.

   (b) Sketch the graph of the example that you developed in part (a). Pay particular attention to the graph in a neighborhood of $x = 3$.

   (c) Repeat parts (a) and (b) for different multiplicities (and different roots).

   (d) Make a conjecture (based on your observations) about the behavior of the graph of a rational function in the neighborhood of a root in terms of the multiplicity of the root.

   (e) Test your conjecture by applying it to a new example that contains roots of both even and odd multiplicities.

   (f) Using your graphs as a guide, develop an analytical proof showing that your conjecture is a theorem. [If you show that your conjecture is false, recycle through parts (d) and (e).]

   (g) Combine your results into a general theorem (it may have more than one part) and include a paragraph explaining your theorem.

Chapter 3

DIFFERENTIATION

3.1 Introduction

In the introduction to Chapter 1, we stated that the ability to analyze and quantify rates of change and to approximate measurements to within a stated degree of accuracy are important in every field of human endeavor. Examples permeate all activities: rate of growth of a baby, discount rate, rate of inflation, rate of exchange, interest rate, velocity (rate of change of distance with respect to time), acceleration (rate of change of velocity with respect to time), cooling rate, rate of decay, absorption rate. In mathematics, the instantaneous rate of change is called the *derivative*.

3.2 Instantaneous Rate of Change

Example 3.2.1 A person drives from point A to point B, a distance of 1.4 miles. Every tenth of a mile the time (in seconds) is recorded. How fast is the person traveling at point B? That is, what is the instantaneous rate of change at point B?

The data is given in the following table.

Time (seconds)	Distance (miles)
0	0
15	0.1
28	0.2
39	0.3
48	0.4
58	0.5
73	0.6
86	0.7
95	0.8
110	0.9
120	1.0
128	1.1
137	1.2
145	1.3
153	1.4

Let us model the situation with a distance function, i.e., let $y = d(t)$, distance traveled during time t. Thus time is the input variable and distance is the output variable. The *average rate of change,* the change in the output (distance) divided by the change in the input (time), will be used to approximate the instantaneous rate of change at point B. That is,

$$r_0 = \frac{d(153) - d(0)}{153 - 0} = \frac{1.4}{153} = 0.00915 \text{ miles/sec (32.94 miles/hr)}$$

Intuitively it seems that averaging over the period starting with the second time reading ($t = 15$) should give a "better" approximation, that is

$$r_1 = \frac{d(153) - d(15)}{153 - 15} = \frac{1.3}{138} = 0.00942 \text{ miles/sec (33.91 miles/hr)}$$

Similarly, averaging over the period starting with the third time reading should give an even "better" approximation.

$$r_2 = \frac{d(153) - d(28)}{153 - 28} = \frac{1.2}{125} = 0.00960 \text{ miles/sec (34.56 miles/hr)}$$

Continuing in this manner, we compute a sequence of 14 average rates with the property that each average rate is a "better" approximation to the instantaneous rate at point B than the preceding average rates.

$$\begin{aligned}
r_3 &= \frac{d(153) - d(39)}{153 - 39} = \frac{1.1}{114} = 0.00965 \text{ miles/sec (34.74 miles/hr)} \\
r_4 &= \frac{d(153) - d(48)}{153 - 48} = \frac{1.0}{105} = 0.00952 \text{ miles/sec (34.29 miles/hr)} \\
r_5 &= \frac{d(153) - d(58)}{153 - 58} = \frac{0.9}{95} = 0.00947 \text{ miles/sec (34.11 miles/hr)}
\end{aligned}$$

etc.

The entire sequence (in miles/hr) is

$$\{32.94, 33.91, 34.56, 34.74, 34.29, 34.11, 36.00, \\ 37.61, 37.24, 41.86, 43.64, 43.20, 45.00, 45.00\}$$

Thus on the basis of the data given, it is reasonable to conclude that the person was probably traveling 45 miles/hr at point B. More data readings, particularly in the time interval between $t = 148$ and $t = 153$, would increase both the accuracy and one's confidence in the final answer. △

Let us now consider an example with infinitely many data points.

Example 3.2.2 Suppose a person throws a rock straight up with an initial velocity of 56 ft/sec. We shall consider that the rock was exactly 6 ft above the ground when it left the person's hand. Assume that the only force acting on the rock after being thrown is the force due to gravity. Suppose also that physical experiments suggests that as long as the rock is in the air, its height $f(t)$ in feet is given by

$$f(t) = 56t - 16t^2 + 6$$

What is the instantaneous rate of change of the height of the rock when $t = 1$?

Now $(f(t) - f(1))/(t-1)$ represents the average rate of change of the height over the interval $[t, 1]$ or $[1, t]$ (depending on whether t is less than or greater than 1). Following the reasoning in the previous example that decreasing the size of the change in the input yields a "better" average approximation, we consider a sequence of t values converging to 1. We will let $t = 1 + 1/n$. Then the nth average rate approximation is

$$r_n = \frac{f(1+1/n) - f(1)}{(1+1/n) - 1}$$

Doing some computations we get

$$\begin{aligned} r_n &= \frac{f(1+1/n) - f(1)}{1/n} \\ &= n\left[56\left(1+\frac{1}{n}\right) - 16\left(1+\frac{1}{n}\right)^2 + 6 - (56 - 16 + 6)\right] \\ &= n\left[\frac{24}{n} - \frac{16}{n^2}\right] \\ &= 24 - \frac{16}{n} \end{aligned}$$

Thus $\{r_n\}$ is a sequence of average rates of change, each of which is a "better" approximation to the instantaneous rate of change of the height at $t = 1$

than the preceding ones. We are tempted to claim that since $\lim_{n\to\infty} r_n = 24$, 24 is the instantaneous rate of change of the height of the rock at $t = 1$. However, recall that it is not enough to show that a particular sequence of average rates converges, but all convergent sequences of average rates must be shown to converge to the same expression. This can be done by replacing $1/n$ by z_n, an arbitrary sequence converging to zero. The instantaneous rate of change of f when $t = 1$ is then given by

$$\begin{aligned} r &= \lim_{n\to\infty} \frac{f(1+z_n) - f(1)}{z_n} \\ &= \lim_{n\to\infty} \frac{56(1+z_n) - 16(1+z_n)^2 + 6 - (56 - 16 + 6)}{z_n} \\ &= \lim_{n\to\infty} (24 - 16z_n) = 24 \text{ ft/sec} \end{aligned}$$

△

The central idea in these two examples for finding the instantaneous rate of change was to generate a sequence of average rates each of which is a "better" approximation to the (unknown) instantaneous rate than all the preceding ones. If *all* such sequences should converge to a common value, then this common value is taken to be the instantaneous rate of change, otherwise the instantaneous rate is not defined.

Notation: Analyzing sequences of approximations is a natural way to develop a concept (e.g., instantaneous rate of change). Thus it is desirable to use notation in the *development process* that emphasizes the sequence aspect [e.g., $\lim_{n\to\infty}(f(a+z_n)-f(a))/z_n$] *for all* sequences z_n converging to 0, $z_n \neq 0$). Once a concept has been established, a more "compact" notation is often adopted. We will use the notation for the limit of a function (Definition 2.4.6) and write

$$\lim_{h\to 0} \frac{f(a+h) - f(a)}{h} \quad \text{for} \quad \lim_{n\to\infty} \frac{f(a+z_n) - f(a)}{z_n} \quad \text{for all } z_n \to 0, z_n \neq 0$$

(The above two limit expressions are equivalent. The left-hand one is sometimes referred to as the "continuous" form, meaning that $h \to 0$ over a continuum of real numbers. The right-hand one is sometimes referred to as the "discrete" form, meaning that $z_n \to 0$ over a discrete set of points.)

Definition 3.2.3 (Instantaneous Rate of Change) *Let* $f : \mathbb{R} \to \mathbb{R}$. *The* instantaneous rate of change *of* f *at a point* a *is defined to be*

$$\lim_{h\to 0} \frac{f(a+h) - f(a)}{h}$$

provided this limit exists.

Example 3.2.4 (Average and Marginal Cost) In economics, instantaneous rate of change is usually referred to by the adjective *marginal.* Suppose the cost of operating a shoe factory is modeled by the simple formula

$$c(t) = 5000 + 150t - t^2$$

Then the average cost during the time interval $[t, t+h]$ is

$$\begin{aligned}
\frac{c(t+h) - c(t)}{h} &= \frac{[5000 + 150(t+h) - (t+h)^2] - [5000 + 150t - t^2]}{h} \\
&= \frac{150h - 2th - h^2}{h} \\
&= 150 - 2t - h
\end{aligned}$$

and the marginal cost at time t is

$$\lim_{h \to 0} \frac{c(t+h) - c(t)}{h} = \lim_{h \to 0}(150 - 2t - h) = 150 - 2t \qquad \triangle$$

To emphasize that the limit in the above definition may not always exist, i.e., a function may not have an instantaneous rate of change at a given point, consider the following example.

Example 3.2.5 We will show that the absolute value function, $f(x) = |x|$, does not have an instantaneous rate of change at the point 0. That is,

$$\lim_{h \to 0} \frac{f(0+h) - f(0)}{h}$$

does not exist.

The procedure is to show that the left and right-hand limits do not agree and thus by Theorem 2.4.13, the limit does not exist.
Thus

$$\lim_{h \to 0-} \frac{f(0+h) - f(0)}{h} = \lim_{h \to 0-} \frac{|h|}{h} = \lim_{h \to 0-} \frac{(-h)}{h} = -1$$

and

$$\lim_{h \to 0+} \frac{f(0+h) - f(0)}{h} = \lim_{h \to 0+} \frac{|h|}{h} = \lim_{h \to 0+} \frac{h}{h} = +1$$

Hence the left- and right-hand limits do not agree, and so by Theorem 2.4.13, the limit does not exist! $\triangle$

Let us now consider the concept of instantaneous rate of change from a geometrical point of view. The question is the same as before: Given a function, how can we find its instantaneous rate of change at a given point p? We shall think of the graph of $y = f(x)$ being traced out by a point moving in the plane according to the relation $y = f(x)$. To be specific, we consider the graph of the quadratic polynomial $y = f(x) = (x-3)^2$ and take (1,4) to be the given point. The lines L_1, L_2, and L_3 in the following figure are called *secant* lines (a secant line is a line that passes through two points of a curve). The basic idea that we will develop, is to consider the segment of the secant line joining the two points on the curve as an approximation to the piece of the curve determined by the two points. And, in general, the closer the points are together, the better the approximation.

Recall from Section 1.5 that a linear (or affine) function is one that models a constant rate of change and the constant rate is given by the slope of the graph, i.e.,

$$\text{slope} = \text{constant rate of change} = \frac{\Delta y}{\Delta x}$$

Thus if a piece of the graph of f containing the point (1,4) was approximated by a segment of a secant line, the slope of the line segment could be used as an approximation to the instantaneous rate of change of f at the point (1,4). Of course, the slope of the line segment is the slope of the secant line. For example, let L_1 in the following figure be the secant line determined by the points (1,4) and (4,1).

$f(x) = (x-3)^2$

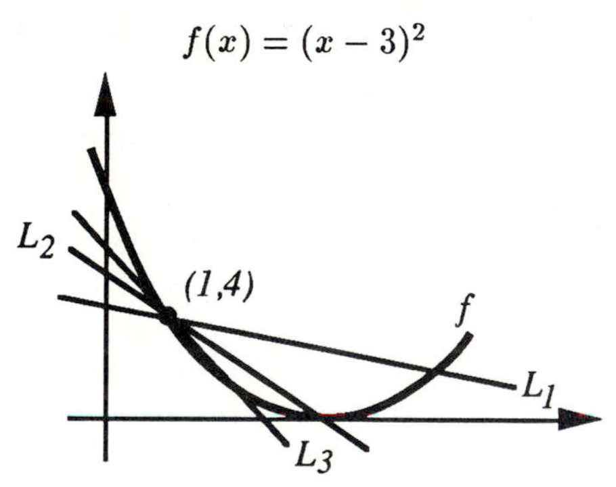

Figure 3.2.6

The slope of L_1,

$$\frac{\Delta y}{\Delta x} = \frac{1-4}{4-1} = -1$$

yields an approximation to the instantaneous rate of change of f at $x = 1$. Can a "better" approximation be found? In the analytic approach illustrated in Examples 3.2.1 and 3.2.2, a "better" (average) approximation was obtained by decreasing the change in the input. The change in the input that determined L_1 was $\Delta x = 4 - 1 = 3$. Thus a "better" approximation would result from using a $\Delta x = 2$. Let L_2 be the secant line determined by the points (1,4) and (3,0). The slope of L_2 is -2, which we take as our second approximation to the instantaneous rate of change of f at $x = 1$. Repeating this process with a $\Delta x = 1$ gives a third secant line, L_3, determined by the points (1,4) and $(2, 1)$. Our third approximation is -3, the slope of L_3.

Continuing this process yields a sequence of secant lines passing through the point (1,4) whose corresponding slopes yield a sequence of approximations to the instantaneous rate of change of f at $x = 1$, provided that rate exists. For example, if the sequence of $\Delta x's$ converging to zero is $\{1/n\}$, then the sequence of slopes of the line segments L_n is

$$\frac{\Delta y}{\Delta x} = \frac{f(1+1/n) - f(1)}{(1+1/n) - 1} = -4 + \frac{1}{n}$$

This suggests that the instantaneous rate of change of f at $x = 1$ is -4. However, this result may be dependent on the particular sequence that was used. Thus we need to replace the particular sequence $\{1/n\}$ with an arbitrary sequence $\{z_n\}$ where $z_n \to 0$ and $z_n \neq 0$. The slope of the line segment L_n is

$$\frac{\Delta y}{\Delta x} = \frac{f(1+z_n) - f(1)}{z_n} = -4 + z_n$$

Since this sequence converges to -4, we conclude that -4 is the instantaneous rate of change of f at $x = 1$.

Note that the expression for the sequence of slopes is exactly the expression obtained in Example 3.2.2 that led to the definition of instantaneous rate of change. Thus our geometric approach leads to the same result as our analytic approach. This should really be no surprise, since the slope of a line represents an average rate of change.

The limiting position for the sequence of secant lines $\{L_n\}$ passing through the point (1,4) has an important geometrical significance. Let m_n be the slope of line L_n. Using the point-slope formula, the equation of L_n is

$$y - 4 = m_n(x - 1)$$

or

$$y = m_n x - (m_n - 4)$$

Since the sequence of slopes $\{m_n\}$ converges, the sequence of secant lines $\{L_n\}$ converges to a unique line passing through the point (1,4). (Why is the "limit" line unique?) This unique line is called the *tangent line* to the graph of the function at the specified point. In our example, $m_n \to -4$ and so $\{L_n\}$ converges to the line $y = -4x + 8$. Thus the line $y = -4x + 8$ is the tangent line to the graph of $y = f(x) = (x - 3)^2$ at the point (1,4).

Definition 3.2.7 (Tangent Line) *The* tangent line *to the graph of a function at the point $(a, f(a))$ is the line passing through the point $(a, f(a))$ with slope equal to $\lim_{h\to 0}(f(a+h) - f(a))/h$, provided this limit exists.*

Given that there exists a tangent line to the graph of f at the point $(a, f(a))$, its equation is

$$y - f(a) = \lim_{h \to 0} \frac{f(a+h) - f(a)}{h}(x - a)$$

The expression $(f(a+h) - f(a))/h$ is often referred to as the *difference quotient.*

Since the instantaneous rate of change of a function at $x = a$ is the limit of the difference quotient which, in turn, is the slope of the line tangent to the graph of f at the point $(a, f(a))$, Example 3.2.5 implies that the graph of the absolute value function does not have a tangent at the point (0,0). This is clear from the graph since the "right-hand tangent line" is $y = x$ and the "left-hand tangent line" is $y = -x$. This example raises an interesting question concerning the relation (if any) between a function being continuous at $x = a$ and the graph of the function having a tangent at the point $(a, f(a))$. Example 3.2.5 shows that continuity does not imply the existence of a tangent, does the existence of a tangent imply continuity?

Section 3.2 Exercises

In Exercises 1 through 4, a particle is moving along a straight line according to the equation given, where d is the distance in feet that the particle has traveled in t seconds. Find (a) the average rate of change from $t = t_1$ to $t = t_2$ and (b) the instantaneous rate of change at $t = t_1$.

1. $d = 5t^2, t_1 = 2, t_2 = 4$

2. $d = t^2 + t, t_1 = 1, t_2 = 5$

3. $d = 13t - 72, t_1 = 4, t_2 = 12$

4. $d = \sqrt{t}, t_1 = 9, t_2 = 16$

5. If the rate of production in a shoe factory is given by $p(t) = 100t + t^2 - t^3/2$, determine the marginal rate at $t = 2$.

6. In Exercise 5 when t is restricted to [0,10], is there a time t_0, $0 < t_0 < 10$, at which the marginal rate of production equals the average rate of production over the time interval [0,10]? If so, find t_0. If not, explain why not.

   In Exercises 7 through 11, find the equation of the tangent line to the graph of the given curve at the indicated point.

7. $y = x^2/2 - 3x + 10, \quad (2,6)$

8. $y = 1/x, \quad (1,1)$

9. $y = 3x^3 - 3, \quad (0, -3)$

10. $y = (x^2 - 2x)/(x + 1), \quad (0,0)$

11. $y = \begin{cases} 2x - 2 & \text{for } x < 1 \\ 0 & \text{for } x = 1 \\ x - 1/x & \text{for } x > 1 \end{cases} \quad (1,0)$

12. Find the points on the graph of $f(x) = x^3$ where the tangent line is parallel to $y = 3x + 2$.

13. Determine constants a, b, c such that the graph of $f(x) = ax^2 + bx + c$ will go through the origin and the tangent to the graph at (1,1) will have a slope of 3.

14. A line is said to be *normal* (perpendicular) to a curve at the point (a, b) if its slope is the negative reciprocal of the slope of the tangent line to the curve at (a, b). Show that the line $2y + x = 3$ is normal to the graph of $y = x^2$ at one point of intersection, but not at the other point of intersection.

15. Find the x-intercept of the normal line to the curve of $f(x) = 2/x$ at the point (1,2). (See Exercise 14.)

16. Find a point on the graph of $f(x) = x^2 + 5$ such that the tangent line to the graph at this point will have an x-intercept of 2.

17. Mark each of the following statements true (T) or false (F).

    (a) There exists a tangent line to the graph of $f(x) = |x|$ at $x = 0$.

    (b) A tangent line may intersect a graph in more than one point.

    (c) There exists a tangent line to the graph of $f(x) = x^{1/2}$ at $x = 0$.

    (d) The slope of the line tangent to the graph of a function f at $x = a$ is the instantaneous rate of change of f at $x = a$.

    (e) It is possible to have two different tangent lines to the graph of a function at the same point.

18. At a certain instant the speedometer of an automobile reads r miles/hr. During the next $\frac{1}{2}$ sec, the automobile travels 30 ft. Estimate r from this information.

19. A metal sphere with radius x is expanding uniformly as a result of being heated. Find:

    (a) The average rate of change of the sphere's volume with respect to the radius as x increases from 2 to 2.01 centimeters.

    (b) The instantaneous rate of change of the sphere's volume with respect to the radius at the instant when $x = 2$ centimeters.

20. A child is sliding along a board. At the end of t sec the child has slid $d(t)$ meters from the starting point, where $d(t) = t^2 + t$. Find the speed (instantaneous rate of change) of the child after exactly 2 sec.

21. A rocket is fired vertically upward and is s feet above the ground t sec after blast off, where $s = 256t - 16t^2$. Find:

    (a) The average rate of change of the distance traveled by the rocket with respect to time during the first 4 sec of flight.

    (b) The instantaneous rate of change of the distance traveled by the rocket with respect to time exactly 4 sec after being fired.

    (c) The time in seconds for the rocket to reach its maximum height.

    (d) The maximum height to which the rocket ascends.

22. The pressure P of a gas depends upon its volume V. The relationship is given by Boyle's law, $P = C/V$, where C is a constant. Suppose that $C = 2000$. Find:

    (a) The average rate of change of P with respect to V as V increases from 100 cubic inches to 125 cubic inches.

    (b) The instantaneous rate of change of P with respect to V at the moment when $V = 100$ cubic inches.

23. Give an example for each of the following statements or explain why no example can exist.

    (a) A nonconstant function defined over [0,1] whose average rate of change is zero.

    (b) A function with the property that every line joining two points on its graph has slope equal to one. (Recall that a line joining two points on the graph is called a secant line.)

    (c) A continuous function f defined at $x = 4$ and two sequences $a_n \to 0, b_n \to 0$ such that

    $$\lim_{n\to\infty} \frac{f(4+a_n) - f(4)}{a_n} \neq \lim_{n\to\infty} \frac{f(4+b_n) - f(4)}{b_n}$$

    (d) A function whose instantaneous rate of change at $x = 2$ is not equal to the slope of the tangent at the point $(2, f(2))$.

    (e) A function whose graph has the property that all secant lines have positive slope.

24. Project: Develop and carry out a rate of change experiment which involves computing a sequence (finite) of average rates of change each of which represents a better approximation to the instantaneous rate at some fixed point than all of the preceding terms of the sequence. (See Example 3.2.1 for ideas.)

3.3 The Derivative

The instantaneous rate of change of a function is called the derivative of the function.

Definition 3.3.1 (Differentiability at a Point) *A function $f : \mathbb{R} \to \mathbb{R}$ is said to be* differentiable at p *if f is defined on an open interval containing p and* $\lim_{h\to 0} \frac{f(p+h)-f(p)}{h}$ *exists.*

The derivative of a function f at a point p in an open interval in its domain is the number denoted by $f'(p) = \lim_{h\to 0} \frac{f(p+h)-f(p)}{h}$.

If the limit of the difference quotient (in the definition) does not exist, then the function f does not have a derivative at the point p. In this case, f is said to be *not differentiable at p*.

The geometric interpretation of derivative is the slope of the tangent line to the curve $y = f(x)$ at the point $(p, f(p))$. (See the definition of the tangent line, Definition 3.2.7, of the previous section.) Thus the graph of a differentiable function has a tangent line at every point and therefore will have no "corners" or "cusps." In fact, a small segment of the graph of a differentiable function when observed on a sufficiently large scale will appear to be just a straight line. (See Exercise 20.)

Notes:

(a) We can define a function $f' : \mathbb{R} \to \mathbb{R}$ by associating with each point p its derivative at p, $f'(p)$. It is important to distinguish between the function f' and the number $f'(p)$ where p is a specified point.

(b) We restrict ourselves to open intervals to avoid the problem of endpoints of intervals.

The *right-hand derivative* at a point is the number obtained by restricting h to be positive in the definition. The *left-hand derivative* is obtained by restricting h to be negative. It follows from Theorem 2.4.13 that a function is differentiable at a point if the right- and left-hand derivatives exist and are equal at the point. If either of the left- or right-hand limits does not exist, or both limits exist but are not equal at a point, then the function does not have a derivative at the point.

Example 3.3.2 We will compute the derivative of $f : \mathbb{R} \to \mathbb{R}$ given by $f(x) = x^2$ at any point x. The difference quotient is

$$\frac{f(x+h) - f(x)}{h} = \frac{x^2 + 2xh + h^2 - x^2}{h} = 2x + h$$

Thus $f'(x) = \lim_{h\to 0}(2x + h) = 2x$, for any real number x. The derivative of $f(x) = x^2$ is $f'(x) = 2x$ for any real number x. △

Example 3.3.3 We will show that

$$g(x) = \begin{cases} 1/x & \text{for } x \leq 1 \\ x & \text{for } x > 1 \end{cases}$$

is not differentiable at $x = 1$.

Computing the left- and right-hand derivatives yields

$$\begin{aligned} \lim_{h \to 0^-} \frac{g(1+h) - g(1)}{h} &= \lim_{h \to 0^-} \frac{\frac{1}{1+h} - 1}{h} \\ &= \lim_{h \to 0^-} \frac{-1}{1+h} \\ &= -1 \end{aligned}$$

$$\begin{aligned} \lim_{h \to 0+} \frac{g(1+h) - g(1)}{h} &= \lim_{h \to 0+} \frac{(1+h) - 1}{h} \\ &= +1 \end{aligned}$$

Since the left- and right-hand derivatives are not equal, g is not differentiable at $x = 0$. Is g continuous at $x = 1$? △

Recall that a function is continuous if it is continuous at every point in its domain. A similar relation holds for differentiability.

Definition 3.3.4 (Differentiable Function) *A function is differentiable if it is differentiable at every point in its domain.*

It is important to realize that in the development of the definition of derivative (instantaneous rate of change, or slope of tangent line) we used a basic procedure in analysis–the use of the approximation process to develop a definition. There are four steps in this process, as outlined in Section 2.2 and below:

Basic Approximation Process

1. Approximate the unknown quantity with a known property.
2. Determine a way to obtain a better approximation.
3. Generate a convergent sequence of approximations such that each is a better approximation than the preceding one.
4. Define the desired property to be the common limit of all possible sequences in part 3. When there is no common limit, the desired property is said not to exist.

This is the fundamental process that will provide the framework for the development of *every* major concept in the calculus.

Notation. The choice of notation is a very important consideration to the ease of learning new material. Simplicity of form and suggestiveness of the intrinsic nature or process involved are usually the two major characteristics of good notation.

Euler (1707-1783), the most prolific writer of mathematics of all times (he published 530 books and papers), introduced the "$f(x)$" functional notation. He is credited with being the first to use Δ (delta) to denote change. Thus Euler expressed the difference quotient

$$\frac{f(x+h)-f(x)}{h} \quad \text{as} \quad \frac{\Delta y}{\Delta x}$$

There are three commonly used notations for derivative that satisfy these two conditions. We shall use them interchangeably.

1. Leibniz (1646-1716), a cofounder of the calculus, had an unusual feeling for mathematical form and the potential of well-devised notation. With respect to the equation $y = f(x)$, Leibniz introduced the notation dy/dx (read "dee y dee x") to represent the *limit* of the difference quotient. Thus we have for $y = f(x)$

   $$\frac{dy}{dx} = f'(x) = \lim_{h\to 0} \frac{f(x+h)-f(x)}{h} = \lim_{h\to 0} \frac{\Delta y}{\Delta x}$$

   This form is sometimes varied to gain clarity in writing. We will, on occasion, write

   $$\frac{df}{dx}(x) \quad \text{or} \quad \frac{df(x)}{dx} \quad \text{or} \quad \frac{d}{dx}f(x)$$

   For instance, if $f(x) = \sin(x)$, we may indicate differentiation by writing $d/dx \sin(x)$.

2. The f' ("f prime") notation was introduced by J.L. Lagrange (1734–1813), the first mathematician to attempt to develop a theory of functions. His notation emphasizes the functional aspect of the derivative. *The derivative of a function is a function* and when evaluated at a prescribed point yields a number. When $y = f(x)$, we may also write y' to denote the derivative.

3. The operator notation D was first used by Arbogast (1759–1803) to indicate that the derivative was obtained by carrying out a certain operation on the function. If $f(x) = \cos(x)$, then Df or $D\cos$ denotes the derivative of f.

We shall illustrate these three notations by expressing the derivatives of the following basic functions in each of the forms.

Example 3.3.5 (Derivatives) (a) *Constant function,* $f(x) = c$.

$$f'(x) = \lim_{h \to 0} \frac{f(x+h) - f(x)}{h} = \lim_{h \to 0} \frac{c - c}{h} = 0$$

In other notations,

$$\frac{d}{dx} f(x) = \frac{d}{dx} c = 0 \quad \text{and } D(c) = 0$$

(b) *Affine function,* $f(x) = mx + b$. (This includes $f(x) = x$.)

$$\begin{aligned} f'(x) &= \lim_{h \to 0} \frac{f(x+h) - f(x)}{h} = \lim_{h \to 0} \frac{(m(x+h) + b) - (mx + b)}{h} \\ &= \lim_{h \to 0} \frac{mh}{h} = m \end{aligned}$$

In other notations,

$$\frac{d}{dx} f(x) = \frac{d}{dx}(mx + b) = D(mx + b) = m$$

(c) *Positive integer power function,* $f(x) = x^n$.

$$f'(x) = \lim_{h \to 0} \frac{f(x+h) - f(x)}{h} = \lim_{h \to 0} \frac{(x+h)^n - x^n}{h}$$

To simplify the quotient, we expand $(x+h)^n$ by the binomial theorem and write

$$(x+h)^n = x^n + nx^{n-1}h + \frac{n(n-1)}{2} x^{n-2}h^2 + \cdots + h^n$$

Thus,

$$\frac{(x+h)^n - x^n}{h} = nx^{n-1} + \frac{n(n-1)}{2} x^{n-2}h + \cdots + h^{n-1}$$

Observe that every term on the right except the first contains h to a positive power. Since the limit of a sum is the sum of the limits, we may evaluate the limit of the quotient by taking the limit of each term on the right and summing the results. Each term on the right approaches 0 as h approaches 0 except for the first term that is independent of h. Hence

$$f'(x) = \lim_{h \to 0} \frac{(x+h)^n - x^n}{h} = nx^{n-1}$$

In other notations,

$$\frac{d}{dx} f(x) = \frac{d}{dx} x^n = D(x^n) = nx^{n-1}$$

(d) The *nth root function,* $f(x) = x^{1/n}$.

$$f'(x) = \lim_{h\to 0} \frac{(x+h)^{1/n} - x^{1/n}}{h}$$

It is not at all obvious how to evaluate this limit. However, let us make the following transformation (which is equivalent to multiplying and dividing by the conjugate):

$$u = (x+h)^{1/n} \quad \text{and} \quad v = x^{1/n}$$

Thus $u^n = x + h$ and $v^n = x$, yielding $h = u^n - v^n$. The difference quotient can now be written as

$$\begin{aligned} \frac{u-v}{u^n - v^n} &= \frac{u-v}{(u-v)(u^{n-1} + u^{n-2}v + u^{n-3}v^2 + \cdots + v^{n-1})} \\ &= \frac{1}{u^{n-1} + u^{n-2}v + u^{n-3}v^2 + \cdots + v^{n-1}} \end{aligned}$$

By the continuity of the n*th* root function, u approaches v as h approaches 0. Thus each of the n terms in the denominator approach v^{n-1} and so

$$\lim_{h\to 0} \frac{u-v}{u^n - v^n} = \frac{1}{nv^{n-1}}$$

Substituting $x^{1/n}$ for v yields

$$f'(x) = \frac{1}{nx^{(1-1/n)}} = \frac{1}{n}x^{(1/n-1)}$$

In other notations,

$$\frac{d}{dx}x^{1/n} = Dx^{1/n} = \frac{1}{n}x^{(1/n-1)}$$

(e) *The sine function,* $f(x) = \sin(x)$.

$$f'(x) = \lim_{h\to 0} \frac{\sin(x+h) - \sin(x)}{h}$$

We now use the trigonometric identity (see the Appendix on trigonometry, equation 9)

$$\sin(a) - \sin(b) = 2\sin\left(\frac{a-b}{2}\right)\cos\left(\frac{a+b}{2}\right)$$

with $a = x + h$ and $b = x$ to obtain

$$f'(x) = \lim_{h\to 0} \frac{\sin(h/2)\cos(x+h/2)}{h/2} = \lim_{h\to 0} \frac{\sin(h/2)}{h/2}\cos\left(x + \frac{h}{2}\right)$$

Since

$$\lim_{h \to 0} \frac{\sin(h/2)}{h/2} = 1$$

and

$$\lim_{h \to 0} \cos(x + \frac{h}{2}) = \cos\left(\lim_{h \to 0}\left(x + \frac{h}{2}\right)\right) = \cos(x)$$

(recall that the cosine function is continuous), we may evaluate the limit of the product as the product of limits to obtain

$$f'(x) = \cos(x)$$

In other notations,

$$\frac{d\sin(x)}{dx} = \frac{d}{dx}\sin(x) = D\sin(x) = \cos(x)$$

(f) *The cosine function,* $f(x) = \cos(x)$.

$$f'(x) = \lim_{h \to 0} \frac{\cos(x+h) - \cos(x)}{h}$$

We use the trigonometric identity

$$\cos(a) - \cos(b) = -2\sin\left(\frac{a-b}{2}\right)\sin\left(\frac{a+b}{2}\right),$$

with $a = x + h$ and $b = x$ to obtain

$$f'(x) = \lim_{h \to 0} -\frac{\sin(h/2)\sin(x + h/2)}{h/2}$$

and thus

$$f'(x) = -\sin(x)$$

(How is the last step obtained?)

In other notations,

$$\frac{d\cos(x)}{dx} = \frac{d}{dx}\cos(x) = D\cos(x) = -\sin(x)$$

Consequently, the derivative of $\sin(x)$ is $\cos(x)$ and the derivative of $\cos(x)$ is $-\sin(x)$.

(g) *The exponential function.* The function $f(x) = b^x$, $b > 0$, will be discussed in detail in Chapter 7. We will introduce its derivative here, assuming that its definition, the laws of exponents, and its derivative at zero, $f'(0)$, exists

$$f'(x) = \lim_{h \to 0} \frac{b^{x+h} - b^x}{h} = \lim_{h \to 0} b^x \frac{b^h - 1}{h} = b^x \lim_{h \to 0} \frac{b^h - 1}{h} = b^x f'(0)$$

Thus if $f(x) = b^x$, $b > 0$,

$$f'(x) = cb^x$$

where the constant c is $f'(0)$.

In the case where $b = e = 2.7182...$ (see the part on exponential functions in Section 1.5), we give a heuristic argument showing that $f'(0) = 1$. Recall that

$$f'(0) = \lim_{h \to 0} \frac{f(0+h) - f(0)}{h} = \lim_{n \to \infty} \frac{f(0+z_n) - f(0)}{z_n}$$

Where z_n is an arbitrary sequence converging to zero.

We approximate the right side of the above equality by letting z_n be the particular sequence $\{1/n\}$ and considering a few values of n. We have

$$\begin{aligned} x_n &= \frac{f(0+1/n) - f(0)}{1/n} \\ &= n(f(\frac{1}{n}) - f(0)) \\ &= n(e^{1/n} - 1) \end{aligned}$$

Now

$$\begin{aligned} x_{10} &= 1.05170918076 \\ x_{10^2} &= 1.05170918076 \\ x_{10^3} &= 1.00050016671 \\ x_{10^4} &= 1.00005000167 \\ x_{10^5} &= 1.00000500002 \\ x_{10^6} &= 1.00000050000 \\ x_{10^7} &= 1.00000005000 \\ x_{10^8} &= 1.00000000500 \end{aligned}$$

It appears that $x_n \to 1$ and this, in turn, suggests that

$$f'(0) = \lim_{n \to \infty} \frac{f(0+z_n) - f(0)}{z_n} = 1$$

That is $f'(0) = 1$ and thus

$$\frac{d}{dx} e^x = e^x$$

This property, that *the function is equal to its derivative*, characterizes the natural exponential function (with base $b = e$). The above heuristic argument is not a proof since we only considered a few values of a particular sequence. A complete proof will be given in Chapter 7. $\triangle$

An important and immediate consequence of the fact that $d/dx\, e^x = e^x > 0$ is that $f(x) = e^x$ is a strictly increasing function (i.e., at every point on the graph of $y = e^x$, the tangent line has a positive slope). Thus e^x is a one-to-one function and hence has an inverse. The inverse function is called the (natural) logarithmic function and is denoted by $\log(x)$. From Section 1.4 we know that the graph of $y = \log(x)$ can be obtained by reflecting the graph of $y = e^x$ in the line $y = x$.

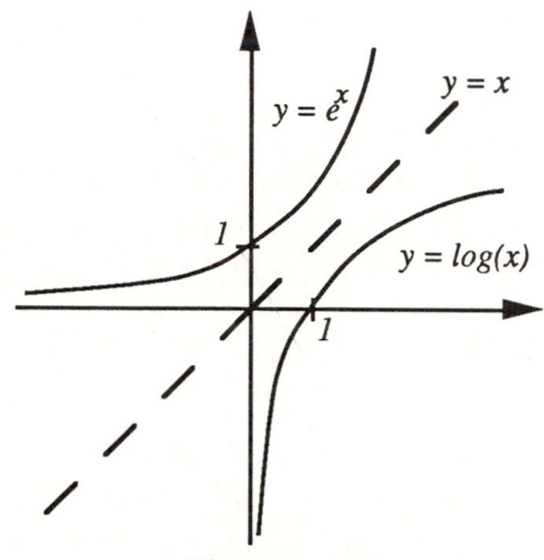

Figure 3.3.6

The question of the differentiability of $\log(x)$ will be taken up in the next section.

We are now in a position to answer the questions raised as to the relation between differentiability and continuity.

Theorem 3.3.7 *If f has a derivative at a point x, then f is continuous at x.*

Proof: Assume that f is differentiable at x. We need to show that $\lim_{h\to 0} f(x+h) = f(x)$. (Recall the definition of continuity.) The idea is to express $f(x+h)$ as $f(x)$ plus some quantity that will go to zero as h approaches 0. We start by multiplying and dividing $f(x+h) - f(x)$ by h. Thus

$$f(x+h) - f(x) = \frac{f(x+h) - f(x)}{h} h$$

and so

$$f(x+h) = f(x) + \frac{f(x+h) - f(x)}{h} h$$

Now taking the limit yields

$$\lim_{h\to 0} f(x+h) = \lim_{h\to 0} \left[f(x) + \frac{f(x+h) - f(x)}{h} h \right]$$

$$\begin{aligned}
&= \lim_{h\to 0} f(x) + \lim_{h\to 0} \frac{f(x+h)-f(x)}{h}\left(\lim_{h\to 0} h\right) \\
&= f(x) + f'(x)(0) \\
&= f(x)
\end{aligned}$$

□

This theorem provides another way of showing that a function is continuous. Every time that we establish that a function is differentiable at a point p, we have automatically shown that the function is continuous at p.

Note: The converse of Theorem 3.3.7 is not true: continuity *does not imply* differentiability. Consider the absolute value function, $f(x) = |x|$. Recall that in Example 3.2.5 we showed that the absolute value function was not differentiable at $x = 0$ by showing that its left- and right-hand derivatives are not equal. However, the absolute value function is continuous at $x = 0$.

The Geometric Interpretation of the Derivative

We repeat for emphasis that the derivative expression for a function is precisely the expression for the slope of the tangent line to the graph of the function. Thus, *f is a differentiable function at p if and only if its graph has a tangent line at p.* Furthermore, the equation of the tangent line at $(p, f(p))$ can be obtained by equating slopes:

$$\text{slope of tangent line} = \frac{\Delta y}{\Delta x} = \text{slope of curve} = f'(p)$$

or

$$\frac{y - f(p)}{x - p} = f'(p)$$

or

$$y = f'(p)(x-p) + f(p)$$

Example 3.3.8 We will find the equation of the tangent line to $y = x^4$ at $x = 2$. Since $f(x) = x^4$, from the above discussion the equation of the tangent line is $y = f'(2)(x-2) + 2^4 = 4(2^3)(x-2) + 16 = 32(x-2) + 16$. △

We can use the intuitive geometry of the tangent line to tell us about the derivative of a function.

Example 3.3.9 Suppose that the graph of f is

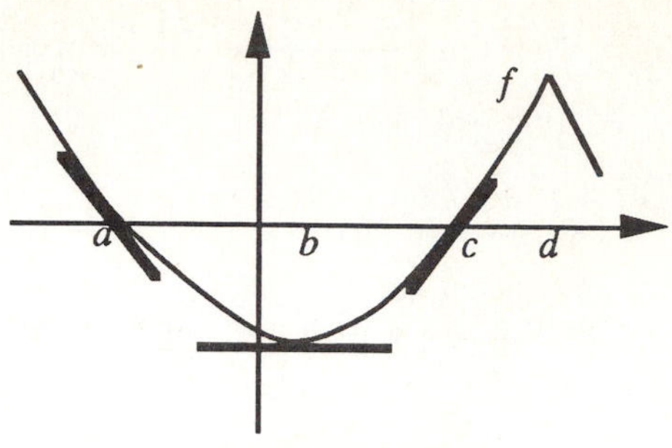

Figure 3.3.10

We first note that f appears to have a tangent line at every point except possibly d, where there appears to be a sharp point. At point a the tangent line slopes downward; thus its slope is negative and $f'(a) < 0$. At point b the curve levels off, the tangent line is horizontal; thus its slope is zero and $f'(b) = 0$. At point c the tangent line slopes upward; thus its slope is positive and $f'(c) > 0$.

If we consider the derivative f' as a function, then we know that it must be negative to the left of b, zero at b, positive between b and d (undefined at d), and negative to the right of d as indicated in the following figure.

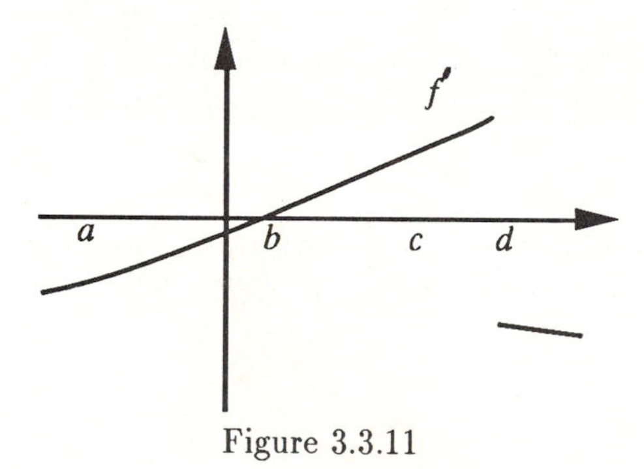

Figure 3.3.11 △

In a later section we will undertake an analytic investigation of the relation between f and f', culminating in the Mean Value Theorem (Theorem 3.6.6).

Section 3.3 Exercises

1. Using the definition of derivative find dy/dx for the functions described by the following relations.

   (a) $y = 10$ **(b)** $y = 4x - 10$ **(c)** $y = x^4$ **(d)** $y = \cos(x)$

   (e) $y = \sin(x)$ **(f)** $y = 1/x$ **(g)** $y = x(x+2)$ **(h)** $y = \sqrt{x}$

2. Find the point on the graph of $y = -x^2$ where the tangent line has an x-intercept of 8.

3. Approximate the graph of $f(x) = x^3$ with a sketch consisting of the points on the graph for $x = -2, -1, 0, 1$, and 2 and short segments of the tangent lines to the graph at those points.

4. Let $f(x) = \begin{cases} x^2 & \text{for } x < 2 \\ 4 & \text{for } x = 2 \\ g(x) & \text{for } x > 2 \end{cases}$
Define a differentiable function g with domain $x > 2$ such that f is continuous, but not differentiable, at $x = 2$.

5. Let f be a differentiable function. Explain the difference (if any) between the derivative of f at $x = 4$, the instantaneous rate of change of f when $x = 4$, and the slope of the line tangent to the graph of f at the point $(4, f(4))$.

6. For each of the following, give an example or explain why no example exists.

   (a) A nonconstant function whose derivative is a constant function.

   (b) A function that is differentiable at $x = 3$, but is not continuous at $x = 3$.

   (c) A function $f : \mathbb{R} \to \mathbb{R}$ that is not differentiable at any point in its domain.

   (d) A function f whose graph does not have a tangent line at the point $(2, f(2))$.

   (e) A function $f : [0, 3] \to \mathbb{R}$ whose right-hand derivative at $x = 0$ is less than the average rate of change of f over any interval $[0, x]$ for $x \in (0, 3]$.

   (f) A differentiable function $f : \mathbb{R} \to \mathbb{R}$ whose left- and right-hand derivatives disagree at $x = 1$.

   (g) A continuous function f defined over the nonnegative real numbers with the property that its average rate of change function is an increasing function.

   (h) A differentiable function $f : \mathbb{R} \to \mathbb{R}$ with the property that f' is a strictly increasing function.

7. Mark each of the following statements true (T) or false (F).

   (a) If $f(x) = g(x) + h(x)$, then $f'(x) = g'(x) + h'(x)$.

   (b) If g and h are differentiable functions and if $f(x) = g(x))/(h(x)$, then $f'(x) = (g'(x))/(h'(x))$.

   (c) If f is a differentiable function and if $f'(3) = 0$, then f has a maximum or minimum value at $x = 3$.

   (d) A sequence is a differentiable function.

(e) A step function is a differentiable function.

(f) The set of continuous functions is a subset of the set of differentiable functions.

8. Sketch and label on the same pair of axes the graphs of $y = f(x)$ and $y = f'(x)$ for (a) $f(x) = x^3$ (b) $f(x) = \sin(x)$ (c) $f(x) = 2^x$.

9. Sketch the graph of a function $f : [-2, 3] \to \mathbb{R}$ such that

$$\begin{array}{lll} f(-2) = 0, & f(-1) = 0, & f(0) = 2, \\ f(1) = 1, & f(2) = 2, & f(3) = 1 \end{array}$$

and the graph of f' is

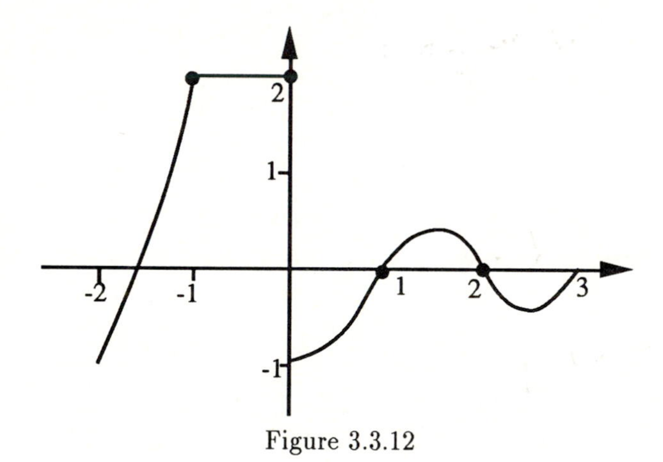

Figure 3.3.12

10. The following sketch indicates the graph of a function $f : \mathbb{R} \to \mathbb{R}$.

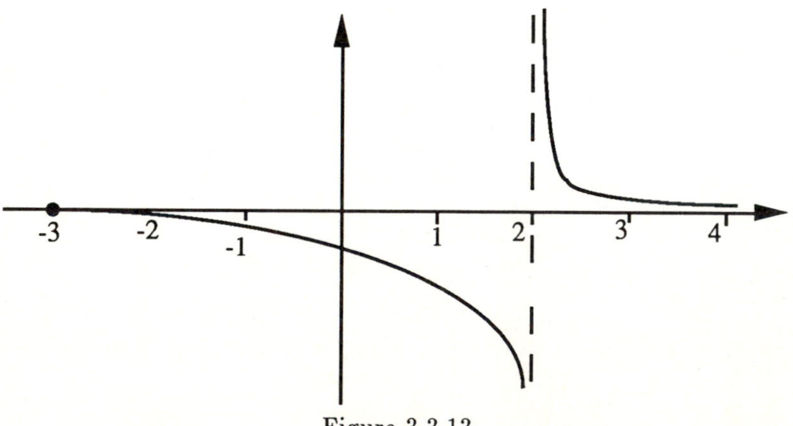

Figure 3.3.13

(a) What is the source of f?
(b) What is the target of f?
(c) What is the natural domain for f?
(d) Is f bounded?Why?
(e) Is f one-to-one? Why?
(f) Is f onto? Why?
(g) Is f monotone? Why?
(h) Is f increasing? Why?
(i) Is f differentiable? Why?
(j) Is f periodic? Why?

11. Given the following graph of f $[f(x) = x^2e^{-x^2}]$

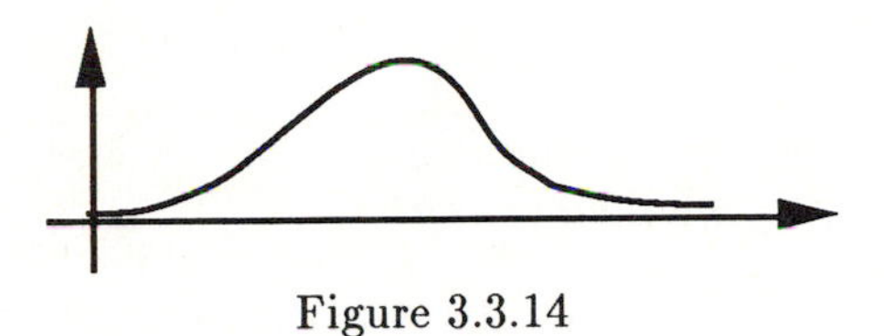

Figure 3.3.14

determine which (if any) of the following graphs is the graph of f'. (The scales are all the same.)

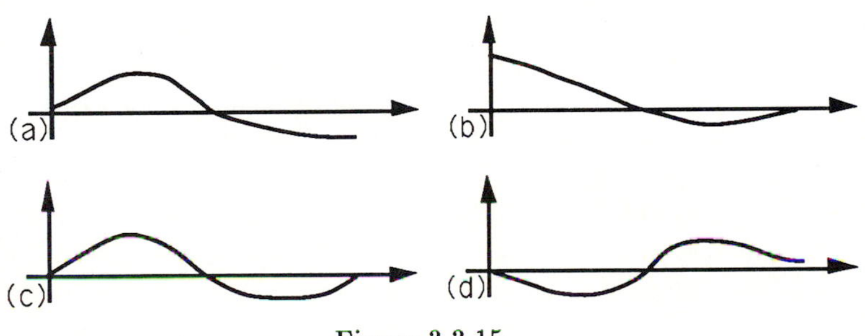

Figure 3.3.15

12. Given the graph of f:

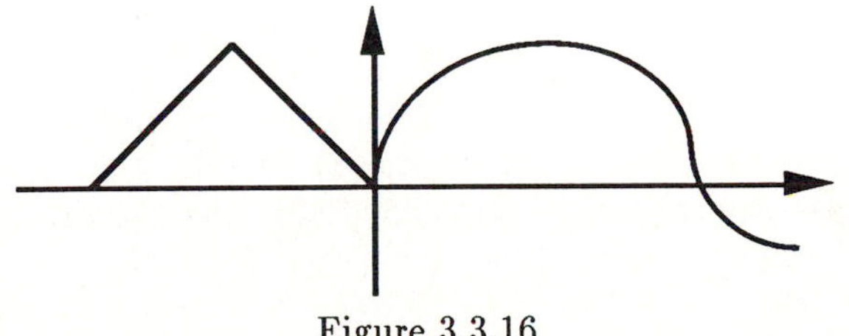

Figure 3.3.16

sketch the graph of f'.

13. Given the graph of f:

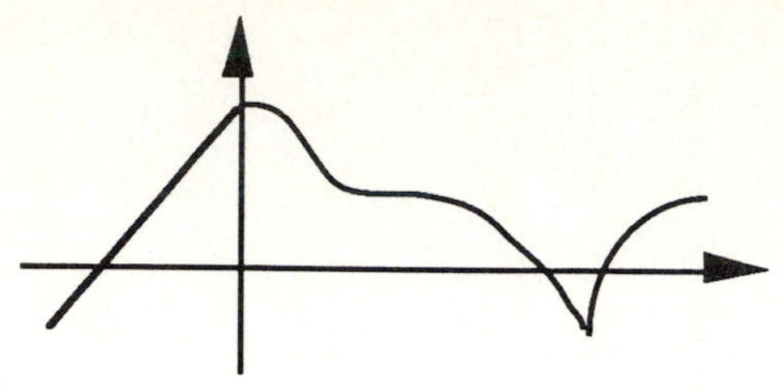

Figure 3.3.17

sketch the graph of f'.

14. What can be said about a function at the x values where the derivative has a maximum or minimum value?

15. (a) Find the equation of the line tangent to the curve $y = x^2$ at (1,1).
    (b) Find the equation of the line tangent to the curve $y = x^3$ at (2,8).
    (c) Find the equation of the line tangent to the curve $y = \sin(x)$ at $(\pi/4, 1/\sqrt{2})$.
    (d) Find the equation of the line tangent to the curve $y = \cos(x)$ at $(1, \cos(1))$.

16. (a) A cow pond needs a new boat ramp. The altitude of the shore is given by $a(x) = x^3$ for x in [2,4]. If the ramp is to be flat and meet the shoreline at $x = 2$, find an equation for the boat ramp.
    (b) A cow's haystack is to be described by $y = x^{1/n}$ for x in [1,2] and some positive integer n. If the slope always exceeds 0.15, what is the smallest integer n that can be used?

17. Consider Theorem 3.3.7
    (a) Is the proof given in the text a direct or indirect proof?
    (b) How does the result $\lim_{h \to 0} f(x + h) = f(x)$ imply that the f is continuous?

18. Using a computer or calculator as a computing tool, determine the derivative of the following functions from the definition. In particular, you should (i) compute the difference quotient, (ii) simplify the difference quotient, (iii) evaluate the limit of the simplified difference quotient, and (iv) check the result of part (iii) using the differentiation routine on your computer or calculator.
    (a) $f : \mathbb{R} \to \mathbb{R}$ defined by $f(x) = \sin(x \cos(x))$.
    (b) $f : \mathbb{R} \to \mathbb{R}$ defined by $f(x) = 1/x(x^2 - 3x)/(2x - x^2)$.
    (c) $f : \mathbb{R} \to \mathbb{R}$ defined by $f(x) = x^2 + 4e^{\cos(x)}$.

19. Following the directions for Exercise 18, evaluate the derivatives (from the definition) of the following functions at the points indicated.

   (a) $f : \mathbb{R} \to \mathbb{R}$ defined by $f(x) = (x^2 - 3x)/(x + 1)$ at $x = 2$.
   (b) $f : \mathbb{R} \to \mathbb{R}$ defined by $f(x) = (3x^2 - 6x)/(4x^5 + 5)$ at $x = 3$.
   (c) $f : \mathbb{R} \to \mathbb{R}$ defined by $f(x) = (e^{3x} - \log(x))^3/(x^3 - 7x)$ at $x = 4$.
   (d) $f : \mathbb{R} \to \mathbb{R}$ defined by $f(0) = 0$ at $x = 0$ and $f(x) = x^2 \sin(1/x)$ at $x \neq 0$.

20. The purpose of this exercise is to illustrate geometrically the difference between a continuous, nondifferentiable function and a differentiable function. In particular, you will see that the graph of a differentiable function when looked at "through a microscope" appears to be a straight line.

   (a) Select your favorite function that is differentiable at $x = 2$. Using a computer plotting routine or a graphing calculator as a microscope, plot the graph of your function and then "zoom" in on a small interval containing the point 2. Continue the zooming until the graph appears to be a straight line.
   (b) Select your favorite function that is continuous, but not differentiable at $x = 2$. Plot the graph and then carry out the zooming process of part (a). Can you obtain a graph that appears to be a straight line? Explain.
   (c) Repeat parts (a) and (b) using different functions.

The following exercise gives a direct geometric argument that $\sin'(x) = \cos(x)$ and $\cos'(x) = -\sin(x)$

21. (Peter Renz) Consider the motion of a point $p(t) = (\cos(t), \sin(t))$ along the rim of the unit circle. If $p(t)$ advances counterclockwise as if carried by a wheel whose rim has a linear velocity of one unit per second, then the magnitude of the velocity of $p(t)$ is one at all times and its direction is perpendicular to the radius from (0,0) to $p(t)$, (i.e., the direction of the velocity is the direction of the tangent line drawn through $p(t)$.) Show that the vertical velocity of $p(t)$ is $\sin'(t)$ and that this must be equal to $\cos(t)$. Likewise show that the horizontal velocity of $p(t)$ is $\cos(t)$ and this must be equal to $-\sin(t)$. Hint: Draw a diagram showing the unit circle, a point $p(t)$ on the rim, the tangent line drawn through $p(t)$, and the vertical and horizontal component forces of the velocity of $p(t)$. The drawings of the vertical and horizontal forces of $p(t)$ should form a right triangle with the tangent line through $p(t)$.

3.4 The Algebra of Derivatives

Having computed the derivative of several basic functions, we are ready to expand our list of examples into classes of functions by developing an algebra of derivatives. Recall how this was done for functions in Chapter 1 and for sequences (Theorem 2.3.1), limits (Theorem 2.4.21), and continuity (Theorem 2.5.10) in Chapter 2.

Theorem 3.4.1 (Algebra of Derivatives) *Let $f, g : \mathbb{R} \to \mathbb{R}$ be two differentiable functions defined on a common interval. The derivative of the sum, difference, product, and quotient of these two functions is given by the following equations:*

(a) $(af)'(x) = af'(x)$

(b) $(f+g)'(x) = f'(x) + g'(x)$

(c) $(f-g)'(x) = f'(x) - g'(x)$

(d) $(fg)'(x) = f(x)g'(x) + f'(x)g(x)$

(e) $(f/g)'(x) = (g(x)f'(x) - f(x)g'(x))/g(x)^2 g(x) \neq 0.$

We shall leave the proofs of parts (a), (b), and (c) to the exercises. The proof of part (d) involves a common transformation in mathematics–*subtracting and adding an expression to carry out a factorization.* We shall sketch the details.

Proof of part (d): The difference quotient for the product $(fg)(x)$ is

$$\frac{f(x+h)g(x+h) - f(x)g(x)}{h}$$

Subtract and add $g(x)f(x+h)$ in the numerator and then factor to obtain the expression

$$\begin{aligned}
&\frac{f(x+h)g(x+h) - f(x)g(x)}{h} \\
&\quad = \frac{f(x+h)g(x+h) - g(x)f(x+h) + g(x)f(x+h) - f(x)g(x)}{h} \\
&\quad = f(x+h)\frac{g(x+h) - g(x)}{h} + \frac{f(x+h) - f(x)}{h}g(x)
\end{aligned}$$

Taking the limit as h approaches 0, the first term on the right approaches $f(x)g'(x)$ and the second term approaches $f'(x)g(x)$. This yields the desired result. □

In taking the limit of the first term on the right, we used the continuity of f to conclude that $f(x+h)$ approaches $f(x)$ as h approaches 0. How do we know that f is continuous?

It is often very helpful to learn to express the parts of Theorem 3.4.1 verbally in the following sing-song fashion:

(a) The derivative of a constant times a function is the constant times the derivative of the function.

(b) The derivative of a sum is the sum of the derivatives.

(c) The derivative of a difference is the difference of the derivatives.

(d) The derivative of a product is the "first times the derivative of the second plus the derivative of the first times the second."

(e) The derivative of a quotient is the "denominator times the derivative of the numerator minus the numerator times the derivative of the denominator all divided by the denominator squared."

We now describe several consequences of Theorem 3.4.1 that illustrate the "power" of having an algebra for a given concept.

1. *Linearity property of derivatives.* Using parts (a) and (b), "the derivative of a sum is the sum of the derivatives," yields for every pair of constants, r and s:

$$(rf + sg)'(x) = rf'(x) + sg'(x)$$

This relationship is referred to as the *linearity property* of derivatives. Note that limits also satisfy a linearity property

$$\lim_{x \to a}(rf + sg)(x) = r \lim_{x \to a} f(x) + s \lim_{x \to a} g(x)$$

as do continuous functions.

2.*Differentiability of polynomials.* The linearity property can be extended to an arbitrary finite sum using summation notation; we have

$$\left(\sum_{k=0}^{n} c_k f_k\right)'(x) = \sum_{k=0}^{n} c_k f_k'(x)$$

where the c_k's are constants and the functions f_k are differentiable. Recall from Chapter 1 that a polynomial is a linear combination of functions of the form x^n, each multiplied by a constant. Thus, for the polynomial

$$p(x) = \sum_{k=0}^{n} c_k x^k$$

we have

$$p'(x) = \sum_{k=1}^{n} c_k k x^{k-1}$$

For example, if $p(x) = 9 - 13x + 7x^2 + 3x^4$, then $p'(x) = -13 + 7 \cdot 2x + 3 \cdot 4x^3$.

In summation notation, this polynomial is written

$$p(x) = \sum_{k=0}^{4} c_k x^k$$

where $c_0 = 9$, $c_1 = -13$, $c_2 = 7$, $c_3 = 0$, and $c_4 = 3$ with

$$p'(x) = \sum_{k=1}^{4} c_k \cdot k x^{k-1}$$

Thus *all polynomials are differentiable functions!*

3. *Differentiability of rational functions.* Recall that a rational function is the quotient of two polynomials with the restriction that the denominator is nonzero. Since polynomials are differentiable functions, part (e) of Theorem 3.4.1 guarantees that *all rational functions are differentiable.*

For example, if

$$h(x) = \frac{3x^5 - x^3 + 2x}{x^2 + 7}$$

then

$$h'(x) = \frac{(x^2+7)(15x^4 - 3x^2 + 2) - (3x^5 - x^3 + 2x)(2x)}{(x^2+7)^2}$$

4. *Power functions are differentiable.* We have previously shown that

$$\frac{d}{dx} x^{1/n} = \frac{1}{n} x^{(1/n-1)} \quad \text{for } x > 0$$

Expressing $x^{2/n}$ as the product $x^{1/n}x^{1/n}$ and using part (d) of Theorem 3.4.1 we have

$$\frac{d}{dx} x^{2/n} = \frac{2}{n} x^{(2/n-1)}$$

We can extend this result and obtain

$$\frac{d}{dx} x^{p/q} = \frac{p}{q} x^{(p/q-1)} \quad \text{for } x > 0$$

Thus, $d/dx\ x^r = rx^{r-1}$ where $x > 0$ and r is a rational number. We will give a proof of this result in the next section.

5. *All trigonometric functions are differentiable* (since both the sine and cosine functions are differentiable).

We summarize the above consequences of Theorem 3.4.1 with the statement: All polynomial functions, rational functions, power functions, and trig functions are differentiable. In addition, the differentiation operation is linear [i.e., $(cf + dg)'(x) = cf'(x) + dg'(x)$]. The following two examples illustrate these results.

Example 3.4.2 We will find the derivative of $\tan(x)$.

We express $\tan(x)$ as the quotient $\sin(x)/\cos(x)$ and then apply part (e) of Theorem 3.4.1 to find the derivative. Thus

$$\begin{aligned} D\tan(x) &= D\left(\frac{\sin(x)}{\cos(x)}\right) = \frac{\cos(x)D\sin(x) - \sin(x)D\cos(x)}{\cos^2(x)} \\ &= \frac{\cos(x)\cos(x) - \sin(x)(-\sin(x))}{\cos^2(x)} \\ &= \frac{1}{\cos^2(x)} \\ &= \sec^2(x) \end{aligned}$$

△

Example 3.4.3 Consider the derivative of

$$f(x) = x^3 + 2x^{2/3}\sin(x) - \frac{3x^4 + 7x}{x^2 - 2}$$

Using the rules for differentiating a sum, product, quotient, power function, and $\sin(x)$, we have

$$f'(x) = 3x^2 + 2x^{\frac{2}{3}}\cos(x) + \frac{4}{3}x^{-1/3}\sin(x) - \frac{(x^2-2)(12x^3+7) - (3x^4+7x)(2x)}{(x^2-2)^2}$$

△

The Chain Rule

Theorem 3.4.1 provides techniques for differentiating functions formed from combining terms using the basic operations of addition, subtraction, multiplication, and division. However, none of these techniques help in finding the derivatives of

$$f(x) = (x^2+1)^{1/2}, \quad g(x) = \sin(x^2+2)$$

or

$$h(x) = (x^4 + 2x)^{100}$$

It is true that the last function could be expanded into a polynomial of degree 400 and then differentiated term by term, but what a job!

The functions in the above paragraph have the common property that each one can be rewritten as a composition of functions whose derivatives can be found using the techniques of Theorem 3.4.1. For example, if $t(x) = x^2 + 1$ and $w(u) = u^{1/2}$, then $f(x) = (x^2+1)^{1/2}$ is expressible as a composition in the form $f(x) = w(t(x))$.

We now develop the technique for differentiating the composition of two differentiable functions. Called the *chain rule,* this technique is a basic rule of differentiation.

The idea behind the chain rule can be illustrated with a compound pulley system of three pulleys such as is often found in machine shops or home workshops where more than one machine is to run off a single motor. In the picture below, we can think of A the pulley on a motor, B a pulley mounted on a "jack shaft" (a transmission shaft), and C a pulley on a band saw. Thus A "drives" B and, in turn, B "drives" C.

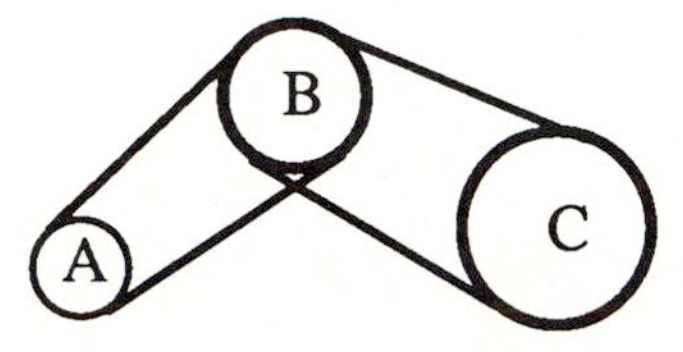

Figure 3.4.4

The interesting rate question is: What is the rate of change of C with respect to A? Abbreviating "with respect to" with "w.r.t.", we get

- Since (rate of change of C w.r.t. B) $= \dfrac{\text{circumference of } C}{\text{circumference of } B}$
- and (rate of change of B w.r.t. A) $= \dfrac{\text{circumference of } B}{\text{circumference of } A}$, then
- (rate of change of C w.r.t. A) $= (\dfrac{\text{circumference of } C}{\text{circumference of } B}) \cdot (\dfrac{\text{circumference of } B}{\text{circumference of } A})$

Note that (rate of change of C w.r.t. A) = (rate of change of C w.r.t. B) times (rate of change B w.r.t. A).

Interpreting the above figure as an abstraction of the pedal, sprocket, and chain system on a bike provides another example of the chain rule. In this case, A represents the "pedal wheel," B the main sprocket, and C the rear wheel sprocket. On a multispeed bike, changing the size of the two sprocket gears (i.e., pulleys B and C) is what produces the different speeds (ratios of pedal revolutions to wheel revolutions).

Another illustration of the chain rule idea is provided by the rate of change of currency between different countries. For example, if one American dollar is worth k French francs ($A = kF$) and one French franc is worth p Italian lira ($F = pI$), what is the rate of change of an American dollar with respect to an Italian lira? Note that the exchange rates are

$$\frac{\Delta A}{\Delta F} = k$$

and

$$\frac{\Delta F}{\Delta I} = p$$

Now if one were to change one American dollar into francs and then change the francs into lira, the resulting number of lira, kp, would represent the exchange rate. That is,

$$\frac{\Delta A}{\Delta I} = \frac{\Delta A}{\Delta F}\frac{\Delta F}{\Delta I} = kp$$

These illustrations suggest the following theorem.

Theorem 3.4.5 (The Chain Rule) *If g is differentiable at x and f is differentiable at $g(x)$, then the composite function $f \circ g$ is differentiable at x and*

$$(f \circ g)'(x) = f'(g(x))g'(x)$$

We can express this in Leibniz's notation by letting $y = f(u)$ and $u = g(x)$. Then the derivative of the composite function $y = f(g(x))$ is given by

$$\frac{dy}{dx} = \frac{dy}{du}\frac{du}{dx}$$

Proof. Since $y = f(u)$ is a differentiable function of u, we may *approximate* dy/du with $\Delta y/\Delta u$ whenever $\Delta u \neq 0$. Thus for $\Delta u \neq 0$, we may write

$$\frac{\Delta y}{\Delta u} = \frac{dy}{du} + ERROR(u)$$

where $ERROR(u)$ approaches zero as Δu approaches zero. Now multiplying both sides by Δu yields

$$\Delta y = \frac{dy}{du}\Delta u + ERROR(u)\Delta u$$

This equation is correct even when $\Delta u = 0$. Dividing both sides of this equation by Δx yields

$$\frac{\Delta y}{\Delta x} = \frac{dy}{du}\frac{\Delta u}{\Delta x} + ERROR(u)\frac{\Delta u}{\Delta x}$$

Since u is a differentiable, and hence continuous, function of x, Δx approaching zero implies that Δu will also approach zero. Thus taking the limit as Δx approaches zero yields

$$\frac{dy}{dx} = \frac{dy}{du}\frac{du}{dx}$$

(How do we know that the term

$$ERROR(u)\frac{\Delta u}{\Delta x}$$

goes to zero as Δx goes to zero?) □

Examples of the Chain Rule:

1. If $y = (x^2+1)^{1/2}$, then define f by $y = f(u) = u^{1/2}$ and define $u = g(x) = x^2+1$. Thus $y = f(g(x))$ and

$$\frac{dy}{dx} = \frac{dy}{du}\frac{du}{dx} = \left(\frac{1}{2}u^{-1/2}\right)(2x) = x(x^2+1)^{-1/2}$$

2. If $y = \sin(x^3+2)$, then define f by $y = f(u) = \sin(u)$ and g by $u = g(x) = x^3+2$. Thus $y = f(g(x))$ and

$$\frac{dy}{dx} = \frac{dy}{du}\frac{du}{dx} = (\cos(u))(3x^2) = 3x^2\cos(x^3+2)$$

3. If $y = (x^4+2x)^{100}$, then define f by $y = f(u) = u^{100}$ and g by $u = g(x) = x^4+2x$. Thus $y = f(g(x))$ and

$$\frac{dy}{dx} = \frac{dy}{du}\frac{du}{dx} = 100u^{99}(4x^3+2) = 100(x^4+2x)^{99}(4x^3+2)$$

4. If $y = (x^2\sin(x)+3x)^3$, define f by $y = f(u) = u^3$ and g by $u = g(x) = x^2\sin(x)+3x$. Thus $y = f(g(x))$ and

$$\begin{aligned}\frac{dy}{dx} &= \frac{dy}{du}\frac{du}{dx}\\ &= 3u^2(x^2\cos(x)+2x\sin(x)+3)\\ &= 3(x^2\sin(x)+3x)^2(x^2\cos(x)+2x\sin(x)+3)\end{aligned}$$

5. Let $y = \sec(x)$. To express this as a composite function, we express $\sec(x)$ as $(1+\tan^2(x))^{1/2}$. Thus $y = (1+\tan^2(x))^{1/2}$. Define f by $y = f(u) = u^{1/2}$ and define g by $u = g(x) = 1+\tan^2(x)$. Then $y = f(g(x))$ and

$$\frac{dy}{du} = \frac{dy}{du}\frac{du}{dx}$$

Treating $\tan^2(x)$ as a product and differentiating accordingly

$$\begin{aligned}\frac{dy}{dy} &= \frac{1}{2}u^{-1/2}(2\tan(x)\sec^2(x))\\ &= (1+\tan^2(x))^{-1/2}\tan(x)\sec^2(x)\\ &= \frac{1}{\sec(x)}\tan(x)\sec^2(x)\\ &= \tan(x)\sec(x)\end{aligned}$$

6. If $y = e^{\sin(x)}$, then define g by $u = g(x) = \sin(x)$ and define f by $y = f(u) = e^u$. Thus $y = f(g(x))$ and

$$\frac{dy}{dx} = \frac{dy}{du}\frac{du}{dx} = e^u\cos(x) = e^{\sin(x)}\cos(x)$$

7. Let $y = \cos^4(5x)$. (Recall that $\cos^4(5x) = (\cos(5x)^4$.) Now y is of the form u^4 where $u = \cos(5x)$. Hence

$$\frac{dy}{dx} = 4\cos^3(5x)\frac{d}{dx}\cos(5x) = 4\cos^3(5x)(-\sin(5x)5)$$

Note that $\cos^4(5x)$ is a composition of three functions and thus the chain rule is applied twice.

Warning: A very common mistake in computing derivatives is to omit the "du/dx" factor required by the chain rule. For example, $d/dx \sin(x) = \cos(x)$, but $d/dx \sin(2x) = \cos(2x)2$ not just $\cos(2x)$. Similarly $d/dx\,(x-6)^2 = 2(x-6)$, but $d/dx\,(3x^2-6)^2 = 2(3x^2-6)(6x)$ not just $2(3x^2-6)$. Likewise $d/dx\, e^x = e^x$, but $d/dx\, e^{\sin(x)} = e^{\sin(x)}\cos(x)$ not just $e^{\sin(x)}$.

Question 3.4.6 Consider the functions $f, g : \mathbb{R} \to \mathbb{R}$ defined by $f(x) = x^2$ and $g(x) = |x|$. Note that g is not a differentiable function.

(a) Is the product $(fg)(x) = f(x)g(x)$ differentiable? If so, does this contradict part (d) of Theorem 3.4.1? Explain.

medskip**(b)** Does the composition of f with g yield a differentiable function? If yes, does this contradict the chain rule? Explain.

◇

The Inverse Function Theorem

As an application of the chain rule we will develop a method to determine the derivative of the inverse function if we know the derivative of the function.

Theorem 3.4.7 (Inverse Function Theorem) *If f is differentiable and strictly monotonic on an interval $I = (a, b)$, then f^{-1} is differentiable and strictly monotonic on $f(I)$, and*

$$(f^{-1})'(f(x)) = \frac{1}{f'(x)}$$

Proof: Since the graph of f is strictly monotonic on I, f^{-1} exists and is nonzero on $f(I)$ (Why?). Since the graph of f on I has a tangent line at every point and the graph of f^{-1} is the reflection of the graph of f in the line $y = x$, the graph of f^{-1} must have a tangent line at every point. Thus f^{-1} is differentiable on $f(I)$.

To find the derivative of f^{-1}, we notice that the composition of f^{-1} and f gives the identity function on I:

$$(f^{-1} \circ f)(x) = f^{-1}(f(x)) = x$$

for all x in I. Now we differentiate both sides, applying the chain rule to the left-hand side:

$$(f^{-1} \circ f)'(x) = (f^{-1})'(f(x))f'(x) = 1 \quad \text{for all } x \in I$$

Dividing both sides by $f'(x)$ gives us the desired result. □

Example 3.4.8 (The Derivative of Arcsin) The sine function is strictly increasing on the interval $I = (-\pi/2, \pi/2)$ and $\sin(I) = (-1, 1)$. By Example 3.3.5 (e) we know that $\sin'(x) = \cos(x)$ on I. Thus by the inverse function theorem sine has an inverse function on $(-1, 1)$, denoted $\arcsin = \sin^{-1}$, and

$$\arcsin'(y) = \frac{1}{\sin'(x)} = \frac{1}{\cos(x)}$$

if $y = \sin(x)$. We want to write both sides of the above equation in terms of the same variable. To do this, we need a relation between y $(= \sin(x))$ and $\cos(x)$. The obvious relationship is the Pythagorean theorem, $\sin^2(x) + \cos^2(x) = 1$, or

$$y^2 + \cos^2(x) = 1$$

So,

$$\arcsin'(y) = \frac{1}{\cos(x)} = \frac{1}{[1 - y^2]^{1/2}}$$

The positive root is used since cosine is positive on I. Rewriting the above equation in terms of the variable x, we have

$$\arcsin'(x) = \frac{1}{[1 - x^2]^{1/2}} \qquad \triangle$$

Example 3.4.9 (Derivative of Logarithm) Given the derivative of the exponential function $f(x) = e^x$ in Example 3.3.5 (g), we can use the inverse function theorem to determine the derivative of $f^{-1} = \log$, which exists since f is strictly monotonic.

$$\log'(f(x)) = \frac{1}{f'(x)} = \frac{1}{f(x)}$$

Let $y = f(x) = e^x$. Then

$$\log'(y) = \frac{1}{y}$$

or, using x as the variable,

$$\log'(x) = \frac{1}{x} \qquad \triangle$$

Example 3.4.10 We will compute the derivative of $f(x) = \log(x^2 + 4x)$. Let $u = x^2 + 4x$ and $g(u) = \log(u)$. Then by the chain rule,

$$\frac{dy}{dx} = \frac{dy}{du}\frac{du}{dx} = \frac{1}{u}\frac{du}{dx} = \frac{1}{x^2 + 4x}(2x + 4) = \frac{2x + 4}{x^2 + 4x} \qquad \triangle$$

Section 3.4 Exercises

In Exercises 1 through 20, find dy/dx for each of the following relations. You need not simplify.

1. $y = 3x^7 - 2\sqrt{x}$
2. $y = (x^2 + 5x)/(x - 2)$
3. $y = x^2 \sin(x) - \tan(x)$
4. $y = 7x^3 - 3x^{2/3} \cos(x)$
5. $y = (3x^2 - 6x)/(4x^5 + 5)$
6. $y = (x^2 - 3x)/(2x - x^2)$
7. $y = \sin(x)/x$
8. $y = x^2 + 4e^{\cos(x)}$
9. $y = \log(x + e^x)$
10. $y = \sin(x^2 + 7x)$
11. $y = \sin(x \cos(x))$
12. $y = [(3x^2 - 6x)/(4x^5 + 5)]^{1399}$
13. $y = \sin(x) \tan(x^3)$
14. $y = \begin{cases} x^3 - 3x & \text{for } x \leq 5 \\ (x^2 - 2)/(x - 4) & \text{for } x > 5 \end{cases}$
15. $y = e^{3x^2} \cos(x)$
16. $y = e^x \cos(x)$
17. $y = \sin^3(e^{2x})$
18. $y = e^{\log(x)} + \log(e^x)$
19. $y = (e^{3x} \log(x))^4$
20. $y = \tan(x^{2)} + (e^{3x} - \log(x))^3$

In Exercises 21 through 24, compute $f'(x_1)$ for the given x_1.

21. $f(x) = 3x^5 - 4x^2$, $x_1 = 1$
22. $f(x) = x^2 \sin(x)$, $x_1 = 2$
23. $f(x) = (x - 1)/x$, $x_1 = 2$
24. $f(x) = (x^2 - 3x)/(x + 1)$, $x_1 = 2$

In Exercises 25 through 28, find the indicated derivatives.

25. $D_t s$ if $s = 16t^2 + 30t + 10$
26. du/dv if $u = 16v^2 + 30v + 10$
27. $D_x y$ if $y = \sin(x) \cos(x)$
28. $D_x y$ if $y = \tan(x) \sec(x)$
29. Determine values for the constants b and d such that the function f is differentiable, where

$$f(x) = \begin{cases} bx^2 + d & \text{for } x \leq 1 \\ 2x^2 - 7 & \text{for } x > 1 \end{cases}$$

30. Determine values for the constants b and d such that the function, f, is differentiable where

$$f(x) = \begin{cases} 3x^5 + 4b & \text{for } x \leq 1 \\ dx^2 - 7 & \text{for } x > 1 \end{cases}$$

31. Mark the following statements true (T) or false (F).

    (a) The function f defined by

    $$f(x) = \begin{cases} 3x - 2 & \text{for } x < 3 \\ 7 & \text{for } x = 3 \\ 10 - x & \text{for } x > 3 \end{cases}$$

    is differentiable at $x = 3$.

    (b) If $f(0) = 0$ and $f'(x) < 0$ for $x > 0$, then f is negative for all $x > 0$.

    (c) If f is a continuous function, then it is a differentiable function.

    (d) If f is a differentiable function, then it is a continuous function.

    (e) If f is a differentiable function, then f' is a continuous function.

32. Compute $f'(1)$ where $f(x) = x^2 \sin(x^3 + 2x)$.

33. Let $f : \mathbb{R} \to \mathbb{R}$ be defined by $f(0) = 0$ and $f(x) = x^2 \sin(1/x)$ for x not equal to zero.

    Using the plotting routine on a computer or graphing calculator, plot the graph of f over an interval containing $x = 0$. Make a conjecture (based on your graph) about the differentiability of f at $x = 0$. [Is f differentiable at $x = 0$? If so, what is the value of $f'(0)$?] You may need to reduce the size of the interval several times in order to get a "clear" picture at $x = 0$. Check your conjecture with an analytical argument. If the results of your conjecture and analytical argument differ, explain which is correct.

34. Follow the instructions in Exercise 33 for the function $f : \mathbb{R} \to \mathbb{R}$ defined by $f(0) = 0$ and $f(x) = x \sin(1/x)$ for x not equal to zero.

35. Consider part (c) of Theorem 3.4.1

    (a) Rewrite part (c) of the theorem in the form of a conditional statement.

    (b) Apply the definition of differentiability to the function $k : \mathbb{R} \to \mathbb{R}$ defined by $k(x) = (f - g)(x) = f(x) - g(x)$.

    (c) Complete the proof of part (c) of Theorem 3.4.1.

36. Consider part (a) of Theorem 3.4.1.

    (a) Rewrite part (a) of the theorem in the form of a conditional statement.

    (b) Define $g : \mathbb{R} \to \mathbb{R}$ by $g(x) = af(x)$. Now apply the definition of differentiability to the function g to show that $g'(x) = (af(x))' = af'(x)$.

37. Consider part (e) of Theorem 3.4.1.

    (a) Consider the quotient $f(x)/g(x)$ as the product $(1/g(x))f(x)$.

(b) To differentiate $1/g(x)$, define a function $k : \mathbb{R} \to \mathbb{R}$ by $k(x) = 1/g(x)$. Then $k(x)g(x) = 1$. Now differentiate this last expression and solve for $k'(x)$.

(c) Complete the proof of part (e) of Theorem 3.4.1.

38. (a) Compute the derivative of arctan. Hint: Recall that $\sec^2(t) = 1 + \tan^2(t)$.

(b) Compute the derivative of arccos.

39. Suppose f and g are differentiable, $h = g^{-1}$, and

$$\begin{array}{lll} f(1) = 2 & f'(1) = 3 & g(2) = 6 \\ g'(2) = -2 & g(3) = 2 & g'(3) = -4 \end{array}$$

Find $h'(f(1))$.

40. Suppose f and g are differentiable, $h = g^{-1}$, and

$$\begin{array}{lll} f(1) = 7 & f'(1) = 3 & g(2) = 5 \\ g'(2) = 9 & g(3) = 7 & g'(3) = 6 \end{array}$$

Find $h'(f(1))$.

41. A cow, who is 5 ft tall, is almost killed when grazed on the head by a moonbeam which also grazed a nearby semicircular hill (upon which the cow *never* grazes) and struck the plain 10 feet away on the other side of the cow. If the hill is 50 ft high, how far away from the cow is its base?

42. Use the inverse function theorem to find the equation of the line tangent to the graph of f^{-1} at the point $(a, f^{-1}(a))$.

(a) $f(x) = x^3 + x + 1$, (3,1)

(b) $f(x) = (x^3 + 8)/(x^5 - 1)$, $(0, -2)$

(c) $f(x) = \cos(x)$, $(1/2, \pi/3)$

43. Prove or disprove that $\lim_{x\to 2} e^x = e^2$.
Hint: Recall that since e^x is differentiable, e^x is continuous.

44. If $f(x) = \log(e^x)$, prove or disprove that $f'(10) = 1$.
Hint: Recall that since log is differentiable, it is continuous.

3.5 Implicit and Higher Derivatives

In (long-range) planning, it is often the rate of change of the rate of change that is important, that is, the derivative of the derivative or the *second derivative.* The second derivative provides information on long-term changes or trends, whereas the first derivative provides information on immediate change.

Notation. If f is a differentiable function whose first derivative is also differentiable, we say that f is twice differentiable and write f'', d^2f/dx^2, $d^2/dx^2 f$, or D^2f. Similarly, if the second derivative of f is differentiable, we say that f is three-times differentiable and write f''', d^3f/dx^3, $d^3/dx^3 f$ or D^3f.

In general, if the $(n-1)$st derivative of f is differentiable, we say that f is n times differentiable and write $f^{(n)}$, d^nf/dx^n, $d^n/dx^n f$, or D^nf. (Note the parentheses about the n in $f^{(n)}$.

Example 3.5.1 (Tree Growth) Suppose a model for the growth of a tree figured over a fall, winter, spring period is modeled by the function $h(t) = t^3 + k$ where k is the height of the tree in the middle of winter.

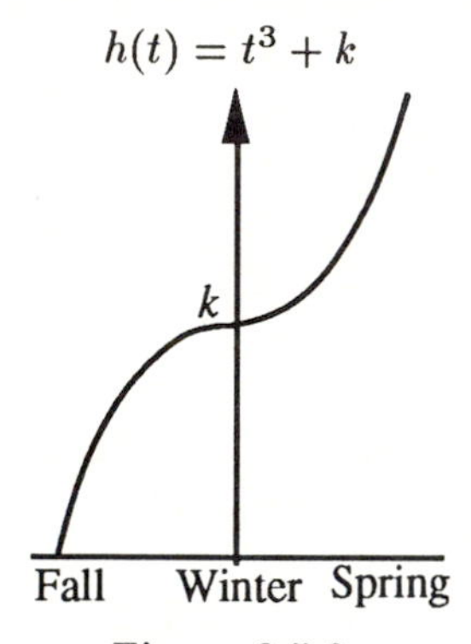

Figure 3.5.2

The graph suggests that the growth rate (first derivative) slows down during the fall, is practically zero during the winter, and then speeds up in the spring. This means the first derivative of h should be positive in the fall and spring and almost zero during the winter; the second derivative of h should be negative during the fall, almost zero during the winter, and positive during the spring. It is easy to verify that this is the case analytically since $h'(t) = 3t^2$ and $h''(t) = 6t$. △

Question 3.5.3 Show that for $f(x) = \sin(x)$, $f^{(4)}(x) = f(x)$ ◇

Definition 3.5.4 (Velocity and Acceleration) *If $y = f(t)$ models the path of an object in motion, then we define the* velocity *of the object to be the first derivative, $f'(t)$, and the* acceleration *of the object to be the second derivative, $f''(t)$.*

Example 3.5.5 A baseball is thrown vertically into the air. Let the height above ground of the ball be given by $f(t) = -15t^2 + 60t + 6$, where $f(t)$ is measured in feet and t in seconds. What is the initial velocity ($t = 0$)? What is the velocity when $t = 3$? What is the acceleration when $t = 4$?

We differentiate $f(t)$ to obtain an expression for the velocity and get $f'(t) = -30t + 60$. Thus, the initial velocity ($t = 0$) is given by $f'(0) = 60$ ft/sec. The velocity when $t = 3$ is $f'(3) = -30$ ft/sec (the negative sign indicates that the ball is falling, toward minus infinity on the coordinate axes). The acceleration is given by the second derivative, that is, $f''(t) = -30$. Thus the acceleration is a constant negative 30 ft/sec/sec for all the time that the ball is in the air. △

We have emphasized the fact that if a function is differentiable, then the function is continuous. However, the differentiability of a function says nothing about either the continuity or differentiability of the first derivative.

Example 3.5.6 (Discontinuous Derivative) This is an example of a function that is differentiable at 0, but whose derivative is not continuous at 0. Let the function f be defined by

$$f(x) = \begin{cases} x^2 \sin(1/x) & \text{for } x \neq 0 \\ 0 & \text{for } x = 0 \end{cases}$$

For x not equal to zero, we compute the derivative (using the Power, Product, and Chain Rules) and obtain

$$f'(x) = -\cos(\frac{1}{x}) + 2x \sin(\frac{1}{x})$$

For x equal to zero, we *must* apply the definition of the derivative. (Why?) Thus,

$$f'(0) = \lim_{h \to 0} \frac{f(0+h) - f(0)}{h} = \lim_{h \to 0} \frac{h^2 \sin(1/h)}{h} = \lim_{h \to 0} h \sin(\frac{1}{h}) = 0$$

(by the "squeeze" technique).

Clearly,

$$f'(x) = \begin{cases} -\cos(1/x) + 2x \sin(1/x) & \text{for } x \neq 0 \\ 0 & \text{for } x = 0 \end{cases}$$

is continuous for all $x \neq 0$. However, when $x = 0$, we see that $\lim_{x \to 0} f'(x)$ does not exist since $\lim_{x \to 0} \cos(1/x)$ does not exist. Thus, f is a differentiable function whose derivative is not continuous! △

Example 3.5.7 (First Derivative Exists, But Not a Second Derivative) This is an example of a function whose first derivative exists and is continuous at 0 (unlike the previous example), but no second derivative exists at 0. Let f be the function defined by

$$f(x) = \begin{cases} x^3 \sin(1/x) & \text{for } x \neq 0 \\ 0 & \text{for } x = 0 \end{cases}$$

The computations involved in finding the derivative are similar to those in the previous example and yield

$$f'(x) = \begin{cases} -x\cos(1/x) + 3x^2\sin(1/x) & \text{for } x \neq 0 \\ 0 & \text{for } x = 0 \end{cases}$$

Clearly f' is continuous for $x \neq 0$. Using the "Squeeze" technique on both terms of $f'(x)$, we can easily show that $\lim_{x\to 0} f'(x) = 0$. Since $f'(0)$ has been computed to be zero, we conclude that f' is continuous at $x = 0$.

Now, applying the definition of derivative to f' at the point $x = 0$, we obtain

$$\begin{aligned} \lim_{h\to 0} \frac{f'(0+h) - f'(0)}{h} &= \lim_{h\to 0} \frac{-h\cos(1/h) + 3h^2\sin(1/h)}{h} \\ &= \lim_{h\to 0} \left(-\cos\left(\frac{1}{h}\right) + 3h\sin\left(\frac{1}{h}\right)\right) \end{aligned}$$

Since this limit does not exist (Why?), f' is not differentiable. △

Implicit Differentiation and Related Rates

The functions that we have considered in our discussion of differentiation have all had the property that one variable could be expressed in terms of the other by a system of one or more equations. Such functions are said to be defined *explicitly.*

For example, a function $f : \mathbb{R} \to \mathbb{R}$ is defined explicitly by the relation $f(x) = x^3 + 7x - 4$. That is, the relation $y = x^3 + 7x - 4$ defines a unique function, $y = f(x)$. On the other hand, the relation $x^2 + y^2 = 4$ may be used to define several functions. Two such functions are:

$$f(x) = (4 - x^2)^{1/2} \text{ and } g(x) = -(4 - x^2)^{1/2}$$

These functions are said to be defined *implicitly* by the relation $x^2 + y^2 = 4$. The graphs of $y = f(x)$ and $y = g(x)$ are the two indicated sections of the graph of $x^2 + y^2 = 4$.

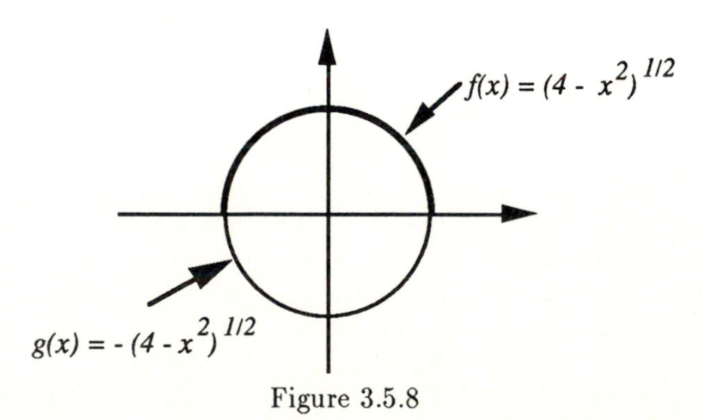

Figure 3.5.8

We illustrate in the next example the interesting and very useful fact that *all* differentiable functions defined implicitly by the same relation have the same derivative expression.

Example 3.5.9 Find the slope of the tangent line to the circle $x^2 + y^2 = 4$ at a point where $y \neq 0$. That is, differentiate $x^2 + y^2 = 4$. However, this object is not a function! What function should be considered: $f(x) = (4 - x^2)^{1/2}$ or $g(x) = -(4 - x^2)^{1/2}$? (The graph of f is the upper semicircle and the graph of g is the lower semicircle.)

Differentiating both f and g using the Chain Rule yields

$$f'(x) = \frac{1}{2}(4 - x^2)^{-1/2}(-2x) = \frac{-x}{f(x)}$$

whenever $f(x) \neq 0$ and

$$g'(x) = -\frac{1}{2}(4 - x^2)^{-1/2}(-2x) = \frac{-x}{g(x)}$$

whenever $g(x) \neq 0$. Now letting y stand for either $f(x)$ or $g(x)$, we obtain $y' = -x/y$ whenever $y \neq 0$ in both equations. △

The common expression for the derivative of the f and g functions in the preceding example (i.e., $y' = -x/y$) can be found by differentiating the original relation, $x^2 + y^2 = 4$, without first explicitly solving for y in terms of $f(x)$ or $g(x)$. Differentiating term by term, assuming that this relation defines y as a differentiable function of x, yields

$$2x + 2yy' = 0$$

where the $2yy'$ is obtained by applying the Chain Rule to y^2. Solving for y' yields the same expression that we had previously, namely $y' = -x/y$ whenever $y \neq 0$.

The method of differentiating an implicit relation to find the derivative of the differentiable functions defined by that relation is called *implicit differentiation*. Note that implicit differentiation is an application of the Chain Rule.

We now use the method of implicit differentiation to prove the power rule for differentiating $f(x) = x^r$ when r is a rational number.

Example 3.5.10 (Delayed Proof of the Power Function Rule) We previously stated that the derivative of $f(x) = x^r$ where $x > 0$ and r is a rational number is $f'(x) = rx^{r-1}$ (consequence 4 of Theorem 3.4.1). We now derive this result as an application of implicit differentiation.

Let $y = x^r$ where $x > 0$ and $r = p/q$, p and q integers, $q \neq 0$. Thus, $y^q = x^p$. Differentiating implicitly, we obtain the expression

$$qy^{q-1}y' = px^{p-1}$$

Solving for y' yields

$$y' = \frac{px^{p-1}}{qy^{q-1}}$$

Now substitute x^r for y and obtain

$$y' = rx^{r-1}$$

(One should write out all the exponent algebra involved in the last step to be convinced of the result.) △

Example 3.5.11 Find dy/dx if $y^2 - x = 0, x \geq 0$.

Solving for y yields two *explicit* equations $y = \sqrt{x}$ and $y = -\sqrt{x}$. For $y = \sqrt{x}$, we have

$$\frac{dy}{dx} = \frac{1}{2\sqrt{x}} = \frac{1}{2y}$$

For $y = -\sqrt{x}$, we have

$$\frac{dy}{dx} = -\frac{1}{2\sqrt{x}} = \frac{1}{2y}$$

In either case, $dy/dx = 1/(2y)$.

Now treating $y^2 - x = 0$ as an implicit relation and differentiating implicitly with respect to x yields

$$\frac{d}{dx}(y^2 - x) = \frac{d}{dx}(0) \quad \text{so } 2y\frac{dy}{dx} - 1 = 0 \quad \text{or} \quad \frac{dy}{dx} = \frac{1}{2y}$$

which is what was obtained before. △

The method of implicit differentiation is often used in solving problems that ask for the rate of change of one variable in terms of the rate of change of a second variable. Such problems are called *related rate* problems. The next few examples illustrate this type of problem.

Example 3.5.12 (A Sliding Ladder) A house painter leans a 26-ft ladder against the side of a house with the foot of the ladder 8 ft from the house. After climbing halfway up the ladder, the painter feels the ladder slipping. Assuming that the ground is level and that the rate at which the foot of the ladder slides away from the house is directly proportional to its distance from the house, how fast is the top of the ladder moving when the foot is 16 ft from the house?

A "good" first step in problem solving is to sketch a figure (whenever possible) representing the situation. Thus we sketch and label a diagram distinguishing between variable and fixed lengths.

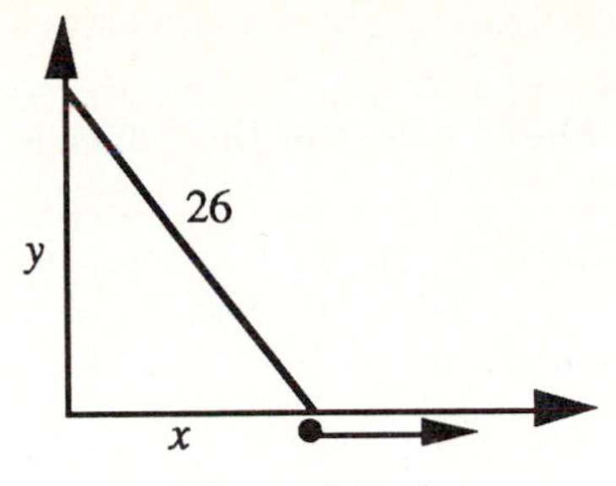

Figure 3.5.13

Warning: Be careful not to label a variable quantity with a constant (nor a constant quantity with a variable). For example, the distance of the foot of the ladder from the wall is variable and thus should be labeled with a variable and *not with the constant 8.*

The problem states that the foot of the ladder slips away from the wall at a rate proportional to its distance from the wall. Thus denoting the distance by x, we have

$$\frac{dx}{dt} = kx \text{ ft/sec} \quad \text{where } k \text{ is the proportionality constant.}$$

We are asked to find

$$\frac{dy}{dt} \text{ when } x = 16$$

The next step is to find an expression relating y to x so that we can differentiate and solve for dy/dt. A good diagram is usually very helpful in suggesting an appropriate equation. In this problem, the desired equation is given by the Pythagorean theorem:

$$x^2 + y^2 = 26^2$$

Since we want to solve for dy/dt, we differentiate implicitly with respect to t and then solve for dy/dt.

$$2x\frac{dx}{dt} + 2y\frac{dy}{dt} = 0 \Longrightarrow \frac{dy}{dt} = -\frac{x}{y}\frac{dx}{dt} = -\frac{kx^2}{y}$$

When $x = 16, y = \sqrt{26^2 - x^2} = \sqrt{26^2 - 16^2} \approx 20.5$. Substituting for x and y yields

$$\frac{dy}{dt} \approx -\frac{256k}{20.5} \approx -12.5k \text{ ft/sec}$$

What is the physical significance of the negative sign?

In order to determine a value for the proportionality constant k, additional information must be given. Suppose we had been told that the foot of the ladder, when 10 ft from the wall, is sliding at 1 ft/sect. Then knowing that $dx/dt = kx$ and $dx/dt = 1$ when $x = 10$, we have $k = 0.1$. In this case, the top of the ladder is moving at ≈ 1.25 ft/sec when the foot of the ladder is 16 ft from the wall. △

Note that there was extraneous information given in the preceding example (e.g., the foot of the ladder was originally 8 ft. from the house). One of the "challenges" in problem solving is discerning what information is relevant to the question.

Example 3.5.14 (Filling a Water Cup) Drinking cups at a water cooler are often shaped in the form of a right-circular cone. The depth of the cup is 4 in and the radius of the top is 3/2 in. If the faucet in the cooler dispenses water at the rate of 30 in^3/min, how fast is the water level rising (in inches per second) when the cup is half-full?

First we sketch and label a diagram showing a water level inside of a cup.

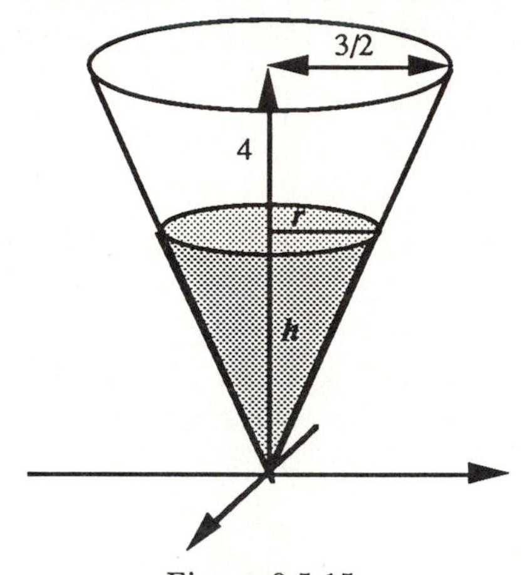

Figure 3.5.15

The volume of a cone of altitude h and base radius r is given by

$$V = \frac{1}{3}\pi r^2 h$$

We are told that the *rate of change* of the volume of water is 30 in^3/min. An (instantaneous) rate of change is a derivative, so this statement is expressed symbolically as

$$\frac{dV}{dt} = 30 \text{ (in}^3\text{/min)}$$

We are asked to find the *rate of change* of the water level, dh/dt, when the cup is half-full. Thus we need to determine the value of h when the cup is half-full. Since V is expressed as a function of two variables, r and h, we try to express r in terms of h in order to express V as a function of h only. One usually refers to the sketch for clues to additional relationships. In this case, note that there are two similar triangles giving the relationship

$$\frac{r}{h} = \frac{3/2}{4} \quad \text{or} \quad r = \frac{3}{8}h$$

Substituting for r in the original expression for V yields

$$V = \frac{3}{64}\pi h^3$$

The total volume of the cup is $\frac{1}{3}\pi(\frac{9}{4})(4) = 3\pi$. Thus when the cup is half-full, $h = 32^{1/3}$. (Why?)

The question is now: What is dh/dt when $h = 32^{1/3}$? We know $V = (3\pi/64)h^3$ and so

$$\frac{dV}{dt} = \frac{9\pi}{64}h^2\frac{dh}{dt}$$

Thus we can solve for dh/dt. We also know $dV/dt = 30 \text{ in}^3/\text{min} = 0.5\text{in}^3/\text{sec}$ (recall that the question asked for the rate of change of h in inches per second). Now substituting for dV/dt and h, we have

$$\frac{dh}{dt} \approx 0.112 \text{ in/sec}$$

△

Example 3.5.16 A spherical balloon is being filled with helium. If the surface area is increasing at a constant rate of 8 in^2/sec, how fast is the volume changing when the radius is 5 in?

The question is about the rate of change of the volume (V) for a particular value of the radius (r), thus we first express volume in terms of the radius:

$$V = \frac{4}{3}\pi r^3$$

Thus

$$\frac{dV}{dt} = 4\pi r^2\frac{dr}{dt}$$

Now we need to find dr/dt. Since we know that the surface area (S) is increasing at the constant rate of 8 sq. inches per second, we can find dr/dt by expressing the surface area in terms of r and differentiating with respect to t.

$$S = 4\pi r^2 \Longrightarrow \frac{dS}{dt} = 8\pi r\frac{dr}{dt}$$

The problem states that $dS/dt = 8 \text{ in}^3/\text{sec}$. Thus

$$\frac{dr}{dt} = \frac{1}{\pi r}$$

Now substituting for r and dr/dt in the expression for dV/dt yields

$$\frac{dV}{dt} = 4r = 20 \text{ in}^3/\text{sec}$$

△

The last three examples are typical *related rate* problems.

Section 3.5 Exercises

In Exercises 1 through 6, find the first and second derivatives of the given functions.

1. $f(x) = 5x^2 - 3/x + 5$ 2. $f(x) = 2x^6 - 4x^2 + 32$

3. $g(x) = x^{1/2} + \tan(4x^3)$ 4. $t(x) = \sin(x^3 + \cos(2x))$

5. $u(x) = [x + (1/x)]^3$ 6. $h(x) = x/\sqrt{x^2+1}$

In the Exercises 7 through 11, find dy/dx for each of the following relations.

7. $y = \csc(x)$ 8. $yx - y^2 = \cos(x)$ 9. $y - 3x - 10xy = 4$

10. $y = (2x + y)^{12}$ 11. $\sin(xy) = 1/2$

12. Let h be a differentiable function such that $h(3) = 2$, $h'(3) = 3$. Let the function f be defined by the relation $f(x) = 4/h(x)$. Compute $f'(3)$.

13. Let the function f be defined by the relation $f(x) = [6 + g(x)]^{3/2}$, where g is a twice differentiable function. If $g(2) = -2$, $g'(2) = 4$, and $g''(2) = 5$, compute $f''(2)$.

14. Find the fiftieth derivative of $f(x) = 2\sin(3x)$.

15. Find the sixtieth derivative of $y = 47x^{49} + 93x^{37} - 28x^{13}$.

16. Assume that you are the production manager of the Left-Right Shoe Company and that $y = d(t)$ is the demand function (number of shoes demanded at time t) for your shoes. For each of the following conditions, explain how you would modify your production rate.

    (a) $d'(t) > 0$ and $d''(t) < 0$ (b) $d'(t) > 0$ and $d''(t) > 0$

    (c) $d'(t) < 0$ and $d''(t) > 0$ (d) $d'(t) < 0$ and $d''(t) < 0$

17. If you, as owner of the Right-Left Shoe Company, are interested in increasing your sales, what is the order of conditions a through d described in the above exercise from most desirable condition to least desirable condition?

18. Suppose you are driving a car and your position as a function of time is given by the relation $y = d(t)$. If $d''(13) = -50$, would your foot be on the brake or the accelerator? Why? Same question if $d''(20) = 40$.

19. Several students at Moose College become ill with a highly infectious strain of measles. Let $y = s(t)$ represent the number of students who have contracted this disease at time t (in days). Give an interpretation for each of the following conditions:

(a) $s'(t) > 0$ and $s''(t) = 0$. **(b)** $s'(t) < 0$ and $s''(t) < 0$.

(c) $s'(t) > 0$ and $s''(t) > 0$. **(d)** $s'(t) > 0$ and $s''(t) < 0$.

(e) $s'(t) = 0$ and $s''(t) > 0$.

20. Let $y = f(t)$ be the number of acres of forest infested with the gypsy moth. Describe each of the following situations by assigning numerical signs (positive, negative, or zero) to the first and second derivatives of f.

    (a) The infestation is spreading, but at a slower rate than in the previous time period.

    (b) The infestation is decreasing, but the number of gypsy moth eggs is much larger than in the past time periods.

    (c) The infestation is decreasing at a faster rate than in the previous time periods.

    (d) The infestation is increasing, but the number of gypsy moth eggs is less than in previous periods.

    (e) The infestation is constant, but the number of gypsy moth eggs is less than in previous periods.

21. Given that

$$\begin{array}{lll} g(4) = 10, & g'(6) = 7, & g'(10) = 2, \\ g'(4) = 13, & f(5) = 1, & f(10) = 2, \\ f'(10) = 6, & f'(5) = 4, & f''(10) = 3, \end{array}$$

compute the derivative of $y = g(f'(x))$ at $x = 10$.

22. Given that

$$g(2) = 6, \quad g'(2) = 4, \quad g''(2) = -3, \quad f'(6) = 10, \quad f''(6) = -4,$$

compute the values the first and second derivatives of y at $x = 2$ where $y = f(g(x))$.

23. If you are driving an automobile up a hill at a velocity of 15 miles per hour (mph) and then suddenly accelerate at a constant rate of 10 mph/sec, how long will it take to reach a speed of 60 mph?

24. A woman who is 5 ft tall is 15 ft from the base of a lamp post that is 12 feet high. If the woman walks directly away from the lamp post at a constant rate of 6 ft/sec, how fast is the tip of her shadow moving? (Assume that it is dark and the light is shining.)

25. A circular plate of metal 2 cm thick expands when heated so that its radius increases at a constant rate of 1 cm/min. At what rate is the surface area of one side of the plate expanding when the radius is exactly 5 cm?

26. Consider a swimming pool 10 m wide and 15 m long. The depth varies uniformly from 1 m at the shallow end to 6 m at the deep end. If water is pumped into the pool at a constant rate of 5 m^3/min, compute how fast the water is rising at the deep end when the depth at the deep end is

    (a) 5 m (b) 3 M.

27. Consider the region between two concentric circles (i.e., a washer) where the radius of the inner circle is increasing at the rate of 4 in/sec and the radius of the outside circle is increasing at the rate of 2 in/sec. Is the area of the region increasing or decreasing, and by how much, when the two radii are 5 in and 9 in?

28. A woman on a dock is pulling in a 1/2 in rope attached to a canoe at the rate of 2 ft/sec. The woman's hands are 3 ft above the level point where the rope is attached to the canoe. How fast is the canoe approaching the dock when the length of rope from her hands to the canoe is 10 ft?

29. Let $y = \sin^{-1}(x)$. Use implicit differentiation to show that $y' = 1/(1 - x^2)^{1/2}$. Hint: By definition, $y = \sin^{-1}(x)$ means that $\sin(y) = x$.

30. Mark each of the following statements true (T) or false (F).

    (a) If f, g, and h are differentiable functions and $y = f(g(h(x)))$, then
    $$\frac{dy}{dx} = f'(g(h(x)))g'(h(x))h'(x).$$
    (b) The equation $x^2 + y^2 = 1$ determines y as a function of x in exactly two ways.
    (c) $D_x \tan(x) = \sec^2(x)$.
    (d) $D_x(\cos(x)\sec(x)) = 0$.
    (e) If $y = \cos^{-1}(x)$, then $y' = 1/(1 - x^2)^{1/2}$.

31. Give an example for each of the following or explain why no example can exist.

    (a) A bounded function whose first and second derivatives are also bounded.
    (b) A negative function having a negative first derivative and a positive second derivative.
    (c) A positive function with a negative first and second derivative.
    (d) A bounded function with an unbounded second derivative.

32. The purpose of this exercise is to distinguish between the local nature of the information given by the first derivative and the global nature of the information given by the second derivative. You should provide a written explanation for each of your answers.

(a) Which is the more important consideration in selecting an investment, the first or second derivative of the performance?

(b) In driving a car, is braking the result of responding to the first or second derivative?

(c) For $f(x) = \sin(x)$, does knowing that $f''(x) = -f(x)$ make it more plausible or less plausible that the sine function is periodic?

(d) Is consuming a triple scoop of ice cream topped with chocolate sauce and nuts a response to the first or second derivative of one's appetite?

(e) Does the phrase "point of diminishing returns" refer to the behavior of the first or of the second derivative of performance?

33. This is an exploratory exercise expanding on Examples 3.5.6 and 3.5.7. Consider the family of functions $f_k : \mathbb{R} \to \mathbb{R}$ defined by

$$f_k(x) = \begin{cases} x^k \sin(1/x) & \text{for } x \neq 0 \\ 0 & \text{for } x = 0 \end{cases}$$

(a) For several successive, positive integer values of k, determine the highest order derivative of f_k that exists and whether or not it is continuous.

(b) Make a conjecture, based on the results of part (a), concerning the highest order derivative of f_k and whether or not it is continuous.

(c) Give an argument verifying your conjecture in part (b).

34. The purpose of this exercise is to *discover* a rule for the nth derivative of a product of two differentiable functions. Recall the product rule for differentiation states that

$$(fg)' = f'g + fg'$$

(a) Using a computer or calculator, compute the expressions for $(fg)''$, $(fg)^{(3)}$, and $(fg)^{(4)}$.

(b) Based on the results of part (a), conjecture a rule for $(fg)^{(n)}$. (It may be necessary to compute more derivatives before the pattern becomes clear.)

(c) Test your conjecture by computing a higher order derivative than was done in part (b) and comparing the result against your conjectured result.

35. Consider the graph of $xy = 1$ in the first quadrant. Prove or disprove that every triangle formed by the x and y axes and a line tangent to this curve has the same area.

3.6 The Mean Value Theorem and Its Applications

The relationship between the *average* rate of change of a function over an interval, a global property, and the *instantaneous* rate of change of the function at an interior point of the interval, a local property, is an example of the "interplay" between discrete and continuous mathematics. These two rates are usually different for most of the points in the interval. (The average rate of rainfall for October does not necessarily tell one how hard it is raining at 2 o'clock on the morning of October 7.) Recall that geometrically the average rate of change is the slope of the secant line joining the endpoints of the graph, whereas the instantaneous or exact rate of change is the slope of the tangent line. Given a differentiable function defined over a fixed interval, is it possible that the average rate would differ from the instantaneous rate at every point? For example, is it possible to average 60 mph on a car trip while always driving slower (faster) than 60 mph? Experience and common sense say "No, there has to be some point where the two rates are the same." The theorem that rigorously establishes this result for "suitably nice" functions is called the Mean Value Theorem. This theorem plays an essential role in the development of several calculus concepts, particularly in higher dimensional situations. The Mean Value Theorem will become a favorite "tool" that we will use throughout the text. In this section we shall study the Mean Value Theorem and some of its applications after first considering a relationship between extreme values and the derivative of a differentiable function.

Theorem 3.6.1 *Let f be a differentiable function defined on an interval I. If f has a local extremum at an* interior *point c of I, then $f'(c) = 0$.*

Proof: Since $f'(c)$ is a real number, it must be positive, negative, or zero. We shall show that $f'(c)$ cannot be positive or negative and, therefore, $f'(c)$ must be zero.

Let $DQ(h) = (f(c+h) - f(c))/h$ be the difference quotient of f at the point c, and define $DQ(0)$ to be $f'(c)$. We shall show that DQ is continuous by noting that since f is a differentiable function, it is continuous and therefore DQ is a continuous function of h for $h \neq 0$. Also, since $\lim_{h\to 0} DQ(h) = f'(c) = DQ(0)$, DQ is also continuous at $h = 0$.

Now, assume that $f'(c)$ is positive. Thus, $DQ(0)$ is positive and by the sign-preserving property of continuous functions, $DQ(h)$ is positive over some interval centered at $h = 0$. Hence both numerator and denominator of $DQ(h)$ must have the same numerical sign for all $h \neq 0$ in this interval. Therefore, $f(c+h) < f(c)$ when $h < 0$ and $f(c+h) > f(c)$ when $h > 0$. However, this contradicts the hypothesis that f takes on a local extremum at the point c. A similar argument can be made if it is assumed that $f'(c)$ is negative. Since $f'(c)$ cannot be positive or negative, we have $f'(c) = 0$. □

This theorem gives us a method for finding extrema of differentiable functions on an open interval:

- find the points of the interval where $f'(x) = 0$
- examine these points individually for extremal values.

We will consider the results of this theorem in detail in Section 3.8.

Theorem 3.6.2 (Rolle's Theorem) *Let f be a real valued function of the real variable x. If*

(a) f is continuous over the closed interval $[a, b]$,

(b) f is differentiable over the open interval (a, b), and

(c) $f(a) = f(b)$,

then there exists at least one number c in (a, b) such that $f'(c) = 0$.

Proof (by Contradiction): We assume the conclusion is false and show that this leads to a logical contradiction of the hypothesis.

Assume that $f'(x) \neq 0$ for all x in (a, b). Since f is continuous over $[a, b]$, we know by the Extreme Value Theorem (Chapter 2) that f must take on its global maximum and global minimum at points in the closed interval. Also, since $f'(x) \neq 0$, f must take on its extreme values at the endpoints. (Why?) However, $f(a) = f(b)$, so the global maximum is equal to the global minimum. Hence, f is a constant function and therefore $f'(x) = 0$ for all x in $[a, b]$. However, this contradicts the assumption that $f'(x) \neq 0$ for all x in $[a, b]$. We must conclude that there exists at least one point c in $[a, b]$ where $f'(c) = 0$. □

The reader's attention is called to the fact that the point c in this theorem must be an interior point of $[a, b]$. Also, there may be more than one point c satisfying the theorem.

As illustrated in the following figure, the geometric significance of Rolle's Theorem is that the graph of f must have a tangent line at some point in the interval that is parallel to the line segment joining the endpoints of the curve.

$f(x) = \cos(x)$

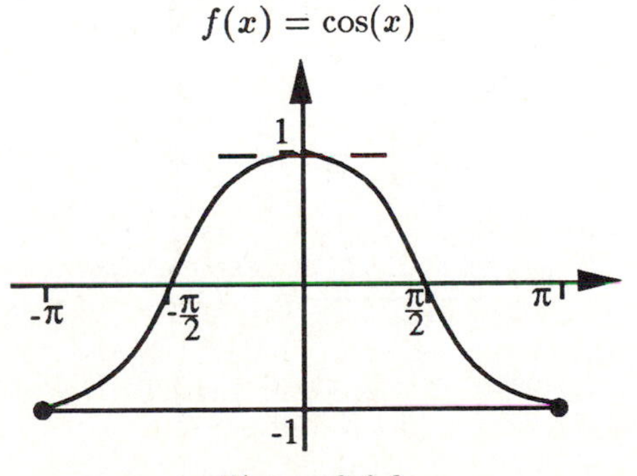

Figure 3.6.3

Example 3.6.4 We show that the polynomial $g(x) = 4x^3 + 3x^2 - 6x + 1$ has a zero between 0 and 1. By inspection we see that if

$$f(x) = x^4 + x^3 - 3x^2 + x$$

then the derivative of $f(x)$ is $g(x)$; that is,

$$f'(x) = 4x^3 + 3x^2 - 6x + 1 = g(x)$$

Since $f(0) = 0$ and $f(1) = 0$, the last part of the hypothesis of Rolle's theorem is satisfied. Note that the first two parts are always satisfied for a polynomial. Therefore, $f'(x)$ has at least one zero between 0 and 1 and hence $g(x)$ does also (since $f'(x) = g(x)$). $\triangle$

If the graph in Figure 3.6.3 were "tipped" as illustrated in the figure below, there would still be interior points of the interval where the tangent line is parallel to the line joining the endpoints of the curve. The Mean Value Theorem, which can be considered as a generalization of Rolle's Theorem, is conveniently interpreted in this geometric fashion.

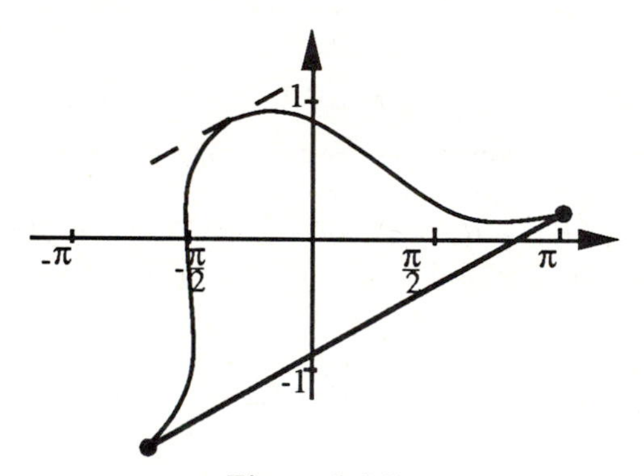

Figure 3.6.5

Theorem 3.6.6 (The Mean Value Theorem) *Let f be a real valued function of the real variable x. If*

(a) f is a continuous function over the closed interval $[a, b]$ and

(b) f is differentiable over the open interval (a, b), then there exists at least one point c in (a, b) for which

$$\frac{f(b) - f(a)}{b - a} = f'(c)$$

Before proving this theorem, we note that the left-hand side of the equation in the conclusion of the theorem is just the slope of the line joining the endpoints of the curve and the right-hand side is the slope of the line tangent to the curve at $x = c$.

The proof will consist of defining a function that satisfies Rolle's theorem and has the property that when its derivative is set equal to zero the conclusion of the Mean Value Theorem is obtained. The inspiration for such a function comes from Figure 3.6.7. We define a function d giving the vertical distance between the line segment joining the endpoints of the curve and the curve itself.

Proof: In the figure below, we sketch the graph of an arbitrary function f which satisfies the hypothesis of the Mean Value Theorem.

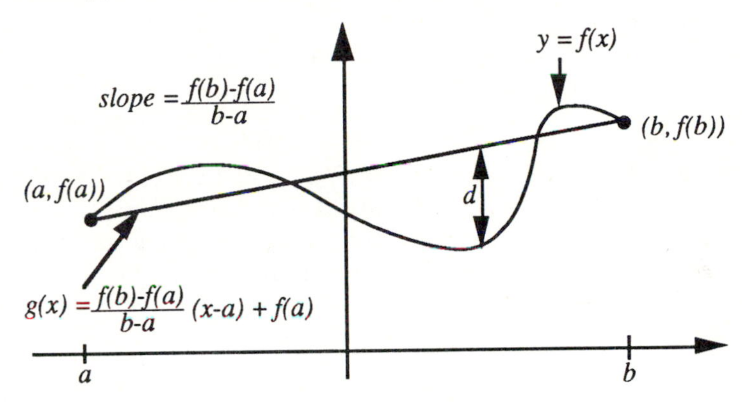

Figure 3.6.7

Let g denote the line segment joining the points $(a, f(a))$ and $(b, f(b))$, the endpoints of the curve. To obtain the equation of g, we substitute the slope of the line segment $(f(b) - f(a))/(b - a)$ and the endpoint $(a, f(a))$ into the point-slope equation for a line. This gives $y - f(a) = [(f(b) - f(a))/(b - a)](x - a)$ or

$$y = g(x) = \frac{f(b) - f(a)}{b - a}(x - a) + f(a)$$

Define the function d by the equation

$$d(x) = f(x) - \left[\frac{f(b) - f(a)}{b - a}(x - a) + f(a)\right]$$

Clearly, d satisfies the hypothesis of Rolle's Theorem, and therefore there is at least one c in (a, b) for which $d'(c) = 0$. Differentiating d, we get

$$d'(x) = f'(x) - \frac{f(b) - f(a)}{b - a}$$

Thus, $d'(c) = 0$ implies that $\frac{f(b)-f(a)}{(b-a)} = f'(c)$ □

Example 3.6.8 We will find a point c in (a, b) where the tangent line to the curve at c is parallel to the line joining $(a, f(a))$ and $(b, f(b))$. This is equivalent to finding a point c where $f'(c)$, the slope of the tangent line, is the same as the slope of the line joining the endpoints, $(f(b) - f(a))/(b - a)$. Suppose $f : [-2, 2] \to \mathbb{R}$ is given by $f(x) = x^3$. Then the slope of the line joining the endpoints is

$$\frac{f(b) - f(a)}{b - a} = \frac{2^3 - (-2)^3}{2 - (-2)} = \frac{16}{4} = 4$$

We want to find c such that

$$f'(c) = 3c^2 = 4.$$

There are two solutions, $c = +\sqrt{4/3}$ and $c = -\sqrt{4/3}$.

$f = x^3$ on $[-2, 2]$

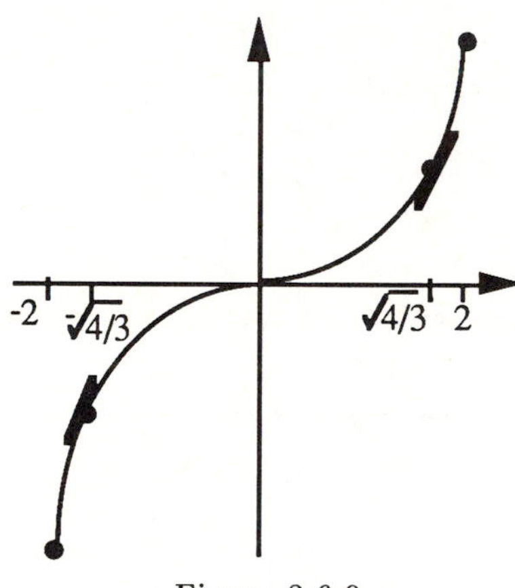

Figure 3.6.9

△

It is often useful to express the conclusion of the Mean Value Theorem in the form

$$f(b) - f(a) = f'(c)(b - a)$$

or

$$f(b) = f(a) + f'(c)(b - a)$$

for some c in (a, b). If we consider the closed interval $[a, x]$, where x is in $[a, b]$, then the latter expression above is

$$f(x) = f(a) + f'(c)(x - a)$$

for some c in (a, x).

Applications of the Mean Value Theorem

The Mean Value Theorem does not give the exact location of the one or more points c where $f'(c) = (f(b) - f(a))/(b - a)$. However, as we shall now show, the existence of such points is extremely useful.

Recall from Chapter 1 the definitions of strictly increasing and strictly decreasing functions.

Theorem 3.6.10 *Assume f is a continuous function on the closed interval $[a, b]$ and differentiable on the open interval (a, b). The following results hold:*

(a) If $f'(x) > 0$ for every x in (a, b), then f is strictly increasing on $[a, b]$.

(b) If $f'(x) < 0$ for every x in (a, b), then f is strictly decreasing on $[a, b]$.

(c) If $f'(x) = 0$ for every x in (a, b), then f is constant on $[a, b]$.

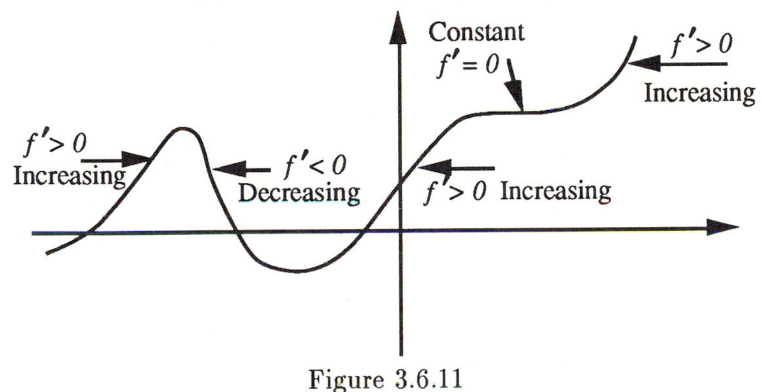

Figure 3.6.11

The proof of this theorem is given in the exercises.

Example 3.6.12 We will find the intervals where

$$f(x) = x^3 - 6x^2 + 9x + 3$$

is monotone.

$$f'(x) = 3x^2 - 12x + 9 = 3(x^2 - 4x + 3) = 3(x - 1)(x - 3).$$

Thus,
If $x < 1$, then $f'(x) > 0$ and f is strictly increasing.

If $1 < x < 3$, then $f'(x) < 0$ and f is strictly decreasing.

If $3 < x$, then $f'(x) > 0$ and f is strictly increasing. △

For our last application of the Mean Value Theorem in this section, we develop a theorem that provides a linear approximation to a differentiable function. One of the main themes in mathematics is the approximation of functions by "nicer" functions. In this section we will continue this theme by approximating by linear (actually affine) functions. Our theorem and its converse will also give an alternative way of defining a derivative that does not involve a difference quotient.

Recall that the conclusion of the Mean Value Theorem can be expressed as

$$(*) \quad f(x) = f(a) + f'(c)(x-a) \qquad \text{for some } c \in (a, x)$$

Suppose f' is a continuous function. Then when x is close to a (and hence c is close to a), $f'(c)$ is close to $f'(a)$. Thus by restricting x to be close to a, we may replace c by a in $(*)$ and obtain a *linear approximation* to $f(x)$, namely,

$$f(x) \approx f(a) + f'(a)(x-a)$$

Note that the right-hand side is just the equation of the tangent line to the graph of f at the point $(a, f(a))$. Thus, geometrically, we are saying that the graph of f can be approximated by the tangent line over a small interval about the point of tangency (see the figure below).

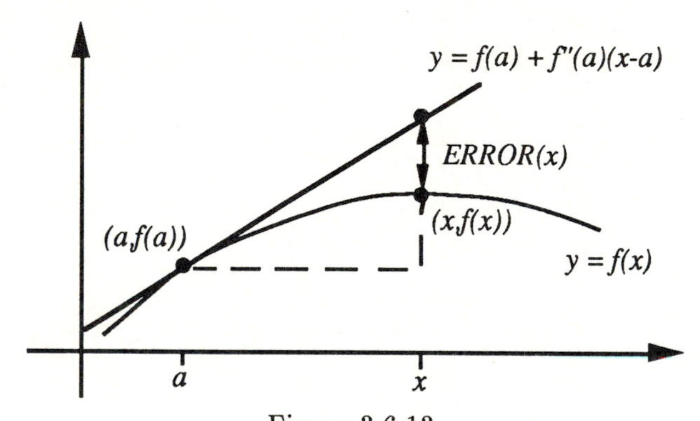

Figure 3.6.13

In Figure 3.6.13, $ERROR$ denotes the error resulting from approximating $f(x)$ by $f(a) + f'(a)(x-a)$, so that

$$f(x) = f(a) + f'(a)(x-a) + ERROR(x)$$

Suppose we *define* a function $ERROR$ by

$$ERROR(x) = f(x) - f(a) - f'(a)(x-a)$$

for any function f which is differentiable at $x = a$. Dividing both sides by $(x-a)$, we get

$$\frac{ERROR(x)}{x-a} = \frac{f(x) - f(a)}{x-a} - f'(a)$$

The first term on the right-hand side is the difference quotient of f and thus

$$\begin{aligned}\lim_{x\to a}\frac{ERROR(x)}{x-a} &= \lim_{x\to a}\left[\frac{f(x)-f(a)}{x-a}-f'(a)\right]\\ &= \lim_{x\to a}\left[\frac{f(x)-f(a)}{x-a}\right]-f'(a)\\ &= f'(a)-f'(a)=0\end{aligned}$$

We have shown that if f is a differentiable function at $x = a$, then there exists a function $ERROR(x)$ such that

$$f(x) = f(a) + f'(a)(x-a) + ERROR(x)$$

where $\lim_{x\to a}\frac{ERROR(x)}{(x-a)} = 0$.

Theorem 3.6.14 (The Linear Approximation Theorem) *If f is a differentiable function over an interval containing $x = a$, then there is a function $ERROR$ with the properties:*

(a) $f(x) = f(a) + f'(a)(x-a) + ERROR(x)$

(b) $\lim_{x\to a} ERROR(x)/(x-a) = 0$

Example 3.6.15 We will use the Linear Approximation Theorem to express $f(x) = x^{1/2}$ near $a = 4$ as the sum of a linear function and an error.

Since $f'(x) = \frac{1}{2}x^{-1/2}$, we have $f'(4) = \frac{1}{4}$ and $f(4) = 2$. Substituting into the linear approximation expression yields

$$x^{1/2} = 2 + \frac{1}{4}(x-4) + ERROR(x)$$

Here the linear function (actually an affine function) is $L(x) = 1 + x/4$.

We can use this result to approximate $\sqrt{3.9}$. Let $f(x) = x^{1/2}$, $a = 4$, and $x = 3.9$. Then

$$\sqrt{3.9} = 2 + \frac{1}{4}(3.9-4) + ERROR(x)$$

and thus the approximation is

$$2 + \frac{-0.1}{4} = 1.975$$

△

The converse of the Linear Approximation Theorem is also a theorem.

Theorem 3.6.16 (Converse of the Linear Approximation Theorem) *Let f be a function defined on an open interval I containing the point $x = a$. If there exists a function $ERROR$ defined on I such that*

(a) $f(x) = f(a) + m(x-a) + ERROR(x)$

(b) $\lim_{x\to a} ERROR(x)/(x-a) = 0$

Then f is differentiable at $x = a$ and $f'(a) = m$.

Proof: To show that f is differentiable, we form the difference quotient and evaluate the limit. Thus, consider

$$\begin{aligned}\lim_{x\to a}\frac{f(x)-f(a)}{x-a} &= \lim_{x\to a}\frac{m(x-a)+f(a)+ERROR(x)-f(a)}{x-a}\\ &= \lim_{x\to a}\left(m+\frac{ERROR(x)}{x-a}\right)=m\end{aligned}$$

Thus f is differentiable at $x = a$ and $f'(a) = m$. □

Note that The Linear Approximation Theorem and its converse taken together yield a characterization of derivative that does not involve a difference quotient. This approach to the derivative can be used in many contexts where the standard definition doesn't make sense. In particular, we will define the derivative this way when we study functions of several variables in the next chapter. We formalize this observation in the next theorem.

Theorem 3.6.17 *The function f is differentiable at $x = a$ and $f'(a) = m$ if and only if there exists a function ERROR such that*

(a) $f(x) = f(a) + m(x-a) + ERROR(x)$ *and*

(b) $\lim_{x\to a} ERROR(x)/(x-a) = 0$

Example 3.6.18 We will show, by the above theorem, that $f(x) = x^2$ is differentiable at $x = 3$ and that $f'(3) = 6$.

We need to define a function *ERROR* satisfying the two conditions of the theorem. To do this, set

$$f(x) = f(3) + m(x-3) + ERROR(x)$$

and solve for $ERROR(x)$. Thus

$$ERROR(x) = f(x) - f(3) - m(x-3) = x^2 - 3^2 - m(x-3)$$

Now determine a value for m such that the second condition in Theorem 3.6.17 is satisfied.

$$\lim_{x\to 3}\frac{ERROR(x)}{x-3} = \lim_{x\to 3}\left(\frac{x^2-3^2}{x-3}-m\right) = \lim_{x\to 3}(x+3-m) = 6-m = 0$$

Thus, $m = 6$ and $f'(3) = 6$. △

The advantage of this approach in establishing the existence of the derivative will become clear when we extend the differentiation concept to functions of more than one variable.

Section 3.6 Exercises

In Exercises 1 through 10, find an explicit numerical value of c such that $a < c < b$ and the tangent line to the graph of the function f at $(c, f(c))$ is parallel to the line joining the points $(a, f(a))$ and $(b, f(b))$. Sketch the graph of f and show the lines.

1. $f(x) = x^2$, $[-1, 3]$
2. $f(x) = x^2 - 4$, $[-2, 2]$
3. $f(x) = x(x^2 - 2)^2$, $[0,2]$
4. $f(x) = x(x - 2)^2$, $[-1, 2]$
5. $f(x) = x^3$, $[-1, 2]$
6. $f(x) = 1/x$, $[1,4]$
7. $f(x) = \sin(x)$, $[\pi/2, 3\pi/2]$
8. $f(x) = \cos(x)$, $[0, \pi]$
9. $f(x) = \sqrt{x}$, $[1,4]$
10. $f(x) = (x - 1)/(x + 1)$, $[0,3]$

In Exercises 11 through 14, find the intervals where each function is monotone (increasing or decreasing). Sketch the graph of the function.

11. $f(x) = 2x^3 - 12x + 4$
12. $f(x) = x + 4/x$
13. $f(x) = \begin{cases} -x & \text{for } x < -2 \\ 2 & \text{for } x = -2 \\ -x^2 + 6 & \text{for } x > -2 \end{cases}$
14. $g(x) = \begin{cases} \cos(x) & \text{for } -\pi < x \leq \pi \\ -\tan(x) & \text{for } \pi < x < 3\pi/2 \end{cases}$
15. For each one of the following, sketch the graph of a function which satisfies the stated condition and, in addition, *does not* satisfy the hypotheses of the Mean Value Theorem.
    (a) A function which is continuous over $[-3, 5]$
    (b) A function which is differentiable over $(-3, 5)$
    (c) A function which is not continuous over $(-3, 5)$
16. Consider the function $f(x) = x^{1/2}$, as a sum of a linear term and an error term.
    (a) Use the Linear Approximation Theorem to express $f(x)$ near $a = 9$ as a sum of a linear term and an error term.
    (b) Compute the numerical value for the *ERROR* in part (a) when $x = 9.06$.
17. Consider the function $f(x) = (x - 1)/(x + 1)$.
    (a) Use the Linear Approximation Theorem to express $f(x)$ near $a = 0$ as a sum of a linear term and an error term.

(b) Compute the numerical value of the *ERROR* in part (a) when $x = .03$.

18. Verify analytically that the function d in the proof of the mean value theorem satisfies the hypothesis of Rolle's Theorem.

19. Mark each of the following statements true (T) or false (F).

    (a) In driving from Kittery to Augusta over the Maine turnpike, there is at least one point where one's speed (instantaneous rate) is equal to one's average rate over the Turnpike.

    (b) It is possible to drive the 100-mile Maine Turnpike in 92 minutes and never drive faster than 65 mph.

    (c) The instantaneous rate of change of a function is zero at the point where the function assumes its maximum value.

    (d) If the instantaneous rate of change of a function f is zero at all points in the domain of f, then f is a constant function.

    (e) The instantaneous rate of change of a linear function is a constant.

    (f) It is possible to have a function whose instantaneous rate of change at every point in its domain is equal to its average rate of change over its domain.

20. Let f be a differentiable function defined over the interval I. Can the instantaneous rate of change of f at every point in the interval I be greater than the average rate of change over interval I? Why?

21. Mark each of the following statements true (T) or false (F).

    (a) If $f(x) = x^3 \sin(1/x)$ for $x \neq 0$ and if $f(0) = 0$, then f' is a continuous function at $x = 0$.

    (b) The function defined in (a) is twice differentiable.

    (c) If f is a twice differentiable function defined for all real numbers and $f''(x)$ is always positive, then f is an unbounded function.

    (d) It is possible to have a function f such that $f'(x) > 0$ and $f''(x) < 0$ for all x.

    (e) If f is a differentiable function defined for all real numbers and $f'(x)$ is always positive, then f is an unbounded function.

    (f) If f and g are two increasing functions, then their product fg is an increasing function.

    (g) If $f'(x) < 1$, for all $a \leq x \leq b$, then $f(d) - f(c) \leq d - c$ for $a \leq c \leq d \leq b$.

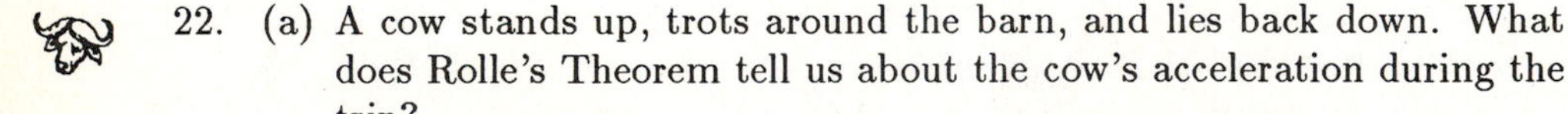

22. (a) A cow stands up, trots around the barn, and lies back down. What does Rolle's Theorem tell us about the cow's acceleration during the trip?

(b) A cow barn's roof is in the shape of the parabola $a(x) = x^2 - 2x + 10$, $-1 \le x \le 0$ near the edge ($x = 0$). Since it is so steep, we wish to attach a straight overhang to the roof which smoothly matches the roofline and is 3 ft long. Use a theorem in this section to find the equation of the overhang (include domain).

23. Prove Theorem 3.6.10, part (a). (Hint: Assume $a \le x < y \le b$. Apply the Mean Value Theorem to the closed interval $[x, y]$ and consider the signs of the terms.)

24. Prove Theorem 3.6.10, part (b). (See the hint for Exercise 23.)

25. Prove Theorem 3.6.10, part (c). (See the hint for Exercise 23.)

Exercises 26 through 28 illustrate some of the important consequences that result when the derivative(s) of a function are bounded. In particular, Exercise 27 presents a different proof of the Mean Value Theorem and Exercise 28 gives a quadratic approximation theorem. After working Exercises 27 and 28 the student should form a conjecture generalizing the results of these two exercises.

26. Complete the proof of the following theorem by justifying each of the steps in the proof outline.

    Theorem: If functions f and g have the property that

    $$f'(x) \le g'(x) \quad \text{for all } x \text{ in } [a, b]$$

    then

    $$f(x) - f(a) \le g(x) - g(a) \quad \text{for all } x \text{ in } [a, b]$$

    Proof (outline): Step 1: Since $0 \le g'(x) - f'(x)$, $(g - f)$ is an increasing function.
    Step 2: $g(a) - f(a) \le g(x) - f(x) \quad$ for all x in $[a, b]$
    Step 3: $f(x) - f(a) \le g(x) - g(a) \quad$ for all x in $[a, b]$

27. Prove the theorem: If the first derivative of f is continuous and bounded on $[a, b]$, say
    $m \le f'(x) \le M$, then $f(x) = f(a) + f'(c)(x - a) \quad$ for some c in $[a, b]$

    Proof (suggestions) Step 1: Since f' is both continuous and bounded, f' has a minimum value, say m, and a maximum value, say M.
    Step 2: (Work with $f'(x) \le M$) Show that $f(x) \le f(a) + M(x - a)$ by applying the theorem of Exercise 26 to the inequality $f'(x) \le M$ (i.e., let $g(x) = Mx$, so $g'(x) = M$).
    Step 3: [Work with $m \le f'(x)$] In a manner analogous to step 1, show that

$f(a) + m(x - a) \leq f(x)$.
Step 4: Combining the results of steps 1 and 2 yields

$$f(a) + m(x - a) \leq f(x) \leq f(a) + M(x - a)$$

Now apply the Intermediate Value Theorem to show that there exists a number k, $m \leq k \leq M$, such that

$$f(x) = f(a) + k(x - a)$$

Since f' is continuous and m and M are the minimum and maximum values of f', there is some point c in $[a, b]$ such that $f'(c) = k$. Thus

$$f(x) = f(a) + f'(c)(x - a)$$

28. Prove the theorem: If the second derivative of f is continuous and bounded on $[a, b]$, say $m \leq f''(x) \leq M$, then

$$f(x) = f(a) + f'(a)(x - a) + \frac{f''(c)}{2}(x - a)^2$$

for some number c in $[a, b]$.

Proof (suggestions) Step 1: Since f'' is both continuous and bounded, f'' has a minimum value, say m, and a maximum value, say M.
Step 2: (work with $f''(x) \leq M$) Since $f''(x)$ is the derivative of $f'(x)$ and M is the derivative of Mx, the theorem of Exercise 26 may be applied to obtained

$$f'(x) - f'(a) \leq Mx - Ma = M(x - a) \quad \text{for all } x \text{ in } [a, b]$$

Since the left side is the derivative of $f(x) - f'(a)x$ and the right side is the derivative of $\frac{1}{2}M(x - a)^2$, the theorem of Exercise 26 may be applied a second time to obtain

$$f(x) - f'(a)x - f(a) + f'(a)a \leq \frac{M}{2}(x - a)^2$$

or

$$f(x) \leq f(a) + f'(a)(x - a) + \frac{M}{2}(x - a)^2$$

Step 3: [Work with $m \leq f'(x)$] In a manner analogous to step 1 show that

$$f(a) + f'(a)(x - a) + \frac{m}{2}(x - a)^2 \leq f(x) \quad \text{for all } x \text{ in } [a, b]$$

Step 4: (See step 4 in Exercise 27.) Use the Intermediate Value Theorem to show that there exists a number c such that

$$f(x) = f(a) + f'(a)(x - a) + \frac{f''(c)}{2}(x - a)^2$$

3.7 Antiderivatives

So far in this chapter we have been concerned with the process of differentiation, i.e., finding f' given a function f. We now consider the opposite process of finding f given f'. This is called *antidifferentiation.* We shall give an introduction to this process here and consider it in more detail in Chapter 6. The following example illustrates how the need for antidifferentiation may arise.

Example 3.7.1 (Speed of a Falling Object) Suppose we drop an object and wish to determine its speed 5 sec later. The acceleration of a falling body near the earth's surface due to the force of gravity is approximately 32 ft/sec/sec downward. Newtonian physics says:

$$F = ma$$

where F is the force acting on the object, m is the mass of the object, and a is the acceleration of the object. If the position of an object at time t is $x(t)$, measured upward from the ground, then the velocity is $v(t) = x'(t)$ and the acceleration is $a(t) = v'(t) = x''(t)$.

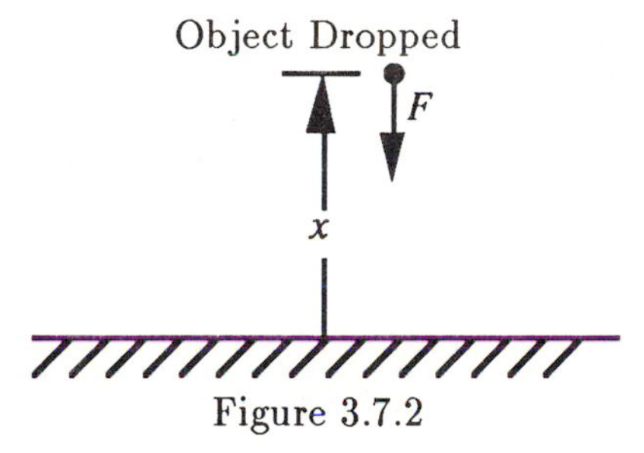

Figure 3.7.2

For an object of mass m dropped from rest at time zero, $v(0) = 0$ and the force due to gravity is $F = -32m$; the sign is negative since the force is acting downward, opposite to the direction of measurement. Using Newton's law we can write:

$$-32m = mv'(t) \quad \text{or} \quad -32 = v'(t)$$

Here we have an equation involving the derivative of the function v' which we wish to solve for the function v. Such an equation, involving derivatives of functions, is called a *differential equation.* In this differential equation we wish to find a function v that satisfies $v'(t) = -32$. Clearly there are many solutions to this equation: if $v(t) = -32t + C$ for an arbitrary constant C, then $v'(t) = -32$. What value should we choose for C? In general, differential equations require additional conditions to uniquely determine a solution. In this case, an initial condition, $v(0) = 0$, gives $0 = -32 \cdot 0 + C$ so that $C = 0$ and $v(t) = -32t$. Thus at time $t = 5$, the velocity is $v(5) = -160$ ft/sec. Notice that the velocity is negative, which says that the object is moving downward, as it should. $\triangle$

In the preceding example, we noted that if two functions differ by a constant, then they have the same derivative. The next theorem shows that the converse also holds.

Theorem 3.7.3 *If f and g are two differentiable functions defined on (a, b) and $f'(x) = g'(x)$ for all x in (a, b), then there exists a constant C such that $f(x) = g(x) + C$ for all x in (a, b).*

The proof which is dependent on a consequence of the Mean Value Theorem is in the exercises.

Note that functions f and g are both defined over the same interval. For emphasis we repeat that Theorem 3.7.3 tells us that if two functions have the same derivative, they differ by a constant.

Definition 3.7.4 (Antiderivative) *An* antiderivative *of a function f is a function F whose derivative is f, $F' = f$.*

We will now find a few antiderivatives and establish an algebra of antiderivatives. We first consider powers of the variable.

Corollary 3.7.5 *If r is a rational number and $f'(x) = ax^r$, for $r \neq -1$, in some interval, then $f(x) = ax^{r+1}/(r+1) + C$, where C is a constant.*

Proof: We know that the derivative of $g(x) = ax^{r+1}/(r+1)$ is ax^r. If f is any function whose derivative is ax^r, then by Theorem 3.7.3 there exists a constant C such that $f(x) = g(x) + C$. □

Note that this corollary implies that there is a family of differentiable functions all of whose derivatives are ax^n. Furthermore, any two members of this family differ only by a constant. The expression for the most general antiderivative must contain an arbitrary constant.

Example 3.7.6 We will find the general antiderivative of $4x^5 - 3x^2 + 13x - 2$.

We need to find a function whose derivative is $4x^5 - 3x^2 + 13x - 2$. Since the derivative of a sum is the sum of derivatives, we may construct a function by applying the preceding corollary to each term separately. This yields

$$f(x) = \frac{4x^6}{6} + C_1 - \frac{3x^3}{3} + C_2 + \frac{13x^2}{2} + C_3 - 2x + C_4$$

Now simplifying and combining all four arbitrary constants into one constant denoted by C yields

$$f(x) = \frac{2x^6}{3} - x^3 + \frac{13x^2}{2} - 2x + C$$

An easy way to check that $f(x)$ is a correct solution is to differentiate it. Note that if the question had included the additional condition that the graph of the function contain the point (1,2), then $f(1)$ would have to have the value 2. Since $f(1) = 25/6 + C$, we would assign C the value $-13/6$, and the unique solution to

$$f'(x) = 4x^5 - 3x^2 + 13x - 2, \quad f(1) = 2$$

is

$$f(x) = \frac{2x^6}{3} - x^3 + \frac{13x^2}{2} - 2x - \frac{13}{6}$$ △

Example 3.7.7 If a particle is moving along a line such that its velocity at time t is $f(t) = 2t + 1$, what is its position at time $t = 5$? Since $d/dt(t^2) = 2t$ and $d/dt(t) = 1$, we see that $d/dt(t^2 + t) = 2t + 1$. By Theorem 3.7.3 any antiderivative of $f(t) = 2t + 1$ differs from $t^2 + t$ by a constant. Thus the general antiderivative of f is $F(t) = t^2 + t + C$. In order to determine C, we need some condition to restrict the possible antiderivatives to a specific one. We will assume that at time 0 the particle is at position 1. Then $1 = F(0) = C$, and the unique solution to the restricted problem is

$$F(t) = t^2 + t + 1$$

and the position at time 5 is $F(5) = 25 + 5 + 1 = 31$. △

Notation: The general antiderivative of a function f is denoted symbolically by $\int f$, or, showing the variable, $\int f(x)dx$.

The elongated s, $\int$, is called the *integral sign.*

The principal subject of Chapter 6 is integration and its relation to differentiation.

Example 3.7.8 $\int(2t + 1)dt = t^2 + t + C$ △

We will develop some rules that will make computing antiderivatives easier. Recall from Theorem 3.4.1 on the algebra of derivatives that $(F + G)' = F' + G'$ and $(aF)' = aF'$ for any constant a. Thus if F is an antiderivative for f, $F' = f$, and G is an antiderivative for g, $G' = g$, then $(F+G)' = F'+G' = f+g$. So $F + G$ is an antiderivative for $f + g$ and the general antiderivative is

$$\int(f + g) = F + G + C$$

Similarly, if F is an antiderivative of f, then (aF) is an antiderivative of af since $(aF)' = aF' = af$. Thus the general antiderivative of af is

$$\int af = aF + C$$

We can rephrase Corollary 3.7.5 as

$$\int ax^n\,dx = a\frac{x^{n+1}}{n+1} + C$$

for any rational number $n \neq -1$.

We summarize these rules with a theorem and a short table of antiderivatives.

Theorem 3.7.9 (Algebra of Antiderivatives)

$$(a) \quad \int (f+g) = \int f + \int g$$

$$(b) \quad \int af = a\int f$$

Some known antiderivatives:

$\int ax^n dx = ax^{n+1}/(n+1) + C, n$ rational, $n \neq -1$ (Corollary 3.7.5)

$\int \cos(x)dx = \sin(x) + C$ [Example 3.3.5 part (e)]

$\int \sin(x)dx = -\cos(x) + C$ [Example 3.3.5 part (f)]

$\int \sec^2(x)dx = \tan(x) + C$ [Example 3.4.2]

(e) $\int e^x dx = e^x + C$ [Example 3.3.5 part (g)]

Example 3.7.10

$$\begin{aligned}\int (x^3 + 4\sin(x) + 3\sqrt{x})dx &= \int x^3 dx + \int 4\sin(x)dx + \int 3\sqrt{x}dx \quad \text{part (a)}\\ &= \int x^3 dx + 4\int \sin(x)dx + 3\int \sqrt{x}dx \quad \text{part (b)}\\ &= \frac{x^4}{4} - 4\cos(x) + 2x^{3/2} + C \qquad \triangle\end{aligned}$$

Example 3.7.11 Suppose the marginal cost (derivative of the cost function) of a certain commodity is

$$c'(x) = e^x - 60x^2 + x$$

where x is the number of units produced. If $c(0) = 70$ [c(0) is called the initial overhead], find the average cost of producing the first 10 units.

The first step is to solve the initial value differential equation. Thus

$$c(x) = \int (e^x - 60x^2 + x)dx = e^x - 20x^3 + \frac{1}{2}x^2 + C$$

Since $c(0) = 70$ and $c(0) = 1 + C, C = 69$. Thus

$$c(x) = e^x - 20x^3 + \frac{1}{2}x^2 + 69$$

The second step is to compute the average cost for producing the first 10 units.

$$\text{av. cost } = \frac{1}{10}c(10) = \frac{1}{10}(e^{10} - 20(10^3) + \frac{1}{2}(10^2) + 69) = 214.55$$ △

Example 3.7.12 Suppose that the object of Example 3.7.1 is thrown upward with an initial velocity of 10 ft/sec from a position 100 ft above the ground. We wish to find the height of the object after 2 sec have elapsed.

Object Thrown Upward

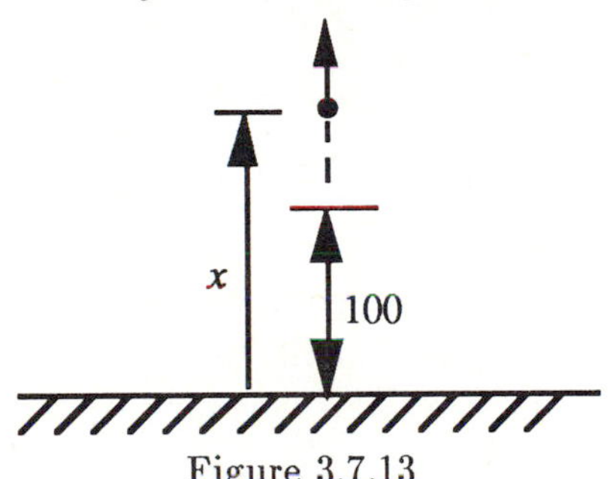

Figure 3.7.13

The basic equations from Example 3.7.1 still hold:

$$-32 = v'(t) \quad \text{and} \quad v(t) = x'(t)$$

where $v(t)$ is the velocity at time t. We could combine the two equations into one:

$$-32 = x''(t)$$

which is an example of a *second-order differential equation* since the derivative appears to the second order. Second-order equations will have two arbitrary constants in the general solution and thus require two conditions [in this case $v(0) = 10$ and $x(0) = 100$] to determine a unique solution. Finding the antiderivative of -32 gives

$$v(t) = \int v'(t)dt = \int -32dt = -32t + C$$

Since the first initial condition is $v(0) = 10$,

$$10 = -32 \cdot 0 + C \quad \text{or } C = 10$$

Thus we now have the equation

$$x'(t) = v(t) = -32t + 10$$

Finding the antiderivative of this polynomial yields

$$x(t) = \int (-32t + 10)dt = -16t^2 + 10t + C$$

Since the second initial condition is $x(0) = 100$, we have

$$100 = x(0) = C$$

and the final equation for the position at time t is

$$x(t) = -16t^2 + 10t + 100$$

Thus the position at time $t = 2$ is

$$x(2) = -16 \cdot 4 + 20 + 100 = 56 \text{ ft “above” the ground}$$

△

We must be careful not to allow our examples to lead us to assume that all functions have antiderivatives. As the following example illustrates, even some very simple looking functions may fail to have antiderivatives.

Example 3.7.14 (A Function with No Antiderivative) Does the following simple function have an antiderivative?

$$f(x) = \begin{cases} 0 & \text{for } x \leq 0 \\ 1 & \text{for } x > 0 \end{cases}$$

Assume that g is an antiderivative of f; that is, $g'(x) = f(x)$. Then at $x = 0$, the left-hand derivative of g would be 0 and the right-hand derivative would be 1. Thus g would not be differentiable at $x = 0$ and hence could not be an antiderivative of f over any interval containing zero. Our assumption has led us to a contradiction and therefore the assumption is impossible. Thus the answer is No, f does not have an antiderivative. △

The difficulty with the function f in the preceding example lies in the fact that it is not continuous. It can be shown (in an advanced calculus course) that every continuous function defined over a closed interval, has an antiderivative over that interval.

Section 3.7 Exercises

Find the general antiderivatives of the given functions in Exercises 1 through 10.

1. $f(x) = 4x^3 + \cos(x) + \sin(x)$
2. $f(x) = 3/x^2 + 5\sin(x)$
3. $f(x) = (x^2 + 1)/x^5$
4. $f(x) = (x^3 + 3x + 4)/x^3$
5. $f(t) = (t + 1)^2$
6. $f(u) = (u^3 + 3u^2 + 3u + 1)/(u + 1)$
7. $f(y) = \sec^2(y) + 1$
8. $f(x) = \tan^2(x)$
9. $f(x) = \sin(2x)$
10. $f(x) = \begin{cases} e^x & \text{for } x \geq 0 \\ 1 + x & \text{for } x < 0 \end{cases}$
11. Find a function h that satisfies all the following conditions:

    (a) $h''(x) = 4x^3 - 9x^2 + 8x - 2$
    (b) $h'(1) = 4$
    (c) $h(0) = 1$

12. Find a function that satisfies all the following conditions

    (a) $g''(x) = x^2 - 6x + 2$
    (b) $g'(2) = 0$
    (c) $g(0) = 5$

13. The marginal revenue (i.e., derivative of the revenue function) of a certain commodity is given by $r'(x) = 10x + \cos(x) - 20$ where x is the number of units sold. If the total revenue is 100 dollars for $x = 10$, find the revenue function.
14. The marginal cost of a commodity is given by $c'(x) = x^2(10 - x) + \sin(x)$. If the initial overhead is 100 [i.e., $c(0) = 100$], find the average cost of producing the first 20 units.
15. Show why it is impossible to find a function g such that $g'(x) = x^2 + x$ on the interval [1,2] with the property that $g(1) = 0$ and $g(2) = 4$.
16. Find the function f given that $f''(x) = 6x^2 - 4$ and $f'(0) = 0 = f(1)$.

17. Find a function g whose x and y intercepts are both equal to 1 and $g''(x) = 12x^2 + x$.

18. Find all functions whose second derivative is cosine and whose graph passes through the origin.

19. Show or give a counterexample to the assertion that for any differentiable function f,

$$\int f(x)f'(x)dx = \frac{1}{2}[f(x)]^2 + C$$

20. Show or give a counterexample to the assertion that

$$\int f(x)g(x)dx = \int f(x)dx \cdot \int g(x)dx$$

21. Find $\int \cos(x)\sin(x)dx$.

22. Find $\int \sec^2(x)\tan(x)dx$.

23. An automobile accelerates at a constant rate for 1 min starting from rest. At the end of this time the velocity is 60 m/sec. How far did the car travel in this time?

24. An object is thrown downward with a velocity of 10 ft/sec. How long does it take the object to travel 300 ft?

25. What rule for finding antiderivatives can you obtain from the chain rule?

26. Using your rule from Exercise 25,

    (a) Find $\int 2x\cos(x^2)dx$.

    (b) Find $\int (u+2)^{-2}du$.

27. Mark each of the following statements true (T) or false (F).

    (a) It is impossible for a function to have a negative first derivative and a positive second derivative.

    (b) If $f'(x) = g(x)$ over an interval I and g is a polynomial of degree 2, then f is a polynomial of degree 3.

    (c) If on the same interval h is an antiderivative of f and k is an antiderivative of g, then hk is an antiderivative of fg.

    (d) If $g'(x) = 0$ for every x in an interval I, then g is a constant function on I.

    (e) If the marginal revenue of a commodity is zero on $(0, \infty)$, then its revenue function is a constant.

(f) A step function has an antiderivative.

28. Consider Theorem 3.7.3.

  (a) Define a differentiable function $k : (a, b) \to \mathbb{R}$ by $k(x) = (f - g)(x)$. Now show that the hypothesis implies that $k'(\mathrm{x}) = 0$.

  (b) Explain why k' $(\mathrm{x}) = 0$ implies that $k(x)$ is a constant by referring to the appropriate theorem in Section 3.6.

  (c) Complete the proof of Theorem 3.7.3. [Hint: Apply Theorem 3.6.10 (c) to $f(x) - g(x)$].

29. Use a computer or a graphing calculator to explore the relationship between an even function and an antiderivative of it. Hint: For each of several examples of even functions, plot both the graph of the function and an antiderivative of it on the same set of axes.

30. Use a computer or a graphing calculator to explore the relationship between an odd function and an antiderivative of it. (See the hint for Exercise 29.)

31. Project. Explore the relationship between even and odd functions and functions and their antiderivatives. In particular consider the following questions

  (a) If f is an odd function whose domain is an interval containing 0 and g is an antiderivative of f, is g an even function? Hint: Differentiate $g(x) - g(-x)$.

  (b) Does there exist an odd function having an antiderivative that is not an even function?

  (c) If f is an even function whose domain is an interval and g is an antiderivative of f, then g is an odd function if and only if $g(0) = 0$.

  You should make up several illustrative examples, state and prove some theorems, or state some conjectures and give heuristic arguments for them.

3.8 Extreme Values of Functions

The determination of maximum or minimum values is a major application of the analysis of rate of change.

Example 3.8.1 (Leaf Raking) A common fall "sport" is raking leaves. Suppose you have raked all the leaves to one side of your yard which is 20 ft long. You now want to rake the leaves into a pile and then carry them to a compost heap at one corner of the yard. The problem is, where should you make the pile in order to minimize your effort? (Note that the pile could be the compost heap itself.)

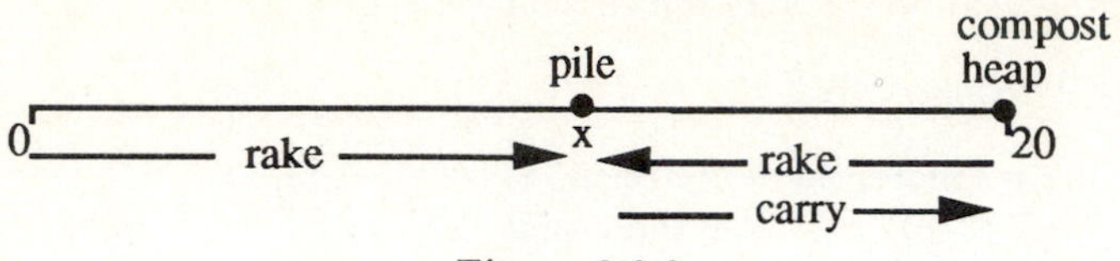

Figure 3.8.2

Suppose the effort exerted in raking is d^2, where d is the distance raked. (The further you rake, the more leaves there are to move and thus more effort is required.) Also suppose the effort in putting the leaves into a basket, lifting, carrying, and then dumping the basket is $4d$, where d is the distance the leaves are carried. (The further you carry the basket, the heavier it seems to get and thus the more mental effort is exerted.)

We can now model the problem with the equation

$$E = \underbrace{x^2 + (20-x)^2}_{\text{raking}} \quad + \underbrace{4(20-x)}_{\text{carrying}}$$

in which x represents the distance raked and E the effort exerted. This equation simplifies to

$$E = 2(x^2 - 22x + 240)$$

whose graph is a parabola opening upward with the vertex at (11,238).

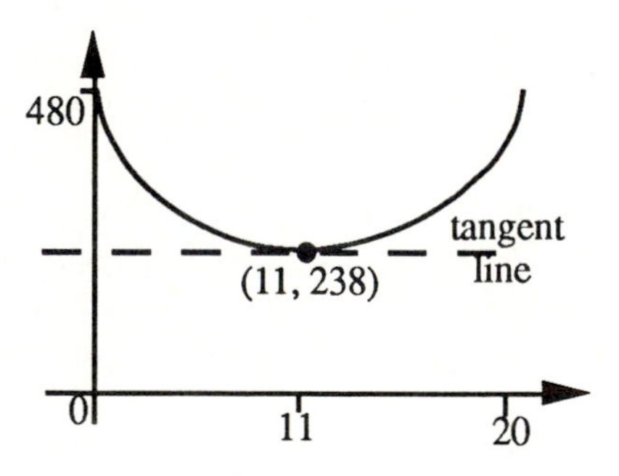

Figure 3.8.3

The graph has a horizontal tangent line at the point $(11, 238)$ and thus the minimum effort exerted will result when the pile is located 11 ft from the starting end (9 ft from the compost heap). △

The previous example is called an *extremal problem*. The objective of such a problem is to determine the minimum or maximum values (extreme values) of a function. Extremal value problems occur frequently in applied mathematics.

Before looking at some additional examples, we need to recognize that there are two different kinds of maxima and minima discussed in the calculus: global (absolute) and local (relative).

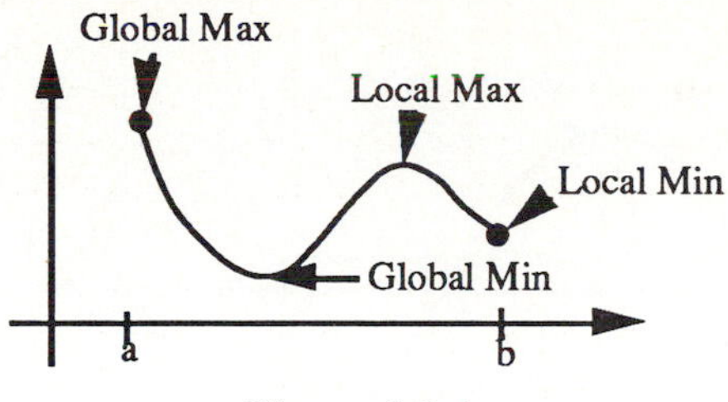

Figure 3.8.4

Definition 3.8.5 (Global Maximum) *If there exists a number c in the domain of the function f with the property that $f(c) \geq f(x)$ for all x in the domain, then $f(c)$ is the* global (absolute) maximum *of f.*

Definition 3.8.6 (Local Maximum) *If there exists a number c and an open interval, I, containing c with the property that $f(c) \geq f(x)$ for all x in the intersection of I and the domain of f, then $f(c)$ is a* local (relative) maximum *of f.*

There are similar definitions for global and local minima of a function. Note that every global maximum of a function is also a local maximum, but the converse need not hold (i.e., a local maximum is not necessarily a global maximum).

Definition 3.8.7 (Extreme Point) *A number that is either a local maximum or minimum is called an* extreme point *or* extremum *of the function.*

Note that every point on the graph of a constant function is a global maximum, local maximum, global minimum, and local minimum.

Recall from Chapter 2 that the Extreme Value Theorem stated that every continuous function defined on a closed bounded interval has both a maximum and a minimum value. Thus, by Theorem 3.3.7, every differentiable function defined on a closed and bounded interval has both a maximum and a minimum value. Observe that one or both of these values may occur at an endpoint of the domain. Is it possible for a function not to have any extremal values?

We approach the problem of determining extrema first from a geometrical view and then analytically.

Example 3.8.8 We will find the extrema of $f(x) = \sin(x)$ for $0 \leq x \leq 2\pi$. Since $f'(x) = \cos(x)$, $f'(x) = 0$ for $x = \pi/2$ and $3\pi/2$. Thus, there are exactly two points on the graph of $f(x) = \sin(x)$ where the tangent line is horizontal, one is a global maximum and the other a global minimum. How could one have identified the extrema analytically without relying on the graph? △

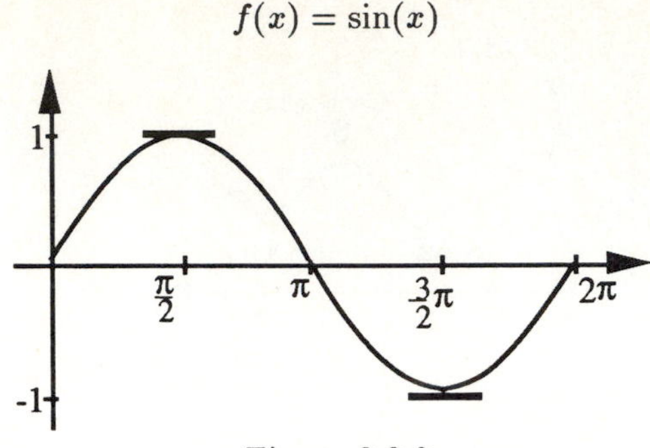

Figure 3.8.9

Example 3.8.10 We will find the extrema for $f(x) = x(1-x)^2$ for $-1 \leq x \leq 2$. Since $f'(x) = (1-x)(1-3x)$, $f'(x) = 0$ implies that $x = 1$ or $\frac{1}{3}$. Thus, there are exactly two points on the graph of f where the tangent line is horizontal. From the figure below we see that one is a local maximum and the other is a local minimum, but neither is a global extremum. The global extrema occur at the endpoints in this example. How could one have identified the extrema analytically without relying on the graph?

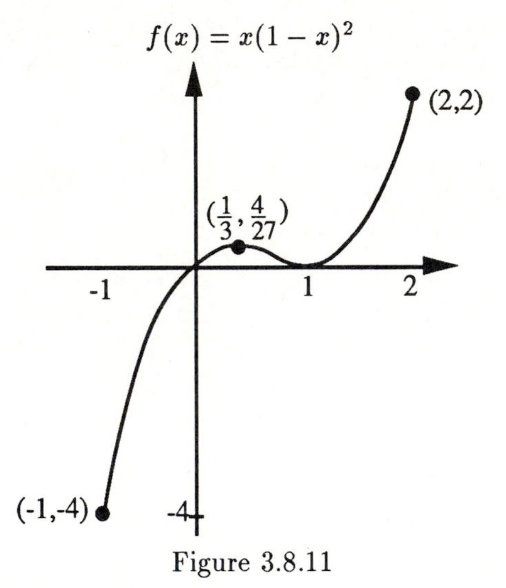

Figure 3.8.11

△

Example 3.8.12 We will find the extrema for $f(x) = x^3+1$. Since $f'(x) = 3x^2$, $f'(x) = 0$ implies that $x = 0$. Thus, there is exactly one point on the graph of f where the tangent line is horizontal. From the following figure, we see that f does not have an extreme value at this point. In fact, f does not have any extreme values at all! How could we have determined this without relying on the graph?

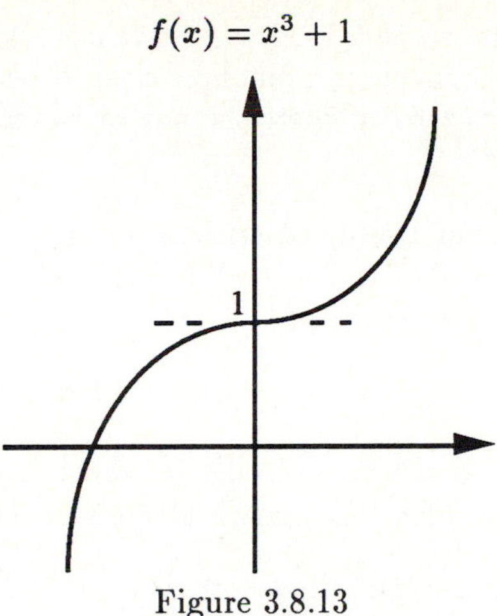

Figure 3.8.13 △

Example 3.8.14 We will find the extrema for

$$g(x) = \begin{cases} x^3 + 1 & \text{for } x < 1 \\ 1/x + 1 & \text{for } x \geq 1 \end{cases}$$

Note that this function is a modification of the function in the previous example. Since

$$g'(x) = \begin{cases} 3x^2 & \text{for } x < 1 \\ -1/x^2 & \text{for } x > 1 \end{cases}$$

then $g'(x) = 0$ only for $x = 0$. We saw in the previous example that even though the tangent line is horizontal at $x = 0$, g does not have an extremum there. However, as we can see from the following figure, g does have a global maximum at $x = 1$. How could one have determined this analytically?

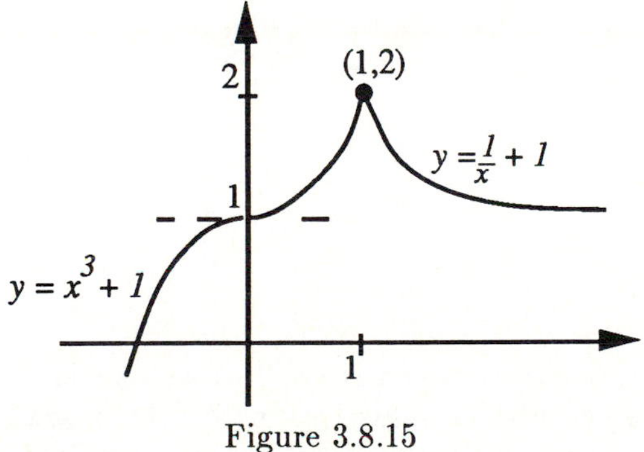

Figure 3.8.15 △

We are now ready to establish several graphical properties of functions by analytical methods. The principal "tool" that we shall use is the Mean Value Theorem.

The graphs in the figures we have used suggest that if a local extremum occurs at an interior point of the domain of a function, then it occurs either where the derivative of the function is zero or where the derivative does not exist. We call these critical points.

Definition 3.8.16 (Critical Point) *A* critical point *of a function is an interior point (in the domain of the function) where the derivative of the function is zero or does not exist (see Figure 3.8.15).*

Note that a function does not necessarily have a local extremum at a critical point (see Figure 3.8.13).

Theorem 3.6.10 provides the tools needed to develop an analytical test to determine if a function has a local extremum at a given critical point.

Theorem 3.8.17 (The First Derivative Test) *Let f be a continuous function on the closed interval $[a, b]$ and differentiable on the open interval (a, b) except possibly at the critical point c.*

(a) *If $f'(x)$ is positive for all $x < c$ and $f'(x)$ is negative for all $x > c$, then f has a local maximum at $x = c$.*

(b) *If $f'(x)$ is negative for all $x < c$ and $f'(x)$ is positive for all $x > c$, then f has a local minimum at $x = c$.*

Proof [part (a)]: By Theorem 3.6.10, we know that f is strictly increasing on $[a, c]$ and strictly decreasing on $[c, b]$. Thus, $f(x) < f(c)$ for all $x \neq c$ in $[a, b]$ and hence f has a local maximum at $x = c$. Part (b) can be proved using a similar argument. The details are left to the exercises. □

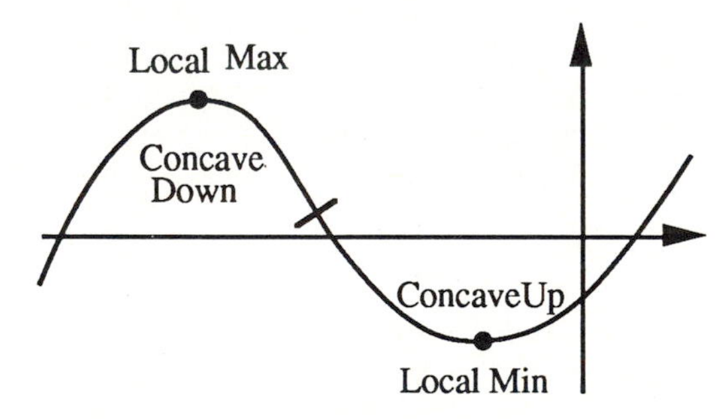

Figure 3.8.18

Looking at Figure 3.8.18, it would seem that we should be able to develop a test for extrema based on the shape of the curve at a point where $f'(x) = 0$. Such is the case. We begin by noting that if f is a twice differentiable function and $f''(x) > 0$ for x in (a, b), then f' is an increasing function over the interval (a, b). This means that as x "moves" from a to b, the slopes of the tangent lines increase. For example, consider the function $f(x) = x^2 - 1$. The graph of this function, along with segments of the tangent lines drawn to the points where

$x = -\frac{3}{2}, -\frac{1}{2}, 0, \frac{1}{2}, \frac{3}{2}$ are sketched below. Note how the tangent lines "turn" in a counterclockwise direction. We shall say that such a curve is concave up.

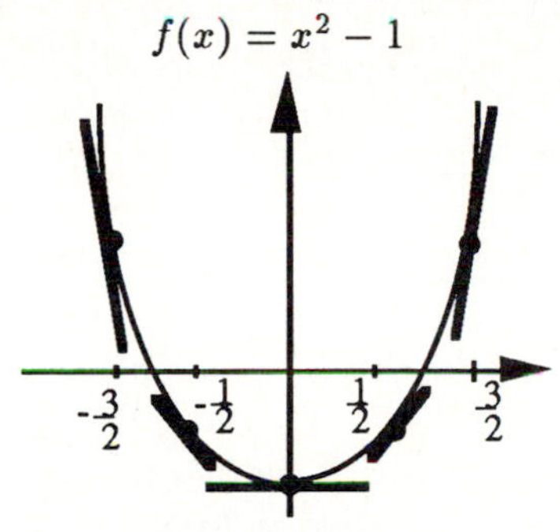

Figure 3.8.19

An analogous situation exists if $f''(x) < 0$ with the tangent lines turning in a clockwise direction.

Question 3.8.20 Suppose that f' has graph

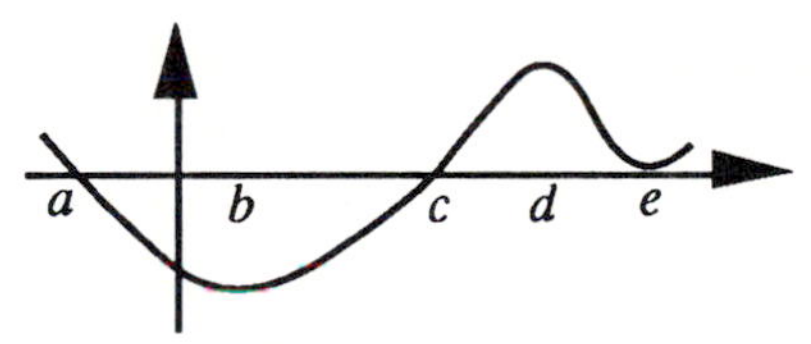

Figure 3.8.21

What are the critical points of f, and which are local minima or local maxima? ◇

Definition 3.8.22 (Concave Functions) *The graph of a twice differentiable function over an open interval* (a, b) *is* concave up *over* (a, b) *if* $f''(x) \geq 0$ *for all* x *in* (a, b) *and* concave down *over* (a, b) *if* $f''(x) \leq 0$ *for all* x *in* (a, b).

A useful mnemonic association for remembering concavity is to think of a curve being concave up as a cup holding water (positive result) and concave down as a cup spilling water (negative result).

Example 3.8.23 The graph of $f(x) = \sin(x)$ is concave up over $(-\pi, 0)$ and concave down over $(0, \pi)$. △

The Second Derivative Test says that if a critical point is in an open interval over which the graph is concave down, then the function has a local maximum at the critical point. Similarly, if a critical point is in an open interval over which the graph is concave up, then the function has a local minimum at the critical point. We state and prove this formally in the following theorem.

Theorem 3.8.24 (The Second Derivative Test) *Let f be a twice differentiable function on an open interval (a, b) and assume that c is a number in the interval such that $f'(c) = 0$.*

(a) *If $f''(c) > 0$, then f has a local minimum at $x = c$.*

(b) *If $f''(c) < 0$, then f has a local maximum at $x = c$.*

(c) *If $f''(c) = 0$, then the test is inconclusive.*

Proof[part (a)]: By the definition of $f''(c)$ and knowing that $f'(c) = 0$, we have

$$f''(c) = \lim_{h \to 0} \frac{f'(c+h) - f'(c)}{h} = \lim_{h \to 0} \frac{f'(c+h)}{h}$$

If this limit is positive, $f'(c+h)$ must have the same sign as h. Thus, $f'(c+h)$ must be negative for h negative and positive for h positive. Hence, f is decreasing for h negative and increasing for h positive. Therefore, by the First Derivative Test f has a local minimum at $x = c$. □

Proofs of parts (b) and (c) are left to the exercises.

Clearly this test does not apply at a critical point where the function is not differentiable.

Curve Sketching and Examples of Extrema Problems

The development of plotting routines for computers and graphing calculators has greatly enhanced the role of graphing in calculus. In particular, graphing is now often used to "guide" the analysis of a problem rather than having a graph be an "end product" of the analysis; e.g., a graph can suggest both the location and nature of extremal values. Thus the availability of technology to produce a graph increases (rather than decreases) the importance of understanding the graphical interpretation of analytical properties of a function and conversely.

Before studying calculus, our curve sketching relied on plotting points and recognizing symmetry. The study of limits enabled us to discuss global properties of graphs using the concept of asymptotes. Continuity, through the Intermediate Value Theorem, provided justification for "joining" points with a curve. Hence, by the end of the last chapter, we were able to sketch graphs "globally" using symmetry, asymptotes, and plotting only intercept points. Now we can refine our sketching still further by using information obtained from differentiation.

Example 3.8.25 We will find all the local extrema for $f(x) = x^3 - 6x^2 + 9x$ and sketch the graph of f.

Differentiating f yields

$$f'(x) = 3x^2 - 12x + 9 = 3(x-3)(x-1)$$

Thus, $x = 1$ and 3 are critical points.

To analyze the situation using the First Derivative Test, we investigate the sign of the derivative at numbers on either side of the critical points. Now, for x

slightly less than one, f' is positive and for numbers slightly more than one, f' is negative. Hence, f has a local maximum at $x = 1$. A similar analysis shows that f has a local minimum at $x = 3$.

To analyze the situation using the Second Derivative Test, we consider the sign of the second derivative evaluated at the critical points. Differentiating f' yields: $f''(x) = 6x - 12 = 6(x - 2)$.

Since $f''(1) = -6 < 0$, f has a local maximum at $x = 1$. Similarly, since $f''(3) = 6 > 0$, f has a local minimum at $x = 3$.

To aid in sketching the graph, we see by inspection that the origin is an intercept and that f is unbounded, going to positive infinity for large positive values of x and to negative infinity for large negative values of x. Looking at the first derivative, we note that f is increasing for $x < 1$ and $x > 3$ (f' is positive) and f is decreasing for $1 < x < 3$ (f' is negative). Also, the graph is concave down for $x < 2$ (f'' negative) and concave up for $x > 2$ (f'' positive). Furthermore, there is a local maximum at (1,4) and a local minimum at (3,0).

We can now easily sketch the graph shown below.

$f(x) = x^3 - 6x^2 + 9x$, x in [0,4]

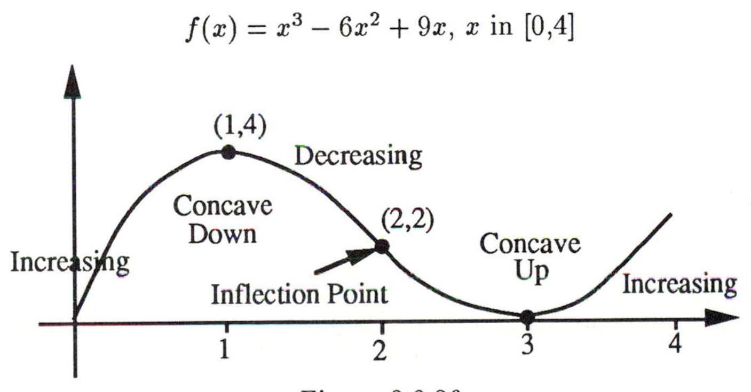

Figure 3.8.26

Just as the critical points give useful information concerning where a function changes direction, the points where the second derivative is zero or does not exist (i.e., critical points of f') give useful information concerning the "trend" of a function. In particular, the point on the curve where the concavity changes is called a *point of inflection* or inflection point. Note that for a twice differentiable function, having $f''(c) = 0$ is a necessary but not sufficient condition for $(c, f(c))$ to be an inflection point. Consider the function $f(x) = x^4$ defined for all real numbers. $f''(x) = 12x^2 = 0$ at the origin, 0. However the concavity does not change at 0 since $f''(x) \geq 0$ for all x and thus the function is concave up everywhere.

Graph of $f(x) = x^4$

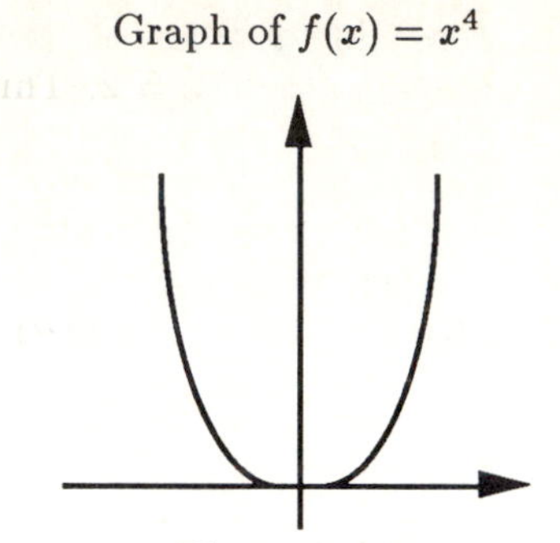

Figure 3.8.27 △

Question 3.8.28 Construct an example of a twice differentiable function for which $f''(3) = 0$, but $(3, f(3))$ is not an inflection point. ◊

The First and Second Derivative Tests apply only at critical points where the first derivative exists and is zero. Thus, to find *all* local extrema, we need to also check those critical points where the derivative does not exist, as well as the endpoints of the curve.

The global maximum is found by taking the largest of all the local extrema, that is, computing the value of f at the endpoints and each of the critical points and then choosing the largest of those values. A similar statement can be made about the global minimum. We illustrate with the following example.

Example 3.8.29 We will find the global maximum and global minimum of

$$f(x) = \begin{cases} x^2 + 2x + 2 & \text{for } -1/2 \leq x \leq 0 \\ x^2 - 2x + 2 & \text{for } 0 < x \leq 2 \end{cases}$$

Since f is continuous over $[-\frac{1}{2}, 2]$, by the extreme value theorem, f will take on both a global maximum and a global minimum at points in the interval. Differentiating to find the critical points, we have

$$f'(x) = \begin{cases} 2x + 2 & \text{for } -1/2 \leq x < 0 \\ 2x - 2 & \text{for } 0 < x < 2 \end{cases}$$

Thus, $f'(1) = 0$ and f is not differentiable at $x = 0$ (note that the left- and right-hand derivatives are not equal). Hence, $x = 0$ and $x = 1$ are critical points. Computing the functional values at the critical points and the endpoints yields

$$\begin{aligned} f(\frac{-1}{2}) &= \left(\frac{-1}{2}\right)^2 + 2\left(\frac{-1}{2}\right) + 2 = \frac{5}{4} \\ f(0) &= 0^2 - 2(0) + 2 = 2 \\ f(1) &= 1^2 - 2(1) + 2 = 1, \text{ and} \\ f(2) &= 2^2 - 2(2) + 2 = 2 \end{aligned}$$

Since the largest of these values is 2 and the smallest is 1, f has a global maximum of 2 that it takes on at $x = 0$ and at $x = 2$. This function has a global minimum of 1 that it takes on at $x = 1$. △

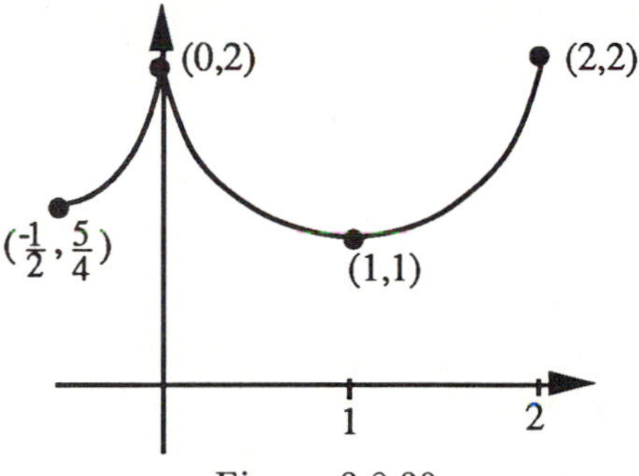

Figure 3.8.30

Many of the early applications of differential calculus involved finding extrema. This area of applications has continued to grow over the past 250 years and is still an area of active interest.

Example 3.8.31 Show that among all rectangles with a given perimeter, the square has the largest area.

First, we sketch and label a picture.

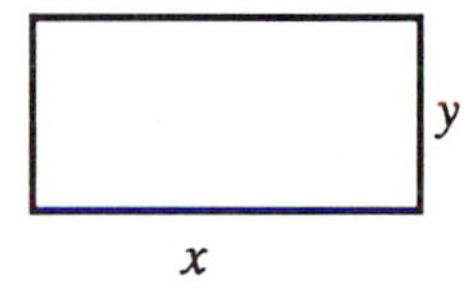

Figure 3.8.32

Let P denote the given perimeter and x and y the lengths of the sides of the rectangle. The area of the rectangle is given by $A = xy$. Although the function we wish to maximize, A, is defined in terms of two variables, x and y, we may rewrite A as a function of a single variable by finding another relationship between x and y and substituting into the original expression for A. Now $P = 2x + 2y$ and so $y = P/2 - x$. Substituting for y in the expression for A yields

$$A = xy = x\left(\frac{P}{2} - x\right) = x\frac{P}{2} - x^2$$

for x in the interval [0,P/2]. Differentiating, since P is constant, we have $A' = P/2 - 2x$ and so $A' = 0$ implies that $x = P/4$ is a critical point.

To determine the (global) maximum for A, we evaluate A at the critical point and at the endpoints. Since A is zero at the endpoints and $A = P^2/16$ at $x = P/4$, the maximum for A occurs at $x = P/4$ and $y = P/2 - P/4 = P/4$. Thus, the rectangle of maximum area is a square. △

Example 3.8.33 A window is to be made in the form of a rectangle surmounted by a semicircle with diameter equal to the base of the rectangle. The rectangular portion is to be of clear glass and the semicircular portion is to be of colored glass that admits only half as much light per square foot as the clear glass. The total perimeter of the window frame is to be a fixed length P. Find, in terms of P, the dimensions of the window that will admit the most light.

First we sketch and label a picture.

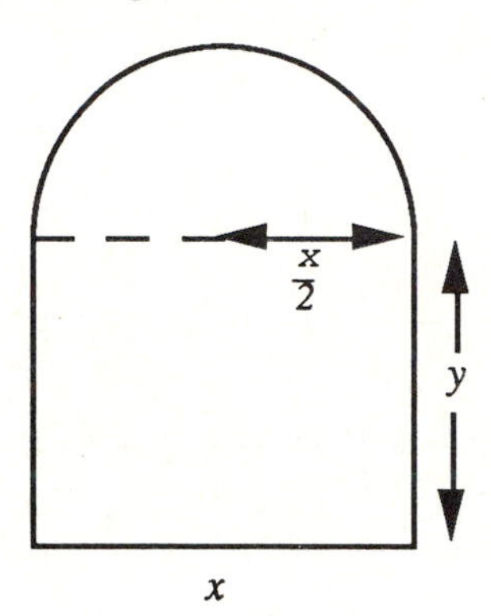

Figure 3.8.34

Now

$$P = (x + 2y) + \pi \frac{x}{2}$$

the first term being the length of the three sides of the rectangle and the second term being the perimeter of the semicircle. To model the amount of light that the window will admit, we define a "light" function L by

$$L(x) = 2xy + \frac{1}{2}\left(\pi \frac{x^2}{4}\right)$$

where the coefficient 2 in the first term represents the fact that the rectangular portion of the window admits twice as much light as the semicircular portion. The second term is the area of the semicircular portion (radius $= x/2$). Solving for y in the expression for P and substituting it into the expression for L, yields

$$L(x) = x\left(P - \left(1 + \frac{\pi}{2}\right)x\right) + \pi \frac{x^2}{8} = Px - \left(1 + 3\frac{\pi}{8}\right)x^2$$

Differentiating L, we obtain

$$L'(x) = P - 2\left(1 + 3\frac{\pi}{8}\right)x$$

Setting $L'(x)$ equal to zero, we find that

$$x = \frac{P}{2(1 + \frac{3}{8}\pi)} = \frac{4P}{8 + 3\pi}$$

is the only critical point.

Using the Second Derivative Test, we differentiate $L'(x)$ to obtain

$$L''(x) = -2\left(1 + 3\frac{\pi}{8}\right) < 0$$

Thus, f has a local maximum value at the critical point. Since L is zero at the endpoints (where $x = 0$ or $y = 0$), the local maximum is a global maximum and therefore the dimensions of the window that will admit the most light are (after some algebra)

$$x = \frac{4P}{8+3\pi} \quad \text{and} \quad y = \left(\frac{P}{2}\right)\frac{4+\pi}{8+3\pi} \qquad \triangle$$

Section 3.8 Exercises

In Exercises 1 through 4, use a computer or graphing calculator to approximate the extremal values (if any exist).

1. $f(x) = x^2 + 3x$
2. $f(x) = x(x^2 - 1)$
3. $f(x) = \begin{cases} -x^2 & \text{for } x \le 0 \\ x^2 & \text{for } x > 0 \end{cases}$
4. $f(x) = \begin{cases} 1/x & \text{for } 0 < x \le 2 \\ x & \text{for } x > 2 \end{cases}$

For each of the functions defined by the relations in Exercises 5 through 9, obtain a graph of the relation using a computer or graphing calculator. Indicate on your sketch where the function is increasing, decreasing, concave up, concave down, has extreme values, and points of inflection. Check your analysis of the graph by analytically verifying the above listed properties.

5. $f(x) = x^3 - 6x^2 + 9x + 1$
6. $f(x) = x^3/3 - x^2 - 3x + 2$
7. $f(x) = 2 - (x - 3)^{2/3}$
8. $f(x) = 2x/(2 + x^2)$
9. $f(x) = (1/4)x^4 - (3/2)x^2 - 1$

For each of the functions defined by the relations in Exercises 10 through 14, determine analytically where the function is increasing, decreasing, concave up, concave down, has extreme values, and points of inflection. Sketch the graph of the function and indicate the above information on your sketch. Check your sketch by plotting the graph using a computer or graphing calculator.

10. $f(x) = x^{4/3} + 4x^{1/3} + 2$

11. $f(x) = x^2(x^2 - 2)$

12. $f(x) = x^2(3 - 2x^2)^{1/2}$

13. $f(x) = 1/(x + 3)$

14. $f(x) = (x^2)/((x - 2)(x + 1))$

In Exercises 15 and 16 assume that the functions are defined on all of $\mathbb{R}$ and are monotone where not shown.

15. If f' has the following graph,

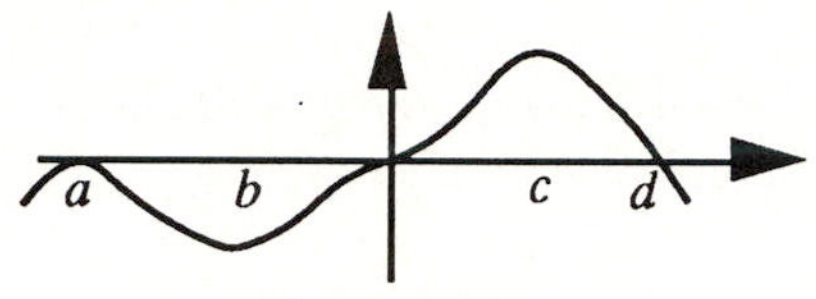

Figure 3.8.35

find the critical points of f and classify them as local maxima, local minima, or neither.

16. If f' has the following graph,

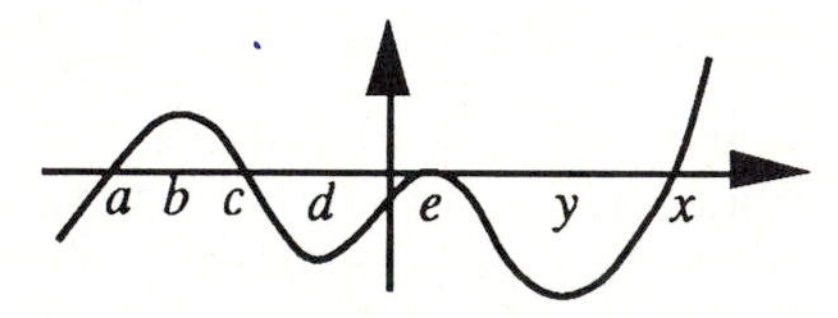

Figure 3.8.36

find the critical points of f and classify them as local maxima, local minima, or neither.

17. Sketch the graph of a function which satisfies all the following conditions

(a) Local minima at $(-1, -2)$, $(1, -5)$, and $(3, -4)$.

(b) Local maxima at $(-3, 2)$, $(0, -1)$, $(2, -3)$, and $(5,0)$.

(c) Global maximum at $(-3, 2)$.

(d) Differentiable for all x except $x = 5$.

18. Sketch the graph of a function which satisfies all the following conditions

(a) f is continuous except at $x = 3$.

(b) f is differentiable except for $x = -1$ and $x = 3$.

(c) f is increasing for $x < -1$ and $x > 3$.

(d) f is concave up for $x < -1$ and $x > 3$.

(e) f is decreasing for $-1 < x < 3$.

(f) $x = 3$ is a vertical asymptote.

19. Find values for a and b such that the function f defined by $f(x) = x^3 + ax^2 + b$ has a local minimum at (2,4).

20. Find values for a, b, c, and d such that the function defined by $f(x) = ax^3 + bx^2 + cx + d$ has a local maximum at (1,1).

21. A closed cylindrical can is to have a volume of 100 in^3. What should the dimensions be in order to minimize the amount of material?

22. A wire k meters long is cut into two parts. One part is then bent into a square and the other into a circle. What should be the lengths of the two pieces if the sum of the two enclosed areas is to be a maximum?

23. A right triangle has sides of length 5, 12, and 13 in. A rectangle is inscribed in the triangle with sides along the two shorter sides of the triangle and the opposite corner on the hypotenuse of the triangle. Find the dimensions of the rectangle in order that its area be as large as possible.

24. Plans for a new supermarket require a floor area of 10,000 ft^2. The supermarket is to be rectangular in shape with three solid brick walls and a fancy all-glass front. If the glass wall costs 1.5 times as much as the brick wall per linear foot, what should be the dimensions of the building so that the cost of materials for the walls is a minimum?

25. A box is to be made from a rectangular sheet of cardboard that measures 10 in by 16 in. Equal squares are cut from the four corners and the sides are then folded up. What are the dimensions of the box if its volume is to be a maximum?

26. Find the point or points on the graph of $f(x) = x^2$ which are closest to the point $(0, a)$. Express your answer as a function of a.

27. Find the dimensions of an isosceles triangle having a perimeter of 40 cm if the area of the region enclosed is to be a maximum.

28. Mark each of the following statements true (T) or false (F).

(a) The function f has an extremum when $x = 3$ provided $f'(3) = 0$.

(b) If f is a twice differentiable function and $f''(2) = 0$, then $(2, f(2))$ is an inflection point.

(c) It is possible for a function to be increasing and have its graph be concave down over the same interval.

(d) A function may have an extremum at an inflection point.

(e) It is possible to have a differentiable function defined for all real numbers which is bounded and whose graph is concave up.

(f) A function may have a local minimum, but no global minimum.

(g) A function may take on an extreme value at a point where the derivative is not equal to zero.

(h) If f is a differentiable function over the interval $a < x < b$, then f takes on its global maximum value at some point c, $a < c < b$.

29. Give an example of a function which has a local minimum that is greater than one of its local maxima.

30. Give an example of a bounded function that has no extrema.

31. Give an example of a continuous function defined over the interval $[-2, 5]$ which has a local minimum at $x = 2$, but the derivative of the function is not zero at $x = 2$.

32. Give an example of a function which has a local minimum at a point of discontinuity.

33. Make a conjecture (an intelligent guess) on how each of the following statements should be completed.

    (a) The extrema of a function may occur ... (three possibilities).

    (b) If $f'(x)$ is positive, then the graph of f is ... .

    (c) If $f'(x)$ is negative, then the graph of f is ... .

34. Consider Theorem 3.8.24 (c).

    (a) Explain what the conclusion of part (c) means.

    (b) Describe the behavior in a neighborhood of the point where the second derivative is zero for each of the functions: $f(x) = x^4$, $g(x) = -x^4$, and $h(x) = x^5$.

    (c) Use the results of part (b) to prove Theorem 3.8.24.

35. Prove Theorem 3.8.17 (b).

36. Prove Theorem 3.8.24 (b).

3.9 L'Hospital's Rule

A property of limits that we studied in Chapter 2 stated that the limit of a quotient is equal to the quotient of the limits, provided the limit of the denominator is not zero. What happens when the limit of the denominator is zero? Does that mean that the limit does not exist or is it just that the quotient rule does not apply? These are the questions that motivate the material in this section.

First, we should note that the limit of a quotient does not exist if the limit of the denominator is zero and the limit of the numerator is not zero. Thus, we shall be concerned with evaluating the limits of quotients in which both numerator and denominator converge to zero. We shall refer to this type of situation as the *zero over zero form* (0/0).

For example, consider the following four problems listed in order of increasing difficulty.

1. $\lim_{x\to 2} \dfrac{x^2-4}{(x-2)x}$

2. $\lim_{h\to 0} \dfrac{2-(4-h)^{1/2}}{h}$

3. $\lim_{x\to 0} \dfrac{\sin(x)}{x}$

4. $\lim_{x\to 0} \dfrac{\cos(x)-\cos(3x)}{\sin(x^2)}$

In Chapter 2, we saw that the 0/0 form in problem 1 could be algebraically resolved by factoring the numerator and then canceling the common $(x-2)$ factor. In problem 2, the algebra was a little more involved because of the square root. However, multiplying and dividing by the conjugate of the numerator [i.e., $2+(4-h)^{1/2}$] transformed the problem into a problem similar to problem 1. That is, where the 0/0 form can be eliminated by canceling. Problem 3 was different from problems 1 and 2 in that no algebraic manipulation would eliminate the 0/0 form. Recall that a geometric argument was used to produce a double inequality to which the "squeeze" technique was applied, thus resolving the 0/0 form. There are no "nice" identities or geometric diagrams that will eliminate the 0/0 form in problem 4. The limit in problem 4 cannot be evaluated with the techniques that have been developed so far. What should we do? A reasonable idea would be to approximate both numerator and denominator functions with their tangents as $x \to 0$. This suggests applying the Linear Approximation Theorem.

Let's attempt to analyze the 0/0 form by transforming the numerator and denominator of the quotient using the linear approximation theorem. Thus we consider

$$\lim_{x\to a} \frac{f(x)}{g(x)} \quad \text{where } \lim_{x\to a} f(x) = f(a) = 0 \text{ and } \lim_{x\to a} g(x) = g(a) = 0.$$

(Note that we are assuming that f and g are continuous at $x = a$.)

Applying the Linear Approximation Theorem to the numerator and denominator separately yields

$$\begin{aligned}\lim_{x\to a}\frac{f(x)}{g(x)} &= \lim_{x\to a}\frac{f(a)+f'(a)(x-a)+ERROR1(x)}{g(a)+g'(a)(x-a)+ERROR2(x)}\\ &= \lim_{x\to a}\frac{f'(a)(x-a)+ERROR1(x)}{g'(a)(x-a)+ERROR2(x)} \quad \text{since } f(a)=g(a)=0\end{aligned}$$

The second result of the Linear Approximation Theorem suggests that both numerator and denominator be divided by $x-a$. This yields

$$\lim_{x\to a}\frac{f(x)}{g(x)} = \lim_{x\to a}\frac{f'(a)+\frac{ERROR1(x)}{x-a}}{g'(a)+\frac{ERROR2(x)}{x-a}}$$

By the Linear Approximation Theorem,

$$\lim_{x\to a}\frac{ERROR1(x)}{x-a}=0$$

and

$$\lim_{x\to a}\frac{ERROR2(x)}{x-a}=0$$

Thus we have

$$\lim_{x\to a}\frac{f(x)}{g(x)}=\frac{f'(a)}{g'(a)}$$

For illustrative purposes, we apply this approach to each of the four problems stated in the beginning of this section.

1. $\lim_{x\to 2} f(x)/g(x) = \lim_{x\to 2}(x^2-4)/((x-2)x) = f'(2)/g'(2) = \frac{4}{2} = 2$
2. $\lim_{h\to 0} f(h)/g(h) = \lim_{h\to 0}(2-(4-h)^{1/2})/h = f'(0)/g'(0) = \frac{1/2}{2} = \frac{1}{4}$
3. $\lim_{x\to 0} f(x)/g(x) = \lim_{x\to 0}\sin(x)/x = f'(0)/g'(0) = \cos(0)/1 = \frac{1}{1} = 1$
4. $\lim_{x\to 0} f(x)/g(x) = (\cos(x)-\cos(3x))/\sin(x^2) = f'(0)/g'(0)$

$$= \frac{-\sin(0)+3\sin(0)}{\cos(0)2(0)} = \frac{0}{0}$$

TROUBLE!

Our approach depended on the existence of $f'(a)/g'(a)$. That is, on $g'(a)\neq 0$. Thus when $g'(a)=0$, we need to "dig deeper." Since the linear approximation theorem was an application of the Mean Value Theorem it makes sense to see what happens when the Mean Value Theorem is used in place of the Linear Approximation Theorem. That is, apply the Mean Value Theorem to both numerator and denominator of the quotient $f(x)/g(x)$. For this to make sense, we need to write $f(x)$ as $f(x) - f(a)$ and $g(x)$ as $g(x) - g(a)$. [Recall that $f(a) = g(a) = 0$.] Applying the Mean Value Theorem to both numerator and denominator yields

$$\frac{f(x) - f(a)}{g(x) - g(a)} = \frac{f'(c)(x-a)}{g'(d)(x-a)} = \frac{f'(c)}{g'(d)}$$

where c and d are between x and a. Now applying the "squeeze" technique, we have

$$\lim_{x \to a} \frac{f(x)}{g(x)} = \lim_{c,d \to a} \frac{f'(c)}{g'(d)}$$

Rewriting the right-hand side as $\lim_{x \to a} f'(x)/g'(x)$ gives the result known as L'Hospital's Rule. Before stating this rule formally, we note that when $f'(x)$ and $g'(x)$ are continuous at $x = a$ and $g'(a) \neq 0$, the result using the Mean Value Theorem is the same as our result using the Linear Approximation Theorem. The advantage of the mean Value Theorem approach is that for a problem such as 4 in which the $\lim_{x \to a} f'(x)/g'(x) = 0/0$, the process may be repeated. For example in problem 4

$$\lim_{x \to 0} \frac{f(x)}{g(x)} = \lim_{x \to 0} \frac{\cos(x) - \cos(3x)}{\sin(x^2)} = \lim_{x \to 0} \frac{-\sin(x) + 3\sin(3x)}{\cos(x^2)2x} = \frac{0}{0}$$

Now repeat the process. Then

$$\lim_{x \to 0} \frac{-\sin(x) + 3\sin(3x)}{\cos(x^2)2x} = \lim_{x \to a} \frac{-\cos(x) + 9\cos(3x)}{-\sin(x^2)4x^2 + 2\cos(x^2)} = \frac{8}{2} = 4$$

We formalize the process illustrated above by stating and proving the famous theorem.

Theorem 3.9.1 (L'Hospital's Rule) *Let f and g have continuous derivatives over the open interval (a, b) such that*

(a) $g(x) \neq 0$ *for* $a < x < b$
(b) $\lim_{x \to a+} f(x) = \lim_{x \to a+} g(x) = 0$ *and*
(c) $g'(x) \neq 0$ *for* $a < x < b$

Then, if $\lim_{x \to a+} f'(x)/g'(x)$ *exists,* $\lim_{x \to a+} f(x)/g(x) = \lim_{x \to a+} f'(x)/g'(x)$.

Proof: Define functions F and G as follows:

$$F(x) = \begin{cases} f(x) & \text{for } a < x < b \\ 0 & \text{for } x = a \end{cases}$$

$$G(x) = \begin{cases} g(x) & \text{for } a < x < b \\ 0 & \text{for } x = a \end{cases}$$

for all values of x in $[a, b)$. Now, F and G are continuous on $[a, b)$ and $F'(x) = f'(x)$, $G'(x) = g'(x)$ for $a < x < b$. Note that F and G are just continuous extensions of f and g to the interval $[a, b)$.

Choose any number x, $a < x < b$, and note that F and G restricted to the closed interval $[a, x]$ satisfy the hypothesis of the Mean Value Theorem. Thus,

$$\frac{F(x) - F(a)}{G(x) - G(a)} = \frac{F'(c)(x-a)}{G'(d)(x-a)} = \frac{F'(c)}{G'(d)}$$

for some c and d between a and x. Since $F(a) = G(a) = 0$, we have

$$\frac{F(x)}{G(x)} = \frac{f(x)}{g(x)} = \frac{f'(c)}{g'(d)}$$

for some c and d between a and x. Now taking limits, the conclusion follows since c and d being squeezed between a and x means that as x approaches a, both c and d are forced to approach a. □

Note that in this theorem x approaches a from the right. A similar theorem can be proved with x approaching a from the left. The theorem is also true if a is replaced by positive or negative infinity.

Example 3.9.2 We will evaluate $\lim_{x \to 0+} x / \sin(x^{1/2})$. As x approaches 0, both the numerator and the denominator approach 0. Since

$$\frac{d \sin(x^{1/2})}{dx} = \frac{\cos(x^{1/2})}{2x^{1/2}}$$

applying L'Hospital's Rule gives

$$\lim_{x \to 0+} \frac{x}{\sin(x^{1/2})} = \lim_{x \to 0+} \frac{2x^{1/2}}{\cos(x^{1/2})} = 0 \qquad \triangle$$

Example 3.9.3 Sometimes it is necessary to transform an expression in order to apply L'Hospital's Rule. Consider $\lim_{x \to 0}(1/(x \sin(x)) - 1/x^2)$. As it stands, the limiting form is $\infty - \infty$. We transform this expression by obtaining a common denominator:

$$\frac{1}{x \sin(x)} - \frac{1}{x^2} = \frac{x - \sin(x)}{x^2 \sin(x)}$$

Now the limiting form is $0/0$ and we can apply L'Hospital's Rule to obtain:

$$\begin{aligned} \lim_{x \to 0} \left(\frac{1}{x \sin(x)} - \frac{1}{x^2} \right) &= \lim_{x \to 0} \frac{x - \sin(x)}{x^2 \sin(x)} \\ &= \lim_{x \to 0} \frac{1 - \cos(x)}{2x \sin(x) + x^2 \cos(x)} \quad \text{(applying the rule)} \end{aligned}$$

$$= \lim_{x\to 0} \frac{\sin(x)}{2\sin(x) + 4x\cos(x) - x^2\sin(x)} \text{ (apply again)}$$

$$= \lim_{x\to 0} \frac{\cos(x)}{6\cos(x) - 6x\sin(x) - x^2\cos(x)} \text{ (and again!)}$$

$$= \frac{1}{6} \qquad \triangle$$

Question 3.9.4 L'Hospital's Rule first appeared in print in 1696 in L'Hospital's book, *Analyse des infiniments petits, pour l'intelligence des lignes courbes*, the *first* textbook written on differential calculus. The discovery of the rule actually belongs to Johann Bernoulli whom L'Hospital employed as a tutor with the provision that he (L'Hospital) could publish results that Bernoulli taught him.

The reader should "step into the pages of history" by working the following problem (stated in today's language) given in L'Hospital's book to illustrate "his" rule.

Evaluate

$$\lim_{x\to a} \frac{\sqrt{2a^3x - x^4} - a\sqrt[3]{a^2x}}{a - \sqrt[4]{ax^3}} \qquad \diamond$$

L'Hospital's Rule also applies when the limit of a quotient yields the "infinity over infinity" form. We will use this result without giving a formal argument justifying it.

Example 3.9.5 We will evaluate the ∞/∞ form $\lim_{x\to 0}(x+\cot(x))/(7+\csc(x))$.

Apply L'Hospital's Rule with

$$f(x) = x + \cot(x) \text{ and } g(x) = 7 + \csc(x)$$

Thus,

$$\lim_{x\to 0} \frac{x + \cot(x)}{7 + \csc(x)} = \lim_{x\to 0} \frac{1 - \csc^2(x)}{-\csc(x)\cot(x)}$$

Note that this is also an ∞/∞ form. However, transforming the cosecant and cotangent functions into sines and cosines and then simplifying yields

$$\lim_{x\to 0} \frac{\sin^2(x) - 1}{-\cos(x)} = 1 \qquad \triangle$$

Any form that can be transformed into either the $0/0$ or ∞/∞ form is called an *indeterminate form.* The term "indeterminate" refers to the fact that two functions which are similar in the sense that they are both converging to zero or both diverging to infinity form ratios that do not behave in any predetermined manner. The $\infty - \infty$ form in Example 3.9.3 is an example of an indeterminate form. The next two examples illustrate two other indeterminate forms. (These examples use properties of the natural exponential and natural logarithm functions that will be formally established in Chapter 7.)

Example 3.9.6 We will evaluate $\lim_{x\to 0+} x\log(x)$.

Since $\lim_{x\to 0+}\log(x) = -\infty$, $\lim_{x\to 0+} x\log(x) = 0(-\infty)$. This form can always be transformed into a 0/0 or ∞/∞ form by rewriting the product as a quotient. That is,

$$\lim_{x\to 0+} x\log(x) = \lim_{x\to 0+} \frac{\log(x)}{1/x} = -\frac{\infty}{\infty}$$

Applying L'Hospital's Rule yields (recall that $d/dx\log(x) = 1/x$)

$$\lim_{x\to 0+} x\log(x) = \lim_{x\to 0+} \frac{\log(x)}{1/x} = \lim_{x\to 0+} \frac{(1/x)}{(-1/x^2)} = \lim_{x\to 0+}(-x) = 0 \qquad \triangle$$

Example 3.9.7 We will evaluate $\lim_{x\to\infty}(1+1/x)^x$. This is an example of the "one to the infinity power" indeterminate form.

When working with an "exponential" indeterminate form, the standard procedure is to form an equation, $y = (1+1/x)^x$, and then take the natural log of both sides. This transforms the problem into the type illustrated in the previous example (i.e., a $0\cdot\infty$ indeterminate form).

$$\log(y) = \log\left(\left(1+\frac{1}{x}\right)^x\right) = x\log\left(1+\frac{1}{x}\right) \quad \text{(logarithm property)}$$

Now

$$\lim_{x\to\infty}\log(y) = \lim_{x\to\infty} x\log(1+\frac{1}{x}) = \infty\cdot 0 \quad (\log(1) = 0)$$

Applying the technique illustrated in the previous example of rewriting a product as a quotient yields

$$\lim_{x\to\infty}\log(y) = \lim_{x\to\infty}\frac{\log(1+1/x)}{1/x} = \frac{0}{0}$$

Applying L'Hospital's Rule gives

$$\lim_{x\to\infty}\log(y) = \lim_{x\to\infty}\frac{(\frac{1}{1+1/x})(-1/x^2)}{-1/x^2} = \lim_{x\to\infty}\frac{1}{1+1/x} = 1$$

Now be careful. Realize that we have only found $\lim_{x\to\infty}\log(y)$, not $\lim_{x\to\infty} y$. There are two very important steps left to do to obtain $\lim_{x\to\infty} y$. The first is to recognize that the natural log function is continuous and therefore

$$\lim_{x\to\infty}\log(y) = \log(\lim_{x\to\infty} y)$$

The second is to recognize that the natural exponential and the natural logarithmic functions are inverses of one another (see Section 1.5). Thus applying the natural exponential function to both sides of

$$\log(\lim_{x\to\infty} y) = 1$$

yields

$$\lim_{x\to\infty} y = e^1 = e \qquad \triangle$$

The result of the last example, $\lim_{x\to\infty}(1 + 1/x)^x = e$, is sometimes used as the definition for e. The next example illustrates a useful application of this limit expression for e.

Example 3.9.8 (Compound Interest) Suppose you invest P dollars in a savings account that pays 8% interest compounded quarterly. The bank divides the 8% into four 2% increments. After the first quarter (3 months), the bank pays 2% interest on the principal P. Thus after three months, your account contains $P(1+.02) = 1.02P$ dollars. After the second quarter, the bank pays 2% interest on the principal $1.02P$. The account now has $1.02P(1+.02) = (1.02)^2P$ dollars. After the third quarter, another 2% interest is paid on the principal $(1.02)^2P$ leaving a new principal of $(1.02)^3P$. Similarly after the fourth quarter, the final 2% interest is paid on the principal $(1.02)^3P$ yielding a balance of $(1.02)^4P$. Writing $(1.02)^4P$ as $(1 + .08/4)^4P$ shows how to generalize this result. If the bank were to pay an interest rate of r and compound x times a year, then after 1 year the initial investment of P dollars would have grown to $(1+r/x)^xP$ dollars. If the bank were to compound your money "continuously" then your investment of P dollars would have grown in 1 year to

$$\lim_{x\to\infty}(1 + \frac{r}{x})^x P = e^r P \quad \text{(using the previous example)}$$

in 2 years to $e^{2r}P$, and in n years to $e^{nr}P$.

Thus one dollar invested at 8% interest compounded quarterly would yield 1.08243 dollars after 1 year. The same dollar invested at 8% compounded continuously would yield 1.08329 dollars after 1 year. △

(Relative) Rates of Growth An important application of L'Hospital's Rule is determining the relative rates of growth of two functions. What does it mean to say that the exponential function $f(x) = e^x$ "grows faster" than a polynomial function? Or, that $f(x) = x$ grows faster than $g(x) = \log(x)$?

The following defines "f grows faster than g" to mean that for *large* x-values, $g(x)$ is negligible when compared to $f(x)$.

Definition 3.9.9 (Rates of Growth) Let f and $g : \mathbb{R} \to \mathbb{R}$. f is said to grow faster than g as $x \to \infty$ if

$$\lim_{x\to\infty} \frac{f(x)}{g(x)} = \infty$$

f grows at the same rate as g as $x \to \infty$ if

$$\lim_{x\to\infty} \frac{f(x)}{g(x)} = L \neq 0 \quad (L \text{ finite and nonzero})$$

and f grows slower than g as $x \to \infty$ if

$$\lim_{x\to\infty} \frac{f(x)}{g(x)} = 0$$

Note that the functions $f(x) = 3x^2 + 117$ and $g(x) = (1/2)x^2 + 789x - 97$ grow at the same rate since

$$\lim_{x\to\infty} \frac{f(x)}{g(x)} = \lim_{x\to\infty} \frac{3x^2 + 117}{(1/2)x^2 + 789x - 97} = 6$$

a finite nonzero limit. In fact, the reader can show that any two polynomials of the same degree will grow at the same rate according to the above definition.

Example 3.9.10 We will show that $f(x) = e^x$ grows faster than $g(x) = x^3 - 207x^2 + 107$ as $x \to \infty$. We apply L'Hospital's Rule three times:

$$\begin{aligned}
\lim_{x\to\infty} \frac{f(x)}{g(x)} &= \lim_{x\to\infty} \frac{e^x}{x^3 - 207x^2 + 107} \\
&= \lim_{x\to\infty} \frac{e^x}{3x^2 - 414x} \\
&= \lim_{x\to\infty} \frac{e^x}{6x - 414} \\
&= \lim_{x\to\infty} \frac{e^x}{6} \\
&= \infty
\end{aligned}$$

$\triangle$

When speaking of rates of growth we will follow the convention of not explicitly stating that the variable goes to ∞

Section 3.9 Exercises

In Exercises 1 through 18, evaluate the limit or show that the limit does not exist.

1. $\lim_{x\to 0} \sin(x)/x$
2. $\lim_{x\to 1}(x^3 - x)/(x^3 - 3x^2 + 3x - 1)$
3. $\lim_{x\to 2}(3x + 1)/(4x - x^2 + 2)$
4. $\lim_{x\to 0}(x - \sin(x))/\cos(x)$
5. $\lim_{x\to 0}(\cos(x) - \cos(x^2))/\sin^2(x)$
6. $\lim_{x\to\infty} \sin(4/x)/\sin(2/x)$
7. $\lim_{x\to 0}(\sin(x) - \tan(x))/(x^3 - 2x^2)$
8. $\lim_{x\to 0} \sin(13x)x$
9. $\lim_{x\to 0}(x^3 - x^2)/\tan(x)$
10. $\lim_{x\to 0} \sin^{-1}(x)/\tan(x)$
11. $\lim_{x\to 0}(e^x/\sin(x) - e^{-x}/\sin(x))$
12. $\lim_{x\to 0}(\cot(x) - \csc(x))$
13. $\lim_{x\to\infty}(x\tan^{-1}(x) + 4)/(x^2 + 7x)$
14. $\lim_{x\to 0}(\csc(x) - x)/(\cot(x) - x)$
15. $\lim_{x\to 0} xe^{-x}$
16. $\lim_{x\to\infty}(\cos(1/x))^x$
17. $\lim_{x\to 2} x^{1/2-x}$
18. $\lim_{x\to 0} e^{-1/x^2}/x$
19. Mark each of the following statements true (T) or false (F).
    (a) If $\lim_{x\to 2} f(x) = 0$ and $\lim_{x\to 2} g(x) = \infty$, then $\lim_{x\to 2} f(x)g(x) = 0$ since "zero times anything is zero."
    (b) If the limits of the functions f and g exist as x approaches a, then
    $$\lim_{x\to a} \frac{f(x)}{g(x)} = \lim_{x\to a} \frac{f'(x)}{g'(x)}$$
    (c) The function g in the statement of L'Hospital's ule may be zero at either or both endpoints of the interval $[a, b]$.
    (d) The functions f and g in L'Hospital's Rule only need to be differentiable rather than C^1.
    (e) If $\lim_{x\to 1} f(x) = 1$ and if $\lim_{x\to 1} g(x) = \infty$, then $\lim_{x\to 1} f(x)^{g(x)} = 1$ since "one raised to any power is one".
20. Give an example for each of the following or show why no example can exist.
    (a) Differentiable functions f and g such that $\lim_{x\to 2} f(x) = \lim_{x\to 2} g(x) = 0$ and $\lim_{x\to 2} f(x)/g(x) = 3$.
    (b) Differentiable functions f and g such that $\lim_{x\to 2} f(x) = \lim_{x\to 2} g(x) = 0$ and $\lim_{x\to 2} f(x)/g(x) = 0$.
    (c) Differentiable functions f and g such that
    $$\lim_{x\to 2} f(x) = \lim_{x\to 2} g(x) = 0 \text{ and } \lim_{x\to 2} \frac{f(x)}{g(x)} = \infty$$

(d) Differentiable functions f and g such that

$$\lim_{x\to\infty} f(x) = \infty, \lim_{x\to\infty} g(x) = -\infty, \text{ and } \lim_{x\to\infty} (f(x) + g(x)) = 6$$

(e) Differentiable functions f and g such that

$$\lim_{x\to 4} f(x) = \lim_{x\to 4} g(x) = \infty \text{ and } \lim_{x\to 4}(f(x) + g(x)) = 10$$

21. Rates of growth.

    (a) Compare the rate of growth of e^x to x^n for any positive integer n.
    (b) Compare the rate of growth of $\log(x)$ to x^n for any n.
    (c) Compare the rate of growth of $\sqrt{3x^2 + 4x}$ to $(\sqrt{x} - 4)^2$. (Hint: Show that both expressions grow at the same rate as x.)

22. Give an example for each of the following.

    (a) A function that grows slower then $f(x) = 1/x$.
    (b) A function that grows at the same rate as $f(x) = \sqrt{x^3 + 6x}$.
    (c) A function that grows faster then $f(x) = x^4 \log(x)$.

23. Use a computer or calculator to do the calculations.

    (a) What interest rate would a bank have to pay for a person to be able to "double" their money in a savings account in n years if interest were compounded yearly (simple interest)?
    (b) If an 8% rate of interest compounded continuously were paid on a savings account, how many years would it take for a person to "double" their money?
    (c) If an 8% rate of interest compounded quarterly were paid on a savings account, how many years would it take for a person to double their money?
    (d) Would \$1000 invested at 6% compounded continuously for 5 years yield as much return as \$1000 invested at 6% compounded yearly for 10 years? Why?
    (e) How much interest is earned when 1 million dollars is invested at 8% interest compounded continuously for one day?

24. Project: How does the expression y^x behave when x and y both approach zero? (Can y be negative?)

25. Project: How many indeterminate forms are there?

3.10 Partial Derivatives

Consider the problem of dividing the line segment $[a, b]$ into three parts so that the product of their lengths is a maximum?

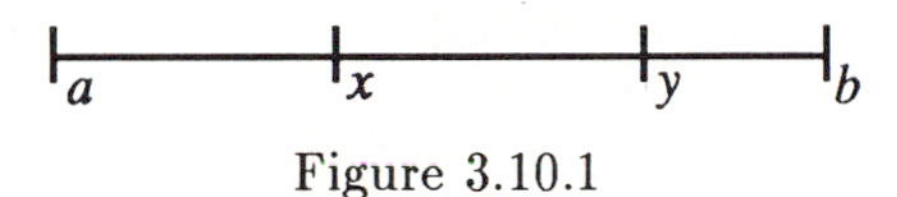

Figure 3.10.1

We will model the problem by expressing the product of the three parts as a function of the two dividing points, say x and y: we have $f : \mathbb{R}^2 \to \mathbb{R}$ defined by

$$\begin{aligned} f(x,y) &= (x-a)(y-x)(b-y) \\ &= x^2y - xy^2 + (b-a)xy - bx^2 + ay^2 - aby + abx \end{aligned}$$

Now we ask at what point or points is the derivative of f equal to zero? That is, at what point or points is the instantaneous rate of change of f equal to zero? This raises a new issue, namely what does it mean to talk about the instantaneous rate of change of a function of two or more variables? For we ask, the instantaneous rate of change of f with respect to what? Do we want to know how f changes as x changes or as y changes or as both x and y change?

The concept of differentiation becomes "richer" very quickly when we extend our thinking to functions of more than one variable, that is to functions $f : \mathbb{R}^n \to \mathbb{R}$. In this section we shall restrict ourselves to examining a special case of differentiating a function of several variables. Namely, the case in which all the variables except one are held constant, and leave the analysis of the general case to Chapter 4.

Now if all the variables except one are held constant, the instantaneous rate of change of the function is just the derivative of the function with respect to the one variable that is allowed to vary. Such a derivative is called a *partial derivative.* Geometrically, the partial derivative of a function $f : \mathbb{R}^n \to \mathbb{R}$ is the slope of the graph of f parallel to an axis. For example, let $f : \mathbb{R}^2 \to \mathbb{R}$. Then the partial derivative of f with respect to the variable y (written $\partial f(x,y)/\partial y$) is the slope of the graph of $z = f(x, y)$ above the point (x, y) in the direction parallel to the y-axis. The set of points on the graph of $z = f(x, y)$ where x is constant is a curve and $\partial f(x,y)/\partial y$ is the slope along this curve.

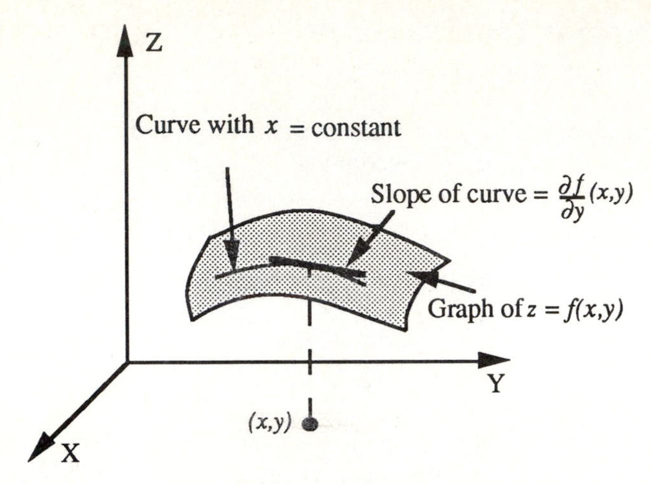

Figure 3.10.2

Analytically the partial derivative is obtained by a direct extension of the definition of the derivative for a single variable: we keep all independent variables except one constant, and then take the limit of the difference quotient.

Definition 3.10.3 (Partial Derivative) *If* $f : \mathbb{R}^2 \to \mathbb{R}$ *then*

$$\frac{\partial f}{\partial y} f(x,y) = \lim_{h \to 0} \frac{f(x, y+h) - f(x,y)}{h}$$

Let us illustrate with the function $f : \mathbb{R}^2 \to \mathbb{R}$ defined by

$$f(x,y) = x^2y^3 - 4x + 7y - 6$$

The *partial derivative of f with respect to x* is the function $\partial f/\partial x\,(x,y)$ obtained by differentiating f with respect to x while holding y constant:

$$\frac{\partial f}{\partial x}(x,y) = 2xy^3 - 4$$

In terms of a limit of a difference quotient, we have

$$\begin{aligned}
\frac{\partial f}{\partial x}(x,y) &= \lim_{h \to 0} \frac{f(x+h,y) - f(x,y)}{h} \\
&= \lim_{h \to 0} \frac{((x+h)^2y^3 - 4(x+h) + 7y - 6) - (x^2y^3 - 4x + 7y - 6)}{h} \\
&= \lim_{h \to 0} \frac{2xhy^3 + h^2y^3 - 4h}{h} \\
&= \lim_{h \to 0} (2xy^3 + hy^3 - 4) \\
&= 2xy^3 - 4
\end{aligned}$$

The *partial derivative of f with respect to y* is the function $\partial f/\partial y(x,y)$ obtained by differentiating f with respect to y while holding x constant. For example, using the above f,

$$\frac{\partial f}{\partial y}(x,y) = 3x^2y^2 + 7$$

Example 3.10.4 Let $f : \mathbb{R}^3 \to \mathbb{R}$ be defined by $f(x,y,z) = x/y + 3\sin(xz)$.

The partial derivative of f with respect to x is

$$\frac{\partial f}{\partial x}(x,y,z) = \frac{1}{y} + 3z\cos(xz)$$

The partial derivative of f with respect to y is

$$\frac{\partial f}{\partial y}(x,y,z) = -\frac{x}{y^2}$$

The partial derivative of f with respect to z is

$$\frac{\partial f}{\partial z}(x,y,z) = 3x\cos(xz) \qquad \triangle$$

Notation. There are various partial derivative notations just as there are for derivatives of a function of a single variable. The most frequently used ones are (for the partial derivative of f with respect to x)

$$\frac{\partial f}{\partial x} = D_x f = D_1 f$$

The subscript one in the last notation refers to the first coordinate variable. Thus the partial derivative of $f(x,y,z)$ with respect to y is

$$D_2 f(x,y,z) = D_y f(x,y,z)$$

and the partial derivative of $f(x,y,z)$ with respect to z is

$$D_3 f(x,y,z) = D_z f(x,y,z)$$

Higher order partial derivatives exist for functions of more than one variable just as higher order derivatives exist for functions of a single variable. However there is more "variety" in the sense that successive derivatives may be taken with respect to any of the variables. For example, consider

$$f(x,y,z) = 2x^2z^3 + y\cos(z)$$

We will find the "second order partial derivative of f first with respect to z and secondly with respect to x".

Notation. we will denote the second-order partial derivative of f first with respect to z and secondly with respect to x in the following ways:

$$\frac{\partial^2 f}{\partial x \partial z}(x,y,z) = D_{1,3}f(x,y,z) = D_{x,z}f(x,y,z) = \frac{\partial}{\partial x}\left(\frac{\partial f}{\partial z}(x,y,z)\right)$$

Now,

$$\frac{\partial f}{\partial z}(x,y,z) = D_z f(x,y,z) = D_3 f(x,y,z) = 6x^2z^2 - y\sin(z)$$

Then

$$\frac{\partial^2 f}{\partial x \partial z}(x,y,z) = D_{1,3}f(x,y,z) = D_{x,z}f(x,y,z) = \frac{\partial}{\partial x}\left(\frac{\partial f}{\partial z}(x,y,z)\right) = 12xz^2$$

Notice that in the "subscript" notations, the partial computed first is with respect to the variable listed closest to the function name. Note also that there are 3 first-order partial derivatives of f, 3^2 second-order partial derivatives of f, 3^3 third-order partial derivatives of f, etc.

We illustrate the partial derivative notation by computing some of the higher order partial derivatives for the function $f : \mathbb{R}^3 \to \mathbb{R}$ defined by $f(x,y,z) = 2x^2z^3 + y\cos(z)$.

$$\frac{\partial^2 f}{\partial z^2}(x,y,z) = D_{z,z}f(x,y,z) = D_{3,3}f(x,y,z) = 12x^2z - y\cos(z)$$
$$\frac{\partial^2 f}{\partial y \partial z}(x,y,z) = D_{y,z}f(x,y,z) = D_{2,3}f(x,y,z) = -\sin(z)$$
$$\frac{\partial^3 f}{\partial y \partial x \partial z}(x,y,z) = D_{y,x,z}f(x,y,z) = D_{2,1,3}f(x,y,z) = 0$$
$$\frac{\partial^3 f}{\partial z \partial x \partial z}(x,y,z) = D_{z,x,z}f(x,y,z) = D_{3,1,3}f(x,y,z) = 24xz$$
$$\frac{\partial^2 f}{\partial y \partial z}(1,3,4) = D_{y,z}f(1,3,4) = D_{2,3}f(1,3,4) = -\sin(4)$$

Example 3.10.5 We will find a function $f : \mathbb{R}^2 \to \mathbb{R}$ such that

(a) $D_{1,2}f(x,y) = x + y$ for all (x,y)

(b) $D_2 f(0,0) = 1$ and

(c) $f(0,0) = 2$

Since $D_{1,2}f(x,y) = x + y$, we can consider the second variable y as fixed and use Theorem 3.7.3 and its Corollary 3.7.5 to obtain an antiderivative with respect to x:

$$D_2 f(x,y) = \frac{x^2}{2} + xy + c(y)$$

where c is a function depending only on y. To see that this is correct, we take the partial derivative of both sides with respect to x, and we obtain the original equation. Now the second equation says that $c(0) = 1$. Since we are looking for only *one* function to satisfy the equations, not *all* functions satisfying the equations, let's assume $c(y)$ is a constant function, 1. This gives

$$D_2 f(x,y) = \frac{x^2}{2} + xy + 1$$

Finding an antiderivative with respect to y, we have

$$f(x,y) = \frac{x^2}{2}y + x\frac{y^2}{2} + y + d(x)$$

where $d(x)$ depends only on x. Using the third initial equation we have $d(0) = 2$. Again, we will assume that d is a constant function to obtain

$$f(x,y) = \frac{x^2}{2}y + x\frac{y^2}{2} + y + 2$$

as *one* function satisfying the three initial equations. We can easily check that this function does satisfy the three conditions. △

Section 3.10 Exercises

In Exercises 1 through 10, compute all first- and second-order partial derivatives.

1. $f(x,y) = 3x^2 + xy$
2. $f(x,y) = \sin(xy)$
3. $f(x,y) = xy/(x-y)$
4. $f(x,y) = xy^3 - x/y$
5. $f(x,y,z) = z + y^2x^3$
6. $f(x,y,z) = 4/xy$
7. $f(x,y,z) = \sin(z)\cos(xy)$
8. $f(x,y,z) = \tan(xy)$
9. $f(x,y,z) = (xy+z)^2$
10. $f(x,y) = x/y + y/x$
11. Construct a function f of two variables such that $D_1 f(0,0)$ exists, but $D_2 f(0,0)$ does not exist.
12. Construct a function f of two variables such that $D_1 f(0,0) \neq D_2 f(0,0)$.
13. From exercises 1 through 10 you may have obtained the impression that for any function f, $D_{x,y}f(x,y) = D_{y,x}f(x,y)$. This is true if the second order partial derivatives $D_{x,y}f$ and $D_{y,x}f$ are continuous functions, but it is not always true as is shown by the following example. Let

$$f(x,y) = \begin{cases} \frac{xy(x^2-y^2)}{(x^2+y^2)} & \text{for } (x,y) \neq (0,0) \\ 0 & \text{for } (x,y) = (0,0) \end{cases}$$

Show that $D_{x,y}f(0,0) = 1$ and $D_{y,x}f(0,0) = -1$.

14. Give an example for each of the following or explain why no such example can exist.

    (a) A function $f : \mathbb{R}^2 \to \mathbb{R}$ such that $D_1 f(1,0) < 0$ and $D_2 f(1,0) > 0$.

    (b) A function $f : \mathbb{R}^2 \to \mathbb{R}$ such that $D_{1,2} f(x,y) = 0$, $D_{2,2} f(x,y) = 0$, and $D_{1,1} f(x,y) = 2$ for all (x,y).

    (c) A function $f : \mathbb{R}^3 \to \mathbb{R}$ such that $D_{1,2,3} f(x,y,z) = D_{1,2} f(x,y,z) = 4$ for all (x,y,z).

15. Find a function $f : \mathbb{R}^2 \to \mathbb{R}$ such that $D_{1,2} f(x,y) = xy + y^2$ for all (x,y), $D_2 f(0,0) = 3$, and $f(0,1) = 6$.

16. Find a function $f : \mathbb{R}^3 \to \mathbb{R}$ such that $D_{2,1,3} f(x,y,z) = x^2 + yz - 2$ for all (x,y,z), $D_{1,3} f(0,0,0) = 10$, $D_3 f(1,0,1) = 20$, and $f(0,0,1) = 10$.

17. [Project] Example 1.5.32 defined the equilibrium point function: $EP : \mathbb{R}^4 \to \mathbb{R}^2$

$$EP(m_s, m_d, c_s, c_d) = \left(\frac{c_d - c_s}{m_s - m_d}, m_s \frac{c_d - c_s}{m_s - m_d} + c_s \right)$$

where m_s is the slope of the supply curve, m_d is the slope of the demand curve, c_s the price intercept of the supply curve, and c_d the price intercept of the demand curve.

The output of the EP function gives the (price, quantity) coordinates of the point of intersection of the supply and demand curves.

   (a) Define a "price equilibrium point" function, $PEP : \mathbb{R}^4 \to \mathbb{R}$, whose output is the price coordinate (first coordinate) in the output of the EP function.

   (b) Define a "quantity equilibrium point" function, $QEP : \mathbb{R}^4 \to \mathbb{R}$, whose output is the quantity coordinate (second coordinate) in the output for the EP function.

   (c) Compute and then give an economic and geometric interpretation of the partial derivatives of your PEP and QEP functions with respect to each of the four input variables.

   (d) Expand your economic analysis of part (c) by computing and interpreting higher order derivatives (e.g., QEP_{m_s,c_d}).

Chapter 4

DERIVATIVES IN SEVERAL VARIABLES

4.1 Vectors

When functions from $\mathbb{R}$ into $\mathbb{R}$ are differentiated, there is only one direction in which the function can change, along the x-axis. A signed number can conveniently represent this change with the sign (positive or negative) denoting the direction and the absolute value of the number denoting the magnitude of the change. However, when functions of two or more variables are differentiated, there are many possible directions to consider. For example, with a function of two variables x and y we can ask about the slope of a line tangent to the graph of a function at a point in the direction 45° counterclockwise from the x-axis. To discuss these possibilities we need to develop a method for easily representing both directions and magnitudes. Directed line segments provide a geometric answer. A given direction and magnitude may be described geometrically by a directed line segment whose direction is the given direction and whose length is the given magnitude.

Directed line segments are called *vectors*. We begin our work with vectors by developing notation and methods of computation. These methods will lead to an *Algebra of Vectors* and the all-important *dot product* operation.

The Geometric Representation of Vectors

Let P and Q be two points in the plane. These two points determine a vector from P to Q, written $\vec{PQ}$, which is the directed line segment from P to Q. The tail of the vector is at P; the head of the vector is at Q. To indicate the direction we use an arrowlike notation, with the point of the arrow at the head of the vector.

Vectors in the Plane

Figure 4.1.1

Since two points determine a line segment in any dimension, we can extend the above concept of geometrical vectors to three or more dimensions (although it is difficult to visualize vectors in more than three dimensions).

Coordinate Representation of Vectors

We are primarily interested in vectors in terms of *magnitude* and *direction*, rather than location. Thus two directed line segments that are "parallel" (in the directed sense) and have the same length represent equivalent vectors. To clarify the idea of "parallel directed line segments," we introduce the notion of "translation" and coordinate representation of vectors.

Consider the plane with the $x-y$ coordinate system. Suppose that a vector $\vec{PQ}$ has tail at $P=(1,2)$ and head at $Q=(4,3)$.

Translation to the Origin

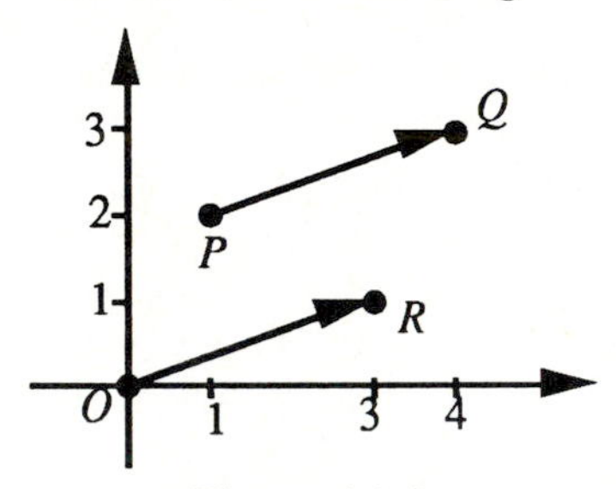

Figure 4.1.2

The change in the x-coordinate between these two points is $\Delta x = 4-1=3$, and the change in the y-coordinate is $\Delta y = 3-2=1$. If the vector $\vec{PQ}$ is translated so that the tail is at the origin $O=(0,0)$ rather than at $P=(1,2)$, then the head will be at $R=(3,1)$. By "translate," we mean to move a vector (i.e., directed line segment) without changing its magnitude or direction. In the above diagram, the vector $\vec{OR}$ is the translation of the vector $\vec{PQ}$ to the origin. Since all translations of a vector have the same magnitude and direction, they are said to be equivalent to one another.

If the tail of a vector is fixed at the origin, then the vector is completely determined by the position of its head. Thus in the plane, a vector (with its tail at the origin) is completely determined by giving the coordinates (a,b) of the point that is the head of the vector. Note that there is a one-to-one correspondence between vectors in the plane with tails at the origin and pairs of real numbers [i.e., the pair of real numbers (3,1) represents a point in the plane and also represents the vector $\vec{OR}$ with its tail at the origin (0,0) and its head at (3,1)]. A similar one-to-one correspondence between vectors and points exists in all dimensions.

Since all vectors with the same magnitude and direction are equivalent, the coordinates (a, b) of the point R may *also* represent all vectors equivalent to $\vec{OR}$. Given a geometric vector, its *coordinate representation* (or *analytic representation) is the pair of real numbers that give the coordinates of the head of the vector when the tail is translated to the origin.* This situation is analogous to the representation of fractions. Consider $\frac{2}{4}$ and $\frac{1}{2}$. They both represent the same number, one-half. Just as the fraction $\frac{1}{2}$ "represents" the fractions $\frac{1}{2}, \frac{2}{4}, \frac{3}{6}$, etc., so does the vector with its tail at the origin "represent" all vectors with the same length and direction.

Consider another example, the vector $\vec{AB}$, where $A = (1, -1)$ and $B = (-3, -3)$. What is the coordinate representation of $\vec{AB}$? Translating $\vec{AB}$ to the origin means that the x-coordinate of both the head and the tail must be reduced by the x-coordinate of the tail, and the y-coordinate of both the head and the tail must be reduced by the y-coordinate of the tail. When we do this, the x-coordinate of the head of the translated vector is $-3 - (1) = -4$ and the y-coordinate of the head of the translated vector is $-3 - (-1) = -2$. The coordinate representation of $\vec{AB}$ is $(-4, -2)$.

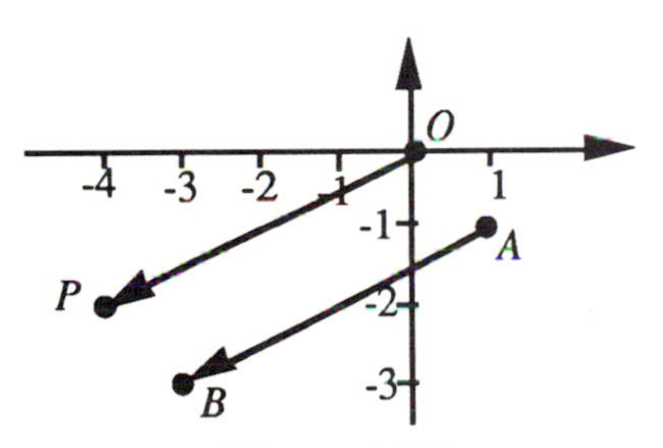

Figure 4.1.3

In general, if a vector in the xy-plane has its tail at $P = (x_1, y_1)$ and its head at $Q = (x_2, y_2)$, then the coordinate representation of the vector $\vec{PQ}$ is $(x_2 - x_1, y_2 - y_1)$. When discussing vectors with tails at the origin, we will suppress the O indicating the position of the tail. Thus the vector $\vec{OR}$ will be written $\vec{R}$.

Question 4.1.4 What is the coordinate representation of the vector with its tail at $(-1, 3)$ and its head at $(2, 0)$? ◇

The same representation of vectors that has been discussed for the plane is used in three or more dimensions. Let $\vec{PQ}$ be a vector in $\mathbb{R}^3$ with its tail at $P = (x_1, y_1, z_1)$ and its head at $Q = (x_2, y_2, z_2)$. When the vector is translated to the origin (0,0,0), we obtain a vector $\vec{OR}$, or simply $\vec{R}$, with head at $(x_2 - x_1, y_2 - y_1, z_2 - z_1)$, so $\vec{R} = (x_2 - x_1, y_2 - y_1, z_2 - z_1)$.

An advantage of representing vectors analytically is the ease in computing the length or magnitude of the vector. (The *magnitude* of a vector is another term for the length of a vector. We shall use both terms.) The length of the vector $\vec{PQ}$ is the distance between the points P and Q as described in Chapter 1. Thus if $P = (x_1, y_1)$ and $Q = (x_2, y_2)$, then the length of the vector $\vec{PQ}$ is $[(x_2 - x_1)^2 + (y_2 - y_1)^2]^{1/2}$. If $\vec{PQ}$ is translated to the origin, then the translated vector $\vec{V}$ has the same length and direction as $\vec{PQ}$. The coordinate representation of $\vec{V}$ is $(x_2 - x_1, y_2 - y_1) = (x, y)$, so the length of $\vec{V}$ is $[x^2 + y^2]^{1/2}$. We will denote the length of a vector $\vec{V}$ by $|\vec{V}|$, similar to the way absolute value represents the magnitude of a real number.

In Figure 4.1.2

$$|\vec{PQ}| = |\vec{OR}| = \sqrt{(\Delta x)^2 + (\Delta y)^2} = \sqrt{3^2 + 1^2} = \sqrt{10}$$

In Figure 4.1.3

$$|\vec{AB}| = |\vec{OP}| = \sqrt{(\Delta x)^2 + (\Delta y)^2} = \sqrt{4^2 + 2^2} = \sqrt{20}$$

Definition 4.1.5 (Length of Vector) *If $\vec{V} = (v_1, v_2, \ldots, v_n)$ is a vector in $\mathbb{R}^n$ with coordinates v_i, then the length of the vector, written $|\vec{V}|$, is the same as the distance of the point $(v_1, v_2, \ldots, v_n)$ to the origin as defined in Chapter 1. Thus*

$$|\vec{V}| = [v_1^2 + v_2^2 + \cdots + v_n^2]^{1/2}$$

Another advantage of representing vectors in the plane by coordinates is in expressing the direction of the vector. We can identify the direction with the slope of the line segment from the tail to the head. (An exception arises when the vector is vertical, since the slope is then ∞ or $-\infty$.) If the vector $\vec{PQ}$ has the analytic representation $\vec{V} = (x, y)$, then $\vec{PQ}$ and $\vec{V}$ have the same direction. The common direction is y/x, the slope of the line segment joining (0,0) with (x, y).

Example 4.1.6 We will find the length (magnitude) and direction of the vector $\vec{AB}$ where $A = (1, -1)$ and $B = (-3, 3)$.

The length is:

$$|\vec{AB}| = [(-3 - 1)^2 + (3 - (-1))^2]^{1/2} = [16 + 16]^{1/2} = 4\sqrt{2}$$

and the direction is $\vec{AB} = \frac{3-(-1)}{-3-1} = -1$. △

Question 4.1.7 What is the magnitude and direction of the vector $\vec{PQ}$ where $P = (1, 2)$ and $Q = (4, 9)$? ◇

The Algebra of Vectors

The primary advantage of representing vectors by tuples of real numbers is that it allows us to extend the algebra of real numbers to an algebra of vectors. Our results will apply to spaces of any number of dimensions, but for the convenience of computation and diagrams we will usually use two or three dimensions in the examples. For vectors in two- or three-dimensional space it is common to represent the coordinates of a vector $\vec{V}$ by (x, y) in two dimensions or (x, y, z) in three dimensions. This notation does not extend well beyond three dimensions, so an alternate notation, subscripting, is also used: $\vec{V} = (v_1, v_2)$.

Let $\vec{A} = (a_1, a_2)$ and $\vec{B} = (b_1, b_2)$ be two vectors in the xy-plane. We can add the vectors $\vec{A}$ and $\vec{B}$ to form $\vec{A}+\vec{B}$ geometrically by translating $\vec{B}$ so that the tail of $\vec{B}$ is at the head of $\vec{A}$. The vector $\vec{A}+\vec{B}$ is the diagonal of the parallelogram determined by $\vec{A}$ and $\vec{B}$. This is called the *Parallelogram Rule.*

Addition of Vectors

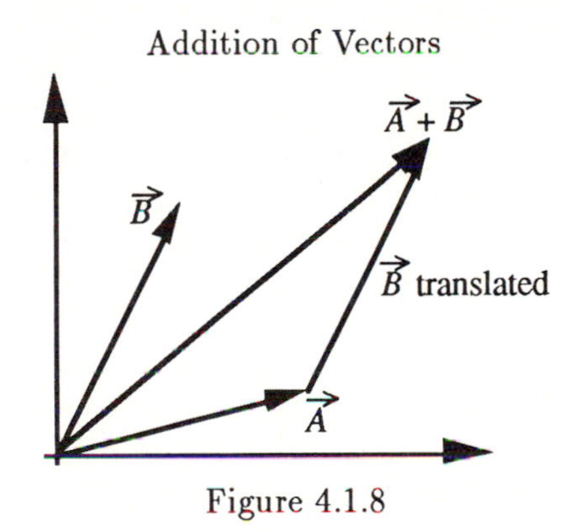

Figure 4.1.8

The head of $\vec{A}+\vec{B}$ is at $(a_1 + b_1, a_2 + b_2)$, so $\vec{A}+\vec{B} = (a_1 + b_1, a_2 + b_2)$. For example, the sum of $\vec{A} = (2, 3)$ and $\vec{B} = (-4, 5)$ is $\vec{A}+\vec{B} = (2 + (-4), 3 + 5) = (-2, 8)$. Notice that vectors are added coordinatewise: *to add two vectors, add the corresponding coordinates.*

The above diagram is familiar to anyone who has paddled a canoe across a lake on a windy day. Let $\vec{A}$ denote the paddling vector and $\vec{B}$ denote the wind vector. Then $\vec{A}+\vec{B}$ represents the resulting canoe vector. Another water example is swimming across a river that has a strong current. Let $\vec{A}$ be the current vector and let $\vec{B}$ be the swimming vector. Then $\vec{A}+\vec{B}$ is the resultant vector of the swimmer.

A vector $\vec{V} = (v_1, v_2)$ can be multiplied by a real number r in the following manner. If the real number r is positive, then $r\vec{V}$ is the vector that has the same direction as $\vec{V}$ and whose length is r times the length of $\vec{V}$. If the real number r is negative, then $r\vec{V}$ will be the vector that has the opposite direction as $\vec{V}$ and whose length is $|r|$ times the length of $\vec{V}$.

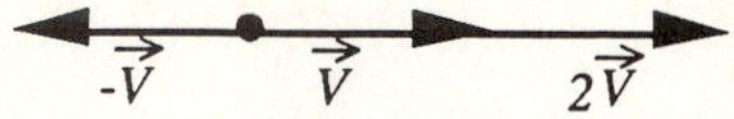

$-\vec{V}$ $\vec{V}$ $2\vec{V}$

Figure 4.1.9

To multiply a vector $\vec{V} = (v_1, v_2)$ by a real number r, the operation of multiplication by r is applied to each coordinate separately, so that $r\vec{V} = (rv_1, rv_2)$. What happens when $r = 0$? Applying the operation, $r\vec{V} = 0\vec{V} = (0,0)$. $\vec{O} = (0,0)$ is a vector of length zero. The concept of direction does not apply to this special vector $\vec{O}$ since the ratio $y/x = 0/0$ is undefined.

Example 4.1.10 If $\vec{V} = (-2,1)$, we will find $3\vec{V}$.
$3\vec{V} = 3 \cdot (-2,1) = (3(-2), 3(1)) = (-6,3)$ △

Question 4.1.11 What is -4 times the vector $\vec{PQ}$ where $P = (-1,1)$ and $Q = (-3,6)$? ◇

We now give the formal definition of how to add vectors and multiply vectors by real numbers. Vectors have both magnitude and direction. Quantities, such as real numbers, which have only magnitude are called *scalars*.

Definition 4.1.12 (Vector Addition and Scalar Multiplication)
Let $A = (a_1, a_2, ..., a_n)$ and $B = (b_1, b_2, ..., b_n)$ be vectors in $\mathbb{R}^n$ and let r be a real number. We define vector addition *by*

$$\vec{A} + \vec{B} = (a_1 + b_1, a_2 + b_2, ..., a_n + b_n)$$

and scalar multiplication *by*

$$r\vec{A} = (ra_1, ra_2, ..., ra_n)$$

Occasionally we will write the scalar on the right:

$$\vec{A}r = r\vec{A}$$

Vector addition combines two vectors to give a third vector. Scalar multiplication combines a real number and a vector to give a new vector. Notice that the operations are coordinatewise; to get the third coordinate of the sum we add the third coordinates of $\vec{A}$ and $\vec{B}$.

Example 4.1.13 Let $\vec{A} = (1,2,3,4)$ and $\vec{B} = (-2,3,0,9)$ be vectors in four dimensional space, then

$$\vec{A} + \vec{B} = (1 + (-2), 2 + 3, 3 + 0, 4 + 9) = (-1, 5, 3, 13) \in \mathbb{R}^4$$

and

$$3\vec{A} = (3 \cdot 1, 3 \cdot 2, 3 \cdot 3, 3 \cdot 4) = (3, 6, 9, 12) \in \mathbb{R}^4$$ △

Subtraction of vectors is defined by combining these two operations:

$$\vec{A} - \vec{B} = \vec{A} + (-1)\vec{B}$$

Question 4.1.14 (a) What is the geometric relation between $\vec{A}$ and $-\vec{A}$?
(b) What is the geometric interpretation of $\vec{B} - \vec{A}$? ◇

Example 4.1.15 If $\vec{A} = (1,2,3)$ and $\vec{B} = (3,-1,2)$, then

$$\begin{aligned} 2\vec{A} - 4\vec{B} &= 2(1,2,3) - 4(3,-1,2) \\ &= (2\cdot 1, 2\cdot 2, 2\cdot 3) - (4\cdot 3, 4\cdot(-1), 4\cdot 2) \\ &= (2,4,6) - (12,-4,8) \\ &= (2,4,6) + (-1)(12,-4,8) = (2,4,6) + (-12,4,-8) \\ &= (-10,8,-2) \end{aligned}$$

△

By using the properties of the arithmetic operations of real numbers, we can obtain similar rules for the algebra of vectors.

Theorem 4.1.16 (Algebra of Vectors) *Let $\vec{A}$, $\vec{B}$, and $\vec{C}$ be vectors, r and s real numbers. Then we have the following rules for the algebra of vectors:*

(a) $\vec{A} + \vec{B} = \vec{B} + \vec{A}$ (commutative law for vector addition)

(b) $(\vec{A} + \vec{B}) + C = \vec{A} + (\vec{B} + \vec{C})$ (associative law for vector addition)

(c) $r(s\vec{A}) = s(r\vec{A})$ (commutative law for scalar multiplication)

(d) $r(s\vec{A}) = (rs)\vec{A}$ (associative law for scalar multiplication)

(e) $(r+s)\vec{A} = r\vec{A} + s\vec{A}$ (distributive law)

(f) $r(\vec{A} + \vec{B}) = r\vec{A} + r\vec{B}$ (distributive law)

(g) $\vec{A} + \vec{O} = \vec{A}$ (a zero element exists)

(h) $0\vec{A} = \vec{O}$

(i) $1\vec{A} = \vec{A}$

(j) $|r\vec{A}| = |r||\vec{A}|$

(k) $|\vec{A} + \vec{B}| \leq |\vec{A}| + |\vec{B}|$ (the Triangle Inequality)

The theorem holds for vectors in any dimensional space $\mathbb{R}^n$. Note that division by a vector is not defined.

The triangle inequality, rule (g), is one of the most frequently used tools in mathematical analysis. It basically says that the shortest distance between two points is a straight line.

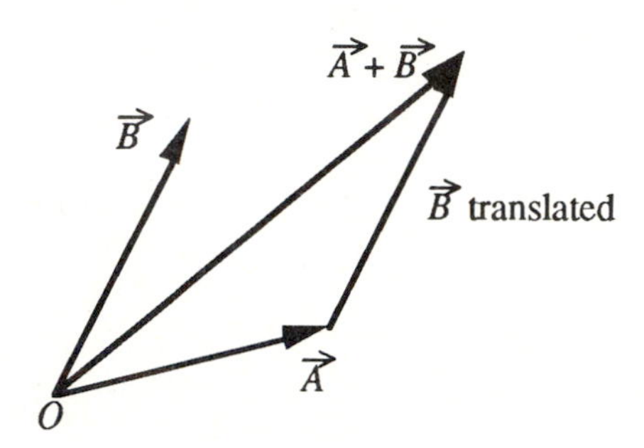

Figure 4.1.17

The distance between O and the head of $\vec{A}+\vec{B}$ is less than the distance from O to the head of $\vec{A}$ plus the distance from the head of $\vec{A}$ along the vector $\vec{B}$ to the head of $\vec{A}+\vec{B}$.

Proof: The proofs of most of these statements are similar. We will give the proof of rule (a). The first thing to do is to reduce the problem to working with real numbers by expressing the vectors analytically in terms of coordinates. Assume that the vectors are in the plane, so that we identify points in the plane, vectors with tails at the origin, and pairs of real numbers in $\mathbb{R}^2$. Let $\vec{A} = (a_1, a_2)$ and $\vec{B} = (b_1, b_2)$. Then

$$\vec{A}+\vec{B} = (a_1 + b_1, a_2 + b_2)$$

Now use the fact that commutativity holds for real numbers:

$$a_1 + b_1 = b_1 + a_1, a_2 + b_2 = b_2 + a_2$$

This gives

$$(a_1 + b_1, a_2 + b_2) = (b_1 + a_1, b_2 + a_2) = \vec{B} + \vec{A} \qquad \square$$

Question 4.1.18 (a) Prove rule (c) of Theorem 4.1.16.

(b) Show that if $|\vec{V}| = 0$, then $\vec{V} = \vec{O}$. $\diamond$

Using the rules of vector algebra, computations involving arbitrary vectors can be reduced to computations with certain natural vectors related to the coordinate system. If we are dealing with three-dimensional space, then there are three special vectors, the coordinate vectors along the coordinate axes:

$$\vec{e_1} = (1,0,0), \ \vec{e_2} = (0,1,0), \quad \text{and} \quad \vec{e_3} = (0,0,1)$$

The vector $\vec{e_1}$ is often called $\vec{i}$, $\vec{e_2}$ is often called $\vec{j}$, and $\vec{e_3}$ is often called $\vec{k}$. The $\vec{i}$, $\vec{j}$, $\vec{k}$ notation is useful in two or three dimensions but does not extend to arbitrarily many dimensions. Both notations will be used. These three vectors are vectors of length 1 (unit length) along the coordinate axes. If $\vec{V} = (v_1, v_2, v_3)$ is any vector in three-dimensional space, then $\vec{V}$ can be written as a *linear combination* of the coordinate vectors:

$$\vec{V} = v_1\vec{e_1} + v_2\vec{e_2} + v_3\vec{e_3}$$

Conversely, any linear combination of the three coordinate vectors $\vec{e_1}$, $\vec{e_2}$, and $\vec{e_3}$ is a vector in three-dimensional space. (Why?)

In the plane we have two coordinate vectors $\vec{e_1} = (1,0)$ and $\vec{e_2} = (0,1)$ along the coordinate axes. These vectors are also denoted $\vec{i} = \vec{e_1}$ along the x-axis and $\vec{j} = \vec{e_2}$ along the y-axis.

Example 4.1.19 We will write the vector $\vec{PQ}$ with $P = (0,2)$ and $Q = (2,1)$ in terms of the coordinate vectors. First we write $\vec{PQ}$ in terms of coordinates by translating the tail to the origin, obtaining $(2-0, 1-2) = (2,-1)$. Now we write

$$(2,-1) = 2(1,0) + (-1)(0,1) = 2\vec{e_1} + (-1)\vec{e_2}$$ △

Example 4.1.20 A *unit vector* is a vector of length 1. Since multiplying a vector by a positive scalar does not change the direction of the vector, we can find a unit vector in the direction of $\vec{V} = (1,2,-1)$ by dividing $\vec{V}$ by its length (i.e., multiplying $\vec{V}$ by the reciprocal of its length). First we find the length of $\vec{V}$:

$$|\vec{V}| = (1^2 + 2^2 + (-1)^2)^{1/2} = (1+4+1)^{1/2} = \sqrt{6}$$

Now we divide $\vec{V}$ by its length, obtaining a unit vector in the direction of $\vec{V}$:

$$\frac{\vec{V}}{|\vec{V}|} = \frac{\vec{V}}{\sqrt{6}} = \left(\frac{1}{\sqrt{6}}, \frac{2}{\sqrt{6}}, \frac{-1}{\sqrt{6}}\right)$$ △

To summarize, we have the following special vectors:

1. If $|\vec{A}| = 1$, then $\vec{A}$ is called a *unit vector.*

2. The $\vec{e_i}$ are called *coordinate vectors.* In three dimensions, the coordinate vectors are also denoted by

$$\begin{aligned} \vec{i} &= \vec{e_1} = (1,0,0) \text{ along the } x\text{-axis} \\ \vec{j} &= \vec{e_2} = (0,1,0) \text{ along the } y\text{-axis} \\ \vec{k} &= \vec{e_3} = (0,0,1) \text{ along the } z\text{-axis} \end{aligned}$$

3. $\vec{O} = (0,0)$ is the *zero vector.*

Question 4.1.21 Show that if $\vec{V}$ is any non-zero vector, then $\frac{\vec{V}}{|\vec{V}|}$ is a unit vector. ◇

The Dot Product of Vectors

Vector addition combines vectors to give vectors and scalar multiplication combines a vector and a real number to give a vector. There is a third way to combine vectors that is useful.

Definition 4.1.22 (Dot Product) *Let $\vec{A} = (a_1, a_2, ..., a_n)$ and $\vec{B} = (b_1, b_2, ..., b_n)$ be two vectors in $\mathbb{R}^n$. The* dot product *(or* inner product*) of $\vec{A}$ and $\vec{B}$ is*

$$\vec{A}\bullet\vec{B} = a_1b_1 + a_2b_2 + \cdots + a_nb_n$$

The dot notation extends the usual notation for the product of two numbers, e.g., $2 \cdot 3 = 6$. Note that the dot product of two vectors combines two vectors to give a real number.

Example 4.1.23 Let $\vec{A} = (1,2)$ and $\vec{B} = (-2,4)$ be two vectors in the plane. Then

$$\vec{A}\bullet\vec{B} = 1 \cdot (-2) + 2 \cdot 4 = -2 + 8 = 6$$

△

Example 4.1.24 We will model buying ice cream using dot products of vectors. Suppose a group buys 10 quarts of ice cream: 1 vanilla, 3 pineapple, 2 blueberry, and 4 strawberry, at unit costs of \$1.50, \$2.00, \$2.25, and \$1.75. Let the "order vector" for the amounts of ice cream be $\vec{A} = (1,3,2,4)$ and the "unit cost vector" be $\vec{C} = (1.50, 2.00, 2.25, 1.75)$. Then the total cost of the ice cream is

$$\vec{A}\bullet\vec{C} = 1 \cdot 1.50 + 3 \cdot 2.00 + 2 \cdot 2.25 + 4 \cdot 1.75 = 19.00$$

△

Since a new way to combine vectors has been introduced, we must establish the algebraic rules the dot product satisfies and see how the dot product relates to the vector sum and scalar product.

Theorem 4.1.25 (Basic Properties of Dot Product) *Let $\vec{A}$, $\vec{B}$, and $\vec{C}$ be vectors; r is a real number. Then the following properties hold:*

(a) $\vec{A}\bullet\vec{B} = \vec{B}\bullet\vec{A}$ (symmetry)

(b) $(\vec{A} + \vec{B})\bullet\vec{C} = \vec{A}\bullet\vec{C} + \vec{B}\bullet\vec{C}$ (distributive law)

(c) $(r\vec{A})\bullet\vec{B} = r(\vec{A}\bullet\vec{B})$ (associative law)

(d) $\vec{O}\bullet\vec{A} = 0$

(e) $|\vec{A}|^2 = \vec{A}\bullet\vec{A}$

Note, again, that division by a vector is not defined.

Proof: The proofs of these results are immediate from the definitions. We will give the proof of property (e). Assume that the vectors are in the plane and that $\vec{A} = (a_1, a_2)$. Then the square of the length of $\vec{A}$ is

$$|\vec{A}|^2 = a_1^2 + a_2^2 = (a_1, a_2) \cdot (a_1, a_2) = \vec{A} \cdot \vec{A} \qquad \square$$

Question 4.1.26 Prove property (a) of Theorem 4.1.25. ◇

Question 4.1.27 Prove that

$$(\vec{A} + \vec{B}) \cdot (\vec{A} + \vec{B}) = \vec{A} \cdot \vec{A} + 2\vec{A} \cdot \vec{B} + \vec{B} \cdot \vec{B} \qquad \diamond$$

Theorem 4.1.28 (Cauchy-Schwarz-Bunyakovsky equality) *Let $\vec{A}$ and $\vec{B}$ be two vectors. Then the* Cauchy-Schwarz-Bunyakovsky Equality *holds:*

$$\vec{A} \cdot \vec{B} = |\vec{A}||\vec{B}| \cos(\theta)$$

where θ is the angle between $\vec{A}$ and $\vec{B}$.

Proof: Notice that two vectors in three or higher dimensional space determine the (two-dimensional) plane in which they lie. This means that we can assume that the initial vectors are in the plane.

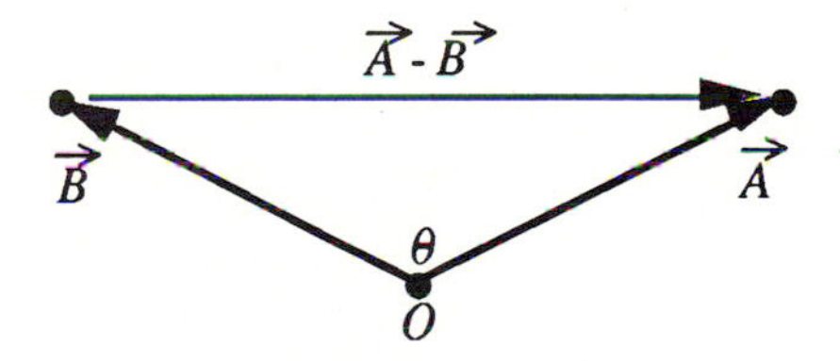

Figure 4.1.29

The law of cosines (see Equation 11 in the Appendix on Trigonometry) and the result in Question 4.1.14 applied to Figure 4.1.29 say that:

$$|\vec{A} - \vec{B}|^2 = |\vec{A}|^2 + |\vec{B}|^2 - 2|\vec{A}||\vec{B}| \cos(\theta)$$

Applying property (e) of Theorem 4.1.25, we have:

$$|\vec{A} - \vec{B}|^2 = (\vec{A} - \vec{B}) \cdot (\vec{A} - \vec{B}) = \vec{A} \cdot \vec{A} - 2\vec{A} \cdot \vec{B} + \vec{B} \cdot \vec{B}$$

(by Question 4.1.27). Substituting into the law of cosines, we have:

$$\vec{A} \cdot \vec{A} - 2\vec{A} \cdot \vec{B} + \vec{B} \cdot \vec{B} = |\vec{A}|^2 + |\vec{B}|^2 - 2|\vec{A}||\vec{B}| \cos(\theta)$$

Again applying property (e) gives:

$$-2\vec{A} \cdot \vec{B} = -2|\vec{A}||\vec{B}| \cos(\theta) \qquad \square$$

Question 4.1.30 Prove the Cauchy-Schwarz-Bunyakovsky *in*equality: For any two vectors $\vec{A}$ and $\vec{B}$, $|\vec{A}\cdot\vec{B}| \leq |\vec{A}||\vec{B}|$. $\diamond$

The Cauchy-Schwarz-Bunyakovsky equality can be used to compute geometric properties of vectors.

Example 4.1.31 Given $\vec{A} = (1,2,3)$ and $\vec{B} = (2,0,1)$, we can compute the cosine of the angle θ between them.

$$\cos(\theta) = \frac{\vec{A}\cdot\vec{B}}{|\vec{A}||\vec{B}|} = \frac{(2+0+3)}{\sqrt{14}\sqrt{5}} = \sqrt{\frac{5}{14}} \qquad \triangle$$

The dot product and the Cauchy-Schwarz-Bunyakovsky equality can be interpreted in terms of the projection of one vector upon another. Consider two vectors $\vec{A}$ and $\vec{B}$ based at the origin.

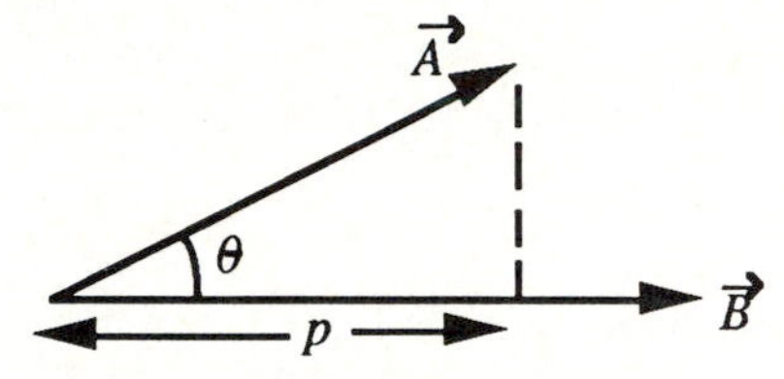

Figure 4.1.32

The length p of the projection of $\vec{A}$ onto $\vec{B}$ is given by the trigonometric relation

$$\cos(\theta) = \frac{p}{|\vec{A}|} \quad \text{or} \quad p = |\vec{A}|\cos(\theta)$$

By the Cauchy-Schwarz-Bunyakovsky equality,

$$\vec{A}\cdot\vec{B} = |\vec{A}||\vec{B}|\cos(\theta) = p|\vec{B}|$$

Thus the length of the projection of $\vec{A}$ onto $\vec{B}$ is

$$p = \frac{\vec{A}\cdot\vec{B}}{|\vec{B}|}$$

The length of the projection of $\vec{A}$ onto $\vec{B}$ is called the *component* of $\vec{A}$ in the direction of $\vec{B}$.

Note that p can be either positive (if $\vec{A}$ and $\vec{B}$ have the same direction) or negative (if they have opposite directions).

Example 4.1.33 The length of the projection of $\vec{A} = (1,2,3)$ onto $\vec{B} = (2,0,3)$ is

$$p = \frac{(1,2,3)\cdot(2,0,3)}{|(2,0,3)|} = \frac{11}{\sqrt{13}}$$

Thus the component of of $\vec{A}$ in the direction of $\vec{B}$ is $\frac{11}{\sqrt{13}}$. $\triangle$

Example 4.1.34 A track meet offers an opportunity to apply vector calculus. A sprinting record is valid provided the wind speed in the direction of the runners does not exceed 2 mph. How do you determine wind speed in a given direction? Since velocity is a vector (it has both magnitude and direction; its length is the *speed*), the component of wind velocity can be determined for a given direction.

A 100-yard sprint is to be run on a straight track in the direction of the vector (2,3). Suppose the wind velocity is (2,1). Under these conditions, would a record sprint be valid?

The wind speed in the direction of the sprint is the length of the projection of (2,1) onto (2,3), i.e., $(4+3)/\sqrt{13} \approx 1.94$. Thus the record would be valid. △

To obtain the projection of $\vec{A}$ in the direction of $\vec{B}$, we take the length of the projection of $\vec{A}$ onto $\vec{B}$ and multiply by a unit vector in the direction of $\vec{B}$, $\vec{B}/|\vec{B}|$. Thus the projection of $\vec{A}$ in the direction of $\vec{B}$ is

$$Proj(\vec{A}, \vec{B}) = \frac{\vec{A} \bullet \vec{B}}{|\vec{B}|} \frac{\vec{B}}{|\vec{B}|} = \frac{\vec{A} \bullet \vec{B}}{\vec{B} \bullet \vec{B}} \vec{B}$$

Example 4.1.35 The projection of $\vec{A} = (1,2,3)$ in the direction of $\vec{B} = (2,0,3)$ is $(11/13)(2,0,3)$. △

Definition 4.1.36 (Geometry of Vectors) *Two vectors are* parallel *if they have the same direction. Two vectors are* antiparallel *if they have opposite directions. Two vectors are* perpendicular *(or* orthogonal*) if the angle between them is $\pi/2$ radians (90°).*

Let $\vec{A}$ and $\vec{B}$ be two nonzero vectors. Then $\vec{A}$ and $\vec{B}$ are parallel if the angle θ between them is 0 radians, i.e., $\cos(\theta) = 1$. Consequently, from the Cauchy-Schwarz-Bunyakovsky equality, two vectors are parallel if and only if $\vec{A} \bullet \vec{B} = |\vec{A}||\vec{B}|$. If $\vec{A}$ and $\vec{B}$ are unit vectors, then they are parallel if and only if $\vec{A} \bullet \vec{B} = 1$.

Similarly, two unit vectors are antiparallel if and only if their dot product is -1.

Two vectors $\vec{A}$ and $\vec{B}$ are perpendicular if and only if the angle θ between them is $\pi/2$, i.e., $\cos(\theta) = 0$. Thus two nonzero vectors $\vec{A}$ and $\vec{B}$ are perpendicular if and only if $\vec{A} \bullet \vec{B} = 0$.

Example 4.1.37 Let $\vec{PQ}$ be the vector with its tail at $P = (0,1)$ and its head at $Q = (2,3)$. Let $\vec{RS}$ be the vector with its tail at $R = (1,-2)$ and its head at $S = (3,-4)$. Are these two vectors perpendicular? First we translate them to the origin (obtain their coordinate representation). The first vector $\vec{PQ}$ is equivalent to $\vec{A} = (2,2)$. The second vector $\vec{RS}$ is equivalent to $\vec{B} = (2,-2)$. $\vec{A} \bullet \vec{B} = 2 \cdot 2 - 2 \cdot 2 = 0$, so $\vec{A}$ and $\vec{B}$ are perpendicular. Since translation of a vector does not change its direction, $\vec{PQ}$ and $\vec{RS}$ are also perpendicular. △

Translating Perpendicular Vectors

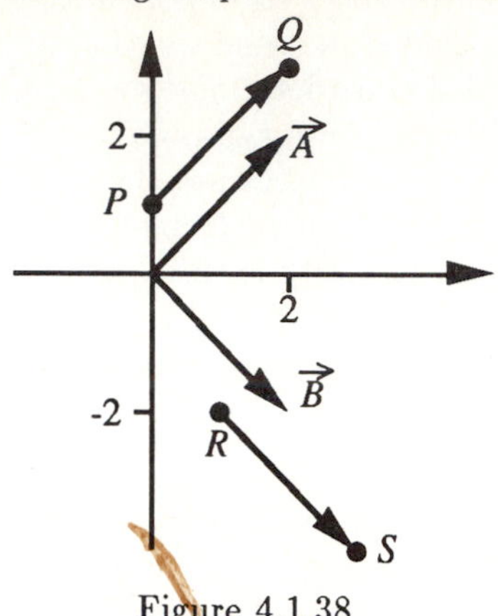

Figure 4.1.38

Example 4.1.39 We will decompose $\vec{A} = (2,4)$ into a sum of two vectors, one parallel to $\vec{B} = (1,1)$ and one perpendicular to $\vec{B}$.

Geometrically this is done by "dropping" a perpendicular from the head of $\vec{A}$ to a line extended along $\vec{B}$ as shown in the following diagram.

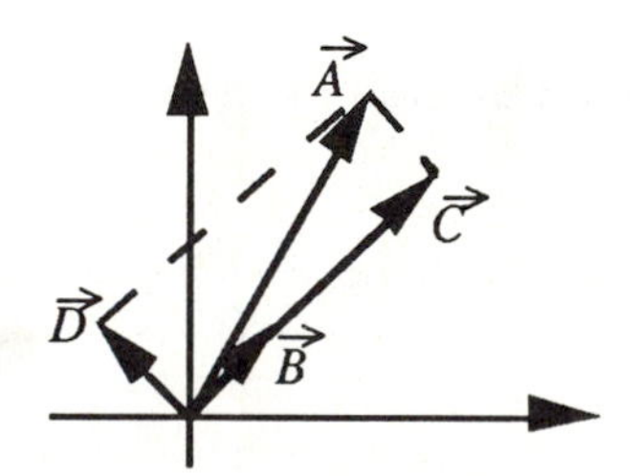

Figure 4.1.40

Since the origin, D, A, and C are the vertices of a parallelogram,

$$\vec{A} = \vec{C} + \vec{CA}$$

(by the Parallelogram Rule) where $\vec{C} = Proj(\vec{A}, \vec{B})$ and $\vec{CA}$ is perpendicular to $\vec{B}$ by construction.

We now solve the problem analytically using the above geometry as a guide.

$$\vec{C} = Proj(\vec{A}, \vec{B}) = \frac{\vec{A} \bullet \vec{B}}{|\vec{B}|} \frac{\vec{B}}{|\vec{B}|} = \frac{\vec{A} \bullet \vec{B}}{\vec{B} \bullet \vec{B}} \vec{B} = (3,3)$$

To find the component of $\vec{A}$ perpendicular to $\vec{B}$, we subtract $Proj(\vec{A}, \vec{B})$ from $\vec{A}$. That is

$$\vec{D} = \vec{A} - Proj(\vec{A}, \vec{B}) = (-1,1)$$

Thus

$$\vec{A} = (2,4) = (3,3) + (-1,1) = \vec{C} + \vec{D}$$

△

Section 4.1 Exercises

1. Where is the head of the vector $\vec{PQ}$ after it is translated to the origin if $P = (1,2)$ and $Q = (3,-1)$?

2. Where is the head of the vector $\vec{PQ}$ after it is translated to the origin if $P = (-2,-3)$ and $Q = (0,0)$?

3. What is the length of the vector $\vec{PQ}$ in $\mathbb{R}^3$ if $P = (1,1,0)$ and $Q = (2,2,3)$?

4. What is the length of the vector $\vec{PQ}$ in $\mathbb{R}^4$ if $P = (0,1,2,3)$ and $Q = (1,2,2,5)$?

5. If $\vec{A} = (1,2,3)$ and $\vec{B} = (2,0,-2)$, what is $3\vec{A} - 2\vec{B}$?

6. If $\vec{A} = (2,3)$ and $\vec{B} = (2,0)$, what is $4\vec{A} + 3\vec{B}$?

7. Write the vector $\vec{PQ}$ where $P = (1,0,1)$ and $Q = (-2,1,1)$ in terms of the coordinate vectors $\vec{i}$, $\vec{j}$, and $\vec{k}$.

8. Write the vector $\vec{PQ}$ where $P = (1,3,4)$ and $Q = (-2,1,0)$ in terms of the coordinate vectors $\vec{i}$, $\vec{j}$, and $\vec{k}$.

9. Find a unit vector in the direction of $\vec{A} = (2,3,4)$.

10. Find a unit vector in the direction of $\vec{A} = (1,2,3,4)$.

11. Prove Theorem 4.1.30.

12. Two vectors having the same length and same direction are said to be equivalent, although they may have different locations. Two fractions having the same numerical value are said to be equivalent, although they may look different (e.g., 1/2 and 7/14). Give two examples of systems in which different elements are defined to be equivalent if they share a common property.

13. What is the angle between $\vec{A} = (1,2,1)$ and $\vec{B} = (-2,3,1)$?

14. What is the angle between $\vec{A} = (1,2)$ and $\vec{B} = (3,1)$?

15. We have discussed one way to multiply vectors, the dot product, which combines two vectors to give a real number. For vectors in $\mathbb{R}^3$ there is another product, called the *vector product* (or *cross product*), which combines two vectors to give a vector. If $\vec{A} = (a_1, a_2, a_3)$ and $\vec{B} = (b_1, b_2, b_3)$, then the cross product is
$$\vec{A} \times \vec{B} = (a_2b_3 - a_3b_2, a_3b_1 - a_1b_3, a_1b_2 - a_2b_1)$$
What is the cross product of $(1,0,0) \times (0,1,0)$?

16. What is the cross product of $(0,1,0) \times (1,0,0)$? Is the cross product commutative, i.e., does $\vec{A} \times \vec{B} = \vec{B} \times \vec{A}$ for all vectors?

17. Show that for any two vectors, $\vec{A} \times \vec{B} = -\vec{B} \times \vec{A}$.

18. Show that the cross product is distributive:
$$\vec{A} \times (\vec{B} + \vec{C}) = \vec{A} \times \vec{B} + \vec{A} \times \vec{C}$$

19. Show that for any three vectors,
$$(\vec{A} \times \vec{B}) \cdot \vec{C} = \vec{A} \cdot (\vec{B} \times \vec{C})$$

20. There is a relationship for the cross product similar to the Cauchy-Schwarz-Bunyakovsky Theorem: $|\vec{A} \times \vec{B}| = |\vec{A}||\vec{B}|\sin(\theta)$ where θ is the angle between the vectors. Using this result, show that two (nonzero) vectors are parallel or antiparallel if and only if their cross product is zero.

21. The cross product has a geometric interpretation. Two vectors $\vec{A}$ and $\vec{B}$ determine a parallelogram.

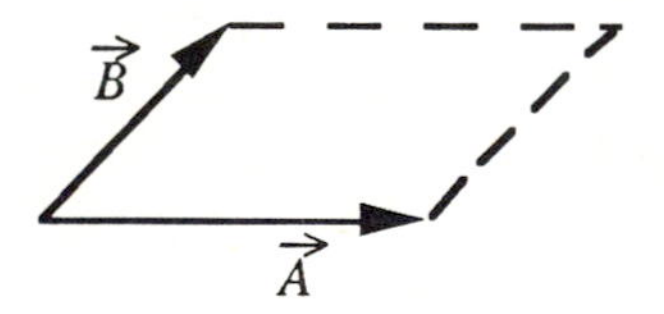

Figure 4.1.1

Show that $|\vec{A} \times \vec{B}|$ is the area of the parallelogram.

22. Prove rules (j) and (k) of Theorem 4.1.16.

23. Prove property (c) of Theorem 4.1.25.

24. (Extension of Example 4.1.39.) Let $Proj(\vec{A}, \vec{B})$ be the (vector) projection of $\vec{A}$ onto $\vec{B}$. Show that the vector $\vec{C} = \vec{A} - Proj(\vec{A}, \vec{B})$ is perpendicular to $\vec{B}$. (Thus given any two vectors $\vec{A}$ and $\vec{B}$, $\vec{A}$ can be decomposed into a component $Proj(\vec{A}, \vec{B})$ parallel to $\vec{B}$ and a component $\vec{A} - Proj(\vec{A}, \vec{B})$ perpendicular to $\vec{B}$.)

25. Let $\vec{V} = (1,2,-4)$. Find the length of the projection of $\vec{V}$ onto

    (a) $\vec{A} = (0,1,0)$

    (b) $\vec{B} = (-2,-2,1)$

    (c) $\vec{C} = (3,-2,4)$

26. Let $\vec{V} = (-5,2,1,3)$. Find a vector $\vec{Z}$ such that the length of the projection of $\vec{V}$ onto $\vec{Z}$ is 4. Is $\vec{Z}$ unique?

27. Let $\vec{V} = (6,1,0,7)$. Find a vector $\vec{Z}$ such that the length of the projection of $\vec{V}$ onto $\vec{Z}$ is zero. Is $\vec{Z}$ unique?

28. With reference to Example 4.1.34 (of the sprint), if the wind maintains the same direction, what is the maximum wind speed which will permit valid records?

29. Let $\vec{V} = (7,2,-1)$. Find a vector $\vec{Z}$ that is perpendicular to $\vec{V}$. Is $\vec{Z}$ unique?

30. State the most general conditions on the vectors (if any) that make the following statements true:

    (a) $\vec{A} \bullet \vec{B} + \vec{C}$ is a vector.

    (b) If $\vec{V}$ is a unit vector, then the vector opposite to $\vec{V}$ has length -1.

    (c) $\vec{A} \bullet \vec{B} = |\vec{A}||\vec{B}|$

31. (a) A cow shed, once a 10 by 10 by 20 box, has developed a lean (along the long side), $\theta°$ from the vertical, and is now a parallelogram when viewed from the side. Find the current height of the shed and the usable volume (that portion where the height is a maximum) as a function of θ.

    (b) A cow is struck by a 5 kg brick thrown by a thoughtless dwarf. The brick strikes descending at an (acute) angle of $\theta°$ from the vertical. If the brick is 20 cm by 8 cm by 2 cm and has a speed of 10 m/sec, find the component of momentum forcing the cow to his knees (recall that momentum is mass times velocity and speed is the length of the velocity vector).

4.2 Derivatives of Vector-Valued Functions

In Chapter 3 we discussed the derivatives of functions with source and target $\mathbb{R}$. In this section the concept of the derivative will be extended to functions from $\mathbb{R}$ into $\mathbb{R}^n$. We will first define the derivative in terms of the difference quotient and then, second, in terms of the Linear Approximation Theorem. Since the two terms in the numerator of the difference quotient, $(f(t+h)-f(t))/h$, are points in $\mathbb{R}^n$, we need to develop an algebra for computing with points in $\mathbb{R}^n$. We will do this by considering functions from $\mathbb{R}$ into $\mathbb{R}^n$ as *vector-valued* functions. Thus we begin our development of differentiation by discussing vector-valued functions and how curves can be parameterized in terms of vector-valued functions.

In the previous section we learned how to identify an n-tuple of real numbers with a vector from the origin to a point in n-dimensional space. Thus a function from $\mathbb{R}$ into $\mathbb{R}^n$ can be considered as a vector-valued function, i.e., the function "inputs" a real number and "outputs" a vector (an n-tuple).

Notation. Sometimes, to stress that a function is vector-valued, we will put an arrow over it: $\vec{f} : \mathbb{R} \to \mathbb{R}^n$. Typically, elementary texts put arrows over vectors, or print the symbol in boldface type, **f**. More advanced texts usually do not, trusting the reader to recall from the definition whether the range is in $\mathbb{R}$ or in a higher dimensional space. We will only occasionally use arrows (or boldface) to stress the fact that a function is vector-valued. Similarly, instead of writing $\vec{v} = (x, y, z)$ we will omit the arrow and write $v = (x, y, z)$. Note the double use of the symbol x: if we have a vector in $\mathbb{R}^n$, we may write $x = (x_1, x_2, ..., x_n)$. Here x is a vector (or point) in $\mathbb{R}^n$. x may also be used for the first coordinate in $\mathbb{R}^2$ or $\mathbb{R}^3$ as above, where $v = (x, y, z)$. It will be clear from the context whether x is a coordinate (a real number) or a vector (an n-tuple of real numbers).

Example 4.2.1 Define the function $SQP : \mathbb{R} \to \mathbb{R}^2$ for all real numbers t by

$$SQP(t) = (t, t^2) = t\vec{e_1} + t^2\vec{e_2} = t\vec{i} + t^2\vec{j}$$

SQP can be considered a vector-valued function (and to stress the fact, write $\overrightarrow{SQP}$ or **SQP**). The value of SQP at t is the vector $t\vec{i} + t^2\vec{j}$. By considering SQP as a vector-valued function, the algebra of vectors developed in the last section can be used in computations. vector-valued functions also have a nice geometric interpretation. If the domain of SQP is considered as time, then SQP gives the position of a particle moving in the plane. $SQP(t)$ is the position at time t.

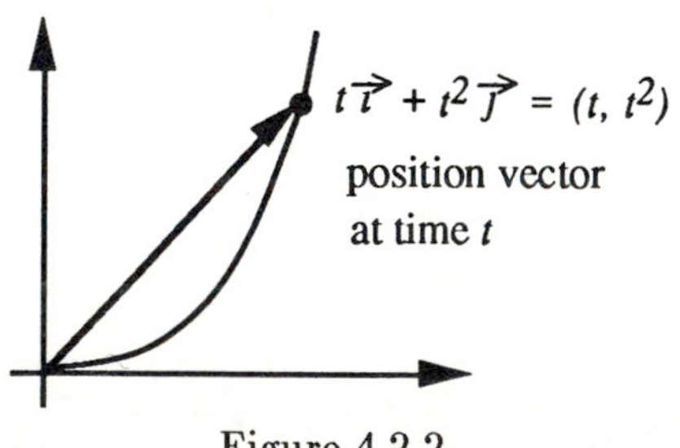

$\triangle$

Figure 4.2.2

Since the dot product of two vectors is a real number, the dot product of two vector-valued functions should be a real valued function. To see that this is the case let $f : \mathbb{R} \to \mathbb{R}^n$ and $g : \mathbb{R} \to \mathbb{R}^n$ be two vector-valued functions. Then a real valued function can be created by taking the dot product of f and g. Define $h : \mathbb{R} \to \mathbb{R}$ by

$$h(t) = (f{\bullet}g)(t) = f(t){\bullet}g(t)$$

Example 4.2.3 We will find the dot product of the two functions $f, g : \mathbb{R} \to \mathbb{R}^2$ given by $f(t) = (t, \sin(t))$ and $g(t) = (\cos(t), 2)$. The dot product $f{\bullet}g : \mathbb{R} \to \mathbb{R}$ is given by

$$f(t){\bullet}g(t) = t \cdot \cos(t) + \sin(t) \cdot 2 = t\cos(t) + 2\sin(t)$$ △

Question 4.2.4 Find the dot product of the two functions $f(t) = (t, t^2, 3t)$ and $g(t) = (\sin(t), 2t, 4)$. ◇

Parameterization

vector-valued functions are often used to describe sets in $\mathbb{R}^n$ as the range of a function. This is called *parameterizing a set as the range of a vector-valued function.*

We illustrate the mechanics of parameterizing a set of points in the following examples. The procedure is to express each of the variables in terms of one variable, called the *parameter.*

Example 4.2.5 The simplest case is when the set to be parameterized is the graph of a function. As an example, we will parameterize the set of points $\{(x, y) : y = x^2\}$.

Let t be the parameterization variable. Set $x = t$. This means that $y = x^2 = t^2$ and the set may be written as $\{(t, t^2) : \text{ for all real t}\}$. This is exactly what is needed to define the vector-valued function $f : \mathbb{R} \to \mathbb{R}^2$ by $f(t) = (t, t^2)$. The range of f is the original set of points. Note that f is just the function SQP, the graph of whose range is pictured in the previous Figure 4.2.1. △

A curve can have many parameterizations. The range of the function $g : \mathbb{R} \to \mathbb{R}^2$ defined by $g(t) = (2t, 4t^2)$ is the same as the range of f in the previous example. If we consider t as time, then replacing t by $2t$ means that the curve is traversed twice as fast.

In general, the simplest way to parameterize the graph of a function $g : \mathbb{R} \to \mathbb{R}$ on an interval $[a, b]$ is to let the parameterization variable be $t = x$. Then the graph is the range of the function $f : [a, b] \to \mathbb{R}$ defined by $f(t) = (t, g(t))$.

Definition 4.2.6 (Parameterization of Curves) *A parameterization of a curve in $\mathbb{R}^n$ is a function $f : I \to \mathbb{R}^n$ where I is an interval in $\mathbb{R}$ and the range $f(I)$ of f is the curve.*

Remember in parameterizing a curve, it is usually helpful to think of the independent variable as time and the range as the path traced out by a moving particle.

Example 4.2.7 (Parameterization of a Directed Line Segment) We will parameterize the directed line segment from point A to point B.

Think of a particle starting at point A and moving along the line segment to point B. The position of the particle at any time can be described by starting at the point A and moving the appropriate distance in the direction toward B (i.e., in the direction of the vector $B - A$). If we let t denote the time the particle has been moving, then the position of the particle P at time t is given by $P(t) = A + t(B - A)$.

Line Parameterization

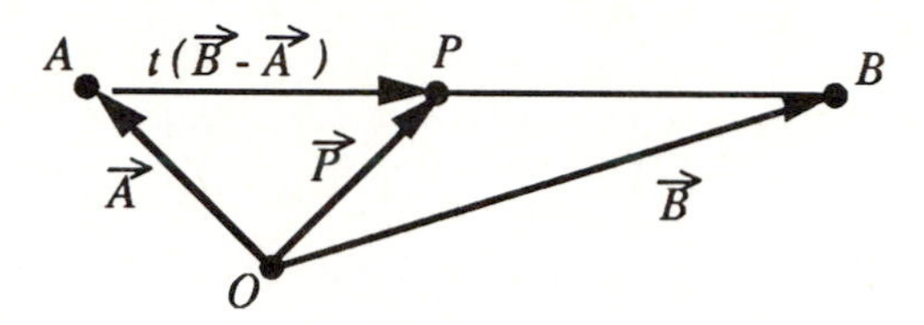

Figure 4.2.8

Note that as t increases from zero to one the entire line segment is traversed, with A being associated with $t = 0$ and B with $t = 1$. Thus if we let t be the parameterization variable, we may parameterize the directed line segment from A to B with the function $f : [0,1] \to \mathbb{R}^2$ defined by $f(t) = P(t) = A+t(B-A)$.△

Example 4.2.9 (Parameterization: Line Segment from (1,4) to (5,2)) If $A = (1,4)$ and $B = (5,2)$, then $f(t) = (1,4) + t((5,2) - (1,4))$. Or, $f(t) = (1,4) + t(4,-2)$.

Note that $f(0) = (1,4) = A$ and $f(1) = (1,4) + (4,-2) = (5,2) = B$. △

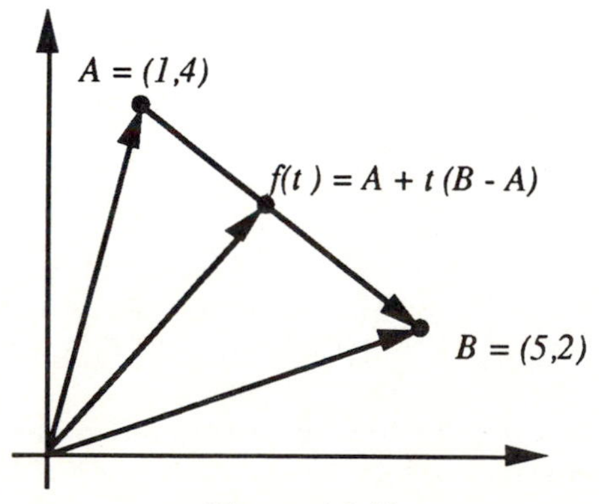

Figure 4.2.10

To parameterize an entire line, rather than just a line segment, we just allow the parameterization variable to take on any real number. Thus, a parameterization of the line determined by two points A and B is $L(t) = A + t(B - A)$, where $L : \mathbb{R} \to \mathbb{R}^n$.

Note that in the expression $L(t) = A + t(B - A)$, $B - A$ is a vector in the direction of the line and A is a point on the line. Thus this expression is the natural generalization of the "point-slope" formula, $y = b + xm$, for a line in the plane.

Example 4.2.11 We will find the parametric form of the straight line determined by (1,2,3) and (2,1,0). Let $A = (1, 2, 3)$ and $B = (2, 1, 0)$. Then by the above discussion a vector in the direction of the line is

$$\vec{AB} = \vec{B} - \vec{A} = (2, 1, 0) - (1, 2, 3) = (1, -1, -3)$$

Let

$$f(t) = \vec{A} + t\vec{AB} = (1, 2, 3) + t(1, -1, -3)$$

Then the range of f is the straight line determined by A and B. △

Example 4.2.12 (A Parameterization of a Circle) How can we parameterize the unit circle of radius 1 centered at the origin (denoted S^1)? We can consider a particle moving around the unit circle, say counterclockwise. The angle is changing; the distance from the center (0,0) remains fixed. So let θ denote the counterclockwise angle between the x-axis and the line joining the origin to (x, y). Then $x = \cos(\theta)$ and $y = \sin(\theta)$. Thus θ is the parameterization variable. To traverse the circle we need θ to range between 0 and 2π radians. We now parameterize the circle by defining a function $CR : \mathbb{R} \to \mathbb{R}^2$ by

$$CR(\theta) = (x, y) = (\cos(\theta), \sin(\theta)), \qquad \text{Domain}(CR) = [0, 2\pi)$$

[Note that 2π is not needed in the domain since it would be sent to the point $(1, 0) = CR(0)$.] △

Example 4.2.13 We will parameterize the curve given by the relation $x + 2y^2 + y = 6$. This is a parabola with its axis parallel to the x-axis. We solve for x in terms of y and obtain $x = 6 - 2y^2 - y$. Thus the curve is $\{(x, y) : x = 6 - 2y^2 - y\}$. Let the parameterization variable be t. Now set $y = t$ and define $f(t) = (6 - 2t^2 - t, t)$ for all real t. The range of f is the desired curve. △

Question 4.2.14 What does the graph of $x + 2y^2 + y = 6$ look like? We know that its a parabola, but where is it's axis of symmetry, and where is its vertex? Hint: see Section 1.6 on transforming the graph of a function. ◇

Example 4.2.15 The function $SPIRAL$ in Example 1.6.18 is a parameterization of a spiral in three-dimensional space. △

Differentiation of vector-valued Functions

In Chapter 2, we showed that a vector-valued function is continuous if and only if each component function is continuous. We will show that a similar result holds for the differentiation of a vector-valued function. The derivative of a vector-valued function can be defined in the same way that the derivative of a real-valued function was defined: the difference quotient still makes sense.

Definition 4.2.16 (Derivative of vector-valued Function) *A vector-valued function f has* derivative $f'(a)$ *at point a in its domain if the limit* $f'(a) = \lim_{h\to 0}(f(a+h) - f(a))/h$ *exists.*

Note that both $f(a)$ and $f'(a)$ are vectors in $\mathbb{R}^n$.

Example 4.2.17 Let $f : \mathbb{R} \to \mathbb{R}^2$ be defined by $f(t) = (t, 2t+3)$ for all t. Then the difference quotient is

$$\begin{aligned} \frac{f(a+h) - f(a)}{h} &= \frac{(a+h, 2a+2h+3) - (a, 2a+3)}{h} = \frac{(h, 2h)}{h} \\ &= \frac{h(1,2)}{h} = (1,2) \end{aligned}$$

Thus

$$f'(a) = \lim_{h\to 0} \frac{f(a+h) - f(a)}{h} = \lim_{h\to 0}(1,2) = (1,2)$$

Note that the derivative $f'(a)$ is a constant vector in $\mathbb{R}^2$. This is analogous to $d(3x+4)/dx$ being the constant 3 in $\mathbb{R}^1 = \mathbb{R}$. △

The problem of computing the derivative of a function $f : \mathbb{R} \to \mathbb{R}^n$ reduces to the problem of computing the derivatives of the coordinate functions. The following theorem formalizes this.

Theorem 4.2.18 *If $f : \mathbb{R} \to \mathbb{R}^n$ has coordinates $f(t) = (f_1(t), f_2(t), ..., f_n(t))$, then f is differentiable at t if and only if every coordinate function f_i is differentiable at t, and in this case*

$$f'(t) = (f_1'(t), f_2'(t), ..., f_n'(t))$$

Proof: Let $f : \mathbb{R} \to \mathbb{R}^2$ for simplicity. As was seen in Chapter 2, Theorem 2.4.15, the limit can be computed coordinate-wise (using coordinate unit vectors $\vec{i}$ and $\vec{j}$). If

$$f(t) = (f_1(t), f_2(t)) = f_1(t)\vec{i} + f_2(t)\vec{j}$$

then

$$\frac{f(t+h) - f(t)}{h} = \frac{f_1(t+h)\vec{i} + f_2(t+h)\vec{j} - f_1(t)\vec{i} - f_2(t)\vec{j}}{h}$$

Taking the limit, we have

$$
\begin{aligned}
f'(t) &= \lim_{h\to 0} \frac{f(t+h)-f(t)}{h} \\
&= \lim_{h\to 0} \frac{f_1(t+h)-f_1(t)}{h}\vec{i} + \lim_{h\to 0} \frac{f_2(t+h)-f_2(t)}{h}\vec{j} \\
&= f_1'(t)\vec{i} + f_2'(t)\vec{j}
\end{aligned}
$$

□

Example 4.2.19 Let $f(x) = (\sin(x), \cos(x), x^2)$. Then taking the derivatives of the component functions we have $f'(x) = (\cos(x), -\sin(x), 2x)$. △

Before giving any more examples, we will discuss the linear approximation approach to differentiation. This approach (which will be essential in later sections) is the higher dimension analogue of Theorem 3.6.17, the Linear Approximation Theorem for differentiating a function $f : \mathbb{R} \to \mathbb{R}$.

Theorem 4.2.20 (Linear Approximation Theorem) *The vector-valued function f is differentiable at a in the domain of f and $f'(a) = v$ if and only if there exists a vector-valued function ERROR such that the following two conditions hold:*

(a) $f(x) = f(a) + v(x-a) + ERROR(x)$

(b) $\lim_{x\to a} \dfrac{|ERROR(x)|}{|x-a|} = 0$

Note that in the case of $f : \mathbb{R} \to \mathbb{R}$, $f'(a) = \lim\limits_{x\to a} \dfrac{f(x)-f(a)}{x-a}$, so that if we define

$$
ERROR(x) = (x-a)\left[\frac{f(x)-f(a)}{x-a} - f'(a)\right]
$$

then the above two conditions hold with $v = f'(x)$.

Example 4.2.21 We can use this theorem to show that the derivative of $f(x) = (3x, 4x^2)$ at $a = 1$ is $f'(1) = (3, 8)$. We need to find a vector-valued function $ERROR : \mathbb{R} \to \mathbb{R}^2$ such that

$$
f(x) = f(1) + (3,8)(x-1) + ERROR(x)
$$

and

$$
\lim_{x\to 1} \frac{|ERROR(x)|}{|x-1|} = 0
$$

The first condition gives

$$\begin{aligned} ERROR(x) &= f(x) - f(1) - (3,8)(x-1) \\ &= (3x, 4x^2) - (3,4) - (3x-3, 8x-8) \\ &= (3x-3-3x+3, 4x^2-4-8x+8) \\ &= (0, 4(x-1)^2) \end{aligned}$$

Thus $\frac{|ERROR(x)|}{|x-1|} = 4|x-1|$ and $\lim_{x\to 1} \frac{|ERROR(x)|}{|x-1|} = 0$.

Therefore the derivative of $f(x)$ at $a = 1$ is $f'(1) = (3,8)$. △

Example 4.2.22 We will find a linear approximation to the function $f(x) = (x^2 + 4, 3\sin(x))$ near $x = 0$. By the Linear Approximation Theorem,

$$f(x) = f(a) + f'(a)(x-a) + ERROR(x)$$

where the error term is small when x is close to $a = 0$. Since $f'(x) = (2x, 3\cos(x))$, $f'(0) = (0,3)$. Thus

$$f(x) \approx f(0) + f'(0)(x-0) = (4,0) + (0,3)x = (4,3x) \quad \text{near } x = 0$$

Note that the range of the approximation function $A(x) = (4, 3x)$ is a straight line (through $(4,0)$ and parallel to the y-axis) which approximates the range of f near $x = 0$. △

Example 4.2.23 Let $f : \mathbb{R} \to \mathbb{R}^2$ be defined by $f(t) = (t^2, t)$. We will

(a) Find a linear approximation to f near $t = 1$

(b) And determine an interval centered at $t = 1$ over which the approximation has an error of at most 0.01.

By the Linear Approximation Theorem, Theorem 4.2.20 (a), a linear approximation to f is

$$g(t) = f(1) + f'(1)(t-1) = (2t-1, t)$$

and

$$ERROR(t) = f(t) - f(1) - f'(1)(t-1) = (t^2 - 2t + 1, 0)$$

Thus

$$|ERROR(t)| = (t-1)^2$$

and

$$|ERROR(t)| \leq 0.01 \quad \text{if and only if} \quad (t-1)^2 \leq 0.01$$

i.e., if and only if $|t-1| \leq 0.1$. Thus $g(t) = (2t-1, t)$ approximates $f(t) = (t^2, t)$ over the interval [0.9,1.1] with an error of at most 0.01. △

In the next several examples we will consider velocity and acceleration as vector-valued functions.

Example 4.2.24 Using $SQP(t) = (t, t^2)$ as an example, we have

$$SQP'(t) = (f_1'(t), f_2'(t)) = (Dt, Dt^2) = (1, 2t) = \vec{\imath} + 2t\vec{\jmath}$$

Recall from Chapter 3 that the derivative was interpreted as velocity. If $SQP(t)$ is interpreted as the position vector at time t, then $SQP'(t)$ is the rate of change of position with time at time t, which is defined as the velocity at time t. When working with a vector-valued function, the velocity is a vector. The length of the velocity vector is the *speed,* a real number. Acceleration is generalized similarly: acceleration is the derivative of velocity. When the velocity is a vector, the acceleration will be a vector. To associate the velocity and acceleration vectors with the particle, the vectors $SQP'(t)$ and $SQP''(t)$ are translated so that the tails are at the position $SQP(t)$ of the particle at time t. $SQP'(t) = (1, 2t)$ and $SQP''(t) = (0, 2)$. Notice that in this example the acceleration is constant, independent of position, just as the acceleration due to gravity is constant at the earth's surface, independent of the position on the earth. △

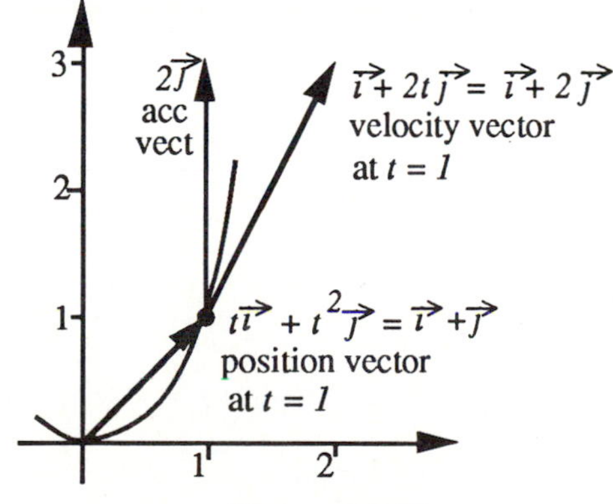

Figure 4.2.25

Example 4.2.26 We will find the velocity and acceleration of a particle moving in $\mathbb{R}^3$ whose position at time t is $f(t) = (3t^2, t, \sin(t))$. The velocity is $f'(t) = (6t, 1, \cos(t))$. The acceleration is $f''(t) = (6, 0, -\sin(t))$. △

Example 4.2.27 We will show that if a particle moves in $\mathbb{R}^3$ with position at time t given by $f(t) = (\sin(t), \cos(t), 2t)$, then the velocity and acceleration vectors are perpendicular.

First we find the velocity vector $f'(t)$ and the acceleration vector $f''(t)$. The velocity vector is $f'(t) = (\cos(t), -\sin(t), 2)$ and the acceleration vector is $f''(t) = (-\sin(t), -\cos(t), 0)$. Recall that two vectors are perpendicular if and only if their dot product is zero. So we compute the dot product:

$$\begin{aligned} f'(t)\bullet f''(t) &= \cos(t)(-\sin(t)) + (-\sin(t)(-\cos(t))) + 2\cdot 0 \\ &= -\cos(t)\sin(t) + \sin(t)\cos(t) + 0 \\ &= 0 \end{aligned}$$

Thus $f'(t)$ and $f''(t)$ are perpendicular.

Geometrically, the path $f(t) = (\sin(t), \cos(t), 2t)$ is a circular spiral. [Note that the sum of squares of the $x = \sin(t)$ and $y = \cos(t)$ coordinates is $x^2+y^2 = 1$, the equation of a circle.] Also, the velocity in the z-direction, $d(2t)/dt = 2$, is constant. For a particle moving in such a manner the velocity and acceleration vectors are always perpendicular. △

Question 4.2.28 What are the velocity and acceleration vectors of a particle at time $t = 1$ if its position at time t is $f(t) = (t^2, 3t, \sin(t))$? ◇

The algebraic properties of the derivative that hold for real valued functions also hold for vector-valued functions. The usual product of real numbers must be replaced by the vector dot product. These properties are collected in the following theorem.

Theorem 4.2.29 (Algebra of vector-valued Functions)
Let f and $g : \mathbb{R} \to \mathbb{R}^n$ and $h : \mathbb{R} \to \mathbb{R}$. Assume that f, g, and h are differentiable where required and that their domains and ranges allow the operations to be performed. Then

(a) $f+g : \mathbb{R} \to \mathbb{R}^n$ and $(f+g)'(t) = f'(t)+g'(t) \in \mathbb{R}^n$ (additivity property)

(b) $hf : \mathbb{R} \to \mathbb{R}^n$ and $(hf)'(t) = h(t)f'(t) + h'(t)f(t) \in \mathbb{R}^n$ (product rule)

(c) $f\bullet g : \mathbb{R} \to \mathbb{R}$ and $(f\bullet g)'(t) = f(t)\bullet g'(t) + f'(t)\bullet g(t) \in \mathbb{R}$ (dot product rule)

(d) $f \circ h : \mathbb{R} \to \mathbb{R}^n$ *and* $(f \circ h)'(t) = f'(h(t))h'(t) \in \mathbb{R}^n$ (chain rule)

Notice that property (b) requires that t be in the intersection of the domains of h and f. Property (d) requires that $h(t)$ be in the domain of f and f be differentiable at $h(t)$.

Proof: These results follow by applying the corresponding results for the derivatives of real-valued functions to the coordinate functions of f and g. As an example we will prove property (c). Suppose that $f, g : \mathbb{R} \to \mathbb{R}^2$. Let $f(t) = (f_1(t), f_2(t))$ and $g(t) = (g_1(t), g_2(t))$.

$$\begin{aligned}
(f \bullet g)'(t) &= (f_1 g_1 + f_2 g_2)'(t) = (f_1 g_1)'(t) + (f_2 g_2)'(t) \\
&= f_1(t) g_1'(t) + f_1'(t) g_1(t) + f_2(t) g_2'(t) + f_2'(t) g_2(t) \\
&= [f_1(t) g_1'(t) + f_2(t) g_2'(t)] + [f_1'(t) g_1(t) + f_2'(t) g_2(t)] \\
&= (f_1(t), f_2(t)) \bullet (g_1'(t), g_2'(t)) + (f_1'(t), f_2'(t)) \bullet (g_1(t), g_2(t)) \\
&= f(t) \bullet g'(t) + f'(t) \bullet g(t)
\end{aligned}$$

□

Example 4.2.30 (a) Suppose that $h : \mathbb{R} \to \mathbb{R}$ has $h(0) = 1$ and $h'(0) = 2$. If $f : \mathbb{R} \to \mathbb{R}^2$ is given by $f(t) = (t, t^2)$, find $(f \circ h)'(0)$. We have

$$(f \circ h)'(t) = f'(h(t))h'(t) = h'(t)f'(h(t)) = h'(t)(1, 2h(t))$$

Letting $t = 0$ we have

$$(f \circ h)'(0) = h'(0)(1, 2h(0)) = 2(1, 2 \cdot 1) = 2(1, 2) = (2, 4)$$

(b) Suppose $f : \mathbb{R} \to \mathbb{R}^2$ satisfies $f(0) = (1, 2)$ and $f'(0) = (2, 3)$; and $g : \mathbb{R} \to \mathbb{R}^2$ satisfies $g(0) = (-1, 1)$ and $g'(0) = (4, 1)$. Find $(f \bullet g)'(0)$. From property (c) above we have

$$(f \bullet g)'(t) = f(t) \bullet g'(t) + f'(t) \bullet g(t)$$

Letting $t = 0$ we have

$$\begin{aligned}
(f \bullet g)'(0) &= f(0) \bullet g'(0) + f'(0) \bullet g(0) = (1, 2) \bullet (4, 1) + (2, 3) \bullet (-1, 1) \\
&= (1 \cdot 4 + 2 \cdot 1) + (2 \cdot (-1) + 3 \cdot 1) = 6 + 1 = 7
\end{aligned}$$

△

Question 4.2.31 Prove property (d) of the theorem. ◇

Example 4.2.32 This example will introduce a little physics. From the beginnings of the Calculus with Newton (1642–1727) and Leibniz in the seventeenth century, physics and mathematics has had a long history of fruitful interaction.

The science of mechanics is concerned with the movement of objects (in three-dimensional space) when a force is applied to them. The movement of objects in $\mathbb{R}^3$ can be modeled by establishing a vector-valued function $POS : \mathbb{R} \to \mathbb{R}^3$ that gives the position $POS(t)$ of the object at time t. We have already introduced two vectors of physical significance, the velocity vector $VEL(t) = POS'(t)$ and the acceleration vector $ACCEL(t) = VEL'(t) = POS''(t)$. The *momentum* of a moving object is defined as the mass m of the object times the velocity of object, $mVEL(t)$. Newton's second law of motion says that if a force is applied to an object, then the rate of change of the momentum of the object is equal to the force applied:

$$FORCE(t) = (mVEL)'(t) = mVEL'(t) = mACCEL(t)$$

i.e., force is equal to mass times acceleration. Thus we see that $FORCE$ is a vector-valued function, since $ACCEL$ is a vector-valued function.

Let's do an example. Suppose that a cannon fires a shell with speed s of 1000 m/s in the direction of the unit vector $(1/\sqrt{2}, 1/\sqrt{3}, 1/\sqrt{6})$. We want to determine where the shell will land. The situation is modeled by assuming a flat segment of the earth corresponding to the $x-y$plane in $\mathbb{R}^3$, with the z-axis upward. The cannon is placed at the origin and pointed in a direction given by a unit vector

$$u = (u_1, u_2, u_3) = (1/\sqrt{2}, 1/\sqrt{3}, 1/\sqrt{6})$$

Let the position of the shell at time t be $POS(t) = (x(t), y(t), z(t))$. Assume that the cannon is fired at time $t = 0$. Since the speed is s in the direction of u, the velocity of the shell when it leaves the cannon is $su = s(u_1, u_2, u_3)$. What forces act on the shell? Gravity alters the path of the shell, and air resistance slows it down. Assume that the air resistance is negligible in order to have a simple model. The force of gravity acts in a vertical direction, accelerating the shell downward at 9.8 m/sec/sec. Thus by Newton's second law,

$$FORCE(t) = mACCEL(t)$$

$FORCE(t) = m(0, 0, -9.8)$, where m is the (unknown) mass of the shell. Since $POS(t)$ is the position at time t,

$$FORCE(t) = mACCEL(t) = mPOS''(t)$$

Equating these expressions for $FORCE(t)$,

$$m(0, 0, -9.8) = mPOS''(t) = m(x''(t), y''(t), z''(t))$$

Dividing both sides of this equation by the mass m we have a differential equation for the position at time t:

$$(x''(t), y''(t), z''(t)) = (0, 0, -9.8)$$

The differential equation is said to be a *second-order* differential equation since the derivative was taken twice. This single equation equating vectors is equivalent to three equations equating real numbers:

$$x''(t) = 0, y''(t) = 0, \quad \text{and} \quad z''(t) = -9.8$$

Using Theorem 3.7.3 (or its Corollary 3.7.5) we have

$$x'(t) = c_1, y'(t) = c_2, \text{ and } z'(t) = -9.8t + c_3$$

The values of the constants can be determined by using the initial condition that the velocity at time $t = 0$ is

$$s(u_1, u_2, u_3) = (x'(0), y'(0), z'(0))$$

This gives

$$\begin{aligned} c_1 &= x'(0) = su_1 = 1000\frac{1}{\sqrt{2}} = \frac{1000}{\sqrt{2}} \\ c_2 &= y'(0) = \frac{1000}{\sqrt{3}} \\ c_3 &= z'(0) = \frac{1000}{\sqrt{6}} \end{aligned}$$

Now the differential equations have been reduced to

$$x'(t) = \frac{1000}{\sqrt{2}}, y'(t) = \frac{1000}{\sqrt{3}}, \text{ and } z'(t) = -9.8t + \frac{1000}{\sqrt{6}}$$

Or, using vector notation to combine the three equations,

$$(x'(t), y'(t), z'(t)) = VEL(t) = \left(\frac{1000}{\sqrt{2}}, \frac{1000}{\sqrt{3}}, -9.8t + \frac{1000}{\sqrt{6}}\right)$$

Applying Corollary 3.7.5 again gives

$$x(t) = \frac{1000t}{\sqrt{2}} + a, y(t) = \frac{1000t}{\sqrt{3}} + b, \quad \text{and} \quad z(t) = -4.9t^2 + \frac{1000t}{\sqrt{6}} + c$$

for some constants a, b, and c. To determine the constants a, b, and c use the second initial condition that at time $t = 0$ the shell is at the origin: $(x(0), y(0), z(0)) = (0, 0, 0)$. This condition gives $a = 0, b = 0, c = 0$ and the equation for position becomes

$$POS(t) = (x(t), y(t), z(t)) = \left(\frac{1000t}{\sqrt{2}}, \frac{1000t}{\sqrt{3}}, -4.9t^2 + \frac{1000t}{\sqrt{6}}\right)$$

(Notice that a second-order differential equation requires two initial conditions to determine the constants.) Now that we have an equation giving the position at any time t, our original question can be answered: where will the shell hit? What condition holds when the shell hits? Clearly, the shell has returned to ground level and so the z-coordinate is zero. Thus we need to determine when

$$z = z(t) = -4.9t^2 + \frac{1000t}{\sqrt{6}} = 0$$

This equation has two solutions for t:

$$t = 0 \text{ (when the shell leaves the cannon)} \quad \text{and} \quad t = \frac{1000}{4.9\sqrt{6}} \approx 83.32$$

At time $t = 83.32$,

$$\begin{aligned} x(83.32) &= 1000\frac{83.32}{\sqrt{2}} \approx 58916.41 \\ y(83.32) &= 1000\frac{83.32}{\sqrt{3}} \approx 48104.80 \\ z(83.32) &= 0 \end{aligned}$$

Thus the shell hits the point on the flat segment of earth modeled by the xy-plane where $x \approx 58916.41$ and $y \approx 48104.80$.

What angle does the shell make with the vertical when it lands? We want to find the angle between the velocity vector $VEC(1000/(4.9\sqrt{6}))$ when the shell lands and a vertical vector, say (0,0,1). We can apply the Cauchy-Schwarz-Bunyakovsky theorem to obtain the relation:

$$VEL\left(\frac{1000}{4.9\sqrt{6}}\right) \bullet (0,0,1) = \left|VEL\left(\frac{1000}{4.9\sqrt{6}}\right)\right| \cos(\theta)$$

where θ is the desired angle between the velocity vector and the positive z-axis. Since

$$VEL\left(\frac{1000}{4.9\sqrt{6}}\right) = \left(\frac{1000}{\sqrt{2}}, \frac{1000}{\sqrt{3}}, -9.8\left(\frac{1000}{4.9\sqrt{6}}\right) + \frac{1000}{\sqrt{6}}\right)$$

we have

$$\left|VEL\left(\frac{1000}{4.9\sqrt{6}}\right)\right| = \left[\left(\frac{1000}{\sqrt{2}}\right)^2 + \left(\frac{1000}{\sqrt{3}}\right)^2 + \left(\frac{-1000}{\sqrt{6}}\right)^2\right]^{1/2} = 1000$$

Thus

$$\cos(\theta) = VEL\left(\frac{1000}{4.9\sqrt{6}}\right) \bullet \frac{(0,0,1)}{1000} = \frac{-1}{\sqrt{6}}.$$

Therefore

$$\theta = 114^\circ$$

Note that the velocity vector on impact is the same as the velocity vector at firing, except that the z-component is negative. This is a result of the symmetry of the physics with air resistance neglected. △

Tangent Vectors

Recall that if $f : \mathbb{R} \to \mathbb{R}$, is a differentiable function, then $f'(t)$ is the slope of the line tangent to the graph of the function at $(t, f(t))$. We will give a similar interpretation for vector valued functions.

If $f : \mathbb{R} \to \mathbb{R}^n$ then $f(a+h) - f(a)$ is a vector from $f(a)$ to $f(a+h)$ (see Question 4.1.14); and $(f(a+h) - f(a))/h$ is a vector with the same direction if $h > 0$. (Why must $h > 0$?) As h approaches 0 through positive values, $(f(a+h) - f(a))/h$ approaches a vector tangent to the curve that is the range of f. Thus $f'(a)$ is a *tangent vector* to the curve that is the range of f at the point $f(a)$.

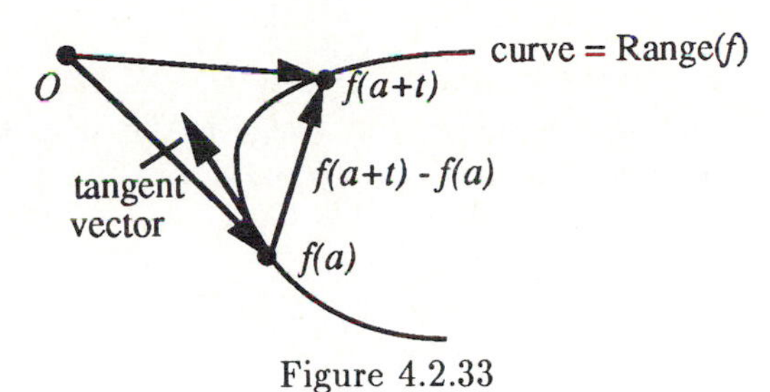

Figure 4.2.33

The vector $f'(a)$ need not be of unit length, but as in Question 4.1.21 a unit vector can be obtained by dividing $f'(a)$ by its length.

Definition 4.2.34 (Unit Tangent Vector) *If $f : \mathbb{R} \to \mathbb{R}^n$ has derivative $f'(t) \neq \vec{O}$ at t, then the* unit tangent vector *to the curve at position $f(t)$ is defined by $T(t) = f'(t)/|f't)|$.*

The unit tangent vector is parallel to the velocity vector, but of unit length. In fact, the unit tangent vector is the velocity, $f'(t)$, divided by the speed, $|f'(t)|$. This shows that the velocity vector points in the direction that the particle is moving, the direction of the unit tangent vector.

Example 4.2.35 Let $SQP(t) = (t, t^2)$. Then $SQP'(t) = (1, 2t)$ is the velocity vector. The unit tangent vector is $T(t) = (1, 2t)/[1 + 4t^2]^{1/2}$. When $t = 1$, the unit tangent vector is $T(1) = (1/\sqrt{5}, 2/\sqrt{5})$. △

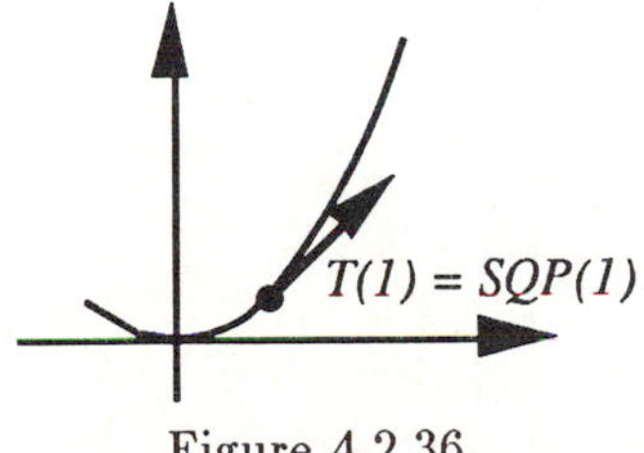

Figure 4.2.36

Example 4.2.37 As a second example of tangent vectors, consider the curve in $\mathbb{R}^3$ whose tangent line is most obvious—a straight line. Consider the straight line determined by two points A and B. As we have seen, the line determined by A and B can be parameterized by $L(t) = A + t(B - A)$, where $L : \mathbb{R} \to \mathbb{R}^3$.

The derivative of the function L is

$$L'(t) = B - A = \vec{AB}$$

which is the vector determining the line, and the unit tangent vector is

$$T(t) = \frac{\vec{AB}}{|\vec{AB}|} = \frac{B - A}{|B - A|} \qquad \triangle$$

Question 4.2.38 Find the function whose range is the straight line through $(1, 2, 1)$ and in the direction of the vector $(3, 2, -1)$. $\diamondsuit$

Question 4.2.39 Determine the unit tangent vector to $f(t) = (3t, t^2, t + 1)$ at $t = 1$. $\diamondsuit$

Example 4.2.40 We will find the parametric form of the line tangent to the curve given parametrically by the range of $f(t) = (4t, 3t + 2, t^3)$ at the point when $t = 1$. When $t = 1$, $f(1) = (4, 5, 1)$. Since $f'(t) = (4, 3, 3t^2)$, a tangent vector to the curve at $t = 1$ is $f'(1) = (4, 3, 3)$. The point $(4, 5, 1)$ and the vector $(4, 3, 3)$ determine the tangent line, i.e., the set of points $\{(x, y, z)\}$ such that

$$(x, y, z) = (4, 5, 1) + r(4, 3, 3)$$

for some r. Thus the tangent line is given parametrically by the range of

$$g(t) = (4, 5, 1) + t(4, 3, 3) \qquad \triangle$$

Section 4.2 Exercises

1. Parameterize the following curves in the plane.

   (a) $\{(x, y) : xy = 1, x > 0\}$
   (b) $\{(x, y) : x + y^2 = 1\}$
   (c) $\{(x, y) : 4x^2 + 9y^2 = 1\}$
   (d) $\{(x, y) : 5x^2 + 2y^2 = 1\}$

2. What curve is described by the given parameterization? Classify the curve, e.g., parabola, ellipse, hyperbola, and convert to rectangular xy-coordinates.

(a) $f(t) = (t^2, t)$ for all real t.

(b) $f(t) = (3\sin(t), 4\cos(t))$ for all real t.

3. Using Definition 4.2.16 show that the derivative of f defined by $f(t) = (1, 2, 3)$ for all t is (0,0,0).

4. Using the Linear Approximation Theorem, Theorem 4.2.20, show that the derivative of f defined by $f(t) = (5, 2)$ for all t is (0,0).

5. Find the derivative of $f(t) = (t, t^2, \cos(t))$.

6. Find the derivative of $f(t) = (3t + 6, \tan(t), (t + 1)/(t + 2))$.

7. Find a linear approximation to $f(t) = (t^3, 3t^2, t)$ at t = 1.

8. Find a linear approximation to $f(t) = (\sin(t), 2t + 1)$ at $t = 0$.

9. Determine the size of the *ERROR* in Example 4.2.22 when $x = 0.1$.

10. Determine the size of the *ERROR* in Example 4.2.22 when $x = 0.2$.

11. Find a linear approximation to $f(t) = (t, 4t^2)$ at $t = 1$ and an interval about $t = 1$ where the error is at most 0.01.

12. Find a linear approximation to $f(t) = (t, 3t^2)$ at $t = 1$ and an interval about $t = 1$ where the error is at most 0.01.

13. What are the velocity and acceleration of a particle at time t if the position at time t is $f(t) = (1, 3t^2, \sin(t))$?

14. What is the velocity of a particle at time t if its position at time t is $f(t) = (\sin(t), \cos(t))$? Sketch the path, showing the velocity vector.

15. If $h'(0) = 3$ and $h(0) = 2$ find $(f \circ h)'(0)$ if $f(t) = (3t, 2t^2 + 1)$.

16. If $h'(1) = -1$ and $h(1) = 2$ find $(f \circ h)'(1)$ if $f(t) = (5t, -3t^2 + t)$.

17. Suppose that the force on a particle at time t is $F(t) = (1, t, t^2)$. If the position at time $t = 0$ is (0,1,2) and the velocity at time $t = 0$ is (2,1,1), find the position vector $P(t) = (x(t), y(t), z(t))$.

18. Find the unit tangent vector to the range of $f(t) = (t, t^2 + 1)$ at $t = 1$.

19. Find the parametric form of the straight line through the point $(1, 2, 3)$ in the direction of $(1, -1, 1)$.

20. Find the parametric form of the straight line through the point $(-1, 0, 3)$ in the direction of $(2, 1, 1)$.

21. Find the parametric form of the tangent line to the curve $f(t) = (t, 3t + 1, t^2)$ when $t = 1$.

22. Find the parametric form of the tangent line to the curve $f(t) = (3t - 1, 1, t^2 + t)$ when $t = 1$.

23. Find the midpoint of the line segment joining $(1, 2, 4)$ and $(-2, 1, -3)$.

24. Find the midpoint of the line segment joining $(1, 2, 4, -4)$ and $(2, 1, -3, 5)$.

25. Does the point $(1, 1, 1)$ lie on the line segment between $(2, 1, -1)$ and $(0, 2, 2)$?

26. Does the point $(1, 0, 1, -2)$ lie on the line segment between $(3, -1, 2, 1)$ and $(-3, 2, -1, -8)$?

27. (a) A cow's head is being approached by a fly, whose position at time t is $f(t) = (10 - t^2, 20 - 3t^2, 7 - t)$ for $t > 0$. When the altitude of the fly is 6 ft, its shoe comes off. Find the position the shoe hits the ground ($z = 0$).

    (b) A cow falls out of an aircraft which has position $(t + 1, 3t^2 + 6, 2t)$ at time t. When the cow falls, the altitude is 100 feet. Where should the burial party gather?

28. Show that if a particle in the plane moves so that the velocity vector is always perpendicular to the position vector, then the particle is moving in a circle.

29. Consider a curve in $\mathbb{R}^2$ parameterized by $V(t) = (f(t), g(t))$, $a \leq t \leq b$.

    (a) Find the slope of the vector from $V(a)$ to $V(b)$, the endpoints of the curve.

    (b) Find the slope of the tangent vector at a point $V(t)$.

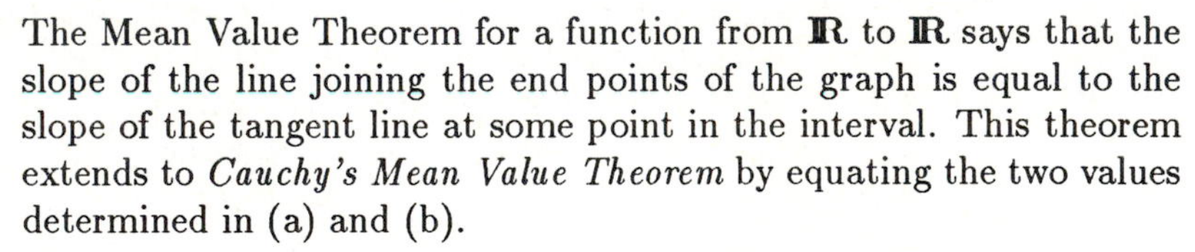

    The Mean Value Theorem for a function from $\mathbb{R}$ to $\mathbb{R}$ says that the slope of the line joining the end points of the graph is equal to the slope of the tangent line at some point in the interval. This theorem extends to *Cauchy's Mean Value Theorem* by equating the two values determined in (a) and (b).

    (c) Formally state Cauchy's Mean Value Theorem.

    (d) Prove Cauchy's Mean Value Theorem by applying Rolle's theorem to the function

$$h(x) = f(x)[g(b) - g(a)] - g(x)[f(b) - f(a)].$$

30. With reference to the physics example, suppose that 100 sec after the first shell is fired (at time $t = 0$), the cannon turns and fires a second shell in the direction of the vector $v = (1, 3, 2)$ with a speed of 500 m/sec. When does the second shell land?

4.3 Directional Derivatives

Our development of differentiation has already undergone three stages. We are now poised for a fourth stage. A derivative represents the instantaneous rate of change of a function, i.e., the instantaneous rate of change of the output with respect to the input. In the first stage, we considered real-valued functions of a single variable, i.e., $f : \mathbb{R} \to \mathbb{R}$. In the second stage (partial derivatives), we considered real-valued functions of several variables, i.e., $f : \mathbb{R}^n \to \mathbb{R}$. However, we restricted our consideration to those situations in which only one of the input variables changed. This effectively reduced stage two to stage one with respect to the variable that changed. The third stage occurred in the previous section where we considered vector-valued functions of a single variable, i.e., $f : \mathbb{R} \to \mathbb{R}^n$. We are now ready to consider real-valued functions of several variables, i.e., $f : \mathbb{R}^n \to \mathbb{R}$, in which we allow combinations of input variables to change simultaneously.

In stage one, an economist studying the rate of change of inflation would assume that inflation was dependent on a single factor. In stage two, the economist could consider inflation dependent on several factors, but could only study the rate of change of inflation by varying one factor at a time. In the stage that we are about to begin, the economist can consider the rate of change of inflation as a result of a linear combination of factors that are changing simultaneously.

As another illustration of differentiating a function of several variables, consider the rate of change of elevation of a person skiing down a mountain. Elevation is defined in terms of longitude and latitude. That is, $elv : \mathbb{R}^2 \to \mathbb{R}$ is a function whose domain is the set of ordered pairs representing longitude and latitude. The rate of change of elevation is dependent, in part, on the *direction* of the skier. How is direction described in $\mathbb{R}^2$? Answer, $\mathbb{R}^2$ is a vector space and thus direction is described by a unit vector. In this section, we shall develop the concept of rate of change of a function in a specified direction.

We consider functions of the form $f : \mathbb{R}^n \to \mathbb{R}$. As usual $\mathbb{R}^2$ and $\mathbb{R}^3$ will be used for most of the examples, although what is said applies to functions defined on $\mathbb{R}^n$ for any n. A function defined on $\mathbb{R}^n$ will be considered to be defined on vectors. To simplify notation, the arrow over the vector will usually be omitted. If $f : \mathbb{R}^2 \to \mathbb{R}$ and $a = (x, y)$ is a vector in $\mathbb{R}^2$ (identified with the point at its head), then

$$f(\vec{a}) = f(a) = f((x, y)) = f(x, y)$$

Notice the simplification of notation that identifies $f((x, y))$ and $f(x, y)$.

Directional Derivatives

Partial derivatives are the slopes of the graph of a function in the directions of the coordinate axes. (See the section on partial derivatives.) If the graph of $f : \mathbb{R}^2 \to \mathbb{R}$, $z = f(x, y)$, is sliced by the vertical yz-plane, $x = 0$, then the section of the graph in this plane has slope $D_y f(0, y)$.

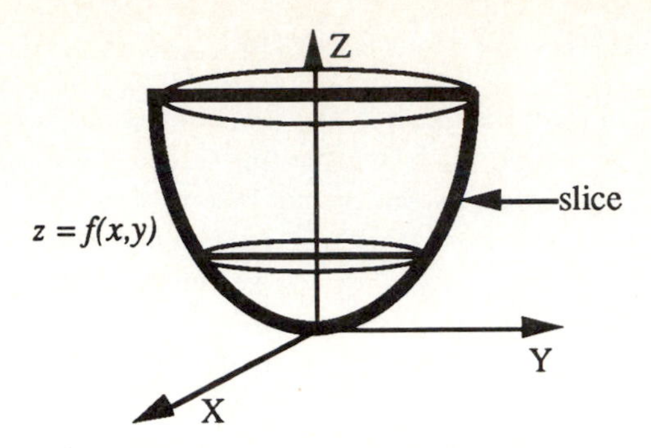

Figure 4.3.1

If the graph of f is sliced by the xz-plane, $y = 0$, then the section of the graph in the xz-plane has slope $D_x f(x, 0)$. If the graph of f is sliced in another direction, does it have a slope? Even if all the partial derivatives exist, the function may not have derivatives in other directions, or even be continuous. Before we can give some examples of this, we must make precise what we mean by the derivative in an arbitrary direction.

Let $f : \mathbb{R}^n \to \mathbb{R}$ be defined at $a \in \mathbb{R}^n$ and let u be a unit vector in $\mathbb{R}^n$. We now develop the concept of the *derivative of f at a in the direction of u*. To move in the direction u from a add a small real multiple $t > 0$ of u to a: $a + tu$ is a point near a in the direction u. (Recall the parametric form of a line in $\mathbb{R}^n$).

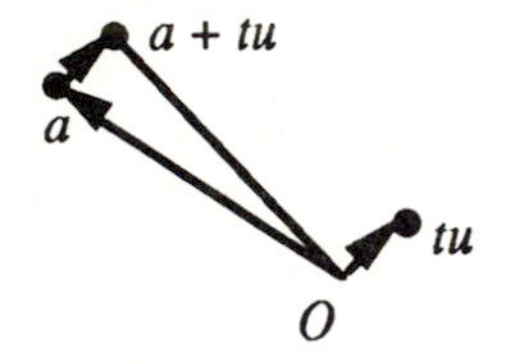

Figure 4.3.2

The average change in f in the direction $a + tu$ is obtained, as in the case of a function of one variable, by forming the difference quotient $(f(a + tu) - f(a))/t$. The limit of this difference quotient (if it exists) is the derivative of f at a in the direction of u.

Definition 4.3.3 (Directional Derivative) *Let $f : \mathbb{R}^n \to \mathbb{R}$ be defined at a. If u is a unit vector in $\mathbb{R}^n$, the* directional derivative *of f at a in the direction u is defined to be*

$$D_u f(a) = \lim_{t \to 0} \frac{f(a + tu) - f(a)}{t}$$

whenever the limit exists.

Warning: The direction vector u *must* be a *unit* vector.

We will show that this definition agrees with the previous definition of partial derivative given in Section 3.10. Let $f : \mathbb{R}^3 \to \mathbb{R}$ and let $a = (x, y, z) \in \mathbb{R}^3$. To take the partial derivative of f in the y direction, a unit vector in the y direction is needed. A unit vector in the y direction is $u = \vec{e_2} = \vec{j} = (0, 1, 0)$. Applying the above definition,

$$\begin{aligned} D_u f(x,y,z) &= \lim_{t\to 0} \frac{f((x,y,z)+t(0,1,0)) - f(x,y,z)}{t} \\ &= \lim_{t\to 0} \frac{f(x,y+t,z) - f(x,y,z)}{t} \\ &= \frac{\partial f}{\partial y}(x,y,z) \end{aligned}$$

The notation $D_u f(x,y)$ is similar to the notation $D_x f(x,y)$ for the partial derivative. It will be clear from the context whether the subscript is a variable or a vector.

We will develop an easy way to compute directional derivatives after the extension of the concept of the derivative to functions of several variables has been completed. Thus only a few examples of computing the directional derivative from the definition will be given.

Example 4.3.4 We will find the directional derivative of $f(x,y) = x + 2y$ in the direction of the unit vector $u = (1/\sqrt{2}, 1/\sqrt{2})$.

$$\begin{aligned} D_u f(x,y) &= \lim_{t\to 0} \frac{f((x,y)+t(1,1)/\sqrt{2}) - f(x,y)}{t} \\ &= \lim_{t\to 0} \frac{f(x+t/\sqrt{2}, y+t/\sqrt{2}) - f(x,y)}{t} \\ &= \lim_{t\to 0} \frac{x+t/\sqrt{2}+2y+2t/\sqrt{2}-x-2y}{t} \\ &= \lim_{t\to 0} \frac{t/\sqrt{2}+2t/\sqrt{2}}{t} = \lim_{t\to 0}(\frac{1}{\sqrt{2}}+\frac{2}{\sqrt{2}}) \\ &= \frac{3}{\sqrt{2}} \end{aligned}$$

△

Can we find directional derivatives from partial derivatives? That is, can a directional derivative be expressed as a linear combination of partial derivatives? For example, are there constants c and d such that the directional derivative in the previous example can be expressed as $c\partial f/\partial x + d\partial f/\partial y$? A very reasonable expectation, but unfortunately not correct as shown by the next example.

Example 4.3.5 (Partial Derivatives vs. Directional Derivatives) Here is an example of a function $f : \mathbb{R}^2 \to \mathbb{R}$ for which all partial derivatives exist, but does not have directional derivatives in all directions. Define $f : \mathbb{R}^2 \to \mathbb{R}$

$$f(x,y) = \begin{cases} x+y & \text{if either } x = 0 \text{ or } y = 0 \\ 1 & \text{otherwise} \end{cases}$$

$$\frac{\partial f}{\partial x}(0,0) = \lim_{t \to 0} \frac{f(0+t,0) - f(0,0)}{t} = \lim_{t \to 0} \frac{t}{t} = 1$$

Similarly,

$$\frac{\partial f}{\partial y}(0,0) = 1$$

To approach $(0,0)$ along the line $y = x$ let $u = (1/\sqrt{2}, 1/\sqrt{2})$, a unit vector in the desired direction. The directional derivative in the direction of u is given by

$$\begin{aligned} D_u f(0,0) &= \lim_{t \to 0} \frac{f((0,0) + t(\frac{1}{\sqrt{2}}, \frac{1}{\sqrt{2}})) - f(0,0)}{t} \\ &= \lim_{t \to 0} \frac{f(\frac{t}{\sqrt{2}}, \frac{t}{\sqrt{2}}) - f(0,0)}{t} \\ &= \lim_{t \to 0} \frac{1}{t} \end{aligned}$$

which doesn't exist!

Notice that f is not continuous at $(0,0)$. △

This last example illustrates that partial derivatives of a function may exist at a point even though the function is not continuous at that point. This is not surprising since partial derivatives only depend on the behavior of the function in directions parallel to the coordinate axes. What is surprising is the fact that the existence of directional derivatives (in all directions) at a point does not imply continuity at that point.

Example 4.3.6 (Directional Derivatives Do Not Imply Continuity) Even if the directional derivatives exist for all directions, the function may not be continuous. Define $f : \mathbb{R}^2 \to \mathbb{R}$ by $f(x,y) = x^2y/(x^4+y^2)$ if $y \neq 0$, and $f(x,y) = 0$ if $y = 0$. In Example 2.4.20 the limit of $f(x,y)$ as (x,y) approached $(0,0)$ was computed. We saw that f is not continuous at $(0,0)$. Now consider the directional derivatives of f. Let $u = (a,b)$ be any unit vector in the plane. The directional derivative of f at $(0,0)$ in the direction (a,b) is

$$\begin{aligned} D_u f(0,0) &= \lim_{t \to 0} \frac{f(ta,tb) - f(0,0)}{t} = \lim_{t \to 0} \frac{(at)^2 b}{(at)^4 + (bt)^2} \\ &= \lim_{t \to 0} \frac{a^2t^2b}{a^4t^4 + b^2t^2} \\ &= \lim_{t \to 0} \frac{a^2 b}{a^4t^2 + b^2} \\ &= \frac{a^2}{b} \end{aligned}$$

if $b \neq 0$. If $b = 0$, then we are computing the partial of f with respect to x and

$$\frac{\partial f}{\partial x}(0,0) = \lim_{t \to 0} \frac{f(t,0) - f(0,0)}{t} = \lim_{t \to 0} \frac{0}{t} = 0$$

Thus the directional derivatives of f exist at $(0,0)$ no matter how $(0,0)$ is approached, but f is not continuous at $(0,0)$. △

Section 4.3 Exercises

1. Compute the directional derivative of $f(x,y) = x^2 - y$ at (x,y) in the direction of the vector $v = (1,1)$.

2. Compute the directional derivative of $f(x,y) = 3x^2 + 2xy$ at (x,y) in the direction of the vector $v = (-1,1)$.

3. Compute the directional derivative of $f(x,y,z) = 2x + 3y - z$ at (x,y,z) in the direction of the vector $v = (0,2,0)$.

4. Compute the directional derivative of $f(x,y,z) = 5x - 2y + z$ at (x,y,z) in the direction of the vector $v = (1,0,1)$.

5. Give an example of a function $f : \mathbb{R}^2 \to \mathbb{R}$ which has all partial derivatives existing at (1,2) but does not have all directional derivatives existing at (1,2).

6. (a) A cow is at coordinates $x = 1$, $y = 1$, $z = 9$ on a hill whose altitude is given by $a(x,y) = 13 - x^2 - 3y^2$. She can go downward in either the direction $u = (0,1)$ or $v = (1/\sqrt{2}, 1/\sqrt{2})$. Which direction has the gentler slope?

   (b) A cow has an insect on her neck. In the neighborhood of the insect at $(1,1,12)$ the altitude is given by $a(x,y) = 2x^4 + 4y^2 + 6$. The insect can either go in the direction $u = (1,0)$ or in direction $v = (1/\sqrt{3}, \sqrt{2/3})$. Which is the quickest way to the head?

4.4 Derivatives of Functions of Several Variables

Recall that for a function of one variable, if the derivative exists at a point, then the function is continuous at the point. The last Examples 4.3.5 and 4.3.6 of directional derivatives show that the situation is more complicated for functions of several variables: the existence of all the directional derivatives at a point does not guarantee continuity of the function at the point. Clearly, more than the concept of directional derivatives is needed in order to discuss derivatives of functions of several variables. In this section we shall develop the concept of derivative without reference to direction.

For functions with domain in the real numbers the derivative was defined in terms of a difference quotient, where we divided by elements in the domain. If the domain is in $\mathbb{R}^n$ with $n > 1$, an attempt to form a difference quotient would mean dividing by a vector, an undefined operation. Thus *the difference quotient approach will not work* for functions whose source is in $\mathbb{R}^n, n > 1$.

The appropriate concept to generalize is the linear approximation approach, Theorem 3.6.17. In fact, all definitions of derivative, in a great many different contexts, generalize the linear approximation approach. Recall that this says that $f : \mathbb{R} \to \mathbb{R}$ is differentiable at the point a with derivative $f'(a)$ if and only if there is a function $ERROR : \mathbb{R} \to \mathbb{R}$ such that

(a) $f(x) = f(a) + f'(a)(x - a) + ERROR(x)$

(b) $\lim_{x \to a} \frac{|ERROR(x)|}{|x-a|} = 0$

This result can be used almost directly as the definition of the derivative of a function of several variables, after making sure that everything makes sense. If $f : \mathbb{R}^n \to \mathbb{R}$, then $f(x)$ and $f(a)$ are real numbers as before. Now x, a, and $x - a$ are vectors in $\mathbb{R}^n$. In order for the term $f'(a)(x - a)$ to give a real number, the product must be changed into a dot product, and $f'(a)$ must be a vector in $\mathbb{R}^n$.

Definition 4.4.1 (Differentiability in Several Variables) *Let $f : \mathbb{R}^n \to \mathbb{R}$. f is* differentiable *at a with derivative $f'(a) = Df(a) \in \mathbb{R}^n$ if there is a function $ERROR : \mathbb{R}^n \to \mathbb{R}$ with the properties:*

(a) $f(x) = f(a) + f'(a) \cdot (x - a) + ERROR(x)$

(b) $\lim_{x \to a} \frac{|ERROR(x)|}{|x-a|} = 0$

Just as in the previous cases, if $\lim_{x \to a} \frac{|ERROR(x)|}{|x-a|} = 0$, then $\lim_{x \to a} ERROR(x) = 0$.

Theorem 4.4.2 *If $f : \mathbb{R}^n \to \mathbb{R}$ is differentiable at a, then*

(a) *f is continuous at a.*

(b) *f has directional derivatives in all directions at a.*

Proof:
(a) We assume the existence of a function ERROR satisfying the theorem. So,

$$f(x) = f(a) + f'(a) \cdot (x - a) + ERROR(x)$$

Thus,

$$\begin{aligned}\lim_{x\to a} f(x) &= \lim_{x\to a}[f(a) + f'(a)\bullet(x-a) + ERROR(x)] \\ &= f(a) + \lim_{x\to a} f'(a)\bullet(x-a) + \lim_{x\to a} ERROR(x) \\ &= f(a) + 0 + 0 = f(a)\end{aligned}$$

(b) Let u be a unit vector in $\mathbb{R}^n$. Let $x = a + tu$ in Definition 4.4.1 (a). Then

$$\begin{aligned}f(a+tu) &= f(a) + f'(a)\bullet(a+tu-a) + ERROR(a+tu) \\ &= f(a) + f'(a)\bullet(tu) + ERROR(a+tu)\end{aligned}$$

$$\begin{aligned}\lim_{t\to 0}\frac{f(a+tu)-f(a)}{t} &= \lim_{t\to 0}\frac{f'(a)\bullet tu + ERROR(a+tu)}{t} \\ &= \lim_{t\to 0}[f'(a)\bullet u + \frac{ERROR(a+tu)}{t}] \\ &= f'(a)\bullet u + \lim_{t\to 0}\frac{ERROR(a+tu)}{t} \\ &= f'(a)\bullet u\end{aligned}$$

Thus the difference quotient exists and $D_u f(a) = f'(a)\bullet u$. □

The last sequence of equalities in the proof of part (b) and Definition 4.3.3 gives an easy way to compute the directional derivative, which is summarized in the next theorem. We reiterate the last line of Theorem 4.4.2 as the following.

Theorem 4.4.3 (Computation of the Directional Derivative)
If $f : \mathbb{R}^n \to \mathbb{R}$ is differentiable at a, then

$$D_u f(a) = f'(a)\bullet u$$

for any unit vector u in $\mathbb{R}^n$.

How is $f'(a)$ determined?

The derivative $f'(a)$ is a vector in $\mathbb{R}^n$, so $f'(a) = (x_1, x_2, \ldots, x_n)$ for coordinates x_i that we wish to find. To find x_1, let $u = (1, 0, 0, \ldots, 0)$. Then

$$x_1 = f'(a)\bullet u = D_u f(a) = D_1 f(a)$$

Similarly,

$$x_2 = D_2 f(a), \ldots, x_n = D_n f(a)$$

Thus,

Theorem 4.4.4 (Computation of Derivative) *If $f : \mathbb{R}^n \to \mathbb{R}$ is differentiable at a, then*

$$f'(a) = (D_1 f(a), D_2 f(a), \ldots, D_n f(a))$$

Example 4.4.5 Let $f(x,y) = x + 2y$ for any (x,y). Then

$$f'(x,y) = \left(\frac{\partial f(x,y)}{\partial x}, \frac{\partial f(x,y)}{\partial y}\right) = (1,2)$$

This gives us an easy way to compute $D_u f$. Let $u = \frac{1}{\sqrt{2}}(1,1)$, so

$$\begin{aligned} D_u f(x,y) &= f'(x,y) \bullet \frac{1}{\sqrt{2}}(1,1) = (1,2) \bullet \frac{1}{\sqrt{2}}(1,1) = \frac{1}{\sqrt{2}} + \frac{2}{\sqrt{2}} \\ &= 3/\sqrt{2} \end{aligned}$$

△

The vector $(D_1 f(a), \ldots, D_n f(a))$ is also called the *gradient* of f at a (grad $f(a)$), and is written in various ways:

$$\nabla(f)(a) = \nabla f(a) = f'(a) = \left(\frac{\partial f(a)}{\partial x_1}, \ldots, \frac{\partial f(a)}{\partial x_n}\right) = (D_1 f(a), \ldots, D_n f(a))$$

Note that $f' = \nabla(f) : \mathbb{R}^n \to \mathbb{R}^n$.

Example 4.4.6 Let $f : \mathbb{R}^2 \to \mathbb{R}$ be defined by $f(x,y) = xy + 3x$. Then

$$f'(x,y) = \nabla f(x,y) = \left(\frac{\partial f}{\partial x}(x,y), \frac{\partial f}{\partial y}(x,y)\right) = (y+3, x)$$

△

In terms of the gradient, we have the relation

$$D_u f(x) = f'(x) \bullet u = \nabla f(x) \bullet u$$

We can now provide a theorem that gives a sufficient condition for a function $f : \mathbb{R}^n \to \mathbb{R}$ to be differentiable at the point $a \in \mathbb{R}^n$. From Example 4.3.6 we know that the existence of partial derivatives at a point does not guarantee continuity at the point, so the existence of partial derivatives at a point cannot guarantee differentiability at the point. However, if the partial derivatives not only exist but are continuous at a point, then the function is differentiable at the point.

Theorem 4.4.7 *If all the partial derivatives of $f : \mathbb{R}^n \to \mathbb{R}$ are continuous at a, then f is differentiable at a and*

$$f'(a) = \nabla f(a) = (D_1 f(a), \ldots, D_n f(a))$$

Proof: The proof uses the Mean Value Theorem for functions of one variable. To simplify our notation, we will consider the case when $f : \mathbb{R}^2 \to \mathbb{R}$. The general case is proved the same way.

How should we proceed? We know what the answer should be: $f'(a) = \nabla f(a)$. Now, $\nabla f(x)$ exists since the partial derivatives are continuous. What we must show is that if ERROR is defined by

$$ERROR(x) = f(x) - [f(a) + \nabla f(a) \bullet (x - a)]$$

then

$$\lim_{t \to a} \frac{|ERROR(x)|}{|x - a|} = 0$$

Let $a = (a_1, a_2)$ and $x = (x_1, x_2)$. Consider the function f of (x_1, x_2) with only the first variable changing. Then we have a real valued function of one variable x_1. The Mean Value Theorem for functions of one variable says that

$$f(x_1, x_2) - f(a_1, x_2) = D_1 f(c_1, x_2)(x_1 - a_1)$$

for some c_1 between x_1 and a_1. Similarly, considering $f(a_1, x_2)$ with only x_2 changing we have:

$$f(a_1, x_2) - f(a_1, a_2) = D_2 f(a_1, c_2)(x_2 - a_2)$$

for some c_2 between x_2 and a_2. This gives

$$\begin{aligned}
ERROR(x) &= f(x) - f(a) - \nabla f(a) \bullet (x - a) = [f(x_1, x_2) - f(a_1, x_2)] \\
&\qquad + [f(a_1, x_2) - f(a_1, a_2)] - \nabla f(a) \bullet (x - a) \\
&\qquad\qquad [\text{adding and subtracting } f(a_1, x_2)] \\
&= D_1 f(c_1, x_2)(x_1 - a_1) + D_2 f(a_1, c_2)(x_2 - a_2) \\
&\qquad - (D_1 f(a_1, a_2), D_2 f(a_1, a_2)) \bullet (x_1 - a_1, x_2 - a_2) \\
&\qquad\qquad [\text{applying the Mean Value Theorem}] \\
&= D_1 f(c_1, x_2)(x_1 - a_1) + D_2 f(a_1, c_2)(x_2 - a_2) \\
&\qquad - D_1 f(a_1, a_2)(x_1 - a_1) - D_2 f(a_1, a_2)(x_2 - a_2) \\
&= [D_1 f(c_1, x_2) - D_1 f(a_1, a_2)](x_1 - a_1) \\
&\qquad + [D_2 f(a_1, c_2) - D_2 f(a_1, a_2)](x_2 - a_2)
\end{aligned}$$

Let $A = [D_1 f(c_1, x_2) - D_1 f(a_1, a_2)]$ and $B = [D_2 f(a_1, c_2) - D_2 f(a_1, a_2)]$. Then from the above equation for $ERROR$,

$$ERROR(x) = A(x_1 - a_1) + B(x_2 - a_2)$$

and

$$\begin{aligned}
\lim_{x \to a} \frac{|ERROR(x)|}{|x-a|} &= \lim_{x \to a} \frac{|A(x_1-a_1)+B(x_2-a_2)|}{|x-a|} \\
&\leq \lim_{x \to a} \left[\frac{|A||(x_1-a_1)|}{|x-a|} + \frac{|B||(x_2-a_2)|}{|x-a|} \right]
\end{aligned}$$

[by the triangle inequality, Theroem 4.1.16 (11)]. So,

$$\lim_{x \to a} \frac{|ERROR(x)|}{|x-a|} \leq \lim_{x \to a} \frac{|A||(x_1-a_1)|}{|x-a|} + \lim_{x \to a} \frac{|B||(x_2-a_2)|}{|x-a|}$$

Note that

$$|x_1-a_1| \leq |x-a| \text{ and } |x_2-a_2| \leq |x-a|$$

since a side of a right triangle ($|x_2-a_2|$) is no longer than the hypotenuse ($|x-a|$) (another version of the triangle inequality). Thus

$$\frac{1}{|x-a|} \leq \frac{1}{|x_1-a_1|} \text{ and } \frac{1}{|x-a|} \leq \frac{1}{|x_2-a_2|}$$

This gives

$$\begin{aligned}
\lim_{x \to a} \frac{|ERROR(x)|}{|x-a|} &\leq \lim_{x \to a} \frac{|A||(x_1-a_1)|}{|x_1-a_1|} + \lim_{x \to a} \frac{|B||(x_2-a_2)|}{|x_2-a_2|} \\
&\leq \lim_{x \to a} |A| + \lim_{x \to a} |B|
\end{aligned}$$

As x approaches a, x_1 approaches a_1 and c_1 approaches a_1 since c_1 is between a_1 and x_1. Thus

$$\lim_{x \to a} |A| = \lim_{x \to a} [D_1 f(c_1, x_2) - D_1 f(a_1, a_2)] = 0$$

since the partial derivative $D_1 f$ is continuous. Similarly

$$\lim_{x \to a} |B| = 0$$

We now have

$$0 \leq \lim_{x \to a} \frac{|ERROR(x)|}{|x-a|} \leq 0$$

Thus

$$\lim_{x \to a} \frac{|ERROR(x)|}{|x-a|} = 0 \qquad \square$$

The set of functions from $\mathbb{R}^n$ into $\mathbb{R}$ that have continuous partial derivatives are important enough to be given a name.

Definition 4.4.8 (C^1 Function) *A function $f : \mathbb{R}^n \to \mathbb{R}$ is a C^1 function (read "cee-one function") if all its first partial derivatives $D_i f$ are continuous functions, $D_i f : \mathbb{R}^n \to \mathbb{R}$.*

This definition is also used for functions of one variable: $f : \mathbb{R} \to \mathbb{R}$ is C^1 if f' is a continuous function.

If a function is continuous, it is said to be a C^0 function. By Theorem 4.4.2 if f is C^1 then f is C^0.

Functions that have continuous higher partial derivatives can also be considered: f is a C^2 function if the second-order partial derivatives are continuous, f is a C^k function if the kth order partial derivatives are continuous. Intuitively, the existence and continuity of derivatives is an indication of smoothness. C^1 functions are smoother than C^0 functions. This gives us a nested, descending hierarchy of smoothness:

$$C^0 \supseteq C^1 \supseteq C^2 \ldots$$

each class of functions containing the next. The class of differentiable functions is between the classes C^0 and C^1: it contains C^1 by Theorem 4.4.7 and is contained in C^0 by Theorem 4.4.2. The intersection of these classes is the class of C^∞ functions, the functions that have derivatives of all orders. The polynomials, trigonometric functions, exponential functions, and their inverses are C^∞ functions where they are defined.

Example 4.4.9 Let us compute a few derivatives.
(a) Let $f(x, y) = x^2y + 3xy$, a polynomial of two variables. Then:

$$\begin{aligned} f'(x,y) &= \nabla f(x,y) = (\frac{\partial f}{\partial x}(x,y), \frac{\partial f}{\partial y}(x,y)) \\ &= (2xy + 3y, x^2 + 3x) \end{aligned}$$

(b) Let $f(x, y, z) = xy^2z^3 + y\cos(x)$. Then:

$$\begin{aligned} f'(x,y,z) &= \nabla f(x,y,z) = (D_x f(x,y,z), D_y f(x,y,z), D_z f(x,y,z)) \\ &= (y^2z^3 - y\sin(x), 2xyz^3 + \cos(x), 3xy^2z^2) \end{aligned}$$ △

Since $f'(a) = \nabla f(a) = (D_1 f(a), ..., D_n f(a))$, the properties of the derivative are easily computed using the corresponding properties of vectors and partial derivatives.

Theorem 4.4.10 (Algebra of Derivatives of Several Variables) *Let $f, g : \mathbb{R}^n \to \mathbb{R}$ be differentiable at $a \in \mathbb{R}^n$. Let r be a real number. Then:*

(a) $(f + g)'(a) = f'(a) + g'(a)$ (additivity of derivative)

(b) $(rf)'(a) = rf'(a)$

(c) $(fg)'(a) = f(a)g'(a) + f'(a)g(a)$

(d) *If $g(a) \neq 0$, then* $\left(\frac{f}{g}\right)'(a) = \frac{g(a)f'(a) - f(a)g'(a)}{g(a)^2}$

Proof: We will only prove (c); the proofs of the others are similar and are left to the exercises. To simplify notation, assume that $f, g : \mathbb{R}^2 \to \mathbb{R}$, with points in $\mathbb{R}^2$ having coordinates (x, y).

By Theorem 4.4.4

$$\begin{aligned}
(fg)'(a) &= \nabla(fg)(a) = (D_x(fg)(a), D_y(fg)(a)) \\
&= (f(a)D_x g(a) + D_x f(a)g(a), f(a)D_y g(a) + D_y f(a)g(a)) \\
&= (f(a)D_x g(a), f(a)D_y g(a)) + (D_x f(a)g(a), D_y f(a)g(a)) \\
&= f(a)(D_x g(a), D_y g(a)) + g(a)(D_x f(a), D_y f(a)) \\
&= f(a)g'(a) + g(a)f'(a)
\end{aligned}$$

□

Example 4.4.11 Let $f(x, y) = (2x + y)/(x - y)$, which is defined for $x \neq y$. By property (d) of the theorem,

$$\begin{aligned}
\nabla f(x, y) &= f'(x, y) = \frac{(x - y)\nabla(2x + y) - (2x + y)\nabla(x - y)}{(x - y)^2} \\
&= \frac{(x - y)(2, 1) - (2x + y)(1, -1)}{(x - y)^2} \\
&= \frac{(2x - 2y - 2x - y, x - y + 2x + y)}{(x - y)^2} \\
&= \frac{(-3y, 3x)}{(x - y)^2} \\
&= \left[\frac{-3}{(x - y)^2}\right](y, -x)
\end{aligned}$$

△

As in the case of functions from the reals into the reals, we can use the derivative to approximate a function near the point where we take the derivative.

Example 4.4.12 As with Theorem 3.6.14, we can use the derivative to approximate functions by "nicer" functions that are easier to work with. We will find an affine function that approximates $f(x, y) = x^2 + y^2$ near $(x, y) = (1, 2)$. From the definition of derivative we have

$$f(x, y) = f(1, 2) + f'(1, 2) \bullet [(x, y) - (1, 2)] + ERROR(x, y)$$

The error function is small when (x, y) is close to $(1, 2)$, so the function near $(1, 2)$ is approximated by the affine function A:

$$\begin{aligned}
A(x, y) &= f(1, 2) + f'(1, 2) \bullet ((x, y) - (1, 2)) \\
&= 5 + (2, 4) \bullet (x - 1, y - 2) \\
&= 5 + 2(x - 1) + 4(y - 2) = 2x + 4y - 5
\end{aligned}$$

The graph of $A : \mathbb{R}^2 \to \mathbb{R}$ is the set of all points (x, y, z) satisfying

$$z = A(x, y) = 2x + 4y - 5$$

This is a plane in $\mathbb{R}^3$ that approximates the surface

$$z = f(x, y) = x^2 + y^2$$

near $(x, y, z) = (1, 2, 5)$. For functions mapping $\mathbb{R}^2$ into $\mathbb{R}$, the graph of the approximating function (given by the Linear Approximation Theorem) is a (tangent) plane.

The *tangent plane* to a surface at a point is that plane which most closely approximates (just touches) the surface at the point.

This is the natural extension of the situation for functions mapping $\mathbb{R}$ into $\mathbb{R}$ where the graph of the approximating function is a tangent line.

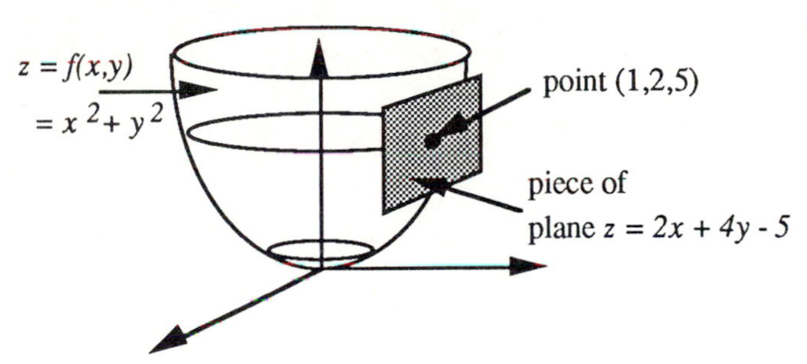

Figure 4.4.13

Compare the above approximation,

$$f(x, y) = f(1, 2) + f'(1, 2)\bullet[(x, y) - (1, 2)] + ERROR(x, y)$$

and related plane

$$z = f(1, 2) + f'(1, 2)\bullet[(x, y) - (1, 2)]$$

to the approximation of a function $g : \mathbb{R} \to \mathbb{R}$ by

$$g(x) = g(p) + g'(p)(x - p) + ERROR(x)$$

using the Linear Approximation Theorem, Theorem 3.6.14, and the related tangent line

$$y = g(p) + g'(p)(x - p)$$

As with functions of one variable, we can ask where the approximating function (A) is sufficiently close to the function (f). Suppose we want to determine the points (x,y) where $ERROR(x,y) \leq 0.01$. Then we must solve for all (x,y) that satisfy

$$ERROR(x,y) = f(x,y) - A(x,y) = (x-1)^2 + (y-2)^2 \leq 0.01$$

The set of points (x,y) satisfying this inequality is a disk of radius 0.1 centered at $(1,2)$. △

The Chain Rule

If $f : \mathbb{R} \to \mathbb{R}^n$ and $g : \mathbb{R}^n \to \mathbb{R}$, then f and g can be composed giving $(g \circ f)(x) = g(f(x))$ for x such that $f(x)$ is in the domain of g. If f is differentiable at a and g is differentiable at $f(a)$, then we can ask whether $(g \circ f)'(a)$ exists. It does, and it equals just what one would hope.

Theorem 4.4.14 (The Chain Rule) *If $f : \mathbb{R} \to \mathbb{R}^n$ is differentiable at a and $g : \mathbb{R}^n \to \mathbb{R}$ is differentiable at $f(a)$, then $g \circ f$ is differentiable at a and*

$$(g \circ f)'(a) = g'(f(a)) \bullet f'(a) \qquad \textit{(chain rule)}$$

Note that since $f : \mathbb{R} \to \mathbb{R}^n$ and $g : \mathbb{R}^n \to \mathbb{R}$, then $g \circ f : \mathbb{R} \to \mathbb{R}$, $f'(a)$ is a vector in $\mathbb{R}^n$ and $g'(f(a))$ is a vector in $\mathbb{R}^n$. This is why the dot product is required in the expression of the chain rule: we must combine two vectors to obtain a real number.

Proof: The linear approximation approach, Theorems 3.6.17 and 4.2.20 and Definition 4.4.1, will be used to obtain the result. Name the variables: $t \in \mathbb{R}$, $x = f(t) \in \mathbb{R}^n$, and $y = g(x) \in \mathbb{R}$. Since we know what the derivative ought to be, we define

$$ERROR(t) = g(f(t)) - g(f(a)) - g'(f(a)) \bullet f'(a)(t-a)$$

We must show that

$$\lim_{t \to a} \frac{|ERROR(t)|}{|t-a|} = 0$$

Since f is differentiable at a, by Theorem 4.2.20 a function $ERROR1 : \mathbb{R} \to \mathbb{R}^n$ exists such that:

(a ′) $f(t) = f(a) + f'(a)(t-a) + ERROR1(t)$

(b ′) $\lim_{t \to a} \frac{|ERROR1(t)|}{|t-a|} = 0$

Since g is differentiable at $f(a)$, by Definition 4.4.1 there is a function $ERROR2 :$ $\mathbb{R} \to \mathbb{R}$ such that:

(a $''$) $g(x) = g(f(a)) + g'(f(a))\bullet(x - f(a)) + ERROR2(x)$

(b $''$) $\lim_{x\to f(a)} \frac{|ERROR2(x)|}{|x-f(a)|} = 0$

Rewriting (a'') by substituting $f(t)$ for x we have

(eq. 1) $g(f(t)) = g(f(a)) + g'(f(a))\bullet[f(t) - f(a)] + ERROR2(f(t))$

Substituting (a') into the right-hand side of (eq. 1) we have:

$$\begin{aligned}
\text{(eq. 2) } g(f(t)) &= g(f(a)) + g'(f(a))\bullet[f'(a)(t - a) \\
&\quad + ERROR1(t)] + ERROR2(f(t)) \\
&= g(f(a)) + g'(f(a))\bullet f'(a)(t - a) \\
&\quad + g'(f(a))\bullet ERROR1(t) + ERROR2(f(t))
\end{aligned}$$

Rewriting the definition of ERROR using this last expression gives

$$ERROR(t) = g'(f(a))\bullet ERROR1(t) + ERROR2(f(t))$$

Now,

$$\begin{aligned}
\lim_{t\to a} \frac{|ERROR(t)|}{|t-a|} &= \lim_{t\to a} \frac{|g'(f(a))\bullet ERROR1(t) + ERROR2(f(t))|}{|t-a|} \\
&\le \lim_{t\to a} \frac{|g'(f(a))\bullet ERROR1(t)|}{|t-a|} + \lim_{t\to a} \frac{|ERROR2(f(t))|}{|t-a|} \\
&\quad \text{(by the Triangle Inequality)} \\
&\le |g'(f(a))| \lim_{t\to a} \frac{|ERROR1(t)|}{|t-a|} + \lim_{t\to a} \frac{|ERROR2(f(t))|}{|t-a|} \\
&\quad \text{(by the Cauchy-Schwarz-Bunyakovsky inequality)} \\
&= 0 + \lim_{t\to a} \frac{|ERROR2(f(t))|}{|t-a|} \\
&= \lim_{t\to a} \frac{|ERROR2(x)|}{|x-f(a)|} \frac{|f(t)-f(a)|}{|t-a|} \\
&= 0 \cdot |f'(a)| = 0 \quad [\text{since as } t \to a,\ x = f(t) \to f(a)]
\end{aligned}$$

□

Example 4.4.15 Suppose $f : \mathbb{R} \to \mathbb{R}^2$ and $g : \mathbb{R}^2 \to \mathbb{R}$ with $f(0) = (1,2)$, $f'(0) = (2,3)$, $g(1,2) = 4$, and $g'(1,2) = (5,6)$.

Then $h = g \circ f$ is differentiable at 0 and

$$h'(0) = g'(f(0))\bullet f'(0) = g'(1,2)\bullet(2,3) = (5,6)\bullet(2,3) = 5 \cdot 2 + 6 \cdot 3 = 28 \quad \triangle$$

Let's restate the chain rule in a simple case. Suppose $f : \mathbb{R} \to \mathbb{R}^2$ is given by $f(t) = (x(t), y(t))$, where x and y are real-valued functions of t. Then

$$f'(a) = (x'(a), y'(a)) = \left(\frac{dx(a)}{dt}, \frac{dy(a)}{dt}\right)$$

$$g'(f(a)) = \nabla g(f(a)) = \left(\frac{\partial g}{\partial x}(f(a)), \frac{\partial g}{\partial y}(f(a))\right)$$

This gives:

$$(g \circ f)'(a) = \frac{\partial g}{\partial x}(f(a))\frac{dx(a)}{dt} + \frac{\partial g}{\partial y}(f(a))\frac{dy(a)}{dt}.$$

Or, more briefly,

$$\frac{dg}{dt} = \frac{\partial g}{\partial x}\frac{dx}{dt} + \frac{\partial g}{\partial y}\frac{dy}{dt}$$

Example 4.4.16 Let $g(x, y) = xy^2 + x\sin(y), f(t) = (3t, t^2)$. Then $x(t) = 3t$, $y(t) = t^2$ and

$$\begin{aligned} \frac{dg}{dt} &= \frac{\partial g}{\partial x}\frac{dx}{dt} + \frac{\partial g}{\partial y}\frac{dy}{dt} \\ &= [y^2 + \sin(y)]3 + [2xy + x\cos(y)]2t \\ &= [t^4 + \sin(t^2)]3 + [2 \cdot 3t^3 + 3t\cos(t^2)]2t \\ &= 15t^4 + 3\sin(t^2) + 6t^2\cos(t^2) \end{aligned}$$

Since $g \circ f(t)$ is a function of t, we have expressed the final answer in terms of t alone. △

Example 4.4.17 Two cars, the first moving 10 m/sec and the second moving 20 m/sec, pass at a (perpendicular) intersection. How fast are the cars separating when they are 100 m apart?

First we model the situation with points moving on the x and y axes of the xy-plane. Let the first car be moving on the positive x-axis and the second car be moving on the positive y-axis.

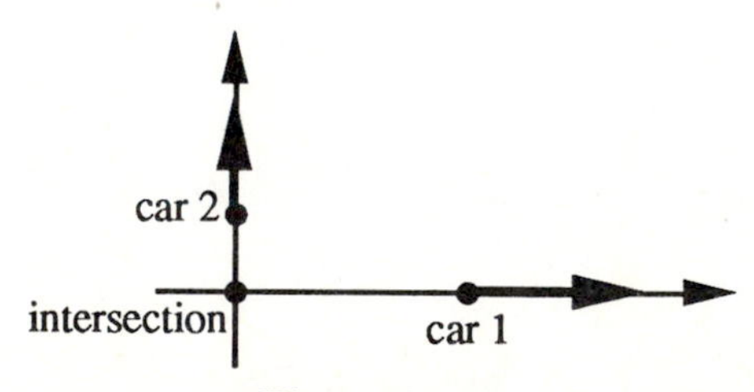

Figure 4.4.18

The positions at time t are $x(t)$ and $y(t)$. The speeds at time t are $x'(t) = 10$ and $y'(t) = 20$. The distance between $x(t)$ and $y(t)$ is

$$g(x(t), y(t)) = (x(t)^2 + y(t)^2)^{1/2}$$

Suppose at time $t = t_0$, $g(x(t_0), y(t_0)) = 100$.

We want dg/dt when $t = t_0$.

$$\begin{aligned}
\frac{dg}{dt}(x(t_0), y(t_0)) &= g'(x(t_0), y(t_0)) \cdot (x'(t_0), y'(t_0)) \\
&= \frac{\partial g}{\partial x}(x(t_0), y(t_0))x'(t_0) + \frac{\partial g}{\partial y}(x(t_0), y(t_0))y'(t_0) \\
&= \frac{x(t_0)}{(x(t_0)^2 + y(t_0)^2)^{1/2}} 10 + \frac{y(t_0)}{(x(t_0)^2 + y(t_0)^2)^{1/2}} 20 \\
&= 10\frac{x(t_0)}{100} + 20\frac{y(t_0)}{100}
\end{aligned}$$

Now we need to find $x(t_0)$ and $y(t_0)$. Since $x'(t) = 10$ and $y'(t) = 20$ for all t, we have from Theorem 3.7.3 or its corollary that $x(t) = 10t+a$ and $y(t) = 20t+b$. At time $t = 0$ the two cars are at the intersection, so $a = b = 0$. At t_0,

$$100^2 = (x(t_0)^2 + y(t_0)^2) = 100t_0^2 + 400t_0^2 = 500t_0^2$$

This gives

$$t_0 = \sqrt{20}, x(t_0) = 10\sqrt{20} \quad \text{and} \quad y(t_0) = 20\sqrt{20}$$

Substituting into the equation for dg/dt, we have

$$\begin{aligned}
\frac{dg}{dt}(x(t_0), y(t_0)) &= \frac{10\sqrt{20}}{100} + \frac{20\sqrt{20}}{100} \\
&= 3\sqrt{20}/10 \approx 1.34 \text{ m/sec}
\end{aligned}$$

△

The chain rule can be extended to the case where $f : \mathbb{R}^m \to \mathbb{R}^n$ and $g : \mathbb{R}^n \to \mathbb{R}$ by considering partial derivatives of f.

In the case where $f : \mathbb{R}^3 \to \mathbb{R}^2$ and $g : \mathbb{R}^2 \to \mathbb{R}$, let's name the variables in $\mathbb{R}^3$ (x, y, z) and the variables in $\mathbb{R}^2$ (u, v), so that u and v depend on (x, y, z) $(u = u(x, y, z)$ and $v = v(x, y, z))$ and

$$f(x, y, z) = (u, v) = (u(x, y, z), v(x, y, z))$$

Then

$$g(f(x, y, z)) = g(u(x, y, z), v(x, y, z))$$

In vector notation, we have

$$\frac{\partial g \circ f}{\partial x}(a) = g'(f(a)) \bullet \left(\frac{\partial u}{\partial x}(a), \frac{\partial v}{\partial x}(a)\right)$$

or

$$\frac{\partial (g \circ f)}{\partial x} = \frac{\partial g}{\partial u}\frac{\partial u}{\partial x} + \frac{\partial g}{\partial v}\frac{\partial v}{\partial x}$$

Example 4.4.19 Suppose $f : \mathbb{R}^3 \to \mathbb{R}^2$, $f = (u, v)$, and $g : \mathbb{R}^2 \to \mathbb{R}$ with

$$\begin{array}{lll} f(1,2,3) = (4,5) & D_1u(1,2,3) = 3 & D_2u(1,2,3) = -2 \\ D_3u(1,2,3) = 4 & D_1v(1,2,3) = 2 & D_2v(1,2,3) = -3 \\ D_3v(1,2,3) = -1 & & \end{array}$$

and

$$g(4,5) = 6 \quad D_1g(4,5) = 1 \text{ and } \quad D_2g(4,5) = -4$$

Let $h = g \circ f : \mathbb{R}^3 \to \mathbb{R}$. Then

$$\begin{aligned} D_1h(1,2,3) &= g'(f(1,2,3)) \bullet (D_1u(1,2,3), D_1v(1,2,3)) \\ &= g'(4,5) \bullet (D_1u(1,2,3), D_1v(1,2,3)) \\ &= (D_1g(4,5), D_2g(4,5)) \bullet (D_1u(1,2,3), D_1v(1,2,3)) \\ &= (1,-4) \bullet (3,2) = 1 \cdot 3 + (-4)2 = -5 \end{aligned}$$

△

Question 4.4.20 What is $D_2h(1,2,3)$? ◇

Example 4.4.21 Let $u(x,y,z) = x + y^2 + x\cos(z)$ and $v(x,y,z) = xy + z$. If $g(u,v) = 2u + 3v$, then

$$\begin{aligned} \frac{\partial g}{\partial z}(u(x,y,z), v(x,y,z)) &= \frac{\partial g}{\partial u}\frac{\partial u}{\partial z} + \frac{\partial g}{\partial v}\frac{\partial v}{\partial z} \\ &= 2[-x\sin(z)] + 3[1] \\ &= 3 - 2x\sin(z) \end{aligned}$$

△

Section 4.4 Exercises

1. Find the gradient of $f(x,y) = 2x\sin(xy)$.
2. Find the gradient of $f(x,y,z) = xyz^2 + y\cos(x)$
3. Find an affine function of two variables which approximates $f(x,y) = x^2 + 3y^2$ near $(x,y) = (1,1)$.

4. Find an affine function which approximates $f(x, y, z) = x^2 + 3y^2 - 5z$ near $(x, y, z) = (1, 0, 1)$.

5. Find the derivative of $f(x(t), y(t))$ with respect to t if $f(x, y) = 2xy^2$, $x(t) = 3t^3 + 1$, and $y(t) = \sin(t)$.

6. Find the derivative of $f(x(t), y(t))$ with respect to t if $f(x, y) = 3x + x^2 y$, $x(t) = 4t^2$, and $y(t) = t + \cos(t)$.

7. If $f(u, v) = u + uv$, $u(x, y) = x^2 \cos(y)$, and $v(x, y) = 3x + y$, then find $D_x f(u(x, y), v(x, y))$.

8. If $f(u, v) = \sin(u^2 v), u(x, y) = x - y$, and $v(x, y) = x^2$, then find $D_x f(u(x, y), v(x, y))$.

9. The volume of a right circular cone is given by $V = \pi r^2 h/3$, where r is the radius of the base and h is the height. At a certain instant the height is 20 cm and is increasing at the rate of 2 cm/sec. At the same instant the radius of the base is 15 cm and is increasing at the rate of 1 cm/sec. At what rate is the volume increasing at that instant?

10. The base b_1 of a trapezoid increases in length at the rate of 3 cm/sec and the base b_2 decreases at the rate of 1 cm/sec. If the altitude is increasing at the rate of 4 cm/sec, how rapidly is the *area* $= ((b_1 + b_2)h)/2$ changing when $b_1 = 20$ cm, $b_2 = 10$ cm, and $h = 5$ cm?

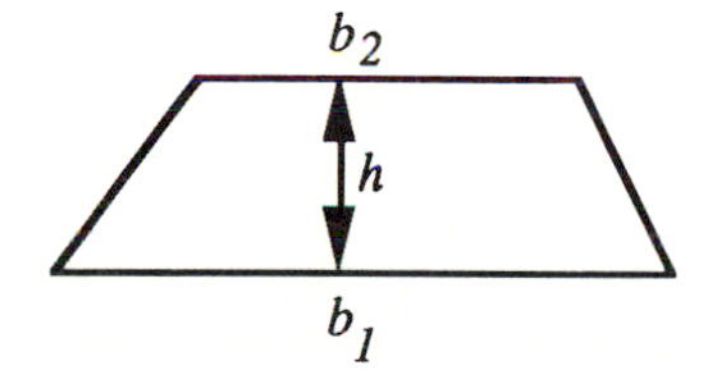

Figure 4.4.22

11. A hot air balloon is traveling westward at 10 m/sec and is rising at the rate of 2 m/sec. At a certain instant the balloon is at an altitude of 1000 m and is 2000 m directly east of an observer. How fast is the distance between the balloon and the observer changing at that instant?

12. If $g(t, x) = f(x + at)$ where $f : \mathbb{R} \to \mathbb{R}$, show that $D_t g(t, x) = a D_x g(t, x)$.

13. We defined the cross product of two vectors in the exercises for Chapter 4 Section 1. There is a closely related concept for functions $f : \mathbb{R}^3 \to \mathbb{R}^3$. Suppose

$$f(x, y, z) = (f_1(x, y, z), f_2(x, y, z), f_3(x, y, z))$$

The *curl* of f is a function from $\mathbb{R}^3$ into $\mathbb{R}^3$ defined by

$$curl(f) = \nabla \times f = (D_2f_3 - D_3f_2, D_3f_1 - D_1f3, D_1f_2 - D_2f_1)$$

If a_i is replaced by D_i in the definition for cross product, the definition of curl is obtained. Compute the curl of $f(x,y,z) = (xyz, x^2z + y, x^2y + z)$

14. Compute the curl of $f(x,y,z) = (\sin(xy), x + y^2, yz)$.

15. The curl satisfies

    (a) $\text{curl}(f + g) = \text{curl}(f) + \text{curl}(g)$

    (b) if h is real valued, then $\text{curl}(hf) = h\text{curl}(f) + h' \bullet f$

    (c) if f is C^2, then $\text{curl}(\text{ grad}f) = 0$.

    Show that (b) holds.

16. Show that property (c) of the curl holds.

17. Let $f : \mathbb{R}^n \to \mathbb{R}^n$ be a smooth function, $f = (f_1, f_2, ..., f_n)$ The *divergence* of f, div(f): $\mathbb{R}^n \to \mathbb{R}$ is defined by

$$\text{div}(f) = \nabla \bullet f = D_1f_1 + D_2f_2 + \cdots + D_nf_n$$

    What is the divergence of $f(x,y) = (xy^2, x\cos(y))$?

18. What is div(f) where $f(x,y,z) = (x + 2y, \cos(z), xy)$?

19. The divergence satisfies

    (a) $\text{div}(f + g) = \text{div}(f) + \text{div}(g)$

    (b) If h is real valued, $\text{div}(hf) = h\text{div}(f) + \text{grad}h \bullet f$.
    Show that (b) holds.

20. Show that if $f : \mathbb{R}^3 \to \mathbb{R}^3$ has C^2 component functions then, $\text{div}(\text{curl}(f)) = 0$.

21. Prove Theorem 4.4.10 part (a).

22. Prove Theorem 4.4.10 part (d).

23. Prove or disprove: If $f : \mathbb{R}^n \to \mathbb{R}$ is differentiable at a, then the directional derivative of f at a in the direction of v is the projection of $f'(a)$ in the direction of v, $Proj(f'(a), v)$.

4.5 Geometric Properties of the Gradient

How do you display three dimensional geometry in two dimensions? For example, how do you indicate elevation on a map? Topographers use contour curves, which are curves of constant elevation, to show elevation. For example, a person hiking along a contour curve always remains at the same level. A contour curve is an example of a *level set* of a function. For studying geometrical properties in a small neighborhood of a point, approximation—a fundamental theme of this text—is the process to use. In particular, we will approximate a surface with a *tangent plane* just as earlier we approximated a curve with a tangent line.

The major emphasis of this section is to develop the concepts of level sets and tangent planes as aids in analyzing the geometrical properties of graphs of functions $f : \mathbb{R}^n \to \mathbb{R}$. The gradient will prove to be the major tool in this task.

Level Sets

Example 4.5.1 (Physical Interpretation of Level Sets)
(a) As noted above, the contour curves on a topographic map are level sets of the elevation function, $elv : \mathbb{R}^2 \to \mathbb{R}$ that maps the latitude and longitude of a point into the elevation (in feet) at that point. Normally contour curves are drawn at 20-ft intervals. That is, the elevation on adjacent curves would differ by 20 ft. The steepness of the terrain being mapped is easily determined by how close the contour curves are together (i.e., the closer the curves are together, the steeper the terrain). The figure below illustrates the level curves, elv $(x, y) = k$ for $k = 100, 110, 120, 130, 140, 150$. Note that the terrain is steep near point A and relatively flat near point B.

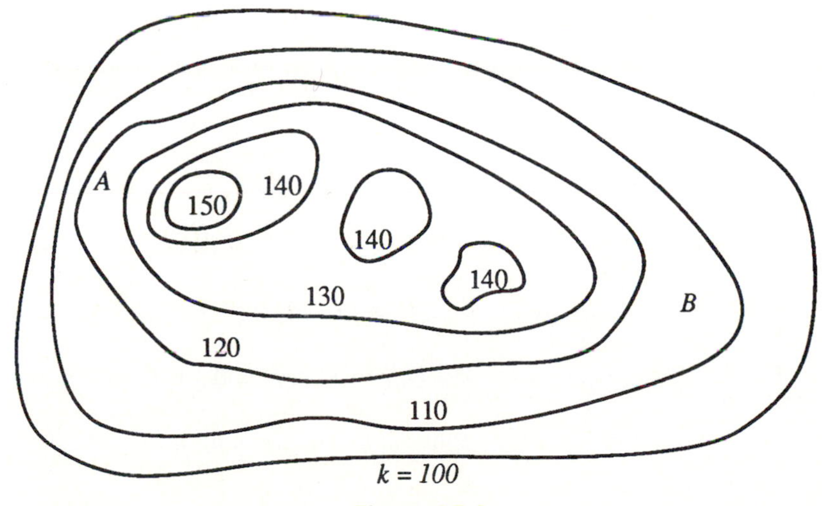

Figure 4.5.2

(b) The contour curves on a navigator's chart are also level curves of the elevation function, only this time the k values are negative to indicate depth below sea level.

(c) The weather map in the local newspaper usually shows isothermal curves along which the temperature is constant. These are identical in appearance to the above elevation contours, since they are level sets of the temperature function, $temp : \mathbb{R}^2 \to \mathbb{R}$, that maps the latitude and longitude of a point into the temperature at that point.

(d) Another weather map example would be the level curves of the barometric pressure function. These level curves are called *isobars.* $\triangle$

The reader should observe that the level sets illustrated above (contour curves, isothermal curves, isobars curves) are subsets in the domain of the respective functions. It is important to understand in the following definition that *level sets of a function are subsets of the domain of the function.*

Definition 4.5.3 (Level Sets) *Let $f : A \to B$ and $c \in B$. The* level set *of f for c is $\{a \in A : f(a) = c\}$.*

We repeat for emphasis, a level set is the set of all points in the *domain* of a function f that are mapped to a fixed value c in B. In fact, the level sets of a function partition the domain of the function into disjoint sets. (Two level sets can never intersect. Why?) If c is not in the range of f, then the level set is empty. Level sets are particularly useful when the function f is real valued.

Example 4.5.4 Let $f : \mathbb{R}^2 \to \mathbb{R}$ be defined by $f(x, y) = x^2 + y^2$. The graph of f is the paraboloid (see Example 1.3.20 in Chapter 1). The level sets of f are circles:
If $c = 9$, then $\{ (x, y) : f(x, y) = 9 \} = \{ (x, y) : x^2 + y^2 = 9 \}$ is the circle of radius $\sqrt{9} = 3$.
If $c = 4$, then $\{ (x, y) : f(x, y) = 4 \} = \{ (x, y) : x^2 + y^2 = 4 \}$ is the circle of radius $\sqrt{4} = 2$.
If $c = 1$, then $\{ (x, y) : f(x, y) = 1 \} = \{ (x, y) : x^2 + y^2 = 1 \}$ is the circle of radius $\sqrt{1} = 1$.
If $c = 0$, then $\{ (x, y) : f(x, y) = 0 \} = \{ (x, y) : x^2 + y^2 = 0 \}$ is the (degenerate) circle of radius $\sqrt{0} = 0$.
If $c = -1$, then $f(x, y) = -1$ has no solution and there is no level set. $\triangle$

Level Sets of $f(x, y) = x^2 + y^2$

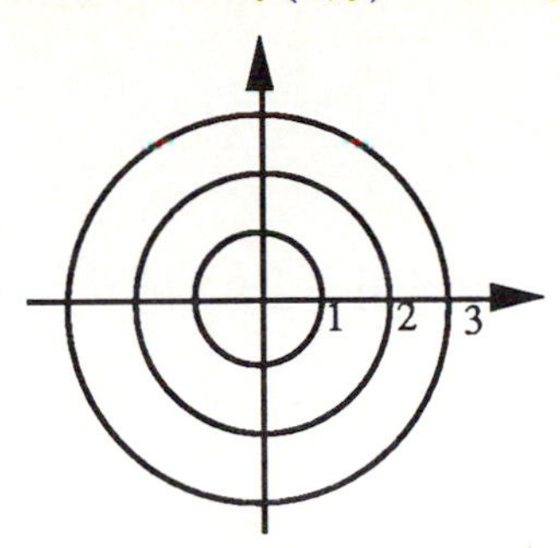

Figure 4.5.5

The word *level* is used because of the interpretation of $f(x, y)$ as the "height" of the point (x, y) on the graph of f. $\{(x, y) : f(x, y) = c\}$ is the set of all points at a "height" of c, and thus all of them are on the same "level."

Example 4.5.6 Consider the function $f(x, y) = x^2+y^2$ of the previous example. The graph of f with the points of heights 9, 4, 1, and 0 is in Figure 4.5.7.

Heights on the Graph of $f(x, y) = x^2 + y^2$

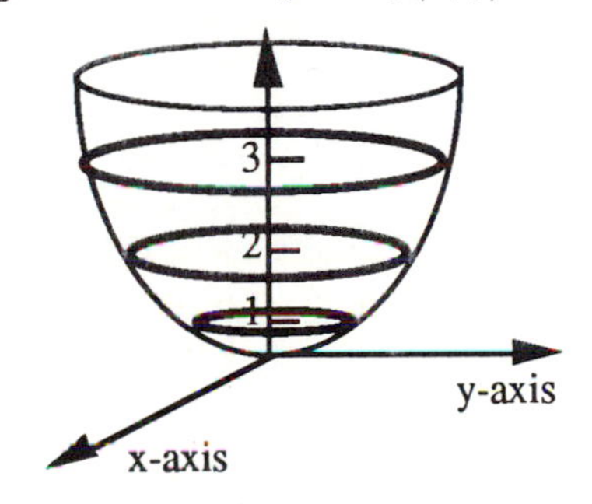

Figure 4.5.7

The level sets of Figure 4.5.5 are the "point" curves of Figure 4.5.7 projected down onto the xy-plane. △

If $c = 0$, then the term *zero set* is often used. If the level set (which is a subset of the domain) is in the plane, then it is often a curve and is called a *level curve.* If the level set is in three-dimensional space, then the set is often a surface and is called a *level surface.*

Example 4.5.8 Consider the parabola $y = x^2$. We have seen that the parabola can be considered as the graph of the function $SQ : \mathbb{R} \to \mathbb{R}$ defined by $SQ(x) = x^2$, or as the range of a parameterizing function

$$SQP : \mathbb{R} \to \mathbb{R}^2, SQP(t) = (t, t^2)$$

We will now consider the parabola as a level set. Define

$$SQL : \mathbb{R}^2 \to \mathbb{R} \text{ by } SQL(x, y) = x^2 - y, \qquad \text{Domain}(SQL) = \mathbb{R}^2$$

Then the level set (or zero set)

$$\{(x, y) : SQL(x, y) = 0\} = \{(x, y) : x^2 - y = 0\}$$

is the parabola $y = x^2$. Since this set is a curve, it is also called a level curve. △

Example 4.5.9 Let $f : \mathbb{R}^2 \to \mathbb{R}$ be defined by $f(x, y) = xy$. The level curves are hyperbolas, $\{(x, y) : xy = c\}$. The zero set

$$\{(x, y) : xy = 0\} = \{(x, y) : x = 0 \text{ or y } = 0\}$$

is the set of all points on the coordinate axes. △

Hyperbolas as Level Sets

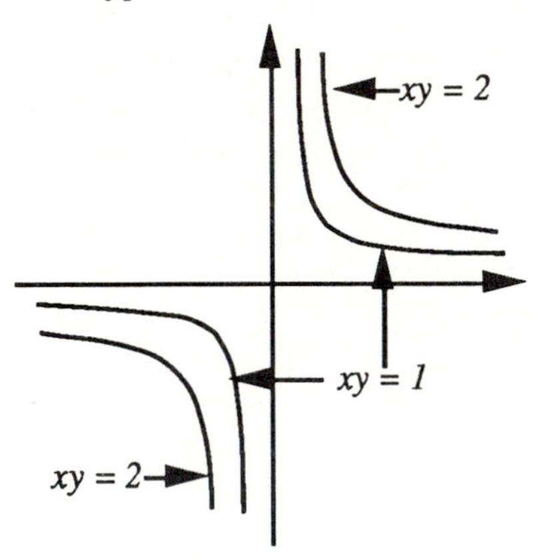

Figure 4.5.10

Level sets are very useful in describing sets in $\mathbb{R}^3$. To sketch sets of points in $\mathbb{R}^3$ given as the level set of a function, it helps to let one of the variables be constant. This is equivalent to looking at a two-dimensional slice of the level set, reducing the visualization problem from three to two dimensions.

Example 4.5.11 Let $S = \{(x, y, z) : x^2/4 + y^2/9 + z^2 = 1\}$, a level surface in $\mathbb{R}^3$. This surface is an ellipsoid centered at the origin $(0, 0, 0)$. If $z = 0$, we have the ellipse $x^2/4 + y^2/9 = 1$ in the xy-plane; if $y = 0$, we have the ellipse $x^2/4 + z^2 = 1$ in the xz-plane; if $x = 0$, we have the ellipse $y^2/9 + z^2 = 1$ in the yz-plane. △

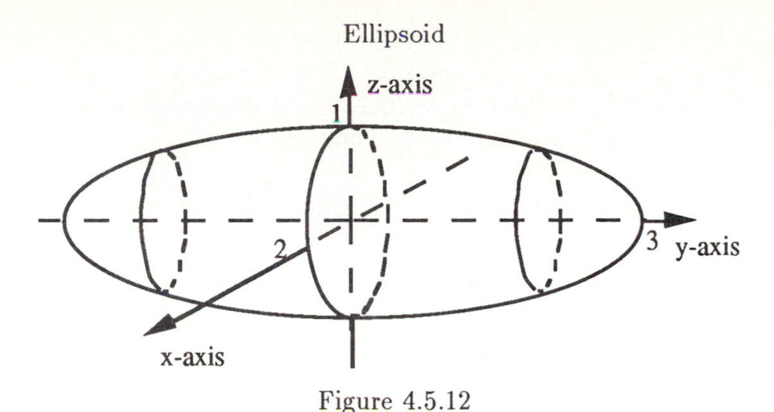

Figure 4.5.12

Tangent Planes

We now consider approximating the graph of a function at a point with a tangent plane similar to how we approximated a curve at a point with a tangent line. Recall that a tangent line was a line passing through a given point (on the curve) in a specified direction given by the derivative. Thus it is reasonable to expect to define a tangent plane as a plane passing through a given point in a specified direction. *How is direction specified for a plane?* We will see that the proper direction to associate with a plane is the direction perpendicular to the plane. First, however, we need to understand the graphical implications of the two vector aspects, magnitude and direction, of the gradient.

Theorem 4.5.13 *If $f : \mathbb{R}^n \to \mathbb{R}$ is differentiable at a, then the maximum value of $D_u f(a)$ is*

$$|f'(a)| = |\nabla f(a)|$$

and occurs when u is parallel to $\nabla f(a)$.

Proof: Let u be any unit vector, $|u| = 1$.

$$D_u f(a) = f'(a) \cdot u = |f'(a)||u| \cos(\theta) = |f'(a)| \cos(\theta)$$

where θ is the angle between $f'(a)$ and u. This product is a maximum when $\cos(\theta) = 1$, that is, when $\theta = 0$. Thus the maximum value of $D_u f(a)$ occurs when u is parallel to $f'(a)$, and that maximum value is $|f'(a)|$. □

This theorem says that $f'(a) = \nabla f(a)$ points in the direction in which f is increasing fastest at a, and that the rate of change of the function in this direction is $|\nabla f(a)|$.

Example 4.5.14 In which direction is the function $f(x,y) = 3x^2 + y^2 + 1$ increasing the most rapidly at $(1,1)$, and what is this maximum rate of increase?

$\nabla f(x,y) = (6x, 2y)$, so $\nabla f(1,1) = (6,2)$. f is increasing most rapidly at $(1,1)$ in the direction of the vector $(6,2)$. The slope is

$$\frac{\Delta y}{\Delta x} = \frac{2}{6} = \frac{1}{3} = \tan(\theta)$$

where θ is measured counterclockwise from the x-axis. Thus $\theta \approx 18.4° \approx 0.32$ radians. Note that the function f is increasing most rapidly at an angle of 0.32 radians from the x-axis, and the rate of increase in this direction is $\sqrt{4+36} = \sqrt{40} \approx 6.32$. △

We now define what is meant for a vector to be perpendicular to a curve. We then use this result to show how the gradient allows us to construct vectors perpendicular to level sets. Recall that a smooth curve has a natural vector associated with it at any point p, the unit tangent vector $T(p)$ at that point.

Definition 4.5.15 *A vector v is perpendicular to a curve at a point p if v is perpendicular to the unit tangent vector of the curve at p.*

Theorem 4.5.16 *If $f : \mathbb{R}^n \to \mathbb{R}$ is a C^1 function and the point p is on the level set $\{x : f(x) = c\}$, then $\nabla f(p)$ is perpendicular to every curve on the level set that passes through p.*

Proof: The range of $g : I \to \mathbb{R}^n$ is a curve in $\mathbb{R}^n$, where I is some interval in $\mathbb{R}$. We want the range of $g, g(I)$, to be in the level set, so we require that $f(g(t)) = c$ for t in I. Assume that $g(t_0) = p$. We want $g(I)$ to have a unit tangent vector at p, so we also require that $g'(t_0) \neq \vec{O}$. We must show that $g'(t_0)$ is perpendicular to $\nabla f(p)$, i.e., that $f'(p) \bullet g'(t_0) = 0$.

The composition $f(g(t)) = c$ and all the derivatives remind us of the chain rule, so let's take the derivative and see what we get.

$$\frac{d}{dt}(f \circ g)(t) = f'(g(t)) \bullet g'(t) = \frac{dc}{dt} = 0$$

Evaluating at $t = t_0$, we obtain

$$f'(p) \bullet g'(t_0) = 0$$

Just what was wanted. □

Example 4.5.17 Let $f : \mathbb{R}^2 \to \mathbb{R}$ be given by $f(x,y) = xy$. The level set $\{(x,y) : f(x,y) = 1\}$ is a hyperbola. A curve in this level set through the point

$(1,1)$ is given by $g(t) = (t, 1/t)$ for $t > 0$. The unit tangent vector to the curve at $(1,1) = g(1)$ is

$$T(1) = \frac{g'(1)}{|g'(1)|} = \frac{(1,-1)}{|g'(1)|}$$

The gradient is $\nabla f(x,y) = (D_x f, D_y f) = (y, x)$, so $\nabla f(1,1) = (1,1)$. As the theorem promises,

$$\nabla f(1,1) \cdot T(1) = (1,1) \cdot \frac{(1,-1)}{|g'(1)|} = \frac{1}{|g'(1)|} - \frac{1}{|g'(1)|} = 0$$

so the gradient is perpendicular to the curve. △

Assume that $f : \mathbb{R}^3 \to \mathbb{R}$ is a C^1 function and that $\nabla f(p) \neq \vec{O}$. A level set $\{(x,y,z) : f(x,y,z) = c\}$ defines a two-dimensional surface in three-dimensional space and $\nabla f(p)$ is perpendicular to this surface at p. Geometrically, a point p and a vector $\nabla f(p)$ determine the plane of all lines through p that are perpendicular to $\nabla f(p)$.

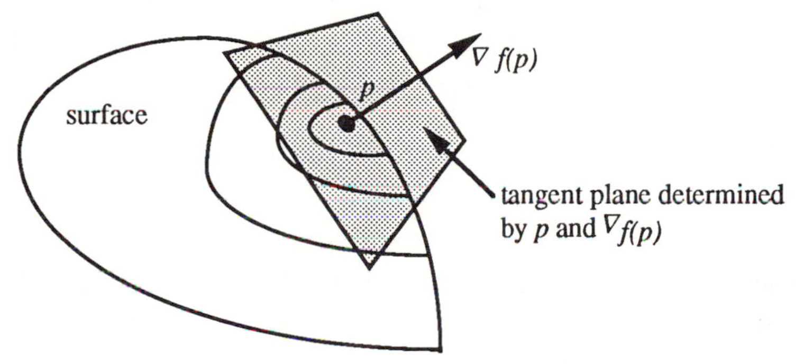

Figure 4.5.18

This plane is the tangent plane to the surface $\{(x,y,z) : f(x,y,z) = c\}$ at p since it just touches the surface at p.

Since it is much easier to work with planes than with arbitrary surfaces, it is often useful to find a tangent plane to a surface and study it rather than the surface itself. (See Example 4.4.12 and its further discussion below in terms of tangent planes.)

The vector $N(p) = \nabla f(p)/|\nabla f(p)|$ is called the *unit normal vector* to the surface $\{(x,y,z) : f(x,y,z) = c\}$ at p.

It is important for the reader to realize that the tangent plane in the following theorem is tangent to a level surface of the function.

Theorem 4.5.19 (The Equation of the Tangent Plane) *Let $f : \mathbb{R}^3 \to \mathbb{R}$ be a C^1 function with $f'(p) \neq \vec{O}$. Then the equation of the tangent plane to the surface $\{x : f(x) = c\}$ at the point p is*

$$(x - p) \cdot f'(p) = 0$$

Proof: We need only note that $x - p$ is a vector from p to x. So x is in the tangent plane to $f(x) = c$ if and only if $x - p$ is perpendicular to $f'(p)$. □

Example 4.5.20 This theorem can be used to find a tangent plane to the graph of a function, f. The procedure is to express the graph of f as a level surface of another function, F, and then apply Theorem 4.5.19 to F.

Let $f(x, y) = 3x^2 + y^2$. The graph of f is all points $\{(x, y, z) : z = 3x^2 + y^2\}$, which is the level surface $\{(x, y, z) : F(x, y, z) = 3x^2 + y^2 - z = 0\}$. $F'(x, y, z) = (6x, 2y, -1)$. When $x = x_0$ and $y = y_0$, then $z_0 = 3x_0^2 + y_0^2$ is on the graph, and the equation of the tangent plane to the graph at (x_0, y_0, z_0) is

$$(x - x_0, y - y_0, z - z_0) \bullet (6x_0, 2y_0, -1) = 0$$

For example, when $x_0 = 1$ and $y_0 = 2$, then $z_0 = 3 + 4 = 7$. The equation of the tangent plane to the graph of f at $(1, 2, 7)$ is

$$(x - 1, y - 2, z - 7) \bullet (6, 4, -1) = 0$$

or,

$$6x + 4y - z = 7$$

△

Example 4.5.21 In this example we will find the equation of the tangent plane to the unit sphere $x^2 + y^2 + z^2 = 1$ at $p = (1/\sqrt{3}, 1/\sqrt{3}, 1\sqrt{3})$.

We first need to express the sphere as a level surface of a function. This is done by defining $f : \mathbb{R}^3 \to \mathbb{R}$ by $f(x, y, z) = x^2 + y^2 + z^2$. The sphere is then the level surface $\{(x, y, z) : f(x, y, z) = 1\}$. Now we apply the previous theorem. Since

$$\nabla f(x, y, z) = (2x, 2y, 2z) \quad \text{or} \quad \nabla f(p) = \left(\frac{2}{\sqrt{3}}, \frac{2}{\sqrt{3}}, \frac{2}{\sqrt{3}}\right)$$

the equation of the tangent plane is

$$\left(x - \frac{1}{\sqrt{3}}, y - \frac{1}{\sqrt{3}}, z - \frac{1}{\sqrt{3}}\right) \bullet \left(\frac{2}{\sqrt{3}}, \frac{2}{\sqrt{3}}, \frac{2}{\sqrt{3}}\right) = 0$$

or,

$$x + y + z = \sqrt{3}$$

△

This example illustrates a third way to describe sets in $\mathbb{R}^n$ in terms of functions. There are three ways to describe a set of points in terms of a function.

1. The set of points may be the *graph* of a function.

2. The set of points may be the *range* of a parameterizing function.

3. The set of points may be a level set in the *domain* of a function.

Example 4.5.22 Consider the approximation to $f(x,y) = x^2 + y^2$ in Example 4.4.12. We saw that the affine function

$$A(x,y) = 2x + 4y - 5$$

approximated f near $(1,2,5)$ and that the graph of A was the plane

$$z = 2x + 4y - 5$$

We now show that this plane is a tangent plane to the graph of f.

The graph of $z = f(x,y)$ is the zero (level) set of the function $F : \mathbb{R}^3 \to \mathbb{R}$ defined by $F(x,y,z) = x^2 + y^2 - z$. That is,

$$\text{graph of } f = \{(x,y,z) : F(x,y,z) = 0\} = \{(x,y,z) : x^2 + y^2 - z = 0.\}$$

This has the normal vector $(2x, 2y, -1)$. At $(1,2,5)$ the normal vector is $(2,4,-1)$ and the equation of the tangent plane is

$$(2,4,-1)\bullet(x-1, y-2, z-5) = 0$$

or

$$z = 2x + 4y - 5$$

Thus if a differentiable function is approximated near p by an affine function A

$$f(x) = A(x) + ERROR(x)$$

where (by the definition of derivative)

$$A(x) = f(p) + f'(p)\bullet(x - p)$$

then the graph of A is the tangent plane to the graph of f. △

Question 4.5.23 Find the equation of the tangent plane to the graph of $f(x,y) = x^2 + 2xy + y$ when $x = 1$ and $y = 0$. ◊

Surfaces and tangent planes are analogous to curves and tangent lines for functions of one variable. These concepts can be extended to functions on $\mathbb{R}^n$ for any n.

Definition 4.5.24 (Hypersurface) *Let $f : \mathbb{R}^n \to \mathbb{R}$ be a C^1 function. A level set $\{x : f(x) = c\}$ defines a $n-1$ dimensional set called a* hypersurface *in $\mathbb{R}^n$. (A hypersurface has one less dimension than the space it is in.) If $f(p) = c$, then $\nabla f(p)$ is a* normal *(i.e. perpendicular) vector to the hypersurface, and $\nabla f(p)/|\nabla f(p)|$ is the* unit normal vector *to the hypersurface.*

If $f : \mathbb{R}^2 \to \mathbb{R}$, then a hypersurface is a curve in the plane; if $f : \mathbb{R}^3 \to \mathbb{R}$, then a hypersurface is a two-dimensional surface in $\mathbb{R}^3$.

Section 4.5 Exercises

1. Find a unit vector in the direction in which the function $f(x, y) = x^2y + 3x$ is increasing most rapidly when $(x, y) = (1, 2)$.

2. Find a unit vector in the direction in which the function $f(x, y) = 2x\cos(y) + 3x$ is increasing most rapidly.

3. Consider the equation $x^2 + y + z^2 = 0$. Describe the shape of the level set slices for each of the following.

(a) $x = 0$	(b) $y = 0$	(c) $z = 0$
(d) $x = 3$	(e) $y = 2$	(f) $z = 1$
(g) $x = -2$	(h) $x = -6$	(i) $y = 10$
(j) $y = -6$	(k) $z = 9$	(l) $z = -4$
(m) $x =$ constant > 0	(n) $y =$ constant > 0	(0) $z =$ constant < 0

4. Consider the equation $z - y^2 + x^2 = 0$. Describe the shape of the level set slices for each of the following:

(a) $x = 0$	(b) $y = 0$	(c) $z = 0$
(d) $x = 2$	(e) $x = 5$	(f) $x = -2$
(g) $y = 2$	(h) $y = -2$	(i) $y = -8$
(j) $z = -3$	(i) $z = -5$	(l) $z = 10$
(m) $x =$ constant > 0	(n) $y =$ constant > 0	(o) $z =$ constant < 0

5. Sketch the level sets:

   (a) $\{(x, y, z) : x^2 + y^2 + z^2 = 1\}$

   (b) $\{(x, y, z) : x^2/4 + y^2/9 + z^2 = 1\}$

6. Sketch the level sets:

   (a) $\{(x, y, z) : x^2 - y^2 - z^2 = 1\}$

   (b) $\{(x, y, z) : x^2 + y^2 - z^2 = 1\}$

7. If the line through $a = (a_1, a_2, a_3)$ in the direction of $v = (v_1, v_2, v_3)$ is given by the range of $f(t) = a + tv$, then the $x, y,$ and z coordinates of the line are given by $x = a_1 + tv_1$, $y = a_2 + tv_2$, and $z = a_3 + tv_3$. Each of these equations can be solved for t and then the different expressions can be equated.

$$\frac{x - a_1}{v_1} = \frac{y - a_2}{v_2} = \frac{z - a_3}{v_3}$$

This allows a line to be expressed without parameterization, i.e., without reference to t. The equalities can be written as

$$\frac{x - a_1}{v_1} = \frac{y - a_2}{v_2}$$

a plane parallel to the z-axis,

$$\frac{y - a_2}{v_2} = \frac{z - a_3}{v_3}$$

a plane parallel to the x-axis, and

$$\frac{x - a_1}{v_1} = \frac{z - a_3}{v_3}$$

a plane parallel to the y-axis. These three planes intersect in the line. Find the above (nonparametric) equations for the line through $(1,2,3)$ in the direction of $(-1,2,1)$.

8. Find the (non-parametric) equations (as in Exercise 7 above) for the line through $(2,0,-1)$ in the direction of $(2,1,3)$.

9. Find the equation of the plane tangent to the surface $x^2 + y^2 - z^2 = 4$ at $(x,y,z) = (2,1,1)$.

10. Find the equation of the plane tangent to the graph of $f(x,y) = x^2 + y^3$ when $(x,y) = (2,1)$.

11. Find the nonparametric equations of the line normal to the surface $z = x^2 + y^2$ at $(x,y) = (1,2)$.

12. Find the nonparametric equations (as in Exercise 7 above) of the line normal to the sphere of radius 3 at $(1,2,\sqrt{2})$.

13. (a) A cow pond needs a new boat ramp (the old one is no longer flat). It is to fit smoothly into the slope which is given as the graph of $f(x,y) = 8 - 4x^2 - 2xy - y^2$ at the point $(1,1,1)$. Find the equation describing the ramp.

    (b) A cow is looking for the quickest way down the slope whose altitude is given by $a(x,y) = 3x^2 + 2xy + 2y^2 + x + 10$. She is at $(1,1,18)$. Which way should the cow go?

14. Show that any line normal to the sphere $x^2 + y^2 + z^2 = r^2$ passes through the center of the sphere.

15. In the exercises for Section 3.2 we stated that a line is normal to a curve in the plane if the slope of the line is the negative reciprocal of the slope of the tangent line to the curve. Suppose $f : \mathbb{R} \to \mathbb{R}$ is a C^1 function. Show that a unit normal vector to the graph of f, as defined in Definition 4.5.24, determines a line which is normal to the graph of f in the sense of the exercises for Section 3.2.

4.6 Extrema of Functions of Several Variables

In Section 3.8, we showed that among all rectangles with a given perimeter, the square has the largest area. What can be said about the three-dimensional version of the question? That is, is it true that among all rectangular shaped boxes with a given surface area, the cube contains the largest volume? It seems reasonable, but how would you show it?

The concepts of finding maxima and minima of functions of several variables are essentially the same as for functions of one variable.

Just as with functions of one variable, the derivative can be used to help find maxima and minima. As in the case of functions of one variable, a local (relative) maximum occurs at a point p if $f(x) \leq f(p)$ for all x near p, i.e., $|x-p|$ is small.

Question 4.6.1 Write out the condition that there is a local minimum at p. ◇

As with differentiable functions of a single variable, the extreme values occur at an interior point (where the derivative is zero) or on the boundary (endpoint). If a local maximum or minimum of a differentiable function occurs at a point p in the interior of a set A, then the slopes at p must be zero in all directions. Expressing this in terms of directional derivatives, if f is a differentiable function and f has a local maximum or minimum at p in the interior of A, then the directional derivatives $D_u f(p) = 0$ for all unit vectors u. Does this mean that we have to compute *all* the directional derivatives? Fortunately not, because of the following result.

Theorem 4.6.2 *If f is a differentiable function, then $D_u f(p) = 0$ for all unit vectors u if and only if $f'(p) = 0$*

Proof: First suppose $f'(p) = 0$. Then

$$|D_u f(p)| = |f'(p) \cdot u| \leq |f'(p)||u| = |f'(p)| = 0$$

Now suppose $D_u f(p) = 0$ for all unit vectors u. In particular, the partial derivatives $D_c f(p) = 0$ for all coordinate vectors c. Thus since for $c = \vec{e_k}$, the kth coordinate vector, $D_c f(p) = D_k f(p)$,

$$f'(p) = (D_1 f(p), ..., D_n f(p)) = 0 \qquad \square$$

This theorem says that all directional derivatives vanish at p if and only if

$$f'(p) = \nabla f(p) = (D_1 f(p), ..., D_n f(p)) = 0$$

Thus all directional derivatives vanish if and only if the n partial derivatives $D_k f(p)$ vanish. If f is a differentiable function and $f'(p) = 0$ then p is called a *critical point* of f, just as in the case of functions of one variable. The value $f(p)$ at a critical point p is called a *critical value.*

To find maxima or minima of f on A we must

1. Find the critical points in the interior of A.
2. Examine the critical points and corresponding critical values for maxima or minima.
3. Examine the boundary of A for maxima or minima.

Example 4.6.3 We will find the maxima and minima of $f : \mathbb{R}^2 \to \mathbb{R}$ defined by $f(x, y) = x^2 + y^2$, with $\text{Domain}(f) = \{(x, y) : x^2 + 2y^2 \leq 4\}$.

We begin by looking for critical points in the interior of the domain. Since $f'(x, y) = (2x, 2y) = (0, 0)$ if and only if $(x, y) = (0, 0)$, there is only one critical point, $(0, 0)$. Since the function f is a sum of squares, it can never be negative. Thus the critical point $(0, 0)$ is a global minimum.

We now consider the boundary. The boundary of the domain is the ellipse $x^2 + 2y^2 = 4$. On this ellipse $x^2 = 4 - 2y^2$, and so

$$f(x, y) = (4 - 2y^2) + y^2 = 4 - y^2$$

The critical points on the ellipse occur when

$$\frac{\partial(4 - y^2)}{\partial x} = 0$$

and

$$\frac{\partial(4 - y^2)}{\partial y} = 0$$

i.e., when $y = 0$, and thus $x^2 = 4$, $x = \pm 2$. Thus we need to examine $(2, 0)$ and $(-2, 0)$. At both of these points the function has the value 4, and this is a maximum since $D_{y,y}(4 - y^2) = -2 < 0$. Thus the function f has a global minimum of 0 at $(0, 0)$ and a global maximum of 4 at $(2, 0)$ and $(-2, 0)$. △

Graph of $f(x, y) = x^2 + y^2$ on Ellipse

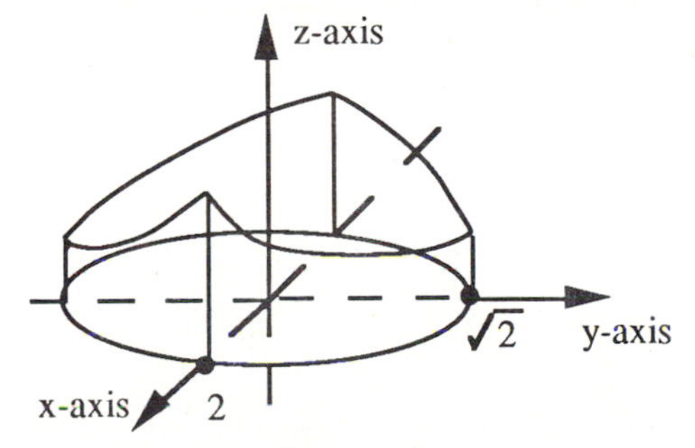

Figure 4.6.4

As with functions of one variable, $f'(p) = 0$ does not necessarily imply that $f(p)$ is a local maximum or minimum. A critical point at which f has neither a local maximum nor minimum is called a *saddle point.*

Example 4.6.5 Consider $f : \mathbb{R}^2 \to \mathbb{R}$ defined over the whole plane by $f(x, y) = y^2 - x^2$. We find the critical points by setting the first-order partial derivatives equal to zero and solving the resulting system of simultaneous equations. Since $\partial f/\partial x(x, y) = -2x = 0$ and $\partial f/\partial y(x, y) = 2y = 0$ give $x = 0$ and $y = 0$, the origin $(0, 0)$ is the only critical point of f on the entire plane. However $(0, 0)$ is neither a local minimum nor a local maximum. $f(0, 0) = 0$. If we consider a point $(x, 0)$ close to $(0, 0)$ on the x-axis, then $f(x, 0) = -x^2 < f(0, 0)$. So f does not have a local minimum at $(0, 0)$. If we consider a point $(0, y)$ close to $(0, 0)$ on the y-axis, then $f(0, y) = y^2 > f(0, 0)$. So f does not have a local maximum at $(0, 0)$. △

Graph of $f(x, y) = y^2 - x^2$

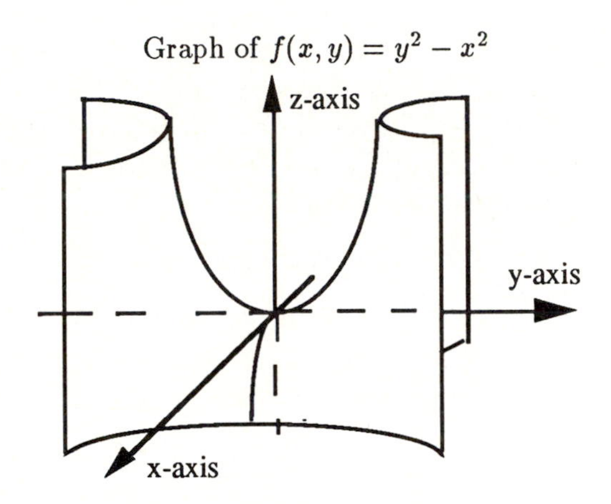

Figure 4.6.6

Example 4.6.7 Find the point on the plane $x + 2y - z = 1$ that is closest to the origin. In this problem it is clear geometrically that a solution must exist.

The distance from (x, y, z) to the origin $(0, 0, 0)$ is

$$d(x, y, z) = \sqrt{x^2 + y^2 + z^2}$$

The minimum of the function $g(x, y, z) = x^2 + y^2 + z^2$ occurs at the same point as the minimum of d (Why?) and avoids the problem of computing with the square root. Since (x, y, z) is on the plane $z = x + 2y - 1$, we must minimize

$$f(x, y) = x^2 + y^2 + (x + 2y - 1)^2 = 2x^2 + 5y^2 + 4xy - 2x - 4y + 1$$

Setting the partials equal to zero we obtain:

$$\begin{aligned} \frac{\partial f}{\partial x}(x, y) &= 4x + 4y - 2 = 0 \\ \frac{\partial f}{\partial y}(x, y) &= 10y + 4x - 4 = 0 \end{aligned}$$

Solving these equations we obtain

$$y = \frac{1}{3} \quad \text{and} \quad x = \frac{1}{6}$$

Since a minimum must exist, and since f must have a critical point at the minimum, and since (1/6,1/3) is the only critical point of f in the plane, we know that the minimum of f occurs at $(\frac{1}{6}, \frac{1}{3})$. Substituting into the equation of the plane, we obtain

$$z = x + 2y - 1 = -\frac{1}{6} \qquad \triangle$$

Just as for functions of one variable (Theorem 3.8.24) there is a second derivative test for determining whether a local maximum, local minimum, or a saddle point occurs at a critical point. The theorem will be stated for functions of two variables, but its proof postponed until Chapter 9.

Theorem 4.6.8 (The Second Derivative Test) *Let $f : \mathbb{R}^2 \to \mathbb{R}$ be a C^2 function with critical point p. Let*

$$DISC = D_{1,1}f(p)D_{2,2}f(p) - (D_{1,2}f(p))^2$$

(a) *If $DISC > 0$ and $D_{1,1}f(p) < 0$, then f has a local maximum at p.*

(b) *If $DISC > 0$ and $D_{1,1}f(p) > 0$, then f has a local minimum at p.*

(c) *If $DISC < 0$, then f has a saddle point at p.*

(d) *If $DISC = 0$, then no conclusion can be drawn.*

The expression

$$DISC = D_{1,1}f(p)D_{2,2}f(p) - (D_{1,2}f(p))^2$$

is called the *discriminant* since it helps to discriminate the type of a critical point.

Example 4.6.9 In Example 4.6.7 we found the critical point of $f(x, y) = 2x^2 + 5y^2 + 4xy - 2x - 4y + 1$ to be $p = (\frac{1}{6}, \frac{1}{3})$. We will apply the above theorem to this problem.

$$\begin{array}{ll}
D_1f(x,y) = 4x + 4y - 2 & D_1f(p) = 0 \\
D_2f(x,y) = 10y + 4x - 4 & D_2f(p) = 0 \\
D_{1,1}f(x,y) = 4 & D_{1,1}f(p) = 4 \\
D_{2,2}f(x,y) = 10 & D_{2,2}f(p) = 10 \\
D_{1,2}f(x,y) = 4 & D_{1,2}f(p) = 4
\end{array}$$

Since

$$DISC = D_{1,1}f(p)D_{2,2}f(p) - (D_{1,2}f(p))^2 = 40 - 16 > 0$$

and $D_{1,1}f(p) = 4 > 0$, part (b) of the theorem tells us that f has a local minimum at $(\frac{1}{6}, \frac{1}{3})$. △

Example 4.6.10 We will find the local maxima, minima, and saddle points for $f(x, y) = x^3 - y^3 + 3xy$.

First we find the critical points by setting the first order partial derivatives equal to zero and solving the resulting system of simultaneous equations.

$$\begin{aligned} D_1f(x,y) &= 3x^2 + 3y = 0 \\ D_2f(x,y) &= -3y^2 + 3x = 0 \end{aligned}$$

Or,

$$\begin{aligned} x^2 + y &= 0 \\ -y^2 + x &= 0 \end{aligned}$$

Substituting $x = y^2$ for x in the first equation gives

$$y^4 + y = 0$$

that has two real solutions, $y = 0$ (whence $x = 0$) and $y = -1$ (whence $x = 1$).

Thus there are two critical points, $p_1 = (0, 0)$ and $p_2 = (1, -1)$. The second-order partials are

$$D_{1,1}f(x,y) = 6x \quad D_{1,2}f(x,y) = D_{2,1}f(x,y) = 3 \quad D_{2,2}f(x,y) = -6y$$

At $p_1 = (0, 0)$,

$$DISC = D_{1,1}f(p_1)D_{2,2}f(p_1) - (D_{1,2}f(p_1))^2 = -9 < 0$$

so by part (c) of the theorem, f has a saddle point at $p = (0, 0)$. At $p_2 = (1, -1)$,

$$DISC = D_{1,1}f(p_2)D_{2,2}f(p_2) - (D_{1,2}f(p_2))^2 = 36 - 9 > 0$$

and so we conclude from part (b) of the theorem that f has a local minimum at $p_2 = (1, -1)$ of $f(p_2) = f(1, -1) = -1$. △

Example 4.6.11 We will find the local maxima, minima, and saddle points of $f(x, y) = \sin(x) + \sin(y) + \sin(x + y)$, x and y in $(0, \pi)$.

First we find the critical points by setting the first order partial derivatives equal to zero and solving the resulting simultaneous equations.

$$\begin{aligned} D_1f(x,y) = \cos(x) + \cos(x+y) &= 0 \\ D_2f(x,y) = \cos(y) + \cos(x+y) &= 0 \end{aligned}$$

Thus $\cos(x) = \cos(y)$ with x and y in $(0, \pi)$. The only solution occurs when $x = y$, since the distance between x and y is less than the period 2π. Substituting $x = y$ into the first equation gives

$$\cos(x) + \cos(2x) = 0$$

or

$\cos(x) + \cos^2(x) - \sin^2(x) = 0$ (Equation 8 in the Appendix on trigonometry)

Substituting $z = \cos(x)$ into the above equation gives a quadratic

$$2z^2 + z - 1 = 0$$

with solutions $z = -1$ and $z = 0.5$. If $\cos(x) = z = -1$, then $x = \pi$, which is outside the specified interval $(0, \pi)$. If $\cos(x) = z = 0.5$, then $x = \pi/3$ and so $y = x = \pi/3$. Thus the only critical point in the rectangle $(0, \pi) \times (0, \pi)$ is $p = (\pi/3, \pi/3)$.

The second-order partials are

$$\begin{aligned} D_{1,1}f(x,y) &= -\sin(x) - \sin(x+y) \\ D_{1,2}f(x,y) &= D_{2,1}f(x,y) = -\sin(x+y) \\ D_{2,2}f(x,y) &= -\sin(y) - \sin(x+y) \end{aligned}$$

At $p = (\pi/3, \pi/3)$ we have

$$D_{1,1}f(p) = -\frac{\sqrt{3}}{2} - \frac{\sqrt{3}}{2} = -\sqrt{3}, \ D_{1,2}f(p) = -\frac{\sqrt{3}}{2}, \ D_{2,2}f(p) = -\sqrt{3}$$

Thus

$$DISC = D_{1,1}f(p)D_{2,2}f(p) - (D_{1,2}f(p))^2 = 3 - 3/4 > 0$$

From part (a) of the theorem, f has a local maximum at $p = (\pi/3, \pi/3)$. △

Example 4.6.12 Consider the functions $f, g, h : \mathbb{R}^2 \to \mathbb{R}$ defined by

$$\begin{aligned} &f(x,y) = x^4 + y^4 \\ &g(x,y) = -(x^4 + y^4) \\ &h(x,y) = x^4 - y^4 \end{aligned}$$

For each of these functions $p = (0,0)$ is the only critical point and at p the discriminant is zero. The function f has a minimum at p, the function g has a maximum at p, and the function h has a saddle point at p (by an argument similar to the one for Example 4.6.5). Thus if the discriminant is zero, anything can happen. △

In the previous examples, if we compute $D_{2,1}f(x,y)$, we obtain the same value as $D_{1,2}f(x,y)$. This is no accident. If f is a C^2 function, then the mixed partial derivatives $D_{1,2}f(x,y)$ and $D_{2,1}f(x,y)$ are equal (see the exercises for Section 3.10).

Question 4.6.13 Answer the question posed at the start of this Section. I.e., is it true that among all rectangular-shaped boxes with a given surface area, the cube contains the largest volume? ◇

Section 4.6 Exercises

1. Find the local maxima, minima, and saddle points of $f(x,y) = x^2 + 2y^2 - x^2y$

2. Find the local maxima, minima, and saddle points of $f(x,y) = 2x^2 - 4xy + y^4 + 2$

3. Find the distance from the point $(-1, 4, 2)$ to the plane $2x - 3y + z = 7$.

4. Find the points on the surface $x^2 - yz = 5$ closest to the origin.

5. (a) A cow's water trough is to be a box with a volume of 32 ft^3. To minimize the cost, the area of sides and bottom should be as small as possible. Find the dimensions of the trough.

   (b) A cow wants to take a piece of luggage (crocodile skin, naturally) on a trip. Airline restrictions state that a cow may carry on one piece of luggage provided the length plus width plus height does not exceed 45 in and the height does not exceed 7 in. How much volume can a piece of carry on luggage occupy?

   (c) A cow pasture is triangular, bounded by an east-west road 10 miles north of the barn, a north-south road 10 miles east of the barn, and a diagonal north-west to south-east road passing next to the barn. The altitude of the pasture is given by $a(x,y) = x^2 - 2xy - y^2 + x + 1$, with coordinates centered on the barn and the x-axis running east-west. Find the lowest point in the pasture.

   (d) A cow may *may* have two pieces of luggage on a flight. Airline restrictions state that two pieces of luggage may be checked without an extra fee provided the length plus width plus height of one piece does not exceed 62 in for one piece and 55 in for the other. What is the maximum volume of luggage that a cow can legally check without paying an extra fee?

6. Let x, y, and z be nonnegative real numbers

(a) Find nonnegative real numbers $x, y,$ and z whose sum is 30 and whose product is a maximum. (If a maximum does not exist, explain why it does not.)

(b) Find nonnegative real numbers whose product is 30 and whose sum is a maximum. (If a maximum does not exist, explain why it does not.)

7. When an experiment is performed a set of data points $(x_1, y_1), \ldots, (x_k, y_k)$ is obtained. The experimenter may wish to draw a straight line through these points. If the points do not lie on a straight line, then the experimenter wants to draw a line which fits the data points as well as possible. Let the line be given by the relation $y = mx + b$. The line predicts that the value corresponding to x_i will be $mx_i + b$ rather than the observed value of y_i. Let $d_i = [y_i - (mx_i + b)]$ be the deviation between the actual and predicted value. The most common method of finding the line is to minimize the sum of the squares of the deviations, the *method of least squares* for fitting a line to experimental data points.

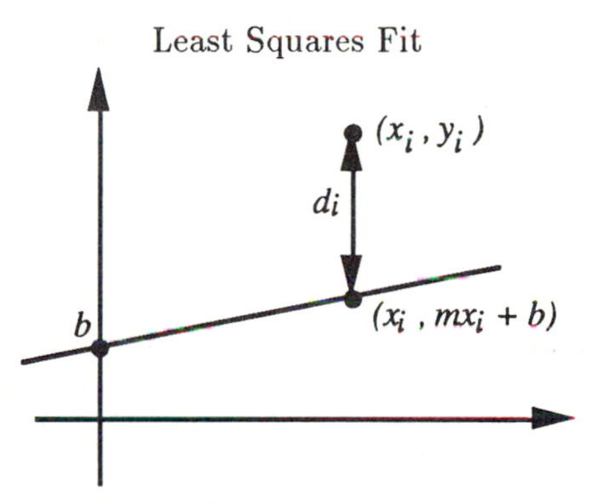

Figure 4.6.14

Find the values of m and b for the least-squares line fitting the data points (1,1), (2,3), (4,3).

8. Find the values of m and b for the least-squares line fitting the data points (1,2), (2,3), (3,3).

9. Show that if two points (x_1, y_1) and (x_2, y_2) are given, then the method of least squares gives the straight line determined by these two points.

10. Given k data points $(x_1, y_1), (x_2, y_2), \ldots, (x_k, y_k)$ find the equations for the least-squares line through these data points.

11. Give an example for each of the following statements or explain why no example can exist.

(a) A function $f : \mathbb{R}^3 \to \mathbb{R}$ that has a critical point at (1,2,3).

(b) A function $f : \mathbb{R}^2 \to \mathbb{R}$ that has a maximum value at every point on the boundary, but no maximum value at an interior point of the domain.

(c) A function $f : \mathbb{R}^2 \to \mathbb{R}$ that has a critical point, but no extreme values.

(d) A differentiable function $f : \mathbb{R}^2 \to \mathbb{R}$ with domain $\{(x, y) : -1 \leq x \leq 1, -1 \leq y \leq 1\}$ that has no critical point on the boundary.

(e) A differentiable function $f : \mathbb{R}^2 \to \mathbb{R}$ that has two or more local maxima, but no local minima. (Note that no such example can exist for functions of a single variable.)

4.7 Lagrange Multipliers

In Example 4.6.7 we considered the problem of finding the point on the plane $x + 2y - z = 1$ that is closest to the origin. This problem can be rephrased as:

Minimize the function $f(x, y, z) = x^2 + y^2 + z^2$ subject to the condition that $x + 2y - z - 1 = 0$.

The condition that $x+2y-z-1 = 0$ is called a *side condition* (or *constraint*), that is, an equation that we want to hold when the function f is minimized. The technique of Lagrange multipliers, developed by the French-Italian mathematician Joseph Louis Lagrange (1734-1813), allows us to find extreme points of a function subject to side conditions.

The geometric properties of the gradient can be used to motivate the result.

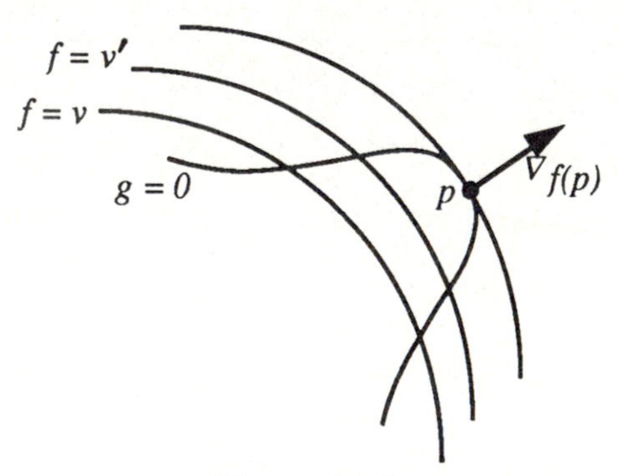

Figure 4.7.1

Let $f : \mathbb{R}^2 \to \mathbb{R}$ be the function to minimize and let $g : \mathbb{R}^2 \to \mathbb{R}$ give the side condition $g(x) = 0$. Consider the side condition $g(x) = 0$ as a level curve. For any value v that f assumes, $\{x : f(x) = v\}$ is a level curve. If x satisfies both $f(x) = v$ and $g(x) = 0$, then x is on the intersection S of these two curves. Suppose that v is not the smallest value that f assumes on the curve $g(x) = 0$. Then there is a smaller value $v' < v$ that f assumes on the curve $g(x) = 0$, and a corresponding intersection S' of $\{x : f(x) = v'\}$ and $\{x : g(x) = 0\}$. Continue decreasing v until the intersection of $\{x : g(x) = 0\}$ and $\{x : f(x) = v\}$ is a single point p. If v is decreased any further, then there is no intersection. At this point p the curves $\{x : f(x) = v\}$ and $\{x : g(x) = 0\}$ are just touching and have the same tangent line. Since the tangent lines are the same, the gradients are multiples of each other. Thus there is a real number r such that

$$\nabla f(p) = r\nabla g(p)$$

Writing this in terms of coordinates gives three equations

$$\begin{aligned}&D_x f(p) = rD_x g(p)\\ &D_y f(p) = rD_y g(p)\\ &g(p) = 0\end{aligned}$$

in three unknowns (x, y, and r). Luckily, these equations are often easy to solve.

This extends to k conditions in the following way:

Theorem 4.7.2 (Lagrange Multipliers) *Let $f, g_1, ..., g_k : \mathbb{R}^n \to \mathbb{R}$ be C^1 functions. Suppose that f has a maximum or minimum at p subject to the conditions*

$$g_1(p) = 0, ..., g_k(p) = 0$$

Then there are real numbers $r_1, r_2, ..., r_k$ such that

$$\nabla f(p) = r_1\nabla g_1(p) + r_2\nabla g_2(p) + \cdots + r_k\nabla g_k(p)$$

We will not prove this result, which requires a substantial amount of analysis, but will consider some examples. Expressed in terms of coordinates, the above theorem leads to $n + k$ equations in $n + k$ unknowns.

Example 4.7.3 First, let's apply the result to Example 4.6.7. Here we want to minimize

$$f(x, y, z) = x^2 + y^2 + z^2$$

subject to

$$g(x, y, z) = x + 2y - z - 1$$

If f has a minimum at p on the level surface $g(x, y, z) = 0$, then there is a real number r such that

$$\nabla f(p) = r\nabla g(p), p = (x, y, z)$$

Since

$$\nabla f(p) = (2x, 2y, 2z)$$

and

$$\nabla g(p) = (1, 2, -1)$$

this gives:

$$2x = r \cdot 1 = r, \quad 2y = r \cdot 2 = 2r, \quad 2z = r(-1) = -r$$

along with the side condition $x + 2y - z - 1 = 0$.

This gives four equations in four unknowns that we can solve by substitution, or, if we have had a course in matrix algebra, by Gaussian elimination.

Substituting the solutions $x = r/2$, $y = r$, and $z = -r/2$ into $g(x, y, z) = 0$ we obtain

$$\frac{r}{2} + 2r + \frac{r}{2} - 1 = 0$$

so $r = \frac{1}{3}$. This gives

$$x = \frac{r}{2} = \frac{1}{6}, \quad y = r = \frac{1}{3}, \quad z = -\frac{r}{2} = -\frac{1}{6}$$

This agrees with the solution found in Example 4.6.7. △

Example 4.7.4 How should the line segment $[a, b]$ be divided into 3 parts so that the product of the lengths of the three parts is a maximum?

Here we are constructing a box in three dimensional space from the three segments and the objective is to maximize the volume.

Let the line segment be divided at $a + x$ and $a + x + y$, so that the lengths of the segments are x, y, and $z = b - a - x - y$. The volume of the box to be maximized is xyz. The side condition is that the sum of the lengths equal the length of the interval: $x + y + z = b - a$. This gives us the following Lagrange multiplier problem:

Maximize $f(x, y, z) = xyz$
Subject to: $g(x, y, z) = x + y + z - (b - a) = 0$

Writing

$$\nabla f(x, y, z) = r\nabla g(x, y, z)$$

in terms of the coordinates we have:

$$yz = r, \qquad xz = r, \qquad xy = r, \qquad \text{and} \qquad x + y + z = b - a$$

Equating the first and second of the equations, we have $yz = r = xz$. If $z = 0$, then the volume is zero, giving a minimum, not a maximum. Thus we may assume $z \neq 0$ and divide by z to obtain $y = x$. Similarly by equating the second two equations, we obtain $z = y$. Thus $x = y = z$. Substituting x for y and z in the level set $x + y + z = b - a$ we obtain $3x = b - a$, or

$$x = y = z = \frac{(b-a)}{3} \qquad \triangle$$

We now consider an example involving two constraints (side conditions).

Example 4.7.5 We will find the maximum and minimum values of $f(x, y, z) = x^2 - 2yz$ on the curve of intersection of the sphere $x^2 + y^2 + z^2 = 4$ and the plane $x + y + z = 0$. Stating this problem as a Lagrange Multiplier problem, gives

$$\begin{aligned} \text{Maximize (minimize) } f(x,y,z) &= x^2 - 2yz \\ \text{Subject to : } g(x,y,z) &= x^2 + y^2 + z^2 - 4 = 0 \\ h(x,y,z) &= x + y + z = 0 \end{aligned}$$

Writing

$$\nabla f(x,y,z) = r_1 \nabla g(x,y,z) + r_2 \nabla h(x,y,z)$$

in terms of coordinates we have:

$$(2x, -2z, -2y) = r_1(2x, 2y, 2z) + r_2(1,1,1)$$

Equating components gives:

$$\begin{aligned} 2x &= 2xr_1 + r_2 \\ -2z &= 2yr_1 + r_2 \\ -2y &= 2zr_1 + r_2 \\ x^2 + y^2 + z^2 &= 4 \\ x + y + z &= 0 \end{aligned}$$

Solving for r_2 in the first three equations and then equating the results gives:

$$r_2 = 2x - 2xr_1 = -2z - 2yr_1 = -2y - 2zr_1.$$

Equating the last two expressions:

$$-2z - 2yr_1 = -2y - 2zr_1$$

which implies

$$z - y = (z - y)r_1$$

and so we have two possibilities:

$$r_1 = 1 \quad \text{or} \quad z = y$$

If the first possibility holds, so that $r_1 = 1$, then: $r_2 = 0$ (by the first component equation) and $z = -y$ (by the second component equation). Thus

$x = 0$ (by the plane constraint)
and so $2z^2 = 4$ (by the sphere constraint).

Thus we have $(x, y, z) = (0, \sqrt{2}, -\sqrt{2})$ or $(0, -\sqrt{2}, \sqrt{2})$
and $f(0, \sqrt{2}, -\sqrt{2}) = f(0, -\sqrt{2}, \sqrt{2}) = 4$.

If the second possibility holds, so that $z = y$, then (by the sphere and plane constraints)

$$\begin{aligned} x^2 + 2z^2 &= 4 \\ x + 2z &= 0 \end{aligned}$$

which gives

$$z^2 = \frac{2}{3}$$

Thus we have

$$(x, y, z) = \left(-2\sqrt{\frac{2}{3}}, \sqrt{\frac{2}{3}}, \sqrt{\frac{2}{3}}\right) \text{ or } \left(2\sqrt{\frac{2}{3}}, -\sqrt{\frac{2}{3}}, -\sqrt{\frac{2}{3}}\right)$$

and $f\left(-2\sqrt{\frac{2}{3}}, \sqrt{\frac{2}{3}}, \sqrt{\frac{2}{3}}\right) = f\left(2\sqrt{\frac{2}{3}}, -\sqrt{\frac{2}{3}}, -\sqrt{\frac{2}{3}}\right) = \frac{4}{3}$.

Thus f assumes its maximum value of 4 at two points and its minimum value of $\frac{4}{3}$ at two points. △

Question 4.7.6 Find the point on the intersection of the two surfaces

$$x + y + z = 1$$

and

$$x^2 + y^2 - z^2 = 0$$

that is closest to the origin. ◇

Section 4.7 Exercises

1. Find the maximum value of $x^2+xy+y^2+yz+z^2$ subject to the condition that (x, y, z) be on the unit sphere.

2. Find the maximum value of $x^2+xy+y^2+yz+z^2$ subject to the condition that (x, y, z) be on the unit sphere and on the plane $x+y+z=0$.

3. Find the dimensions of the open rectangular box of volume 32 cm^3 which requires the least amount of material for its construction.

4. Find the maximum value of $\sin(x)\sin(y)$ where x and y are the acute angles of a right triangle.

5. A closed rectangular box with a volume of 16 cm^3 is made from two kinds of materials. The top and bottom are made of material costing 10 cents/cm^2. The sides are made from a material costing 5 cents/cm^2. Find the dimensions of the box so as to minimize the cost.

6. A manufacturer makes two models of widget, cheap and expensive. The cheap widget costs $20 to make and the expensive one costs $30 to make. Market research suggests that if the cheap widget is priced at x dollars and the expensive widget is priced at y dollars, then $250(y-x)$ cheap widgets will be sold and $10000+250(x-2y)$ expensive widgets will be sold. How should the widgets be priced to maximize the profit?

7. (a) A cow food storage bin (with a bottom) is in the form of a right circular cylinder with a hemisphere on top. The volume is to be 1000 ft^3. If the cost of the hemisphere is twice the cost of the cylinder per square foot, find the dimensions of the storage bin that will minimize the cost.

   (b) A cow is attached to a post by a chain of length 10. The temperature in the cow barn is $T(x, y) = 2x^2-3xy+4y^2+3x+10$, with coordinates centered on the post. Where should the cow go to find the warmest place to rest?

   (c) A cow is attacked by a swarm of flies which maintain a distance of 100 inches from her head, wary of her lashing tail. The probability that a fly is struck is inversely proportional to $p(x, y, z) = (x-10)^2 + (y-20)^2 + (z-30)^2 + 10$, in coordinates centered on the cow's head. Where should the flies go to minimize their chance of being struck?

   (d) A cow, wearing a spherical diving helmet, is looking at a fish in the direction of the vector (1,1,2). What are the coordinates of the point on the helmet closest to the fish?

8. Use a computer program for solving systems of equations or a computer algebra system for solving the system of equations in the following exercise.

Find the minimum of the function $f : \mathbb{R}^4 \to \mathbb{R}$ defined by

$$f(x, y, z, t) = 2x^2 + y^2 + z^2 + 2t^2$$

subject to the conditions:

$$x + y + z - t = 1, \quad 2x + y - z + 2t = 2, \quad \textit{and} \quad x - y + z - t = 4$$

Chapter 5

INTEGRATION

5.1 Introduction

The origins of integration theory lie within some of the oldest problems in mathematics, those of determining areas, volumes, and arc lengths. Although these types of problems probably first arose in connection with agriculture, the first recorded developments are attributed to the fifth century B.C. Greek philosopher Antiphon in his work on the famous problem of "squaring the circle." This problem, which was listed as Problem 48 in the Rhind Papyrus (1700 B.C.), is to construct a square (using only a straight edge and compass) having an area equal to that of a given circle. It is interesting to note that the problem remained unsolved until the nineteenth century, when it was finally established, as a consequence of the work of Ferdinand Lindemann (1852–1939), that the desired construction was impossible. Antiphon's approach was to approximate the given circle by a regular inscribed polygon. Noting that the approximation could be improved by doubling the number of sides of the polygon, he proposed that by successively doubling the number of sides the region between the circle and inscribed polygon would eventually be exhausted. The problem would then have been reduced to constructing a square with an area equal to that of a given regular polygon, a construction that was known.

Antiphon's idea was criticized and rejected on the grounds that magnitudes are infinitely divisible and therefore the region between the circle and the inscribed polygon could never be exhausted. His approach, however, prepared the way for Eudoxus (ca. 370 B.C.) to establish the famous *Method of Exhaustion.* This method, which accepts the hypothesis of the infinite divisibility of magnitudes, is based on the following proposition.

> If from any magnitude there be subtracted a part not less than its half, from the remainder another part, not less than its half, and so on, there will remain a magnitude less than any preassigned magnitude of the same kind.

We shall show by the Method of Exhaustion that Antiphon's approach was a valid way to determine the area of a circle, although it requires knowledge of convergence for the final step.

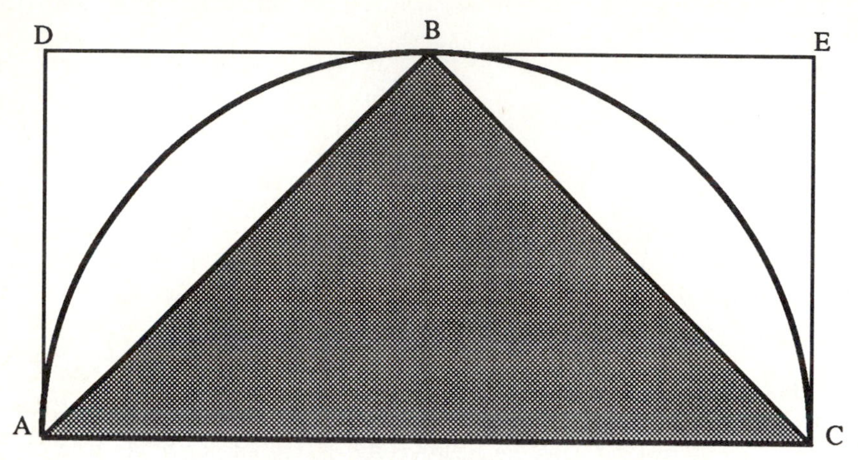

Figure 5.1.1

In the figure, arc ABC is an arc of a circle, chord AC is the edge of an inscribed polygon P, chords AB and BC are edges of the inscribed polygon obtained by doubling the number of sides of P, and DE is tangent to the circle at point B. Since the area of the shaded triangle ABC is half that of the rectangle $ADEC$, it is greater than half the area of the circular region ABC. Thus, by doubling the number of sides of the regular inscribed polygon, the area of the polygon is increased by more than half the difference between the area of the circle and the inscribed polygon. Therefore, by the Method of Exhaustion, Antiphon's process yields a way of approximating the area of a given circle by the area of a regular inscribed polygon in such a way that the resulting error can be made less than any predetermined positive number.

In order to better appreciate the problem of determining the area of a circle as well as to gain a "feeling" for the origins of the number π, do the calculations in the following question.

Question 5.1.2 Let C be a circle of radius r. Let a_{2n} denote the area of a regular polygon of $2n$ sides inscribed in C, and let A_{2n} denote the area of a regular polygon of $2n$ sides circumscribed about C. The algebraic expressions for a_{2n} and A_{2n} are defined by the recursive sequences (which we will not justify; this can be a project for the reader!):

$$a_4 = 2r^2 \quad A_4 = 4r^2$$

$$a_{2n} = \sqrt{a_n A_n} \quad A_{2n} = \frac{2a_{2n}A_n}{a_{2n} + A_n}$$

Verify that the expressions for the base cases a_4 and A_4 are correct and then compute a_N and A_N for $N = 4, 8, ..., 256$. $\diamond$

The Greek Method of Exhaustion provided a process for rigorously proving or disproving a conjectured formula for area, volume, or arc length, but it did not provide the formula. Archimedes (287–212 B.C.), one of the greatest mathematicians of all time, relied on his engineering skills to develop a process for conjecturing formulas for areas and volumes that he could then check by the Method of Exhaustion. The idea in Archimedes' method is to subdivide the region or solid in question into thin parallel strips or layers and then hang these strips or layers at the end of a balance rod so as to be in equilibrium with a known quantity at the other end of the rod. For us, the important idea in Archimedes' Method of Equilibrium is that of partitioning a region into thin strips and then considering the area of the region as the summation of the areas of these strips. We shall encounter numerous and widespread applications of this process of partitioning and then summing. One such application is illustrated in the next question.

Question 5.1.3 The "New Home Sales" function, NHS, maps the day of the year to the number of private homes sold that day. The graph of NHS is displayed in the figure below. Note that NHS(May 1,1882) = 375,000 and NHS(April 1,1883) = 625,000.

Approximate how many private homes were sold during 1882. Give an upper bound for the error associated with your approximation. ◊

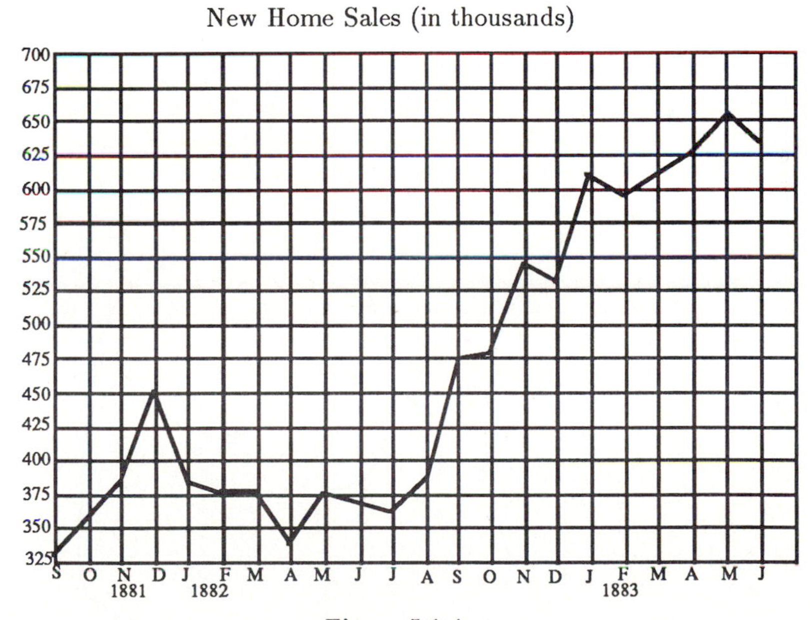

Figure 5.1.4

Major progress in the theory of integration had to wait almost 1800 years until the relationship between integration and differentiation became known, as expressed in the Fundamental Theorem of Calculus.

We begin our development of the theory of integration by combining the approximation concept of Antiphon with Archimedes' idea of subdividing a region into thin strips to formulate a definition of area. We shall follow the *Basic Approximation Process.* (The process that was followed in developing the derivative concept.)

1. Approximate the unknown quantity with a known property.
2. Determine a way to obtain a better approximation.
3. Generate a convergent sequence of approximations such that each is a better approximation than the preceding ones.
4. Define the desired property to be the common limit of all possible sequences in part (3). When there is no common limit, the desired property is said not to exist.

Note: Unless stated otherwise, all the bounded functions in this chapter will be required to have the property that over any closed interval in the domain the maximum value is equal to the least upper bound and the minimum value is equal to the greatest lower bound. (In more advanced courses in analysis, this restriction is removed.)

Section 5.1 Exercises

In Exercises 1 and 2, find three numerical approximations to the area of the specified region such that the second approximation is more accurate than the first, and the third approximation is more accurate than the second. Also, determine an error bound for each approximation.

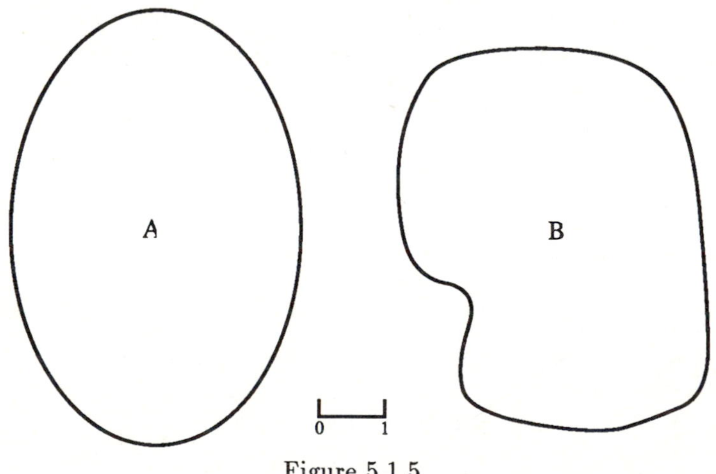

Figure 5.1.5

1. Region A.

2. Region B.

3. Approximate the number of new homes sold from January 1, 1883, through May 31, 1883 (refer to Figure 5.1.4). Give an error bound for your approximation.

5.2 Area

The "region under the graph" of a nonnegative function $f : [a, b] \to \mathbb{R}$ is called the *ordinate set* of f. That is, the ordinate set of f is the region bounded by the graph of f, the x-axis, and the vertical lines $x = a$, $x = b$ [the shaded region in Figure 5.2.2(a)]. The purpose of this section is to develop a definition for the area of an ordinate set. The process, of course, will be the Basic Approximation Process outlined in the previous section. (We shall explicitly enumerate each of the four steps.)

In developing a method or technique, it is often helpful to begin with a simple example in which the result is already known. This allows the reader to focus on the technique being developed as well as providing a check on the "answer." Thus we shall begin our development with a function whose ordinate set is a trapezoid.

Example 5.2.1 We will find the area, if it exists, of the ordinate set of the function $f(x) = x + 2$ over $[1, 5]$. This set is the shaded region in Figure 5.2.2 on the left.

Step 1: Approximate the unknown quantity with a known property.

The unknown quantity is the area of the ordinate set. We shall give two approximations by "known" properties, the areas of the rectangles $ABCD$ and $AEFD$ as shown Figure 5.2.2(b).

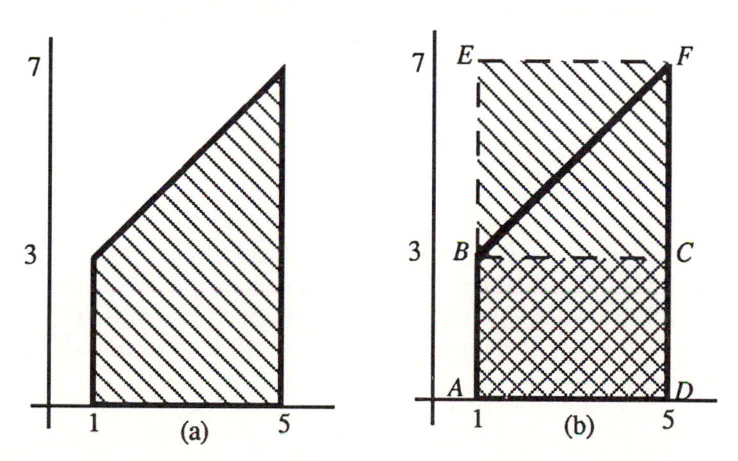

Figure 5.2.2

Geometrically, it is clear that the area of the inscribed rectangle $ABCD$ is a lower approximation to the "area" of the ordinate set and the area of the circumscribed rectangle $AEFD$ is an upper approximation. That is (assuming that the ordinate set has an area A), $12 \leq A \leq 28$. Furthermore, the error associated with either the lower or upper approximation is less than

$$\text{(upper approximation - lower approximation)} = 16 \text{ square units.}$$

Step 2: Determine a way to obtain a better approximation.

A better approximation is one with less error. We subdivide the interval $[1,5]$ into two equal subintervals $[1,3]$ and $[3,5]$ and construct inscribed and circumscribed rectangles over each of these intervals as shown in Figure 5.2.3.

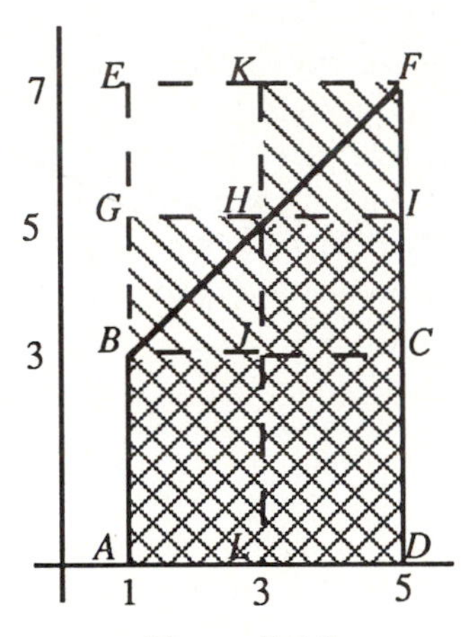

Figure 5.2.3

Let the lower approximation be the sum of the areas of the two inscribed rectangles $ABJL$ and $LHID$ and let the upper approximation be the sum of the two circumscribed rectangles $AGHL$ and $LKFD$ (rectangles are labeled clockwise from the lower left corner). We now have $16 \leq A \leq 24$, and the error for each approximation is bounded above by $24 - 16 = 8$ square units. Note that the second lower approximation is larger than the first lower approximation (i.e., $16 > 12$) and the second upper approximation is less than the first upper approximation (i.e., $24 < 28$). Geometrically, the improvements in the approximations result from the inclusion of rectangle $JHIC$ in the second lower approximation and the exclusion of rectangle $GEKH$ in the second upper approximation.

Let us illustrate the second step in the Basic Approximation Process again by subdividing the interval $[1,5]$ into three equal subintervals. Figure 5.2.4 shows the situation with the inscribed and circumscribed rectangles. A little calculation yields a lower approximation of

$$(3)\left(\frac{4}{3}\right) + \left(\frac{13}{3}\right)\left(\frac{4}{3}\right) + \left(\frac{17}{3}\right)\left(\frac{4}{3}\right) = \frac{52}{3}$$

an upper approximation of

$$\left(\frac{13}{3}\right)\left(\frac{4}{3}\right)+\left(\frac{17}{3}\right)\left(\frac{4}{3}\right)+(7)\left(\frac{4}{3}\right)=\frac{68}{3}$$

and an error bound of $\frac{16}{3}$.

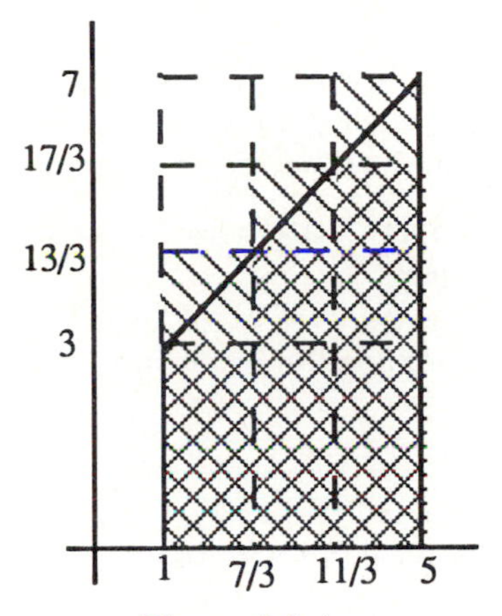

Figure 5.2.4

Note that again the lower approximation has increased and the upper approximation has decreased. It is clear that continuing this approximation process with inscribed and circumscribed rectangles will yield an increasing sequence of lower approximations and a decreasing sequence of upper approximations. With just three pairs of approximations the sequences (finite) are

$$12 < 16 < \frac{52}{3} < A < \frac{68}{3} < 24 < 28$$

Step 3: Generate a convergent sequence of approximations such that each is a better approximation than the preceding one.

We shall define two sequences: an increasing sequence of lower approximations and a decreasing sequence of upper approximations. Furthermore, every lower approximation will be less than or equal to every upper approximation. To form the nth terms in the sequences, the original interval [1,5] is subdivided into n equal subintervals and both an inscribed and a circumscribed rectangle are formed over each subinterval. The nth *lower approximation* is defined to be the sum of the areas of the n inscribed rectangles and the nth *upper approximation* is defined to be the sum of the areas of the n circumscribed rectangles. Before developing the algebraic expressions for the nth terms for each of these sequences, we make three observations:

1. The length of the base for each of the inscribed and circumscribed rectangles is the length of the original interval divided by the number of subin-

tervals. Thus, if there are n subintervals, the length of each is $(5-1)/n = 4/n$.

2. The height of an inscribed rectangle is the minimum value obtained by the function on the base of that rectangle.
3. The height of a circumscribed rectangle is the maximum value obtained by the function on the base of that rectangle.

Notation. We now develop notation for the sequences of upper and lower approximations that are suggestive of the process involved. Denoting upper by U, lower by L, function by f, and number of equal subintervals by n, we have

$UA(f,n)$ = the upper approximation to the area of the ordinate set of f taken over n equal subintervals.

$LA(f,n)$ = the lower approximation to the area of the ordinate set of f taken over n equal subintervals.

Also, we denote the minimum of the function f over the kth subrectangle by $(minf)_k$ and the maximum value by $(maxf)_k$. Using these notations, the third approximation obtained above can be written:

$$\begin{aligned} LA(f,3) &= (minf)_1\left(\frac{4}{3}\right)+(minf)_2\left(\frac{4}{3}\right)+(minf)_3\left(\frac{4}{3}\right) \\ UA(f,3) &= (maxf)_1\left(\frac{4}{3}\right)+(maxf)_2\left(\frac{4}{3}\right)+(maxf)_3\left(\frac{4}{3}\right) \end{aligned}$$

Using summation notation, these expressions are

$$LA(f,3)=\sum_{k=1}^{3}(minf)_k\left(\frac{4}{3}\right)$$

and

$$UA(f,3)=\sum_{k=1}^{3}(maxf)_k\left(\frac{4}{3}\right)$$

The power and beauty of the summation notation is illustrated by the ease in which the expressions for the nth approximations can be obtained by merely replacing the 3 (the number of subintervals) by n. That is,

$$LA(f,n)=\sum_{k=1}^{n}(minf)_k\left(\frac{4}{n}\right)$$

and

$$UA(f,n)=\sum_{k=1}^{n}(maxf)_k\left(\frac{4}{n}\right)$$

The following observation is now used: The minimum value of an *increasing* function over a closed interval occurs at the left-hand endpoint and the maximum value occurs at the right-hand endpoint of the interval.

Note that the left-hand endpoint of the kth subinterval is $1+(k-1)4/n$ and the right-hand endpoint is $1+k(4/n)$. The "1" in both of these expressions is the left-hand endpoint of the original interval $[1,5]$. Since the function in question, $f(x)=x+2$, is increasing,

$$(min f)_k = f\left(1+(k-1)\left(\frac{4}{n}\right)\right) = \left(1+(k-1)\left(\frac{4}{n}\right)\right)+2 = 3+(k-1)\left(\frac{4}{n}\right)$$

and

$$(max f)_k = f\left(1+k\left(\frac{4}{n}\right)\right) = \left(1+k\left(\frac{4}{n}\right)\right)+2 = 3+k\left(\frac{4}{n}\right)$$

Now, substituting these expressions into the summation expressions yields:

$$LA(f,n) = \sum_{k=1}^{n}\left(3+(k-1)\left(\frac{4}{n}\right)\right)\left(\frac{4}{n}\right)$$

and

$$UA(f,n) = \sum_{k=1}^{n}\left(3+k\left(\frac{4}{n}\right)\right)\left(\frac{4}{n}\right)$$

The following formulas will often be used in summation evaluations. These formulas can be proved using the technique of mathematical induction.

1. $\sum_{k=1}^{n} c = cn$
2. $\sum_{k=1}^{n} ca_k = c\sum_{k=1}^{n} a_k$ for any constant c
3. $\sum_{k=1}^{n}(a_k+b_k) = \sum_{k=1}^{n} a_k + \sum_{k=1}^{n} b_k$
4. $\sum_{k=1}^{n} k = n(n+1)/2$
5. $\sum_{k=1}^{n} k^2 = n(n+1)(2n+1)/6$
6. $\sum_{k=1}^{n} k^3 = n^2(n+1)^2/4$

We will use formulas 1 through 4 to evaluate

$$LA(f,n) = \sum_{k=1}^{n}\left[\left(3+(k-1)\left(\frac{4}{n}\right)\right)\frac{4}{n}\right]$$

$$
\begin{aligned}
LA(f,n) &= \frac{4}{n}\sum_{k=1}^{n}\left[3+(k-1)\frac{4}{n}\right] \\
&= \frac{4}{n}\left[\sum_{k=1}^{n}3+\sum_{k=1}^{n}(k-1)\left(\frac{4}{n}\right)\right] \\
&= \frac{4}{n}\left[3n+\frac{4}{n}\sum_{k=1}^{n}(k-1)\right] \\
&= \frac{4}{n}\left[3n+\frac{4}{n}\left(\sum_{k=1}^{n}k-\sum_{k=1}^{n}1\right)\right] \\
&= \frac{4}{n}\left[3n+\frac{4}{n}\left(\frac{(n+1)n}{2}-n\right)\right] = 20-\frac{8}{n}
\end{aligned}
$$

Similar calculations for $UA(f,n)$ yield:

$$
\begin{aligned}
UA(f,n) &= \sum_{k=1}^{n}\left(3+k\frac{4}{n}\right)\frac{4}{n} \\
&= \frac{4}{n}\left(3n+\frac{4}{n}\sum_{k=1}^{n}k\right) \\
&= \frac{4}{n}\left(3n+\frac{4}{n}\frac{n(n+1)}{2}\right) \\
&= 20+\frac{8}{n}
\end{aligned}
$$

We have defined two convergent sequences that bound the unknown quantity, the area of the ordinate set, in the following manner:

$$LA(f,n) = 20-\frac{8}{n} < A < 20+\frac{8}{n} = UA(f,n)$$

Let us now form a third sequence of approximations, called an *arbitrary sequence,* denoted by $AA(f,n)$, where the interval $[1,5]$ is divided into n equal subintervals as before and a rectangle is constructed over each of the subintervals. Let the height of the kth rectangle be f(x_k^*) where x_k^* represents an arbitrary point in the kth subinterval. Since

$$(minf)_k \le f(x_k^*) \le (maxf)_k$$

we have

$$LA(f,n) \le AA(f,n) \le UA(f,n)$$

Thus, by the Squeeze Theorem (Chapter 2), the sequences $LA(f,n)$, $AA(f,n)$, and $UA(f,n)$ all converge to the same value (20 square units).

Step 4: Define the desired property to be the common limit of all possible sequences in step 3. When there is no common limit, the desired property is said not to exist.

Define the *area* of the ordinate set of $f(x) = x + 2$ over $[1, 5]$ to be

$$\lim_{n\to\infty} LA(f,n) = \lim_{n\to\infty} UA(f,n) = 20 \text{ "square" units}$$

Note that the result of 20 square units agrees with the area computed using the formula

$$\text{Area of trapezoid } = \frac{1}{2}h(L + R)$$

where h is the length of the base and L and R are the lengths of the left and right sides (see Figure 5.2.2). In this problem, $h = 5 - 1 = 4$, $L = 3$, and $R = 7$. Thus the area is $\frac{1}{2}(4)(3 + 7) = 20$. This suggests that the procedure of defining area as the limit of a sequence of approximations is a reasonable one. △

We now look at a slightly more complicated example.

Example 5.2.5 We will find the area, if it exists, of the ordinate set for $f(x) = -x^2 + 4$ over $[-2, 1]$. In the previous example, we saw that if the function is monotonic, it is easy to determine the maximum and minimum values. In this example, since f is not monotonic over $[-2, 1]$, we start by subdividing $[-2, 1]$ into subintervals such that f is monotonic over each subinterval $[-2, 0]$ and $[0, 1]$. We have divided the problem into two subproblems.

Problem A: Find the area, if it exists, of the ordinate set for

$$f(x) = -x^2 + 4, x \in [-2, 0]$$

Problem B: Find the area, if it exists, of the ordinate set for

$$f(x) = -x^2 + 4, x \in [0, 1]$$

Figure 5.2.6 shows the ordinate sets for these two problems.

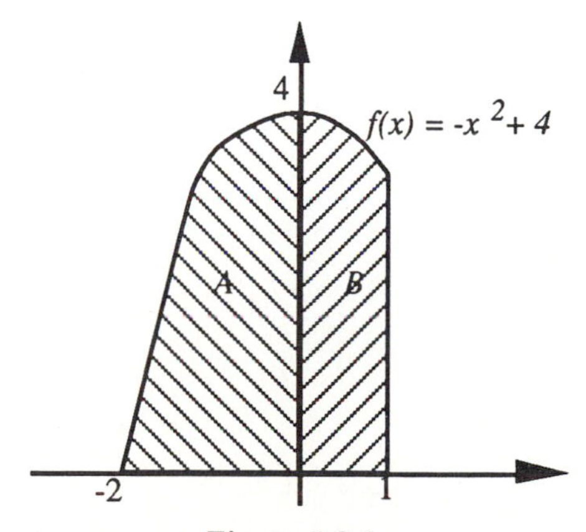

Figure 5.2.6

We shall work problems A and B separately and then combine the results. Although our thinking will follow the four steps in the Basic Approximation Process, we shall combine steps 1, 2, and 3 and write down the expressions for $LA(f,n)$ and $UA(f,n)$ with only a few preliminary comments.

Problem A: We start by subdividing $[-2,0]$ into n equal subintervals, each of length $2/n$. Since f is monotonically *increasing* over $[-2,0]$, the minimum value of f over the kth subinterval will occur at the left-hand endpoint and the maximum value at the right-hand endpoint. Thus,

$$(minf)_k = f\left(-2+(k-1)\frac{2}{n}\right)$$

and

$$(maxf)_k = f\left(-2+k\frac{2}{n}\right)$$

Hence,

$$\begin{aligned}
LA(f,n) &= \sum_{k=1}^{n}(minf)_k\left(\frac{2}{n}\right)\\
&= \sum_{k=1}^{n} f\left(-2+(k-1)\frac{2}{n}\right)\left(\frac{2}{n}\right)\\
&= \sum_{k=1}^{n}\left[-\left(-2+(k-1)\frac{2}{n}\right)^2+4\right]\left(\frac{2}{n}\right)\\
&= \left(\frac{2}{n}\right)\sum_{k=1}^{n}\left[-\left(4-\frac{8(k-1)}{n}+(k-1)^2\frac{4}{n^2}\right)+4\right]\\
&= \left(\frac{2}{n}\right)\sum_{k=1}^{n}\left[\left(\frac{-8}{n}-\frac{4}{n^2}\right)+\left(\frac{8}{n}+\frac{8}{n^2}\right)k-\frac{4k^2}{n^2}\right]\\
&= \left(\frac{2}{n}\right)\left[\left(\frac{-8}{n}-\frac{4}{n^2}\right)\sum_{k=1}^{n}1+\left(\frac{8}{n}+\frac{8}{n^2}\right)\sum_{k=1}^{n}k-\frac{4}{n^2}\sum_{k=1}^{n}k^2\right]\\
&= \left(\frac{2}{n}\right)\left[\left(\frac{-8}{n}-\frac{4}{n^2}\right)n+\left(\frac{8}{n}+\frac{8}{n^2}\right)\frac{n(n+1)}{2}-\frac{4}{n^2}\frac{n(n+1)(2n+1)}{6}\right]\\
&= \frac{16}{3}-\frac{4}{n}-\frac{4}{3n^2}
\end{aligned}$$

In a similar manner,

$$UA(f,n) = \sum_{k=1}^{n}(maxf)_k\left(\frac{2}{n}\right)$$

$$
\begin{aligned}
&= \sum_{k=1}^{n} f\left(-2+k\frac{2}{n}\right)\left(\frac{2}{n}\right) \\
&= \sum_{k=1}^{n}\left[-\left(-2+k\frac{2}{n}\right)^2+4\right]\left(\frac{2}{n}\right) \\
&= \left(\frac{2}{n}\right)\sum_{k=1}^{n}\left[-\left(4-\frac{8k}{n}+\frac{4k^2}{n^2}\right)+4\right] \\
&= \left(\frac{2}{n}\right)\left[\frac{8}{n}\sum_{k=1}^{n}k-\frac{4}{n^2}\sum_{k=1}^{n}k^2\right] \\
&= \left(\frac{2}{n}\right)\left[\left(\frac{8}{n}\right)\frac{n(n+1)}{2}-\left(\frac{4}{n^2}\right)\frac{n(n+1)(2n+1)}{6}\right] \\
&= \frac{16}{3}+\frac{4}{n}-\frac{4}{3n^2}
\end{aligned}
$$

Since $\lim_{n\to\infty} LA(f,n) = \lim_{n\to\infty} UA(f,n) = \frac{16}{3}$, the area of the ordinate set in Problem A exists and is $\frac{16}{3}$ square units.

Problem B: We start by subdividing $[0,1]$ into n equal subintervals each of length $1/n$. Since f is monotonically *decreasing* over $[0,1]$, the minimum value of f over the kth subinterval occurs at the right-hand endpoint and the maximum value occurs at the left-hand endpoint of the subinterval. Thus,

$$
\begin{aligned}
LA(f,n) &= \sum_{k=1}^{n} f\left(0+\frac{k}{n}\right)\left(\frac{1}{n}\right) \\
&= \sum_{k=1}^{n}\left(-\frac{k^2}{n^2}+4\right)\left(\frac{1}{n}\right) \\
&= \left(\frac{1}{n}\right)\left[-\left(\frac{1}{n^2}\right)\frac{n(n+1)(2n+1)}{6}+4n\right] \\
&= \frac{11}{3}-\frac{1}{2n}-\frac{1}{6n^2}
\end{aligned}
$$

and

$$
\begin{aligned}
UA(f,n) &= \sum_{k=1}^{n} f\left(0+\frac{k-1}{n}\right)\left(\frac{1}{n}\right) \\
&= \sum_{k=1}^{n}\left[-\frac{(k-1)^2}{n^2}+4\right]\left(\frac{1}{n}\right)
\end{aligned}
$$

Now set $t = k-1$, obtaining

$$
\sum_{k=1}^{n} -\frac{(k-1)^2}{n^2} = \sum_{t=1}^{n-1} -\frac{t^2}{n^2} = -\frac{1}{n^2}\frac{(n-1)n(2n-1)}{6}
$$

Thus,

$$\begin{aligned} UA(f,n) &= \left(\frac{1}{n}\right)\left[-\left(\frac{1}{n^2}\right)\frac{(n-1)n(2n-1)}{6}+4n\right] \\ &= \frac{11}{3}+\frac{1}{2n}-\frac{1}{6n^2} \end{aligned}$$

Since $\lim_{n\to\infty} LA(f,n) = \lim_{n\to\infty} UA(f,n) = \frac{11}{3}$, the area of the ordinate set in Problem B exists and is $\frac{11}{3}$ square units.

The answer to the original problem is thus $\frac{16}{3}+\frac{11}{3} = 9$ square units. Note that we could not have obtained this result geometrically. △

As was noted in Chapter 2, most physical applications (as well as many theoretical ones) rely on approximate values instead of exact values. This is certainly true about "area type" problems. For instance, in determining the surface area of a wall to be painted one usually is interested in an upper approximation in which the error will be less than what can be covered with 1 quart of paint. Or, in buying grass seed one usually is interested in an approximation of the area to be seeded in which the error will be less than that which could be seeded with 1 pound of seed. Thus, determining a bound on the error associated with a certain approximation is a very important consideration. The following theorem provides an easily obtained error bound for an upper or lower approximation to the area of the ordinate set of a monotonic function.

Theorem 5.2.7 *If f is a nonnegative monotonic function over $[a,b]$, then an error bound for an upper or lower approximation of the ordinate set of f is*

$$UA(f,n) - LA(f,n) = |f(b)-f(a)|\frac{b-a}{n}$$

Proof: Let us assume that f is monotonically increasing over $[a,b]$. Consider a subdivision (partition) of $[a,b]$ into n subintervals, each of length $(b-a)/n$. Since f is increasing, the minimum of f over the kth subinterval occurs at the left endpoint and the maximum value of f occurs at the right endpoint. That is,

$$(minf)_k = f\left(a+(k-1)\frac{b-a}{n}\right)$$

and

$$(maxf)_k = f\left(a+\frac{k(b-a)}{n}\right)$$

Now, expressing the upper and lower approximations in summation form and subtracting one from the other yields:

$$\begin{aligned} UA(f,n) - LA(f,n) &= \sum_{k=1}^{n} f\left(a + \frac{k(b-a)}{n}\right)\frac{b-a}{n} \\ &\quad - \sum_{k=1}^{n} f\left(a + (k-1)\frac{b-a}{n}\right)\frac{b-a}{n} \\ &= \frac{b-a}{n}\sum_{k=1}^{n}[f\left(a + k\frac{b-a}{n}\right) \\ &\quad - f\left(a + (k-1)\frac{b-a}{n}\right)] \\ &= \frac{b-a}{n}[f(b) - f(a)] \end{aligned}$$

To better understand the mechanics in obtaining this last step, the reader is encouraged to choose a value for n, say $n = 7$, and expand the summation. Note that when the expression is simplified, only the first and last terms are left. This reflects the fact that the maximum value of f over the kth subinterval is equal to the minimum value of f over the ($k+1$)st subinterval.

A similar argument can be given in the case where f is monotonically decreasing. □

We are now ready to formalize the intuitive ideas on area that we have been developing, make some observations, and raise some questions.

Definition 5.2.8 (Area of an Ordinate Set) *The* area *of the ordinate set of a nonnegative function f over the interval $[a,b]$ is the common limit of the two sequences: the sequence of lower approximations $\{LA(f,n)\}$ and the sequence of upper approximations $\{UA(f,n)\}$ provided that these sequences have a common limit.*

The area of the ordinate set will be denoted by $\int_a^b f(x)dx$, read as "integral of f from a to b" (see below for a discussion of symbolism).

If $\lim_{n\to\infty} LA(f,n) \neq \lim_{n\to\infty} UA(f,n)$, then the ordinate set is said not to have an area.

The proof of the following theorem is left as an exercise.

Theorem 5.2.9 *The ordinate set of a nonnegative monotone function has an area.*

Not all ordinate sets have an area in the sense defined above as the next example shows.

Example 5.2.10 (An Ordinate Set That Has No Area) Consider the ordinate set for the function $f : [0,1] \to \mathbb{R}$ defined by

$$f(x) = \begin{cases} 1 & \text{if } x \text{ is a rational number} \\ 0 & \text{if } x \text{ is an irrational number} \end{cases}$$

(We shall use without proof the fact that between any two rational numbers there is an irrational number and between any two irrational numbers there is a rational number.)

We start by subdividing the interval $[0,1]$ into n equal subintervals each of length $1/n$. Since each subinterval contains both rational and irrational numbers, $(minf)_k = 0$ and $(maxf)_k = 1$. Thus,

$$LA(f,n) = \sum_{k=1}^{n}(minf)_k\left(\frac{1}{n}\right) = \sum_{k=1}^{n}0\left(\frac{1}{n}\right) = 0$$

and

$$UA(f,n) = \sum_{k=1}^{n}(maxf)_k\left(\frac{1}{n}\right) = \sum_{k=1}^{n}1\left(\frac{1}{n}\right) = \left(\frac{1}{n}\right)\sum_{k=1}^{n}1 = \left(\frac{1}{n}\right)n = 1$$

Since $\lim_{n\to\infty} LA(f,n) = 0$ and $\lim_{n\to\infty} UA(f,n) = 1$, the ordinate set for f over $[0,1]$ *does not have an area in the defined sense!* △

Definition 5.2.11 (Area of a Region Bounded by Two Functions) *The area of a region in the plane bounded by the graphs of the functions f and g and by the lines $x = a$ and $x = b$ is the area of the ordinate set of $h : [a,b] \to \mathbb{R}$, where $h(x) = |f(x) - g(x)|$.*

We will give an example showing that this definition is consistent with our intuitive notions of area.

Example 5.2.12 Consider the functions f, g, and h in Figure 5.2.13.

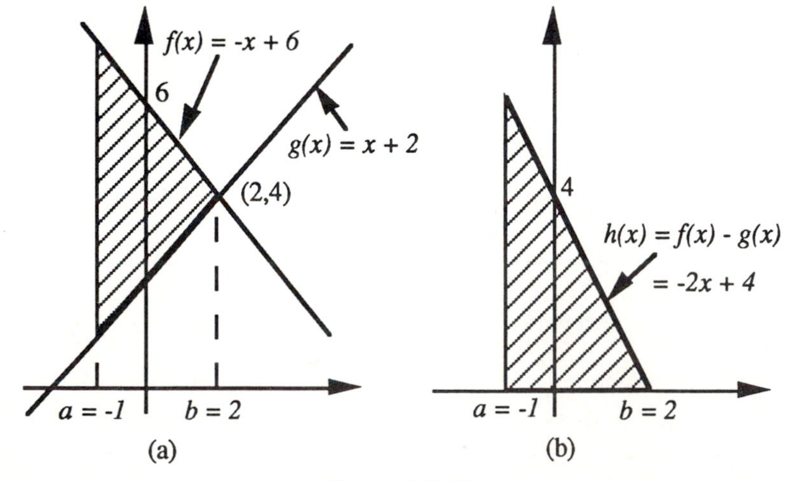

Figure 5.2.13

Figure 5.2.13(a) shows the area of a triangle with (vertical) base $7 - 1 = 6$ and (horizontal) height $2 - (-1) = 3$. Thus the area is $(\frac{1}{2})6 \cdot 3 = 9$. Figure 5.2.13(b) shows the area of a triangle with base $2 - (-1) = 3$ and height 6. Thus the area is again 9. △

The importance of good notation in learning new material was emphasized earlier (see Chapter 3). In particular, it was noted that simplicity of form and suggestiveness of the intrinsic processes involved in the concept are usually the two major characteristics of good notation. Leibniz, who introduced the dy/dx notation for differentiation, is responsible for the notation $\int_a^b f(x)dx$. The symbol $\int$ is called an *integral sign*. Written as a slight modification of a capital S, it suggests the summation process. The expression $f(x)$ is called the *integrand*. The numbers a and b are called the *lower and upper limits of integration*, respectively. They are the endpoints of the interval $[a, b]$ over which the ordinate set is defined. This interval is also referred to as the *path of integration* or the *domain of the integrand*. The domain of the integrand along with the integrand defines a function f whose graph bounds the ordinate set from above (recall that f is nonnegative). Finally the dx (change in x) suggest the width of a subinterval determined by a subdivision of $[a, b]$. Thus, the expressions $f(x)$ and dx written as a product, $f(x)dx$, suggests the area of a rectangle (height times width). Hence, the whole expression

$$\int_a^b f(x)dx$$

suggests summing up the areas of rectangles whose bases are the subintervals defined by a subdivision of the interval $[a, b]$. The reader should compare the corresponding parts and form of the Riemann Sum (to be defined in Section 5.3),

$$\sum_{k=1}^{n} f(x_k)\Delta x_k$$

with the integral

$$\int_a^b f(x)dx$$

We conclude this section with some observations and questions.

Observations:

1. Not every ordinate set has an area (see Example 5.2.10).

2. The critical points in the domain of a C^1 function (a function that has a continuous derivative) divide the domain into subintervals such that the function is monotonic over each subinterval.

3. The interval $[a,b]$ is bounded and the function f, whose graph bounds the ordinate set from above, is a bounded function. [If f were not bounded, we could not talk about $(max f)_k$.]

Questions:

1. Can the area concept be generalized to functions that assume both positive and negative values? For example, if given a profit function P for a business, then total profit from time a to time b could be described as the area of the ordinate set of P over the interval $[a,b]$. What happens if the business loses money for part of the time and therefore P takes on negative as well as positive values?

2. Is it possible to generalize the definition of "area of an ordinate set" to the case where the underlying interval is unbounded (infinite)?

3. Is it possible to generalize the definition of "area of an ordinate set" to the case where the function is unbounded?

4. Can the upper limit be less than the lower limit?

We shall develop answers to these questions (as well as others) in the next few sections.

Section 5.2 Exercises

In Exercises 1 through 6, evaluate the indicated summations. The use of calculators or computers is strongly encouraged.

1. $\sum_{k=1}^{10} 3k$
2. $\sum_{k=1}^{10}(k+1)^3$
3. $\sum_{k=2}^{10}(k^2-1)/(k-1)$
4. $\sum_{k=1}^{8}(2k^3-3k+1)$
5. $\sum_{k=1}^{10}(k+1)(k+2)$
6. $\sum_{k=3}^{3}(k^4+10)$
7. Indicate by shading, the ordinate set of $f(x)=\sin(x)$ on the interval $[0,\pi]$.
8. Indicate by shading, the ordinate set of $h(x)=f(x)-g(x)$ over $[-1,1]$, where $f(x)=-x^2+1$ and $g(x)=x^2-1$.

In Exercises 9 through 18, sketch the ordinate set and indicate at least one "representative" rectangle.

9. $UA(f,6)$ for $f(x)=x^3$ over $[0,3]$.
10. Compute $LA(f,4)$ and $UA(f,4)$ for $f(x)=\cos(x)$ over $[0,\pi/2]$.
11. Compute $LA(h,6)$ and $UA(h,6)$ for $h(x)=x^2$ over $[-1,2]$.
12. Compute $LA(f,n)$ and $UA(f,n)$ for $f(x)=\sqrt{1-x^2}$ over $[0,1]$ and $n=1,2,3,4,5$. By analyzing your results, determine the smallest value of n such that $UA(f,n)-LA(f,n)<0.1$.

13. Compute $UA(h, 5)$ for $h(x) = f(x) - g(x)$ over $[-1, 2]$, where $f(x) = -x + 4$ and $g(x) = x$. Sketch the region bounded by the graphs of f and g over $[-1, 2]$ as well as the ordinate set of h.

14. Compute $LA(h, 5)$ for $h(x) = f(x) - g(x)$ over $[-3/2, 1]$, where $f(x) = -x^2 + 4$ and $g(x) = 3x$. Sketch the region bounded by the graphs of f and g over $[-3/2, 1]$ as well as the ordinate set of h.

15. Find the area of the ordinate set of $h(x) = f(x) - g(x)$ over $[-1, 2]$, where $f(x) = -x + 4$ and $g(x) = x$, by evaluating the limit of an appropriate sum.

16. Find the area of the ordinate set of $h(x) = f(x) - g(x)$ over $[-3/2, 1]$, where $f(x) = -x^2 + 4$ and $g(x) = 3x$, by evaluating the limit of an appropriate sum.

17. Let $h = f - g$ be defined over $[0, 1]$ by $h(x) = f(x) - g(x)$, where $f(x) = x^2 - 2x + 2$ and $g(x) = x^2$.

    (a) Compute $LA(h, 5)$.

    (b) Find the area of the ordinate set of h by evaluating an appropriate sum.

18. Find the area (by the definition) of the region bounded by the graphs of $g(x) = 1/2x + 3$ and $f(x) = |x|$. Sketch the region.

19. Find the area of the region bounded by the graphs of $f(x) = -x + 6$, $g(x) = 3x$, and $h(x) = 2x$. Sketch the region.

In Exercises 20 through 23, evaluate the integrals by the definition.

20. $\int_{-1}^{2}(x + 1)dx$ 21. $\int_0^1 |x|dx$ 22. $\int_0^2 x^2 dx$ 23. $\int_1^2 x^3 dx$

24. Prove or Disprove: A monotonically increasing function whose domain is the interval $[a, b]$ assumes its minimum value at $x = a$ and its maximum value at $x = b$.

25. Give an example for each of the following or explain why no example can exist.

    (a) A discontinuous function defined over $[-2, 3]$ whose ordinate set has an area of 12.

    (b) Two different functions f and g defined over $[-1, 3]$ each of whose ordinate sets has an area of 7.

    (c) A function defined over $[1, 3]$ whose ordinate set has an area of negative four.

(d) A strictly increasing function defined over $[3,7]$ such that $UA(f,6)$ approximates the area of the ordinate set of f with an $ERROR < \frac{1}{10}$.

(e) A monotone function f defined over $[1,5]$ and an integer n such that if $[1,5]$ is subdivided into n equal subintervals then
$UA(f,n) - LA(f,n) = |f(5) - f(1)|(4/n)$ (see Theorem 5.2.7).

26. Prove Theorem 5.2.7 for a non-negative monotone decreasing function.

27. Prove Theorem 5.2.9. Hint: Use Definition 5.2.8, Theorem 5.2.7, and the squeeze theorem.

5.3 The Riemann Integral

In this section we shall generalize the concept that was developed to answer the question: How is area defined? The process of generalizing from a particular instance to a more general case is very characteristic of mathematics. The typical situation is this: a mathematical concept involving certain restrictions is developed in order to answer a given question. Once established, the concept is then generalized by relaxing one or more of the restrictions. In the situation under discussion, the "given question" was: How is the area of an ordinate set defined? The answer that was developed depended on the following restrictions:

1. The function in question was bounded.

2. The interval involved was bounded.

3. The interval was subdivided into equal subintervals.

4. The function was nonnegative.

5. In the approximating sums, the function value was either the minimum or the maximum value of the function.

We shall continue to require the first two restrictions (they will be removed in Chapter 6, where the integral concept will be further generalized), but will eliminate the other three.

In the last section we subdivided the interval $[a,b]$ into equal subintervals, however it is sometimes necessary to work with an arbitrary subdivision, or partition. The formal definitions are as follows.

Definition 5.3.1 (Partition of an Interval) *A partition P of an interval $[a,b]$ is a finite set of points, $P = \{x_0, x_1, x_2, \ldots, x_n\}$ where $a = x_0 < x_1 < x_2 < \ldots < x_n = b$.*

A partition P is usually denoted by $P = \{a = x_0, x_1, x_2, \ldots x_n = b\}$.

The *norm* of a partition P, denoted by $|P|$, is the length of the longest subinterval defined by P. That is,

$$|P| = max\{|x_j - x_{j-1}| : j = 1, 2, 3, \ldots, n\}$$

Definition 5.3.2 (Regular Partition of an Interval) *A* regular *partition is a partition that subdivides the interval $[a, b]$ into subintervals of equal length.*

Note that the norm of a regular partition of $[a, b]$ is $(b - a)/n$, where n is the number of subintervals. The subdivisions of intervals that were made in examples in the last section were all regular partitions.

The following definition is attributed to G.R.B. Riemann (1826–1866).

Definition 5.3.3 (Riemann Sum) *Let f be a bounded function defined over $[a, b]$, let $P = \{x_0, x_1, \ldots, x_n\}$ be a partition of $[a, b]$, and let $\Delta x_k = x_k - x_{k-1}$ denote the change in x over the kth subinterval, $[x_{k-1}, x_k]$.*

For any set of numbers $\{w_k\}$, $x_{k-1} \leq w_k \leq x_k$, $k = 1, 2, \ldots, n$, the Riemann Sum (RS) *of f with respect to the partition P and the set of points $\{w_k\}$ is defined by*

$$RS(f, P, \{w_k\}) = \sum_{k=1}^{n} f(w_k)\Delta x_k$$

Definition of Riemann Sum

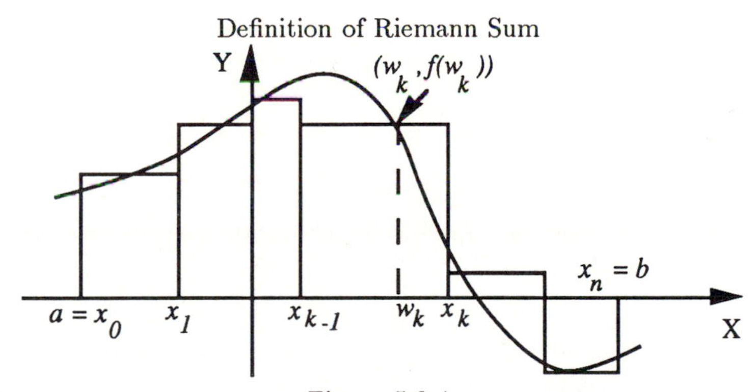

Figure 5.3.4

Note that when P is chosen to be a regular partition, $\Delta x_k = (b - a)/n$. If in addition, w_k is chosen such that $f(w_k) = (minf)_k$, then $RS(f, P, \{w_k\}) = LA(f, n)$. Similarly, if P is a regular partition and w_k is chosen such that $f(w_k) = (maxf)_k$, then $RS(f, P, \{w_k\}) = UA(f, n)$. In any case, $LA(f, n) \leq RS(f, P, w_k) \leq UA(f, n)$.

Example 5.3.5 Given the function $f : [-1, 3] \rightarrow \mathbb{R}$, defined by $f(x) = (x - 1)^2 - 2$, with partition $P = \{-1, 1/2, 1, 7/4, 5/2, 3\}$ and $\{w_k\} = \{0, 1, 1, 2, 3\}$, find $RS(f, P, \{w_k\})$.

The situation is illustrated in Figure 5.3.6.

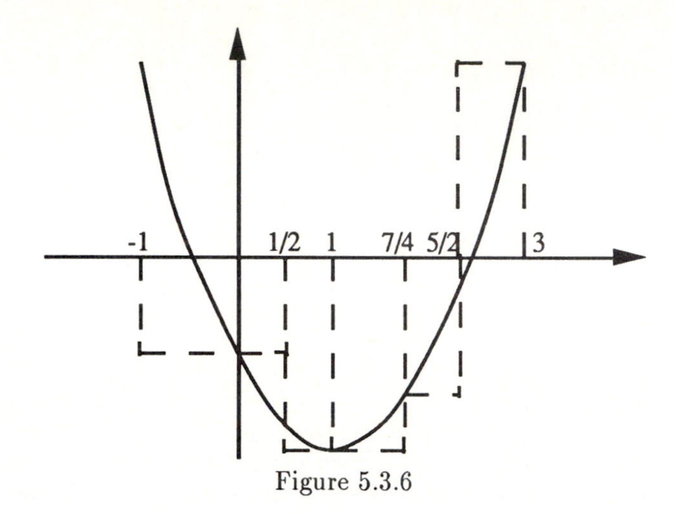

Figure 5.3.6

$$\begin{aligned} RS(f,P,(\{w_k\}) &= \sum_{k=1}^{5} f(w_k)\Delta x_k \\ &= f(0)\left(\frac{3}{2}\right)+f(1)\left(\frac{1}{2}\right)+f(1)\left(\frac{3}{4}\right)+f(2)\left(\frac{3}{4}\right)+f(3)\left(\frac{1}{2}\right) \\ &= (-1)\left(\frac{3}{2}\right)+(-2)\left(\frac{1}{2}\right)+(-2)\left(\frac{3}{4}\right)+(-1)\left(\frac{3}{4}\right)+(2)\left(\frac{1}{2}\right) \\ &= \frac{-15}{4} \end{aligned}$$

Note that the answer is the sum of the areas of the rectangles lying above the x-axis and the negative of the sum of the areas of the rectangles lying below the x-axis. △

In the following definition $UA(f,P)$ and $LA(f,P)$ will denote the upper and lower Riemann Sums associated with the partition P. That is, if $P=\{x_0,\ldots,x_n\}$, we define

$$UA(f,P)=\sum_{k=1}^{n}(max f)_k\Delta x_k$$

and

$$LA(f,P)=\sum_{k=1}^{n}(min f)_k\Delta x_k$$

Definition 5.3.7 (Refinement of a Partition) *A partition P' is a* refinement *of a partition P if every point in P is in P'. P' is said to be* finer than P.

Note that if P' is a refinement of P, then

$$LA(f,P) \leq LA(f,P') \text{ and } UA(f,P') \leq UA(f,P)$$

as is illustrated in Figure 5.3.8 where $P = \{a,b,c,d\}$ and $P' = \{a,t,b,c,d\}$.

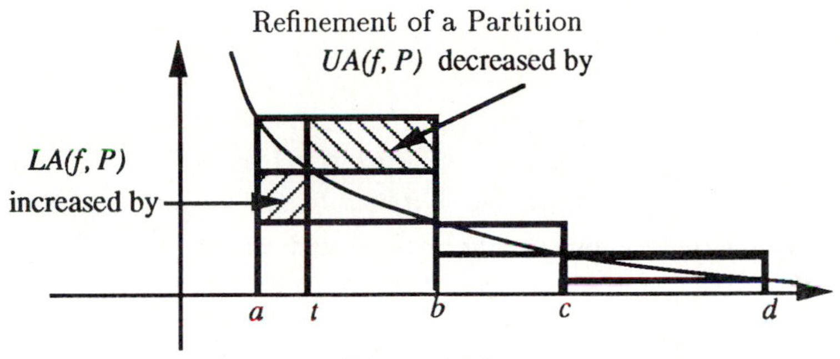

Figure 5.3.8

Definition 5.3.9 (The Riemann Integral) *Let $f : [a,b] \to \mathbb{R}$. The* Riemann Integral of f from a to b *is the unique real number (if such exists), denoted by*

$$\int_a^b f(x)dx$$

having the property that for every partition P *of* $[a,b]$

$$LA(f,P) \leq \int_a^b f(x)dx \leq UA(f,P)$$

If such a unique (i.e., one and only one) number exists, then we say that f is *integrable* over $[a,b]$. Otherwise, we say that f is *not integrable* over $[a,b]$.

Now for some observations:

1. The function in Example 5.2.10 ($f(x) = 1$ if x is rational in $[0,1]$, $f(x) = 0$ if x is irrational in $[0,1]$) is *not integrable* over $[0,1]$ (since *every* real number r, $0 \leq r \leq 1$, satisfies the condition $LA(f,P) \leq r \leq UA(f,P)$ for any partition P of $[0,1]$, there is no *unique* r).

2. A function f is integrable over $[a,b]$ if and only if given any accuracy, $ACC > 0$, there exists a number h such that for every partition P with $|P| < h$, $|UA(f,P) - LA(f,P)| < ACC$.

3. There are infinitely many partitions with norms less than a given number h (as in observation 2), and there are infinitely many Riemann Sums associated with a given partition (there are infinitely many choices for each w_k). Consequently if f is integrable over $[a, b]$, then given any $ACC > 0$ and a suitably small h there are infinitely many Riemann Sums, $RS(f, P, \{w_k\})$, that approximate $\int_a^b f(x)dx$ with accuracy less than ACC.

4. A function f is integrable over $[a, b]$ if and only if

$$\int_a^b f(x)dx = \lim_{n\to\infty} UA(f, n) = \lim_{n\to\infty} LA(f, n)$$

5. The *symbol* used for the variable of integration ("x" in the above definition) does not matter: $\int_a^b f(x)dx$ and $\int_a^b f(t)dt$ both refer to the same number. In fact, we could eliminate the symbol and write $\int_a^b f$. This variable, whose form does not matter, only the set it ranges over, is called a *dummy variable*.

The second observation suggests a type of question that could be asked concerning the integral of a function. Namely, approximate a given integral to within a given accuracy. (Recall questions of the type: approximate $\sqrt{2}$ with error less than 0.01.)

Example 5.3.10 Suppose we wish to approximate $\int_0^1 x^2dx$ with $ACC \leq 0.01$. Let P be a regular partition of n subintervals. Since $f(x) = x^2$ is monotonically increasing on $[0, 1]$,

$$\begin{aligned} |\int_0^1 x^2dx - RS(f, P, \{w_i\})| &\leq (UA(f, n) - LA(f, n)) \\ &\leq (f(1) - f(0))\frac{1-0}{n} \\ &= \frac{1}{n} \end{aligned}$$

Setting $1/n = 0.01$ and solving for n, we have $n = 100$. Thus, one acceptable answer (note that there are infinitely many acceptable answers) is to approximate the integral with an upper approximation taken over a regular partition with 100 subintervals. That is,

$$\begin{aligned} UA(f, 100) &= \sum_{k=1}^{100} (maxf)_k \left(\frac{1}{100}\right) \\ &= \sum_{k=1}^{100} f\left(0 + k\left(\frac{1}{100}\right)\right)\left(\frac{1}{100}\right) \end{aligned}$$

$$= \frac{1}{100}\sum_{k=1}^{100}\left(\frac{k}{100}\right)^2$$
$$= \left(\frac{1}{100}\right)^3 \frac{(100)(101)(201)}{6}$$
$$= 0.33835$$

We will later see that $\int_0^1 x^2 dx = \frac{1}{3}$, and hence the error is approximately $0.00502 < 0.01$, as was required. △

An immediate consequence of the Riemann Integral definition is that if f is integrable over $[a, b]$ and if f is nonnegative, then a geometric interpretation of $\int_a^b f(x)dx$ is that of the area of the ordinate set of the function defined by the integrand and its domain.

Example 5.3.11 We will evaluate $\int_2^4 (3x + 1)dx$.

The function defined by the integrand and the path of integration is $f : [2, 4] \to \mathbb{R}$ where $f(x) = 3x + 1$. The ordinate set of f is the trapezoid shown in Figure 5.3.12.

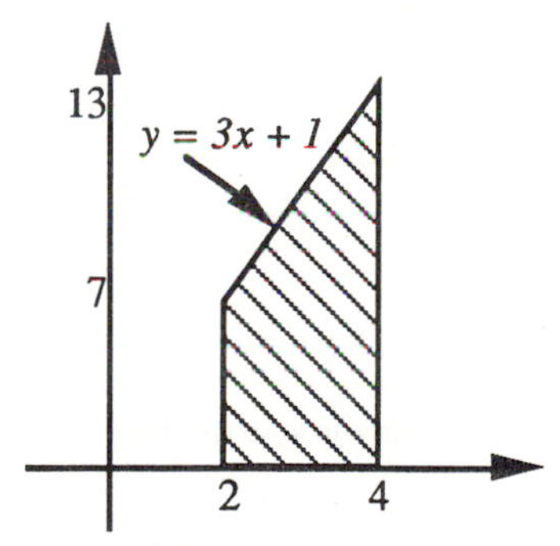

Figure 5.3.12

Thus the value of the integral is the area of the trapezoid that is one-half the length of the base times the sum of the lengths of the two sides. That is,

$$\int_2^4 (3x + 1)dx = \frac{4-2}{2}(7 + 13) = 20$$

△

Example 5.3.13 We will evaluate $\int_{-1}^1 \sqrt{1 - x^2}dx$. The function defined by the integrand and the path of integration is $f : [-1, 1] \to \mathbb{R}$ where $f(x) = \sqrt{1 - x^2}$. The ordinate set of f is the upper half of the unit disk centered at the origin.

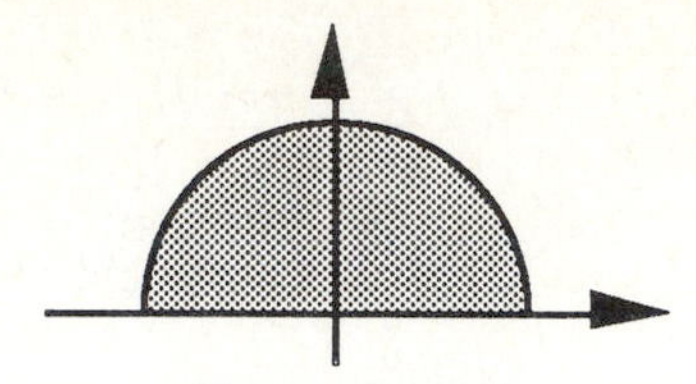

Figure 5.3.14

Since the area of the unit circle is π,

$$\int_{-1}^{1} \sqrt{1-x^2}dx = \frac{1}{2}\pi \qquad \triangle$$

The determination of necessary and sufficient conditions for a function to be integrable over an interval $[a, b]$ is a very difficult problem. The crux of the difficulty concerns the set of points of discontinuity of the function. We have already evaluated the integrals of several functions, and each of these functions was continuous. However, being continuous at every point is not a necessary condition for integrability. A little thought would easily convince us that a step function is integrable, although it is not continuous. Example 5.2.10, on the other hand, suggests that integrability must impose some restrictions on the set of points of discontinuity, because the function in that example, which was discontinuous everywhere, was not integrable.

We shall define the term *piecewise continuous*, state without proof an existence theorem, make some observations, and then leave further analysis of the problem of integrability to an advanced calculus course.

Definition 5.3.15 (Piecewise Continuous Function) *A function f is said to be* piecewise continuous *over an interval $[a, b]$ if the interval can be partitioned,*

$$a = x_0 < x_1 < x_2 < \ldots < x_n = b$$

such that f is continuous over each open interval (x_{k-1}, x_k), $k = 1, \ldots, n$.

As an example, a step function having a finite number of different values is piecewise continuous.

Figure 5.3.16 illustrates the graph of a piecewise continuous function.

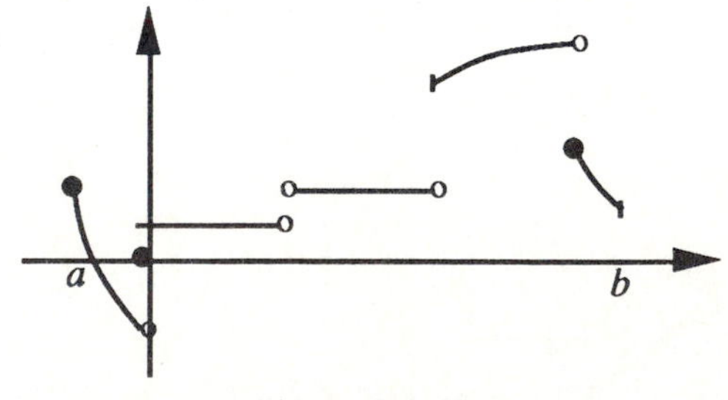

Figure 5.3.16

Theorem 5.3.17 (Integral Existence Theorem) *If f is a bounded piecewise continuous function over $[a,b]$, then f is integrable over $[a,b]$.*

The following observations will be important in our development of the integral. Make sure that you understand their relationship to the previous theorem.

1. Most of the functions that we have studied are piecewise continuous and thus are integrable. That is, polynomial functions, rational functions, trigonometric functions, root or power functions, absolute value functions, exponential functions, logarithmic functions as well as combinations of these are all integrable when considered over suitable intervals.

2. As a consequence of Theorem 5.3.17, we know that the limit of any sequence of Riemann Sums of a (bounded) piecewise continuous function exists and is unique, independent of how the set of points $\{w_i\}$ is chosen and independent of the sequence of partitions as long as the corresponding sequence of norms of these partitions converge to zero. For example, we know that the sine function is continuous and therefore, $\int_1^4 \sin^{2/3}(x)\ \mathrm{dx}$ exists. Furthermore,

$$\int_1^4 \sin^{2/3}(x)dx = \lim_{n\to\infty} \sum_{k=1}^{n} \sin^{2/3}\left(1 + k\frac{3}{n}\right)\left(\frac{3}{n}\right)$$

   The right-hand side is the limit of a sequence of Riemann Sums with the point w_k chosen to be the right-hand endpoint of the kth subinterval and a regular partition with norm $3/n$. We do not know how to evaluate this limit even though Theorem 5.3.17 tells us that this limit exists!

   An existence theorem (such as Theorem 5.3.17) guarantees the existence of a result, but usually does not provide any algorithm for obtaining the result.

3. If the definition of an integrable function is changed at one point, then the new function is still integrable, and the integral is unchanged. This is because the "area" under a single point on the graph of a function is zero.

Further study of the evaluation of an integral (in contrast to the existence or approximation) will have to wait until we have studied the Fundamental Theorem of Calculus in Chapter 6.

Section 5.3 Exercises

1. Let $f(x) = x^2 - x$, $P = \{0, \frac{1}{2}, \frac{3}{4}, 1, \frac{3}{2}\}$, and $\{w_k\} = \{\frac{1}{4}, \frac{1}{2}, 1, 1\}$. Compute $RS(f, P, \{w_k\})$.

2. Let $f : [-3, 2] \to \mathbb{R}$, where $f(x) = FLOOR(x)$. If $P = \{-3, -1, 0, 1, 2\}$ and $\{w_k\} = \{-\frac{5}{2}, -\frac{1}{2}, \frac{1}{2}, \frac{3}{2}\}$, then compute $RS(f, P, \{w_k\})$.

3. Let $f : [-3, 7] \to \mathbb{R}$, where $f(x) = |\cos(x)|$. If $P = \{-3, -2, 0, \frac{1}{2}, 5, 7\}$ and $\{w_k\} = \{-2, -1, 0, 2, 6\}$, then compute $RS(f, P, \{w_k\})$.

4. Approximate $\int_0^1 x dx$ to within an accuracy of $ACC = 0.01$ by using Riemann Sums.

5. Approximate $\int_0^{\pi/2} \cos(x) dx$ to within an accuracy of $ACC = 0.1$ by using Riemann Sums.

6. Approximate $\int_{-1}^1 -x^2 dx$ to within an accuracy of $ACC = 0.1$ by using Riemann Sums.

7. Approximate the area of the region bounded by the graphs of $y = x^2$ and $y = x$ to within an accuracy of $ACC = 0.1$ by using Riemann Sums.

8. Find three different partitions P_1, P_2, and P_3 such that for $k = 1, 2, 3$,

$$|RS(f, P_k, \{w_k\}) - \int_0^2 f(x) dx| < 0.1$$

where $f(x) = -x^2$ and $f(w_k) = (min f)_k$, for all k.

In Exercises 9 through 15, sketch the graph of the ordinate set of the integrand and then evaluate the integral by geometrically computing the area of the ordinate set.

9. $\int_1^6 f(x)dx$ where $f(x) = FLOOR(x)$

10. $\int_0^4 x dx$

11. $\int_{-2}^1 (-4x + 6) dx$

12. $\int_0^2 \sqrt{4 - x^2} dx$

13. $\int_0^1 (\sqrt{1 - x^2} + 1) dx$

14. $\int_1^3 [(2x + 2) - (3x - 1)] dx$

15. $\int_0^1 (\sqrt{1 - x^2} - x) dx$

16. Which answers are *unreasonable*? $\int_1^2 (x^2 + \sin(x^2 + 2)) dx =$

    (a) 1.9 (b) 0 (c) 3.9 (d) -2 (e) 1.3

17. Which answers are *unreasonable*? $\int_0^{2/3} \sqrt{1 + \cos(x)} dx =$
    (a) 8 (b) 0 (c) 1 (d) -2 (e).8

18. Which answer *best* approximates the integral? $\int_{-1}^1 \sqrt{1 + t^2} dt =$
    (a) -1 (b) 1 (c) 4 (d) 3 (e) $\frac{3}{2}$

19. Which answer *best* approximates the integral of the function f that is graphed in the following figure? $\int_0^2 f(x) dx =$
    (a) -1 (b) 0 (c) 2 (d) 3 (e) $\frac{5}{2}$

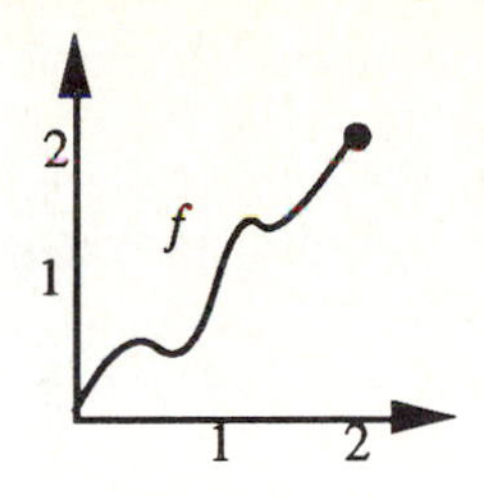

Figure 5.3.18

20. Which answer *best* approximates the integral? $\int_1^3 1/t dt =$
(a) -1 (b) 1 (c) 2 (d) $\frac{3}{2}$ (e) $\frac{1}{2}$

21. Which answer *best* approximates the integral? $\int_0^{2\pi} \sin(t)dt =$
(a) -1 (b) -2 (c) 0 (d) 1 (e) π

22. Which answer *best* approximates the integral? $\int_0^4 |f(x)|dx =$
(a) -1 (b) 0 (c) 2 (d) 4 (e) -2

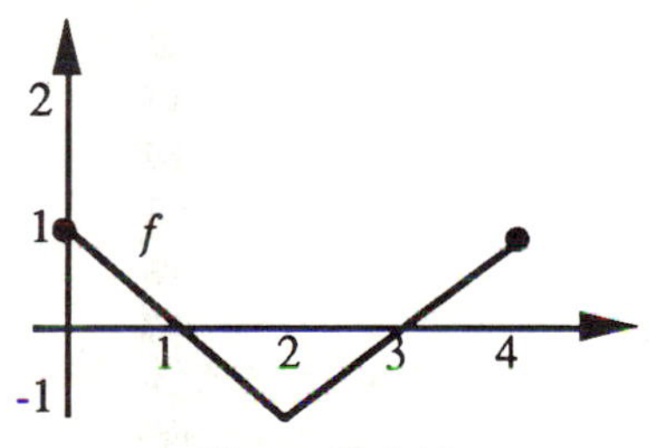

Figure 5.3.19

23. Mark each of the following T (true) or F (false).

(a) A differentiable function is a piecewise continuous function.

(b) If $\int_a^b f(x)dx > 0$, then $f(x) > 0$ for all x in $[a, b]$.

(c) All polynomial functions are integrable functions.

(d) If f is an integrable function, then f is a continuous function.

(e) The integral of a quotient is the quotient of integrals.

24. Give an example for each of the following or explain why no example can exist.

(a) A function f defined over $[-2, 3]$ such that $\int_{-2}^3 f(x)dx = -7$.

(b) A step function g defined over $[0, 1]$ that approximates $f(x) = x^2$ with $ACC = 0.01$ in the sense that $|\int_0^1 f(x)dx - \int_0^1 g(x)dx| \leq 0.01$.

(c) A function f defined over $[-2,3]$ such that $\int_{-2}^{3} f(x)dx = 0$ and $\int_{-2}^{3} |f(x)|dx = 5$.

(d) A function defined over $[2,5]$ such that $\int_{2}^{5} |f(x)|dx < \int_{2}^{5} f(x)dx$.

(e) Functions f and g defined over $[-1,4]$ such that $\int_{-1}^{4} f(g(x))dx > \int_{-1}^{4} g(f(x))dx$.

(f) A function f defined over $[0,1]$ such that $\int_{0}^{1} f(x)dx = 0$, $f(x) \geq 0$ for all x, and $f(0) = f(1) = 1$.

25. We know that if a partition P' is finer than a partition P, then $UA(f, P') \leq UA(f, P)$ and $UL(f, P') \geq UL(f, P)$. Find a function f on an interval $[a, b]$ and partitions P' and P of $[a, b]$ such that P' has more points than P but $UA(f, P') > UA(f, P)$ and $LA(f, P') < LA(f, P)$. (Clearly P' will not be a refinement of P.)

26. In this exercise you will experimentally investigate the error analysis in approximating the Riemann integral using Riemann Sums. From Theorem 5.2.7 we know that for a monotonic function the error in approximating an ordinate set by upper and lower sums is proportional to $1/n$. If the error is inversely proportional to the number of subintervals, then doubling the number of subintervals should (approximately) halve the error. Does this actually happen?

(a) Let the interval be [0,1] and $f(x) = x^2$. The precise value of the definite integral is $\int_{0}^{1} x^2 dx = \frac{1}{3}$. Find the errors in the approximation to $\int_{0}^{1} x^2 dx$ using Riemann Sums with uniform subintervals of lengths $\Delta x_k = 1/2^n$ for $n = 2, 3, 4, 5$. Choose w_k to be the left endpoint of each subinterval. How does the error change with the change in the subinterval size? That is, if the subinterval is halved, how does the error change?

(b) Repeat (a) with w_k the right endpoint of the subinterval. Do you expect any significant change?

(c) Repeat (a) with w_k the midpoint of each subinterval.

(d) From your results in (a) through (c), form a conjecture about how the error changes as n changes and as the choice of w_k changes.

(e) Verify your conjecture with another function, such as $f(x) = \cos(x)$ on $[0, \pi/2]$, whose exact integral is 1.

27. Approximate the surface area of Johnson Pond to within an accuracy of ACC = 10,000 ft^2.

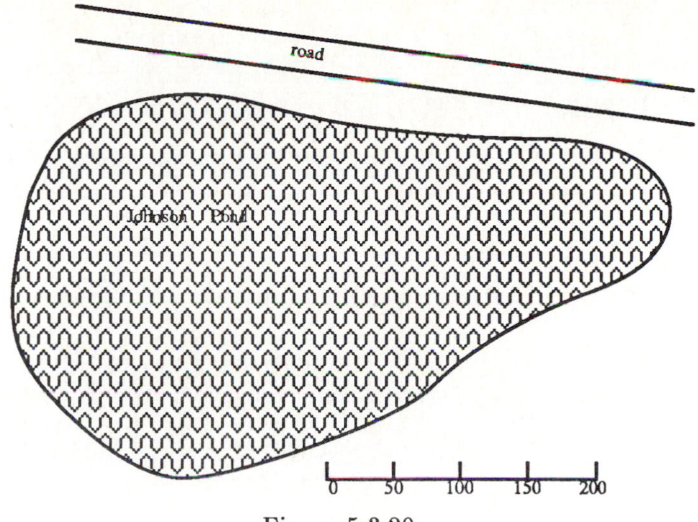

Figure 5.3.20

5.4 Numerical Integration

The formulation of a Riemann Sum is a numerical integration technique, that is, a method for numerically approximating an integral. Most functions cannot be exactly integrated, even using the methods that will be discussed in later sections. In practice, when a function must be integrated some numerical integration technique (applied with the aid of a computer) is usually used. In the method of Riemann Sums the interval in question is partitioned into subintervals and over each subinterval the function is approximated by a constant function, a polynomial of degree zero. The integral of the function over a subinterval is then approximated by the integral of the corresponding constant function, which is just the area of a rectangle. The Riemann Sum approximation to the integral is the sum of the areas of these rectangles.

The establishment of an error bound is an important consideration of any approximation technique. For Riemann Sum approximations, if f is monotone then the error bound is proportional to $1/n$ where n is the number of subintervals (see Theorem 5.2.7). Thus doubling the number of subintervals will reduce the error bound by a factor of $\frac{1}{2}$. This is also true (for nonmonotonic functions) if the derivative of f is bounded.

Theorem 5.4.1 (Error Bound for Riemann Sums) *Let A be an approximation to $\int_a^b f(x)dx$ using Riemann sums with n equal subintervals. If $|f'(x)| \leq B$ for all x in $[a,b]$, then*

$$|A - \int_a^b f(x)dx| \leq \frac{B(b-a)^2}{n}$$

Proof: Let the partition be $P = \{x_0, \ldots, x_n\}$. Since $LA(f,P) \leq \int_a^b f(x)dx \leq UA(f,P)$ and $LA(f,P) \leq A \leq UA(f,P)$ by the definitions,

$$\begin{aligned}
|\int_a^b f(x)dx - A| &\leq UA(f,P) - LA(f,P) \\
&= \sum_{k=1}^{n}(maxf)_k \Delta x_k - \sum_{k=1}^{n}(minf)_k \Delta x_k \\
&= \sum_{k=1}^{n}[(maxf)_k - (minf)_k]\Delta x_k \\
&= \sum_{k=1}^{n}[(maxf)_k - (minf)_k]\frac{b-a}{n} \\
&= \frac{b-a}{n}\sum_{k=1}^{n}[(maxf)_k - (minf)_k]
\end{aligned}$$

We can apply the Mean Value Theorem at the points in $[x_{k-1}, x_k]$ where $(maxf)_k$ and $(minf)_k$ are attained, giving

$$[(maxf)_k - (minf)_k] \leq f'(c_k)\Delta x_k = f'(c_k)\frac{b-a}{n}$$

for some c_k in $[x_{k-1}, x_k]$. Thus,

$$\begin{aligned}
|\int_a^b f(x)dx - A| &\leq \frac{b-a}{n}\sum_{k=1}^{n} f'(c_k)\frac{b-a}{n} \leq \frac{(b-a)^2}{n^2}\sum_{k=1}^{n} B \\
&= \frac{(b-a)^2 B}{n^2} n \\
&= \frac{B(b-a)^2}{n}
\end{aligned}$$

□

Example 5.4.2 We will approximate $\int_1^2 \sin(x^3)dx$ to within 0.01 using Riemann Sums. $f(x) = \sin(x^3)$ is not monotonic over [1,2] (as a plot shows) but it is differentiable for all x. Thus we can apply Theorem 5.4.1 to determine the maximum number of subintervals n required. Since $f'(x) = 3x^2\cos(x^3)$, $|f'(x)| \leq 3 \cdot 4 \cdot 1 = 12$ on $[1,2]$. Thus we want $12/n \leq 0.01$ or $n \geq 1200$. Error bound estimates are always conservative: this large an n guarantees an accuracy of 0.01, although a smaller number of subintervals may also give the desired accuracy. Numerical computation shows that an approximation to $\int_1^2 \sin(x^3)dx$ of 0.22 with an accuracy of 0.01 occurs using only 10 subintervals. △

It is possible to obtain approximations in which the error decreases more rapidly as the number of subintervals is increased by approximating the function over each subinterval with a polynomial of degree greater than zero. In this section, we shall develop the approximation formulas with error bounds for the cases where the approximating polynomial has degree one (Trapezoidal Rule) and degree two (Simpson's Rule). In Chapter 9, we will consider approximations using polynomials of arbitrarily high degree.

Trapezoidal Approximations

Consider the function $f : [2,4] \to \mathbb{R}$, defined by $f(x) = \sqrt{x^3 - 1}$, whose graph is sketched below along with the straight line segment joining the endpoints, $(2, f(2))$ and $(4, f(4))$. The straight line segment is the graph of an approximating polynomial P of degree one. (Note that here P is a polynomial, not a partition—there are only so many letters available, so some must do double duty.)

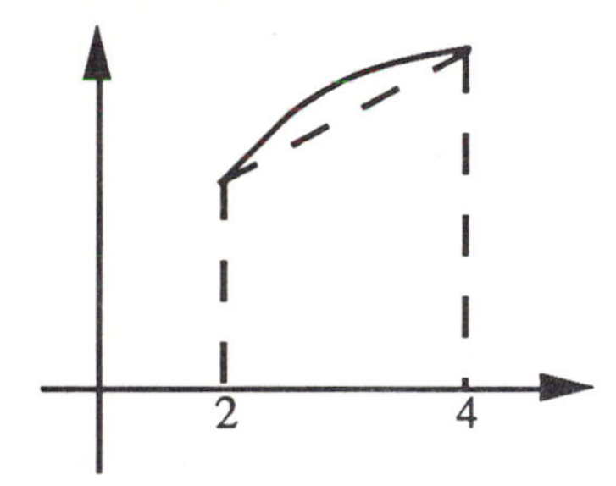

Figure 5.4.3

The integral of P is just the area of the trapezoid with vertices $(2,0)$, $(2, f(2))$, $(4,0)$, and $(4, f(4))$.

Since the area of a trapezoid is $h/2(L + R)$, where h is the length of the base and L and R are the lengths of the left and right sides, the integral of the approximating polynomial is

$$\int_2^4 P(x)dx = (1.0)[f(2) + f(4)] = 2.645751 + 7.937254 = 10.583005$$

In general, for any continuous function $f : [a,b] \to \mathbb{R}$, we can estimate the integral over $[a,b]$ as

$$\int_a^b f(x)dx \approx \frac{h}{2}[f(a) + f(b)]$$

where $h = b - a$ is the length of the interval.

It is important, as always when using an approximation, to consider bounds on the error associated with the approximation. For the trapezoidal approximation $(b-a)/2[f(a)+f(b)]$ of the integral of a twice differentiable function f over $[a,b]$, the following result can be proved.

Theorem 5.4.4 (Single Trapezoid Error Estimate) *Let $f : [a,b] \to \mathbb{R}$ and suppose that there is a positive number B such that $|f''(x)| \leq B$ for all $a \leq x \leq b$.*

Then the error of a trapezoidal approximation of $\int_a^b f(x)dx$ is no more than $B(b-a)^3/12$. That is, $|\int_a^b f(x)dx - h/2[f(a)+f(b)]| \leq Bh^3/12$, where $h = b-a$.

Example 5.4.5 Consider the integral just estimated. It can be shown, using the techniques for finding extrema of Chapter 3, that the maximum of $|f''|$ on $[2,4]$ occurs at 2.732051 and is, to six decimal places, 0.393320. Thus the bound B is 0.393320, and an error bound for the approximation 10.583005 of $\int_2^4 \sqrt{x^3-1}dx$ is

$$\frac{Bh^3}{12} = 0.393320\left(\frac{2^3}{12}\right) = 0.262213$$

Thus, the value of $\int_2^4 \sqrt{x^3-1}dx$ is within the interval:

$$10.320792 \leq \int_2^4 \sqrt{x^3-1}dx \leq 10.845218$$

This estimate is not too bad, but how can it be improved? △

Consider the factors controlling the error. An important consideration is h, the width of the interval. If h is large, h^3 will be very large. Consider $h = 10$, for example! On the other hand, if h is very small, h^3 will be very, very small. An obvious way to improve the method is to break the interval $[a,b]$ into many smaller intervals, compute the trapezoidal approximation on each of the subintervals, and then sum the results. This method is formalized in the following rule.

Definition 5.4.6 (The Trapezoidal Approximation) *If f is a continuous function on $[a,b]$ and if $P = \{a = x_0, x_1, ..., x_n = b\}$ is a regular partition of $[a,b]$, then the* Trapezoidal Approximation (Rule) *to $\int_a^b f(x)dx$ is $h/2[f(x_0) + 2f(x_1) + \cdots + 2f(x_{n-1}) + f(x_n)]$, where $h = (b-a)/n$ is the length of each subinterval.*

Before considering an error bound or looking at any examples, let us see how this approximation is related to the original trapezoidal approximation.

Essentially, the above formula is derived in the following manner.

1. Partition the interval $[a,b]$ into n subintervals, each of length $h = (b-a)/n$.

2. Use the trapezoidal approximation to find the area of the trapezoid over *each subinterval.* The areas of these trapezoids are

$$\begin{aligned}
A_1 &= \frac{h}{2}[f(x_0) + f(x_1)] \\
A_2 &= \frac{h}{2}[f(x_1) + f(x_2)] \\
A_3 &= \frac{h}{2}[f(x_2) + f(x_3)] \\
&\vdots \\
A_{n-1} &= \frac{h}{2}[f(x_{n-2})] + f(x_{n-1})] \\
A_n &= \frac{h}{2}[f(x_{n-1})] + f(x_n)]
\end{aligned}$$

3. Find the sum of all areas. Note, that each of the terms, $f(x_k)$ is found in two area computations with the exception of $f(x_0)$ and $f(x_n)$. Hence, the sequence $1, 2, 2, ..., 2, 2, 1$ is in the formula.

Trapezoidal Rule

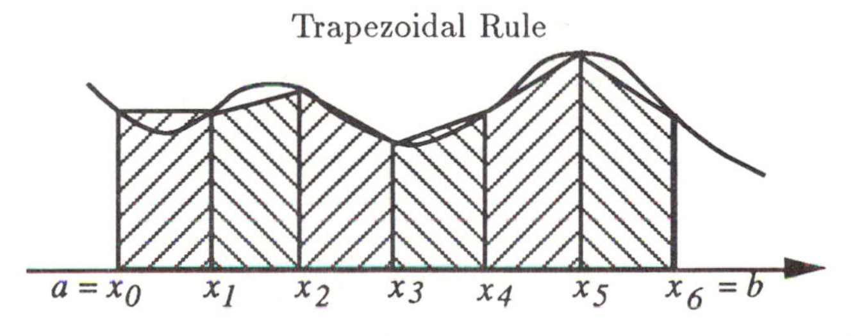

Figure 5.4.7

The proof of the following error bound is similar to that for Riemann Sums, and is omitted.

Theorem 5.4.8 (Error Bound for the Trapezoidal Rule) *Let A be an approximation of $\int_a^b f(x)dx$ using the Trapezoidal Rule with n equal subintervals. If $|f''(x)| \leq B$ for all x in $[a, b]$, then*

$$|A - \int_a^b f(x)dx| \leq \frac{B(b-a)^3}{12n^2}$$

The first feature of the error bound to notice is that as n increases, the size of the error decreases. Furthermore, the decrease in the error is proportional to the square of n. This means, for example, that if n is doubled that the size of the error will decrease by a factor of $\frac{1}{4}$. Since, over any interval, B and $(b - a)$ are fixed, we may choose n large enough to reduce the error to any desired level. When sums of areas of rectangles (Riemann Sums) are used to approximate the integral the error is proportional to $1/n$ (see Theorems 5.2.7 and 5.4.1) while with the trapezoidal method, the error is proportional to $1/n^2$.

Example 5.4.9 We will approximate $\int_2^4 \sqrt{x^3 - 1}dx$ by using the Trapezoidal Rule with $n = 4$. (Note that this is the same integral that was approximated in the previous example with $n = 1$.)

We have $h = (4 - 2)/4 = 0.5$ and the formula gives

$$A = 0.25[f(2) + 2f(2.5) + 2f(3) + 2f(3.5) + f(4)]$$

The pertinent numerical information, accurate to six decimal places, obtained by using a hand calculator, is given in the following table.

x_k	$f(x_k)$	c	$cf(x_k)$
2.0	2.645751	1	2.645751
2.5	3.824265	2	7.648529
3.0	5.099020	2	10.198039
3.5	6.471090	2	12.942179
4.0	7.937254	1	7.937254

summing the values in the right-hand column and multiplying by 0.25 yields

$$\int_2^4 \sqrt{x^3 - 1}dx \approx 10.342938$$

Again, we need to determine an error bound for the approximation. Recall that $|f''|$ attains a maximum value of 0.393320 at $x = 2.732051$. Therefore,

$$|A - \int_2^4 \sqrt{x^3 - 1}dx| \leq \frac{0.393320(2^3)}{12(4)^2} = 0.0163883$$

The error is diminished from the previous approximation (Example 5.4.5) by a factor of 16. Therefore, we know that

$$10.326550 \leq \int_2^4 \sqrt{x^3 - 1}dx \leq 10.359326 \qquad \triangle$$

Example 5.4.10 Using a regular partition, determine how large n (the number of subintervals) must be in order to approximate $\int_{-\pi/4}^{\pi/4} \tan(x)dx$ with $ACC = \frac{1}{10}$ using a Riemann Sum. Using the Trapezoidal Rule.

Since $f(x) = \tan(x), -\pi/4 \leq x \leq \pi/4$, is monotonic, by Theorem 5.2.7 the error using a Riemann Sum approximation is

$$\left|f\left(\frac{\pi}{4}\right) - f\left(-\frac{\pi}{4}\right)\right| \left(\frac{\pi}{2}\right) \left(\frac{1}{n}\right) = 2\left(\frac{\pi}{2}\right)\frac{1}{n} = \frac{\pi}{n}$$

Thus to obtain an accuracy of $\frac{1}{10}$, we set $\pi/4 < \frac{1}{10}$, or $n \geq 32$.

To obtain the error bound for the Trapezoidal Rule, we first need to find an upperbound B for $|f''(x)|$ over $[-\pi/4, \pi/4]$. Since a maximum value is an upper bound, we employ the methods of Chapter 3. That is, set $f'''(x) = 0$ to find the critical values of f''.

$$\begin{aligned} f(x) &= \tan(x) \\ f'(x) &= \sec^2(x) \\ f''(x) &= 2\sec^2(x)\tan(x) \\ f'''(x) &= 2\sec^4(x) + 4\sec^2(x)\tan^2(x) \end{aligned}$$

Since $f'''(x) > 0$, $f''(x)$ is increasing. Hence $f''(\pi/4) = 4$ is the maximum value of $|f''|$ over $[-\pi/4, \pi/4]$ and the error bound is

$$\frac{4(\pi/2)^3}{12n^2} \approx \frac{1.29}{n^2} < \frac{1}{10} \quad \text{or} \quad n^2 > 12.9 \quad \text{or} \quad n \geq 4$$

Thus the Trapezoidal Rule is a great improvement over Riemann Sums, since at most 2 intervals rather than 32 intervals are required to obtain an accuracy of $\frac{1}{10}$. △

Since first-degree (trapezoidal) approximations are usually more accurate than Riemann Sums, we can hope for further improvement by considering approximating polynomials of degree two.

Parabolic Approximations

If f is a continuous function over $[a, b]$, then the *parabolic approximation of the integral of f* is

$$\int_a^b f(x)dx \approx \frac{h}{3}[f(a) + 4f(m) + f(b)]$$

where $m = (a+b)/2$ is the midpoint between a and b and $h = (b-a)/2$ is the length of the intervals $[a, m]$ and $[m, b]$.

The idea behind the parabolic approximation is to approximate the graph of $y = f(x)$ with a polynomial of degree two and use the integral of this polynomial as an approximation to the integral of f over $[a, b]$. The name "parabolic" comes from the fact that the graph of a polynomial of degree two is a parabola.

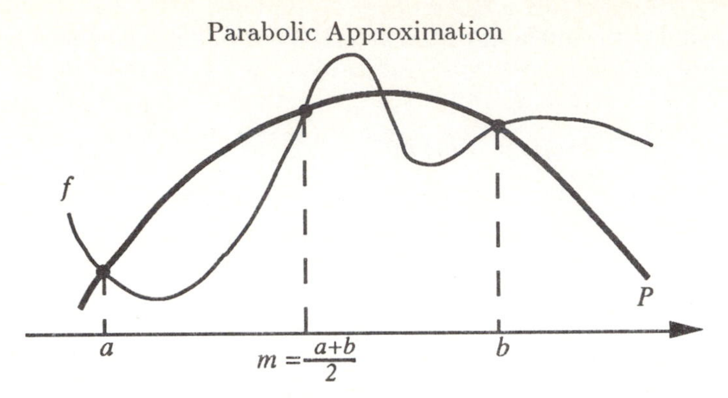

Figure 5.4.11

To derive this approximation, let f be defined on $[a, b]$ and let $m = (a + b)/2$. Construct the polynomial P of degree two whose graph passes through the three points $(a, f(a))$, $(m, f(m))$, and $(b, f(b))$. (The method to construct this *Lagrange interpolating polynomial* is described in the exercises at the end of this section.)

$$P(x) = \frac{(x-b)(x-m)}{(a-b)(a-m)}f(a) + \frac{(x-a)(x-m)}{(b-a)(b-m)}f(b) + \frac{(x-a)(x-b)}{(m-a)(m-b)}f(m)$$

Direct substitution shows that $P(a) = f(a)$, $P(m) = f(m)$, and $P(b) = f(b)$.

Letting $h = (b-a)/2$ be the length of each of the two subintervals, the integral of P can be shown to be

$$\int_a^b P(x)dx = \frac{h}{3}[f(a) + 4f(m) + f(b)]$$

using the limiting process of the previous section.

This is a parabolic approximation of the integral of f over the interval $[a, b]$ with midpoint m.

As always when working with an approximation, it is important to determine an error bound. If f is a fourth differentiable function, there is an error expression for the parabolic approximation that is similar to the error expression for the Trapezoidal Rule.

Theorem 5.4.12 (Single Parabola Error Estimate) *Let $f : [a, b] \to \mathbb{R}$ and suppose that $|f^{(4)}(x)| \leq B$ for all $a \leq x \leq b$.*

Then the error of a parabolic approximation of $\int_a^b f(x)dx$ is no more than $B\,(b-a)h^4/180$ where $h = (b-a)/2$. That is,

$$\left|\int_a^b f(x)dx - \frac{h}{3}[f(a) + 4f(m) + f(b)]\right| \leq B\frac{(b-a)^5}{2880}$$

Compare the above error bound with that of the trapezoidal approximation. The term $b-a$ is raised to the fifth rather than the third power, and hence the approximation is more sensitive to this quantity. For example, if the width of the interval is halved, the error bound is reduced by a factor of $\frac{1}{32}$.

Example 5.4.13 We will estimate $\int_2^4 \sqrt{x^3-1}dx$ using a parabolic approximation. Note that this is the same integral that was approximated using the Trapezoidal Rule in the previous example. We will also determine a bound for the error of the approximation.

In this example we have $a=2$, $b=4$, $m=3$ and $h=1$. The values of f, to six decimal places, are

$$f(2)=2.645751 \quad f(3)=5.099020 \quad f(4)=7.937254$$

Substituting into the approximation formula we get

$$A = 10.326362$$

To determine a bound for the error, we graph $|f^{(4)}|$ and note that it is monotonic, attaining its maximum at $x=2$ of $B=|f^{(4)}(2)|=1.150422$. Therefore,

$$|A - \int_2^4 \sqrt{x^3-1}dx| \leq 1.150422\left(\frac{2}{180}\right) = 0.012783$$

△

We can do even better. The improved performance of parabolic approximation is due, in part, to our choice of examples. An obvious way to improve the approximation is to partition the interval $[a,b]$ into subintervals (as was done in the Riemann Sum and trapezoidal cases) and to do parabolic approximation on *each subinterval.* The result of this effort is known as Simpson's Rule.

Definition 5.4.14 (Simpson's Rule) *If f is a continuous function on $[a,b]$ and if $P=\{a=x_0,x_1,\ldots,x_n=b\}$ is a regular partition of $[a,b]$ with n even, then the* Simpson's Rule approximation *to $\int_a^b f(x)dx$ is*

$$\frac{h}{3}[f(x_0)+4f(x_1)+2f(x_2)+4f(x_3)+\cdots+2f(x_{n-2})+4f(x_{n-1})+f(x_n)]$$

where $h=\frac{(b-a)}{n}$.

Note that the number of subintervals in Simpson's Rule *must be even.*

Simpson's Rule is obtained by piecing together parabolic approximations on pairs of consecutive subintervals (similar to what was done in developing the Trapezoidal Rule). Suppose $n=8$. Then there are 8 subintervals of length $h=(b-a)/8$. On pairs of consecutive subintervals, apply the parabolic approximation to obtain

$$\begin{array}{ll} \text{for } \{x_0, x_1, x_2\} & h/3[f(x_0) + 4f(x_1) + f(x_2)] \\ \text{for } \{x_2, x_3, x_4\} & h/3[f(x_2) + 4f(x_3) + f(x_4)] \\ \text{for } \{x_4, x_5, x_6\} & h/3[f(x_4) + 4f(x_5) + f(x_6)] \\ \text{for } \{x_6, x_7, x_8\} & h/3[f(x_6) + 4f(x_7) + f(x_8)] \end{array}$$

Adding these approximations together gives

$$\frac{h}{3}[f(x_0) + 4f(x_1) + 2f(x_2) + 4f(x_3) + 2f(x_4) + 4f(x_5) + 2f(x_6) + 4f(x_7) + f(x_8)]$$

Simpson's Rule

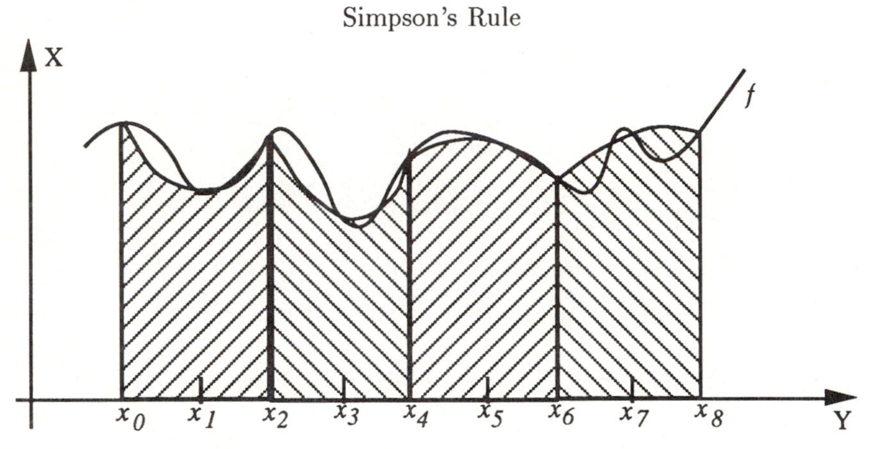

Figure 5.4.15

Example 5.4.16 We will use Simpson's Rule with $n = 4$ [note that there will be 4 subintervals and $h = (4-2)/4 = 1/2$] to approximate $\int_2^4 \sqrt{x^3 - 1}dx$. (See Example 5.4.13.)

The pertinent numerical information, accurate to six decimal places and obtained by using a hand calculator, is given in the following table.

x_k	$f(x_k)$
2.0	2.645751
2.5	3.824265
3.0	5.099020
3.5	6.471090
4.0	7.937254

Thus the approximation is

$$A = \frac{1/2}{3}[f(2) + 4f(2.5) + 2f(3) + 4f(3.5) + f(4)] = 10.327077$$

△

Theorem 5.4.17 (Error Bound for Simpson's Rule) *Let A be an approximation of $\int_a^b f(x)dx$ using Simpson's Rule with n subintervals and $h = (b-a)/n$. If $|f^{(4)}(x)| \leq B$ for all x in $[a, b]$ then*

$$|A - \int_a^b f(x)dx| \leq \frac{B(b-a)^5}{180n^4}$$

This estimate is very similar to the one for parabolic approximation. The basic difference is the factor of n^4 in the denominator. This means that if we double the number of subintervals then the error bound will decrease by a factor of $\frac{1}{16}$. Recall that with Riemann Sums the error bound decreased by a factor of $\frac{1}{2}$ when the number of subintervals was doubled, and that with the Trapezoidal Rule the error bound decreased by a factor of $\frac{1}{4}$ when the number of subintervals was doubled.

Example 5.4.18 We will estimate the error in the Simpson's Rule approximation in Example 5.4.16. Since we know from Example 5.4.13 that $|f^{(4)}(x)| \leq 1.150422$ for x in $[2, 4]$, we get

$$|A - \int_2^4 \sqrt{x^3 - 1}dx| \leq \frac{1.150422(2)^5}{180(4)^4} = 0.000799 \qquad \triangle$$

Example 5.4.19 In the above examples on Simpson's Rule we determined an approximation and its error bound given the number of subintervals used. In applications, the analysis usually starts with a maximum allowable error for the approximation, then the number of subintervals is determined (from the maximum allowable error), and finally the approximation is computed. For example, if we wish to approximate $\int_0^3 \sin(x)dx$ with an error of at most 0.0001, how many subintervals should we use to obtain the desired accuracy? The error estimate says that if n subintervals are used and $|f^{(4)}(x)| \leq B$ on $[0, 3]$, then the error will be at most $3^5B/(180n^4)$. Therefore we need:

$$3^5B/180n^4 \leq 10^{-4}$$

For the sine function, $|\sin^{(4)}(x)| \leq 1$, thus we can take $B = 1$ and solve the inequality

$$\frac{3^5}{180n^4} \leq 10^{-4}$$

for n. This requires that

$$n^4 \geq \frac{10^4 \cdot 243}{180} = 13,500 \quad \text{or} \quad n \geq 11$$

or $n = 12$ since n must be even in Simpson's Rule. Thus if we use 12 subintervals, the error in our approximation will be at most 0.0001. $\triangle$

In previous examples the extrema of $f^{(4)}$ occurred at endpoints of the interval and were relatively easy to estimate. In practice the extrema will be more difficult to estimate.

Example 5.4.20 Suppose we want to compute $\int_1^4 \sin(x^2)dx$ with an accuracy of at least 10^{-4} using Simpson's Rule.

First we need to determine the number of intervals, n, which is required to obtain the desired accuracy. That is, we want to find an n such that

$$\frac{B(b-a)^5}{180n^4} \leq 10^{-4} \quad \text{for all } x \text{ in } [1,4].$$

To solve this inequality, we need to find B, a bound for $|f^{(4)}(x)|$ on [1,4]. We can compute $f^{(4)}$ by hand [or using the symbolic capabilities of a computer algebra system (CAS)] to obtain

$$f^{(4)}(x) = 16\sin(x^2)x^4 - 48\cos(x^2)x^2 - 12\sin(x^2).$$

We need to estimate the maximum of $|f^{(4)}|$ on [1,4]. There are several methods we might use.

1. We can make a rough estimate of B by bounding the individual terms:

$$\begin{aligned} |f^{(4)}(x)| &\leq |16\sin(x^2)x^4 - 48\cos(x^2)x^2 - 12\sin(x^2)| \\ &\leq 16x^4 + 48x^2 + 12 \\ &\leq 4096 + 768 + 12 = 4876 \end{aligned}$$

   on [1,4]. This would require n subintervals, where

$$\frac{4876(4-1)^5}{180n^4} \leq 10^{-4} \quad \text{or} \quad n^4 \geq \frac{4876 \times 3^5 \times 10^4}{180}.$$

   Calculating, we obtain $n \geq 90.07$, or 92 subintervals, since n must be even.

2. We can try to estimate the maximum by using calculus. We know that since $f^{(4)}$ is a differentiable function, its derivative,

$$f^{(5)}(x) = 32\cos(x^2)x^5 + 160\sin(x^2)x^3 - 120\cos(x^2)x$$

   will vanish at a point in (1,4) where $f^{(4)}$ has a maximum. However, it is difficult to find a zero of $f^{(5)}$. A good idea would be to graph $f^{(4)}$ to see what it looks like.

Graph of $f^{(4)}$

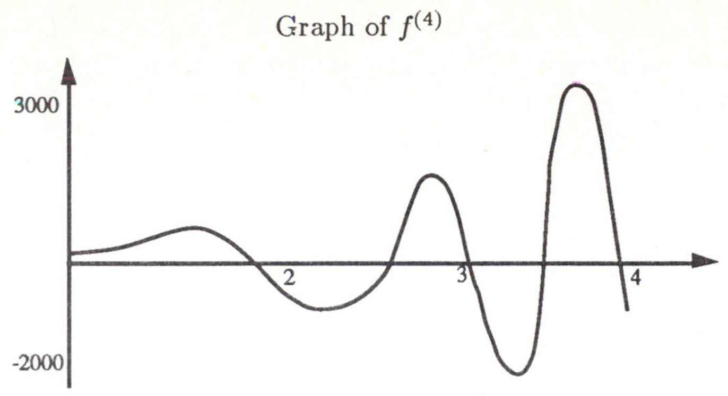

Figure 5.4.21

3. We may be able to obtain a sufficiently good estimate of the bound B from the graph alone. We see that $f^{(4)}$ has several local extrema in (1,4), that $|f^{(4)}|$ attains its maximum on [1,4] at a point between $x = 3.5$ and $x = 4$, and that the maximum appears to be less than 3500. If we need a more precise answer, we can use the bisection algorithm to solve for a zero of $f^{(5)}$ in [3.5,4] with a small error, for example 10^{-6}. Using a computer we obtain an approximation for the desired zero of $f^{(5)}$ of 3.804649831. The value of $f^{(4)}$ at this point is

$$f^{(4)}(3.804649831) \approx 3382.$$

Thus we have found a bound B for $|f^{(4)}|$ on [1,4] of 3500, or, more precisely, 3382.

We can now compute the number of intervals, n, which are required to guarantee the desired accuracy of 0.0001. We need to solve

$$\frac{3382(4-1)^5}{180n^4} \leq 10^{-4} \quad \text{or} \quad n^4 \geq \frac{10^4 \times 3382 \times 3^5}{180}.$$

Computing we obtain $n \geq 82.2$. Since we need an even number of subintervals, we will use 84. Notice that in this case the rough estimate of 92 was not far from the more refined estimate of 84. This is because in the rough estimate the first term dominated and there was little cancellation between terms. In other cases, the rough estimate might be substantially larger than a more refined estimate.

We can now apply Simpson's Rule to the function $f(x) = \sin(x^2)$ on [1,4] obtaining 0.436870, or 0.4369 to four significant digits.

Note that the number of subintervals used, 84, is based not on the specific function, $f(x) = \sin(x^2)$, but on the bound for the fourth derivative. Eighty-four subintervals are guaranteed to work for any function with the same bound. Thus the number of subintervals is likely to be conservative; fewer subintervals may give the desired accuracy. For this function, if we use 42 subintervals, then the value returned is 0.436943, which is 0.4369 using four significant digits. $\triangle$

Question 5.4.22 What is the maximum number of subintervals needed to approximate $\int_0^1 \sin(x^3)dx$ with an accuracy of 0.01 using the t Trapezoid Rule and Simpson's rule? ◇

Section 5.4 Exercises

In Exercises 1 through 4, approximate the integrals using (a) the Trapezoidal Rule and (b) Simpson's Rule, with the indicated number n of subintervals.

1. $\int_{-1}^{0} \sqrt{1+x^2}dx$ for $n = 4$
2. $\int_{-1}^{1} \sin(\pi x)dx$ for $n = 4$
3. $\int_0^{\pi} \sqrt{1+\cos(x)}dx$ for $n = 2$
4. $\int_1^2 \sin(x)/x dx$ for $n = 6$
5. Estimate the error in each of the methods employed in Exercise 2.
6. Estimate the error in each of the methods employed in Exercise 4.
7. Consider the integral $\int_0^2 (x^3 + x - 3)dx$. Determine how many subintervals are required in order to approximate this integral using the Trapezoidal Rule if the approximation is to be accurate to within 0.001.
8. Consider the integral $\int_{-1}^{1} (x^2 - x - 3\sin(\pi x))dx$. Determine how many subintervals are required in order to approximate this integral using Simpson's Rule if the approximation is to be accurate to within 0.001.
9. Consider the integral $\int_1^3 1/x dx$. Determine how many subintervals are required in order to approximate this integral using Simpson's Rule if the approximation is to be accurate to within 0.01.
10. Consider the integral $\int_1^3 1/x dx$. Determine how many subintervals are required in order to approximate this integral using the Trapezoidal Rule if the approximation is to be accurate to within 0.01.
11. Based on the error estimate, what is the minimum number of subintervals in a partition of $[0, 1]$ needed to approximate $\int_0^1 \sin(x^2+1)dx$ with $ACC = 0.01$ using

    (a) Riemann Sum? **(b)** Trapezoidal Rule? **(c)** Simpson's Rule?

12. The estimated number of intervals in the previous exercises are conservative estimates; the actual number of subintervals required may be substantially fewer, as we have seen in several examples. The following can be done for each of the three numerical integration methods.

    (a) Do Exercise 11 to determine an upper bound on the number of subintervals required for the approximation to be accurate to within 0.01.

(b) Using a calculator or computer, determine an estimate, A, of $\int_0^1 \sin(x^2+1)dx$ that is accurate to within 0.005.

(c) Reduce the number of subintervals (found in step (a)) and calculate a new estimate, E, based on this reduced number of subintervals. If $|A - E| \leq 0.005$, then $|E - \int_0^1 \sin(x^2+1)dx| \leq 0.01$ (by the triangle inequality). Thus E is accurate to within 0.01. Estimate the minimum number of subintervals required to obtain the desired accuracy of 0.01.

13. Based on the error estimate, how many subintervals in a partition of $[0,5]$ are needed to approximate $\int_0^5 \sin(\cos(x))dx$ with $ACC = 0.01$ using

(a) Riemann Sum? (b) Trapezoidal Rule? (c) Simpson's Rule?

14. Geometrically it can be easily shown that $\int_0^1 \sqrt{1-x^2}dx = \pi/4$. Is it possible to determine how many subintervals are required to approximate this integral with accuracy of 0.001 using the Trapezoidal Rule? Why or why not?

15. Give an example for each of the following or show why no example can exist.

(a) An integral that can be evaluated exactly using the Trapezoidal Rule.

(b) An integral that can be evaluated exactly using Simpson's Rule.

(c) A function whose integral over $[1,6]$ can be approximately more accurately with the t Trapezoidal Rule than with Simpson's Rule when the number of subintervals used is the same in both approximations.

(d) An integrable function for which the error bound theorem for the Trapezoidal Rule does not apply.

(e) An integrable function for which the error bound theorem for Simpson's Rule does not apply.

16. A cow barn needs a new door. Door material costs 8 ft^2, and the shape of the door is a 5 ft by 5 ft square surmounted by the curve

$$y = x(5-x), \quad 0 \leq x \leq 5.$$

What is the cost of the material for the door?

17. Show that a polynomial of degree three can be integrated exactly by using Simpson's Rule. (See exercises 15(b) and 16.)

18. (Lagrange interpolation). This exercise introduces the method of obtaining the *Lagrange interpolating polynomial,* developed by J. L. Lagrange. Suppose we are given n points in the plane, $(x_1, y_1), (x_2, y_2), \ldots, (x_n, y_n)$, with no two x-coordinates the same. Then there is a *unique* polynomial of degree at most $n-1$ which passes through these n points.

(a) Find the unique polynomial of degree zero which passes through the point (x_1, y_1)

(b) Find the unique polynomial of degree (at most) one which passes through the two points (x_1, y_1) and (x_2, y_2), $x_1 < x_2$.

It is easy to write down the polynomial P in Lagrange's form. Consider the first y-value, y_1. We want $P(x_1) = y_1$. To do this we construct the term $y_1((x - x_2)(x - x_3) \cdots (x - x_n) \ / \ (x_1 - x_2)(x_1 - x_3) \cdots (x_1 - x_n)$. Notice that when $x = x_1$, the value of the term is y_1, and when $x = x_k$ for $k \neq 1$ the value of the term is zero. We construct a similar term for each of the remaining $(x_2, y_2), \ldots, (x_n, y_n)$ and add them together to form P.

$$\begin{aligned} P(x) &= y_1 \frac{(x - x_2)(x - x_3) \cdots (x - x_n)}{(x_1 - x_2)(x_1 - x_3) \cdots (x_1 - x_n)} \\ &\quad + y_2 \frac{(x - x_1)(x - x_3) \cdots (x - x_n)}{(x_2 - x_1)(x_2 - x_3) \cdots (x_2 - x_n)} \\ &\quad + \cdots + y_n \frac{(x - x_1)(x - x_2) \cdots (x - x_{n-1})}{(x_n - x_1)(x_n - x_2) \cdots (x_n - x_{n-1})} \end{aligned}$$

P is the sum of n terms each of degree $n - 1$, so P is of degree $n - 1$. For any point, x_k, the kth term in P is of the form

$$y_k \frac{(x - x_1)(x - x_2) \cdots (x - x_{k-1})(x - x_{k+1}) \cdots (x - x_n)}{(x_k - x_1)(x_k - x_2) \cdots (x_k - x_{k-1})(x_k - x_{k+1}) \cdots (x_k - x_n)}$$

so in $P(x_k)$ every term except the kth is zero, and the value of the kth is y_k. Thus P is a polynomial of degree $n - 1$ which passes through the n points.

(c) Find the Lagrange interpolating polynomial for the three points (1,2), (2,3), and (3,3).

(d) Write down the Lagrange interpolating polynomial for the four points $(x_1, y_1), (x_2, y_2), (x_3, y_3), (x_4, y_4)$.

19. In the following exercises give a complete report: the method used, the theoretically estimated number of subintervals required, the approximate integral value, and an experimental estimate of the minimum number of subintervals required for the given accuracy in estimating $\int_a^b f(x)dx$. (See Exercise 12 above.)

(a) $f(x) = 1/(1 + x^3)$, interval [1, 2], error 0.01.

(b) $f(x) = \sin(x^2 + 3x + 1)$, interval [1,2], error 0.01.

5.5 Properties of the Riemann Integral

Among the results of this section will be those pertaining to the integral of the sum or difference of integrable functions. However, noticeably absent will be any type of property dealing with the product or quotient of integrable functions, since no simple theorems on integrals of products exist. Thus we shall not talk about an "Algebra" of integrals as we did with the concepts of limit, continuity, and derivative.

Most of the proofs of the results in this section are very straightforward, requiring only an application of the definition of the Riemann Integral along with some properties of summation. The reader is urged to illustrate geometrically the theorems and proofs before studying the analytical proofs.

Theorem 5.5.1 *If c is a constant, then $\int_a^b c\,dx = c(b-a)$.*

Proof: Let $f : [a,b] \to \mathbb{R}$, be defined by $f(x) = c$.
Let $P = \{a = x_0, x_1, ..., x_n = b\}$ be any partition of $[a,b]$. Then

$$LA(f,P) = \sum_{k=1}^{n} c\Delta x_k = c\sum_{k=1}^{n} \Delta x_k = c(b-a)$$

Similarly, $UA(f,P) = c(b-a)$. By the definition of the Riemann Integral, $f(x) = c$ is integrable with integral $c(b-a)$. □

(What is a geometrical interpretation of the previous theorem?) Notice that

$$\int_a^b c\,dx = c(b-a) = -c(a-b) = -\int_b^a c\,dx$$

Thus, interchanging the order of the limits of integration changes the sign of the integral of a constant function. That a similar result holds for any integrable function can be seen by defining the Δx in a regular partition as follows

$$\Delta x = \frac{(\text{upper limit} - \text{lower limit})}{n}$$

rather than as $(b-a)/n$.

We state the result formally as a theorem.

Theorem 5.5.2 *If f is an integrable function, then $\int_a^b f(x)dx = -\int_b^a f(x)dx$.*

The next two theorems yield a linearity property for integrable functions similar to the linearity property that was established for limits, continuous functions, and differentiable functions. This linearity property will be used to establish the result that all polynomial functions are integrable in the same manner in which the linearity property for derivatives was used to establish that all polynomials are differentiable.

Theorem 5.5.3 *If f is an integrable function over $[a,b]$ and c is a constant, then $\int_a^b cf(x)dx = c\int_a^b f(x)dx$.*

The proof of this theorem is left to the reader with the comment that the second summation formula in Section 5.2 yields the following result for Riemann Sums: $RS(cf, P, \{w_k\}) = (c)RS(f, P, \{w_k\})$.

Theorem 5.5.4 *If f and g are integrable functions over $[a,b]$, then*

$$\int_a^b (f(x) + g(x))dx = \int_a^b f(x)dx + \int_a^b g(x)dx$$

The proof of this theorem is also left to the reader with the comment that the third summation formula in Section 5.2 yields

$$RS((f+g), P, \{w_k\}) = RS(f, P, \{w_k\}) + RS(g, P, \{w_k\})$$

Combining the above two results yields the following important theorem.

Theorem 5.5.5 (Linearity of Integration) *For any two constants c and d and any two integrable functions $f, g : [a,b] \to \mathbb{R}$,*

$$\int_a^b (cf(x) + dg(x))dx = c\int_a^b f(x)dx + d\int_a^b g(x)dx$$

Since the linearity property can be extended to an arbitrary finite sum, we have the following extension of Theorem 5.5.5.

Theorem 5.5.6 *If $c_1, c_2, ..., c_n$ are constants and $f_1, f_2, ..., f_n$ are all integrable functions with domain $[a,b]$, then*

$$\int_a^b \sum_{k=1}^{n} c_k f_k(x)dx = \sum_{k=1}^{n} c_k \int_a^b f_k(x)dx$$

Theorem 5.5.4 expressed the additive property of an integral with respect to the integrand. The next theorem shows that an integral is also additive with respect to its path of integration.

Theorem 5.5.7 *If f is an integrable function defined on an interval containing a, b, and c, then*

$$\int_a^b f(x)dx = \int_a^c f(x)dx + \int_c^b f(x)dx$$

Note that the point c is not restricted to lie between points a and b.

Theorem 5.5.7 essentially says that "the whole is equal to the sum of its parts." The reader is urged to sketch pictures illustrating Theorem 5.5.7 for all possible arrangements of the points a, b, and c before studying the following proof.

Proof: We shall first consider the case where $a < c < b$. To prove the theorem we need to show that for any partition P of $[a, b]$,

$$LA(f, P) \leq \int_a^c f(x)dx + \int_c^b f(x)dx \leq UA(f, P)$$

Let P' be the partition of $[a, b]$ consisting of P and $\{c\}$. Since P' is a refinement of P,

$$LA(f, P) \leq LA(f, P') \quad \text{and} \quad UA(f, P') \leq UA(f, P)$$

Now we obtain partitions of $[a, c]$ and $[c, b]$ by letting P_1 be all points of P' in $[a, c]$ and P_2 be all points of P' in $[c, b]$. This gives

$$LA(f, P') = LA(f, P_1) + LA(f, P_2) \quad \text{and} \quad UA(f, P') = UA(f, P_1) + UA(f, P_2)$$

Since

$$LA(f, P_1) \leq \int_a^c f(x)dx \leq UA(f, P_1)$$

and

$$LA(f, P_2) \leq \int_c^b f(x)dx \leq UA(f, P_2)$$

we can add these inequalities to obtain

$$LA(f, P_1) + LA(f, P_2) \leq \int_a^c f(x)dx + \int_c^b f(x)dx \leq UA(f, P_1) + UA(f, P_2)$$

Thus

$$LA(f, P') \leq \int_a^c f(x)dx + \int_c^b f(x)dx \leq UA(f, P')$$

and so

$$LA(f, P) \leq \int_a^c f(x)dx + \int_c^b f(x)dx \leq UA(f, P)$$

Now consider the case where $c < a < b$. Using Theorem 5.5.2, replace $\int_a^c f(x)dx$ by $-\int_c^a f(x)dx$ in the conclusion of Theorem 5.5.7 and obtain (after transposing)

$$\int_c^b f(x)dx = \int_c^a f(x)dx + \int_a^b f(x)dx$$

This is just the first case considered since $c < a < b$. A similar argument can be made for the case where $a < b < c$. □

Question 5.5.8 Integrate $\int_{-1}^{4} FLOOR(t)dt$. ◇

The last result in this section is concerned with the question of how to determine an average value for a continuous function over a given domain. For example, several pollution control standards are expressed in terms of averages expressed over a fixed period of time. If emissions are monitored on a continuous basis, how does one determine from a continuous printout whether pollution control standards are being satisfied?

Another example involves the diagnosis and treatment of heart failure. A catheter inserted in the pulmonary artery measures pressures in the left ventricle of the heart, which are displayed in the form of a continuous curve by means of an oscilloscope. The pertinent information is the average pressure over a fixed period. How does one obtain the average?

Another example involves earth tides. Just as the gravitational effect of the sun and moon cause ocean tides, there is a tidal effect on land masses, termed earth tides. A gravity meter at a stationary site is able to detect changes in the earth's gravitational field of one part in 10^8. Gravity surveys are performed to detect small changes in the local gravitational field due to heavier or lighter rocks below. In order to process these data, the effect of earth tides must first be removed. This is done by averaging gravity measurements over a period of time. Since the gravity function is a continuous function of time, the problem is that of averaging a continuous function over a time interval. How is this done?

Let us apply the Basic Approximation Process to the question of obtaining an average value for a continuous function. To be specific, we consider the continuous function f defined over the interval $[0, 1]$. Let

$$av_1 = f(1)$$

be a first approximation to the average value of f. Let

$$av_2 = \frac{f(1/2) + f(1)}{2}$$

be a second approximation (i.e., average two function values). Let

$$av_3 = \frac{f(1/3) + f(2/3) + f(1)}{3}$$

be a third approximation. Continuing in this manner, we have

$$av_n = \frac{f(1/n) + f(2/n) + f(3/n) + \cdots + f(1)}{n}$$

as the nth approximation. Expressing this sequence of approximations in summation notation gives

$$av_n = \frac{1}{n}\sum_{k=1}^{n} f\left(\frac{k}{n}\right) = \sum_{k=1}^{n} f\left(\frac{k}{n}\right)\frac{1}{n}$$

This is just a Riemann Sum! Now since f is continuous, f is integrable and hence the sequence of approximations converges to $\int_0^1 f(x)dx$! The average value of a continuous function is given by the integral of the function. If the interval is $[a, b]$ rather than $[0, 1]$, then the $1/n$ factor in the above sequence is replaced by $(b - a)/n$:

$$av_n = \sum_{k=1}^{n} f\left(a + k\frac{b-a}{n}\right)\frac{1}{n} = \frac{1}{b-a}\sum_{k=1}^{n} f\left(a + k\frac{b-a}{n}\right)\frac{b-a}{n}$$

Theorem 5.5.9 (Integral Mean Value Theorem) *If f is a continuous function over $[a, b]$, then there exists a number c , $a < c < b$, such that*

$$f(c) = \frac{1}{b-a}\int_a^b f(x)dx$$

Proof: Since f is continuous over a *closed bounded interval* $[a, b]$, the Extreme Value Theorem (Chapter 2) insures that f will attain its minimum value m and its maximum value M on $[a, b]$. Thus $m \le f(x) \le M$ for all x in $[a, b]$, and by the Comparison Theorem (Corollary 5.5.15 in the exercises to this section),

$$\int_a^b m dx \le \int_a^b f(x)dx \le \int_a^b M dx$$

Or,

$$m(b-a) \le \int_a^b f(x)dx \le M(b-a)$$

Now, divide all three terms by $(b - a)$ and apply the Intermediate Value Theorem (Chapter 2). That is, since f is continuous over $[a, b]$ and

$$m \le \frac{1}{b-a}\int_a^b f(x)dx \le M$$

there exists a number c in the open interval (a, b) such that

$$f(c) = \frac{1}{b-a}\int_a^b f(x)dx$$

□

Definition 5.5.10 (Average Value of a Function) *The* average (mean) value *of an integrable function f over an interval $[a, b]$ is*

$$\frac{1}{b-a}\int_a^b f(x)dx$$

Notice that the Mean Value Theorem for Integrals states that a continuous function "takes on" its average or mean value at some point in the interval $[a, b]$. The conclusion of this theorem can be written as

$$f(c)(b - a) = \int_a^b f(x)dx$$

The left-hand side can be interpreted as the area of a rectangle of height $f(c)$ and width $(b - a)$. Thus, one way to interpret the Mean Value Theorem for Integrals is to equate the value of an integral of f over an interval $[a, b]$ with the area of a rectangle with base $[a, b]$ and height equal to the average value of f over $[a, b]$. The following figure illustrates this interpretation.

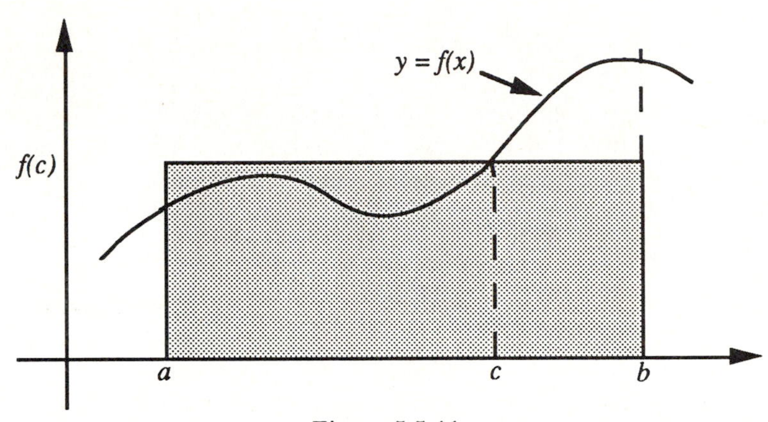

Figure 5.5.11

Question 5.5.12 If the average value of f over $[a, b]$ is greater than the average value of g over $[a, b]$, is it necessarily the case that $f(x) > g(x)$ for all x in $[a, b]$? Why? ◊

Section 5.5 Exercises

In Exercises 1 and 2 assume that f and g are continuous functions on the interval $[c, e]$, $c \leq d \leq e$, $\int_c^e f(t)dt = R$, $\int_c^d g(t)dt = S$, and that $\int_c^e g(t)dt = W$.

1. Find the following in terms of R, S, and W.

   (a) $\int_e^d g(x)dx$ (b) $\int_e^c (f - g)(t)dt$ (c) $\int_e^c (5 - f(t))dt$

2. Find the following in terms of R, S, and W.

   (a) $\int_d^e g(t)dt$ (b) The average value of f over $[c, e]$.

In Exercises 3 through 8, assume

$$\int_a^b x\,dx = \frac{b^2 - a^2}{2}, \int_a^b x^2 dx = \frac{b^3 - a^3}{3}, \text{ and } \int_a^b x^3 dx = \frac{b^4 - a^4}{4}.$$

3. Find the average value of $f(x) = x^2 + 4x$ over the interval $[-2, 4]$.

4. Find the average value of $f(x) = |x|$ over the interval $[-1, 3]$.

5. Find the average value of f over $[0, 4]$, where

$$f(x) = \begin{cases} -x^2 & \text{for } 0 \le x < 2 \\ 2x - 8 & \text{for } 2 \le x < 4 \end{cases}$$

6. Find the average value of f over $[-1, 4]$, where

$$f(x) = \begin{cases} x^3 & \text{for } -1 \le x < 0 \\ x & \text{for } 0 \le x < 2 \\ 2 & \text{for } 2 \le x \le 4 \end{cases}$$

7. Find the average value of $f(x) = (x+1)^2$ over the interval $[-1, 3]$.

8. Find the average value of $f(x) = x(x+1)^2$ over the interval $[0, 2]$.

9. Give an example of each of the following or explain why no example can exist.

   (a) A function f such that $\int_{-1}^{1} f(x)dx = 0$, but $f(x) \neq 0$ for all x in $[-1, 1]$.

   (b) A function f with $f(x) \neq f(-x)$ and $\int_{-1}^{0} f(x)dx = \int_0^1 f(x)dx$.

   (c) A function $f : [0, 10] \to \mathbb{R}$ such that $\int_0^{10} f(x)dx = \sum_{k=1}^{10} f(k)$.

   (d) A function f such that $|\int_{-2}^{1} f(x)dx| \neq \int_{-2}^{1} f(x)dx$.

   (e) Functions f and g such that $(\int_2^5 f(x)dx) \,/\, (\int_2^5 g(x)dx) \neq \int_2^5 f(x)/g(x)\mathbf{dx}$.

   (f) Functions f and g such that

$$\int_{-2}^{1} f(x)g(x)dx \neq \left(\int_{-2}^{1} f(x)dx\right)\left(\int_{-2}^{1} g(x)dx\right)$$

   (g) A function $f : [-1, 3] \to \mathbb{R}$ whose average value is not equal to any $f(c)$, for c in $[-1, 3]$.

   (h) An interval $[a, b]$ such that the average value of $f(x) = x^2$ over $[a, b]$ is 3.

In Exercises 10 through 18, prove the given statement or provide a counterexample.

10. If f is continuous over $[a,b]$ then $|\int_a^b f(x)dx| \leq \int_a^b |f(x)|dx$.

11. If $\int_a^b f(x)dx \geq 0$, then $f(x) \geq 0$ for all x in $[a,b]$.

12. If f is a continuous function, $f(x) \geq 0$, and $\int_a^b f(x)dx = 0$, then $f(x) = 0$ for all x in $[a,b]$.

13. If $f(x) \geq g(x)$ for all x in $[a,b]$, then $\int_a^b (f(x) - g(x))dx \geq 0$.

14. If $\int_a^b (f(x) - g(x))dx \geq 0$, then $f(x) \geq g(x)$ for all x in $[a,b]$.

15. If $\int_a^b f(x)dx \geq \int_a^b g(x)dx$, then $\int_a^b (f(x) - g(x))dx \geq 0$.

16. The average value of a constant function defined over an interval $[a,b]$ is the constant value of the function.

17. If f and g are integrable functions and $f(x) \geq g(x)$ for all x in $[a,b]$, then the average value of f is greater than or equal to the average value of g over the interval.

18. The average value of $f(x) = \sin(x)$ over the interval $[-\pi/2, \pi/2]$ is 0.

19. Prove Theorem 5.5.3.

20. Prove Theorem 5.5.4.

21. Prove Theorem 5.5.7 for the case where $a < b < c$.

22. Let $A = \{2,5,3,7,8\}$. Define $f : [0,5] \to \mathbb{R}$ by

$$f(x) = \begin{cases} 2 & \text{for } 0 \leq x < 1 \\ 5 & \text{for } 1 \leq x < 2 \\ 3 & \text{for } 2 \leq x < 3 \\ 7 & \text{for } 3 \leq x < 4 \\ 8 & \text{for } 4 \leq x \leq 5. \end{cases}$$

Show that the average value of the elements of A is the average value of f.

23. The following chart shows the amount of money that an orphanage received from 1879 through 1883. Find the average (approximate) dollar amount of gifts per year received by this orphanage from 1879 through 1882.

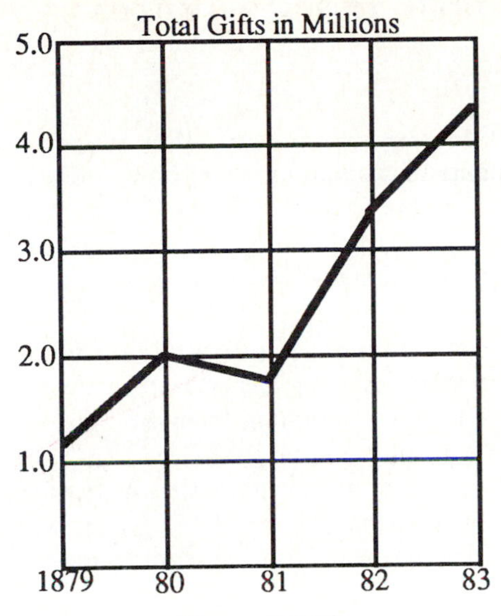

Figure 5.5.13

24. Project: Discuss whether or not an integral is a continuous function. That is if f is integrable over $[a, b]$, is $F(x) = \int_a^x f(t)dt$, $a \le x \le b$, a continuous function?
    Hint Sketch the graph of the integral of;
    (a) A differentiable function.
    (b) A continuous but nondifferentiable function.
    (c) A discontinuous function (i.e., a step function).

25. The following theorem and its corollary provide important comparison results.

    Theorem 5.5.14 *If f is an integrable function on $[a, b]$ and if $f(x) \ge 0$ for all x in $[a, b]$, then*

$$\int_a^b f(x)dx \ge 0$$

    Prove this theorem. Hint: Consider $LA(f, P)$ for any partition P.

26. **Corollary 5.5.15 (Comparison Theorem)** *If f and g are integrable functions over $[a, b]$ and $f(x) \le g(x)$ for all x in $[a, b]$, then*

$$\int_a^b f(x)dx \le \int_a^b g(x)dx$$

    Prove this result. Hint: Apply Exercise 25 to the function $g - f$.

5.6 Arc Length and Revolutes

How is the length of a curve determined? How do you determine how far an aircraft travels from takeoff to landing? Or, how is the formula ($C = 2\pi r$) for the circumference of a circle of radius r derived?

Recall that *a curve in n-dimensional space is described parametrically by the range of a vector-valued function* $f : A \to \mathbb{R}^n$. In developing an expression for arc length, we shall follow the Basic Approximation Process just as we did in the development of the area concept. In this section our emphasis will be on using approximations rather than on obtaining formulas.

Let us consider the curve C described as the range of the vector-valued function $f : [a, b] \to \mathbb{R}^2$ defined by $f(t) = (f_1(t), f_2(t))$, where f_1 and f_2 are C^1 functions (i.e., both f_1 and f_2 have continuous first derivatives).

The first step in the Basic Approximation Process is to approximate C with a line segment whose length we can determine. One possibility is to consider the secant line joining the endpoints of C, whose equation is

$$y = f(a) + x(f(b) - f(a)) \quad \text{for } 0 \leq x \leq 1$$

[Note that this is the equation of a straight line. When $x = 0$, we are at $f(a)$ and when $x = 1$, we are at $f(b)$; see Section 4.2].

The second step is to determine a way to obtain a better approximation. Geometrically it seems "obvious" that a polygonal path with vertices on C would give a better approximation than the above secant line. We define such a polygonal path by first forming a partition

$$P = \{a = t_0, t_1, ..., t_n = b\}$$

of $[a, b]$ and then by using the images of the partition points as the vertices of the polygonal path. Let PP_n denote such a polygonal path having n chords and $n + 1$ vertices.

Polygonal Path Approximating Arc Length

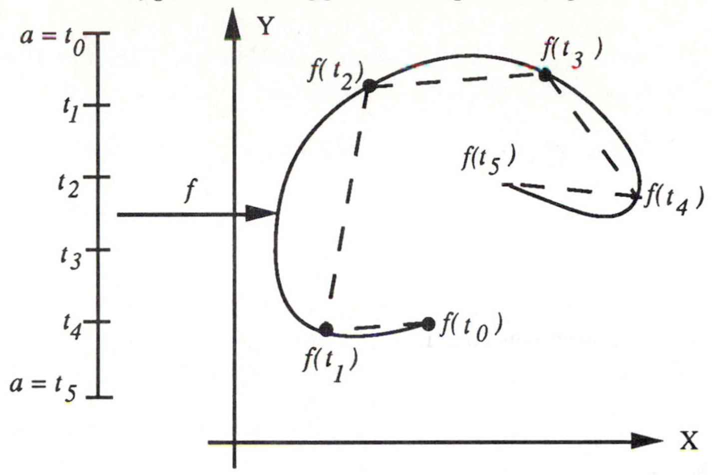

Figure 5.6.1

Since a chord in PP_n is determined by two vectors, say $(f_1(t_k), f_2(t_k))$ and $(f_1(t_{k-1}), f_2(t_{k-1}))$, the length of that chord is $|f(t_k) - f(t_{k-1})|$. Hence, the length of PP_n is

$$\begin{aligned} |PP_n| &= \sum_{k=1}^{n} |f(t_k) - f(t_{k-1})| \\ &= \sum_{k=1}^{n} \sqrt{[f_1(t_k) - f_1(t_{k-1})]^2 + [f_2(t_k) - f_2(t_{k-1})]^2} \end{aligned}$$

We now use the fact that f_1 and f_2 are differentiable functions (recall that they are, in fact, C^1 functions) and therefore we may apply the Mean Value Theorem to each of the two terms of the square root in the above summation. Hence, there exists numbers c_k and d_k in (t_{k-1}, t_k) such that

$$f_1(t_k) - f_1(t_{k-1}) = (t_k - t_{k-1})f_1'(c_k) = \Delta t_k f_1'(c_k)$$

and

$$f_2(t_k) - f_2(t_{k-1}) = (t_k - t_{k-1})f_2'(d_k) = \Delta t_k f_2'(d_k)$$

where $\Delta t_k = t_k - t_{k-1}$. The length of PP_n may be written using summation notation as

$$|PP_n| = \sum_{k=1}^{n} \sqrt{[f_1'(c_k)]^2 + [f_2'(d_k)]^2}\ \Delta t_k$$

Step three in the Basic Approximation Process (generate a convergent sequence of approximations such that each is a better approximation than the preceding ones) is satisfied by a sequence of polygonal paths $\{PP_n\}$.

Step four (define the desired property to be the common limit of all possible sequences in step three) shows how arc length should be defined. Since both f_1 and f_2 are C^1 functions, their derivatives are continuous. Therefore the limit of the above sum exists and is given by the integral

$$\int_a^b \sqrt{[f_1'(t)]^2 + [f_2'(t)]^2}\, dt = \int_a^b \sqrt{f'(t)\cdot f'(t)}\, dt = \int_a^b |f'(t)|dt$$

(Note the use of the dot product.) We are now in a position to define the length of a curve.

Definition 5.6.2 (Arc Length) *Let C be a curve which is the range of $f : [a,b] \to \mathbb{R}^n$ and let f have components*

$$f(t) = (f_1(t), f_2(t), \ldots, f_n(t))$$

where each component function, f_i , is a C^1 function. The arc length of C is given by

$$\int_a^b \sqrt{(f_1'(t))^2 + \cdots + (f_n'(t))^2}\, dt = \int_a^b |f'(t)|dt$$

This expression can be obtained intuitively in the following manner. For simplicity, suppose $f : [a,b] \to \mathbb{R}^2$ is given by $f(t) = (f_1(t), f_2(t))$.

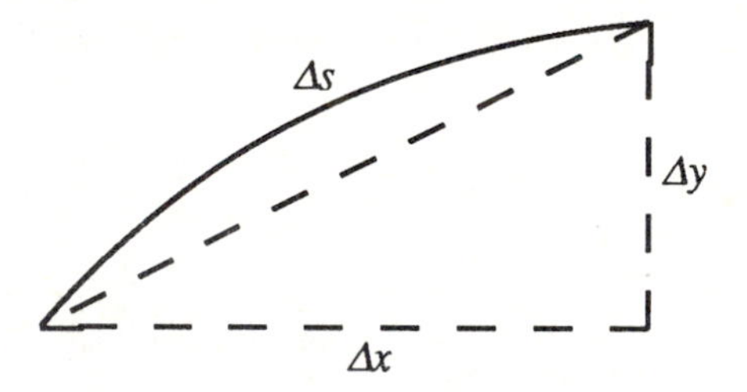

Figure 5.6.3

Consider a small segment of the curve of length Δs. By the Pythagorean theorem,

$$(\Delta s)^2 \approx (\Delta x)^2 + (\Delta y)^2$$

or, writing Δs in terms of the variable t,

$$\begin{aligned}(\Delta s) &\approx \sqrt{(\Delta x)^2 + (\Delta y)^2} = \sqrt{\left(\frac{\Delta x}{\Delta t}\right)^2 + \left(\frac{\Delta y}{\Delta t}\right)^2}\,\Delta t \\ &\approx \sqrt{f_1'(t)^2 + f_2'(t)^2}\,\Delta t\end{aligned}$$

Summing these up and passing to the integral as the limit, we have

$$s = \int_a^b \sqrt{f_1'(t)^2 + f_2'(t)^2}\, dt = \int_a^b |f'(t)|dt$$

the formula in the preceding definition.

Example 5.6.4 We will use the definition of arc length to find the length of the circumference of a circle of radius r. It is often useful to test a new procedure on problems with a known answer.

Let the vector valued function $f : [0, 2\pi] \to \mathbb{R}^2$ be defined by $f(t) = (r\cos(t), r\sin(t))$. The range of f is the desired circle. The length of the circumference is

$$\int_0^{2\pi} \sqrt{[-r\sin(t)]^2 + [r\cos(t)]^2}\, dt = \int_0^{2\pi} r dt = r\int_0^{2\pi} dt = 2\pi r \qquad \triangle$$

Since this example gave the correct answer, we will try the method on something new.

Example 5.6.5 We will find the length of the spiral described by the function $f : [0, 10] \to \mathbb{R}^3$, where $f(t) = (\cos(t), \sin(t), t)$.

Since the three coordinate functions are all C^1 functions, the length of the spiral is given by

$$\int_0^{10} \sqrt{[-\sin(t)]^2 + [\cos(t)]^2 + [1]^2}\, dt = \int_0^{10} \sqrt{2} dt = 10\sqrt{2} \qquad \triangle$$

Surface Area of Revolution

The determination of surface area in three-dimensional space is an interesting, challenging, and important problem in applied mathematics, especially when the surface is irregular. Some interesting questions one could consider are: What percentage of the surface area of your car is glass? Do all 6 oz cups have the same surface area? If not, what is the shape of a 6 oz cup having minimum surface area? How much paint is required to paint your favorite flag pole?

We shall consider a special type of surface, namely, a surface formed by revolving a curve about a line (axis) that does not intersect the curve. Such a surface is called a *surface of revolution.* The surfaces of baseball bats, water tanks, domes, footballs, pipes, drinking glasses, catsup bottles, and rolling pins are examples of surfaces of revolution. Additional examples are called for in the exercises. Since the development of the integral expression for the surface area of a surface of revolution parallels very closely the development of the integral expression for the length of a curve, we shall leave most of the details to the reader. The only significant difference in the two developments is that instead of using the length of a chord to approximate the length of an arc, we shall use the lateral surface area of a frustum of a right circular cone to approximate the area of a surface generated by revolving an arc about a line. The frustum of a right circular cone is illustrated below. (Think of a cheerleader's megaphone.)

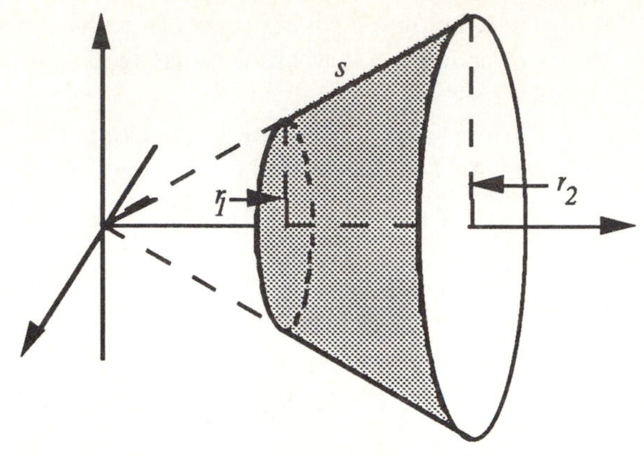

Figure 5.6.6

The lateral surface area of the frustum of a cone is given by the formula

$$A = 2\pi \frac{r_1 + r_2}{2} s$$

where r_1 and r_2 are the radii of the two ends of the cone and s is the slant height of the cone.

Consider a curve C described by the function $f : [a, b] \to \mathbb{R}^2$, and a polygonal path approximation PP_n of C as described in the first part of this section. Let S be the surface of revolution generated by revolving C about a line L, the axis of revolution. Each chord of the polygonal path PP_n generates a frustum of a cone having L as its axis. The sum of the lateral surface areas of the frustums of these cones will approximate the area of the surface of revolution, just as the length of the polygonal path approximated the length of the curve C.

The points $f(t_0), f(t_1), \ldots, f(t_n)$ partition the curve into segments $C_1, \ldots, C_n$.

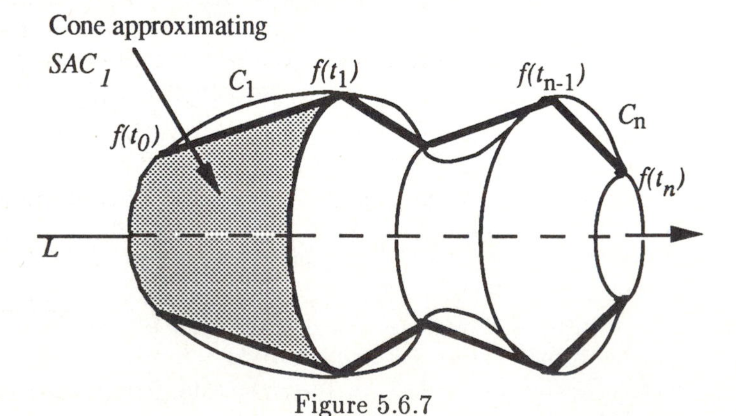

Figure 5.6.7

Let SAC_1 denote the area of the surface generated by the arc C_1 revolved around the x-axis. Thus, SAC_1 can now be approximated by

$$2\pi \text{ (radius) } \Delta s = 2\pi |f_2(t_1^*)| \sqrt{[f_1'(c_1)]^2 + [f_2'(d_1)]^2} \, \Delta t_1$$

where the points t_1^*, c_1, d_1 are in (t_0, t_1). Hence, the area of the entire surface, SAC, generated by revolving the curve C around the x-axis can be approximated by a Riemann Sum. That is, when the norm of the partition is sufficiently small, then $c_i \approx d_i \approx t_i$ and

$$SAC \approx \sum_{k=1}^{n} SAC_k \approx 2\pi \sum_{k=1}^{n} |f_2(t_k^*)| \sqrt{[f'_1(t_k)]^2 + [f'_2(t_k)]^2} \, \Delta t_k$$

The details of the convergence of the sequence of (approximating) Riemann Sums are beyond the scope of this text. We shall merely note that requiring the component functions, f_1 and f_2, to be C^1 functions is sufficient to guarantee convergence and the limit is expressed by

$$SAC = \int_a^b 2\pi |f_2(t)| \sqrt{[f_1'(t)]^2 + [f_2'(t)]^2} \, dt$$

As before, this formula can be obtained intuitively by considering a small segment of curve Δs rotated about the x-axis.

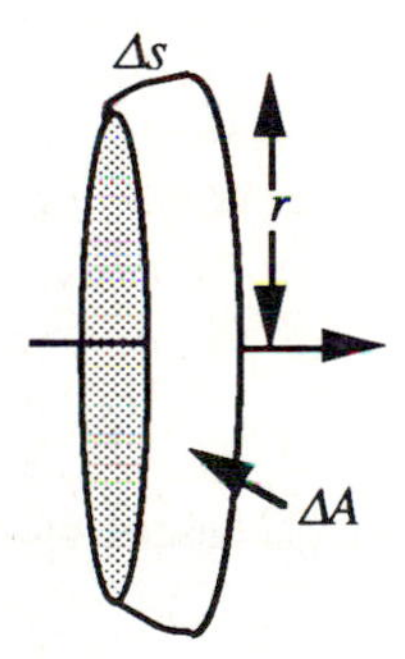

Figure 5.6.8

The piece of surface area is approximated by

$$\Delta A \approx 2\pi r \Delta s \approx 2\pi |f_2(t)| \sqrt{(f_1'(t))^2 + (f_2'(t))^2} \, \Delta t$$

When we sum and pass to the limit we obtain

$$A = \int_a^b 2\pi |f_2(t)| \sqrt{f_1'(t)^2 + f_2'(t)^2} \, dt$$

the same equation as above.

Again, let us check this method on a case where the answer is known, the surface of a cylinder. (The surface area of a cylinder of length L and radius R is $2\pi RL$.)

Example 5.6.9 We will find the (outside) surface area of a cylinder that is 10 ft long and has a radius of 4 ft. (The answer should be 80π.)

Consider the surface of the cylinder as having been generated by revolving the line segment $y = 4$, x in $[0, 10]$, around the x-axis.

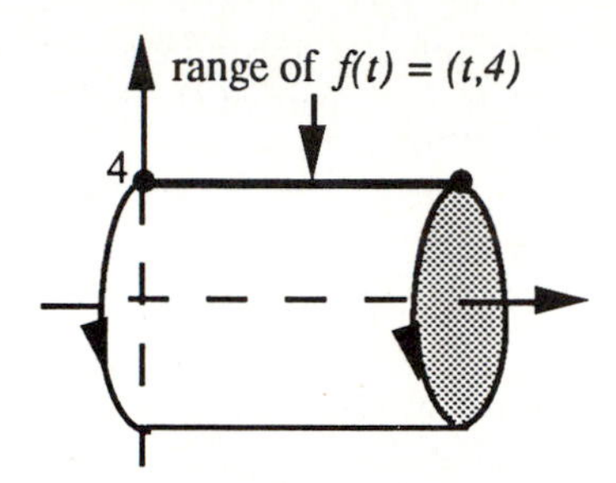

Figure 5.6.10

We will express the line segment as a curve and substitute into the integral expression for surface area.

Define the curve C by $f : [0, 10] \to \mathbb{R}^2$, where $f(t) = (t, 4)$. The surface area of the cylinder is

$$SAC = 2\pi \int_0^{10} (4)\sqrt{[1]^2 + [0]^2}\, dt = 8\pi \int_0^{10} dt = 80\pi \qquad \triangle$$

Let us consider another example from our past experience.

Example 5.6.11 We will find the surface area of a sphere of radius r. (The answer should be $4\pi r^2$.) Consider the surface of the sphere to be generated by revolving the semicircle C described by $f : [0, \pi] \to \mathbb{R}^2$, where $f(t) = (r\cos(t), r\sin(t))$, about the x-axis.

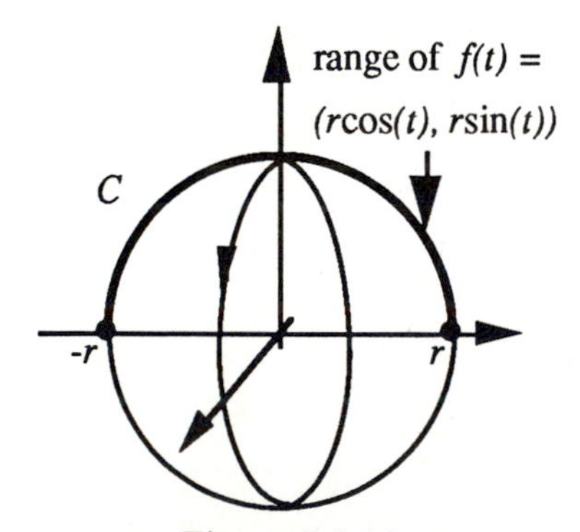

Figure 5.6.12

So,

$$SAC = 2\pi \int_0^{\pi} |r\sin(t)|\sqrt{[-r\sin(t)]^2 + [r\cos(t)]^2}\, dt = 2\pi r^2 \int_0^{\pi} \sin(t)dt$$

[The absolute value signs were removed since $\sin(t) \geq 0$ when $0 \leq t \leq \pi$.] In Chapter 6, we will see that the value of this integral is equal to 2 and thus the surface area of a sphere of radius r is $4\pi r^2$ △

The restriction in the definition of surface area that the generating curve not cross the x-axis is not a serious one. A curve can be partitioned into subcurves that do not cross the x-axis and the entire surface area represented as the sum of the areas of the surfaces generated by the subcurves. Note that this requires using the additivity property of the Riemann Integral.

A frequent special case occurs when the curve is the graph of a function $f : \mathbb{R} \to \mathbb{R}$, i.e., $y = f(x), a \leq x \leq b$. In this case, the curve is described parametrically by $g(x) = (x, y) = (x, f(x))$ for $a \leq x \leq b$.

Question 5.6.13 Find an integral expression for the length of the graph of the function f on the interval $[a, b]$ ◊

In the following examples, we will use the intuitive approach to obtaining an expression for the desired quantity.

Example 5.6.14 In this example we will rotate about an axis other than the x-axis or y-axis to find the surface area obtained by rotating the graph of $y = x^2$ for $0 \leq x \leq 1$ about the line $x = 2$.

Rotation about $x = 2$

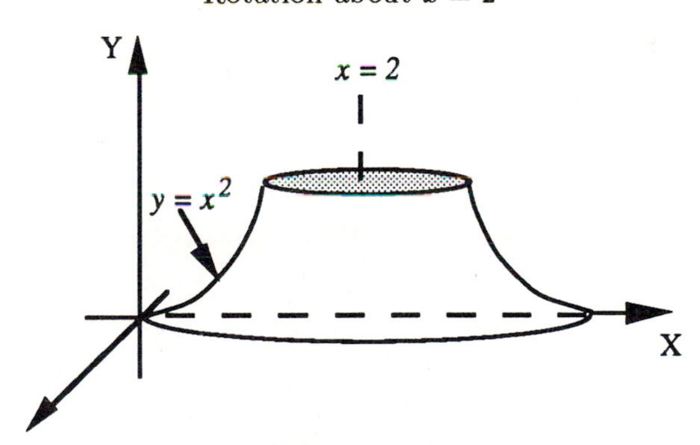

Figure 5.6.15

An approximation to the length of a piece of arc along the curve is

$$\Delta s = \sqrt{(\Delta x)^2 + (\Delta y)^2} = \sqrt{1 + \left(\frac{\Delta y}{\Delta x}\right)^2}\, \Delta x \approx \sqrt{1 + (2x)^2}\, \Delta x$$

so the approximate area is

$$\Delta A = 2\pi r \Delta s = 2\pi(2 - x)\Delta s \approx 2\pi(2 - x)\sqrt{1 + (2x)^2}\, \Delta x$$

Adding these up gives

$$A = \int_0^1 2\pi(2-x)\sqrt{1+4x^2}\,dx$$

The integral can now be approximated numerically. We will use Simpson's Rule to obtain an error of at most 10^{-5}. A bound for the fourth derivative of the integrand on $[0,1]$ is 603. (This bound can be found by using a computer to compute and then plot the fourth derivative.) Thus the number of subintervals needed is n, where

$$n^4 \geq \frac{100000 \cdot 603}{2880}$$

or $n > 12$. With $n = 14$ subintervals, Simpson's Rule gives 13.25454. △

Volumes of Revolution

How can you determine the volume of air in an auditorium with a sloping floor? How can you compute the volume of a catsup bottle? How can you find the volume of sand in a sand pile? Or, how can you determine the volume of water in a pond? How is the volume of a three-dimensional region defined?

In this section, we shall develop approximations for the volume of revolution in a manner analogous to what was done for surfaces of revolution. Of course, the central procedure is the Basic Approximation Process.

In the last section, we considered a special type of surface, namely a surface of revolution formed by revolving a curve around an axis. Let us now consider the three-dimensional solid that is obtained when the region between a curve and an axis is revolved around the axis. Such a solid is called a *solid of revolution.* For example, a solid cylinder can be described as a solid of revolution, as can a baseball bat or an Indian teepee. How is the volume of this special type of solid determined? Slicing the solid perpendicularly to the axis of revolution yields pieces whose volumes are easy to approximate by circular disks. The volume V of the solid of revolution can then be approximated by summing up the volumes of the (approximating) disks. Since the volume of a circular disk of radius r and thickness h is $\pi r^2 h$,

$$V \approx \sum_{k=1}^{n} \pi r_k^2 h_k$$

where r_k and h_k are the radius and thickness, respectively, of the kth circular disk.

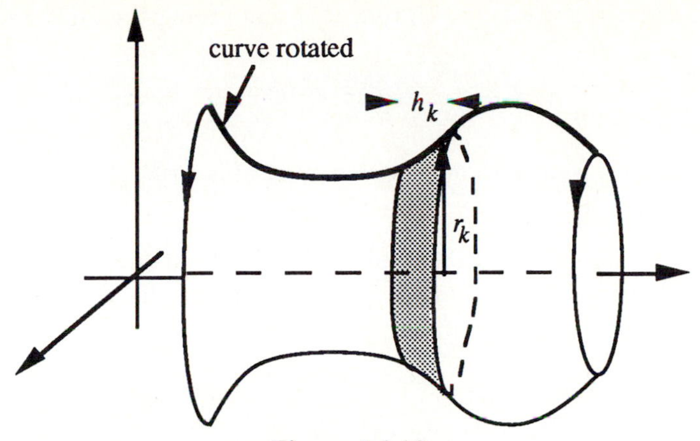

Figure 5.6.16

It seems clear, at least geometrically, that the thinner the slices, the more accurate is the approximation. This strongly suggests the possibility of forming a sequence of approximating Riemann Sums.

As usual, as a check on the method, we will first apply it to a case where the answer is known.

Example 5.6.17 We will find the volume of a sphere of radius R, the answer should be $\frac{4}{3}\pi R^3$. We will rotate the top half of a circle of radius R centered at the origin about the x-axis. The equation of the curve is

$$y = \sqrt{R^2 - x^2} \quad \text{for } -R \leq x \leq R$$

The Sphere as Volume of Revolution

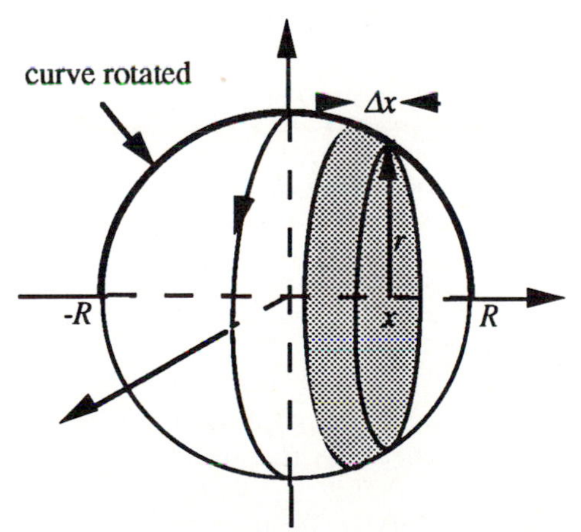

Figure 5.6.18

A disk of thickness Δx has volume

$$\Delta V = \pi r^2 \Delta x \approx \pi(\sqrt{R^2 - x^2})^2 \Delta x \approx \pi(R^2 - x^2)\Delta x$$

Summing up the disks from $x = -R$ to $x = R$ and taking the limit we have

$$V = \int_{-R}^{R} \pi(R^2 - x^2)dx$$

In the next chapter, we will show that this integral is $\frac{4}{3}\pi R^3$. $\triangle$

Example 5.6.19 We will find the volume of the "herald trumpet" obtained by revolving the region over the interval $[1, 5]$ and under the graph of $y = 1/x$ around the x-axis.

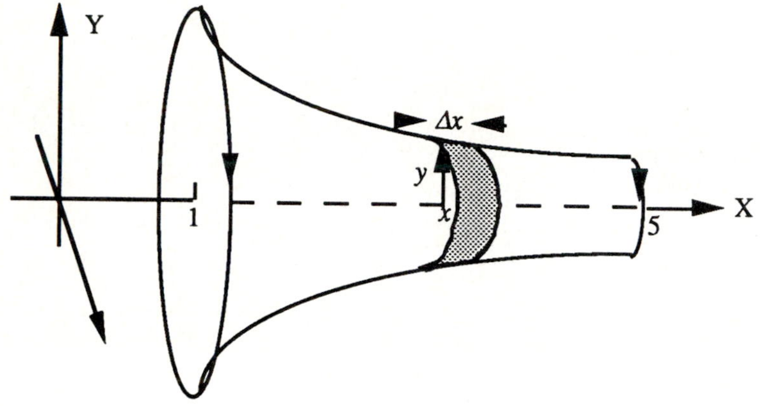

Figure 5.6.20

A disk of thickness Δx has volume

$$\Delta V \approx \pi y^2 \Delta x \approx \pi \frac{1}{x^2} \Delta x$$

Summing the disks and taking the limit gives

$$V = \int_{1}^{5} \pi \frac{1}{x^2} dx$$

In the next chapter, we will see that this integral has the value $\frac{4}{5}\pi$, or we can use the approximation methods of this chapter to obtain the integral. $\triangle$

Example 5.6.21 We will find an integral expression for the volume of revolution obtained by revolving the region between the curve

$$f(t) = (t^2, \sin(t)) \quad \text{for } 0 \leq t \leq 1$$

and the x-axis about the x-axis. The approximate element of volume is

$$\Delta V \approx \pi r^2 \Delta x$$

Expressing this in terms of the variable t gives

$$\Delta V \approx \pi r^2 \frac{\Delta x}{\Delta t} \Delta t \approx \pi y^2 x' dt = \pi \sin^2(t) 2t dt$$

Summing and passing to the limit gives

$$V = \int_0^1 \pi \sin^2(t) 2t dt \qquad \triangle$$

An alternative method to using disks to find the volume of revolution is to use cylinders. We will give a brief intuitive development of the cylinder (shell) method. The details, which are very similar to those in the disk method, will be left to the student.

Example 5.6.22 (Cylinders of Revolution) As usual, our first example with a new method will be a problem with a known answer. We will find the volume of the cone formed by rotating the triangular region bounded by the graphs of

$$y = f(x) = 2 - \frac{2}{3}x \quad \text{for } y = 2, x = 3$$

about the line $x = 3$.

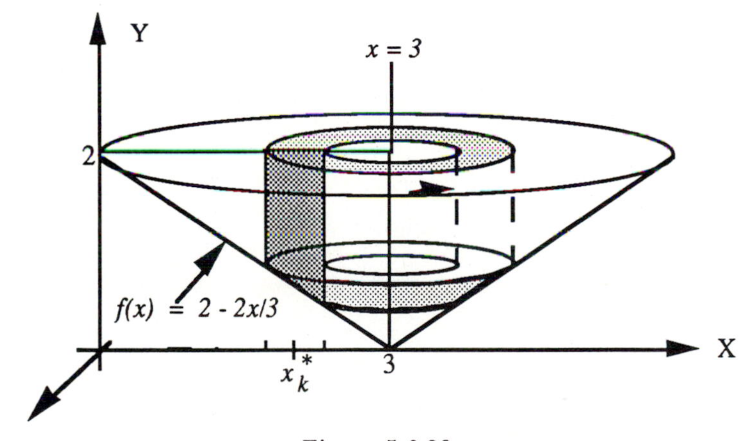

Figure 5.6.23

Since the solid is a cone, the volume is $\frac{1}{3}\pi r^2 h$ where r is the radius of the base, 3, and h is the height, 2. Thus the volume, which can also be obtained by using disks, is 6π.

A partition of [0,3] yields a partition of the generating region into vertical strips (draw a vertical line through each partition point). Each vertical strip (darkly shaded above) generates a cylinder and the resulting set of cylinders partitions the solid. The volume of the cylinder obtained from the strip between x_{k-1} and x_k containing x_k^* is approximately the area of the vertical strip [height times width $= (2 - f(x_k^*))(x_k - x_{k-1})$] times the circumference ($2\pi r$, where r is the distance of x_k^* from the axis of rotation, so $r = 3 - x_k^*$). Letting ΔV_k denote the volume of the kth cylinder, we have

$$\Delta V_k \approx 2\pi(3 - x_k^*)(2 - f(x_k^*))(x_k - x_{k-1})$$

The volume V of the solid is the sum of the volumes of the cylinders. Thus

$$V \approx \sum_{k=1}^{n} \Delta V_k \approx 2\pi \sum_{k=1}^{n} (3 - x_k^*)(2 - f(x_k^*))\Delta x_k$$

is a Riemann Sum! Therefore if f is continuous (as it surely is), the limit of the sum exists as the norm of the partition goes to zero and the volume of the solid is

$$V = 2\pi \int_0^3 (3 - x)(2 - f(x))dx = \frac{4}{3}\pi \int_0^3 (3 - x)x dx = 6\pi$$

(We will learn how to compute the integral in the next chapter.) △

Example 5.6.24 We will find the volume of the solid obtained by revolving the (generating) region bounded by

$$y = x\sin(x^2) + 4 \quad \text{for } 0 \le x \le 3$$

and the x-axis (the shaded region in Figure 5.6.25) about the line $x = 5$.

$$y = x\sin(x^2) + 4$$

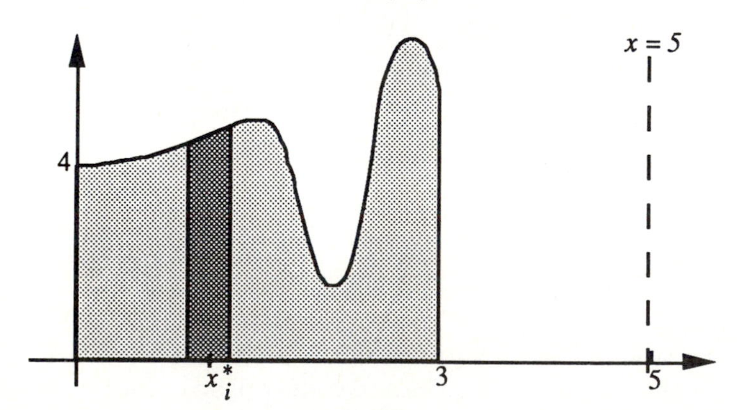

Figure 5.6.25

Again, a vertical strip (darkly shaded) in the generating region generates a cylinder when rotated about the axis $x = 5$.

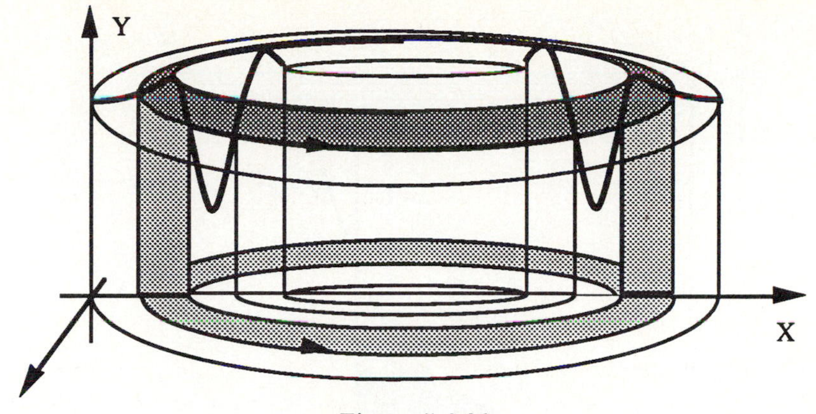

Figure 5.6.26

The volume V of the solid is the sum of the volumes of the cylinders. Thus

$$V \approx \sum_{k=1}^{n} \Delta V_k = \sum_{k=1}^{n} 2\pi r h \Delta x_k = 2\pi \sum_{k=1}^{n} (5 - x_k^*) f(x_k^*) \Delta x_k$$

Taking the limit we obtain the volume of the solid:

$$V = 2\pi \int_0^3 (5-x) f(x) dx = 2\pi \int_0^3 (5-x)(x \sin(x^2) + 4) dx = 283.1405$$

using Simpson's Rule with $n = 20$ (error $< .000059$). △

The reader should be able (based on the past developments of integrals) to write a rigorous development of a "cylinder" definition of volume. This development should result in the formula

$$V = 2\pi \int_a^b r(x) h(x) dx$$

where $r(x)$ is the distance of x from the axis of rotation and $h(x)$ is the height of the strip.

Example 5.6.27 We will find the volume when the region bounded by the curve $y = x^2$, the x-axis, and the line $x = \frac{1}{2}$ is rotated about the axis $y = \frac{3}{4}$.

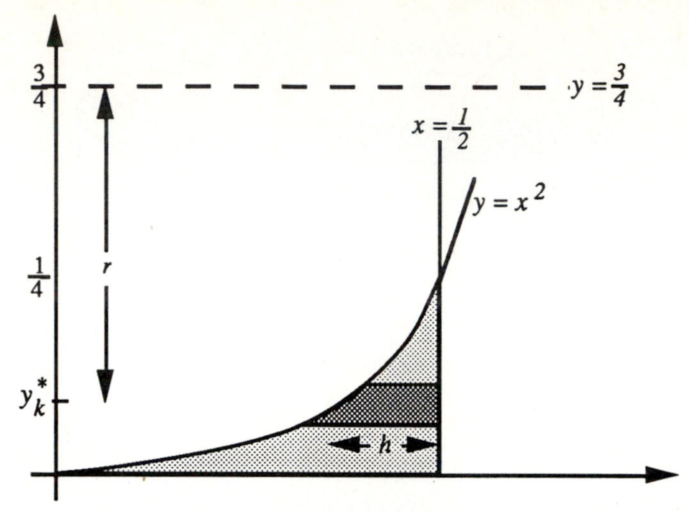

Figure 5.6.28

We will generate a cylinder by rotating a horizontal strip about the axis of rotation. Since we will integrate with respect to y, all expressions are put in terms of y:

$$\Delta V_k \approx 2\pi r h \Delta y_k \approx 2\pi\left(\frac{3}{4} - y_k^*\right)\left(\frac{1}{2} - x_k^*\right)\Delta y_k = 2\pi\left(\frac{3}{4} - y_k^*\right)\left(\frac{1}{2} - \sqrt{y_k^*}\right)\Delta y_k$$

Taking the limit we have

$$V = 2\pi\int_0^{\frac{1}{4}}\left(\frac{3}{4} - y\right)\left(\frac{1}{2} - \sqrt{y}\right)dy$$

(which we will be able to integrate after studying the next chapter). $\triangle$

Question 5.6.29 Example 5.6.27 can also be done using disks (with holes in the center). Develop an integral expression for the volume using disks with holes. $\diamond$

Section 5.6 Exercises

In Exercises 1 through 6, define a function $f : A \to \mathbb{R}^2$ where A is a subset of $\mathbb{R}$ and the range of f is the subset of $\mathbb{R}^2$ satisfying the given relation.

1. $y = x$ 2. $y = x^2 + \cos(x)$ 3. $y^2 + x^2 = 1$

4. $y = 1/x$ 5. $x + 4 = \sin(y), 2 \le y \le 5$ 6. $y^2 = x + 7$

In Exercises 7 through 12, sketch the graph of C and express the arc length of C in terms of an integral.

7. C is the curve described by $f : [-1,3] \to \mathbb{R}^2$, where $f(t) = (t^2+3, 4t)$.

8. C is the curve described by $f : [2,3] \to \mathbb{R}^2$, where $f(t) = (t\cos(t), 2)$.

9. C is the curve described by $f : [0,2] \to \mathbb{R}^3$, where $f(t) = (t, t, 0)$.

10. C is the curve described by $f : [-1,1] \to \mathbb{R}^3$, where $f(t) = (0, 1, t)$.

11. C is the curve described by $f : [-1,3] \to \mathbb{R}^2$, where $f(t) = (t, \sin(t))$.

12. C is the curve described by $f : [1,3] \to \mathbb{R}^3$, where $f(t) = (\cos(t), \sin(t), t)$.

13. Write an integral expression for the length of the graph of $f : \mathbb{R} \to \mathbb{R}$ on the interval $[a, b]$.

14. Write an integral expression for the length of the curve parameterized by $f(t) = (f_1(t), \ldots, f_n(t))$ for $a \leq t \leq b$.

In Exercises 15 through 20, find a numerical approximation to the length of the curve C by computing the length of the polygonal path defined by the given partition P.

15. C is the curve described by $f : [0, 2\pi] \to \mathbb{R}^2$, where $f(t) = (t, \cos(t))$ and $P = \{0, \pi/2, \pi, 3\pi/2, 2\pi\}$.

16. C is the curve described by $f : [-1, 2] \to \mathbb{R}^2$, where $f(t) = (t, |t|)$ and $P = \{-1, 1, 2\}$.

17. C is the curve described by $f : [0, 2\pi] \to \mathbb{R}^2$, where $f(t) = (\cos(t), \sin(t))$ and $P = \{0, \pi/4, \pi/2, 3\pi/2, 2\pi\}$. Sketch the curve and the approximating polygonal path.

18. C is the curve described by $f : [0, 2] \to \mathbb{R}^3$, where $f(t) = (t, t^2, t)$ and $P = \{0, \frac{1}{2}, 1, \frac{3}{2}, 2\}$.

19. C is the spiral described by $f : [0, 2\pi] \to \mathbb{R}^3$, where $f(t) = (\cos(t), \sin(t), t)$ and $P = \{0, \pi/2, \pi, 2\pi\}$.

20. C is the circle described by $f : [0, 2\pi] \to \mathbb{R}^3$, where $f(t) = (\cos(t), \sin(t), 4)$ and $P = \{0, \pi, 3\pi/2, 5\pi/4, 2\pi\}$.

21. Compute the (finite) sequence of approximations, $|PP_1|, |PP_2|, |PP_3|$ to the length of the curve described by $f : [1, 5] \to \mathbb{R}^2$, where $f(t) = (t, 1/t)$. Let each of the three partitions, P_1, P_2, and P_3 be a regular partition.

22. Let S be the surface generated by revolving the curve C described by $f : [1, 5] \to \mathbb{R}^2$, where $f(t) = (t, 1/t)$ around the x-axis. Find two numerical approximations to the surface area of S by computing the surface areas of the surfaces generated by revolving the polygonal paths PP_1 and PP_2 around the x-axis. Let $P_1 = \{1, 5\}$ and $P_2 = \{1, 3, 5\}$.

23. Prove that the lateral surface area of a right circular cone with base radius r and height h is $\pi r\sqrt{h^2+r^2}$

24. Let C be the curve described by $f : [1,3] \to \mathbb{R}^2$, where $f(t) = (t, t^{1/2})$. Write an integral for the area of the surface generated by revolving C around the following lines. Then approximate the integral with an error of at most 0.01.

    (a) line $y = 4$ (b) line $y = -3$ (c) line $x = -1$ (d) line $x = 4$

25. Let C_1 be the curve described by $f : [-1,1] \to \mathbb{R}^2$, where $f(t) = (t, t^2)$ and let C_2 be the curve described by $g : [-1,1] \to \mathbb{R}^2$, where $g(t) = (t, t^2 - 1)$. Let S_1 and S_2 be the respective surfaces generated by revolving the curves C_1 and C_2 around the x-axis. Which of the two surfaces, S_1 or S_2, has the larger surface area? Why?

In Exercises 26 through 32, find an integral expression for the volume of revolution that is obtained by revolving the region between the x-axis and the specified curve about the x-axis.

26. $y = \sqrt{x}$, $0 \le x \le 1$

27. $y = \tan(x)$, $0 \le x \le \pi/4$

28. $y = x^3$, $1 \le x \le 3$

29. The range of $f : [0,2] \to R^2$, where $f(t) = (t, t^2)$.

30. The range of $f : [-\pi, 0] \to \mathbb{R}^2$, where $f(t) = (t, \cos(t))$.

31. The range of $f : [-1,2] \to \mathbb{R}^2$, where $f(t) = (t,t)$.

32. The range of $f : [0, 2\pi] \to \mathbb{R}^2$, where $f(t) = (t, \sin(t))$.

33. Find an integral expression for the volume of revolution obtained by revolving the region between the x-axis and the graph of $x = y^2$ for $0 \le x \le 2$ about the y-axis.

34. Consider the "herald trumpet" obtained by revolving the region between the x-axis and the range of $f : [1,2] \to \mathbb{R}^2$, where $f(t) = (t, 1/t)$ about the x-axis (see Example 5.6.19). Which is larger, the volume of the trumpet or the surface area of the trumpet? Why?

35. Write an integral for the volume between the x-axis and the curve

    $y = 1 + \sin(x), 0 \le x \le 5$ when rotated about the line $x = 8$.

36. Write an integral for the volume between the x-axis and the curve

    $y = \cos(x), -\pi/2 \le x \le \pi/2$ when rotated about the line $x = 4$.

37. Use Simpson's Rule to approximate the volume of Exercise 33 to within 0.001.

38. Use Simpson's Rule to approximate the volume of Exercise 34 to within 0.001.

39. Give a rigorous development of the "cylinder" definition of volume of revolution detailing each step in the Basic Approximation Process.

Chapter 6

EVALUATION OF INTEGRALS

6.1 Fundamental Theorem of Calculus

The Fundamental Theorem of Calculus expresses the extremely gratifying fact that the two basic operations of Calculus, differentiation and integration, are, for a suitable class of functions, inverse operations. This situation is analogous to squaring and extracting a square root as inverse operations. As another example, the exponential and logarithmic operations are inverses of one another. Although Issac Barrow (1630–1677) investigated the Fundamental Theorem of Calculus geometrically, its significance was not realized until approximately 20 years later when Newton and Leibniz, working independently of one another, developed the Calculus. This theorem, which provides an analytical approach to the evaluation of integrals, freed integration for certain classes of functions from a strong reliance on geometry and summation formulas.

To understand the Fundamental Theorem, it is important to think of an integral as a function. Recall that a function $f : A \to B$ is a rule that assigns to every element in A a unique element in B. Now, consider an integrable function f defined over $[a, b]$. If x is a number in the interval $[a, b]$, then $\int_a^x f(t)dt$ is a real number. Thus, we may define a function $F : [a, b] \to \mathbb{R}$ by assigning to each element x in $[a, b]$ the real number $\int_a^x f(t)dt$.

Example 6.1.1 Consider the function $f(t) = t$ on $[3, 5]$. Then $F(x) = \int_3^x t dt$.

Integral as Function

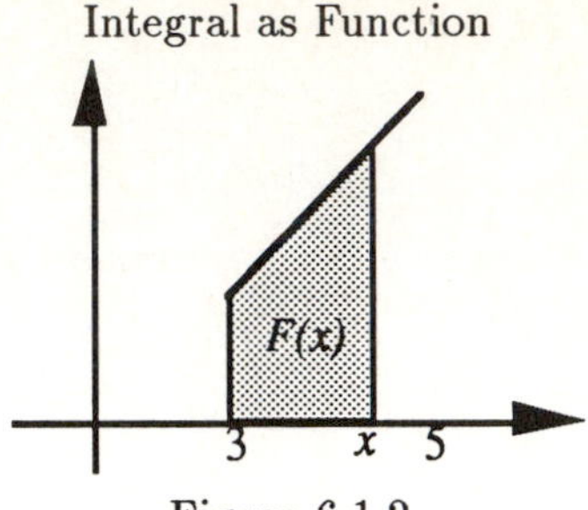

Figure 6.1.2

$F(x)$ is the area under the curve from 3 to x. The area of this trapezoid can easily be computed geometrically to be $(x^2 - 3^2)/2$ [i.e., the area of the trapezoid is one-half the length of the base times the sum of the lengths of the sides, or $1/2(x-3)(x+3)$]. Thus in this case $F(x)$ can be given by an explicit formula $F(x) = (x^2 - 3^2)/2$. △

When considered as functions, it makes sense to ask the same questions about integrals that we have asked in analyzing other functions. For example, given an integral we may ask: Is it continuous? Is it differentiable? Is it one-to-one? Is it onto? Is it increasing? Is its graph concave upward? What are its extreme values? And so on. The Fundamental Theorem will be our major tool in finding answers to these questions.

Before stating and proving the Fundamental Theorem, let us consider a simple example.

Example 6.1.3 Consider the C^1 function defined by $f(t) = t$ on $[0, 1]$. As in the previous example, the function $F(x) = \int_0^x f(t)dt$ can easily be computed (as the area of a triangle) to be $F(x) = x^2/2$. Now,

$$\frac{d}{dx}\int_0^x f(t)dt = \frac{d}{dx}\int_0^x t dt = \frac{d}{dx}\left(\frac{x^2}{2}\right) = x = f(x)$$

and

$$\int_0^x \frac{d}{dt}f(t)dt = \int_0^x \frac{d}{dt}t dt = \int_0^x 1 dt = x = f(x)$$

Hence,

$$\frac{d}{dx}\int_0^x f(t)dt = f(x) = \int_0^x \frac{d}{dt}f(t)dt$$

That is, integrating f first and then differentiating yields the same result, the function f itself, as does differentiating f first and then integrating. Thus integration and differentiation are inverse operations of each other. △

The reader should review Chapter 3, Section 7, on antiderivatives (particularly the definition of "antiderivative" and Theorem 3.7.3) before studying the *Fundamental Theorem of Calculus.*

Theorem 6.1.4 (The Fundamental Theorem of Calculus) *Let f be a continuous function on a closed interval $[a,b]$.*

Part I: *If $F : [a,b] \to \mathbb{R}$ is defined as $F(x) = \int_a^x f(t)dt$, then F is an antiderivative of f on $[a,b]$. (I.e. $F'(x) = f(x)$.)*

Part II: *If F is any antiderivative of f, then $\int_a^b f(x)dx = F(b) - F(a)$.*

Proof: We shall prove Part I by applying the definition of derivative to the function F and showing that $F'(x) = f(x)$ for all x in $[a,b]$. Let x and $x+h$ be numbers in $[a,b]$. In the following argument, it is understood that whenever $x = a$, the limits in question will be right-handed limits, and whenever $x = b$, the limits will be left-handed limits. By the definition of derivative, we have

$$\begin{aligned}\lim_{h\to 0} \frac{F(x+h) - F(x)}{h} &= \lim_{h\to 0} \frac{1}{h}\left[\int_a^{x+h} f(t)dt - \int_a^x f(t)dt\right] \\ &= \lim_{h\to 0} \frac{1}{h}\int_x^{x+h} f(t)dt\end{aligned}$$

(Why is the last equality true?) Since, by the Mean Value Theorem for Integrals, there exists a number c between x and $x+h$ such that

$$\int_x^{x+h} f(x)dx = hf(c)$$

we have

$$\lim_{h\to 0} \frac{F(x+h) - F(x)}{h} = \lim_{h\to 0} \frac{1}{h}[hf(c)] = \lim_{h\to 0} f(c)$$

Since f is continuous over $[a,b]$ and c is between x and $x+h$, $\lim_{h\to 0} f(c) = f(x)$. Thus, $F'(x) = f(x)$. □

Note that the hypothesis of continuity was needed in the application of the Mean Value Theorem for Integrals as well as in the evaluation of $\lim_{h\to 0} f(c)$.

We begin the proof of Part II by recalling that two antiderivatives for the same function differ by a constant. Let F be any antiderivative of f. Since Part I states that $\int_a^x f(t)dt$ is an antiderivative of f, there is a number C such that

$$\int_a^x f(t)dt - F(x) = C$$

Substituting a for x and observing that $\int_a^a f(t)dt = 0$, we have

$$0 - F(a) = C \quad \text{or} \quad F(a) = -C$$

Therefore,

$$\int_a^x f(t)dt = F(x) - F(a)$$

for all x in $[a, b]$. The desired result now follows by substituting b for x. That is,

$$\int_a^b f(t)dt = F(b) - F(a) \qquad \square$$

Notation: We shall follow the standard convention of denoting $F(b) - F(a)$ by the expression $F(x)]_a^b$. Thus, Part II of the Fundamental Theorem of Calculus may be expressed as:

If f is continuous on $[a, b]$ and F is *any* antiderivative of f, then

$$\int_a^b f(t)dt = F(t)]_a^b = F(b) - F(a)$$

Let us now make some observations concerning the Fundamental Theorem.

1. *Terminology.* The word *integral* is used in two ways. In Chapter 5 we introduced the integral $\int_a^b f(x)dx$, which defines a real number when we are given f, a, and b. [If a limit of integration, e.g., b, is considered as a possible variable, then $F(b) = \int_a^b$ defines a function F whose value at b is the number $\int_a^b f$.]

   The word integral is also used for the antiderivative $\int f(x)dx$, introduced in Chapter 3. One sometimes expresses $\int x dx = x^2/2 + C$ as "the integral of x is $x^2/2 + C$" rather than "the antiderivative of x is $x^2/2 + C$." Here the arbitrary *constant of integration* C should be included since $\int f(x)dx$ is the *general* antiderivative of F, and any two antiderivatives differ by a constant.

   The term *definite integral* is sometimes used for the integral $\int_a^b f(x)dx$, where there are limits of integration. The phrase *indefinite integral* is sometimes used for the antiderivative $\int f(x)dx$, where there are no limits of integration.

2. Theorem 5.5.2 implies that Part II of the Fundamental Theorem of Calculus also holds in the case where $a \geq b$. The argument is

   $$\int_a^b f(t)dt = -\int_b^a f(t)dt = -[F(a) - F(b)] = F(b) - F(a)$$

   If $a = b$, it is clear that $\int_a^b f(t)dt = 0$.

Let us consider several examples, first using Part II.

Example 6.1.5 We will evaluate $\int_1^4 (3x^2+4)dx$. Since x^3+4x is an antiderivative of $3x^2+4$,

$$\int_1^4 (3x^2+4)dx = (x^3+4x)]_1^4 = (64+16)-(1+4) = 75 \qquad \triangle$$

Notice that if an arbitrary constant had been added as an additional term in the above antiderivative, the same result would have been obtained. That is,

$$\int_1^4 (3x^4+4)dx = (x^3+4x+C)]_1^4 = (64+16+C)-(1+4+C) = 75$$

Thus, including a constant of integration in the evaluation of a definite integral does not affect the result and we need not add a constant of integration when evaluating a definite integral.

Example 6.1.6 We will evaluate $\int_0^2 e^x dx$. Since e^x is its own antiderivative (this result will be proved in Chapter 7),

$$\int_0^2 e^x dx = e^x]_0^2 = e^2 - 1 \qquad \triangle$$

Example 6.1.7 We will evaluate $\int \cos(3x)dx$. Since $\sin(3x)/3$ is an antiderivative of $\cos(3x)$,

$$\int \cos(3x)dx = \frac{\sin(3x)}{3} + C \qquad \triangle$$

Note that since the integral is indefinite, an arbitrary constant of integration is therefore included in the solution.

Example 6.1.8 We will evaluate $\int_{-1}^4 |x-1|dx$. By the definition of the absolute value function and Theorem 5.5.7,

$$\int_{-1}^4 |x-1|dx = \int_{-1}^1 -(x-1)dx + \int_1^4 (x-1)dx$$

Evaluating these integrals we get

$$\begin{aligned}\int_{-1}^4 |x-1|dx &= \left(\frac{-x^2}{2}+x\right)\bigg]_{-1}^1 + \left(\frac{x^2}{2}-x\right)\bigg]_1^4 \\ &= \left(\frac{-1}{2}+1\right) - \left(\frac{-1}{2}-1\right) + \left(\frac{16}{2}-4\right) - \left(\frac{1}{2}-1\right) \\ &= \frac{13}{2}\end{aligned} \qquad \triangle$$

Example 6.1.9 We will integrate $\int_{1/2}^{x} FLOOR(t)dt$ for x in $(2,3)$. Note that $FLOOR$ is *not* a continuous function. Thus, we may not apply the Fundamental Theorem *until* the integral is transformed using the definition of the FLOOR function and Theorem 5.5.4. Therefore, for $2 < x < 3$, we have

$$\begin{aligned} \int_{1/2}^{x} FLOOR(t)dt &= \int_{1/2}^{1} (0)dt + \int_{1}^{2} (1)dt + \int_{2}^{x} (2)dt \\ &= 0 + (2-1) + 2\int_{2}^{x} (1)dt = 1 + 2(x-2) = 2x - 3 \end{aligned}$$ △

Question 6.1.10 Sketch the graph of the function $F : [-1,4] \rightarrow \mathbb{R}$ defined by $F(x) = \int_{-4}^{x} FLOOR(t)dt$. (The graph illustrates that the integral of a discontinuous function is continuous.) ◊

Example 6.1.11 We will find the average value of the continuous function f defined by $f(x) = x^3$ over the interval $[-1,2]$. From the Mean Value Theorem for integrals, the average value is $\frac{1}{3}\int_{-1}^{2} x^3 dx$. Since an antiderivative of x^3 is $x^4/4$, we have, applying the Fundamental Theorem of Calculus, Part II,

$$\frac{1}{3}\int_{-1}^{2} x^3 dx = \frac{1}{3}\left(\frac{x^4}{4}\right)\Bigg]_{-1}^{2} = \frac{15}{12}$$ △

Question 6.1.12 Find the average value of $f : [-2,4] \rightarrow \mathbb{R}$ defined by $f(x) = |x-2|$. ◊

Now for some examples using Part I of the Fundamental Theorem.

Example 6.1.13 Recall that the first derivative of a distance function, $DIST$, yields a velocity function, VEL, and the second derivative yields an acceleration function, $ACCEL$. Thus we have

$$\frac{d}{dt}\int_{0}^{t} VEL = \frac{d}{dt}DIST = VEL = \int_{0}^{t} ACCEL = \int_{0}^{t} \frac{d}{dt}VEL$$

Thus

$$\frac{d}{dt}\int_{0}^{t} VEL = \int_{0}^{t} \frac{d}{dt}VEL$$ △

Example 6.1.14 We will determine the interval(s) on which the set of functions (antiderivatives) $\int x^2 - 4dx$ are increasing. Thus we wish to determine when the derivative of $\int x^2 - 4dx$ is positive i.e., when

$$D_x \int x^2 - 4dx > 0$$

or when

$$(x-2)(x+2) = x^2 - 4 > 0$$

The solution is $x < -2$ or $x > 2$. △

Question 6.1.15 What role does the constant of integration play in the preceding example? Explain your answer. ◇

The next examples illustrate the necessity of having the limits of integration in the proper form before applying the Fundamental Theorem, Part I.

Example 6.1.16 We will differentiate $\int_2^{3x} r^2 dr$ with respect to r. Note that the upper limit is not just x and thus we cannot immediately apply Part I of the Fundamental Theorem. What we have is a composite function and therefore we must use the chain rule to differentiate. So let $u = 3x$ and, using the chain rule, differentiate $\int_2^u r^2 dr$ with respect to x.

$$D_x \int_2^{3x} r^2 dr = D_u \int_2^u r^2 dr \frac{du}{dx} = u^2 \frac{du}{dx} = (u^2)(3) = (3x)^2(3) = 27x^2$$

Notice that the factor u^2 comes from applying Part I of the Fundamental Theorem and the factor 3 comes from differentiating u. △

Example 6.1.17 We will differentiate $\int_x^{x^2}(t+3)dt$. It will be necessary to make three transformations before applying the Fundamental Theorem. The first transformation is to apply Theorem 5.5.7:

$$\int_x^{x^2}(t+3)dt = \int_x^c (t+3)dt + \int_c^{x^2}(t+3)dt$$

where c is any real number. The second transformation is to apply Theorem 5.5.2 to interchange the limits of integration on the first integral on the right side. The third transformation is to set $u = x^2$ and substitute for x^2 so that the upper limit is a single letter in the second integral on the right-hand side. Thus,

$$\int_x^{x^2}(t+3)dt = -\int_c^x (t+3)dt + \int_c^u (t+3)dt$$

where $u = x^2$. The integrals on the right side are now in the form to which we may apply the Fundamental Theorem:

$$\begin{aligned} D_x \int_x^{x^2} (t+3)dt &= -D_x \int_c^x (t+3)dt + D_x \int_c^u (t+3)dt \\ &= -(x+3) + (u+3)\frac{du}{dx} = -(x+3) + (x^2+3)(2x) \\ &= 2x^3 + 5x - 3 \end{aligned}$$

△

The major point of the last two examples was to emphasize that the Fundamental Theorem, Part I, can only be applied to integrals that are in the form: $\int_a^x f(t)dt$, where f is a continuous function over the interval $[a, x]$. Thus if an integral fails to be in this form, it must first be transformed into the correct form before the Fundamental Theorem is applied.

Question 6.1.18 Prove: If f is an integrable function, then $F(x) = \int_c^x f(x)dx$ is a continuous function. Remember to consider the case where f is not continuous as well as the case where f is continuous. (In Question 6.1.10 we saw an example of when the integral of a discontinuous function was continuous. In this Question, you are to prove that the integral of a function is *always* continuous. Hint: Do a direct computation from the definition of derivative.) ◇

Section 6.1 Exercises

In Exercises 1 through 8, find an explicit formula for $F(x) = \int_a^x f(t)dt$, for x in the interval $[a, b]$. Check to see if $F'(x) = f(x)$.

1. $f(x) = 3$, [1,3]
2. $f(x) = 2$, [0,3]
3. $f(x) = 2x$, [0,3]
4. $f(x) = 5x$, [0,1]
5. $f(x) = 3x - 2$, [1,2]
6. $f(x) = -2x + 1$, $[-1, 0]$
7. $f(x) = -x + 1$, $[-3, -1]$
8. $f(x) = x + 3$, [0,3]

In Exercises 9 through 16, evaluate the integrals in two ways. First, by computing the areas; second, by using the Fundamental Theorem. Then compare the answers.

9. $\int_0^3 4dx$
10. $\int_{-2}^4 3dx$
11. $\int_0^2 2xdx$
12. $\int_{-2}^0 -2xdx$
13. $\int_{-1}^0 -xdx$
14. $\int_{-2}^0 -3xdx$
15. $\int_0^4 (2x+4)dx$
16. $\int_1^2 (-2x+4)dx$

In Exercises 17 through 32, evaluate the definite integrals using Part II of the Fundamental Theorem.

17. $\int_{-2}^{2} 3x^2dx$

18. $\int_1^3 6x^2dx$

19. $\int_{-2}^{2} 4x^3dx$

20. $\int_1^3 4x^3dx$

21. $\int_{-2}^{0}(3x^2 - 2x + 1)dx$

22. $\int_0^2(-3x^2 + 2x - 4)dx$

23. $\int_1^4 \sqrt{x}dx$

24. $\int_1^8 x^{-13}dx$

25. $\int_0^\pi \sin(t)dt$

26. $\int_0^{\pi/2} \cos(t)dt$

27. $\int_1^4 3/x^2dx$

28. $\int_1^3(2x - 1/x^3)dx$

29. $\int_{-3}^{0}(t + 2)(t - 3)dt$

30. $\int_{-1}^{1}(t^2 + 1)(t - 1)dt$

31. $\int_0^1(u + 1)^3du$

32. $\int_0^1(u - 1)^3du$

In Exercises 33 through 41, calculate the derivatives.

33. $D_x \int_1^{x^2} 1/tdt$

34. $D_x \int_0^{1+x^2} 1/\sqrt{2t + 5}dt$

35. $D_x \int_0^{x^3} 1/\sqrt{1 + t}dt$

36. $D_x \int_x^a f(t)dt$

37. $D_x \int_x^{x^2} 1/tdt$

38. $D_x \int_{1-x}^{1+x}(t - 1)/tdt$

39. $D_x \int_2^x |t - 1|dt, 3 < x < 5$

40. $D_x \int_x^2 |1 - t|dt, 3 < x < 5$

41. $D_x \int_2^5(x^2 + 3x)dx$

For Exercises 42 through 49, assume that $\int_a^b f(x)dx = 0$, where f is continuous and $a < b$. Tell whether each of statements 42 through 49 is true or false. If a statement is true, explain why. If it is false, give a counterexample.

42. $f(x) = 0$ for all x in $[a, b]$.

43. $f(x) = 0$ for at least one x in $[a, b]$.

44. $|\int_a^b f(x)dx| = 0$.

45. $\int_a^b |f(x)|dx = 0$.

46. $\int_a^b[f(x)]^2dx = 0$.

47. $\int_a^b[f(x) + 1]dx = b - a$.

48. $\int_a^b[f(x) + g(x)]dx = \int_a^b g(x)dx$ for all integrable functions g.

49. $D_x \int_a^b f(x)dx = 0$.

50. Mark each of the following statements true(T) or false(F).

    (a) $\int \sec^2(x)dx = \tan(x) + c$
    (b) $\int (2x\tan(x) + x^2\sec^2(x))dx = x^2\tan(x) + c$
    (c) $\int (x\cos(2x) + \sin(2x))dx = x\sin(x) + c$
    (d) $\int \frac{1}{4+x^2}dx = 1/2\tan^{-1}(x/2) + c$
    (e) $\int \sin(\cos(x))dx = -\cos^2(x)\sin(x) + c$

51. Give three different examples of a function $f : [1,5] \to \mathbb{R}$ such that $\int_1^5 f(x)dx = 8$.

52. Project. Develop a class presentation on Phillip Straffin's article "Using Integrals to Evaluate Voting Power" in the *Two Year College Mathematics Journal*, volume 10, number 3 (June 1979), pages 179–181.

53. Answer the following questions for each of the functions listed below: (1) Is f continuous? (2) Is f differentiable? (3) Is f one-to-one? (4) Is f onto? (5) Where is the graph of f concave up? (6) What are the extreme values of f? (If the integral cannot be evaluated in closed form, find a numerical approximation accurate to 0.01.) Now, (7), sketch the graph of f.

    (a) $f : (0,\infty) \to \mathbb{R}$, $f(x) = \int_1^x \sqrt{t^3+1}dt$.
    (b) $f : (0,\infty) \to \mathbb{R}$, $f(x) = 10e^{-x}\int_1^x \log(t)dt$.
    (c) $f : (0,\infty) \to \mathbb{R}$, $f(x) = 10\int_2^x e^{-t}\log(t)dt$.
    (d) $f : [0,4\pi] \to \mathbb{R}$, $f(x) = \int_\pi^x \sqrt{t}\sin(t)dt$.

54. Write out the proof requested in Question 6.1.18.

55. Let $f : [1,3] \to \mathbb{R}$ be defined by $f(x) = \int_1^{x^2} \sqrt{t}\sin(t)dt$.

    (a) Analyze f by determining critical numbers, extremal values, concavity, and inflection points. (Hint: What can you tell from the graph of f'?)
    (b) Sketch the graph of f. Use $ACC = .01$ for any numerical approximation of a value of f.

6.2 Integration by Parts

A basic technique in every branch of mathematics is that of applying a transformation to a given expression to transform it (change it) into an expression that is easier to work with, better understood, or in a standard form. The reader has already used numerous transformations, although the transformations were not necessarily labeled as such. Examples of this abound: the reduction of a fraction to lowest terms, the expression of a multiplication problem as an addition problem by a logarithm transform, the application of L'Hospital's Rule in a limit problem, the expression of a function as a composite function, the expression of a quadratic as the sum or difference of squares using the completing the square transformation, etc. Howard Eves, a noted historian of mathematics, says that the most used technique in mathematics is that of "transform–solve–(re)transform."

Evaluating an integral by Part II of the Fundamental Theorem of Calculus requires knowing an explicit formula for an antiderivative of the integrand. To find such a formula usually involves transforming the integrand into a "standard" form. Integration (i.e., integrand) transformations for special classes of functions will be the topic of this section and the next. The reader should realize that although the Fundamental Theorem assures us that every bounded continuous function defined over a bounded interval has an antiderivative [i.e., $\int_a^x f(t)dt$], there are no universal algorithms for finding antiderivatives.

In this section we shall consider an important transformation: integration by parts, based on the differentiation formula for products.

If f and g are C^1 functions, then

$$\int_a^b (fg)'(x)dx = f(x)g(x)]_a^b$$

but

$$\int_a^b (fg)'(x)dx = \int_a^b [f'(x)g(x) + f(x)g'(x)]dx$$

Thus,

$$\int_a^b f'(x)g(x)dx = f(x)g(x)]_a^b - \int_a^b f(x)g'(x)dx$$

More briefly, integration by parts transforms one integral into another integral, which may be easier to work with. In simplified notation we have

$$\int f'g = fg - \int fg'$$

Example 6.2.1 We will use the integration by parts transformation to integrate $\int x\cos(x)dx$.

To do this we need to express $x\cos(x)$ as a product of two functions, one of which is a derivative. We can express either x or $\cos(x)$ as a derivative. We will consider both possibilities.

(a) Consider x as a derivative. We need to find a function f such that $f'(x) = x$. Clearly, $f(x) = x^2/2$ is such a function. Thus

$$\int x\cos(x)dx \quad = \quad \int (\frac{x^2}{2})'\cos(x)dx$$

$$
\begin{aligned}
&= \frac{x^2}{2}\cos(x) - \int \frac{x^2}{2}\cos'(x)dx \\
&= \frac{x^2}{2}\cos(x) + \int \frac{x^2}{2}\sin(x)dx
\end{aligned}
$$

The transformed integral on the right-hand side is more complicated than the original integral, so this was not a good choice.

(b) Consider $\cos(x)$ as the derivative. We need to find a function f such that $f'(x) = \cos(x)$. Such a function is $f(x) = \sin(x)$, and, with $g(x) = x$, we have

$$
\begin{aligned}
\int x\cos(x)dx &= \int \cos(x)x dx = \int \sin'(x)\cdot x dx \\
&= \sin(x)\cdot x - \int \sin(x)\cdot x' dx \\
&= x\sin(x) - \int \sin(x)dx \\
&= x\sin(x) + \cos(x) + C
\end{aligned}
$$

$\triangle$

In applying integration by parts there are often several choices of which part of the integrand to consider as a derivative. If one choice does not work, another choice may work, and practice will give you experience in making a good choice.

Before considering additional examples using integration by parts, we will consider some notation.

The Differential Notation

Consider $f : \mathbb{R} \to \mathbb{R}$ as a function of the variable x. We have commented before on the many heuristic advantages of Leibniz' differential notation, dy/dx. Let us now consider dy/dx as a quotient. From the difference quotient for the derivative we have

$$
f'(x) = \frac{dy}{dx} \approx \frac{\Delta y}{\Delta x} = \frac{\Delta f}{\Delta x} \quad \text{or,} \quad \Delta f \approx f'(x)\Delta x
$$

This says that the change in f, Δf, is approximately equal to $f'(x)$ times the change in x, Δx. It is useful to write $df = f'(x)dx$ where df is thought of as the "small" change in f resulting from the "small" change dx in x. The symbols "dx" and "df" are called *differentials.* Symbolically we can obtain $df = f'(x)dx$ from $df/dx = f'(x)$ simply by multiplying both sides by dx.

There are several ways to give a precise meaning to dx and the operations on it (in terms of "infinitesimals" or "differential forms") but we are unable to do so at this level of mathematical development (see Section 8.4 for an introduction to the idea of differential forms). However the notation and its computational advantages are so useful that we will use it without a precise definition.

Although most manipulations with differentials lead to valid results, there are some warnings.

1. Do not have a differential in the denominator without a differential in the numerator. For example, the expression

$$\frac{1}{df} = \frac{1}{f'(x)dx}$$

is undesirable.

2. If one term has a differential factor, then every term should have a differential factor. For example, do not use

$$df = g(x)$$

where the differential appears only in one term.

The differential notation can be used with functions of several variables, where it is most useful. If $f : \mathbb{R}^2 \to \mathbb{R}$, $x = x(t)$, $y = y(t)$, then

$$\frac{df}{dt} = \frac{\partial f}{\partial x}\frac{dx}{dt} + \frac{\partial f}{\partial y}\frac{dy}{dt}$$

Multiplying by dt we obtain an expression involving differentials,

$$df = \frac{\partial f}{\partial x}dx + \frac{\partial f}{\partial y}dy$$

We can find the differential of an expression by considering each variable symbol as a function (of t) and applying the rules for differentiation. We then multiply through by dt to obtain a differential expression. After a little practice, we can just pretend that a dt exists and write the differential expression directly.

Example 6.2.2 We will find the differential of $3x^2y$. The variable symbols are x and y; we will consider them as functions of t and obtain

$$\frac{d(3x^2y)}{dt} = \frac{\partial(3x^2y)}{\partial x}\frac{dx}{dt} + \frac{\partial(3x^2y)}{\partial y}\frac{dy}{dt} = 6xy\frac{dx}{dt} + 3x^2\frac{dy}{dt}$$

Multiplying through by dt we obtain a differential expression

$$d(3x^2y) = 6xydx + 3x^2dy$$

△

The notation of differentials can be used with integration by parts. In particular, if u and v are functions of the same variable, the differential of the product, uv, is the differential expression

$$d(uv) = udv + vdu$$

Integrating both sides of this equality yields

$$\int d(uv) = \int udv + \int vdu$$

Solving for $\int udv$ and noting that $\int d(uv) = uv$ gives us integration by parts:

$$\int udv = uv - \int vdu$$

We do not have a constant of integration, since one is implied by the indefinite integral on the right-hand side.

In using "parts," we must be able to differentiate our selection for u and be able to integrate our selection for dv. We also note that the purpose is to "simplify" the function that needs to be integrated. Thus, we should compare the expression for vdu against that for udv. If the integrand vdu is more complicated than udv, we should investigate reversing the selections for u and dv.

Example 6.2.3 We will integrate $\int x\sqrt{x+1}dx$. Applying the parts transformation

$$u = \sqrt{x+1}, \quad dv = xdx, \quad du = \frac{dx}{\sqrt{x+1}}, \quad v = \frac{x^2}{2}$$

we have

$$\int x\sqrt{x+1}dx = \frac{x^2}{2}\sqrt{x+1} - \int \frac{x^2}{4\sqrt{x+1}}dx$$

Note that $vdu = x^2/(2\sqrt{x+1})dx$ is more complicated than $udv = x\sqrt{x+1}dx$. Thus, we should consider reversing the selections for u and dv. That is, letting

$$u = x, \quad dv = \sqrt{x+1}dx, \quad du = dx, \quad v = \frac{2}{3}(x+1)^{3/2}$$

we now have

$$\begin{aligned} \int x\sqrt{x+1}dx &= \frac{2}{3}x(x+1)^{3/2} - \frac{2}{3}\int (x+1)^{3/2}dx \\ &= \frac{2}{3}x(x+1)^{3/2} - \frac{4}{15}(x+1)^{5/2} + C \end{aligned}$$ △

The presence of an exponential factor in an integrand often suggests using a parts transformation.

Example 6.2.4 We will integrate $\int xe^x dx$.

The presence of the e^x factor suggests a parts transformation. Furthermore since e^x is its own antiderivative (see Section 7.3), the e^x factor behaves the same way whether it is chosen to be the u factor or the dv factor. Thus in selecting the partition of the integrand, we give primary consideration to the nonexponential factor. So let

$$u = x, \quad dv = e^x, \quad du = dx, \quad v = e^x$$

Thus

$$\int xe^x dx = xe^x - \int e^x dx = xe^x - e^x + C$$ △

It is sometimes necessary to apply Parts two or more times in the same exercise. After applying Parts, the reader should always be alert to the possibility of combining the $\int v du$ term with the original $\int u dv$ term. Further work on applying the parts transformations to integrals involving exponential functions will be considered in Chapter 7.

Section 6.2 Exercises

Compute the differentials of the following expressions.

1. $\sin(x)\cos(y)$ 2. $\tan(x) + cot(y)$ 3. $x^2\sin(y) + 2y^{3/2}$

4. $xy\sin(z) + x/y$ 5. $xyz/(x+y+z)$ 6. $\sin(xyz) + xy^2 + y/z$

Compute the integrals.

7. $\int x\sin(3x)dx$ 8. $\int x\sqrt{x+5}dx$ 9. $\int x\sin(5x)dx$

10. $\int x\cos(4x)dx$ 11. $\int \sin^2(x)dx$ 12. $\int \cos^2(x)dx$

13. $\int x^2\sin(x)dx$ 14. $\int x^3\cos(x)dx$ 15. $\int_1^2 x(x-1)^5dx$

16. $\int_2^3 x^2(x-1)^5dx$ 17. $\int \sin(2x)\sin(3x)dx$ (Hint: Parts twice)

18. $\int \sin(x)\cos(2x)dx$ 19. $\int x^2e^x dx$ 20. $\int x\log(x)dx$

21. $\int_0^1 xe^x dx$ 22. $\int(xe^x + x\sin(x))dx$

In Exercises 23 through 26 determine the error when the integral is approximated using Simpson's Rule with 10 subintervals. Consider using a computer for these problems.

23. $\int_0^2 x(x-1)^{10}dx$ 24. $\int_1^2 x^2\cos(x)dx$

25. $\int_0^4 \sin(x)\cos^2(x)dx$ 26. $\int_2^4 \sin^2(x)dx$

27. The object of this exercise is to develop some of the integration formulas found in tables. This will be done by integrating (with the help of a computer algebra system) a number of examples and looking for patterns. In this exercise, we want a formula for $\int x^n\log(x)dx$, where n is a positive integer.

    (a) Find $\int x\log(x)dx$.
    (b) Find $\int x^2\log(x)dx$
    (c) Find $\int x^3\log(x)dx$

(d) From the results of (a), (b), and (c) (or further examples if needed), conjecture a formula for $\int x^n \log(x)dx$.

(e) Verify your conjecture by differentiation.

28. The object of this exercise is to develop some of the integration formulas found in tables. This will be done by integrating (with the help of a computer algebra system) a number of examples and looking for patterns. In this exercise, we want a formula for $\int x \log(x^n)dx$, where n is a positive integer.

(a) Find $\int x \log(x)dx$.

(b) Find $\int x \log(x^2)dx$

(c) Find $\int x \log(x^3)dx$

(d) From the results of (a), (b), and (c) (or further examples if needed), conjecture a formula for $\int x \log(x^n)dx$.

(e) Verify your conjecture by differentiation.

6.3 Change of Variable Transformations

When we are asked to evaluate an integral, we have four choices: use a table of integrals, use a numerical integration technique (Chapter 5), apply the Fundamental Theorem of Calculus (find an antiderivative), or use a packaged program in a computer or hand calculator. As we have mentioned, most continuous functions do not have antiderivatives that can be expressed in useful forms. Computer algebra systems, which are coming into widespread use, can compute antiderivatives about as well as a mathematician. Thus we put less emphasis on the computation of antiderivatives than was common in the past.

If we wish to use the Fundamental Theorem to compute an antiderivative, we must usually transform the integrand into a standard form for which we know an antiderivative or can find one in a book of integral tables. The integrals that we can evaluate (using what we have studied) are linear combinations of those in the following list. That is, the integrand is a sum of constants times the following integrands:

$$\int x^r dx = \frac{x^{r+1}}{r+1} + C,\ r \neq -1$$

$$\int \sin(x)dx = -\cos(x) + C$$

$$\int \cos(x)dx = \sin(x) + C$$

$$\int \sec^2(x)dx = \tan(x) + C$$

$$\int \csc^2(x)dx = -\cot(x) + C$$

$\int \frac{1}{a^2+x^2} dx = \frac{1}{a} \tan^{-1}(\frac{x}{a}) + C$

$\int \frac{1}{\sqrt{a^2-x^2}} dx = \sin^{-1}(\frac{x}{a}) + C$

$\int e^x dx = e^x + C$

Thus, at present an integrand is in standard form if and only if it is a linear combination of

$$x^r \quad \sin(x) \quad \cos(x) \quad \sec^2(x) \quad \csc^2(x) \quad \frac{1}{a^2+x^2} \quad \frac{1}{\sqrt{a^2-x^2}} \quad e^x$$

where r is a rational number different from -1. Therefore, to integrate a function by the Fundamental Theorem, we must transform the function into standard form. We shall call a transformation *applicable* (to the problem in question) if it transforms the integrand into an expression that is "closer" to standard form. We would like to get the standard form in one transformation, but sometimes it may require several.

Once a transformation is obtained, the next step is to check it. The differential notation provides an easy way to mechanically check a transformation to see if it is applicable to the exercise in question. The checking procedure is to substitute for x and dx in the expression $f(x)dx$. If the resulting expression can be written in standard form (or closer to it), then the transformation is applicable and the integral can be evaluated (or the process continued). The reader should realize that if there is an applicable transformation for a given exercise, there may be more than one.

We shall consider two forms of change of variable transformations: algebraic and trigonometric. No attempt is made to be exhaustive in this study of integral transformations. The reader should be aware that there are many specialized integral transformations (which we will not consider) as well as extensive tables of integrals. Most integrable functions do not have an explicit formula for an antiderivative, and so evaluating an integral by finding an antiderivative is the special technique in contrast to the general technique of numerical integration.

Algebraic Transformations

Example 6.3.1 We will evaluate $\int_2^6 \cos(3x)dx$. The integrand is not in a standard form because of the argument of $3x$ in the cosine function. To transform the integrand into standard form, we let $u = 3x$. Thus, the variable is changed from x to u and $du = 3dx$ or $\frac{1}{3}du = dx$. Since the new variable of integration is u (rather than x), we need to change the limits of integration so that they are limits on u (rather than x). This is done by applying the transformation ($u = 3x$). Thus, $u = 6$ when $x = 2$, and $u = 18$ when $x = 6$. The transformed integral is now, after substitution,

$$\int_6^{18} \cos(u)\frac{1}{3}du = \frac{1}{3}\int_6^{18} \cos(u)du$$

which is in standard form. Thus, $u = 3x$ is an applicable transformation and

$$\int_2^6 \cos(3x)dx = \frac{1}{3}\int_6^{18} \cos(u)du = \left.\frac{\sin(u)}{3}\right]_6^{18} = \frac{\sin(18) - \sin(6)}{3} \qquad \triangle$$

Example 6.3.2 We will integrate $\int x/\cos^2(x^2)dx$. This integral is not in standard form because of the quotient and the argument x^2 of the cosine function. Thus, we need a transformation.

The xdx combination in the integrand is the differential of $x^2/2$. This suggests considering the transformation $u = x^2$, $du = 2xdx$. Checking this transformation, we see that $xdx = \frac{1}{2}du$, and upon substitution we have

$$\int \frac{1}{\cos^2(u)}\frac{1}{2}du = \frac{1}{2}\int \sec^2(u)du$$

which is an integral in standard form. Thus,

$$\int \frac{x}{\cos^2(x^2)}dx = \int \frac{1}{2}\sec^2(u)du = \frac{1}{2}\tan(u) + C$$

Now, we need to (re)transform to express our answer in terms of the original variable of integration, x. Thus substituting x^2 for u yields our answer of $\frac{1}{2}\tan(x^2) + C$. $\triangle$

Example 6.3.3 We will integrate $\int(x^2 + 1)\sqrt{x-2}dx$. This integral is not in standard form. Thus we need a transformation. A clue to choosing a transformation is to think of the radical term as a quantity raised to a power. This would suggest the transformation $u = x - 2$, so that the radical term can be expressed as $u^{1/2}$. Applying this transformation and substituting for x and dx in the integrand yields

$$\int(x^2 + 1)\sqrt{x-2}dx = \int((u+2)^2 + 1)u^{1/2}du = \int(u^{5/2} + 4u^{3/2} + 5u^{1/2})du$$

This last integral is in standard form since the integrand is a linear combination of terms of the form u^r where r is a rational number different from -1. Therefore, it can be integrated giving an expression in the variable u which then needs to be (re)transformed into an expression in x, the original variable of integration.

The reader can check that the transformation $u^2 = x - 2$ also applicable for this problem. $\triangle$

Example 6.3.4 We will integrate $\int(x^2 + 2x)/(x^2 + 2x + 1)dx$. This integral is not in standard form because of the quotient. Thus, we need a transformation. Since the degree of the numerator (2) is greater than or equal to the degree of the denominator (2), the integrand is an *improper* rational function. Such a function should be transformed into a polynomial plus a *proper rational function* (with the degree of the numerator less than the degree of the denominator) by dividing the denominator into the numerator. Carrying out this transformation and then using the linearity of integration yields

$$\int \frac{x^2+2x}{x^2+2x+1}dx = \int (1 - \frac{1}{(x+1)^2})dx = \int 1dx - \int \frac{1}{(x+1)^2}dx$$

Now transforming the second integral by setting $u = x+1$, $du = dx$, yields

$$\int \frac{x^2+2x}{x^2+2x+1}dx = \int 1dx - \int u^{-2}du = x - \frac{u^{-1}}{-1} + C = x + \frac{1}{x+1} + C \qquad \triangle$$

This example illustrates the use of the following rule:

> If the integrand is an improper rational function, divide the denominator into the numerator to obtain a polynomial plus a proper rational fraction.

Example 6.3.5 We will integrate $\int 1/(x^2+2x+2)dx$. This integral is not in standard form since the denominator of the integrand is not written as the sum of two squares. Since the numerator is 1 and the denominator is a quadratic, we think of an inverse tangent. This suggests that we apply the transformation of completing the square to the denominator. Completing the square in the denominator yields

$$\int \frac{1}{(x+1)^2+1}dx$$

which is almost in standard form. Applying the transformation $u = x+1, du = dx$ yields

$$\int \frac{1}{u^2+1}du = \tan^{-1}(u) + C = \tan^{-1}(x+1) + C \qquad \triangle$$

Example 6.3.6 We will integrate $\int \sin^{-1}(x)dx$. Because of the inverse sine function, the integrand is not in standard form. Since the "difficulty" appears to be in the inverse function, we look for a transformation that will eliminate the inverse aspect. Noting that composing an inverse function with the function itself yields just the argument of the function, we consider the transformation $x = \sin(u)$. Substituting for x and dx in the original integral yields

$$\int \sin^{-1}(x)dx = \int u\cos(u)du$$

This is just the integral that was evaluated in the previous section using a parts transformation.

Another approach to integrating $\int \sin^{-1}(x)dx$ is to initially consider a parts transformation. By inspection, it is clear that we can differentiate $\sin^{-1}(x)$ and integrate the differential dx. Hence, we consider the transformation

$$u = \sin^{-1}(x), \quad dv = dx, \quad du = \frac{1}{\sqrt{1-x^2}}dx, \quad v = x$$

So, we have

$$\int \sin^{-1}(x)dx = x\sin^{-1}(x) - \int \frac{x}{\sqrt{1-x^2}}dx$$

Employing the change of variable transformation $t = 1 - x^2$ in the integral on the right-hand side of the expression yields

$$\begin{aligned}\int \sin^{-1}(x)dx &= x\sin^{-1}(x) - \int t^{-1/2}\frac{1}{2}dt = x\sin^{-1}(x) + t^{1/2} + C \\ &= x\sin^{-1}(x) + \sqrt{1-x^2} + C\end{aligned}$$

△

Trigonometric Transformations

The following standard trigonometric identities (transformations) are frequently utilized in evaluating integrals by the Fundamental Theorem:

$$\sin^2(x) + \cos^2(x) = 1$$

$$\tan^2(x) + 1 = \sec^2(x)$$

$$1 + \cot^2(x) = \csc^2(x)$$

$$\sin(2x) = 2\sin(x)\cos(x)$$

$$\cos(2x) = 2\cos^2(x) - 1 = 1 - 2\sin^2(x) = \cos^2(x) - \sin^2(x)$$

$$\cos^2(x) = \tfrac{1+\cos(2x)}{2} \qquad \sin^2(x) = \tfrac{1-\cos(2x)}{2}$$

$$\sin(x)\cos(y) = \tfrac{1}{2}[\sin(x+y) + \sin(x-y)]$$

$$\cos(x)\cos(y) = \tfrac{1}{2}[\cos(x+y) + \cos(x-y)]$$

$$\sin(x)\sin(y) = \tfrac{1}{2}[\cos(x-y) - \cos(x+y)]$$

Powers of Sine and Cosine

Any integral of the form

$$\int \sin^r(x)\cos(x)dx \quad \text{for} r \neq -1,\ r \text{ rational}$$

can be expressed in standard form by applying the transformation $u = \sin(x)$. In a similar fashion, any integral of the form

$$\int \cos^r(x)\sin(x)dx \quad \text{for} r \neq -1,\ r \text{ rational}$$

can be expressed in standard form by employing the transformation $u = \cos(x)$.

Example 6.3.7 We will integrate $\int \sin^{2/3}(x)\cos(x)dx$. Let $u = \sin(x)$, $du = \cos(x)dx$. The transformed integral is now, after substitution,

$$\int u^{2/3}du = \frac{3}{5}u^{5/3} + C = \frac{3}{5}\sin^{5/3}(x) + C$$

△

Integrals of the form $\int \sin^m(x)\cos^n(x)dx$ where m and n are nonnegative integers constitute a subclass of integrals whose evaluations depend on using the trigonometric identities in the above list. The following example illustrates the method to use when either m or n is a positive odd integer.

Example 6.3.8 We will integrate $\int \sin^3(x)\cos^2(x)dx$. Transforming the integrand by using the appropriate identities from the above list yields

$$\begin{aligned} \int \sin^3(x)\cos^2(x)dx &= \int \sin^2(x)\cos^2(x)\sin(x)dx \\ &= \int (1-\cos^2(x))\cos^2(x)\sin(x)dx \\ &= \int \cos^2(x)\sin(x)dx - \int \cos^4(x)\sin(x)dx \end{aligned}$$

Applying the transformation $u = \cos(x)$ and integrating yields

$$\int \sin^3(x)\cos^2(x)dx = \frac{-1}{3}\cos^3(x) + \frac{1}{5}\cos^5(x) + C$$

△

The next example illustrates the method to use when m and n are both even integers.

Example 6.3.9 We will integrate $\int \cos^4(x)\sin^2(x)dx$. The idea behind the following transformations is to use the identities

$$\cos^2(x) = \frac{1}{2}(1 + \cos(2x))$$

and

$$\sin^2(x) = \frac{1}{2}(1 - \cos(2x))$$

to express the integrand as a linear combination of cosine terms raised to *odd* powers. We get

$$\begin{aligned}
\int \cos^4(x)\sin^2(x)dx &= \int (\cos^2(x))^2 \sin^2(x)dx \\
&= \int \left(\frac{1}{4}\right)[1+\cos(2x)]^2 \left(\frac{1}{2}\right)[1-\cos(2x)]\,dx \\
&= \frac{1}{8}\int [1+\cos(2x) - \cos^2(2x) - \cos^3(2x)]dx \\
&= \frac{1}{8}\int [1+\cos(2x) - \frac{1+\cos(4x)}{2} \\
&\qquad - \cos^2(2x)\cos(2x)]dx \\
&= \frac{1}{8}\int [\frac{1}{2} + \cos(2x) - \frac{1}{2}\cos(4x) \\
&\qquad - (1-\sin^2(2x))\cos(2x)]dx \\
&= \frac{1}{8}\int \left[\frac{1}{2} - \frac{1}{2}\cos(4x) + \sin^2(2x)\cos(2x)\right] dx \\
&= \frac{x}{16} - \frac{\sin(4x)}{64} + \frac{\sin^3(2x)}{48} + C
\end{aligned}$$

How is this last step obtained? △

Question 6.3.10 The reader should experiment with integrals of this subclass by considering instances where m and n are both even integers, where one exponent is an even integer and the other is an odd integer, where both exponents are odd integers, etc. Can any general type of statement be made if one or the other of the exponents is negative or rational? ◊

Question 6.3.11 Does the subclass of integrals of the form

$$\int \sin^m(x)\cos^n(x)dx$$

contain all integrals whose integrands are products of trigonometric functions? For example, are

$$\int \tan^3(x)\csc^7(x)dx \quad \text{and} \quad \int \cot^2(x)\sec^{-2}(x)\sin^4(x)dx$$

in this subclass? ◊

Substitutions into Quadratics

Another large subclass of integrals that often involve trigonometric substitutions for their evaluations are those whose integrands contain a quadratic expression raised to a power. The idea is to transform the quadratic expression into a single term by using an identity that is equivalent to the fundamental trigonometric identity

$$\sin^2(x) + \cos^2(x) = 1$$

Since every quadratic expression can be transformed into the sum or difference of squares using the completing the square transform, there are just three possibilities that need to be considered.

Term in the Integrand	Trigonometric Transformation	Transformed Term
$a^2 - x^2$	$x = a\sin(u)$	$a^2\cos^2(u)$
$a^2 + x^2$	$x = a\tan(u)$	$a^2\sec^2(u)$
$x^2 - a^2$	$x = a\sec(u)$	$a^2\tan^2(u)$

In each of these transformations the value u is restricted so that the corresponding inverse trigonometric function is defined.

Example 6.3.12 We will integrate $\int \sqrt{1-x^2}dx$. The integrand is not in standard form since the radicand is a binomial and thus it is not in the form of the power function, x^r. The fact that the radicand is close to the standard form of the power function suggests that we look for a way to transform it into the power function form. That is, find a transformation that will express $1-x^2$ as a single term.

Consider the transformation $x = \sin(u)$, $dx = \cos(u)du$. We will restrict u to $[-\pi/2, \pi/2]$ so that $\cos(u)$ will be positive. Substituting into the integrand yields

$$\begin{aligned}\int \sqrt{1-x^2}dx &= \int \sqrt{1-\sin^2(u)}\cos(u)du \\ &= \int \sqrt{\cos^2(u)}\cos(u)du = \int \cos(u)\cos(u)du\end{aligned}$$

[here we use the fact that $\cos(u) \geq 0$]

$$\begin{aligned} &= \int \frac{1}{2}(1+\cos(2u))du \\ &= \frac{1}{2}(u + \frac{1}{2}\sin(2u)) + C \\ &= \frac{1}{2}(u + \sin(u)\cos(u)) + C \\ &= \frac{1}{2}\sin^{-1}(x) + \frac{1}{2}x\sqrt{1-x^2} + C\end{aligned}$$

How is this last equality obtained? △

Example 6.3.13 We will integrate $\int 1/(1+x^2)dx$. Even though the integrand is in standard form and we know that the integral is $\tan^{-1}(x)+C$, let us arrive at this result by employing an appropriate trigonometric transformation. Let $x=\tan(u)$, $dx=\sec^2(u)du$ and hence (substituting)

$$\begin{aligned}\int \frac{1}{1+x^2}dx &= \int \frac{1}{1+\tan^2(u)}\sec^2(u)du \\ &= \int \frac{1}{\sec^2(u)}\sec^2(u)du \\ &= \int du = u+C = \tan^{-1}(x)+C\end{aligned}$$

The transformation $x=\tan(u)$ determines a *reference triangle* which serves to define all six trigonometric functions in terms of x. Recall that the angle u is restricted so that the inverse tangent function is defined, that is, $-\pi/2 < u < \pi/2$.

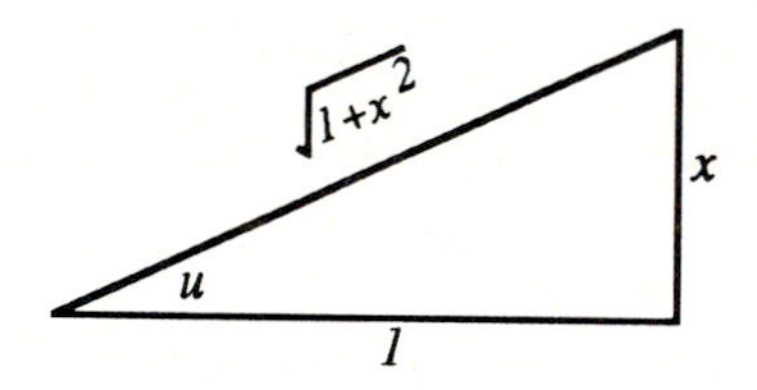

Figure 6.3.14

Thus,

$$\sin(u) = \frac{x}{\sqrt{1+x^2}}$$

and

$$\csc(u) = \frac{\sqrt{1+x^2}}{x}$$

The reader should practice drawing the reference triangles and evaluating the trigonometric functions in terms of x for transformations

$$x=\sin(u) \quad \text{for } \frac{-\pi}{2} \le u \le \frac{\pi}{2} \quad \text{and} \quad x=\sec(u) \quad \text{for } 0 \le u \le \pi$$ △

Recall that every quadratic expression can be transformed into the sum or difference of squares by the technique of completing the square. It is often desirable to carry out such a transformation before using a trigonometric substitution.

Example 6.3.15 We will integrate $\int 1/\sqrt{-x^2+4x-2}dx$. First transform the quadratic expression into a difference of squares by completing the square:

$$-x^2+4x-2 = 2-(x-2)^2$$

Thus,

$$\begin{aligned}\int 1/\sqrt{-x^2+4x-2}dx &= \int \frac{1}{\sqrt{2-(x-2)^2}}dx \\ &= \int \frac{1}{\sqrt{2}} \frac{1}{\sqrt{1-\frac{1}{2}(x-2)^2}}dx\end{aligned}$$

Use the transformation

$$\frac{1}{\sqrt{2}(x-2)} = \sin(u) \quad \frac{1}{\sqrt{2}}dx = \cos(u)du$$

where $-\frac{\pi}{2} \le u \le \frac{\pi}{2}$ as before. Thus,

$$\begin{aligned}\int \frac{1}{\sqrt{-x^2+4x-2}}dx &= \int \frac{1}{\sqrt{1-\sin^2(u)}}\cos(u)du \\ &= \int du = u + C = \sin^{-1}\left(\frac{x-2}{\sqrt{2}}\right) + C\end{aligned}$$

△

Example 6.3.16 We will evaluate $\int_2^4 \sqrt{x^2-4}/x dx$. We shall first consider the indefinite integral to obtain an antiderivative of the integrand and then apply Part II of the Fundamental Theorem of Calculus to evaluate the given integral. The term $\sqrt{x^2-4}$ suggests using the transformation $x = 2\sec(u), dx = 2\sec(u)\tan(u)du$. (Why?)

Substituting, we get

$$\begin{aligned}\int \frac{\sqrt{x^2-4}}{x}dx &= \int \frac{\sqrt{4\sec^2(u)-4}}{2\sec(u)} 2\sec(u)\tan(u)du \\ &= 2\int \sqrt{\tan^2(u)}\tan(u)du \\ &= 2\int |\tan(u)|\tan(u)du\end{aligned}$$

Since the original definite integral restricts x to be in the interval $[2,4]$ and $x = 2\sec(u)$, we have that $1 \le \sec(u) \le 2$. Thus, u is an angle in the first quadrant and $\tan(u)$ is nonnegative. Therefore, we may remove the absolute value signs and get

$$\int \frac{\sqrt{x^2-4}}{x}dx = 2\int \tan^2(u)du = 2\int (\sec^2(u)-1)du = 2\tan(u) - 2u + C$$

The reference triangle determined by the transformation $x = 2\sec(u)$ is given below.

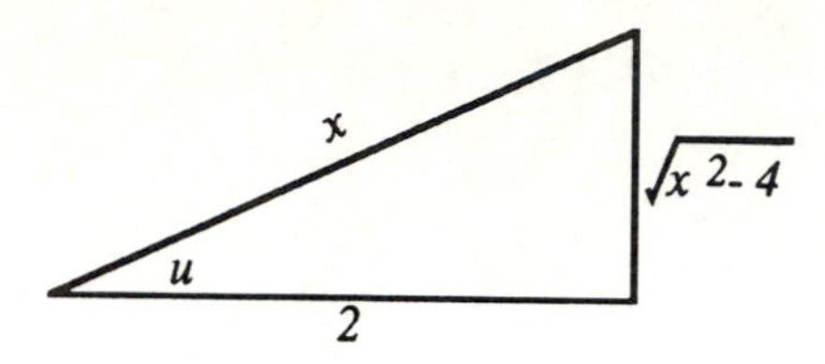

Figure 6.3.17

Finally,

$$\int \frac{\sqrt{x^2-4}}{x}dx = 2(\tan(u) - u) + C = 2\left[\sqrt{\left(\frac{x}{2}\right)^2 - 1} - \sec^{-1}\left(\frac{x}{2}\right)\right] + C$$

We have now found an antiderivative for the integrand of the given problem. The last step in the problem is to apply Part II of the Fundamental Theorem and simplify. Thus,

$$\begin{aligned}\int_2^4 \frac{\sqrt{x^2-4}}{x}dx &= 2\left[\sqrt{\left(\frac{x}{2}\right)^2 - 1} - \sec^{-1}\left(\frac{x}{2}\right)\right]_2^4 \\ &= 2[\sqrt{3} - \sec^{-1}(2) - 0 + \sec^{-1}(1)] \\ &= 2\sqrt{3} - \frac{2}{3}\pi\end{aligned}$$

△

Section 6.3 Exercises

Compute the integrals.

1. $\int (x-1)^5 dx$
2. $\int x(x^2+1)^4 dx$
3. $\int \cos(x^{1/2})/x^{1/2} dx$
4. $\int \cos(x)/(1+\sin^2(x)) dx$
5. $\int x/(x^4+1) dx$
6. $\int 1/(3+2x)^2 dx$
7. $\int \sqrt{2x+1} dx$
8. $\int (5x^2+20x-24)/(x+5)^{1/2} dx$
9. $\int (x^5-8x^3)/(x^2-4)^{1/2} dx$
10. $\int 1/(2x+5)(2x-3)^{1/2} dx$
11. $\int_3^5 x\sqrt{x^2-9} dx$
12. $\int_1^5 (x+3)/(2x-1)^{1/2} dx$
13. $\int_0^1 (x+3)/(x+1)^{1/2} dx$
14. $\int_1^2 (6-x)^{-3} dx$
15. $\int x^2/(x^2+1) dx$
16. $\int (x^2-1)/(x+1) dx$

17. $\int (x^3 + 2x)dx$

18. $\int (x^2 + 2x)/(x^2 + 6x + 1)dx$

19. $\int 1/(x^2 - 2x + 2)dx$

20. $\int 4/(x^2 + 4x + 12)dx$

21. $\int x \tan^{-1}(x)dx$

22. $\int \cos^{-1} x dx$

23. $\int x/(ax + b)^{3/2}dx$

24. $\int x/(ax + b)^{1/2}dx$

25. $\int \tan(x)/\cos(x)dx$

26. $\int \sin^2(x)dx$

27. $\int \cos^2(x)dx$

28. $\int \sin^2(x)\cos^5(x)dx$

29. $\int \sin^5(x)dx$

30. $\int \sin^2(x)\cos^2(x)dx$

31. $\int \sin^3(3x)dx$

32. $\int \cos^5(x/2)dx$

33. $\int (\sin(x + 1))^5 dx$

34. $\int \sqrt{\sin(x)}\cos(x)dx$

35. $\int \sin(x)\cos(2x)dx$

36. $\int \sin(2x)\cos(x)dx$

37. $\int \sin(2x)\cos(3x)dx$

38. $\int \sin(2x)\sin(4x)dx$

39. $\int \sqrt{9 - x^2}dx$

40. $\int \sqrt{x^2 - 9}/4x dx$

41. $\int 1/(x^2 + 2x + 5)dx$

42. $\int 1/(x^2 + 3x + 5)dx$

43. $\int \tan^6(x)dx$

44. $\int \sec^4(x)dx$

45. $\int \sqrt{4 - x^2}/x^2 dx$

46. Let $B_n = \int_0^{\pi/2} \sin^{2n}(x)dx$, n a positive integer.

(a) Compute B_n for successive odd values of n starting with $n = 1$ and continuing until a pattern is recognized. (Hint: Try to express B_n in terms of B_k for some $k < n$.)

(b) Compute B_n for successive even values of n starting with $n = 2$ and continuing until a pattern is recognized.

(c) Conjecture a relation expressing B_n in terms of B_k for a value of $k < n$. (Such a relation is called *recursive*, and is discussed in Chapter 1.)

(d) Verify your conjecture by computing B_n using a Parts transformation. (Hint: express $\sin^n(x)$ as $\sin^{n-1}(x)\sin(x)$.)

(e) Find a numerical expression for B_n in terms of factorials. [Hint: Find two expressions, one for n odd and one for n even.]

47. Mark each of the following statements true (T) or false (F). [Recall that $d\log(x)/dx = 1/x$.]

(a) $\int \cot^2(x)dx = -\cot(x) - x + C$.

(b) $\int \tan^{-1}(x)dx = x\tan^{-1}(x) - 1/2\log(1+x^2) + C$.

(c) $\int (x^2-a^2)^{-1/2}dx = \log|(x+(x^2-a^2)^{1/2}| + C$.

(d) $\int \sec^3(x)dx = \frac{1}{2}\sec(x)\tan(x) + \frac{1}{2}\log|\sec(x)+\tan(x)| + C$.

(e) $\int x^2\sin(x)dx = 2x\sin(x) - x\cos(x) + C$

(f) $\int \log(x)dx = x\log(x) - x + C$.

48. The object of this exercise is to develop some of the integration formulas found in tables. This will be done by integrating (with the help of a computer algebra system) a number of examples and looking for patterns. In this exercise, we want a formula for $\int \sin^n(x)dx$, where n is a positive integer.

(a) Find $\int \sin(x)dx$.

(b) Find $\int \sin^2(x)dx$

(c) Find $\int \sin^3(x)dx$

(d) Find $\int \sin^4(x)dx$

(e) From the results of (a) through (d) (and further examples if needed), conjecture a formula for $\int \sin^n(x)dx$.

(f) Verify your conjecture by differentiation.

6.4 Improper Integrals

In the development of the integral (Sections 5.1 to 5.3), we required that both the integrand and the path of integration be bounded. We are now prepared to generalize the integral concept by removing both of these restrictions. When either the integrand is unbounded or the path of integration is unbounded the integral is called *improper.*

The need for such extensions arises in practical settings as well as theoretical ones. We will begin by removing the restriction that the path of integration be bounded. Consider the following problem.

Example 6.4.1 We wish to find the area between the curve $y = 1/x^2$. the x-axis, and the line $x = 1$.

Unbounded Region with Finite Area

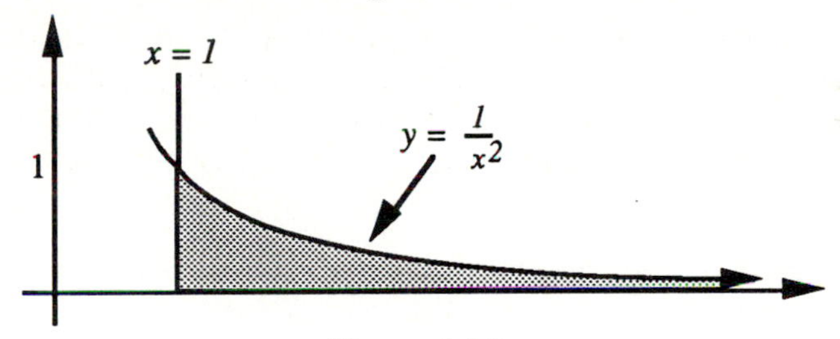

Figure 6.4.2

How can we determine an area for an unbounded region? Naturally, we use an approximation process. If we truncate the region at $x = b$, then we have an approximation to the region. We know that the area of the truncated region is

$$\int_1^b \frac{1}{x^2}dx = \left.\frac{-1}{x}\right]_1^b = 1 - \frac{1}{b}$$

We now let $b \to \infty$ so that the truncated regions approach the unbounded region. The limit of the areas of the truncated regions, $\lim_{b\to\infty}(1 - 1/b) = 1$, is a natural definition for the area in question. $\triangle$

Following the above example, the generalization of the integral concept to unbounded paths of integration is developed in the natural way. That is, by applying the Basic Approximation Process. Let $f : \mathbb{R} \to \mathbb{R}$ be a function integrable over any bounded interval. Now consider $\int_a^b f(x)dx$ to be an approximation to $\int_a^\infty f(x)dx$.

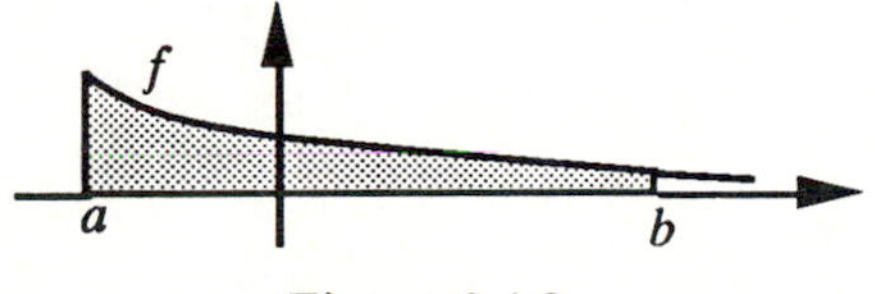

Figure 6.4.3

Since the accuracy of the approximation is improved by increasing the value of b, we choose an arbitrary sequence $\{b_n\}$ diverging to ∞ and consider the behavior of the approximating sequence $\{\int_a^{b_n} f(x)dx\}$. We state a definition based on the convergence or divergence of the approximating sequence.

Definition 6.4.4 *The* improper integral $\int_a^\infty f(x)dx$ converges *and has the value* $\lim_{n\to\infty} \int_a^{b_n} f(x)dx$*, provided the limit exists and is the same for all sequences* $\{b_n\}$ *diverging to* ∞*, otherwise the improper integral is said to* diverge.

Recall from Chapter 2 that the limit in the above definition is equivalent to $\lim_{b\to\infty} \int_a^b f(x)dx$. A similar definition exists for the improper integral $\int_{-\infty}^a f(x)dx$.

Example 6.4.5 Does $\int_1^\infty 1/\sqrt{x}dx$ converge? Evaluating the integral over the bounded interval $[1, b]$ yields

$$\int_1^b \frac{1}{\sqrt{x}}dx = 2\sqrt{x}]_1^b = 2(\sqrt{b} - 1)$$

Since $\lim_{b\to\infty} 2(\sqrt{b} - 1)$ diverges to infinity, the improper integral diverges. $\triangle$

A class of improper integrals that will be *very important* to us in Chapter 9 are those of the form $\int_1^\infty 1/x^p dx$. For what values of p will this integral converge? The above two examples showed that this integral converges for $p = 2$ and diverges for $p = \frac{1}{2}$. Can a more general statement be made? The answer is obtained by evaluating the (proper) integral, $\int_1^b 1/x^p dx$ and then asking for what values of p does the limit exist as $b \to \infty$. Evaluating, we have when $p = 1$,

$$\int_1^b 1/x^p dx = \log(b) \quad \text{(proved in Chapter 7)}$$

and when $p \neq 1$,

$$\int_1^b \frac{1}{x^p}dx = \int_1^b x^{-p}dx = \left.\frac{x^{-p+1}}{-p+1}\right]_1^b = \frac{1}{-p+1}\left(\frac{1}{b^{p-1}-1}\right)$$

Since

$$\lim_{b\to\infty} \frac{1}{-p+1}\left(\frac{1}{b^{p-1}} - 1\right) = \frac{1}{p-1} \quad \text{for } p > 1$$

and does not exist if $p < 1$, we have the result

$$\int_1^\infty \frac{1}{x^p}dx = \frac{1}{p-1} \quad \text{for } p > 1$$

and

$$\int_1^\infty \frac{1}{x^p}dx \text{ diverges, for } p \leq 1$$

The reader's attention is called to the fact that *comparison arguments* are characteristic of analysis. We have utilized the process of comparison several times without formally labeling it as such. The Squeeze Theorem is a primary example of a comparison argument. Another example is showing that a sequence $\{x_n\}$ converges to L by comparing $|x_n - L|$ to an arbitrarily small number. The source of comparison arguments for establishing convergence or divergence is the result (see Chapter 2) that a monotonic sequence converges if and only if it is bounded.

A "natural" extension of the comparison Theorem 5.5.15 yields the following theorem.

Theorem 6.4.6 (Comparison Test for Improper Integrals) *Assume that* $0 < f(x) \leq g(x)$ *for all* $x > a$.

(a) If $\int_a^\infty g(x)dx$ *converges, then* $\int_a^\infty f(x)dx$ *converges.*

(b) If $\int_a^\infty f(x)dx$ *diverges, then* $\int_a^\infty g(x)dx$ *diverges.*

Example 6.4.7 Does $\int_1^\infty x/\sqrt{x^6+1}dx$ converge or diverge? Since $x/\sqrt{x^6+1} < x/x^{6/2} = 1/x^2$ and $\int_1^\infty 1/x^2dx$ converges, $\int_1^\infty x/\sqrt{x^6+1}dx$ converges. △

Question 6.4.8 (a) Develop a definition for improper integrals of the form $\int_{-\infty}^\infty f(x)dx$. Does $\int_{-\infty}^\infty e^{-x/2}dx$, converge or diverge by your definition? [A graphical analysis indicates that $\int_{-\infty}^\infty e^{-x/2}dx$ diverges. (How?)].

(b) Make up and analyze two examples illustrating your definition. ◇

We now remove the restriction (from Chapter 5) that the integrand be bounded. (For example, we will now consider problems such as $\int_0^1 1/x^2dx$, where the integrand is unbounded on the interval of integration.) The process for developing the definition of convergence for an improper integral in which the integrand is unbounded over a finite interval is (naturally) the Basic Approximation Process. Since the details are so similar to those in the development of the definition of $\int_a^\infty f(x)dx$, we shall omit them and just state the definition.

Definition 6.4.9 *Let* $f : [c,d) \to \mathbb{R}$ *be unbounded as* x *approaches* d. *The* improper integral $\int_c^d f(x)dx$ converges *and has the value* $\lim_{b\to d}\int_c^b f(x)dx$ *provided this limit exists, otherwise,* $\int_c^d f(x)dx$ *is said to* diverge.

An analogous definition holds when the function is unbounded at the left endpoint c.

Example 6.4.10 Let $f : (0,1] \to \mathbb{R}$ be defined by $f(x) = 1/x^2$. Note that f is an unbounded function. The integral $\int_0^1 f(x)dx$ diverges since

$$\int_b^1 \frac{1}{x^2}dx = -\frac{1}{x}\Big]_b^1 = -1 + \frac{1}{b}$$

and $\lim_{b\to 0}(-1 + 1/b)$ does not exist. △

Example 6.4.11 Let $f : (0,1] \to \mathbb{R}$ be defined by $f(x) = 1/\sqrt{x}$. Again f is an unbounded function. The integral $\int_0^1 f(x)dx$ converges and has value 2 since

$$\int_b^1 \frac{1}{\sqrt{x}}dx = 2\sqrt{x}]_b^1 = 2(1 - \sqrt{b})$$

and $\lim_{b\to 0} 2(1 - \sqrt{b}) = 2$. △

Question 6.4.12 Show that $\int_0^1 1/x^p dx$ converges and has value $1/(1-p)$ for $p < 1$ and diverges for $p \geq 1$. ◇

Section 6.4 Exercises

In Exercises 1 through 10, evaluate or show divergence.

1. $\int_0^\infty \sin(x)dx$
2. $\int_{-\infty}^0 \frac{1}{(x-1)^3}$ dx
3. $\int_{-\infty}^\infty \frac{1}{1+x^2}dx$
4. $\int_{-1}^1 \frac{1}{x^2}dx$
5. $\int_0^2 \frac{1}{(x-1)^2}dx$
6. $\int_1^\infty \frac{x}{(x^2+1)^{1/2}}dx$
7. $\int_1^\infty \int_0^1 \frac{x}{y^2}dxdy$
8. $\int_1^\infty \int_1^\infty \frac{xy}{(x^2+y^2)^{1/2}}dxdy$
9. $\int_0^2 \int_1^2 (r^2-1)^{-3/2} r dr d\theta$
10. $\int_0^\pi \int_1^\infty r^{-2}\cos(\theta) dr d\theta$.

In Exercises 11 through 16, determine convergence or divergence without evaluating the integrals.

11. $\int_1^\infty (1+x^3)^{-1/2}dx$
12. $\int_{-\infty}^\infty (1+x^5)^{-1/6}dx$
13. $\int_1^\infty \frac{\sin^2(2x)}{x^2}dx$
14. $\int_2^\infty 4(x^2-1)^{-3/2}dx$
15. $\int_0^\infty \frac{1}{1+x^4}dx$
16. $\int_0^\infty \frac{1}{\tan(x)}dx$
17. Find the length of the curve given by the range of $C : [0,a] \to \mathbb{R}^2$, defined by $C(t) = (t, (a^{2/3} - t^{2/3})^{3/2})$.

18. For what values of p will $\int_0^\infty x^{-p}\,dx$ converge?

19. Give an example for each of the following statements or explain why no example can exist.

    (a) A divergent improper integral whose integrand contains a trigonometric function.

    (b) A convergent improper integral whose integrand contains a trigonometric function.

    (c) A convergent improper double integral.

    (d) A divergent improper triple integral.

20. Mark each of the following statements true (T) or false (F).

    (a) $\int_1^\infty \frac{\sin(x)}{x^2}dx$ is convergent.

    (b) $\int_{-\infty}^{+\infty} \frac{1}{x^2}dx$ is convergent.

    (c) $\int_{-\infty}^{+\infty} x\,dx = 0$.

    (d) If $\int_1^\infty f(x)dx$ converges, then $\lim_{x\to\infty} f(x) = 0$.

    (e) If $\lim_{x\to\infty} f(x) = 0$, then $\int_1^\infty f(x)dx$ converges.

21. Prove or disprove that an improper integral that can be expressed as a sum of improper integrals converges if and only if each of the summands converge.

22. [Project] (Question 6.4.8 should be answered before (or as part of) this project.) Develop a class presentation comparing an improper integral of the form $\int_{-\infty}^{\infty} f(x)dx$, defined in Question 6.4.8, and its Cauchy Principal Value defined below. Give several examples to illustrate parts (a) and (b) below.

    The *Cauchy Principal Value* of the improper integral $\int_{-\infty}^{\infty} f(x)dx$ is defined by

    $$(P)\int_{-\infty}^{\infty} f(x)dx = \lim_{b\to\infty}\int_{-b}^{b} f(x)dx,$$

    provided that the limit exists.

    (a) Show that whenever $\int_{-\infty}^{\infty} f(x)dx = L$, then $(P)\int_{-b}^{b} f(x)dx = L$. Hint: Use Question 6.4.8 and the definition of improper integral.

    (b) Show that the converse to part (a) is false. (I.e., give an example of a divergent improper integral of the form $\int_{-\infty}^{\infty} f(x)dx$ whose Cauchy Principal Value exists.) Hint: Consider an odd function.

Chapter 7

THE LOG AND EXPONENTIAL FUNCTIONS

7.1 Introduction

A major goal of a college calculus course is to aid the students in their development of analytical reasoning skills. The study and analysis of functions is the vehicle by which we hope to promote this growth. In this chapter, we shall illustrate how to use the calculus (definitions, theorems, procedures) to analyze the logarithmic transformation. In the process, we shall develop two of the most useful functions in science, the natural logarithm and the natural exponential. Recall from your previous study of common logarithms that the logarithm function transforms a multiplication problem into an addition problem. Thus it transforms a complicated computational problem into a comparatively simple one. Laplace (1749–1827) described the computational advantages of logarithms saying "by shortening the labors doubled the life of the astronomer" (Eves, *An Introduction To The History Of Mathematics*, Fourth Edition (1967), Holt, Rinehart, and Winston, p 250). The development of logarithms by John Napier (1550-1617), a Scottish mathematician, ranks as one of the great labor-saving devices in history of computing.

The logarithm transformation can be stated in terms of the functional relation

$$f(ab) = f(a) + f(b) \qquad \text{(Eq.7.1)}$$

We know that there is at least one function that satisfies (Eq. 7.1), namely, the logarithm to the base 10 (common logarithm). There may be others. The goal of the next section is to carefully analyze this functional relation and to learn all that we can about functions that satisfy it. Throughout the analysis, calculus will be an invaluable tool.

The reader is encouraged to verify all of the results in the first two sections.

7.2 The Logarithm: An Application of Calculus

We begin our analysis of Eq.7.1 with the assumption that some function $f : \mathbb{R} \to \mathbb{R}$ satisfies Eq.7.1. In an attempt to develop a feeling for f, we shall assign some numerical values to a and b and see what happens.

1. Let $a = 0$ and let b be any value at all in A, the domain of f. Substituting these values into Eq.7.1 yields

$$f(0) = f(0) + f(b) \quad \text{or} \quad f(b) = 0$$

for all b in A. Thus, if 0 is in the domain of a function that satisfies Eq.7.1, then that function must be identically 0. Since this is the trivial solution, we shall restrict A so that it does not contain 0.

2. Let $a = 1$ and let b be any value in A. Substituting into Eq.7.1 yields

$$f(b) = f(1) + f(b)$$

Therefore, $f(1) = 0$.

3. Let $a = -1$ and let $b = -1$. Again, substituting these values into Eq.7.1 yields

$$f(1) = f(-1) + f(-1)$$

In step 2 we learned that $f(1) = 0$. Therefore, $f(-1) = 0$.

4. Let $a = -1$ and let b be any value in A. Substituting into Eq.7.1 yields

$$f(-b) = f(-1) + f(b)$$

Since $f(-1) = 0$, $f(-b) = f(b)$ for all b in A. So every function that satisfies Eq.7.1 can be extended to an even function. (The graph of an even function is symmetrical about the y-axis.)

5. Let $a = 1/b$, where b is any value in A. (Recall that 0 is not in A.) Substituting into Eq.7.1 yields

$$f(1) = f\left(\frac{1}{b}\right) + f(b)$$

Since $f(1) = 0$, $f(1/b) = -f(b)$. Therefore,

$$f(\frac{1}{b}) = -f(b) \quad \text{and} \quad f(\frac{a}{b}) = f(a) - f(b)$$

6. Let $b = a$. Substituting into Eq.7.1 yields

$$f(a^2) = f(a) + f(a)$$

Therefore, $f(a^2) = 2f(a)$. This result can be extended to

$$f(a^n) = nf(a) \text{ and } f(a^{-n}) = -nf(a)$$

for any positive integer n.

All the above statements are true for every function that satisfies the functional equation Eq.7.1.

Question 7.2.1 Show that any function f which satisfies Eq.7.1, has the property that $f(x^{p/q}) = p/qf(x)$, where p and q are integers and $q \neq 0$. ◇

We shall now begin to use the calculus in our analysis. In particular, let us assume that f is a differentiable function that is not identically zero and see what happens! (We shall continue to restrict A to not contain zero.) The reader should realize that at this point we do not know if there are any differentiable functions satisfying Eq.7.1 that are not identically zero.

Let us apply the definition of derivative to the function f at a point $x \neq 0$:

$$f'(x) = \lim_{h \to 0} \frac{f(x+h) - f(x)}{h}$$

Since f satisfies Eq.7.1, by property 5 we know that

$$f(x+h) - f(x) = f(\frac{x+h}{x})$$

and we have

$$f'(x) = \lim_{h \to 0} \frac{f(1+\frac{h}{x})}{h}$$

The right-hand side suggests a difference quotient at 1. The numerator is in the correct form, $f(1+u) - f(1)$, since $f(1) = 0$. However the denominator should be the same as u, $u = h/x$. We apply a transformation to put the denominator in the form h/x. Multiplying and dividing by x we get $f'(x) = \lim_{h \to 0}(1/x)\frac{f(1+h/x)}{h/x}$. Apply the transformation $t = h/x$ and note that t approaches 0 as h approaches 0. So,

$$f'(x) = \lim_{t \to 0} \left(\frac{1}{x}\right) \frac{f(1+t)}{t} = \frac{1}{x} \lim_{t \to 0} \frac{f(1+t)}{t}$$

Now this last limit is just the definition of $f'(1)$! To verify this, set $x = 1$ in the definition of derivative and recall that $f(1) = 0$. Thus, if f is any differentiable function satisfying Eq.7.1, then

$$f'(x) = c\frac{1}{x}$$

where the constant c is $f'(1)$.

Can this constant ever be 0? If the constant is 0, then $f'(x) = 0$ for all x. From our work in differentiation, we know that a function whose derivative is identically 0 must be a constant function. However, if f is a constant function, then $f(x) = f(1)$. Since $f(1) = 0$, f would be the function identically equal to 0. However, f was assumed not to be identically 0, and therefore the constant $f'(1)$ is not 0. We state this result again (this time in words) because it is so important.

> If f is any differentiable function that satisfies Eq.7.1, then $f'(x)$ is equal to a constant times $1/x$.

The reader should remember, however, that we do not know if there exists a differentiable function that satisfies Eq.7.1 *other than the zero function.*

We now apply our knowledge of integration. Let us integrate the expression for $f'(x)$ to obtain an expression for $f(x)$. Since $f'(x)$ is continuous over any interval that does not contain zero, Part II of the Fundamental Theorem of Calculus says that for $x > 0$,

$$\int_1^x f'(t)dt = f(x) - f(1) = f(x)$$

since $f(1) = 0$. Therefore,

$$f(x) = c\int_1^x \frac{1}{t}dt \quad \text{for} \quad x > 0$$

Now, we make an observation that illustrates both the *beauty and strength* of the calculus: Since the integrand function, $g : (0, \infty) \to \mathbb{R}$ given by $g(x) = 1/x$, is a continuous function, the existence theorem for the Riemann Integral asserts that $\int_1^x 1/t dt$ exists! Consequently,

$$f(x) = c\int_1^x \frac{1}{t}dt \quad \text{for} \quad x > 0$$

defines a function that is a candidate for a non-trivial, differentiable solution of Eq.7.1. We need to check to see that

$$f(x) = c\int_1^x \frac{1}{t}dt \quad \text{for} \quad x > 0$$

is in fact a solution. That is, is the following true?

$$f(ab) = c\int_1^{ab} \frac{1}{t}dt = f(a) + f(b) = c\int_1^a \frac{1}{t}dt + c\int_1^b \frac{1}{t}dt$$

Recall from Chapter 5 that

$$\int_1^{ab} \frac{1}{t}dt = \int_1^a \frac{1}{t}dt + \int_a^{ab} \frac{1}{t}dt$$

The problem has now been reduced to showing that the second integral on the right-hand side can be transformed into $\int_1^b 1/t dt$. Since such a transformation must map the upper limit ab into b and the lower limit a into 1, we try setting $t = au$ and $dt = adu$ in the second integral on the right-hand side. Then

$$\int_a^{ab} \frac{1}{t} dt = \int_1^b \frac{1}{au} adu = \int_1^b \frac{1}{u} du$$

Thus

$$\int_1^{ab} \frac{1}{t} dt = \int_1^a \frac{1}{t} dt + \int_1^b \frac{1}{u} du$$

and hence

$$f(x) = c \int_1^x \frac{1}{t} dt \quad \text{for} \quad x > 0$$

is a solution of Eq.7.1. Consequently, the function we have so diligently been seeking actually exists.

Hence, we have shown the existence of a differentiable function that satisfies Eq.7.1 and have defined it in terms of an integral. Given any function that satisfies Eq.7.1, any constant times that function also satisfies the equation. (See the exercise at the end of this section.) Therefore, *any* constant times $\int_1^x 1/t dt$, $x > 0$ satisfies the logarithmic transformation Eq.7.1! Since the simplest such function that is not identically zero is the one in which the constant is chosen to be 1, we shall define this function to be the natural logarithm (naturally!) and denote it by log. (The natural logarithm is sometimes denoted by "ln" in elementary texts.)

Definition 7.2.2 (Natural Logarithm) *The natural logarithm,* $\log : (0, \infty) \to \mathbb{R}$ *is defined by*

$$\log(x) = \int_1^x \frac{1}{t} dt \quad \textit{for} \quad x > 0$$

We have followed the standard convention in defining the domain of the natural logarithm function to be the positive real numbers. We could also have chosen the domain to be the negative real numbers since we have shown that any solution f of Eq.7.1 can be extended to a function with the property that $f(x) = f(-x)$.

Note that we have shown that every logarithm function is equal to a constant times the natural logarithm. For example, the common (base 10) logarithm function is a constant times log and the logarithm to the base 2 is a constant times log (these constants are not necessarily the same). In the next section, we shall derive a expression for the constant associated with a given logarithm.

Geometrically, $\log(x)$ for $x > 1$ represents the area of the shaded portion in the following figure. Thus, if $x > 1$, $\log(x)$ is positive and if $0 < x < 1$, $\log x$ is negative.

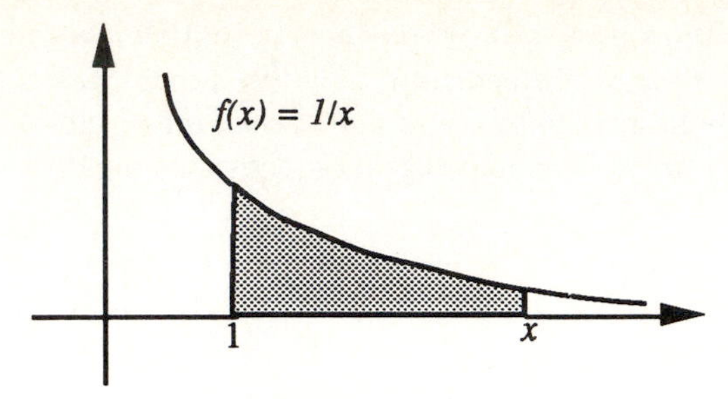

Figure 7.2.3

What properties does the natural logarithm function have? We have already shown that $\log(1) = 0$ and that for p and q integers, $q \neq 0$,

$$\log(x^{p/q}) = \frac{p}{q} \log(x)$$

(see Question 7.2.1).

By Part I of the Fundamental Theorem of Calculus,

$$\frac{d}{dx} \log(x) = \frac{d}{dx} \int_1^x \frac{1}{t} dt = \frac{1}{x} > 0$$

This result has several implications which we list.

1. Since log is a differentiable function, log is a continuous function.
2. Since the derivative of the log function is positive, the log function is strictly increasing, one-to-one, and has an inverse function.
3. Since the log function is strictly increasing and $\log(1) = 0$, the graph of the log function has exactly one x intercept. This intercept is at $x = 1$.
4. Since the log function is strictly increasing, $\log(x) > 0$ for $x > 1$.
5. Since $\log(2) > 0$ and $\log(x^n) = n \log(x)$, $\lim_{x \to \infty} \log(x) = \infty$.
6. Since $\log(x^{-n}) = -n \log(x)$, $\lim_{x \to 0} \log(x) = -\infty$.
7. As a consequence of implications 5 and 6, the range of the log function is all of $\mathbb{R}$.
8. The y-axis is a vertical asymptote for the graph of log.

What does the graph of the log function look like? Since

$$\frac{d^2}{dx^2} \log(x) = -\frac{1}{x^2} < 0$$

the graph of $\log(x)$ is *concave down* for all $x > 0$. Thus

1. The graph of log is unbroken (log is a continuous function).

2. The graph of log is concave down.

3. The graph of log has an x intercept at (1,0).

4. The graph of log has tangents with an arbitrarily steep slope as x approaches 0.

5. The graph of log has tangents with an arbitrarily small positive slope as x approaches ∞.

To help get a "fix" on the position of the graph of the log function, let us compare the graph of the log function to the graph of a known function. For example, let us compare $\log(x)$ to the identity function $f(x) = x$ by asking if $\log(x) < x \quad \text{for} \quad x > 0$. If the answer is yes, then we know that the graph of $\log(x)$ lies below the graph of $f(x) = x$. Let us rephrase this question by comparing the difference of the two functions to zero. That is, by asking:

$$\text{Is } g(x) = x - \log(x) > 0 \text{ for } x > 0?$$

One way to answer this question is to determine the minimum value of g, for if the minimum value is positive, then $g(x) > 0$ for all $x > 0$. Let us find the minimum value (using the calculus). We begin by finding the critical values of g. That is, we differentiate g and set the derivative equal to 0. Since

$$g'(x) = 1 - \frac{1}{x} = 0$$

yields $x = 1$, $x = 1$ is the only critical point. Applying the second derivative test, we have

$$g''(x) = \frac{1}{x^2} > 0$$

and thus g has its minimum value at $x = 1$. Since $g(1) = 1$, $g(x) > 0$ and thus $x > \log(x)$ for $x > 0$. Hence, the graph of $\log(x)$ lies below the graph of $f(x) = x$. We are now in a position to plot a few points and then sketch the graph of the log function. [Note that $\log(2) \approx 0.7$ and thus $\log(2^5) \approx 3.5$ and $\log(2^{10}) \approx 7$.]

Graph of the Natural Logarithm Function

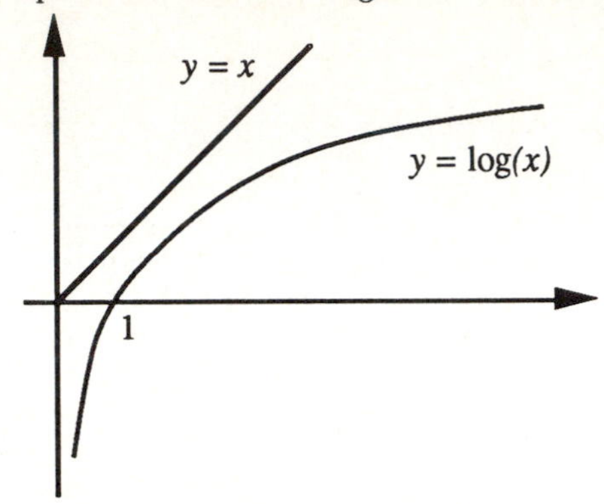

Figure 7.2.4

Question 7.2.5 Using the Trapezoidal Rule and a regular partition with four subintervals, find a numerical approximation of log(3). ◇

When the natural logarithm function is composed with another differentiable function, say u, the derivative of the resulting expression is (using the Chain Rule)

$$\frac{d}{dx}\log(u) = \left(\frac{1}{u}\right)\frac{du}{dx}$$

Example 7.2.6 We will differentiate $\log(x^2 + 3x)$. Using the transformation $u = x^2 + 3x$, yields

$$\frac{d}{dx}\log(x^2 + 3x) = \frac{d}{dx}\log(u) = \left(\frac{1}{u}\right)\frac{du}{dx} = \left(\frac{1}{u}\right)(2x + 3) = \frac{1}{x^2 + 3x}(2x + 3) \quad \triangle$$

Since the logarithm transform a product into a sum, the natural logarithm provides a useful transformation for simplifying the differentiation of a product of differentiable functions. The technique is called *logarithmic differentiation.* Let y be the product of differentiable functions, say $y = u(x)v(x)w(x)$. We want to compute dy/dx. Transform the product into a sum by applying the natural logarithm to y. That is,

$$\log(y) = \log(uvw) = \log(u) + \log(v) + \log(w)$$

In order to simplify notation, we are writing $u(x)$ as u, $v(x)$ as v, and $w(x)$ as w. Now we differentiate using the Chain Rule. Thus,

$$\left(\frac{1}{y}\right)\frac{dy}{dx} = \left(\frac{1}{u}\right)\frac{du}{dx} + \left(\frac{1}{v}\right)\frac{dv}{dx} + \left(\frac{1}{w}\right)\frac{dw}{dx}$$

Solving for dy/dx yields

$$\frac{dy}{dx} = y\left[\left(\frac{1}{u}\right)\frac{du}{dx} + \left(\frac{1}{v}\right)\frac{dv}{dx} + \left(\frac{1}{w}\right)\frac{dw}{dx}\right]$$

Finally, substituting for y on the right-hand side yields

$$\frac{dy}{dx} = uvw\left[\left(\frac{1}{u}\right)\frac{du}{dx} + \left(\frac{1}{v}\right)\frac{dv}{dx} + \left(\frac{1}{w}\right)\frac{dw}{dx}\right]$$

Example 7.2.7 We will find dy/dx for $y = (x^2+3)\cos^3(x)\log(x)$. Applying the logarithmic transformation, we have

$$\begin{aligned}\log(y) &= \log[(x^2+3)\cos^3(x)\log(x)] \\ &= \log(x^2+3) + \log[\cos^3(x)] + \log[\log(x)]\end{aligned}$$

Differentiating using the Chain Rule yields

$$\left(\frac{1}{y}\right)\frac{dy}{dx} = \frac{1}{x^2+3}(2x) + \frac{1}{\cos^3(x)}[-3\cos^2(x)\sin(x)] + \frac{1}{\log(x)}\left(\frac{1}{x}\right)$$

Hence,

$$\frac{dy}{dx} = y\left[\frac{2x}{x^2+3} - \frac{3\cos^2(x)\sin(x)}{\cos^3(x)} + \frac{1}{x\log(x)}\right]$$

or

$$\frac{dy}{dx} = (x^2+3)\cos^3(x)\log(x)\left[\frac{2x}{x^2+3} - 3\tan(x) + \frac{1}{x\log(x)}\right] \qquad \triangle$$

Question 7.2.8 Verify the following relationships for the natural logarithm.

(a) $\log(f(x)g(x)) = \log(f(x)) + \log(g(x))$.

(b) $\log(f(x)/g(x)) = \log(f(x)) - \log(g(x))$.

(c) $\log(f(x)^{p/q}) = p/q\log(f(x))$, for p, q integers. $\diamond$

The reader is encouraged to review the development of the definition of the natural logarithm and the derivation of its derivative. In earlier chapters, we had informally defined the logarithm as the inverse of the (undefined) exponential function and stated its derivative (without proof).

Section 7.2 Exercises

In Exercises 1 through 16, find dy/dx.

1. $y = \log(3x^2 + 4)$
2. $y = \log[x \cos^2(x)]$
3. $y = \log(x^3 + 3x)^3$
4. $y = \log[\sec(x) + \tan^{-1}(x)]$
5. $y = \log[x^2 \log(\sin(x) + 4)]$
6. $y = x^2 \log(2 + x^2)$
7. $y = 3x^2 - 4 + \log[(x^2 + 6)/(x + 10)]$
8. $y = \log[x/(x + 7) + 2x]^3$
9. $y = x^2 \cos(x)/(x + 2)$
10. $y = x^2 \sqrt{\tan^2(x) + 7}$
11. $y = \log(xy)$
12. $\log(y) - \tan(x + y) = 7$
13. $y = \sin(x) \log(y^2)$
14. $x/y = \log(1/y)$
15. $y = \int_2^{\log(x)} (t^3 + 3)dt$
16. $y = \int_{\log(x)}^3 \log(t)dt$
17. Show that if f satisfies the functional relation (Eq.7.1), then so does any constant multiple of f.
18. Sketch the graphs of the functions $f(x) = \log(x^n)$ for $n = -2, 1, 2$ on the same set of axes.
19. Prove or disprove that $\log(x^{p/q}) = (\log(x))^{p/q}$.
20. Prove or disprove the statement:

    Given any positive number m, there exists a number N such that the graph of $f(x) = \log(x)$ lies under the graph of $g(x) = mx$ for all $x > N$.

21. For each of the following integrals, determine the values of x for which the integral is improper and then determine whether the integral converges or diverges. Explain your thinking in complete sentences.
    (a) $\int_0^\infty \log(x)/(x - 1)$ (b) $\int_0^1 1/\sqrt{x} \log(x)dx$

    (c) $\int_0^1 \log(x)/\sqrt{x}dx$ (d) $\int_0^1 [x \log(1/x)]^{1/3}dx$
22. Project. Analyze the functional relation $f(a)f(b) = f(a + b)$.
23. Project. Analyze the functional relation $f(a + b) = f(a) + f(b)$.

7.3 The Natural Exponential Function

[Some of the results of this section have been stated (without proof) and used in previous sections. The reader is encouraged to pay particular attention to the definition of the exponential function and the derivation of its derivative.]

Recall that the log function is one-to-one, and therefore it has an inverse function. In this section we shall analyze the inverse function of the natural logarithm.

The inverse of a logarithmic function is an exponential function, denoted by exp. Recall, that for the base 10 (common) logarithm

$$\log_{10}(x) = y \quad \text{if and only if} \quad 10^y = x$$

Since x is the inverse of $\log_{10}(x)$ and $x = 10^y = \exp_{10}(y)$, $\exp_{10}(y)$ is the inverse of $\log_{10}(x)$. For the logarithm to the base 2,

$$\log_2(x) = y \quad \text{if and only if} \quad 2^y = x$$

Thus, $\exp_2(y) = 2^y$ is the inverse of $\log_2(x)$.

To express the inverse of the natural logarithm function as an exponential, we need to first determine the base of the natural logarithm. The base of a logarithm is the unique number in the domain that is mapped into the number 1 by the logarithm function. To see that such a number exists, we recall from the previous section that the range of log is $\mathbb{R}$. The fact that log is a one-to-one function guarantees that this number is unique. This number was labeled e" by Leonhard Euler (1707–1783) and is given to 15 decimal places as

$$e = 2.718281828459045$$

Thus, $\log(e) = 1$.

To show that this result is consistent with the limit expression obtained for e in Section 3.9 [$e = \lim_{x\to\infty}(1 + 1/x)^x$], we will compute $\log(e)$ using the limit expression.

$$\begin{aligned}
\log(e) &= \log[\lim_{x\to\infty}(1 + 1/x)^x] \\
&= \lim_{x\to\infty} \log[(1 + 1/x)^x] \quad \text{(continuity of the log function permits the} \\
&\qquad\qquad \text{interchange of the limit and log operations; see Section 2.5)} \\
&= \lim_{x\to\infty} x\log(x + 1/x) \quad (\infty \cdot 0 \text{ indeterminate form}) \\
&= \lim_{x\to\infty} \frac{log(1 + 1/x)}{1/x} \quad \text{(the } \tfrac{0}{0} \text{ indeterminate form)} \\
&= \lim_{x\to\infty} \frac{-1/x^2}{(1 + 1/x)(-1/x^2)} \quad \text{(L'Hospital's Rule)} \\
&= 1
\end{aligned}$$

Definition 7.3.1 (Exponential Function) *The* natural exponential function *is the function* $\exp : \mathbb{R} \to (0, \infty)$ *defined by* $\exp = \log^{-1}$.

Thus $\log(x) = y$ if and only if $\exp(y) = x$

The expression $\exp(x)$ is often denoted by e^x and referred to as *the exponential function.* The reader should note that since logarithmic and exponential functions are inverses of each other,

$$\log[\exp(x)] = x \quad \text{or} \quad \log(e^x) = x$$

for all x and

$$\exp[\log(x)] = x \quad \text{or} \quad e^{\log(x)} = x$$

for all $x > 0$. (Why only for $x > 0$?)

The graph of $y = e^x$ can be obtained by reflecting the graph of $y = \log(x)$ in the line $y = x$.

The Exponential Function

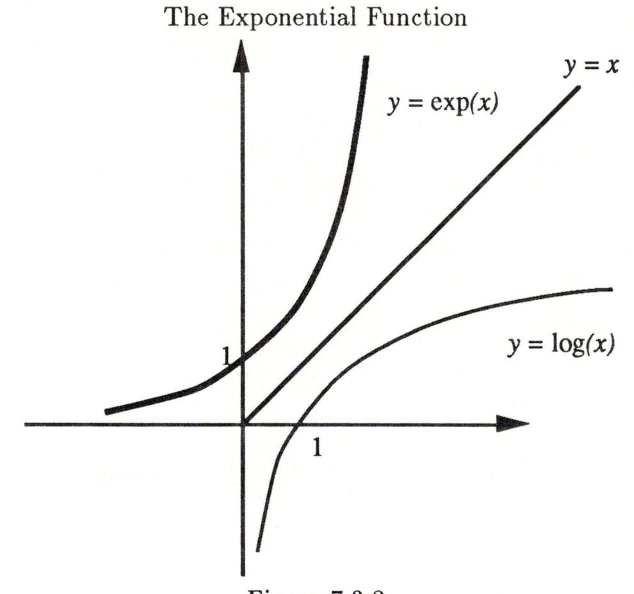

Figure 7.3.2

The standard properties of exponents can be shown to apply to e^x. For example, to show that $e^x e^y = e^{x+y}$, write

$$e^x e^y = \exp[\log(e^x e^y)] = \exp[\log(e^x) + \log(e^y)] = \exp(x + y) = e^{x+y}$$

Question 7.3.3 Show that $\exp(x)/\exp(y) = \exp(x - y)$. ◇

Let us now consider differentiating the exponential function. Since exp is an inverse function, we can use the Inverse Function Theorem,

$$(f^{-1})'(f(x)) = \frac{1}{f'(x)}$$

With $y = f(x) = \log(x)$ and $f^{-1}(y) = \exp(y) = x$ we have

$$\exp'(y) = (f^{-1})'(f(x)) = \frac{1}{f'(x)} = \frac{1}{\log'(x)} = x = \exp(y)$$

Thus, $\exp'(x) = \exp(x)$; *the exponential function is its own derivative!* The importance of this last result is underscored by the fact that only the functions that are constant multiples of the natural exponential function have the property of being equal to their own derivative. Since exp is its own derivative, it is its own antiderivative and thus

$$\int e^x \, dx = e^x + C$$

When the exponential function is composed with another differentiable function, say u, and the composition is differentiated using the Chain Rule, we have

$$\frac{d}{dx} e^u = e^u \frac{du}{dx}$$

Example 7.3.4 We will differentiate e^{2x} with respect to x. We will express e^{2x} as a composite function by applying the transformation, $u = 2x$. Now, differentiating by the Chain Rule yields

$$\frac{d}{dx} e^{2x} = \frac{d}{dx} e^u = e^u \frac{du}{dx} = e^u(2) = 2e^{2x}$$

△

Example 7.3.5 We will differentiate $\exp[\cos(x)]$ with respect to x. We express $\exp[\cos(x)]$ as a composite function by applying the transformation $u = \cos(x)$. Now, differentiating by the Chain Rule yields

$$\begin{aligned} \frac{d}{dx} \exp[\cos(x)] &= \frac{d}{dx} \exp(u) = \exp(u) \frac{du}{dx} = \exp(u)[-\sin(x)] \\ &= -\exp[\cos(x)] \sin(x) \end{aligned}$$

△

The natural exponential function exp is used to define the expression a^b when a and b are both real numbers, $a > 0$. Recall that if n and m are positive integers, and if x and y are real numbers, then we use the following conventions:

x^n is the product of n x 's
x^{-n} is the product of n $1/x$'s
$x^{n/m} = y$ if and only if $x^n = y^m$

How is the general exponential expression defined when the exponent is an irrational number? For example, how is 2^π defined? Since the domain of the natural exponential function is $\mathbb{R}$, e^x is defined for every real number. However, e is the *only* real number that we know (at this point) that has the property of being defined when raised to a power. (I.e., we know that e^x is defined, but we don't know that a^x is defined.) We shall use the inverse relationship between the exponential and log functions to define the general exponential form a^b. Since the exponential and log functions are inverses of each other,

$$a^b = \exp[\log(a^b)] = \exp[b\log(a)] = e^{b\log(a)}$$

Note also that since $a > 0$, $\log(a)$ is defined and since the domain of exp is $\mathbb{R}$, $\exp[b\log(a)]$ is defined. Thus we use the equality

$$a^b = \exp[b\log(a)]$$

to define a^b.

Definition 7.3.6 (Exponentials) *If a and b are real numbers, $a > 0$, then* $a^b = \exp[\log(a^b)] = \exp[b\log(a)] = e^{b\log(a)}$.

We call the transformation expressing a^b as $\exp[b\log(a)]$ *the exponential transformation.* The next few examples will illustrate the "power" of this transformation.

Example 7.3.7 We will differentiate $f(x) = 2^x$. The procedure is to apply the exponential transformation to 2^x and then differentiate using the Chain Rule. That is,

$$f'(x) = \frac{d}{dx}e^{x\log(2)} = e^{x\log(2)}\log(2) = 2^x\log(2)$$

Thus, $f'(x) = 2^x\log(2)$. △

Example 7.3.8 We will differentiate the expression x^x with respect to x. The procedure is to apply the exponential transformation and then differentiate using the Chain Rule. That is,

$$\frac{d}{dx}x^x = \frac{d}{dx}e^{x\log(x)} = e^{x\log(x)}\frac{d}{dx}x\log(x) = x^x[1+\log(x)]$$ △

Note how the application of the exponential transformation simplified the differentiation in the above examples.

The fact that e^x is equal to its derivative (i.e., its rate of change) is the reason that exponential functions are used to model change in situations where the change is proportional to the amount present. The growth of an interest account is a classic example of such a situation.

Example 7.3.9 If an amount of money P (principal) is invested for t years at an interest rate of r compounded once a year (simple interest), the amount of money A after t years is given by the formula

$$A = P(1+r)^t$$

If interest is compounded x times a year, the expression for A is given by

$$A = P\left(1 + \frac{r}{x}\right)^{tx}$$

We will find the expression for A when interest is compounded continuously. That is, evaluate $\lim_{x\to\infty} P(1 + r/x)^{tx}$.

Since the principal P is a constant,

$$\lim_{x\to\infty} P\left(1 + \frac{r}{x}\right)^{tx} = P \lim_{x\to\infty} \left(1 + \frac{r}{x}\right)^{tx}$$

Apply the exponential transformation and get

$$\lim_{x\to\infty} P\left(1 + \frac{r}{x}\right)^{tx} = P \lim_{x\to\infty} \exp\left[tx \log\left(1 + \frac{r}{x}\right)\right]$$

Since exp is a continuous function, evaluating the limit and evaluating the function may be interchanged so that

$$\lim_{x\to\infty} P\left(1 + \frac{r}{x}\right)^{tx} = P \exp\left[\lim_{x\to\infty} tx \log\left(1 + \frac{r}{x}\right)\right]$$

Since the limit on the right side yields the $0 \cdot \infty$ indeterminate form, the limit cannot be evaluated as a product of limits. Thus we rewrite (transform) the product as a quotient to obtain the indeterminate form of $0/0$ to which we may apply L'Hospital's Rule. Thus,

$$\lim_{x\to\infty} tx \log\left(1 + \frac{r}{x}\right) = \lim_{x\to\infty} \frac{\log(1 + r/x)}{1/tx}$$

Applying L'Hospital's Rule,

$$\lim_{x\to\infty} tx \log(1 + \frac{r}{x}) = \lim_{x\to\infty} \frac{\frac{1}{1+r/x}\left(\frac{-r}{x^2}\right)}{-1/tx^2} = \lim_{x\to\infty} \left(\frac{1}{1 + r/x}\right)(rt) = rt$$

Now, substituting rt for the limit in the exponent, yields $A = P\exp(rt) = Pe^{rt}$.

An interesting consequence of the above calculations is found when one considers the limit expression with $P = r = t = 1$. That is,

$$\lim_{x\to\infty} \left(1 + \frac{1}{x}\right)^{x} = e$$

It would be a very useful exercise for the reader to verify the above result directly (using the exponential transform and L'Hospital's Rule) rather than just substituting 1 for P, r, and t.

The formula derived above, $A = Pe^{rt}$, is the basic one used in solving problems concerning continuously compounded interest, where

P equals amount invested, $\quad r$ equals interest rate

t equals number of years, A equals amount of money after t years.

Three common problems are

1. How much interest is earned when P dollars is invested at an interest rate of r compounded continuously for t years? (Answer: $A - P$.)

2. How long does it take for P dollars to double when invested at an interest rate of r compounded continuously? (Answer: Solve for t when $A = 2P$.)

3. What is the present value of B dollars t years from now when invested at an interest rate of r compounded continuously? (Answer: Solve for P when $A = B$.) △

We conclude this section by using the exponential transformation to develop the relation between logarithms to different bases. Recall, in the previous section we showed that every solution of the logarithmic transformation Eq.7.1 was equal to a constant times the natural logarithm. We shall now derive the expression for this constant. Consider the logarithm to the base a,

$$\log_a(x) = y \quad \text{if and only if} \quad x = a^y$$

Applying the exponential transformation to $x = a^y$ yields

$$x = \exp[y \log(a)]$$

Thus, $\log(x) = \log(\exp[y \log(a)]) = y \log(a)$. Now solving for y yields,

$$y = \frac{1}{\log(a)} \log(x)$$

So

$$\log_a(x) = \frac{1}{\log(a)} \log(x)$$

Thus, the constant for the logarithm to the base a is $\frac{1}{\log(a)}$.

Section 7.3 Exercises

In Exercises 1 through 10, find dy/dx

1. $y = xe^{x^2}$
2. $y = \log(x^x)$
3. $y = e^x \tan(e^x)$
4. $y = xe^x / \log[\sin(x)]$
5. $y = 4^x + x^2$
6. $y = x^{\log(x)} \log(x)^x$
7. $xy = 2^{\sec(x)}$
8. $x = y^y$
9. $y = \log[e^{2x}/(x+2)]$
10. $y = \exp(\cos(x)) + \log(x^2)$

In Exercises 11 through 16, evaluate the limits.

11. $\lim_{x\to\infty}(1+x)^{1/x}$
12. $\lim_{x\to\infty} x^{1/\log(x)}$
13. $\lim_{x\to 1}[x/\log(x) - 1/x\log(x)]$
14. $\lim_{x\to 0}(1+2x)^{3/x}$
15. $\lim_{x\to 0}[1+\tan(x)]^{1/x}$
16. $\lim_{x\to 0}(1+x)^{\cot(x)}$
17. State and verify properties for the natural exponential function that correspond to the 10 properties that are stated for the natural logarithm function in Section 7.2.

In Exercises 18 through 20, find the indicated derivatives by applying the definition of derivative (see Chapter 3).

18. Find $h'(2)$ for $h(x) = e^{-x}$ given that $d/dx \exp(0) = 1$.
19. Find $g'(1)$ for $g(x) = e^{2x}$.
20. Find $f'(2)$ for $f(x) = \log(2x+1)$ given that $d/dx \log(1) = 1$.
21. Approximate $\int_0^2 e^x dx$ by using both the Trapezoidal Rule and Simpson's Rule with four equal subintervals.
22. Sketch the graphs of $f(x) = e^{nx}$ for $n = -2, 1, 2$ on the same set of axes.

In Exercises 23 through 26, assume that interest is compounded continuously.

23. How long does it take to double \$500 when it is invested at 8% interest? [$\log(2) \approx 0.6931$]
24. What is the present value of a \$500 bond that matures in 10 years if the interest rate is 6%?
25. Assume that you are paid only once a year (on December 31). If your present salary is \$25,000 and you receive a 10% raise for next year, will your purchasing power increase assuming a 5% inflation rate? Why?
26. If Mrs. Jones is paid her yearly salary of \$40,000 on December 31, how much of a salary increase would she have to receive in order to maintain her purchasing power? Assume an inflation rate of 4%.
27. Prove or disprove that $e^{ar} = (e^a)^r$.
28. Prove or disprove that the equation $e^x = \log(x)$ has no solution. (Hint: Consider the graphs of the two functions.)
29. Prove or disprove that the following improper integrals converge.

    (a) $\int_0^\infty \sin(x)\log(x)dx$ **(b)** $\int_2^\infty 1/[x\log(x)^k]dx$ for $k > 1$

    (c) $\int_{-\infty}^0 e^{-x}\log(|x|)dx$ **(d)** $\int_0^\infty xe^{-x}\log(x)dx$

30. Sketch the graph of $f(x) = e^x / \log(x)$, $x > 0$. Use a calculator or computer to generate a set of points on the curve. What can you say about f being increasing or decreasing, convexity, extreme values, asymptotes, and intercepts?

7.4 Integration of Logs and Exponentials

The exponential function and the (natural) log function allow us to greatly expand the family of functions that can be integrated using the methods of Sections 6.1 through 6.4. We can add the form $1/u$ to our list of standard forms, which now consists of linear combinations of u^r, $\sin(u)$, $\cos(u)$, $\sec^2(u)$, $\csc^2(u)$, $1/(a^2 \pm u^2)$, $1/\sqrt{a^2 \pm u^2}$, and $1/u$.

Let us consider some examples illustrating the integration of functions involving the natural exponential.

Example 7.4.1 We will evaluate $\int_{-1}^{4} 2^x\,dx$. The integrand is not in standard form and thus we need a transformation. The form of the integrand being a general exponential suggests using the exponential transformation in an attempt to transform the integrand into the standard form e^u. So, applying the exponential transformation, we have

$$\int_{-1}^{4} 2^x\,dx = \int_{-1}^{4} \exp[x\log(2)]dx$$

Since we want the exponent of e to be u, we apply the transformation $u = x\log(2)$, $du = \log(2)dx$ and the (transformed) limits are $u = -\log(2)$ and $u = 4\log(2)$. We now have

$$\begin{aligned}
\int_{-1}^{4} 2^x\,dx &= \int_{-\log(2)}^{4\log(2)} \exp(u)\frac{1}{\log(2)}du \\
&= \frac{1}{\log(2)}\exp(u)\Big]_{-\log(2)}^{4\log(2)} \\
&= \frac{1}{\log(2)}[\exp(4\log(2)) - \exp(-\log(2))] \\
&= \frac{1}{\log(2)}[\exp(\log(2^4)) - \exp(\log(2^{-1}))] \\
&= \frac{1}{\log(2)}\left(2^4 - \frac{1}{2}\right) \\
&= \frac{31}{2\log(2)}
\end{aligned}$$

△

An interesting situation occurs in the integration of a function that is a product of the exponential and a function, such as the sine function, which is periodic with respect to integration or differentiation.

Example 7.4.2 We will evaluate $\int e^x \sin(x)dx$. The integrand is not in standard form, but it is a product in which one factor is the exponential function. Thus, we consider a parts transformation. It is not obvious that there is any advantage in selecting the $\sin(x)$ factor (the nonexponential factor) to be to be the "u" term as compared to the dv term. So, let us see what happens when

$$\begin{array}{ll} u = \sin(x) & dv = e^x dx \\ du = \cos(x)dx & v = e^x \end{array}$$

Applying the parts transformation yields

$$\int e^x \sin(x)dx = e^x \sin(x) - \int e^x \cos(x)dx$$

The integral on the right side appears to be of the same level of difficulty as the original integral. This suggests that we apply the parts transformation again. So

$$\begin{array}{ll} u = \cos(x) & dv = e^x dx \\ du = -\sin(x)dx & v = e^x \end{array}$$

Thus,

$$\int e^x \sin(x)dx = e^x \sin(x) - \left[e^x \cos(x) + \int e^x \sin(x)dx\right]$$

The integral on the right side is the same as the original integral. Have we caught ourselves in a revolving door? Not really, for the *integral term* on the right side has the opposite sign of the *integral term* on the left side and hence transposing the integral term on the right side yields

$$2\int e^x \sin(x)dx = e^x \sin(x) - e^x \cos(x)$$

and thus

$$\int e^x \sin(x)dx = \frac{1}{2}\left[e^x \sin(x) - e^x \cos(x)\right] + C \qquad \triangle$$

In the application of the parts transformation, one should be alert to the possibility that the $\int vdu$ term may be the same as the original. If this happens and their numerical signs are different, then the two terms can be combined as was done in the previous example. If the numerical signs are the same, then one has either made a mistake in the mechanics of differentiating the u term or integrating the dv term or has chosen an incorrect partition of the integrand.

We now consider some of the ramifications of being able to integrate the $1/u$ form.

Since the natural logarithm function is only defined over the positive real numbers, the indefinite integral of $1/u$ is defined by

$$\int \frac{1}{u} du = \log(|u|) + C$$

(Here u cannot be zero.)

Note the absolute values are necessary to guarantee that the argument of the natural logarithm function is positive. Let us verify that $\log(|u|)$ is an antiderivative of $1/u$. That is, differentiate $\log(|u|)$ with respect to u, $u \neq 0$.

$$\frac{d}{du}\log(|u|) = \frac{1}{|u|}\frac{d}{du}|u|$$

Now, if $u > 0$ we get

$$\frac{1}{|u|}\frac{d}{du}|u| = \frac{1}{u}\frac{d}{du}(u) = \frac{1}{u}$$

Alternatively, if $u < 0$, we get

$$\frac{1}{|u|}\frac{d}{du}|u| = \frac{1}{-u}\frac{d}{du}(-u) = \frac{1}{u}$$

The existence of the integral $\int 1/udu$ removes the restriction that had been placed on integrating the power function. Recall, that in Chapter 6, x^r was listed as an integrand in standard form provided that $r \neq -1$. The expression for the antiderivative of the power function may now be written as

$$\int u^r du = \begin{cases} u^{r+1}/(r+1) + C & \text{for } r \neq -1 \\ \log(|u|) & \text{for } r = -1 \end{cases}$$

The presence of $\log(u)$ as a factor in an integrand often suggests that an integration by parts be considered. (Recall that a similar observation was made in Section 6.2 concerning the exponential function.) Since we know how to differentiate $\log(u)$ and integrating $\log(u)$ requires an application by parts (try it), the choice is usually made to let the $\log(u)$ factor be the u term in the parts partition of the integrand. This situation is illustrated in the next example.

Example 7.4.3 We will evaluate $\int x\log(x)dx$. The integrand is not in standard form and therefore we look for a transformation. Since $\log(x)$ is a factor of the integrand, we consider a parts transformation with $\log(x)$ being chosen for the u term. That is,

$$\begin{array}{ll} u = \log(x) & dv = xdx \\ du = 1/xdx & v = x^2/2 \end{array}$$

So

$$\int x\log(x)dx = \frac{x^2}{2}\log(x) - \int \frac{x}{2}dx = \frac{x^2}{2}\log(x) - \frac{x^2}{4} + C \qquad \triangle$$

The natural logarithm allows us to extend the scope of integrals that can be evaluated using trigonometric transformations. For example, the integrals $\int \sec(u)du$ and $\int \sec^3(u)du$ often occur when a trigonometric transformation is applied to expressions of the form $\sqrt{a^2+x^2}$ or $\sqrt{a^2-x^2}$. To see how this happens, consider $\int \sqrt{a^2+x^2}dx$.

Applying the trigonometric transformation $x = a\tan(u)$, $dx = a\sec^2(u)du$, we obtain $\int a^2\sec^3(u)du$.

Another example is the integral $\int \sqrt{x^2-a^2}dx$. Applying the trigonometric transformation $x = a\sec(u)$, $dx = a\sec(u)\tan(u)du$, we obtain

$$\begin{aligned}\int a^2\sec(u)\tan^2(u)du &= \int a^2\sec(u)[\sec^2(u)-1]du \\ &= \int a^2\sec^3(u)du - \int a^2\sec(u)du\end{aligned}$$

Integrating the $\sec(x)$ function involves a special transformation. The idea is to transform $\sec(x)$ into a du/u form by multiplying and dividing $\sec(x)$ by $(\sec(x)+\tan(x))$. Here are the details.

$$\int \sec(x)dx = \int \sec(x)\frac{\sec(x)+\tan(x)}{\sec(x)+\tan(x)}dx = \int \frac{\sec^2(x)+\sec(x)\tan(x)}{\sec(x)+\tan(x)}dx$$

Now apply the transformation $u = \sec(x)+\tan(x)$. Note that $dx = [\sec(u)\tan(u)+\sec^2(u)]du$. Thus, after substitution, the resulting integral is

$$\int \frac{1}{u}du = \log(|u|)+C = \log(|\sec(x)+\tan(x)|)+C$$

Thus,

$$\int \sec(x)dx = \log(|\sec(x)+\tan(x)|)+C$$

The parts transform applied to $\int \sec^3(x)dx$ yields an expression involving $\int \sec(x)dx$. Let us illustrate the procedure. It is clear that in a parts transform if the u term is chosen to be $\sec^3(x)$ and thus the dv term is dx, then the resulting integral of vdu is considerably more complicated than the original integral. Thus, that particular choice of partitioning the integrand is discarded. Another possibility is to break up $\sec^3(x)$ and consider it as a product of $\sec(x)\sec^2(x)$. It appears that the simplest choice (in terms of calculations) is to let $u = \sec(x)$ and $dv = \sec^2(x)dx$. So let

$$\begin{aligned} u &= \sec(x) & dv &= \sec^2(x)dx \\ du &= \sec(x)\tan(x)dx & v &= \tan(x)\end{aligned}$$

Thus,

$$\begin{aligned}\int \sec^3(x)dx &= \sec(x)\tan(x) - \int \sec(x)\tan^2(x)dx \\ &= \sec(x)\tan(x) - \int [\sec^3(x) - \sec(x)]dx \\ &= \sec(x)\tan(x) - \int \sec^3(x)dx + \int \sec(x)dx\end{aligned}$$

Since the first integral on the right side is the same as the original integral, it can be transposed to the left side. Thus,

$$2\int \sec^3(x)dx = \sec(x)\tan(x) + \int \sec(x)dx$$

or

$$\begin{aligned}\int \sec^3(x) &= \frac{1}{2}\left[\sec(x)\tan(x) + \int \sec(x)dx\right] \\ &= \frac{1}{2}\left[\sec(x)\tan(x) + \log|\sec(x) + \tan(u)|\,\right] + C\end{aligned}$$

Partial Fraction Transformation

The ability to integrate the power function without any restrictions on the exponent is the key element in developing a universal method for integrating a rational function. The following theorem is important for this analysis.

Theorem 7.4.4 (The Fundamental Theorem of Algebra) *Every polynomial with real coefficients can be factored into a product of linear and irreducible quadratic factors.*

A quadratic factor is *irreducible* if it can not be expressed as the product of two linear factors with real coefficients. For example, $x^2 + 1$ is irreducible, whereas $x^2 - 1$ is reducible.

The idea behind the partial fraction transformation is to reverse the process of summing rational functions by the method of finding the lowest common denominator. So, given a rational function, factor the denominator and then consider how it could have been formed through the process of finding the lowest common denominator of a sum of partial fractions.

Example 7.4.5 We will decompose $3x/(x^2 - 1)$ into partial fractions.

The first step is to *always* check that the rational function is a proper rational function (i.e., the degree of the numerator is less than the degree of the denominator). If the rational function is not proper (i.e., it is improper), divide the denominator into the numerator to obtain a polynomial plus a proper rational function. Since the rational function in this example is proper, we proceed to factor the denominator into linear or irreducible quadratic factors. Factoring yields

$$\frac{3x}{x^2-1} = \frac{3x}{(x-1)(x+1)}$$

The most general case in which the right-hand side could represent a sum of *proper* rational functions is

$$\frac{3x}{(x-1)(x+1)} = \frac{A}{x-1} + \frac{B}{x+1}$$

where both A and B are constrained to be constants (so that the rational functions are proper). Summing the functions on the right-hand side gives

$$\frac{3x}{x^2-1} = \frac{A(x+1)+B(x-1)}{(x-1)(x+1)}$$

Thus for equality to hold

$$3x = A(x+1) + B(x-1)$$

or

$$3x = (A+B)x + (A-B)$$

So, $3 = A+B$ and $0 = A-B$. (Recall that two polynomials are equal provided coefficients of corresponding terms are equal.) Therefore,

$$A = \frac{3}{2} \quad \text{and} \quad B = \frac{3}{2}$$

We have now derived the following decomposition:

$$\frac{3x}{x^2-1} = \left(\frac{3}{2}\right)\frac{1}{x-1} + \left(\frac{3}{2}\right)\frac{1}{x+1}$$

Suppose the original problem had been to evaluate the integral of $3x/(x^2-1)$. We would have

$$\begin{aligned}
\int \frac{3x}{x^2-1}dx &= \frac{3}{2}\int \frac{1}{x-1}dx + \frac{3}{2}\int \frac{1}{x+1}dx \\
&= \frac{3}{2}\log(|x-1|) + \frac{3}{2}\log(|x+1|) + C \\
&= \frac{3}{2}[\log(|x-1|) + \log(|x+1|)] + C \\
&= \frac{3}{2}\log(|x-1||x+1|) + C \\
&= \log(|x^2-1|)^{3/2} + C
\end{aligned}$$

△

We now consider an example that involves an irreducible quadratic factor.

Example 7.4.6 We will evaluate $\int (x^4+3)/(x^3+2x)dx$.

We first observe that the integrand is an *improper* rational function. Thus, dividing the denominator into the numerator yields the integral

$$\int \left(x - \frac{2x^2-3}{x^3+2x} \right) dx$$

The next step is to decompose the rational function in the integrand into a sum of *proper* rational functions. So

$$\frac{2x^2-3}{x^3+2x} = \frac{2x^2-3}{x(x^2+2)} = \frac{A}{x} + \frac{Bx+C}{x^2+2}$$

The numerator in the second term is linear to provide the most general proper rational function.

$$\frac{2x^2-3}{x^3+2x} = \frac{A(x^2+2)+(Bx+C)x}{x(x^2+2)}$$

Therefore,

$$2x^2-3 = A(x^2+2)+(Bx+C)x$$

or $2 = A+B$, $0 = C$ and $-3 = 2A$. Solving yields

$$A = -\frac{3}{2} \quad B = \frac{7}{2} \quad C = 0$$

Thus,

$$\begin{aligned} \int \frac{x^4+3}{x^3+2x} dx &= \int \left(x + \frac{3}{2}\frac{1}{x} - \frac{7}{2}\frac{x}{x^2+2} \right) dx \\ &= \frac{1}{2}x^2 + \frac{3}{2}\log(|x|) - \frac{7}{4}\log(x^2+2) + C \end{aligned}$$

The reader should understand why absolute values were used in one of the logarithm expressions, but not in the other. The reader should also check the integration details for the last term and should also understand why the numerator in the second term in the partial fraction decomposition above is $Bx+C$ rather than just the constant C. In particular, show that there are no values for A and C such that $(2x^2-3)/[x(x^2+2)] = A/x + C/(x^2+2)$. △

It is certainly possible that the denominator of a rational function may contain repeated linear or quadratic factors. It can be shown without too much difficulty that these factors yield the partial fraction decompositions illustrated below.

$$\frac{1}{(x-a)^2} = \frac{A}{x-a} + \frac{B}{(x-a)^2}$$

$$\frac{1}{(x^2+1)^2} = \frac{Ax+B}{x^2+1} + \frac{Cx+D}{(x^2+1)^2}$$

The reader is reminded that in treating the subject of integral transformations, the emphasis has been on the transformation process rather than attempting to present an exhaustive treatment of transformations.

Section 7.4 Exercises

In Exercises 1 through 40, find the integral.

1. $\int e^{3x+2}dx$
2. $\int e^{-4x}dx$
3. $\int x3^{x^2}dx$
4. $\int e^x/(e^x+4^2)dx$
5. $\int 4^{-2x}dx$
6. $\int \sqrt{1-x^2}dx$
7. $\int (e^x+3)/e^x dx$
8. $\int e^{\cos(x)}\sin(x)dx$
9. $\int \log(x)/x dx$
10. $\int x^2/(x^3-1)dx$
11. $\int (x^2+7)/x dx$
12. $\int \log[\log(x)]/x\log(x)dx$
13. $\int xe^{-2x}dx$
14. $\int x^2\log(x)dx$
15. $\int x\sec^2(x)dx$
16. $\int x^3\cos(x^2)dx$
17. $\int \log(x)dx$
18. $\int \sin^{-1}(x)dx$
19. $\int dx/x^2-16$
20. $\int dx/(x^3-x)$
21. $\int x+1/[x(x-2)]dx$
22. $\int (3x-5)/(x^2-x-2)dx$
23. $\int (5x-4)/(x^2-x-3))dx$
24. $\int (3x^2+4x+2)/x(x+1)^2dx$
25. $\int x^2/(x^2-x-6)dx$
26. $\int (x^2-2)/(x-2)^2dx$
27. $\int dx/[(x-1)(x^2+4)]$
28. $\int (x^3+5x^2-21)/(x^2+3x)dx$
29. $\int (x+3)/[x(x^2+1)]dx$
30. $\int (5x^2+20)/(x+1)(x^2+4)dx$
31. $\int (x^2+1)^{3/2}dx$
32. $\int x/\sqrt{x^2-4}dx$
33. $\int x^2\sin^2(x)dx$
34. $\int x\sec(x)\tan(x)dx$
35. $\int xe^{3x}dx$
36. $\int \sqrt{x^2-16}/x dx$
37. $\int x^2/\sqrt{x^2+4}dx$

38. $\int (3x^3 + 2x - 2)/(x^4 + 2x^2)dx$

39. Find the length of the curve C defined by $C(t) = (1, \log(1 - t^2))$, $0 \le t \le 1/3$.

40. Find the area of the region bounded by the graphs of $f(x) = e^x$, $g(x) = \log(x)$, $x = 1$ and $x = 3$.

41. Find the equation of the line tangent to the graph of $x = y^y$ at $(1, 1)$.

42. Prove or disprove that $f(x) = e^{-4x} + e^{4x}$ is a constant function.

43. Define the *hyperbolic sine function* by $\sinh(x) = e^x - e^{-x}/2$ for all real x. Define the *hyperbolic cosine function* by $\cosh(x) = e^x + e^{-x}/2$ for all real x.

    (a) Show $\cosh^2(x) - \sinh^2(x) = 1$
    (b) Find $\sinh'(x)$.
    (c) Find $\cosh'(x)$.
    (d) Sketch the graph of sinh.
    (e) Sketch the graph of cosh.

44. Use the hyperbolic sine and cosine functions to parameterize the standard parabola $x^2 - y^2 = 1$.

45. Give the natural definitions for

    (a) the *hyperbolic tangent,* tanh,
    (b) the *hyperbolic cotangent,* coth,
    (c) the *hyperbolic secant,* sech,
    (d) the *hyperbolic cosecant,* csch.

46. (a) Find $\tanh'(x)$. (b) Find $\coth'(x)$.
    (c) Find $\operatorname{sech}'(x)$. (d) Find $\operatorname{csch}'(x)$.

47. (a) Find $\int \sinh(x)dx$. (b) Find $\int \cosh(x)dx$.
    (c) Find $\int \operatorname{sech}^2(x)dx$. (d) Find $\int \tanh(x)dx$.

48. Let $B_n = \int_0^\infty x^n e^{-x} dx$ where n is a non-negative integer.

    (a) Compute the values of B_n for successive values of n starting with $n = 0$ and continuing until you recognize a pattern.
    (b) Conjecture the value of B_n.
    (c) Test your conjecture using a value of n different from those used in part a.
    (d) Prove (or disprove) your conjecture. (Hint: Try integrating by parts.)

7.5 Applications

The purpose of this section is to provide the reader with an appreciation of the logarithmic and exponential functions and their wide applicability. There are four parts to this section: the first describes a model that is applicable in many population studies; the second attempts to provide the reader with a sense of the exponential growth rate; the third illustrates compounding interest in the field of finance; and the fourth shows the fundamental role of the exponential function in the solution of nth order linear homogeneous differential equations.

Logistic Equation

In this part, we will develop an equation, the logistic equation, for modeling the growth of a population when there are ecological factors (e.g., space) that bound the size of the population. Ecologists call such a bound the "carrying capacity of the environment." Such populations (usually) have the following characteristics:

1. When the population is very small, the growth rate is small.

2. When the population reaches a certain size. the growth rate becomes very large,

3. When the population is very large, the growth rate is again very small.

The growth curve for such a population resembles an S-shaped curve as illustrated below.

Logistic Curve

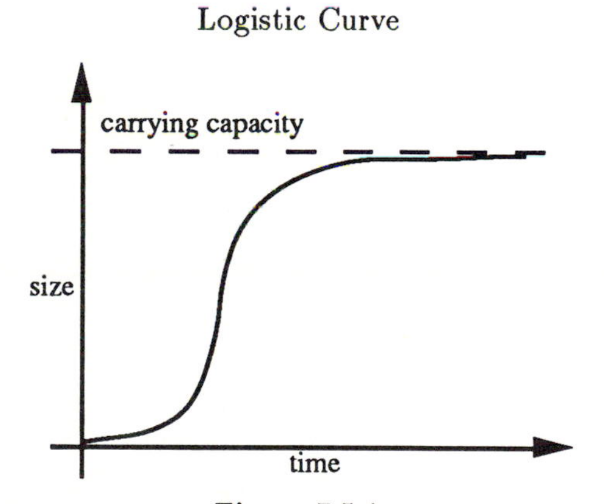

Figure 7.5.1

We develop an elementary mathematical model for this type of population growth based on the two general assumptions:

1. The growth rate for the initial period is proportional to the size of the population.

2. The growth rate for the later period is proportional to the difference between the carrying capacity and the size of the population.

If we describe population size as a function p of time, then assumption 1 is $p'(t) = Ap(t)$ for "small" values of t and some proportionality constant A; assumption 2 is $p'(t) = B(C - p(t))$ for "large" values of t, some proportionality constant B, and the bound or carrying capacity of the environment C.

Now we need to find a function that will satisfy both of these assumptions. Although there are several possibilities, we will require p to satisfy

$$p'(t) = kp(t)[C - p(t)]$$

based on the following reasoning. When $p(t)$ is small, the $C - p(t)$ factor is approximately C and thus $p'(t) \approx kCp(t)$, which is assumption 1 with $A = kC$. When $p(t)$ is large, $p(t)$ is approximately C and thus $p'(t) \approx kC[C - p(t)]$ which is assumption 2 with $B = kC$.

The next step is to find an explicit expression for $p(t)$. We would like to transform the equation for $p'(t)$ so as to be able to antidifferentiate both sides and then solve for $p(t)$. The key to the transformation is recognizing that the left side is just the derivative of $p(t)$ and some of the right side factors contain $p(t)$. Thus dividing both sides by the right side factors containing $p(t)$ yields a left side that is almost a du/u form and the right-side is a constant. That is,

$$\frac{p'(t)}{p(t)(C - p(t))} = k$$

Applying the partial fraction transformation to the left side (the reader should supply the details) yields

$$\frac{1}{C}\left[\frac{p'(t)}{C - p(t)} + \frac{p'(t)}{p(t)}\right] = k$$

We can now antidifferentiate both sides (and multiply by a negative one) getting

$$\log[C - p(t)] - \log p(t) = -Ckt + D$$

where D is a constant of integration, or

$$\log\left[\frac{C - p(t)}{p(t)}\right] = -Ckt + D$$

or

$$\frac{C - p(t)}{p(t)} = e^{-Ckt+D} = e^{D}e^{-Ckt}$$

or

$$p(t) = \frac{C}{1 + Ee^{-Ckt}}$$

where $E = e^{D}$.

Note that D can be determined when given the size of the population at a fixed time, say $t = 0$ (i.e., the initial size).

This last equation, $p(t) = C/(1 + Ee^{-Ckt})$, is called the *logistic equation*. Although a very elementary model, it gives a good "fit" to the growth data for unicellular organisms and small multicellular organisms.

Question 7.5.2 Criticize this model by naming some important growth factors that are not taken into account. Also find another equation that satisfies assumptions 1 and 2. ◇

The Amazing Exponential Growth

In this part, we will attempt to gain a feeling for the amazing growth of exponential functions. We will then compare several types of functions with respect to size and rates of growth. Our illustrations begin with the exponential sequence $\{2^n\}$ and then expand to the smooth function $f : \mathbb{R} \to \mathbb{R}$ defined by $f(x) = 2^x$.

Example 7.5.3 (Doubling on a Checkerboard) Consider an 8 by 8 checkerboard. Number the 64 blocks and then place 2 pennies on the first block, 4 on the second block, 8 on the third block, and so on doubling the number of pennies as you move from one block to the next. The last block will have 2^{64} pennies. Estimate answers to the following questions:

1. How much money in terms of dollars is on the sixty-fourth block? (Would you believe $184,467,440,737,095,516.16? Check it out.) On the sixty-third block?
2. If a penny is $\frac{1}{16}$ of an inch thick, how tall is the stack of pennies on the sixty-fourth block? Give an answer in feet and in miles. Compare the height of the pennies to the distance: to the moon ($\approx$ 248,560 miles), to the sun ($\approx$ 93,210,000 miles), to Proxima Centuri ($\approx$ 4 light years) the closest star to Earth (light travels approximately 186,000 miles/sec.).
3. There are approximately 1600 blueberries per quart and a quart weighs approximately 1.5 lbs. If the pennies were blueberries, how many tons of blueberries would be on the sixty-fourth block?
4. Twenty-four pounds of blueberries ($\frac{1}{2}$ bushel) occupies a volume of approximately 0.69 ft^3. What are the dimensions of a cubical box that is large enough to hold all of the blueberries on the sixty-fourth block? (Would you believe such a box would measure over 15 miles/side?)

Consider modifying the above questions to ask for rates of growth (of the sum of money, height, weight, volume). That is, suppose you plotted the points $(n, 2^n)$ for $n = 1, 2, ..., 64$ and then passed a smooth curve through the points. What would be the slope of the tangent to the curve for different values of n? How does the slope change as a function of n? Is it true that given any positive number B, the rate of growth of 2^x will eventually be greater than B? If so, can you determine a minimum value of x (in terms of B) for which this will be true?

Now let us consider the log function to the base 2, $\log_2$, the inverse function of $f(x) = 2^x$. What can be said about the rate of growth of $\log_2$? What is the relationship between the slope of the tangent line to the graph of $y = 2^x$ and the slope of the tangent line to the graph of $y = \log_2(x)$? Will the rate of change of the log function (any base) ever be zero? Why? Given any (small) positive number b, will the rate of change of $\log_2(x)$ be less than b for sufficiently large values of x? If so, how large must x be for this to be true? △

Question 7.5.4 How do the polynomial functions compare with exponential and logarithmic functions for large values of x? (Assume that the coefficient of the highest degree term is positive.) Apply L'Hospital's Rule to the limit (as x approaches infinity) of the quotient of an exponential function divided by an arbitrary polynomial. What does the conclusion tell you? Repeat the L'Hospital's Rule argument with a quotient of a logarithmic function and a polynomial. What does the conclusion tell you?

How does the growth rate of the sine and cosine functions compare with that of the logarithmic functions, polynomials, and exponential functions? How does the growth rate of the logarithmic function compare with the growth rate of $f(x) = -e^{-x}$? ◇

Annuities and Bonds

An *annuity* is a (finite) sequence of payments. We will consider annuities of equal payments at fixed periods of time (e.g., an annuity may pay $10,000 per year for 20 years). There are two basic questions of interest:

1. What is the value of an annuity?

2. What is the present value of an annuity?

We will determine an answer to question 1 and then use the results of Section 7.3 to obtain the answer to question 2.

The *value of an annuity* is determined in the following manner. Each annuity payment is invested for the rest of the annuity period. The sum of of money realized in these investments is defined to be the value of the annuity.

Example 7.5.5 (Value of an Annuity) We will determine the value of an annuity that pays $10,000 at the end of the year for 20 years when the payments are invested at an interest rate of 6%.

We will sum up the 20 values obtained by investing the annuity payments, starting with the last payment. Since the last payment is received at the end of the annuity period, there is no time left to invest it and therefore the value of the last payment is just $10,000. The next to last payment is invested for 1 year at 6% interest. Thus its value is $\$10,000(1 + .06)$. The second to last payment is invested for 2 years and thus its value is $\$10,000(1+.06)^2$. Continuing in this manner gives us the value S of the annuity:

$$S = \$10,000 + \$10,000(1.06) + \$10,000(1.06)^2 + \cdots + \$10,000(1.06)^{19} \qquad \triangle$$

The expression for the value S of the annuity is an example of a *finite geometric series.* A finite geometric series is a finite sum of terms arranged so that the first term is one and each of the other terms is a constant multiple c of the preceding term. Thus a finite geometric series has the form

$$1 + c + c^2 + \cdots + c^n$$

A nice compact formula for a finite geometric series is derived as follows. The sum S of a finite geometric series is obtained by multiplying each term of the series by the constant c, subtracting this series from the original, and then solving for S. That is,

$$\begin{aligned} S - cS &= (1 + c + c^2 + \cdots + c^n) - c(1 + c + c^2 + \cdots + c^n) \\ &= 1 - c^{n+1} \end{aligned}$$

Thus

$$S = \frac{1 - c^{n+1}}{1 - c}$$

We can now express the value of the annuity in the above example in the form:

$$S = \$10,000 \frac{1 - 1.06^{20}}{-.06} = \$367,855.92$$

Let us now abstract from the above example and the geometric series formula to obtain the formula for the value of an annuity paying Y dollars per year for n years assuming an investment interest rate of r. Since $c = 1 + r$, we have

$$S = Y\frac{1 - (1+r)^n}{1 - (1+r)} = Y\frac{(1+r)^n - 1}{r}$$

If there are k payments a year (instead of just one), then

$$S = \frac{Y}{k}\frac{[(1 + r/k)^{kn} - 1]}{r/k} = Y\frac{(1 + r/k)^{kn} - 1}{r}$$

The exponential function comes into the picture when the interest is compounded continuously. Since

$$\lim_{k\to\infty}\left(1 + \frac{r}{k}\right)^{kn} = e^{rn}$$

the value of an annuity with k payments per year and interest compounded continuously is

$$S = Y\frac{e^{rn} - 1}{r}$$

Example 7.5.6 How much money would a person have to invest to set up the annuity described in the previous example?

This is a "present value" problem (see Section 7.3). We solve for P in the expression

$$\$367,855.92 = P(1.06)^{20}$$

which yields $P = \$114,699.21$. △

Let us abstract this result as we did for annuity value. The *present value of an annuity* that pays Y dollars per year for n years assuming an investment interest rate of r is

$$P = S(1+r)^{-n} = Y\frac{(1+r)^n - 1}{r}(1+r)^{-n}$$

If the payments are made k times per year, then

$$P = \frac{Y}{k}\frac{(1+r/k)^{kn} - 1}{r/k}(1+r/k)^{-kn} = Y\frac{1-(1+r/k)^{-kn}}{r}$$

If the interest is compounded continuously, then

$$P = Y\frac{1-e^{-rn}}{r}$$

Question 7.5.7 Illustrate the results of compounding interest by reworking the previous two examples first with interest compounded quarterly, second with interest compounded monthly, and third with interest compounded continuously. ◇

Example 7.5.8 We want to buy a house that would require our taking a 30-year, \$50,000 mortgage at an interest rate of 8% compounded monthly. What would be the monthly payments?

The \$50,000 is the present value of a 30-year annuity with interest compounded monthly at an investment rate of 8%. That is,

$$\$50,000 = Y\frac{1-(1+.08/12)^{-360}}{0.08}$$

where Y is the yearly payment. Hence the monthly payment is $Y/12 = \$366.45$.

What would the monthly payment be if the interest were compounded continuously? △

A *bond* is a contract for the issuing agent (business, town) to pay the buyer a fixed series of equal payments over the time period of the bond and to pay the "face amount" of the bond at the end of the time period. Usually the fixed payments are made once or twice a year. The question of what is a "fair" price for a particular bond is just a present value problem as is illustrated in the next example.

Example 7.5.9 (Bonds) What would be a "fair" price to pay for a $10,000, 20-year bond with semiannual payments of $100 assuming an interest rate of 8%?

Note that the bond consists of the value of a 20-year annuity with semiannual payments of $100 with an interest rate of 8% plus a single payment of $10,000 after 20 years. The present value of this is

$$Y\frac{1-(1+.08/2)^{-40}}{0.08} + \$10,000(1+0.08/2)^{-40}$$

where Y is the yearly payment; thus $Y = \$200$. The fair price of this bond is thus $4,062.17. (The reader should check the arithmetic with a calculator.)

To verify that this figure is correct, the reader should compute the value of a 20-year annuity of semiannual payments of $100 with an interest rate of 8%. This value plus the $10,000 face value of the bond should equal $\$4,062.17(1.04)^{40}$, the amount that would be realized by investing the fair value of the bond for 20 years at 8% interest compounded semi-annually. △

Differential Equations

A differential equation is simply an equation containing a derivative. We will consider differential equations of the form

$$a_n y^{(n)} + a_{n-1}y^{(n-1)} + \cdots + a_1 y' + a_0 y = 0$$

where the a_k are constants. These are called nth order linear homogeneous differential equations with constant coefficients. The order is that of the highest order derivative in the equation. The following is a third order linear homogeneous differential equation (y is a function of x):

$$\frac{d^3y}{dx^3} - 6\frac{dy}{dx} = 7y$$

The purpose of this section is to illustrate the role of the natural exponential function in the solution of nth order linear homogeneous differential equations with constant coefficients. No attempt will be made to be either rigorous or exhaustive in our treatment. These aspects of the study will be left to a course in differential equations. The simplest linear homogeneous differential equation is

$$\frac{dy}{dx} = ay$$

where a is constant. To solve this equation, we treat dy/dx as the quotient of differentials. This allows us to write

$$dy = aydx$$

or

$$\frac{dy}{y} = adx$$

Thus (antidifferentiating both sides)

$$\log(y) = ax + C$$

and so

$$y = e^{ax+C} = e^C e^{ax}$$

Note that $e^C = y(0)$, the "initial" value, and $a = y'(0)/y(0)$. (This method of solving differential equations is called *separation of variables*: all terms containing y are put on one side and all terms containing dx are placed on the other side.)

We have shown that the solution to the differential equation $dy/dx = ay$ is the exponential function $y = y(0)e^{ax}$. This verifies the remark in Section 7.3 that the exponential function is the only non-zero function whose derivative is equal to the function itself, as well as the remark in Section 1.1 that the rate of change of the exponential function is proportional to the function.

Understanding this last statement, i.e.,

$$y(x) = e^{ax} \quad \text{if and only if} \quad y'(x) = ae^{ax}$$

is the key to solving nth-order linear homogeneous differential equations.

We will demonstrate how to solve a second order linear homogeneous differential equation with constant coefficients. This will illustrate the general method.

Example 7.5.10 Consider the differential equation $y'' - 6y' + 5y = 0$ where $y(0) = 2$ and $y'(0) = 4$.

We transform the problem into one of finding the zeros of a polynomial by assuming a solution of the form $y = e^{ax}$ and substituting into the differential equation. The steps are:

(a) $a^2e^{ax} - 6ae^{ax} + 5e^{ax} = 0$ (substitute e^{ax} for y)

(b) $e^{ax}(a^2 - 6a + 5) = 0$ (factor out e^{ax})

(c) $a^2 - 6a + 5 = 0$ (since $e^{ax} \neq 0$)

(d) $(a - 5)(a - 1) = 0$

This gives $a = 1$ and $a = 5$, so $y = e^{5x}$ and $y = e^x$ are two solutions to the differential equation. The general solution is obtained by forming a linear combination of these two solutions. That is,

$$y = c_1 e^{5x} + c_2 e^x$$

The question is now how do we determine values for the coefficients in order to obtain a particular solution? The answer is to form a system of two equations in the two unknowns c_1 and c_2 from the initial conditions, $y(0) = 2$, and $y'(0) = 4$. That is,

$$c_1 + c_2 = 2 \quad [\text{from } y(0) = 2]$$

and

$$5c_1 + c_2 = 4 \quad [\text{from } y'(0) = 4]$$

Thus $c_1 = \frac{1}{2}$ and $c_2 = \frac{3}{2}$ and so the solution of the differential equation is

$$y(x) = \frac{1}{2}e^{5x} + \frac{3}{2}e^x$$

△

The relative ease with which we solved the above example reflects more on the choice of the example than on the type of problem. Note that an nth degree linear homogeneous differential equation with constant coefficients yields an nth-degree polynomial in step (c). This is called the *characteristic polynomial.* In general, we cannot find the zeros of an nth-degree polynomial for $n > 4$. Therefore a numerical approximation is often required (we might try the Bisection Algorithm). Repeated zeros and complex roots also supply additional "wrinkles" to the general problem. Further difficulties may be encountered in determining the coefficients for the particular solution. Since the n zeros of the characteristic equation yield n terms in the general solution, there are n coefficients that need to be determined by solving a system of n linear equations. For large n, solving a system of n linear equations in n unknowns will usually require some special techniques studied in a linear algebra course.

Section 7.5 Exercises

1. Fill in all of the missing calculations in the derivation of the logistic curve.

2. Show that for positive constants C, E, and k, the graph of the logistic equation is an S curve.

3. Find the point where the growth rate of the logistic curve is a maximum.

4. Consider the logistic equation $p(t) = 1000/(1 + 4e^{-t})$.

   (a) Approximate the initial size of the population.

   (b) Approximate the carrying capacity of the environment.

(c) How long does it take for the population to reach half of its maximum size?

5. Repeat the checkerboard illustration starting with three pennies rather than two.

6. How does the growth rate of $y = n^n$ compare with that of $y = 2^x$? (Hint: Use the exponential transformation.)

7. How does the growth rate of $y = n!$ compare to that of $y = n^n$? To $y = 2^x$? (Hint: Use *Stirling's formula* $n! \approx \sqrt{2\pi n} n^n e^{-n}$ to approximate the factorial.)

8. Given any (small) positive number b, will the rate of change of $\log_2(x)$ be less than b for sufficiently large values of x? If so, how large must x be for this to be true?

9. Compare the growth rates of $\log_2(x)$ and $\log_2[\log_2(x)]$.

10. Order (from small to large) the following expressions according to their rates of growth

$$x^n,\ \log_2(x),\ x^x,\ 2^x,\ |\sin(x)|,\ FLOOR(x)!,\ \log_2[\log_2(x)].$$

11. What is the value of a 15-year annuity with quarterly payments of \$1,000 if the investment interest rate is 7% compounded continuously?

12. What is the present value of the annuity in Exercise 11?

13. Determine the monthly payments required for a 25-year \$60,000 mortgage with interest compounded continuously at an interest rate of 9%?

14. What is the largest 30-year home mortgage that a person can obtain if the interest is compounded monthly at 8% and if the monthly payments cannot exceed \$300? What is the answer if the interest is compounded continuously?

15. What would be the quarterly payments on a 10-year annuity established with \$10,000 if the investment interest rate is 6% compounded quarterly, or compounded continuously?

16. A person invests \$10,000 at 6% interest compounded continuously for 10 years and then uses the accumulated value to establish a 10-year monthly annuity with an investment rate of 8% compounded continuously. What would be the monthly payments?

17. What is the "fair" market price for a \$2500, 10-year bond with yearly payments of \$400 if the interest rate is fixed at 7% compounded continuously? What is the value of the bond?

18. If you could earn 8% compounded continuously, which of the following would you prefer:

    (a) \$1,000 right now

    (b) \$500 now and \$650 at the end of 5 years

    (c) \$500 now, \$200 after two years, and \$450 after 4 years

    (d) an annuity which pays \$150 per year for 10 years

19. Find a solution for $y'' - 3y' - 4y = 0$ given that $y(0) = 2$ and $y'(0) = -3$.

20. Find a solution for $y'' - 5y' - 14y = 0$ given that $y(0) = 5$ and $y'(0) = 9$.

21. Find a solution for $y''' - 3y'' - y' + 3y = 0$ given that $y(0) = 2$, $y'(0) = 4$, and $y''(0) = 6$.

22. For each of the following statements, give an example or show that no example can exist.

    (a) A nonhomogeneous third-order linear differential equation.

    (b) A nonlinear homogeneous differential equation.

    (c) A nonlinear, nonhomogeneous differential equation.

    (d) A second-order linear homogeneous differential equation having $y(x) = 12e^{2x} - 4e^{-2x}$ for its solution.

    (e) A fourth-order linear homogeneous differential equation whose characteristic equation has zeros: 2, 2, −2, −2.

23. The logistic equation depends upon three parameters C, E, and k. Determine how each of these parameters affects the shape of the curve; i.e., if these parameters are changed, how is the curve changed? You may find it useful to graph the curve with several values of C, E, and k.

    Exercises 24 through 26 are on the separation of variables, introduced at the beginning of the subsection on differential equations.

24. Solve the differential equation

$$\frac{dy}{dx} = \frac{x}{y}$$

by separating the variables

$$ydy = xdx$$

and antidifferentiating

$$\int ydy = \int xdx.$$

Check your solution for y by differentiating.

25. Solve the logistic equation

$$p'(t) = kp(t)(C - p(t))$$

by writing in dp/dt notation

$$\frac{dp}{dt} = kp(C - p)$$

and separating variables.

26. Solve by separating variables:

(a) $dy/dx = (1 + y)/(1 + x)$ (b) $dy/dx = x^2/y$

(c) $dy/dx = (x - e^{-x})/(y + e^y)$ (d) $dy/dx = x/(y\sqrt{1 - x^2})$

Chapter 8

INTEGRALS IN SEVERAL VARIABLES

8.1 Line Integrals

In this section, we expand on the concept of arc length that was developed in Section 5.6 by considering the idea of integrating a function along a curve. The vector approach is used since it extends easily from the two dimensional case to the n-dimensional case.

As well as providing a very useful application, the concept of work, force times distance, motivates our development of line integrals. Let us begin by recalling a few "work" situations. In each situation, the reader should identify the "force" and the curve along which an object is moving. Consider paddling a canoe across a lake, a fish swimming up a river, jogging, launching a spaceship, sailing, pushing a car, or skiing.

In some situations, such as in downhill skiing, the force (gravity) is constant while in other situations, such as in sailing, the force (wind) is variable. The complexity of the situation is realized when we consider that the magnitude of a force depends, in part, on the direction from which it is being applied. (Ask a sailor to explain "tacking.") For example, in pushing a large box across the floor, a 50 lb force F exerted at an angle of $60°$ from the floor would have the same effect as a 25 lb force exerted parallel to the floor. That is, the component of the force F in the horizontal direction is $50\cos(\pi/3) = 25$. The component of F in the vertical direction is $50\sin(\pi/3)$.

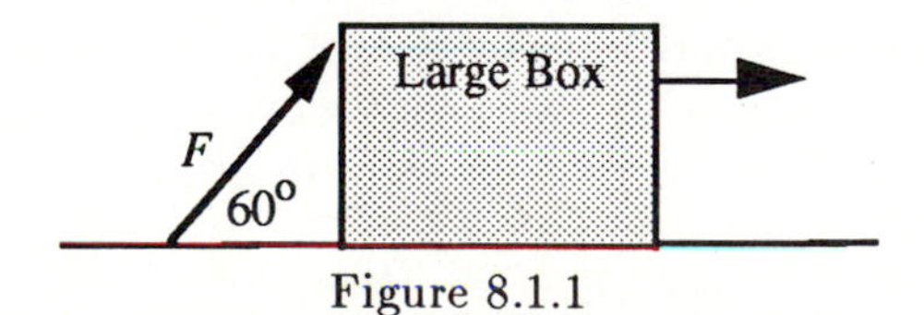

Figure 8.1.1

Notice that in the above illustration if the box moved 10 ft, then the amount of work done would be $50\cos(\pi/3)(10) = 250$ foot-pounds (ft-lb). The point that is being made is that in a work situation one needs to consider the *component of the force in a given direction*, not just the force.

What is meant by the phrase "component of force in a given direction"? Recall four things from Chapter 4:

1. The definition of the dot product of two vectors.

2. Direction is given by a *unit* vector. For example, the direction of the x-axis in the plane is given by the vector $(1, 0)$. The direction of the x-axis in three-dimensional space is given by the vector $(1, 0, 0)$. In general, the direction of vector D is given by $D/|D|$.

3. Force and velocity are vectors. For example, a weather forecaster describes a wind by giving both its magnitude (miles per hour) and its direction.

4. The angle between vectors is described by the Cauchy-Schwarz-Bunyakovsky equality:

$$F{\bullet}D = |F||D|\cos(\theta)$$

where θ is the angle between the vectors F and D.

This last fact gives us the information we need to make a definition. Since $|F|\cos(\theta) = |F|F{\bullet}D/|F||D| = F{\bullet}D/|D| = F{\bullet}D/|D|$, we have the following definition.

Definition 8.1.2 *The* component of force F in the direction D *is* $F{\bullet}\frac{D}{|D|}$.

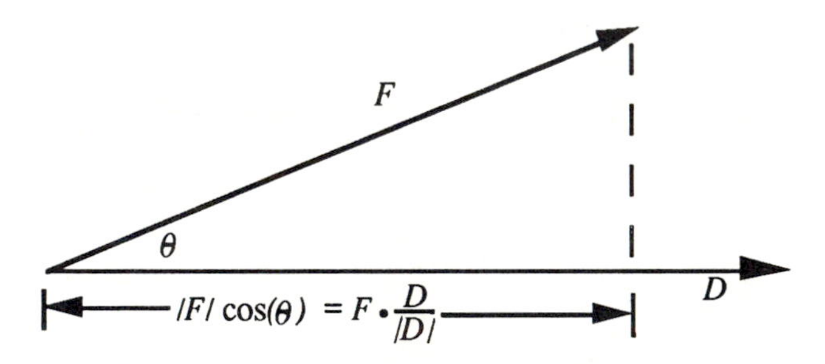

Figure 8.1.3

Technically, *work done by a force acting on an object* is defined to be the component of the force in the direction of motion of the object times the distance the object moves. Thus, the work done by a constant force F applied to an object that moves in a straight line D is the component of F in the direction of the vector $D/|D|$ times the distance the object moves, $|D|$. That is,

$$W = F{\bullet}\frac{D}{|D|}(|D|) \quad \text{or} \quad W = F{\bullet}D$$

We are now ready to develop the concept of a line integral by generalizing the definition of work to a variable force applied to an object moving along a curve. The generalization will be carried out in two steps.

Step 1. Consider F to be a force in $\mathbb{R}$ described by a function $F : \mathbb{R} \to \mathbb{R}$ acting in the direction along the x-axis from $x = a$ to $x = b$. The work W done by F in moving an object from a to b would be $F(b - a)$, provided F is a constant force. If F is not constant, W can be approximated by partitioning $[a, b]$ into subintervals and assuming that F is approximately constant over each subinterval. This process yields

$$W \approx \sum_{i=1}^{n} F(x_i)\Delta x_i$$

which is a Riemann Sum. Hence, if F is continuous, then by the existence theorem for Riemann integrals (Theorem 5.3.17),

$$W = \int_a^b F(x)dx$$

Step 2. Consider F to be a variable force in $\mathbb{R}^n$ acting on an object moving along a curve D in $\mathbb{R}^n$. Suppose $F : \mathbb{R}^n \to \mathbb{R}^n$ and the curve D is parameterized as the range of a C^1 vector-valued function $C : [a, b] \to \mathbb{R}^n$. Notice that F is restricted to D by forming the composition of functions, $F(C(t))$. We want to describe the amount of work done in moving an object from $C(a)$ to $C(b)$ along the curve D. The process is, of course, the Basic Approximation Process that underlies the development of *every* integral concept. We shall omit many of the details since they are similar to those that have been illustrated in the developments of other integral concepts (area, length of curve, volume, etc.).

We briefly sketch the basic ideas involved in formulating a Riemann Sum approximation for work.

1. Partition the curve D into subcurves D_i by the partition $P_n = \{a = t_0, t_1, ..., t_n = b\}$ of $[a, b]$, so that $D_i = C([t_{i-1}, t_i])$.

2. Replace each subcurve D_i with the chord joining its endpoints. Assume F is constant over each chord. (This will allow us to reduce the general case to step 1.)

3. Approximate the work done by F applied to an object moving along the subcurve D_i with the work done by the "assumed constant" force F applied to an object moving along the chord joining the endpoints of D_i. (Since the curve D is the *range* of the function C, the vector expression for the chord joining the endpoints of the subcurve D_i is $C(t_i) - C(t_{i-1})$). Hence, we have

$$W \approx \sum_{i=1}^{n} F(C(t_i)) \cdot (C(t_i) - C(t_{i-1}))$$

4. Apply the Mean Value Theorem for derivatives to $C(t_i) - C(t_{i-1})$ and obtain

$$C(t_i) - C(t_{i-1}) = C'(t_i^*)\Delta t_i$$

for some t_i^* in $[t_{i-1}, t_i]$. Note that the application of the Mean Value Theorem uses the C^1 property of the function C.

The Riemann Sum approximation is thus

$$W \approx \sum_{i=1}^{n} F(C(t_i)) \cdot C'(t_i^*)\Delta t_i$$

Furthermore, if F is continuous, then

$$W = \int_a^b F(C(t)) \cdot C'(t)dt$$

Since integration is additive with respect to the path of integration (Theorem 5.5.7), the curve D does not have to be C^1, but can be made up of a finite number of C^1 curves that are joined together. Such a curve is called a *piecewise C^1 curve.*

Abstracting the above result, that is, removing the interpretation of work, we can state the definition for the line integral.

Definition 8.1.4 (Line Integral) *Let D be a piecewise C^1 curve parameterized by $C : [a, b] \to \mathbb{R}^n$ and let F be a continuous function $F : A \to \mathbb{R}^n$, where A is a subset of $\mathbb{R}^n$ containing the curve D. The* integral of F along D, *denoted by $\int_D F$, is*

$$\int_D F = \int_a^b F(C(t)) \cdot C'(t)dt$$

Example 8.1.5 We will find the work done by a force $F : \mathbb{R}^3 \to \mathbb{R}^3$, defined by $F(x, y, z) = (xy, yz, x)$, in moving an object along the curve D described by the range of $C : [0, 2] \to \mathbb{R}^3$, where $C(t) = (t, t, t^2)$.

Since F is continuous and C is C^1,

$$\begin{aligned} W &= \int_D F = \int_0^2 F(C(t)) \cdot C'(t)dt \\ &= \int_0^2 (t^2, t^3, t) \cdot (1, 1, 2t)dt = \int_0^2 (t^2 + t^3 + 2t^2)dt = \left(t^3 + \frac{t^4}{4}\right)\Big]_0^2 \\ &= 12 \end{aligned}$$

△

Example 8.1.6 We will evaluate the integral of $F : \mathbb{R}^2 \to \mathbb{R}^2$ defined by $F(x, y) = (2x, y)$, over the edges of the triangle with vertices $\{(1,1), (3,-2), (5,4)\}$ in a counterclockwise direction.

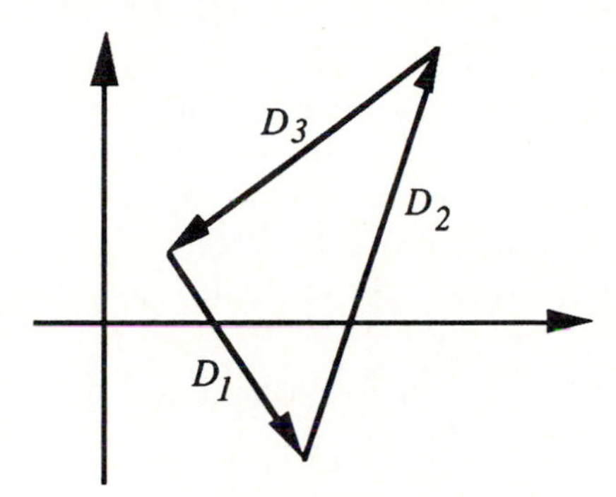

Figure 8.1.7

Let D be the piecewise C^1 curve consisting of C^1 pieces D_1, D_2, and D_3. Recall from Chapter 4 that one way to parameterize a line segment joining points A and B in the plane is to define the function $f : [0,1] \to \mathbb{R}^2$ by $f(t) = A + t(B - A)$. We now parameterize the three sides of the triangle D_1, D_2, and D_3 with the functions C_1, C_2, and C_3 defined below. In each case, the domain is the interval $[0,1]$.

$$\begin{aligned}
C_1(t) &= (1,1) + t(2,-3) = (1 + 2t, 1 - 3t) \\
C_2(t) &= (3,-2) + t(2,6) = (3 + 2t, -2 + 6t) \\
C_3(t) &= (5,4) + t(-4,-3) = (5 - 4t, 4 - 3t)
\end{aligned}$$

By Theorem 5.5.5, the integral of F along D is

$$\begin{aligned}
\int_D F(C(t)) \cdot C'(t)dt &= \int_{D_1} F(C(t)) \cdot C'(t)dt + \int_{D_2} F(C(t)) \cdot C'(t)dt \\
&\quad + \int_{D_3} F(C(t)) \cdot C'(t)dt
\end{aligned}$$

Thus,

$$\begin{aligned}
\int_D F &= \int_0^1 (2 + 4t, 1 - 3t) \cdot (2,-3)dt + \int_0^1 (6 + 4t, -2 + 6t) \cdot (2,6)dt \\
&\quad + \int_0^1 (10 - 8t, 4 - 3t) \cdot (-4,-3)dt = \int_0^1 (-51 + 102t)dt \\
&= 0
\end{aligned}$$

△

The reader should note that there is always a direction associated with a curve. If D is a curve, then $-D$ will denote the curve passing through the same points as D does, but with the opposite direction (orientation). The following is an analogue of Theorem 5.5.2.

Theorem 8.1.8 $\int_D F = -\int_{-D} F$.

We show that the integral expression for the length of a curve that was derived in Section 5.6 is, in fact, a special case of a line integral.

Let the curve D be described by the range of the C^1 function C. Consider F to be a unit vector force ($|F| = 1$) in the direction of the curve D. Now, the component of F in the direction of D is, by definition, $F(C(t)) \cdot C'(t) = |F(C(t))||C'(t)|\cos(\theta)$. Since F is a unit vector, $|F(C(t))| = 1$, and since the direction of F is the direction of D, $\theta = 0$. Hence, $\cos(\theta) = 1$ and

$$\int_D F = \int_a^b F(C(t)) \cdot C'(t)dt = \int_a^b |F(C(t))||C'(t)|\cos(\theta)dt = \int_a^b |C'(t)|dt$$

(See Definition 5.6.2).

Notation. There is a popular notation for line integrals using the differential notation of Section 6.2. Suppose $F : \mathbb{R}^n \to \mathbb{R}^n$ has components $F = (f_1, f_2, ..., f_n)$ and that the curve $D = C([a, b])$ is given by C with coordinates $C(t) = (x_1(t), x_2(t), ..., x_n(t))$. Then

$$\begin{aligned}
\int_D F &= \int_a^b F(x_1(t), \ldots, x_n(t)) \cdot \left(\frac{dx_1}{dt}, \frac{dx_2}{dt}, \ldots, \frac{dx_n}{dt}\right) dt \\
&= \int_a^b \left(f_1\frac{dx_1}{dt} + f_2\frac{dx_2}{dt} + \cdots + f_n\frac{dx_n}{dt}\right) dt \\
&= \int_a^b (f_1 dx_1 + f_2 dx_2 + \cdots + f_n dx_n)
\end{aligned}$$

Thus, if $F(x, y) = (f_1, f_2)$ and $C(t) = (x(t), y(t))$, then

$$\int_{C([a,b])} F = \int_a^b f_1 dx + f_2 dy$$

We have considered the integral of a vector-valued function F along a curve D. Does the value depend on the choice of parameterization of D, or do all parameterizations (traveling along D in the same direction) give the same value?

As an example, let D be the parabolic segment $\{x, x^2\} : -1 \leq x \leq 1$. One parameterization is $C_1(t) = (t, t^2)$ for t in $[-1, 1]$. Another parameterization is $C_2(x) = (2x - 1, (2x - 1)^2)$ for x in $[0, 1]$. Both parameterizations traverse D from $(-1, 1)$ to $(1, 1)$. C_1 and C_2 have different domains, and, if we consider the parameter as time, then C_1 takes twice as long as C_2 to traverse the curve. Notice that we can express one parameter in terms of the other:

Let $\quad t = f(x) = 2x - 1 \quad$ where $\quad f : [0, 1] \to [-1, 1]$.

Then $\quad C_2(t) = C_1(f(t))$.

With this example in mind, we can now prove the theorem.

Theorem 8.1.9 (Independence of Parameterization) *The line integral is independent of the parameterization of the curve.*

Proof: Suppose that we have two parameterizations of D. Let $C_1 : [a,b] \to D$ and $C_2 : [c,d] \to D$ traverse D in the same direction. Then there is a differentiable function $f : [c,d] \to [a,b]$ where $a = f(c)$ and $b = f(d)$, and $C_2(t) = C_1(f(t))$. If F is a C^1 function on D, then

$$\begin{aligned}
\int_{C_1[a,b]} F &= \int_a^b F(C_1(t) \bullet C_1'(t) dt \\
&= \int_c^d F(C_1(f(x))) \bullet C_1'(f(x)) f'(x) dx \\
&= \int_c^d F(C_2(x)) \bullet C_2'(x) dx \\
&= \int_{C_2[c,d]} F
\end{aligned}$$

Thus the line integral is independent of the choice of parameterization. □

Section 8.1 Exercises

In Exercises 1 through 10, find the line integral of F over the given curve.

1. $F : \mathbb{R}^2 \to \mathbb{R}^2$ defined by $F(x,y) = (y, \sin(xy))$ and the curve is the interval $[-2,5]$ on the x-axis.

2. $F : \mathbb{R}^2 \to \mathbb{R}^2$ defined by $F(x,y) = (x^2 y \cos(xy), y)$ and the curve is the interval $[1,4]$ on the y-axis.

3. $F : \mathbb{R}^2 \to \mathbb{R}^2$ defined by $F(x,y) = (x, x+y)$ and the curve is the unit circle oriented in the counterclockwise direction.

4. $F : \mathbb{R}^2 \to \mathbb{R}^2$ defined by $F(x,y) = (x-y, y)$ and the curve is the boundary of the rectangle with vertices $\{(0,0),(4,0),(4,2),(0,2)\}$ oriented in the clockwise direction.

5. $F : \mathbb{R}^3 \to \mathbb{R}^3$ defined by $F(x,y,z) = (x,y,z)$ and $C : [1,2] \to \mathbb{R}^3$, defined by $C(t) = (t, t^2, 0)$.

6. $F : \mathbb{R}^3 \to \mathbb{R}^3$ defined by $F(x,y,z) = (2x-y, y-z, 2y)$ and $C : [0,2\pi] \to \mathbb{R}^3$, defined by $C(t) = (\cos(t), \sin(t), 3t)$.

7. $F : \mathbb{R}^3 \to \mathbb{R}^3$ defined by $F(x, y, z) = (x^2z, -yx^2, 3xz)$ and the curve is the straight line from $(0, 0, 0)$ to $(2, 3, 4)$.

8. $F : \mathbb{R}^3 \to \mathbb{R}^3$ defined by $F(x, y, z) = (x, -z, 2y)$ and the curve is the straight line segment from $(0, 0, 0)$ to $(0, 2, 0)$.

9. $F : \mathbb{R}^3 \to \mathbb{R}^3$ defined by $F(x, y, z) = (x, y^2, 2z)$ and $C : [0, 2\pi] \to \mathbb{R}^3$, defined by $C(t) = (\cos(t), \sin(t), t)$.

10. $F : \mathbb{R}^4 \to \mathbb{R}^4$ defined by $F(x, y, z, w) = (x, y, 0, 1)$ and C is the "space triangle" with vertices $A(0, 1, 1, 0)$, $B(1, 0, 1, 0)$, and $D(1, 1, 0, 1)$, oriented from A to B to D to A.

11. Give an example for each of the following statements or explain why no example can exist.

    (a) A nonzero function $F : \mathbb{R}^2 \to \mathbb{R}^2$ such that the integral of F around the unit circle is zero.

    (b) A curve D passing through the points $(1, 0)$ and $(0, 1)$ such that $\int_D F = 0$, where $F : \mathbb{R}^2 \to \mathbb{R}^2$ is defined by $F(x, y) = (x, y)$.

    (c) A function $F : \mathbb{R}^2 \to \mathbb{R}^2$ such that $\int_D F = 1$, where D is the straight line segment joining $(0, 0)$ and $(1, 1)$.

    (d) A function $F : \mathbb{R}^2 \to \mathbb{R}^2$ such that $\int_D F = \int_E F$ where D is the straight line segment joining $(0, 0)$ and $(1, 1)$ and E is the path parallel to the coordinate axes, joining $(0, 0)$ to $(0, 1)$ to $(1, 1)$.

8.2 The Fundamental Theorem for Line Integrals

In this section the Fundamental Theorem of Calculus will be used to motivate the Fundamental Theorem for Line Integrals. We begin by recalling the statement of the Fundamental Theorem (Part II) from Section 6.1.

If F is any antiderivative of f, then $\int_a^b f(t)dt = F(b) - F(a)$.

Now we consider $F : \mathbb{R}^n \to \mathbb{R}^n$ and ask: Does it make any sense to talk about an "antiderivative" of F? That is, given a function $F : \mathbb{R}^n \to \mathbb{R}^n$, does there exist a function g such that F is equal to the derivative of g? Note that the derivative of g would have to be a function from $\mathbb{R}^n$ into $\mathbb{R}^n$. The only functions we know that have this property are those that map $\mathbb{R}^n$ into $\mathbb{R}$.

For example, if $g : \mathbb{R}^2 \to \mathbb{R}$, then $g'(x, y) = \nabla g(x, y) = (\partial g/\partial x, \partial g/\partial y)$ which is a mapping from $\mathbb{R}^2$ into $\mathbb{R}^2$.

Thus an "antiderivative" for $F : \mathbb{R}^n \to \mathbb{R}^n$ is a function $g : \mathbb{R}^n \to \mathbb{R}$ with the property that $F = \nabla g$. If such a function g exists, F is said to be conservative and g is called a potential function for F ("antiderivative" and "potential function" are equivalent).

Definition 8.2.1 *A vector valued function F defined over an open rectangle in $\mathbb{R}^n$ is said to be* conservative *if there exists a function g with the property that $F = \nabla g$. Such a function g is called a* potential function *for F.*

Example 8.2.2 The function $g : \mathbb{R}^3 \to \mathbb{R}$, defined by $g(x, y, z) = xy + xz^2$, is a potential function for $F : \mathbb{R}^3 \to \mathbb{R}^3$ defined by $F(x, y, z) = (y + z^2, x, 2xz)$, since $F = \nabla g$. △

We are now ready to state and prove the Fundamental Theorem for Line Integrals.

Theorem 8.2.3 (The Fundamental Theorem for Line Integrals) *Let $C : [a, b] \to \mathbb{R}^n$, $D = C([a, b])$ be a piecewise C^1 curve, and $F : A \to \mathbb{R}^n$ be a continuous function, where A is an open rectangle in $\mathbb{R}^n$ containing the curve D. If g is a potential function for F, then*

$$\int_D F = g(C(b)) - g(C(a))$$

Proof: The key to the proof is to recall the form of the expression for the derivative of a composite function (the Chain Rule, Section 4.4). That is,

$$\frac{d}{dt} g(C(t)) = \nabla g \bullet C'(t)$$

Thus,

$$\begin{aligned} \int_D F &= \int_a^b F(C(t)) \bullet C'(t) dt = \int_a^b \nabla g(C(t)) \bullet C'(t) dt \\ &= \int_a^b dg(C(t)) = g(C(b)) - g(C(a)) \end{aligned}$$ □

A conservative function $F : A \to \mathbb{R}^n$ where A is an open rectangle in $\mathbb{R}^n$ is also called a *path independent function.* This expression reflects the fact that the line integral of a conservative function over a curve connecting two points is independent of the curve (path) since it depends only on the endpoints.

A closed curve is one whose two endpoints coincide. Thus, if F is a conservative (path independent) function and D is a piecewise C^1 closed curve in the domain of F, then $\int_D F = 0$.

Example 8.2.4 Let $F : \mathbb{R}^2 \to \mathbb{R}^2$ be defined by $F(x, y) = (2xy, x^2)$ and let the curve D be the range of $C : [0, 2\pi] \to \mathbb{R}^2$ $C(t) = (\cos(t), \sin(t))$. We will compute $\int_D F$.

$$\begin{aligned}
\int_D F &= \int_0^{2\pi} F(C(t))\bullet C'(t)dt \\
&= \int_0^{2\pi} (2\cos(t)\sin(t), \cos^2(t))\bullet(-\sin(t), \cos(t))dt \\
&= \int_0^{2\pi} [-2\sin^2(t)\cos(t) + \cos^3(t)]dt \\
&= \left[\frac{-2\sin^3(t)}{3} + \sin(t) - \frac{\sin^3(t)}{3}\right]_0^{2\pi} \\
&= 0
\end{aligned}$$

Note that if $g : \mathbb{R}^2 \to \mathbb{R}$ is defined by $g(x, y) = x^2y + 7$, then

$$\nabla g = \left(\frac{\partial g}{\partial x}, \frac{\partial g}{\partial y}\right) = (2xy, x^2) = F$$

Thus g is a potential function for F and therefore by the previous theorem

$$\int_D F = g(C(2\pi)) - g(C(0)) = 0$$

since D is a closed curve. △

As a further illustration of the Fundamental Theorem for Line Integrals, we shall change the curve in the previous example and then compute $\int_D F$ by the definition of line integral and by the Fundamental Theorem.

Example 8.2.5 Let $F : \mathbb{R}^n \to \mathbb{R}^n$ be defined by $F(x, y) = (2xy, x^2)$ and let the curve D in the range of $C : [-1, 2] \to \mathbb{R}^2$ be defined by $C(t) = (t, t^2)$. We will compute $\int_D F$ in two ways.

1. (By definition of line integral)

$$\begin{aligned}
\int_D F &= \int_{-1}^{2} F(C(t))\bullet C'(t)dt = \int_{-1}^{2} (2t^3, t^2)\bullet(1, 2t)dt \\
&= \int_{-1}^{2} (2t^3 + 2t^3)dt = \int_{-1}^{2} 4t^3 dt \\
&= t^4]_{-1}^{2} = 16 - 1 = 15
\end{aligned}$$

2. (By Fundamental Theorem for Line Integrals) Recall that $g : \mathbb{R}^2 \to \mathbb{R}$ defined by $g(x, y) = x^2y + 7$ is a potential function for F. (Verify that $\nabla g = F$.)

$$\int_D F = g(C(2)) - g(C(-1)) = g((2,4)) - g((-1,1)) = 23 - 8 = 15 \qquad \triangle$$

These previous two examples and theorem (should) raise two important questions concerning a function $F : \mathbb{R}^n \to \mathbb{R}^n$.

1. How can one determine whether F has a potential function?

2. If one knows that F has a potential function, how can that function be determined?

To develop a feeling for question 1, we shall assume that g is a potential function for F and ask: What must be true of g?

Let $F : \mathbb{R}^2 \to \mathbb{R}^2$, and let $g : \mathbb{R}^2 \to \mathbb{R}$. If g is a potential function for F, then

$$F = (f_1, f_2) = \left(\frac{\partial g}{\partial x}, \frac{\partial g}{\partial y}\right) = \nabla g$$

Thus $\partial g/\partial x = f_1$ and $\partial g/\partial y = f_2$. The problem is now to determine under what conditions is it possible to recover g from these last two equations when f_1 and f_2 are known.

The key is to recall (from Section 3.10) that if g has continuous second partial derivatives (i.e., g is a C^2 function), then

$$\frac{\partial^2 g}{\partial y \partial x} = \frac{\partial^2 g}{\partial x \partial y}$$

Thus if g does have continuous second partial derivatives, then

$$\frac{\partial f_1}{\partial y} = \frac{\partial^2 g}{\partial y \partial x} = \frac{\partial^2 g}{\partial x \partial y} = \frac{\partial f_2}{\partial x}$$

Hence a necessary condition for $F = (f_1, f_2)$ to have a potential function is that

$$\frac{\partial f_1}{\partial y} = \frac{\partial f_2}{\partial x}$$

We state this result formally as a theorem.

Theorem 8.2.6 *If the C^1 function $F = (f_1, f_2)$ has a potential function, then $\partial f_1/\partial y = \partial f_2/\partial x$. (In operator notation: $D_2 f_1 = D_1 f_2$.)*

Proof: Let g be a potential function for F. Thus,

$$F = \nabla g = \left(\frac{\partial g}{\partial x}, \frac{\partial g}{\partial y}\right)$$

Therefore,

$$f_1 = \frac{\partial g}{\partial x} \quad \text{and} \quad f_2 = \frac{\partial g}{\partial x}$$

Since F has continuous partial derivatives, g has continuous second partial derivatives. Thus the mixed partial derivatives are equal and we can write:

$$\frac{\partial f_1}{\partial y} = \frac{\partial^2 g}{\partial y \partial x} = \frac{\partial^2 g}{\partial x \partial y} = \frac{\partial f_2}{\partial x} \qquad \square$$

This theorem provides an easy test of whether a potential function exists for the function F.

Test for (Non)existence of a Potential Function

If $F = (f_1, f_2)$ and $\partial f_1/\partial y \neq \partial f_2/\partial x$, then F does not have a potential function.

Example 8.2.7 $F(x, y) = (xy, x^2)$ does not have a potential function since $\partial f_1/\partial y = x \neq 2x = \partial f_2/\partial x$. △

The converse of this Test can be stated as a theorem by placing suitable constraints on the domain of F.

Theorem 8.2.8 *If f_1 and f_2 are C^1 functions defined over an open rectangle A in $\mathbb{R}^2$ and $\partial f_1/\partial y = \partial f_2/\partial x$ on A, then $F = (f_1, f_2)$ has a potential function.*

Proof: We are seeking an antiderivative of F. Recall from the Fundamental Theorem of Calculus, Theorem6.1.4, Part I, that if $f : \mathbb{R} \to \mathbb{R}$ is continuous, then an antiderivative of f in a interval is obtained by choosing a point p in the interval and defining the antiderivative at x to be the integral of f from p to x: if $g(x) = \int_p^x f(t)dt$, then $g'(x) = f(x)$. The idea of this proof is the same: to take any point p in the set A and define the potential function at any other point (x, y) by integrating F along the straight line from p to (x, y).

Let $p = (a, b)$ be any fixed point in the rectangle and (x, y) any other point in the set. The line segment L joining (a, b) to (x, y) lies within A. We can parameterize the line segment by

$$C(t) = (a, b) + t[(x, y) - (a, b)] = (a + t(x - a), b + t(y - b))$$

Notice that when $t = 0$, we are at the point $C(0) = (a, b)$, and when $t = 1$, we are at the point $C(1) = (x, y)$. The potential function for F is defined by

$$\begin{aligned} g(x, y) &= \int_L F = \int_0^1 F(C(t)) \cdot C'(t) dt = \int_0^1 F(C(t)) \cdot (x - a, y - b) dt \\ &= \int_0^1 [(x - a) f_1(C(t)) + (y - b) f_2(C(t))] dt \end{aligned}$$

Now we only need compute $\partial g/\partial x$ and $\partial g/\partial y$ and see that we get f_1 and f_2. We will show that $\partial g/\partial x = f_1$ and leave the other, which is completely analogous, to the reader. We take the partial derivative, moving the partial derivative operator inside the integral sign:

$$\begin{aligned}
\frac{\partial g}{\partial x} &= \int_0^1 \left[\frac{\partial}{\partial x}(x-a)f_1(C(t)) + \frac{\partial}{\partial x}(y-b)f_2(C(t))\right] dt \\
&= \int_0^1 f_1(C(t)) + (x-a)\frac{\partial f_1}{\partial x}(C(t)) + (y-b)\frac{\partial f_2}{\partial x}(C(t))dt \\
&= \int_0^1 f_1(C(t)) + (x-a)\left[D_1f_1(C(t))\frac{\partial}{\partial x}(a+t(x-a))\right. \\
&\quad \left.+D_2f_1(C(t))\frac{\partial}{\partial x}(b+t(y-b))\right](y-b)\left[D_1f_2(C(t))\frac{\partial}{\partial x}(a+t(x-a))\right. \\
&\quad \left.+D_2f_2(C(t))\frac{\partial}{\partial x}(b+t(y-b)\right] dt \\
&= \int_0^1 f_1(C(t)) + (x-a)D_1f_1(C(t))t + (y-b)D_1f_2(C(t))tdt \\
&= \int_0^1 f_1(C(t)) + (x-a)D_1f_1(C(t))t + (y-b)D_2f_1(C(t))tdt \\
&\quad \text{(since } D_1f_2 = D_2f_1 \text{ by assumption)}
\end{aligned}$$

Now we need to evaluate the integrand. We show that it is the derivative of $h(x,y,t) = tf_1(C(t))$ with respect to t:

$$\frac{dh}{dt} = f_1(C(t)) + t[D_1f_1(C(t))(x-a) + D_2f_1(C(t))(y-b)]$$

Thus,

$$\frac{\partial g}{\partial x} = \int_0^1 \frac{dh}{dt}dt = h(1) - h(0) = f_1(C(1)) = f_1(x,y) \qquad \square$$

The integration method of the preceding theorem can be used to find an antiderivative. An alternative approach is used in the following example. [It is equivalent to integrating from p to (x,y) along lines parallel to the coordinate axes rather than along a straight line from p to (x,y).]

Example 8.2.9 Let $F(x,y) = (2xy, x^2+3y^2)$. We will determine if F has a potential function and, if it does, find one. Since

$$\frac{\partial f_1}{\partial y} = \frac{\partial(2xy)}{\partial y} = 2x$$

and

$$\frac{\partial f_2}{\partial x} = \frac{\partial(x^2+3y^2)}{\partial x} = 2x$$

and F is defined over all of $\mathbb{R}^2$ (which is an open rectangle), F has a potential function. Our task is to find (construct) a potential function for F. If g is such a function, then

$$F = \nabla g = \left(\frac{\partial g}{\partial x}, \frac{\partial g}{\partial y}\right)$$

or

$$\frac{\partial g}{\partial x} = 2xy \quad \text{and} \quad \frac{\partial g}{\partial y} = x^2 + 3y^2$$

Note that we can obtain an expression for g by integrating $\partial g/\partial x$ (partially) with respect to x, holding y fixed. That is,

$$g(x, y) = \int \frac{\partial g}{\partial x} dx = \int 2xy dx = x^2 y + k(y)$$

where $k(y)$ is the "constant" of integration. We shall find an expression for $k(y)$ and substitute this expression into the expression for g. The steps are:

1. Differentiate g partially with respect to y: $\partial g/\partial y = x^2 + k'(y)$.
2. Set $\partial g/\partial y = f_2$: $x^2 + k'(y) = x^2 + 3y^2$.
3. Solve for $k'(y)$: $k'(y) = 3y^2$.
4. Integrate $k'(y)$ to obtain $k(y)$: $k(y) = y^3$.
5. Substitute for $k(y)$ in step 1: $g(x, y) = x^2 y + y^3$.

The constant of integration C for $k(y)$ can be omitted since the question asked for a potential function, not for the whole family of potential functions. △

Example 8.2.10 Let $F(x, y) = (y\cos(xy), x\cos(xy))$. We will determine if F has a potential function and, if it does, find one. Since

$$\frac{\partial f_1}{\partial y} = \frac{\partial y\cos(xy)}{\partial y} = \cos(xy) - xy\sin(xy)$$

and

$$\frac{\partial f_2}{\partial x} = \frac{\partial x\cos(xy)}{\partial x} = \cos(xy) - xy\sin(xy)$$

and the domain of F is all of $\mathbb{R}^2$, F has a potential function.

Let g be a potential function for F. Thus, $F = \nabla g = (\partial g/\partial x, \partial g/\partial y)$. So

$$\frac{\partial g}{\partial x} = y\cos(xy) \quad \text{and} \quad \frac{\partial g}{\partial y} = x\cos(xy)$$

We can obtain an expression for g by integrating $\partial g/\partial x$ (partially) with respect to x, holding y fixed. Thus,

$$g(x,y) = \int \frac{\partial g}{\partial x} dx = \int y\cos(xy)dx = \sin(xy) + k(y)$$

where $k(y)$ acts as the constant of integration. We now obtain an expression for $k'(y)$ by differentiating g partially with respect to y and setting it equal to f_2. Thus,

$$\frac{\partial g}{\partial y} = x\cos(xy) + k'(y) = f_2 = x\cos(xy) \quad \text{and, therefore,} \quad k'(y) = 0$$

Integrating $k'(y)$ yields $k(y) = C$ and, hence, $g(x,y) = \sin(xy)$. △

The last two theorems and the Test have n-dimensional analogues that are obtained by considering all possible pairs of first partial derivatives.

Example 8.2.11 We will find a potential function for $F : \mathbb{R}^3 \to \mathbb{R}^3$ where $F(x,y,z) = (3x^2yz, x^3z, x^3y + 7)$, if one exists.

We begin by testing for existence:

$$\frac{\partial f_1}{\partial y} = \frac{\partial f_2}{\partial x} = 3x^2z \qquad \frac{\partial f_1}{\partial z} = \frac{\partial f_3}{\partial x} = 3x^2y \qquad \frac{\partial f_2}{\partial z} = \frac{\partial f_3}{\partial y} = x^3$$

Thus there exists a potential function g. That is, $F = \nabla g$ or $(f_1, f_2, f_3) = (\partial g/\partial x, \partial g/\partial y, \partial g/\partial z)$. Hence, $g(x,y,z) = \int 3x^2yz dx + k(y,z) = x^3yz + k(y,z)$. Since $\partial g/\partial y = f_2 = x^3z$ and $\partial g/\partial y = x^3z + \partial k/\partial y$, then $\partial k/\partial y = 0$ and so k is a function of z alone. That is, $k(y,z) = h(z)$.

Now to find h (and hence k), we note that $\partial g/\partial z = f_3$ and also $\partial g/\partial z = x^3y + h'(z)$.

Thus $h'(z) = 7$ and $h(z) = 7z$. (No constant of integration is added since the question only asked for a potential function, not the whole family of potential functions.)

Therefore $g(x,y,z) = x^3yz + 7z$.

As a check, we compute ∇g: $\nabla g = (3x^2yz, x^3z, x^3y + 7) = F$. △

Question 8.2.12 State a three-dimensional analogue of Theorems 8.2.6 and 8.2.8. ◇

Theorem 8.2.13 *Let $C : [a,b] \to \mathbb{R}^2$ parameterize a closed piecewise C^1 curve D and $F : A \to \mathbb{R}^2$ be a C^1 function, where A is an open rectangle in $\mathbb{R}^2$ containing the curve D. If $F = (f_1, f_2)$ has $\partial f_1/\partial y = \partial f_2/\partial x$, then $\int_D F = 0$.*

Example 8.2.14 We will integrate $F(x, y) = (3x^2y, x^3)$ around D, the edges of the triangle with vertices $(1,1), (3,-2), (5,4)$, in a counterclockwise direction. Note that $\partial 3x^2y/\partial y = 3x^2 = \partial x^3/\partial x$ on the open rectangle $(0.5, 5.5)\times(-2.5, 4.5)$ containing the triangle. Thus F is a conservative function on D and D is a piecewise C^1 closed curve, so $\int_D F = 0$. △

Example 8.2.15 If the set A is not an open rectangle, then the conclusion of the Theorem 8.2.8 may not hold. Let $A = \{(x, y) : (x, y) \neq (0,0)\}$. Let $F : A \to \mathbf{R}^2$ be defined by $F(x, y) = (f_1(x, y), f_2(x, y))$, where $f_1(x, y) = -y/(x^2 + y^2)$ and $f_2(x, y) = x/(x^2 + y^2)$. Let $C(t) = (\cos(2\pi t), \sin(2\pi t))$ on $[0, 1]$. The range $D = C([0,1])$ of C is the unit circle. Note that

$$\frac{\partial f_1}{\partial y}(x, y) = \frac{y^2 - x^2}{(x^2 + y^2)^2}$$

and

$$\frac{\partial f_2}{\partial x}(x, y) = \frac{y^2 - x^2}{(x^2 + y^2)^2}$$

so $\partial f_1/\partial y = \partial f_2/\partial x$. Let us compute:

$$\begin{aligned} \int_D F &= \int_0^1 F(C(t)) \cdot C'(t) dt \\ &= \int_0^1 (-\sin(2\pi t), \cos(2\pi t)) \cdot (-2\pi \sin(2\pi t), 2\pi \cos(2\pi t)) dt \\ &= \int_0^1 2\pi[\sin^2(2\pi t) + \cos^2(2\pi t)] dt \\ &= \int_0^1 2\pi dt = 2\pi \end{aligned}$$

△

In this example the set was not an open rectangle and the conclusion of the theorem did not hold. The condition of an open rectangle guarantees that $\int_D F = 0$, but is a stronger condition than is actually required.

Section 8.2 Exercises

1. For each of the following, determine whether or not the function defined on all of $\mathbf{R}^2$ has a potential function or not. If it does, then find the potential function g satisfying the given condition.

   (a) $F(x, y) = (x^2 + y, x + y^2), \quad g(0,0) = 2.$

   (b) $F(x, y) = (xy, xy), \quad g(1,1) = 1.$

   (c) $F(x, y) = (x(x^2 + y^2), y(x^2 + y^2)), \quad g(0,0) = 1.$

2. Compute the line integral of the function F on the given curve.

   (a) $F(x,y) = (x,y)$, $C(t) = (a\cos(t), b\sin(t)), t \in [0,2\pi]$.

   (b) $F(x,y) = (x+y, y)$, on the curve of part (a).

   (c) $F(x,y) = (\cos(\pi y), -\pi x \sin(\pi y))$, $C(t) = (t^2, -t^3), t \in [0,1]$.

   (d) $F(x,y) = (3x^2y + 2, x^3 + 4y^3)$, on the straight line from $(0,0)$ to $(1,1)$.

   (e) $F(x,y) = (y^2 + 2xy, x^2 + 2xy)$, on the straight line from $(-1,2)$ to $(3,1)$.

   (f) $F(x,y) = (x^2 + x\sin(xy), y\sin(xy))$, on the curve of part (a) where t is in $[0,\pi]$.

   (g) The function of Example 8.2.15 on the triangle with vertices $(1,0)$, $(2,0)$, and $(1,1)$.

3. Let $F : \mathbb{R}^2 \to \mathbb{R}^2$ be defined by $f(x,y) = (x^2 + y^2, 3x^2y)$. Find the integral of F along the curve D, where D is the graph of the parabola $y = x^2$ from $(-1,1)$ to $(2,4)$.

4. Evaluate $\int_B (x^2 + 3y)dx - (x + y^3)dy$ where B is the boundary of the triangle with vertices $(0,0)$, $(1,0)$ and $(1,1)$ oriented in a counterclockwise direction.

5. Show that if f and g are continuous functions defined on all real numbers, then $F(x,y) = (f(x), g(y))$ is a conservative function.

6. Find a potential function for F of Exercise 5.

7. Mark each of the following statements true (T) or false (F).

   (a) A path independent function is a continuous function.

   (b) If f has a potential function, then f is a path independent function.

   (c) If $D_1f_1 = D_2f_2$, then f has a potential function.

   (d) A constant function is a path independent function.

   (e) A function is conservative if and only if it has a potential function.

   (f) A piecewise C^1 curve is a connected curve.

   (g) If g is a potential function for both F and H, then $F = H$.

8. Give an example for each of the following statements or explain why no example can exist.

   (a) A path independent function whose integral over a curve joining $(1,1)$ and $(0,3)$ is equal to 4.

   (b) A function having a potential function $g(x,y,z) = x^3y^2 + 3xz^2 - 7yz + 4$.

(c) A linear function $F : \mathbb{R}^3 \to \mathbb{R}^3$ that does not have a potential function.

(d) A conservative function $F : \mathbb{R}^2 \to \mathbb{R}^2$, $F(x, y) = (f_1(x, y), f_2(x, y))$, where f_1 and f_2 are not differentiable.

(e) A conservative function $F : \mathbb{R}^2 \to \mathbb{R}^2$ such that $\int_D F = 2\pi$ where D is the unit circle.

8.3 Multiple and Iterated Integrals

How can you determine the volume of air in an auditorium with a sloping floor? How can you compute the volume of a catsup bottle? How can you find the volume of sand in a sand pile? Or how can you determine the volume of water in a pond? How is the volume of a three-dimensional region defined?

In this section, we shall develop the concept of volume in a manner analogous to what was done for the concept of area. Of course, the central procedure is the Basic Approximation Process. A key to the development of a volume concept is to reconsider the development of the area concept and ask what aspects of it can be extended to a higher dimension. In particular, let us consider three aspects.

1. Consider the region in question. The two-dimensional region in the area concept was the ordinate set of a bounded, nonnegative function defined over an interval, i.e., $f : [a, b] \to \mathbb{R}$. Thus, let us start the development of a volume concept by restricting the three-dimensional region to be the three-dimensional ordinate set of a bounded, nonnegative function defined over a rectangle, i.e., $f : [a, b] \times [c, d] \to \mathbb{R}$. Recall from Chapter 1 that a rectangle, expressed as a Cartesian product of the form $[a, b] \times [c, d]$, is the generalization of an interval to two dimensions.

2. Next, develop a partition of the domain of the function. In the area problem, the interval $[a, b]$ was partitioned into subintervals by a partition $P = \{a = x_0, x_1, \ldots, x_n = b\}$. This idea can be extended to give a partition of the rectangle $[a, b] \times [c, d]$ into subrectangles by forming the Cartesian product of a partition P of $[a, b]$ with a partition Q of $[c, d]$. This is illustrated in the following figure where $[a, b] = [0, 3]$, $[c, d] = [0, 2]$, $P = \{0, 1, 2, 3\}$, and $Q = \{0, 1, 2\}$.

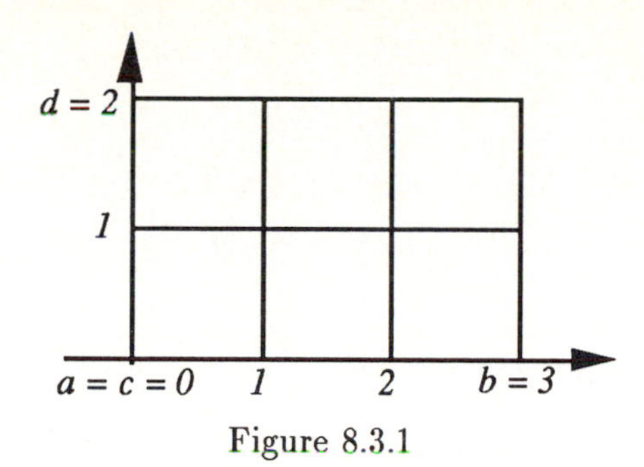

Figure 8.3.1

3. Compute the volumes of approximating subregions. The approximating subregions in the area problem were rectangles or 2-cells. The base of a 2-cell was a subinterval in the partitioned domain and the height was obtained by evaluating the function at some particular point in the base. Thus, in three dimensions, let us consider an approximating subregion to be a 3-cell, where a 3-cell is defined to be a parallelepiped whose base is a subrectangle in the partitioned domain and whose height is the function evaluated at some particular point in the base.

We are now in a position to work through the four steps of the Basic Approximation Process to formulate a definition of volume. Since the details are so similar to those in the development of the definition of area, we shall omit most of the them and illustrate only a few of the key steps.

Consider the three-dimensional ordinate set of a bounded, nonnegative function $f : [a,b] \times [c,d] \to \mathbb{R}$. Let $P_n = \{a = x_0, x_1, \ldots, x_n = b\}$ and $Q_m = \{c = y_0, y_1, \ldots, y_m = d\}$. $P \times Q$ partitions the domain of f into $n \cdot m$ subrectangles R_{ij} where

$$R_{ij} = [x_{i-1}, x_i] \times [y_{j-1}, y_j] \quad \text{for} \quad 1 \le i \le n \text{ and } 1 \le j \le m$$

The volume of an *(i,j)*th 3-cell is the area of R_{ij} times $f(x_i^*, y_j^*)$, where (x_i^*, y_j^*) is any point in R_{ij}.

The sum of the volumes of the 3-cells is expressed in terms of a double Riemann Sum:

$$\sum_{j=1}^{m}\sum_{i=1}^{n} f(x_i^*, y_j^*)(x_i - x_{i-1})(y_j - y_{j-1})$$

We shall illustrate the expansion of this Riemann Sum in the case where $n = 3$ and $m = 2$. The "inside" summation is expanded first yielding

$$\sum_{j=1}^{2}[f(x_1^*, y_j^*)(x_1 - x_0)(y_j - y_{j-1}) \\ + f(x_2^*, y_j^*)(x_2 - x_1)(y_j - y_{j-1}) \\ + f(x_3^*, y_j^*)(x_3 - x_2)(y_j - y_{j-1})]$$

Now, expanding the "outside" sum and recalling that the summation of a sum is the sum of the summation yields

$$[f(x_1^*, y_1^*)(x_1 - x_0)(y_1 - y_0) + f(x_2^*, y_1^*)(x_2 - x_1)(y_1 - y_0) \\ + f(x_3^*, y_1^*)(x_3 - x_2)(y_1 - y_0) + f(x_1^*, y_2^*)(x_1 - x_0)(y_2 - y_1) \\ + f(x_2^*, y_2^*)(x_2 - x_1)(y_2 - y_1) + f(x_3^*, y_2^*)(x_3 - x_2)(y_2 - y_1)]$$

Note that there are six 3-cells corresponding to the six subrectangles in the previous figure. Lower and upper approximating double Riemann Sums are formed in exactly the same manner as they were in the development of area. That is,

$$LA(f, P, Q) = \sum_{j=1}^{m}\sum_{i=1}^{n}(minf)_{ij}\Delta x_i \Delta y_j$$

and

$$UA(f, P, Q) = \sum_{j=1}^{m}\sum_{i=1}^{n}(maxf)_{ij}\Delta x_i \Delta y_j$$

where $(minf)_{ij}$ represents the minimum value of f over R_{ij} and $(maxf)_{ij}$ represents the maximum value of f over R_{ij}.

Recall that $|P|$ denotes the norm of the partition P, that is, the length of the largest subinterval defined by P. Now consider two sequences of partitions, $\{P_n\}$ and $\{Q_m\}$, with the properties

$$|P_{n+1}| < |P_n|, \quad \lim_{n\to\infty} |P_n| = 0 \quad \text{and} \quad |Q_{m+1}| < |Q_m|, \quad \lim_{m\to\infty} |Q_m| = 0$$

Note that

$$\begin{aligned} LA(f, P_n, Q_m) &\leq \sum_{j=1}^{m}\sum_{i=1}^{n} f(x_i^*, y_j^*)(x_i - x_{i-1})(y_j - y_{j-1}) \\ &\leq UA(f, P_n, Q_m) \end{aligned}$$

As either n or m increases the sequence of lower approximating sums $\{LA(f, P_n, Q_m)\}$ increases and the sequence of upper approximating sums $\{UA(f, P_n, Q_m)\}$ decreases since the partition becomes finer. The volume of the ordinate set is now obtained by taking the limits of these two sequences as $|P_n|$ and $|Q_m|$ both go to zero. Formally, then we have the following definition.

Definition 8.3.2 (Volume) *Let $f : [a,b] \times [c,d] \to \mathbb{R}$ be bounded and nonnegative. Let $\{P_n\}$ and $\{Q_m\}$ be two arbitrary sequences of partitions of $[a,b]$ and $[c,d]$ respectively with the properties that $\lim_{n\to\infty} |P_n| = 0$ and $\lim_{m\to\infty} |Q_m| = 0$.*

The volume of the ordinate set of f *is the common limit of $\{LA(f,P_n,Q_m)\}$ and $\{UA(f,P_n,Q_m)\}$ as the norms, $|P_n|$ and $|Q_m|$, approach zero, provided this common limit exists. If the common limit does not exist, then the ordinate set is said not to have a volume.*

We formulate the definition for a double integral by dropping the requirement that the function in the development of the volume concept be nonnegative (recall that this was exactly the process that was followed in defining the "single" integral).

Definition 8.3.3 (Double Integral) *Let $f : [a,b] \times [c,d] \to \mathbb{R}$. The* Riemann Integral of f *over $[a,b] \times [c,d]$ is the unique real number, denoted by $\int_c^d \int_a^b f(x,y)dxdy$, having the property that for* every *partition P of $[a,b]$ and partition Q of $[c,d]$*

$$LA(f,P,Q) \leq \int_c^d \int_a^b f(x,y)dxdy \leq UA(f,P,Q)$$

Thus, the double integral of a bounded, nonnegative function defined over a rectangle yields the volume of the ordinate set of that function.

Notation: If R denotes the rectangle $[a,b] \times [c,d]$, an alternative notation for the double integral of f over R is

$$\iint_R f(x,y)dxdy$$

The expression $dxdy$ is also denoted by dA. Thus, the integral may be written as $\iint_D Rf(x,y)dA$.

Example 8.3.4 As usual, we start with a problem whose answer is known. We will find the volume of the ordinate set of the constant function $f : [1,5] \times [2,4] \to \mathbb{R}$ defined by $f(x,y) = 10$.

Let R denote the domain of f. Since f is a constant function, $LA(f,P_n,Q_m) = UA(f,P_n,Q_m)$ and thus

$$\begin{aligned}\text{Volume} &= \iint_R 10dxdy = \sum_{j=1}^{m}\sum_{i=1}^{n}(10)\Delta x_i \Delta y_j \\ &= 10\sum_{j=1}^{m}\sum_{i=1}^{n}\Delta x_i \Delta y_j = 10(5-1)(4-2) = 80\end{aligned}$$

△

The definition of double integration can be generalized by replacing the requirement that the domain of the function be a rectangle with the requirement that the domain be bounded.

Let $f : A \to \mathbb{R}$, where A is a bounded set in $\mathbb{R}^2$. By the definition of boundedness, A is contained within a rectangle R. Now extend f to all of R by defining it to be zero outside of A: $g : R \to \mathbb{R}$ is defined by

$$g(x,y) = \begin{cases} f(x,y) & \text{for } (x,y) \in A \\ 0 & \text{otherwise} \end{cases}$$

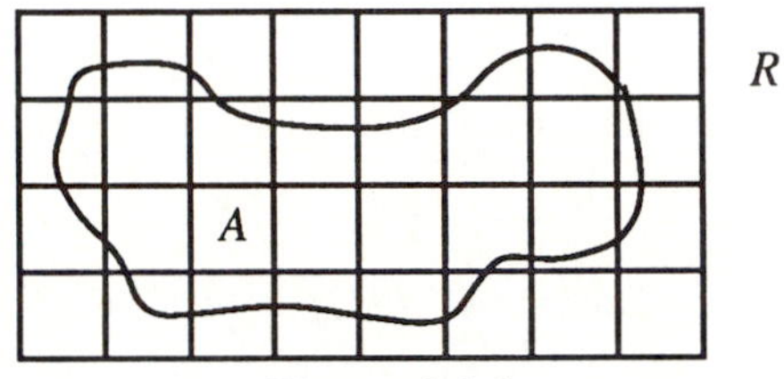

Figure 8.3.5

We define the double integral of f over A to be the double integral of g over R. That is,

$$\iint_A f(x,y)dxdy = \iint_R g(x,y)dxdy$$

the double integral of g over R.

The process of extending the integral concept from a single integral to a double integral may be repeated (with some obvious changes) to obtain the definition of a multiple integral for a bounded function $f : A \to \mathbb{R}$ where A is a bounded subset of $\mathbb{R}^n$.

The idea of a double integral arose very naturally in the context of defining the concept of volume. Applications of triple integrals occur in the determination of mass and moments of three-dimensional objects. Profit or cost functions for business often involve n-fold integrals. Situations that involve summing values of an n-dimensional function over a region are usually modeled with an n-fold integral.

Multiple integrals can be evaluated by iterated single integrals, e.g.,

$$\iint_R f(x,y)dxdy = \int_c^d \left[\int_a^b f(x,y)dx \right] dy$$

In evaluating an n-fold iterated integral, the inside integral is evaluated first. This will yield an (n-1)-fold iterated integral. The (new) inside integral is then evaluated leaving an (n-2)-fold iterated integral. This process continues until all n integrals have been evaluated. Thus, the evaluation of an n-fold iterated integral consists of the evaluation of n "single" integrals. It is important to realize when integrating that all variables are held constant except for the variable of integration under consideration.

The most important aspect of evaluating or setting up an iterated integral is to understand that the role of the limits of integration is to describe the region (region of integration) over which the integrand function is being integrated. For example, in the integral $\int_2^8 f(x)dx$, the variable of integration x ranges from 2 to 8 ($2 \leq x \leq 8$). Every definite integral has *one* variable of integration and that variable of integration ranges from the lower limit to the upper limit of integration.

Example 8.3.6

$$\begin{aligned}
\int_1^2 \int_{-1}^3 2xydxdy &= \int_1^2 \left(\int_{-1}^3 2xydx\right) dy \\
&= \int_1^2 (x^2y]_{-1}^3)dy = \int_1^2 (9y - y)dy \\
&= 4y^2]_1^2 = 12
\end{aligned}$$

Since x is the variable of integration of the inside integral, the range of x is $-1 \leq x \leq 3$. The variable of integration of the "outside" integral is y and so the range of y is $1 \leq y \leq 2$. Hence the region of integration is the rectangle in the xy-plane described by the double inequalities: $-1 \leq x \leq 3$ and $1 \leq y \leq 2$. $\triangle$

Example 8.3.7

$$\begin{aligned}
\int_1^2 \int_y^{y^2} 2xydxdy &= \int_1^2 \left(\int_y^{y^2} 2xydx\right) dy = \int_1^2 x^2y]_y^{y^2} dy = \int_1^2 (y^5 - y^3)dy \\
&= \frac{27}{4}
\end{aligned}$$

In this example the region of integration is described analytically by:

$$y \leq x \leq y^2 \quad \text{and} \quad 1 \leq y \leq 2$$

and geometrically by the shaded region in the following figure. $\triangle$

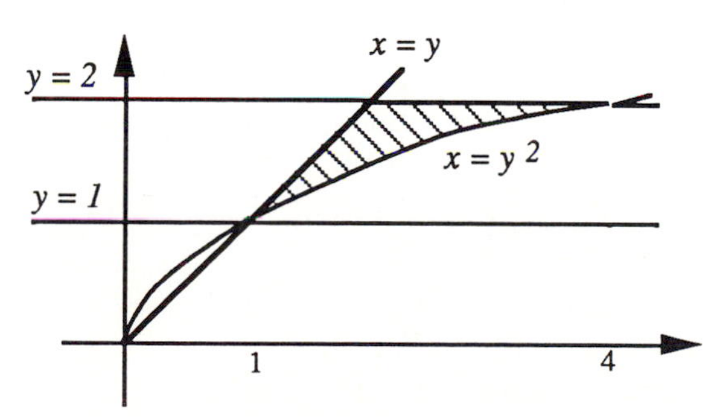

Figure 8.3.8

In Chapter 5, we developed the concept of area of an ordinate set of a bounded function f over an interval $[a,b]$ as $\int_a^b f(x)dx$. An important observation to make is that the area of a planar region is equal to the volume of the three-dimensional region that has the planar region as a base and a height of one. Thus the area of the ordinate set of f over $[a,b]$ can be computed by evaluating the double integral of the function that is identically one over the ordinate set. That is,

$$\text{Area} = \int_a^b f(x)dx = \int_a^b \int_0^{f(x)} (1)dydx$$

Example 8.3.9 We will find the area of the ordinate set of $f(x) = x^2 + 1$ over $[-2,3]$ in two ways: first, by means of a single integral and second, by a double integral.

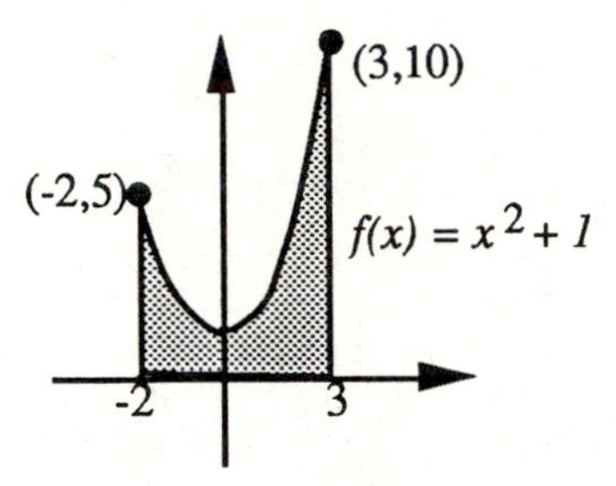

Figure 8.3.10

$$\text{Area} = \int_{-2}^{3} (x^2+1)dx = \frac{1}{3}x^3 + x\Big]_{-2}^{3} = \frac{50}{3}$$

$$\text{Area} = \int_{-2}^{3} \int_0^{x^2+1} (1)dydx = \int_{-2}^{3} (x^2+1)dx = \frac{1}{3}x^3 + x\Big]_{-2}^{3} = \frac{50}{3}$$

(Note that the region of integration, the ordinate set of f, is described by the inequalities: $-2 \le x \le 3$ and $0 \le y \le x^2 + 1$.) △

To further illustrate that any single integral can be expressed as a double integral with integrand one, we compute the area of the region bounded by the graphs of two integrable functions f and g over $[a,b]$, where $f(x) \le g(x)$. In Chapter 5, this area was defined to be the area of the ordinate set of g minus the area of the ordinate set of f. That is,

$$\begin{aligned} \text{Area} &= \int_a^b g(x)dx - \int_a^b f(x)dx = \int_a^b [g(x) - f(x)]dx \\ &= \int_a^b \int_{f(x)}^{g(x)} (1)dydx \end{aligned}$$

Question 8.3.11 Use a repeated integral to find the area of the circle, $x^2 + y^2 = 1$. ◇

We shall now extend this concept of area to volume and express the volume of the ordinate set of an integrable function g defined over a bounded region in the plane as the triple integral of $f(x, y, z) = 1$, defined over the three-dimensional ordinate set of g.

Example 8.3.12 We will express the volume of the ordinate set of $g(x, y) = xy$ over the region bounded by the graphs of $y = x^2$, $x = 4$, and the x-axis as a double integral and then as a triple integral.

We first sketch the region in the plane.

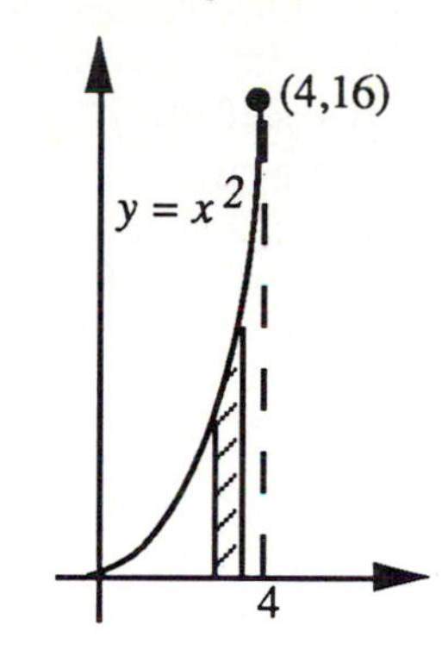

Figure 8.3.13

Integrating first "over a vertical strip" we have

$$\text{Volume} = \int_0^4 \int_0^{x^2} g(x, y)\,dy\,dx$$

For the triple integral expression, we start by sketching the ordinate set of g in three dimensions.

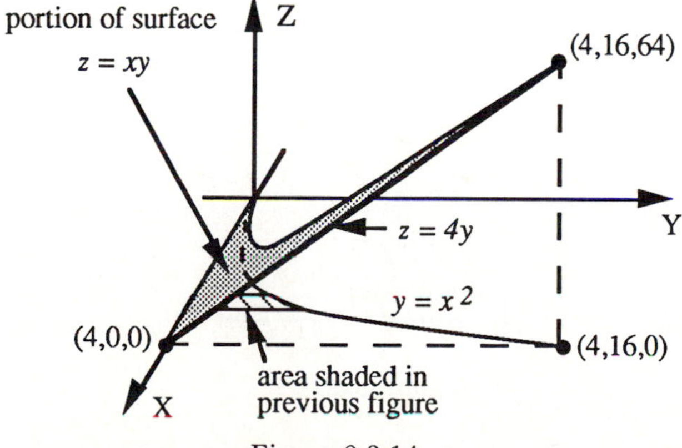

Figure 8.3.14

Integrating first over a vertical parallelepiped (i.e., with respect to z) and then over the region in the xy-plane as we did above yields

$$\text{Volume} = \int_0^4 \int_0^{x^2} \int_0^{xy} (1)dzdydx \qquad \triangle$$

We once again call the reader's attention to the fact that the limits of integration describe the region of integration. Thus in the previous example, the region of integration is described by the inequalities $0 \leq z \leq xy$, $0 \leq y \leq x^2$, and $0 \leq x \leq 4$.

Determining the proper limits is often difficult, particularly when the number of integrals is greater than two. The difficulties can be lessened considerably by a carefully drawn sketch of the region of integration. Consider integrating $f(x, y)$ over the triangular region with vertices $(0, 0)$, $(3, 0)$, and $(3, 3)$. Geometrically, one can think of integrating first with respect to x and second with respect to y [i.e., $\int_0^3 \int_y^3 f(x, y)dxdy$] as integrating $f(x, y)$ over the horizontal strip shown in the next figure as the strip moves from the bottom to the top of the region.

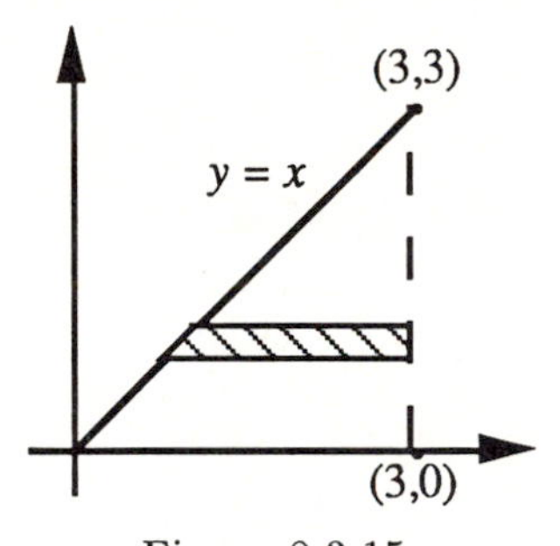

Figure 8.3.15

In a similar manner, one can think of integrating first with respect to y and second with respect to x [i.e., $\int_0^3 \int_0^x f(x, y)dydx$] as integrating $f(x, y)$ over the vertical strip shown in the next figure as the strip moves from the left end to the right end of the region.

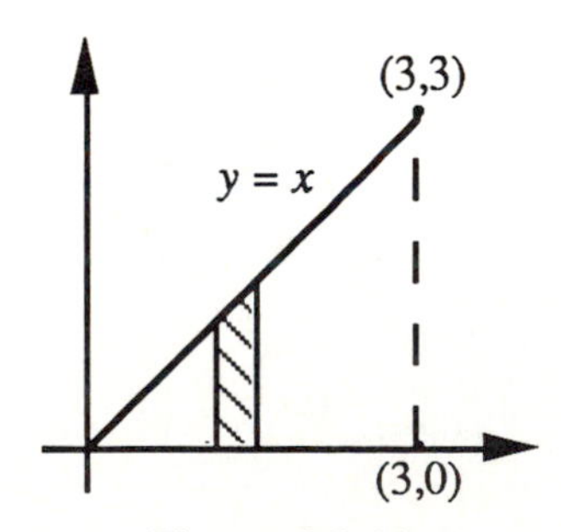

Figure 8.3.16

Thus, $\iint_R f(x, y)dA = \int_0^3 \int_y^3 f(x, y)dxdy = \int_0^3 \int_0^x f(x, y)dydx$.

Although theoretically a repeated integral of a function gives the same value for every order of integration, it is often the case that the integral is easier to evaluate in one order than another because of the form of the integrand or the shape of the region. We will give two examples of repeated integrals where the evaluation is relatively easy for one order of integration and more difficult for the other order. The difficulty will be due to the form of the integrand in the first example and the region of integration in the second example.

Example 8.3.17 We will set up the integral for the function defined by

$$f(x,y) = y\sqrt{x^3 - 2}$$

over the triangular region R illustrated in Figure 8.3.16.

We can evaluate the integral first with respect to y and second with respect to x, but not in the reverse order. That is, we can evaluate

$$\int_0^3 \int_0^x y\sqrt{x^3 - 1}dydx$$

but we do not know how to evaluate (except by numerical approximation)

$$\int_0^3 \int_y^3 y\sqrt{x^3 - 2}dxdy$$

△

Example 8.3.18 We will set up the integrals of $f(x,y) = xy$ over the region bounded by the graphs of $y = -x$ and $x = y^2 - 2$.

We first sketch the graphs in order to visualize the region of integration.

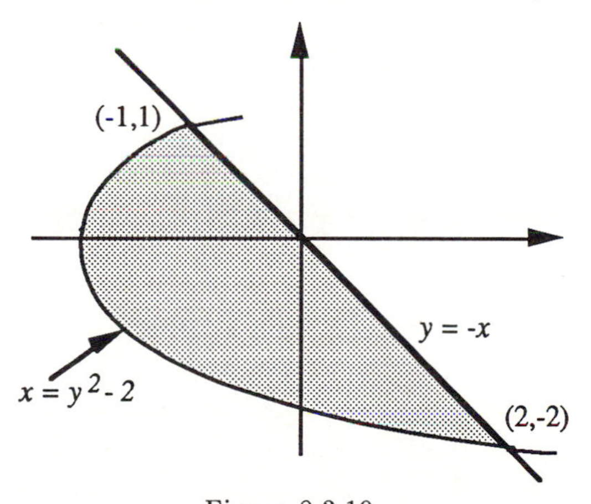

Figure 8.3.19

We can evaluate the double integral over the region as a single repeated integral if we integrate first with respect to x and second with respect to y. That is,

$$\int_{-2}^1 \int_{y^2-2}^{-y} xydxdy$$

Interchanging the order of integration, however, would require that we partition the region into two regions and thus have to evaluate two repeated integrals. That is,

$$\int_{-2}^{-1}\int_{-(x+2)^{1/2}}^{(x+2)^{1/2}} xydydx + \int_{-1}^{2}\int_{-(x+2)^{1/2}}^{-x} xydydx \qquad \triangle$$

Section 8.3 Exercises

In Exercises 1 through 16, evaluate the iterated integrals.

1. $\int_0^1 \int_0^2 (x+3)dydx$
2. $\int_1^3 \int_{-1}^1 (2x-4y)dydx$
3. $\int_2^4 \int_0^1 x^2ydxdy$
4. $\int_{-2}^0 \int_{-1}^2 (x^2+y^2)dxdy$
5. $\int_0^2 \int_0^1 y\sin(x)dydx$
6. $\int_0^3 \int_0^1 \sqrt{y+x^2}dxdy$
7. $\int_{-1}^2 \int_2^4 (2x^2y+3xy^2)dxdy$
8. $\int_{-1}^0 \int_2^5 dxdy$
9. $\int_4^6 \int_{-3}^7 dydx$
10. $\int_3^4 \int_1^2 1/(x+y)^2dydx$
11. $\int_{-1}^1 \int_0^2 \int_0^1 (x^2+y^2+z^2)dxdydz$
12. $\int_0^2 \int_{-1}^{y^2} \int_1^z yzdxdzdy$
13. $\int_0^{\pi} \int_0^1 \int_0^{x^2} x\cos(y)dzdxdy$
14. $\int_0^3 \int_0^{\sqrt{9-z^2}} \int_0^x xydydxdz$
15. $\int_0^2 \int_0^{\sqrt{4-x^2}} \int_{-5+x^2+y^2}^{3-x^2-y^2} xdzdydx$
16. $\int_{1/3}^{1/2} \int_0^{\pi} \int_0^1 zx\sin(xy)dzdydx$

In Exercises 17 and 18, find the area of the ordinate set of f over the interval in two ways: first, by means of a single integral and, second, by using a double integral.

17. $f:[-2,2]\to \mathbb{R}$, $f(x)=2x+6$
18. $f:[0,\pi]\to \mathbb{R}$, $f(x)=\sin(x)$

In Exercises 19 through 22, use a double integral to find the volume.

19. The volume of the ordinate set of $f(x,y) = 2x+y$ over the rectangle $R=[3,5]\times[1,2]$.
20. The volume of the ordinate set of $z=f(x,y)=2$ over the interior of the circle $x^2+y^2=1$.
21. The volume of the solid in the first octant enclosed by the surface $f(x,y)=x^2$ and the planes $x=2$, $y=3$.
22. The volume of the solid enclosed by $x^2+y^2+z^2=4$.

In Exercises 23 and 24, express the integral as an equivalent integral with the order of integration reversed.

23. $\int_0^2 \int_0^{\sqrt{x}} f(x,y)dydx$ 24. $\int_0^4 \int_{2y}^8 f(x,y)dxdy$

25. Use a triple integral to find the volume of the solid enclosed between the cylinder $x^2 + y^2 = 9$ and the planes $z = 1$ and $x + z = 5$.

26. Use a triple integral to find the volume of the solid enclosed by the paraboloids $z = x^2 + y^2$ and $z = 4 - (x^2 + y^2)$.

In Exercises 27 through 30, sketch the solid whose volume is given by the integral.

27. $\int_0^3 \int_{x^2}^9 \int_0^2 dzdydx$ 28. $\int_0^2 \int_0^{2-y} \int_0^{2-x-y} dzdxdy$

29. $\int_0^1 \int_0^{\sqrt{1-x^2}} \int_0^2 dydzdx$ 30. $\int_{-2}^2 \int_0^{4-y^2} \int_0^2 dxdzdy$

31. Give an example of each of the following statements or explain why no example exists.

    (a) A nonconstant function $f : \mathbb{R}^2 \to \mathbb{R}$ and a region D in the plane such that

$$\iint_D f(x,y)dA = 0$$

    (b) A single integral that cannot be expressed as a double integral.

    (c) A nonconstant function $f : \mathbb{R}^3 \to \mathbb{R}$ and a region D in $\mathbb{R}^3$ such that

$$\iiint_D f(x,y,z)dV = 1$$

32. Let T be a right triangle with legs of length 5 and 10 units. Find the average distance of points in T to the longer leg.

33. Set up a triple integral expression for the volume of the polar cap that is formed by the plane $z = 1$ and the solid sphere $x^2 + y^2 + z^2 \leq 4$.

34. Project. Develop a concept of "equilibrium point." Consider both discrete and continuous situations. For example, your concept should encompass the balancing point for one- and two-dimensional seesaws, center of mass for uniform and nonuniform rods, plates, and solids, political compromise point, etc.

8.4 Differential Forms and Green's Theorem

In this section we will extend the Fundamental Theorem of Calculus to integration over two-dimensional sets in the plane.

Recall that the Fundamental Theorem of Calculus for functions from the reals to the reals states that if $f : \mathbb{R} \to \mathbb{R}$ is C^1, then

$$\int_a^b f'(x)dx = f(b) - f(a)$$

This result was extended to functions defined on curves in Theorem 8.2.3: if $f : \mathbb{R}^n \to \mathbb{R}$ is defined on a curve $C : [a, b] \to \mathbb{R}^n$, then

$$\int_{C([a,b])} \nabla f = f(C(b)) - f(C(a))$$

Both of these theorems relate the integral over a line (a one-dimensional set) of the derivative of a function to the values of the function at the boundary (a zero-dimensional set of two points). To see how this can generalize, we need to examine two aspects of these results, orientation and differentials.

Orientation

The integral depends on how we traverse the domain of integration. If we integrate from b to a, then $\int_b^a f(x)dx = -\int_a^b f(x)dx$. Similarly, if we reverse the orientation of the curve in a line integral, the sign of the integral changes.

Now we will consider a bounded domain D in the plane and its boundary, denoted ∂D. We will assume that the boundary is either smooth (like a circle) or piecewise smooth (like a square). Here "smooth" means that the curve has a unique tangent line at each point. We will first consider the case when D is a square. The boundary ∂D is piecewise smooth, consisting of four straight line segments. We will orient ∂D counterclockwise, as in Figure 8.4.1(a). Notice that if the boundary is traversed in the counterclockwise orientation, then the interior of D is always on the left-hand side.

Domain and Oriented Boundary

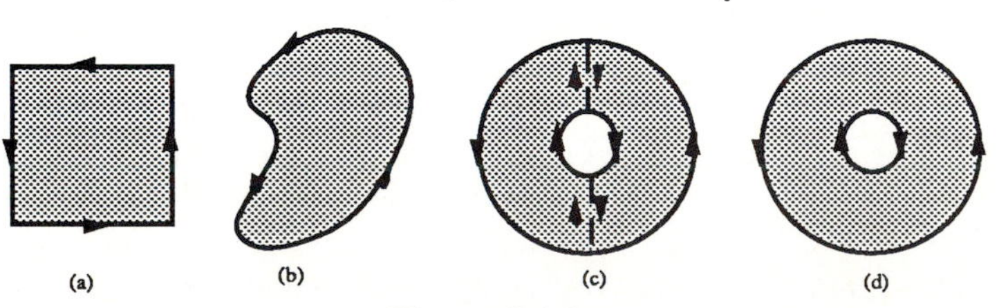

Figure 8.4.1

If D has no holes, then we can again orient the boundary ∂D counterclockwise as in Figure 8.4.1(b). If D has a hole, as in part (c), then we can consider D as constructed of pieces without holes, and orient each of the pieces. Notice that on each of these pieces, the direction traveled cancels out except on the boundary of D, leaving us with part (d). (If the domain is divided into more than two subdomains, then the final result is the same, since the pieces of the subboundary crossing the interior of the domain will cancel out.) Again, as we walk counterclockwise around the boundary, the interior of D is on our left.

In the Fundamental Theorem of Calculus, $\int_a^b f'(x)dx = f(b) - f(a)$, the set $D = [a, b]$ and $\partial D = \{a, b\}$. When we evaluate f on ∂D, we subtract to reflect the orientation: $f(b) - f(a)$. The fundamental theorem can now be written

$$\int_D f'(x)dx = \int_{\partial D} f$$

where we *define* integration over a zero-dimensional set of two points to be

$$\int_{\partial D} f = \int_{\{a,b\}} f = f(b) - f(a)$$

Differential Forms

Using the notation of differentials (see Section 6.2) so that $df = f'(x)dx$, we can write the above form of the fundamental theorem as $\int_D df = \int_{\partial D} f$. The differential dx can be thought of as an infinitesimally small change in x. Since we integrate in the positive direction, from a to b, if $a < b$, we can consider dx as a positively oriented infinitesimally small change in x.

In order to generalize the above equation, we need to develop an algebra of differentials.

A *0-differential form* is a C^1 function.

A *1-differential form* for the variables x, y, and z is an expression of the form $fdx + gdy + hdz$ where f, g, and h are smooth functions of the three variables. Here, by "smooth" we mean sufficiently differentiable, e.g., a C^1 function. If the variables are $x_1, \ldots, x_n$, then a 1-differentiable form is an expression of the form $f_1dx_1 + \cdots + f_ndx_n$ where each f_i is a smooth function of n variables.

A *2-differential form* for the variables x, y, and z is an expression of the form $fdx \wedge dy + gdx \wedge dz + hdy \wedge dz$ where f, g, and h are smooth functions of the three variables. If there are n variables $x_1, \ldots, x_n$, then a 2-differentiable form is an expression of the form $\sum_{i,j} f_{i,j}dx_i \wedge dx_j$, where each $f_{i,j}$ is a smooth function of n variables. (Using the rules for wedge product below, we can always rewrite such an expression as $\sum_{i<j} f_{i,j}dx_i \wedge dx_j$.)

The $dx_i \wedge dx_j$ is the product of 1-differential forms dx and dz. Called the *wedge product,* this expression can be thought of as an infinitesimally small oriented piece of the plane.

Wedge Product

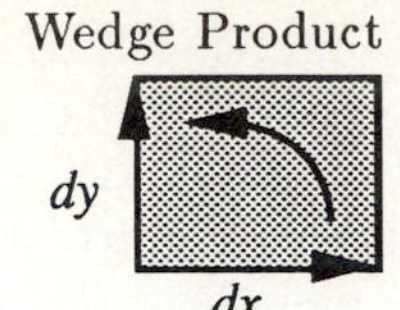

Figure 8.4.2

Since it is oriented, $dx \wedge dy = -dy \wedge dx$. The wedge product obeys all the usual laws of algebra except that it is *not* commutative. In the following, α, β, and γ are differential forms (traditionally denoted by lowercase Greek letters) and f is a real-valued function.

Theorem 8.4.3 (Rules for Wedge Product)

(a) $\alpha \wedge (\beta \wedge \gamma) = (\alpha \wedge \beta) \wedge \gamma$ (associative law)
(b) $\alpha \wedge (\beta + \gamma) = (\alpha \wedge \beta) + (\alpha \wedge \gamma)$ (distributive law)
(c) $\alpha \wedge (f\beta) = (f\alpha) \wedge \beta = f(\alpha \wedge \beta)$ (real values factor out)
(d) $dx \wedge dy = -dy \wedge dx$ (anti-commutivity)

Example 8.4.4 We will exhibit a distributive property and the fact that $dx \wedge dx = 0$

$$\begin{aligned}
\text{(a) } dy \wedge (fdx + gdz) &= dy \wedge fdx + dy \wedge gdz && \text{[by rule (b)]} \\
&= fdy \wedge dx + gdy \wedge dz && \text{[by rule (c)]} \\
&= -fdx \wedge dy + gdy \wedge dz && \text{[by rule (d)]}
\end{aligned}$$

(b) $dx \wedge dx = -dx \wedge dx$ [by rule (d)], thus $dx \wedge dx = 0$ △

An n-differential form is sometimes just called an *n-form*. The n refers to the number of differentials in each term of the expression. Each term must have the same number of differentials; e.g.,

$$3dx \wedge dy + 2dz$$

is not a 2-form, since it is the sum of a 2-form and a 1-form.

The Exterior Derivative

The final step is to differentiate a differential form. Differential forms are usually indicated by lowercase Greek letters. We will explain how to take the *exterior derivative* d of a differential form α giving $d\alpha$.

If α is a 0-differential form, i.e., a function, then we use the differential. For example, if $\alpha = f : \mathbb{R}^3 \to \mathbb{R}$ with variables x, y, and z, then

$$d\alpha = df = \frac{\partial f}{\partial x}dx + \frac{\partial f}{\partial y}dy + \frac{\partial f}{\partial z}dz$$

For other differential forms, we require d to be linear and first apply d to the function part as above, and then apply the rules of wedge product.

Example 8.4.5 Let α be a 1-form with two variables, x and y. Then $\alpha = fdx + gdy$ for some $f, g : \mathbb{R}^2 \to \mathbb{R}$.

$$\begin{aligned}
d\alpha &= d(fdx + gdy) = (df) \wedge dx + (dg) \wedge dy \quad \text{(by linearity)} \\
&= \left(\frac{\partial f}{\partial x}dx + \frac{\partial f}{\partial y}dy\right) \wedge dx + \left(\frac{\partial g}{\partial x}dx + \frac{\partial g}{\partial y}dy\right) \wedge dy \quad \text{(function differential)} \\
&= \frac{\partial f}{\partial x}dx \wedge dx + \frac{\partial f}{\partial y}dy \wedge dx + \frac{\partial g}{\partial x}dx \wedge dy + \frac{\partial g}{\partial y}dy \wedge dy \quad \text{(by distributivity)} \\
&= \frac{\partial f}{\partial y}dy \wedge dx + \frac{\partial g}{\partial x}dx \wedge dy \quad (\text{since } dx \wedge dx = 0 \text{ and } dy \wedge dy = 0) \\
&= \left(\frac{\partial g}{\partial x} - \frac{\partial f}{\partial y}\right) dx \wedge dy \quad (\text{since } dy \wedge dx = -dx \wedge dy)
\end{aligned}$$

△

Example 8.4.6 If α is a 1-differential form with three variables x, y, and z so that $\alpha = fdx + gdy + hdz$ for some f, g, and $h : \mathbb{R}^3 \to \mathbb{R}$ in the variables x, y, and z, then

$$d\alpha = (df) \wedge dx + (dg) \wedge dy + (dh) \wedge dz$$

where df, dg, and dh are the differentials of functions:

$$\begin{aligned}
df &= \frac{\partial f}{\partial x}dx + \frac{\partial f}{\partial y}dy + \frac{\partial f}{\partial z}dz \\
dg &= \frac{\partial g}{\partial x}dx + \frac{\partial g}{\partial y}dy + \frac{\partial g}{\partial z}dz \\
dh &= \frac{\partial h}{\partial x}dx + \frac{\partial h}{\partial y}dy + \frac{\partial h}{\partial z}dz
\end{aligned}$$

Thus, substituting into the equation for $d\alpha$,

$$\begin{aligned}
d\alpha &= \left(\frac{\partial f}{\partial x}dx + \frac{\partial f}{\partial y}dy + \frac{\partial f}{\partial z}dz\right) \wedge dx + \left(\frac{\partial g}{\partial x}dx + \frac{\partial g}{\partial y}dy + \frac{\partial g}{\partial z}dz\right) \wedge dy \\
&\quad + \left(\frac{\partial h}{\partial x}dx + \frac{\partial h}{\partial y}dy + \frac{\partial h}{\partial z}dz\right) \wedge dz \\
&= \frac{\partial f}{\partial x}dx \wedge dx + \frac{\partial f}{\partial y}dy \wedge dx + \frac{\partial f}{\partial z}dz \wedge dx + \frac{\partial g}{\partial x}dx \wedge dy + \frac{\partial g}{\partial y}dy \wedge dy \\
&\quad + \frac{\partial g}{\partial z}dz \wedge dy + \frac{\partial h}{\partial x}dx \wedge dz + \frac{\partial h}{\partial y}dy \wedge dz + \frac{\partial h}{\partial z}dz \wedge dz \\
&= \left(\frac{\partial f}{\partial y} - \frac{\partial g}{\partial x}\right) dy \wedge dx + \left(\frac{\partial f}{\partial z} - \frac{\partial h}{\partial x}\right) dz \wedge dx + \left(\frac{\partial g}{\partial z} - \frac{\partial h}{\partial y}\right) dz \wedge dy
\end{aligned}$$

△

Example 8.4.7 ($d^2 = 0$)

$$ddx = d(1dx) = d(1) \wedge dx = 0dx \wedge dx = 0$$

and

$$dd(x^3) = d(3x^2dx) = 6xdx \wedge dx = 0 \qquad \triangle$$

Integration of Differential Forms

Let α be an n-differential form. The α is integrated over an n-dimensional smooth oriented domain. We will look at several cases for small n.

Case $n = 0$. If α is a 0-differential form (a function), then α is integrated over a zero-dimensional domain D, which is just a pair of points $D = \{a, b\}$. We *define* $\int_D \alpha$ to be evaluation at the endpoints:

$$\int_D \alpha = \alpha(b) - \alpha(a)$$

Case $n = 1$. If α is a 1-differential form on an interval $D = [a, b]$ with one variable x, then $\alpha = f(x)dx$ for some smooth function f and

$$\int_D \alpha = \int_a^b \alpha = \int_a^b f(x)dx$$

(Note that this is the Riemann Integral defined in Section 5.3.) If α is a 1-differential form on a curve $D = C([a, b])$, $C : [a, b] \to \mathbb{R}^3$ with variables x, y, and z [so that $C(t) = (x(t), y(t), z(t))$], then $\alpha = fdx + gdy + hdz$ for some smooth functions f, g, h and

$$\begin{aligned}
\int_D \alpha &= \int_{C([a,b])} \alpha = \int_{C([a,b])} fdx + gdy + hdz \\
&= \int_a^b f\frac{dx}{dt}dt + g\frac{dy}{dt}dt + h\frac{dz}{dt}dt = \int_a^b (f, g, h)\bullet(x'(t), y'(t), z'(t))dt \\
&= \int_a^b F(C(t))\bullet C'(t)dt = \int_{C([a,b])} F
\end{aligned}$$

where $F : \mathbb{R}^3 \to \mathbb{R}^3$ is defined by $F(x, y, z) = (f(x, y, z), g(x, y, z), h(x, y, z))$. This is precisely the line integral defined in Definition 8.1.4.

Case $n = 2$. If α is a 2-differential form on a two-dimensional domain D in the plane, then $\alpha = fdx \wedge dy = -fdy \wedge dx$. To integrate this, we always use the orientation $dx \wedge dy$ and turn it into a multiple integral:

$$\int_D \alpha = \int_D fdx \wedge dy = \iint_D fdxdy$$

Generalized Stokes' Theorem

The theorem which gives us the various versions of the Fundamental Theorem of Calculus is called the *Generalized Stokes' Theorem.*

Theorem 8.4.8 (Generalized Stokes' Theorem) *Assume α is an $(n-1)$-form on an oriented n-dimensional domain D with piecewise smooth boundary. Then*

$$\int_D d\alpha = \int_{\partial D} \alpha$$

Note that since α is an $(n-1)$-differential form, it must be integrated over an $(n-1)$-dimensional domain. If D is an n-dimensional domain, then the boundary of D, ∂D, is $(n-1)$-dimensional. Note also that since α is an $(n-1)$-differential form, $d\alpha$ is an n-differential form.

This theorem holds for any dimension n; however we will only consider the case for D a two-dimensional domain in the plane. Other cases are complicated by the necessity of defining the orientation.

Green's Theorem in the Plane

Let D be a domain in the plane with a piecewise smooth boundary oriented (as above) counterclockwise. Let $f, g : \mathbb{R}^2 \to \mathbb{R}$ be smooth functions defined on D with variables x and y. Then we can construct a 1-differential form α from f and g: $\alpha = f dx + g dy$. We will integrate α around the boundary of D. Suppose that the boundary is given by the smooth function $C : [a, b] \to \mathbb{R}^2$ [thus D can have no hole, as in Figure 8.4.9(a)].

Green's Theorem

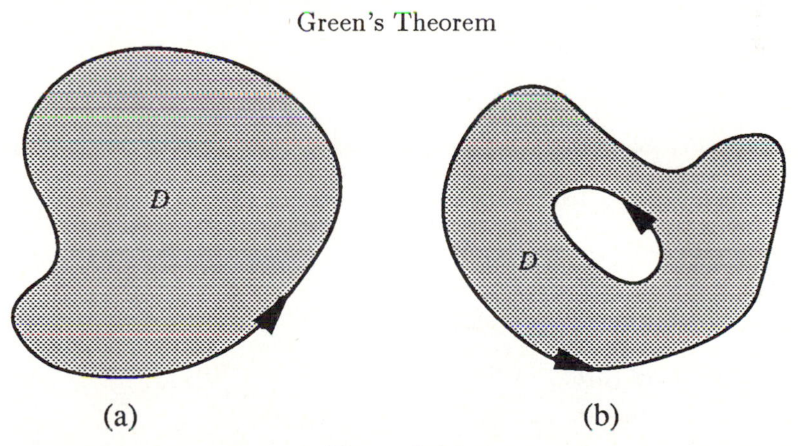

Figure 8.4.9

Stokes' Theorem above says that $\int_D d\alpha = \int_{\partial D} \alpha$. Let $F = (f, g) : \mathbb{R}^2 \to \mathbb{R}^2$ as before. Then the right-hand side gives

$$\int_{\partial D} \alpha = \int_{\partial D} f dx + g dy = \int_{\partial D} F(C(t)) \cdot C'(t) dt = \int_{\partial D} F$$

and the left-hand side gives

$$\begin{aligned}\int_D d\alpha &= \int_D d(fdx + gdy) = \int_D \left(\frac{\partial g}{\partial x} - \frac{\partial f}{\partial y}\right) dx \wedge dy \\ &= \int_D \int \left(\frac{\partial g}{\partial x} - \frac{\partial f}{\partial y}\right) dxdy\end{aligned}$$

by Example 8.4.5. Thus we have:

$$\int_{C([a,b])} F = \int_a^b F(C(t))\cdot C'(t)dt = \int_{\partial D} fdx + gdy = \int_D \int \left(\frac{\partial g}{\partial x} - \frac{\partial f}{\partial y}\right) dxdy$$

Theorem 8.4.10 (Green's Theorem) *If ∂D is the counterclockwise boundary of a domain D and $F = (f, g)$ is a C^1 function on D, then*

$$\int_{\partial D} F = \int_{\partial D} fdx + gdy = \iint_D \left(\frac{\partial g}{\partial x} - \frac{\partial f}{\partial y}\right) dxdy$$

If the domain D has a hole, as in Figure 8.4.9 (b), then from the discussion on orientation, we must use two curves C_1 and C_2 oriented as indicated. This gives:

$$\int_{C_1([a,b])} F + \int_{C_2([a,b])} F = \iint_D \left(\frac{\partial g}{\partial x} - \frac{\partial f}{\partial y}\right) dxdy$$

Example 8.4.11 We will evaluate the integral of the function $F(x, y) = (y, -2x)$ about the unit circle C traversed in the counterclockwise direction, $\int_C F$. Green's Theorem is applied with $f(x, y) = y$, $g(x, y) = -2x$, and D the unit disk, $C = \partial D$. Thus

$$\begin{aligned}\int_C F &= \int_D \int \left(\frac{\partial g}{\partial x} - \frac{\partial f}{\partial y}\right) dxdy = \iint_D (-2-1)dxdy \\ &= -3\iint_D dxdy = -3area(D) = -3\pi\end{aligned}$$ △

Example 8.4.12 We will evaluate the integral of the function $F(x, y) = (y^2 + x^3, 3x^4)$ about the boundary of the unit square $D = [0, 1] \times [0, 1]$ in the counterclockwise direction. By Green's Theorem,

$$\begin{aligned}\int_B F &= \iint_D \left(\frac{\partial g}{\partial x} - \frac{\partial f}{\partial y}\right) dxdy = \int_{[0,1]\times[0,1]} \int (12x^3 - 2y)dxdy \\ &= \int_0^1 \left(\int_0^1 12x^3 - 2ydx\right) dy = \int_0^1 (3-2y)dy = 3 - 1 = 2\end{aligned}$$ △

To appreciate the usefulness of Green's Theorem, you should try evaluating the line integrals in the above two examples using Definition 8.1.4 rather than Green's Theorem.

Corollary 8.4.13 (Area of a Region) *Let D be a region in $\mathbb{R}^2$ with its boundary oriented in a counterclockwise direction. The area of D is*

$$Area(D) = \frac{1}{2}\int_{\partial D} xdy - ydx = \frac{1}{2}\int_{\partial D} F$$

where $F(x,y) = (-y,x)$ on D.

Proof: Applying Green's Theorem to the function $F(x,y)$ we have

$$\begin{aligned}\int_{\partial D} xdy - ydx &= \int_{\partial D} F = \iint_D \left(\frac{\partial g}{\partial x} - \frac{\partial f}{\partial y}\right) dxdy \\ &= \iint_D [1-(-1)]dxdy = 2\text{ area}(D)\end{aligned}$$

□

Example 8.4.14 We will find the area contained within the ellipse $x^2/a^2 + y^2/b^2 = 1$. Here the ellipse is in standard form, with axes of half-lengths $a > 0$ and $b > 0$.

Ellipse in Standard Form

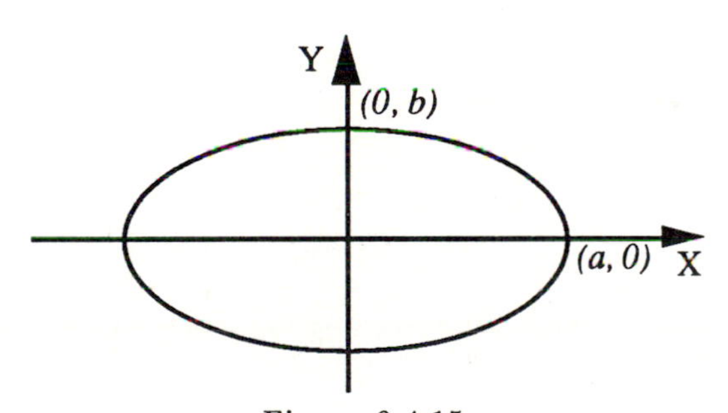

Figure 8.4.15

First we will parameterize the ellipse by $c(t) = (a\cos(t), b\sin(t)), 0 \le t \le 2\pi$, so that $c([0,2\pi])$ is the ellipse. By the above corollary,

$$\begin{aligned}\text{Area} &= \frac{1}{2}\int_{\partial D} F = \frac{1}{2}\int_0^{2\pi} (-y,x)\bullet c'(t)dt \\ &= \frac{1}{2}\int_0^{2\pi} (-b\sin(t), a\cos(t))\bullet(-a\sin(t), b\cos(t))dt \\ &= \frac{1}{2}\int_0^{2\pi} [ab\sin^2(t) + ab\cos^2(t)]dt = \frac{1}{2}\int_0^{2\pi} abdt = ab\pi\end{aligned}$$

△

Section 8.4 Exercises

1. Simplify each of the following differential forms. Write all forms so that the variables are in alphabetical order, e.g., $dx \wedge dy$ rather than $dy \wedge dx$.

   (a) $(dx + dy) \wedge dx$

   (b) $(dy - 2dz) \wedge dy$

   (c) $(2dx + 3dy + 4dz) \wedge (dx \wedge dy)$

   (d) $(dz \wedge dx) \wedge (2xdx - 4dy + 3dz)$

2. Compute the derivative of each of the following forms.

   (a) $\alpha = xydx + dy$

   (b) $\alpha = \sin(xy)dx$

   (c) $\alpha = x\sin(x + z)dx \wedge dy + dx \wedge dz$

   (d) $\alpha = \cos(x + 2y + z)dx \wedge dz + xdx \wedge dy$

3. Extend Example 8.4.7 by showing that for any C^2 function $f : \mathbb{R}^3 \to \mathbb{R}$, $ddf = 0$.

4. Extend Example 8.4.7 and Exercise 3 by showing that for any 1-form

$$\alpha = f(x, y, z)dx + g(x, y, z)dy + h(x, y, z)dz$$

   we have $dd\alpha = 0$.

5. Extend Example 8.4.7 and Exercises 3 and 4 by showing that for any 2-form

$$\alpha = f(x, y, z)dx \wedge dy + g(x, y, z)dy \wedge dz + h(x, y, z)dx \wedge dz$$

   we have $dd\alpha = 0$.

6. If C is a closed curve and $F(x, y) = (y, x)$, what is $\int_C F$?

In Exercises 7 through 10 below, find the line integral using Green's Theorem (and also by using Definition 8.1.4) where B is the (counterclockwise) boundary of the rectangle determined by the lines $x = 1$, $x = 4$, $y = 2$, and $y = 3$.

7. $\int_B (y^2 + x)dx + x^2dy$

8. $\int_B (x^2 - 3x)dx + (x^2 - 2y)dy$

9. $\int_B \cos(x)\sin(y)dx + \sin(x)\cos(y)dy$

10. $\int_B y\sin(x)dx + 2x\cos(y)\sin(x)dy$

11. Find the area inside the ellipse $3x^2 + 4y^2 = 8$.

12. Find the area inside the curve $f(t) = (\cos(t), \sin^3(t)))$ for $0 \le t \le 2\pi$.

13. Find the area inside the curve $f(t) = (\cos^3(t), \sin^3(t))$ for $0 \le t \le 2\pi$.

14. Find the area bounded by the curves $y = x^3$ and $y = \sqrt{x}$ using Green's Theorem.

15. Why cannot Green's Theorem be applied to $F(x, y) = (1/(x^2 + y^2), x + y)$ on the unit disk D?

16. If $f : \mathbb{R}^n \to \mathbb{R}$ is a C^2 function, then the *Laplacian of* f, $\nabla^2 f$, is defined by

$$\nabla^2 f(x) = \frac{\partial^2 f}{\partial {x_1}^2} + \cdots + \frac{\partial^2 f}{\partial {x_n}^2}$$

Show that $\nabla^2 f = \text{div}(\text{grad}(f)) = \nabla(\nabla f) = \nabla \bullet \nabla f$.

17. If D is a domain in the plane and f and g are C^2 functions on D, then prove Green's First Identity:

$$\iint_D f\nabla^2 g dx dy = -\iint_D \nabla f \bullet \nabla g dx dy + \int_{\partial D} f\frac{\partial g}{\partial x}dy - f\frac{\partial g}{\partial y}dx$$

18. If D is a domain in the plane and f and g are C^2 functions on D, then prove Green's Second Identity:

$$\iint_D (f\nabla^2 g - g\nabla^2 f) dx dy = \int_{\partial D} \left(f\frac{\partial g}{\partial x} - g\frac{\partial f}{\partial x}\right) dy - \left(f\frac{\partial g}{\partial y} - g\frac{\partial f}{\partial y}\right) dx$$

19. A function $f : \mathbb{R}^2 \to \mathbb{R}$ with second derivatives is called *harmonic* on a domain D if for every (x, y) in D, $\nabla^2 f(x, y) = 0$. Show that if a C^2 function f is harmonic on D, then $\int_{\partial D}(\partial f/\partial y)dx - (\partial f/\partial x)dy = 0$.

20. Prove that if f is harmonic on D and f vanishes on the boundary of D, then

    (a) $\nabla f = 0$ everywhere in D.

    (b) $f = 0$ everywhere in D.

21. Give an example of

    (a) A 0-form α with $d\alpha = x dx$.

    (b) A 1-form α with $d\alpha = dx \wedge dy$.

    (c) A 2-form α with $d\alpha = dx \wedge dy \wedge dz$.

22. Show that there is no 1-form α (of variables x and y) such that $d\alpha = y dx$. (Hint: Consider Exercises 3, 4, and 5 above.)

8.5 Change of Coordinates

The difficulty associated with determining the limits of integration for a multiple integral is usually a reflection of the difficulty one has in describing the curves that bound the region in question. For example, what would the limits of integration be for the double integral expression of the area contained inside the "four-petal rose" whose equation is $(x^2 + y^2)^{3/2} = x^2 - y^2$?

Four-Petal Curve

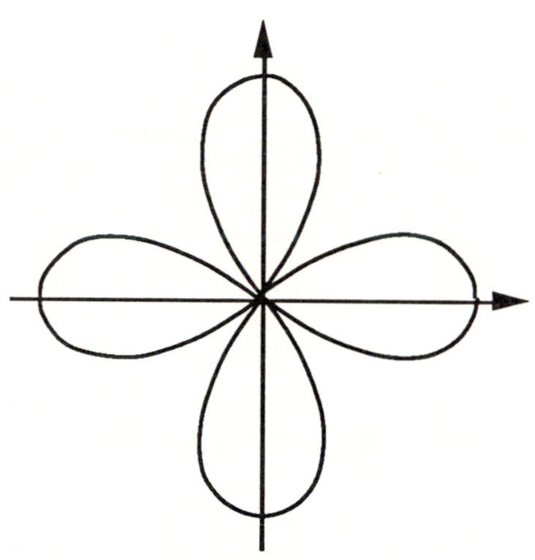

Figure 8.5.1

The difficulties in working with the equation for the four-petal rose may reflect more the limitations of the rectangular coordinate system rather than inherent problems in the shape being described. Thus we consider an alternative system, called the *Polar Coordinate* system. The basis of this system consists of a fixed point, called the polar origin, and a fixed ray, called the polar axis, emanating from the polar origin. In the polar system a point is located by its distance from the (polar) origin and the angle, measured in a counterclockwise direction, formed by the (polar) axis and a ray from the origin through the point. (With the rectangular xy-coordinate system, a point is uniquely described by its signed distances from the axes.) We shall identify the polar origin with the rectangular origin and the polar axis with the positive x-axis. Thus the polar coordinates of a point p, (r, θ), are:

r—the distance from p to the origin (r is called the *radial distance*).

θ—the angle measured counterclockwise from the polar axis (x-axis) to the ray from the origin through p (θ is called the *polar angle*).

The angle is usually measured in radians, where 2π radians equals 360°. Thus the point with xy-coordinates $(0, 1)$ has $r\theta$-coordinates $(1, \pi/2)$.

Polar Coordinates

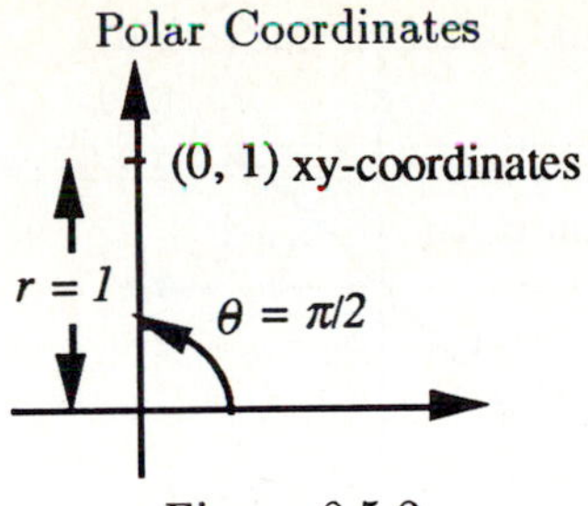

Figure 8.5.2

Rectangular xy-coordinates form a rectangular grid on the plane.

Grid Lines for Rectangular Coordinates

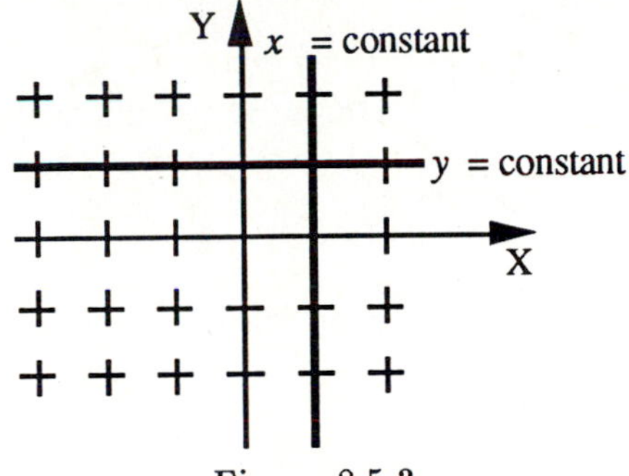

Figure 8.5.3

In polar coordinates the grid lines are formed by lines with r = constant (circles about the origin) and lines with θ = constant (rays from the origin).

Grid Lines for Polar Coordinates

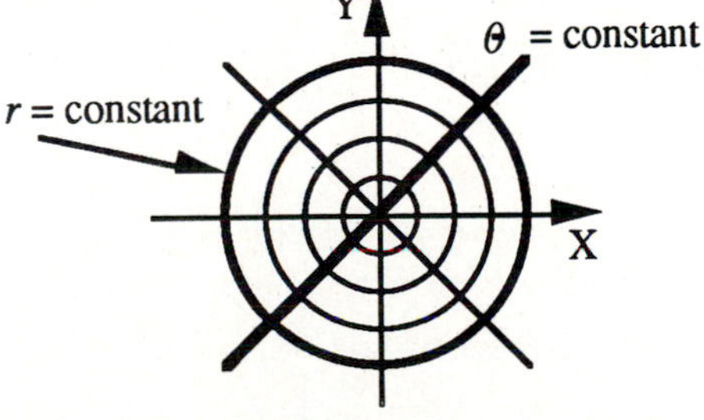

Figure 8.5.4

The value of r is allowed to be negative. To locate a point with a negative r, first go to the correct angle, then go the distance $|r|$ in the opposite direction through the origin.

Points in Polar Coordinates

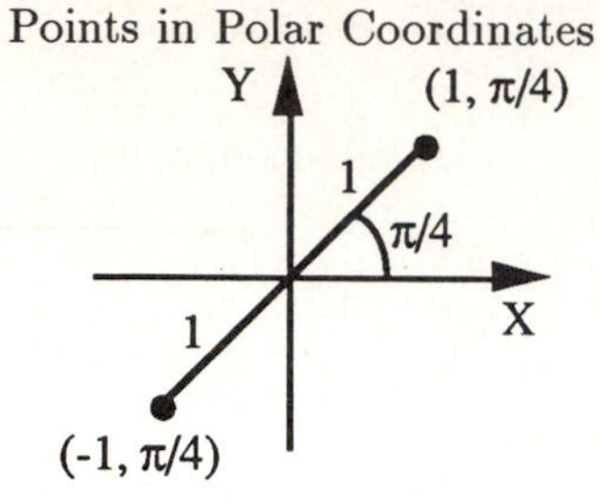

Figure 8.5.5

In rectangular coordinates there is a one-to-one correspondence between points in the plane and pairs of real numbers (x, y). This is not true when we use polar coordinates, since (r, θ) and $(r, \theta + 2\pi n)$ correspond to the same point in the plane for any integer n. In addition, (r, θ) and $(-r, \theta + \pi)$ correspond to the same point in the plane.

Example 8.5.6 We will find the rectangular coordinates of the point whose polar coordinates are given by $(-2, -\pi/4)$. Since the angle is negative, we move clockwise rather than counterclockwise. Then we go two units in the opposite direction through the origin.

The point with polar coordinates $(-2, -\pi/4)$

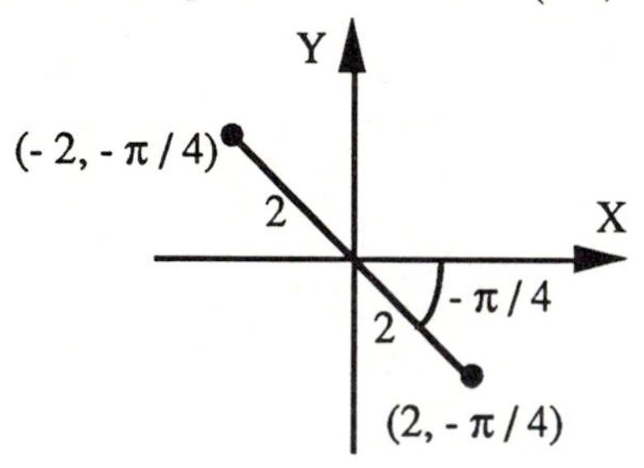

Figure 8.5.7

The x-coordinate is thus $-2\cos(\pi/4) = -\sqrt{2}$ and the y-coordinate is $\sqrt{2}$. △

Example 8.5.8 Express the point with rectangular coordinates $(-2, -1)$ in polar coordinates. First we locate the point.

The Point with Rectangular Coordinates $(-2, -1)$

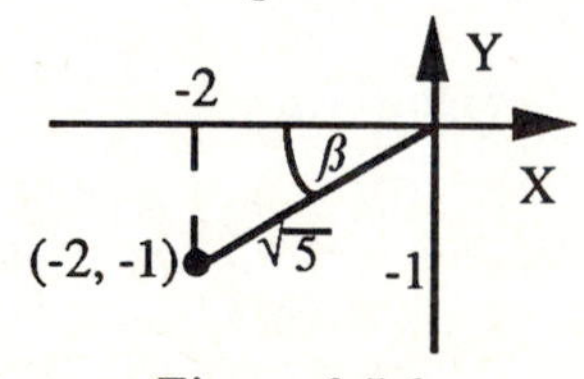

Figure 8.5.9

We use the Pythagorean theorem to find the distance of the point $(-2, -1)$ from the origin: $r^2 = (-2)^2 + (-1)^2 = 5$, so the distance to the origin is $\sqrt{5}$. The angle β has $\tan(\beta) = 1/2$, so $\beta \approx 0.47$ radians (from a table or calculator). The polar coordinate angle θ is thus $\pi + 0.47$ radians. The polar coordinates of the point are $(\sqrt{5},\ 0.47 + \pi)$, where we have $0 \leq \theta < 2\pi$. We could use negative angles and write the polar coordinates as $r = \sqrt{5}$ and $\theta = -\pi + 0.47$ radians. We could also express the polar coordinates of the point using a negative value for r. This would give $(-\sqrt{5},\ 0.47)$ and $(-\sqrt{5},\ 2\pi - 0.47)$. △

This last example suggests transformation relations for converting from rectangular xy-coordinates to polar coordinates and conversely. Namely for $0 \leq \theta \leq 2\pi$

$$x = r\cos(\theta),\ \ y = r\sin(\theta)$$

and

$$r^2 = x^2 + y^2,\ \tan(\theta) = \frac{y}{x}(\text{ or }\ \cot(\theta) = \frac{x}{y})$$

(In polar coordinates, the origin has coordinates $(0, \theta)$ where θ can be any angle.)

Using polar coordinates we can sketch many interesting curves in the plane much more easily than with rectangular coordinates. For example, the unit circle is $r = 1$ rather than $x^2 + y^2 = 1$ as with rectangular coordinates. On the other hand transforming the equation $x + y^2 = 1$ to polar coordinates by substituting $x = r\cos(\theta)$ and $y = r\sin(\theta)$ gives $r\cos(\theta) + r^2\sin^2(\theta) = 1$. This is not an improvement over the rectangular form. Generally the use of polar coordinates simplifies the expressions for curves with a lot of circular symmetry.

To aid in developing intuition with respect to the polar coordinate system, we shall "merge" the terminology of the rectangular and polar systems. By identifying the two origins and the polar axis with the positive x-axis, we may easily shift from one system to the other as well as using terms such as quadrants and reference angles. To sketch a curve, begin with $\theta = 0$ on the x-axis and compute the value of r. Now let θ increase and using our knowledge of the trigonometric functions (continuity, increasing, decreasing, symmetry), note what r does. It is often helpful to use known reference angles such as $\theta = 0,\ \pi/6,\ \pi/4,\ \pi/3$, and $\pi/2$ (and the corresponding angles in the other quadrants) to compute "check" points for the curve.

Question 8.5.10 Sketch the curve satisfying the relation $\theta = \pi/4$. ◇

Example 8.5.11 (Spiral of Archimedes) Another nice curve that can be easily defined using polar coordinates is the spiral of Archimedes, $r = \theta, \theta \geq 0$. What does this look like? As θ increases from 0, the radius is strictly increasing. Thus a spiral is traced out. △

The Spiral $r = \theta$

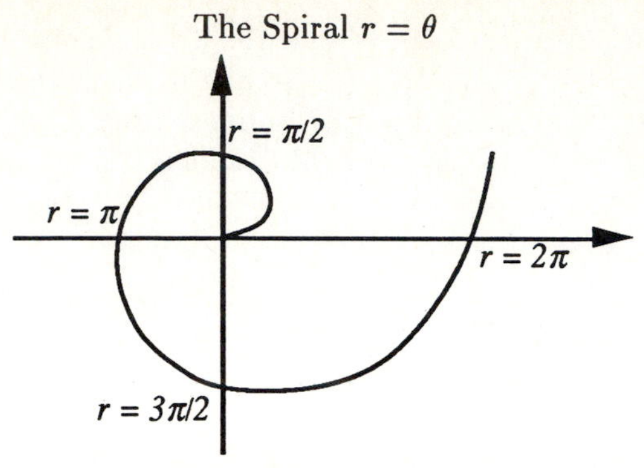

Figure 8.5.12

Example 8.5.13 (Circles) We will sketch the curve $r = \cos(\theta)$. Note that $\cos(\theta) = \cos(-\theta)$. Thus the sketch will be symmetric with respect to the polar axis (x-axis), and hence it is only necessary to consider θ ranging from 0 to π.

When $\theta = 0$, $r = 1$. As θ increases from 0 to $\pi/2$, r decreases from 1 to 0. This gives a curve in the first quadrant passing through the points (in rectangular coordinates) (1,0), $(1/\sqrt{2}, \pi/4)$, (0,0). As θ increases from $\pi/2$ to π, r decreases from 0 to -1. Since r is negative, this portion of the curve will lie in the fourth quadrant. Now using symmetry with respect to the polar axis, we obtain the following sketch.

$r = \cos(\theta)$

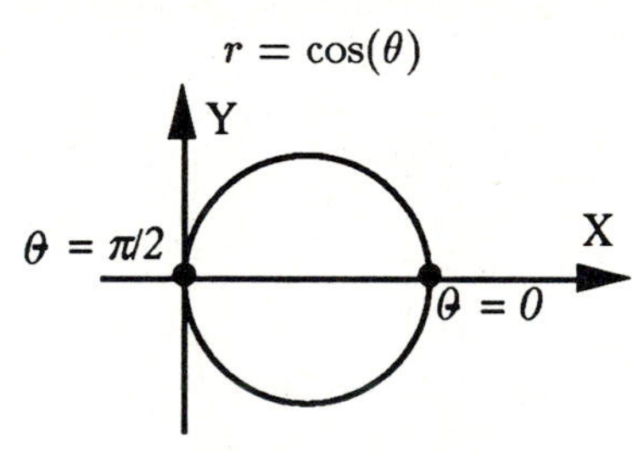

Figure 8.5.14

To verify that the graph is a circle, we transform $r = \cos(\theta)$ into the more familiar rectangular coordinate system. Multiplying both sides by r, we obtain

$$r^2 = r\cos(\theta)$$

which yields

$$x^2 + y^2 = x$$

or (transposing and completing the square)

$$\left(x - \frac{1}{2}\right)^2 + y^2 = \frac{1}{4}$$

This is the equation of a circle of radius $\frac{1}{2}$ centered at $(\frac{1}{2}, 0)$. △

We are now ready to show how to obtain the graph of the four-petal rose.

Example 8.5.15 (A Four-Petal Rose) Sketch the graph of $(x^2 + y^2)^{3/2} = x^2 - y^2$. Transforming this expression into polar coordinates yields $r = \cos(2\theta)$. (How?) Now when $\theta = 0$, $r = 1$ and when θ increases from 0 to $\pi/4$, r decreases from 1 to 0. This gives half of a "leaf." As θ continues from $\pi/4$ to $\pi/2$, r decreases from 0 to -1. Since r is negative, the curve is in the third quadrant. For θ from $\pi/2$ to $3\pi/4$, r goes from -1 to 0 and the curve is in the fourth quadrant. We have now finished one and a half leaves. The figure continues until an entire four-petal rose is traced out. △

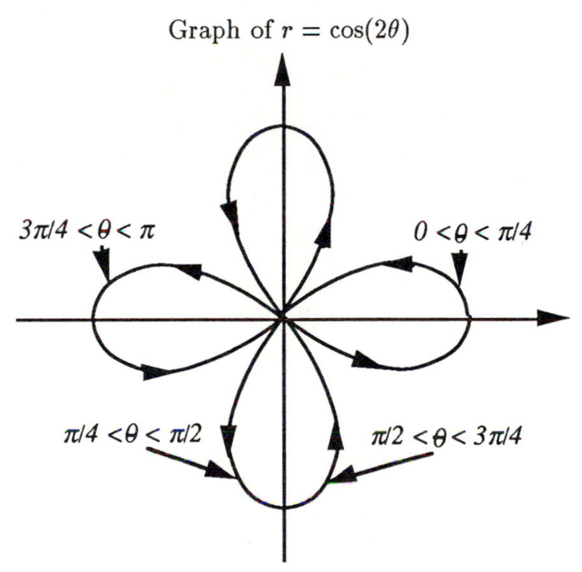

Figure 8.5.16

We now return to the question posed at the beginning of this section, how can we find the area enclosed by the four-petal rose? Clearly working with the rectangular coordinate description of the curve, $(x^2 + y^2)^{3/2} = x^2 - y^2$, could easily lead into an exercise in frustration. Is there a "polar" expression for area? Yes; we will obtain an expression for area in polar coordinates in two ways: first following the method used in developing the area as a double integral; second by substituting the transformation formulas $[x = r\cos(\theta)$ and $y = r\sin(\theta)]$ into the integral expression and using the rules for differential forms.

Area in Polar Coordinates: Development from Geometry

We begin by giving a brief review of the process involved in expressing the area A of region S as a double integral, $A = \iint_S (1)dxdy$.

We start by enclosing S in a rectangle, say $R = [a, b] \times [c, d]$, and then partition R with a rectangular grid $P \times Q$, where P is a partition of $[a, b]$ and

Q is a partition of $[c, d]$. A lower approximation to the area of S is the sum of the areas of the grid rectangles that are completely contained inside of S and an upper approximation is the sum of the grid rectangles that contain at least one point of S. The situation is illustrated in the following figure. (The sum of the areas of the shaded rectangles is a lower approximation of A.)

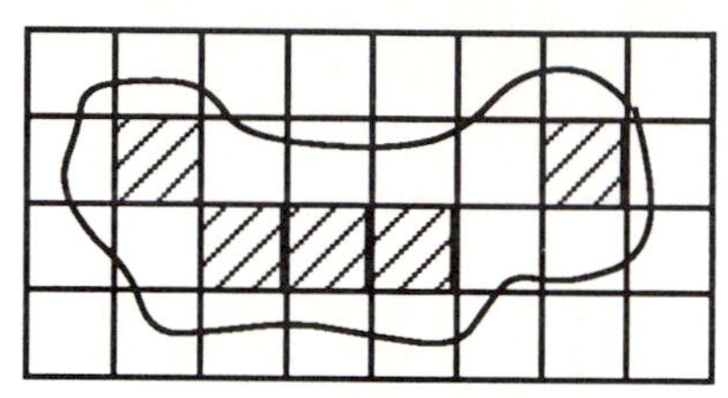

Figure 8.5.17

The lower and upper approximations can be improved by increasing the number of grid lines (thus reducing the sizes of the grid rectangles). This is accomplished by replacing the partitions P and Q with partitions of smaller norms. Thus, the sequences of partitions $\{P_n\}$ and $\{Q_m\}$ with the properties that $\lim_{n\to\infty} |P_n| = 0$ and $\lim_{m\to\infty} |Q_m| = 0$ will, in turn, yield an increasing sequence of lower approximations, $\{LA(1, P_n, Q_m)\}$, and a decreasing sequence of upper approximations, $\{UA(1, P_n, Q_m)\}$, both of which converge to a common value that is defined to be the area of region S.

The analogous procedure when S is described using polar coordinates would be to enclose S in a circle (rather than a rectangle) C of radius r, and then partition C with a polar grid, $P \times Q$, where P is a partition of $[0, r]$ and Q is a partition of $[0, 2\pi]$.

We shall refer to the regions in the polar grid as *polar rectangles* and denote the (i,j)th one by $PR(i, j)$ (P for polar and R for rectangle). Finding the area of a polar rectangle is a bit more involved than finding the area of a rectangular rectangle. We shall refer to the labeling in the following diagram in developing the expression for the area of the polar rectangle $PR(i, j)$.

Polar Rectangle

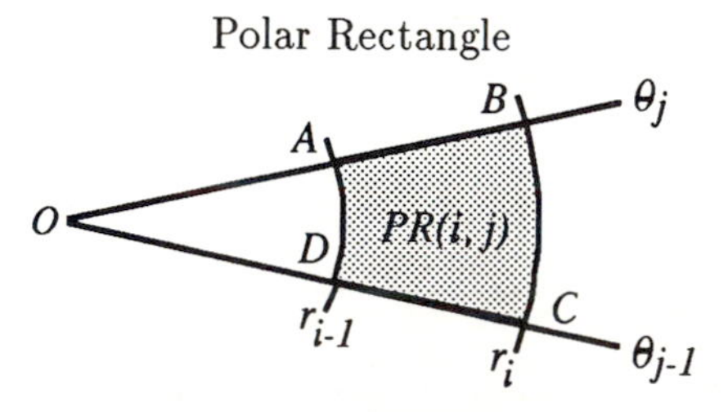

Figure 8.5.18

The area of $PR(i, j)$ is the area A_1 of the sector OBC minus the area A_2 of the sector OAD. Note that the ratio of the central angle of a sector to 2π is the same as the ratio of the area of the sector to the area of the circle containing it,

πr^2. Thus, the area of sector OBC is

$$\frac{A_1}{\pi {r_i}^2} = \frac{\theta_j - \theta_{j-1}}{2\pi}$$

which yields

$$A_1 = \frac{r_i^2}{2}(\theta_j - \theta_{j-1})$$

In a similar manner, the expression for the area of sector OAD is

$$A_2 = \frac{r_{i-1}^2}{2}(\theta_j - \theta_{j-1})$$

Thus, the area A of the polar rectangle is

$$\begin{aligned} A &= \frac{r_i^2 - r_{i-1}^2}{2}(\theta_j - \theta_{j-1}) = \frac{r_i + r_{i-1}}{2}(r_i - r_{i-1})(\theta_j - \theta_{j-1}) \\ &= \frac{r_i + r_{i-1}}{2}\Delta r_i \Delta\theta_j = r_i^* \Delta r_i \Delta\theta_j \end{aligned}$$

where r_i^* is in the interval $[r_{i-1}, r_i]$.

The remaining steps in the development are now identical to those that were sketched out using rectangular coordinates. That is, define an increasing sequence of lower approximations and a decreasing sequence of upper approximations and show that they converge to a common value. This common value is then defined to be the area of region S and is given by the double integral

$$A = \iint_S r dr d\theta$$

The limits of integration are, of course, described in polar coordinates.

This gives us the *polar change of coordinates transformation formula for integrals:*

$$\iint_D f(x, y) dx dy = \iint_S f(r\cos(\theta), r\sin(\theta)) r dr d\theta$$

where S is the region in the $r\theta$-coordinates corresponding to the region D in the xy-coordinates.

Area in Polar Coordinates: Development from Differential Forms

Suppose we have a double integral in rectangular coordinates, $\iint_D f(x, y) dx dy$. Recall the interpretation of the integral using the wedge product of differential forms:

$$\iint_D f(x, y) dx dy = \int_D f(x, y) dx \wedge dy$$

Now substitute $x = r\cos(\theta)$ and $y = r\sin(\theta)$ into the expression. We compute $dx \wedge dy$:

$$\begin{aligned}
dx \wedge dy &= d(r\cos(\theta)) \wedge d(r\sin(\theta)) \\
&= \left[\frac{\partial}{\partial r} r\cos(\theta)dr + \frac{\partial}{\partial \theta} r\cos(\theta)d\theta\right] \wedge \left[\frac{\partial}{\partial r} r\sin(\theta)dr + \frac{\partial}{\partial \theta} r\sin(\theta)d\theta\right] \\
&= [\cos(\theta)dr - r\sin(\theta)d\theta] \wedge [\sin(\theta)dr + r\cos(\theta)d\theta] \\
&= \cos(\theta)\sin(\theta)dr \wedge dr + \cos(\theta)r\cos(\theta)dr \wedge d\theta \\
&\quad -r\sin^2(\theta)d\theta \wedge dr - r\sin(\theta)r\cos(\theta)d\theta \wedge d\theta \\
&= r\cos^2(\theta)dr \wedge d\theta - r\sin^2(\theta)d\theta \wedge dr \\
&= r\cos^2(\theta)dr \wedge d\theta + r\sin^2(\theta)dr \wedge d\theta \\
&= rdr \wedge d\theta
\end{aligned}$$

Thus we have:

$$\begin{aligned}
\iint_D f(x,y)dxdy &= \int_D f(x,y)dx \wedge dy = \int_S f(r\cos(\theta), r\sin(\theta))rdr \wedge d\theta \\
&= \iint_S f(r\cos(\theta), r\sin(\theta))rdrd\theta
\end{aligned}$$

where S is the region D expressed in polar coordinates.

The above computation of $dx \wedge dy$ is long, but it is entirely mechanical. (In fact, some computer algebra systems can compute with wedge products and differential forms.) Contrast this with the thought required in the geometrical development. A theme in mathematics is to *study a problem, understand it, and then develop an algebra which allows thought to be replaced by mechanical computation.* The computational method makes for easier applications and allows thought to be concentrated on problems that are not yet understood.

Example 8.5.19 We will find the area enclosed by the circle $x^2 + y^2 = 4$. The polar equation for this circle is $r = 2$. Thus the region enclosed by the circle is described by the inequalities $0 \leq r \leq 2$ and $0 \leq \theta \leq 2\pi$. The area of this region is

$$\int_0^{2\pi} \int_0^2 rdrd\theta = 4\pi$$

△

In order to better appreciate the above calculations, the reader should compute the area inside the circle with a double integral using rectangular coordinates.

Example 8.5.20 (Cardioid) We will find the area of the region enclosed by the cardioid ("car-dee-oid") curve $r = 1 - \cos(\theta)$. First we sketch a picture. Note that the region of integration is described by the inequalities:

$$0 \leq r \leq 1 - \cos(\theta) \quad \text{and} \quad 0 \leq \theta \leq 2\pi$$

Also note that the sketch will be symmetric with respect to the x-axis (since θ only appears in the cosine function). Plot some reference points to "tie down" the sketch, say for $\theta = 0, \pi/2$, and π. The sketch can now be drawn and then checked by plotting reference points for $\theta = \pi/4$ and $\theta = 3\pi/4$.

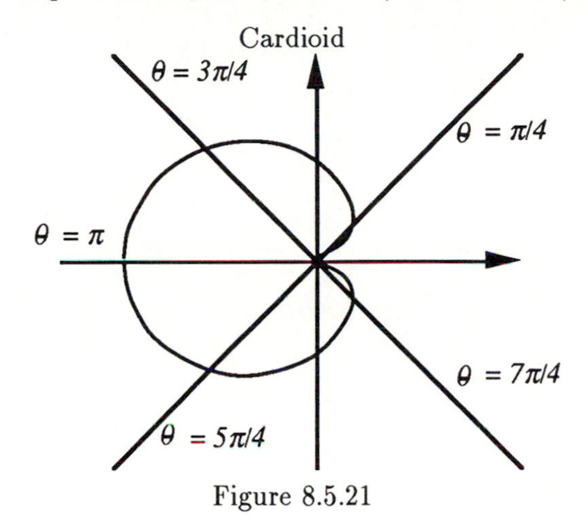

Figure 8.5.21

Thus, the double integral expression for the area is

$$\begin{aligned}\int_0^{2\pi}\int_0^{1-\cos(\theta)} r dr d\theta &= \frac{1}{2}\int_0^{2\pi}[1-2\cos(\theta)+\cos^2(\theta)]d\theta \\ &= \frac{1}{2}\int_0^{2\pi}[1-2\cos(\theta)+\frac{1}{2}+\frac{1}{2}\cos(2\theta)]d\theta \\ &= \frac{3}{4}\theta]_0^{2\pi} = 3\frac{\pi}{2}\end{aligned}$$

The reader should fill in the missing steps. △

Example 8.5.22 (Limaçon) Set up the integral expression for the area of the region enclosed inside the small loop of the limaçon ("lee-ma-sohn") $r = 1 + 2\sin(\theta)$. First we draw a sketch (see the next figure). Note that the sketch will be symmetric with respect to the "vertical" axis, $\theta = \pi/2$ (since θ appears only in the sine function). Plot some reference points to "tie down" the sketch, say for $\theta = 0$, $\pi/2$, and $3\pi/2$. Since r is positive for $\theta = \pi$ and negative for $\theta = 3\pi/2$, r must be zero for some value of θ in the third quadrant. (Why?) It is important to find this value because it will tell when the curve passes through the origin. Setting $r = 0$ and solving for θ yields $\theta = 7\pi/6$. The graph can now be easily sketched starting with $\theta = 0$ and letting θ increase to 2π. Thus the region of integration is described by the inequalities $0 \le r \le 1 + 2\sin(\theta)$ and $7\pi/6 \le \theta \le 11\pi/6$.

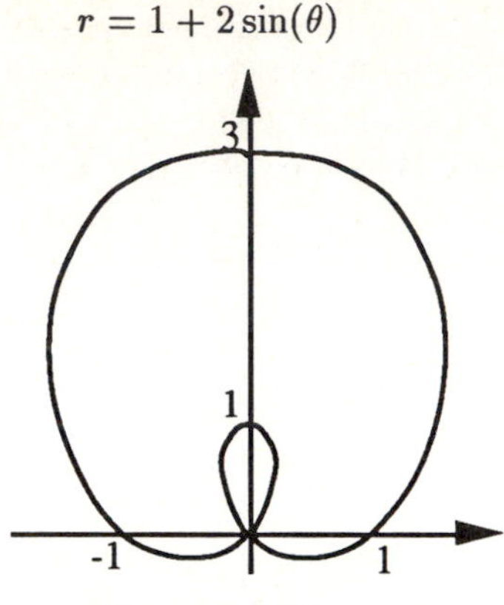

Figure 8.5.23

Hence:

$$\text{Area} = \int_{7\pi/6}^{11\pi/6} \int_0^{1+2\sin(\theta)} r\,dr\,d\theta$$

or, making use of symmetry,

$$\text{Area} = 2\int_{7\pi/6}^{3\pi/2} \int_0^{1+2\sin(\theta)} r\,dr\,d\theta$$

△

Cylindrical Coordinates

The cylindrical coordinate system is the extension of the polar system to three dimensions. The third coordinate is the "usual" third coordinate in the rectangular system. Thus, a point in three dimensions can be represented by (x, y, z) in the rectangular coordinate system and by (r, θ, z) in the cylindrical coordinate system. The transformation formulas from the rectangular system to the cylindrical system are

$$x = r\cos(\theta), \quad y = r\sin(\theta), \quad \text{and } z = z$$

where $r^2 = x^2 + y^2$ and $\theta = \tan^{-1}(y/x)$. For example, if the point P has rectangular coordinates of $(2, 2, 3)$, it will have cylindrical coordinates of $(2\sqrt{2}, \pi/4, 3)$. In a similar fashion, if the point Q has cylindrical coordinates of $(2, \pi/3, -2)$, it will have rectangular coordinates of $(1, \sqrt{3}, -2)$.

In the rectangular system, the graph of $x = c$, where c is a consta[illegible] plane perpendicular to the x-axis. Likewise, the graph $y = c$ is a plane perpendicular to the y-axis and the graph of $z = c$ is a plane perpendicular to the z-axis. In the cylindrical system, however, the graph of $r = c$ is a cylinder about the z-axis, the graph of $\theta = c$ is a plane containing the z-axis, and the graph of $z = c$ is a plane perpendicular to the z-axis.

The formal development of a triple integral of a function f using cylindrical coordinates follows the four steps of the Basic Approximation Process (of course). Since we have illustrated this process several times, we shall merely comment on the partitioning of the region of integration and then leave the rest of the details to the reader. The region of integration is partitioned with respect to each of the three coordinates into "wedges." The "base" of a wedge is a polar rectangle and the height is Δz.

Wedge

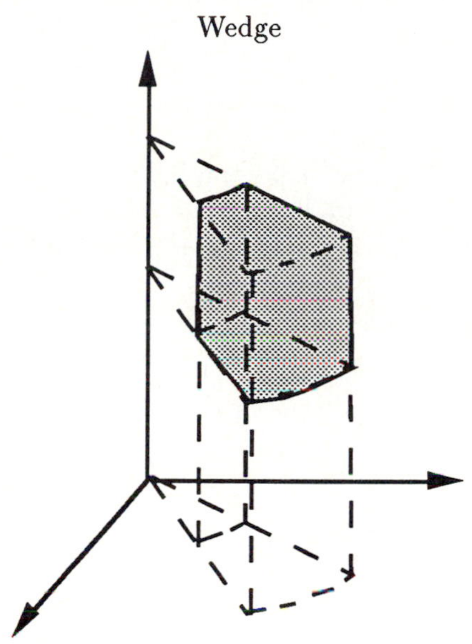

Figure 8.5.24

The volume of a wedge is $r_i^* \Delta r_i \Delta\theta_j \Delta z_k$. The resulting triple Riemann Sum has the form

$$\sum_{k=1}^{p}\sum_{j=1}^{m}\sum_{i=1}^{n} f(r_i, \theta_j, z_k) r_i^* \Delta r_i \Delta\theta_j \Delta z_k$$

By considering minimum and maximum values of f over each of the wedges, sequences of lower and upper approximations can be formulated. If these sequences converge to a common value, this value is defined to be the triple integral of f over D and is denoted by

$$\iiint_D f(r, \theta, z) r dr d\theta dz$$

Note that if f is a continuous function, the convergence of the sequences of lower and upper approximations to a common value is guaranteed.

Thus we have the *cylindrical change of coordinates transformation for integrals:*

$$\iint_D f(x, y, z) dx dy dz = \iint_S f(r\cos(\theta), r\sin(\theta), z) r dr d\theta dz$$

where S is the region in $r\theta z$-coordinates corresponding to D in xyz-coordinates.

Example 8.5.25 As usual, our first example will be a problem whose solution is known, so as to confirm our method. We will find the volume contained inside the cone of height b and base radius a, which is $\frac{1}{3}\pi a^2 b$. Think of the cone being oriented with the vertex at the origin and the axis of the cone being along the z-axis. Cutting a cone perpendicular to its axis produces a circular edge. Thus, a level surface determined by fixing a value for z between 0 and b is a circle. Hence, the equation for the cone must be of the form $x^2 + y^2 = c^2 z^2$, where c is a constant to be determined. At the base (or top) of the cone, $z = b$ and the radius of the circle is a. Thus, we have $x^2 + y^2 = c^2 b^2 = a^2$. Hence, $c = a/b$ and the equation of the cone in rectangular coordinates is $x^2 + y^2 = (a/b)^2 z^2$ and in cylindrical coordinates it is $r = (a/b)z$.

The region of integration is described by the inequalities:

$$0 \le r \le (\frac{a}{b})z, \quad 0 \le \theta \le 2\pi, \quad \text{and} \quad 0 \le z \le b$$

The sketch below pictures the cone in only the first octant.

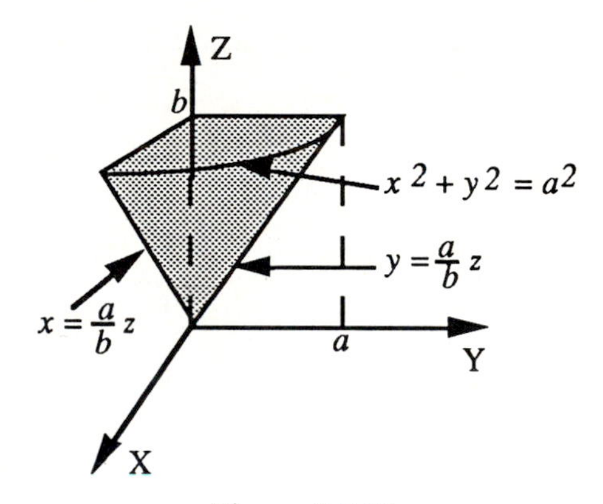

Figure 8.5.26

The volume of the cone is

$$\int_0^b \int_0^{2\pi} \int_0^{a/bz} (1) r dr d\theta dz = \int_0^b \int_0^{2\pi} \frac{a^2 z^2}{2b^2} d\theta dz = \int_0^b \pi \frac{a^2 z^2}{b^2} dz = \frac{1}{3}\pi a^2 b \qquad \triangle$$

Example 8.5.27 We will set up the expression for the triple integral of the function $f(x, y, z) = x^2$ over that portion of the cylinder $x^2 + y^2 \le 16$ lying between the planes $z = 1$ and $z = 3$.

Since the domain of f is a portion of a cylinder, we will use cylindrical coordinates. We first sketch the domain in the first octant.

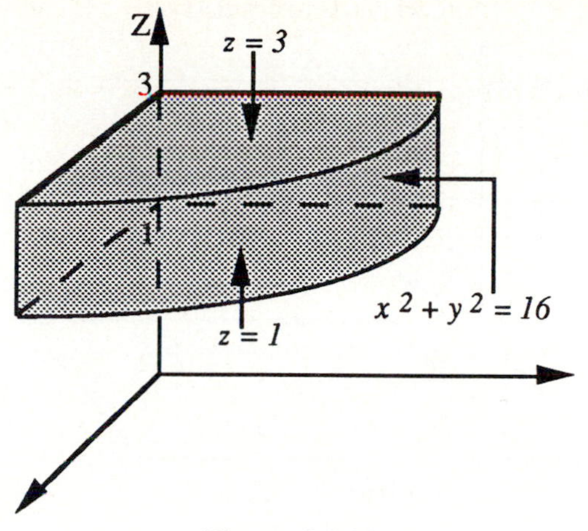

Figure 8.5.28

The region of integration is described by the inequalities:

$$0 \leq r \leq 4, 0 \leq \theta \leq 2\pi \text{ and } 1 \leq z \leq 3$$

Thus we have $\int_1^3 \int_0^{2\pi} \int_0^4 r^2 \cos^2(\theta) r dr d\theta dz$. △

Spherical Coordinates

Spherical coordinates are used to describe sets in $\mathbb{R}^3$ which have a lot of spherical symmetry. A point with rectangular coordinates (x, y, z) has spherical coordinates (ρ, θ, ϕ) where:

ρ is the distance of (x, y, z) from the origin $(0, 0, 0)$.

θ is, as in the cylindrical coordinate system, the angle between the positive x-axis and $(x, y, 0)$.

ϕ is the angle between the positive z-axis and the line from the origin $(0, 0, 0)$ to (x, y, z).

Spherical Coordinates

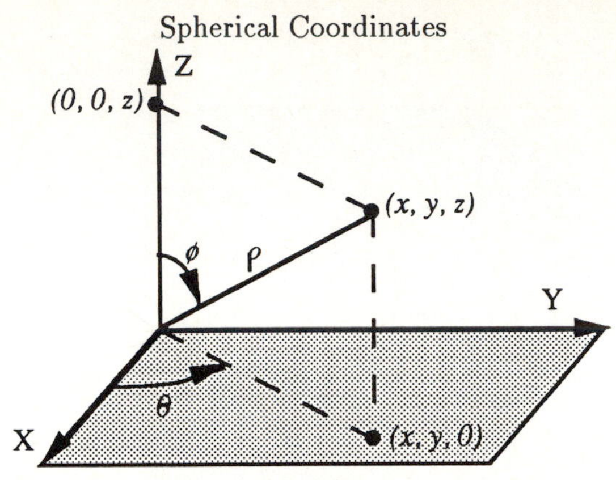

Figure 8.5.29

We can obtain the relations between (x, y, z) and (ρ, θ, ϕ) from trigonometry. The cylindrical coordinate r, the distance between $(0,0,0)$ and $(x, y, 0)$, satisfies

$$r = \sqrt{x^2 + y^2} = \rho \sin(\phi)$$

and the z coordinate satisfies $z = \rho \cos(\phi)$. Thus

$$x = r \cos(\theta) = \rho \sin(\phi) \cos(\theta)$$

and

$$y = r \sin(\theta) = \rho \sin(\phi) \sin(\theta)$$

Thus we have the transformations between rectangular and spherical coordinates:

$$\begin{array}{ll} x = \rho \sin(\phi) \cos(\theta) & \rho = \sqrt{x^2 + y^2 + z^2} \\ y = \rho \sin(\phi) \sin(\theta) & \theta = \tan^{-1}(y/x), x \geq 0 \\ z = \rho \cos(\phi) & \phi = \cos^{-1}(z/\sqrt{x^2 + y^2 + z^2}) \end{array}$$

Here the distance from the origin, ρ, satisfies $\rho \geq 0$; the angle θ satisfies $0 \leq \theta \leq 2\pi$; and the angle ϕ satisfies $0 \leq \phi \leq \pi$.

Example 8.5.30 The sphere $x^2 + y^2 + z^2 = 9$ in spherical coordinates is $\rho = 3$, since the sphere consists of all points three units from the origin. The equation $\rho = 3$ may be obtained algebraically by substituting into the rectangular coordinate equation $x^2 + y^2 + z^2 = 9$. △

Example 8.5.31 The cone that makes an angle of $\pi/3$ with the positive z-axis is described in spherical coordinates by $\phi = \pi/3$. △

Example 8.5.32 The xy-plane is described in spherical coordinates by $\phi = \pi/2$. $\triangle$

We will now derive the relation between integrals in rectangular coordinates and spherical coordinates.

$$\iiint_D f(x,y,z)dxdydz = \int_D f(x,y,z)dx \wedge dy \wedge dz$$

Substituting the change of coordinates into the differential form, we have,

$$\begin{aligned}
dx \wedge dy \wedge dz &= \left[\frac{\partial}{\partial \rho}[\rho \sin(\phi)\cos(\theta)]d\rho + \frac{\partial}{\partial \phi}[\rho \sin(\phi)\cos(\theta)]d\phi \right.\\
&\quad \left. + \frac{\partial}{\partial \theta}[\rho \sin(\phi)\cos(\theta)]d\theta\right] \wedge \left[\frac{\partial}{\partial \rho}[\rho \sin(\phi)\sin(\theta)]d\rho \right.\\
&\quad \left. + \frac{\partial}{\partial \phi}[\rho \sin(\phi)\sin(\theta)]d\phi + \frac{\partial}{\partial \theta}[\rho \sin(\phi)\sin(\theta)\ d\theta\right]\\
&\quad \wedge \left[\frac{\partial}{\partial \rho}[\rho \cos(\phi)]d\rho + \frac{\partial}{\partial \phi}[\rho \cos(\phi)]d\phi + \frac{\partial}{\partial \theta}[\rho \cos(\phi)]d\theta\right]\\
&= [\sin(\phi)\cos(\theta)d\rho + \rho\cos(\phi)\cos(\theta)d\phi - \rho\sin(\phi)\sin(\theta)d\theta]\\
&\quad \wedge[\sin(\phi)\sin(\theta)d\rho + \rho\cos(\phi)\sin(\theta)d\phi + \rho\sin(\phi)\cos(\theta)d\theta]\\
&\quad \wedge[\cos(\phi)d\rho - \rho\sin(\phi)d\phi - 0]\\
&= \rho^2\sin(\phi)\cos^2(\phi)\cos^2(\theta)d\phi \wedge d\theta \wedge d\rho\\
&\quad -\rho^2\sin(\phi)\cos^2(\phi)\sin^2(\theta)d\theta \wedge d\phi \wedge d\rho\\
&\quad -\rho^2\sin^3(\phi)\cos^2(\theta)d\rho \wedge d\theta \wedge d\phi\\
&\quad +\rho^2\sin^3(\phi)\sin^2(\theta)d\theta \wedge d\rho \wedge d\phi\\
&= \rho^2\sin(\phi)\{\cos^2(\phi)[\cos^2(\theta) + \sin^2(\theta)]\\
&\quad + \sin^2(\phi)[\cos^2(\theta) + \sin^2(\theta)]\}d\rho \wedge d\phi \wedge d\theta\\
&= \rho^2\sin(\phi)d\rho \wedge d\phi \wedge d\theta
\end{aligned}$$

Thus,

$$\begin{aligned}
&\iiint_D f(x,y,z)dxdydz = \int_D f(x,y,z)dx \wedge dy \wedge dz\\
&\quad = \int_S f(\rho\sin(\phi)\cos(\theta), \rho\sin(\phi)\sin(\theta), \rho\cos(\phi))\rho^2\sin(\phi)d\rho \wedge d\phi \wedge d\theta\\
&\quad = \iiint_S f(\rho\sin(\phi)\cos(\theta), \rho\sin(\phi)\sin(\theta), \rho\cos(\phi))\rho^2\sin(\phi)d\rho d\phi d\theta
\end{aligned}$$

This gives us the *spherical change of coordinates transformation for integrals:*

$$\begin{aligned}
&\iiint_D f(x,y,z)dxdydz\\
&\quad = \iiint_S f(\rho\sin(\phi)\cos(\theta), \rho\sin(\phi)\sin(\theta), \rho\cos(\phi))\rho^2\sin(\phi)d\rho d\phi d\theta
\end{aligned}$$

where S is the region in $\rho\phi\theta$-coordinates corresponding to D in xyz-coordinates.

Example 8.5.33 As a check on our result, we will find the volume of D, a sphere of radius r. We know the answer should be $4\pi r^3/3$. The equation of a sphere of radius r is $\rho = r$.

$$\begin{aligned}
\text{Volume} &= \iiint_D (1)dxdydz = \iiint_S \rho^2 \sin(\phi)d\rho d\phi d\theta \\
&= \int_0^{2\pi}\int_0^{\pi}\int_0^{r} \rho^2 \sin(\phi)d\rho d\phi d\theta \\
&= \int_0^{2\pi}\int_0^{\pi} \frac{r^3}{3}\sin(\phi)d\phi d\theta = \int_0^{2\pi} \frac{r^3}{3}(-\cos(\phi)]_0^{\pi})d\theta \\
&= \int_0^{2\pi} \frac{2r^3}{3}d\theta \\
&= \frac{4\pi r^3}{3}
\end{aligned}$$

△

Example 8.5.34 We will find the volume of the region D bounded by the cone $z = \sqrt{x^2+y^2}$ and the sphere $x^2+y^2+z^2 = 9$. The region is shaped like an ice cream cone. Note that the cone intersects the plane $y = 0$ in the lines $z = x$ and $z = -x$.

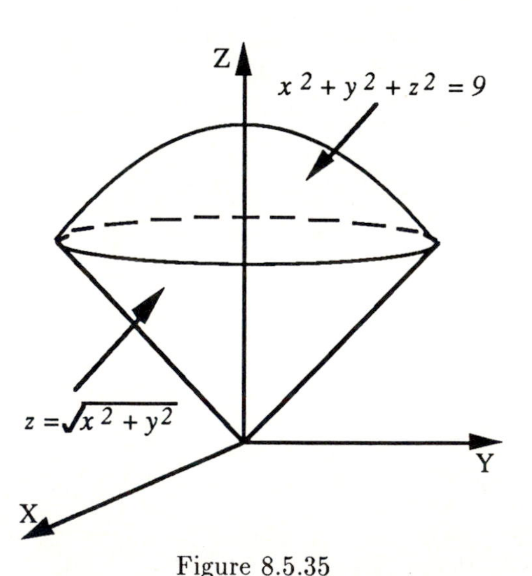

Figure 8.5.35

In spherical coordinates, the region is bounded by $\rho = 3$ and $\phi = \pi/4$. Thus,

$$\text{Volume} = \iiint_D (1)dxdydz = \int_0^{2\pi}\int_0^{\frac{\pi}{4}}\int_0^{3} \rho^2 \sin(\phi)d\rho d\phi d\theta$$

$$= \int_0^{2\pi}\int_0^{\frac{\pi}{4}} \frac{3^3}{3}\sin(\phi)d\phi d\theta = \int_0^{2\pi} 9(-\cos(\phi)]_0^{\frac{\pi}{4}})d\theta$$

$$= \int_0^{2\pi} 9\left[1-\cos(\frac{\pi}{4})\right]d\theta = 18\pi\left(1-\frac{1}{\sqrt{2}}\right) \qquad \triangle$$

Section 8.5 Exercises

In Exercises 1 through 10, sketch the indicated region in the plane and then find the area of the region.

1. Inside $r = 2$.
2. Inside $r = \sin(\theta)$.
3. Inside both $r = \cos(\theta)$ and $r = \sin(\theta)$.
4. Inside $r = 3 + 2\sin(\theta)$.
5. Inside $r^2 = 9\sin(2\theta)$.
6. Inside $r = 2 - 2\cos(\theta)$ and outside $r = 3$.
7. Inside $r = 2[1 + \cos(\theta)]$.
8. Inside both $r = 1 - \cos(\theta)$ and $r = 1 + \cos(\theta)$.
9. Inside one petal of $r = \sin(3\theta)$.
10. Inside the region bounded by $r = \theta$ and $\theta = 0$.

In Exercises 11 through 16, evaluate the integrals.

11. $\int_0^{\pi}\int_0^2 (6r - 3/2r^3)drd\theta$
12. $\int_0^{\pi}\int_0^{\sec(\theta)} r^3\sin^2(\theta)drd\theta$
13. $\int_{-a}^{a}\int_0^{\sqrt{a^2-x^2}} (x^2+y^2)^{3/2}dydx$
14. $\int_1^2\int_0^{\sqrt{4-x^2}} (x^2+y^2)^{-1/2}dydx$
15. $\iint_D \cos(x^2+y^2)dxdy$, where D is the region inside the circle of radius 2 in the first quadrant.
16. $\iint_D (x^3+y^3)dxdy$, where D is the region enclosed by the circle $x^2+y^2=1$.
17. Find the volume of the solid bounded by the paraboloid $z = 4 - x^2 - y^2$ and the xy-plane. Evaluate the double integral using both rectangular and cylindrical coordinates.
18. Find the volume of the solid bounded by the paraboloid $z = r^2$, the cylinder $r = \sin(\theta)$, and the xy-plane.

19. Find the volume of the solid that is bounded by the cone $z = r$, the plane $z = 0$, and the cylinder $r = 2\sin(\theta)$.

20. Find the volume of the cone $z = r$ that lies between the planes $z = 2$ and $z = 5$.

21. Find both an upper and lower Riemann Sum approximation to the volume of the cone $z = r, 0 \le r \le 2$, using the partitions $P = \{0, 1, 2\}$ and $Q = \{0, \pi/4, \pi/2, \pi, 2\pi\}$.

22. Using polar coordinates, approximate the area of the unit circle by forming a Riemann Sum using partitions $P = \{0, \frac{1}{4}, \frac{1}{2}, \frac{3}{4}, 1\}$ and $Q = \{0, \frac{\pi}{4}, \frac{\pi}{2}, \frac{3\pi}{4}, \pi, \frac{3\pi}{2}, 2\pi\}$.

23. Find the volume of the region bounded by the sphere $\rho = 4$ and the cone $\phi = \frac{\pi}{3}$

24. Find the volume of the region bounded by the spheres $\rho = 4$, $\rho = 2$, and the cone $\phi = \pi/3$.

25. Find the volume of the region bounded by the cone $z^2 = x^2 + y^2$ the sphere $x^2 + y^2 + z^2 = 2az$.

26. Derive the formula for change of variables into cylindrical coordinates by means of differentials.

27. Evaluate the integral $\iiint_D 1/(x^2 + y^2 + z^2)^{3/2} dxdydz$ where D is the region bounded by the spheres $x^2 + y^2 + z^2 = a^2$ and $x^2 + y^2 + z^2 = b^2$ where $a < b$.

28. Integrate $\sqrt{x^2 + y^2 + z^2} e^{-(x^2+y^2+z^2)}$ over the region D of Exercise 27.

29. (a) Graph the region D bounded by the lines $y = 1 - x$, $x = 0$, and $y = 0$.

    (b) Using the change of coordinates $u = x + y$ and $y = uv$ show that the corresponding region in the uv-plane is $S = [0, 1] \times [0, 1]$.

    (c) Find the change of coordinates formula, expressing $dx \wedge dy$ in terms of $du \wedge dv$.

30. Show that $\int_0^1 \int_0^{1-x} e^{y/(x+y)} dydx = (e - 1)/2$ by using the change of coordinates $u = x + y$ and $y = uv$. (See Exercise 29.)

31. Project. Describe the graphs (in polar coordinates) of the following relations

    (a) $r = \sin(n\theta)$ (b) $r = \cos(n\theta)$ for various values of n.
    How can you describe the graph as a function of n?

32. Project. Describe the graphs (in polar coordinates) of the relations

(a) $r^2 = a^2 \cos(n\theta)$ (b) $r^2 = -a^2 \cos(n\theta)$

(c) $r^2 = a^2 \sin(n\theta)$ (d) $r^2 = -a^2 \sin(n\theta)$

How does the value of a affect the graph? How does the value of n affect the graph?

Chapter 9

SERIES

9.1 Introduction

As stated in the preface, calculus is the study and analysis of functions. We have seen functions in a variety of contexts and have analyzed functions from several points of view.

For any function f, the following questions might be asked.

1. How do we *actually compute* $f(x)$ for a given x?
2. What are the source and target, domain and range of f?
3. Is f continuous?
4. Is f differentiable? If it is, how do we differentiate f?
5. Is f integrable? If it is, how do we integrate f?
6. For what values of x is $f(x) = 0$?

We have spent much of our efforts in this text on questions 2 through 6. Question 1 is much more basic and may appear to be far simpler to answer. However, these basic questions are often extremely difficult to handle!

Let us concentrate on question (1). Given a function $f : A \to B$ and a number x in A, how do we compute $f(x)$?

Example 9.1.1 Let f be the polynomial given as $f(x) = x^2 - x + 2$. If we want to compute $f(3)$, for example, it is an easy task. We find $f(3) = 8$ as 3 times 3, minus 3, plus 2. The operations involved in this computation are multiplication, subtraction, and addition. Of course, these operations are three of the four basic arithmetic operations: addition, subtraction, multiplication, and division. *All computations involving real numbers are based on these four basic operations.* $\triangle$

Example 9.1.2 Consider a slightly more complicated problem: Evaluate the function $f : [0, \infty) \to \mathbb{R}$ at $x = 3$ where f is defined by $f(x) = \sqrt{x}$. If we think of f as an input/output machine, we wish to know the output when the input is 3. As a practical matter, the only operations that the "machine" can really perform are the four basic operations. There is an algorithm for computing a sequence of approximations for the square roots of integers. The difficulty is that the algorithm requires an *infinite* number of steps to recover $\sqrt{3}$ exactly. △

Example 9.1.3 Our problem is even worse if we wish to compute cos(1). Using basic operations, how is this computation accomplished? The cosine function is differentiable (it has derivatives of all orders) and is bounded. It is as "nice" a function as you could encounter, but how does one find the cosine of a number? (How does a calculator find the cosine of a number?) △

The goal of this chapter is to develop techniques that will enable us to easily approximate values for the variety of functions that appear in applications of mathematics. It is natural to begin this process by considering polynomial approximations to functions, since polynomials can be evaluated exactly (see Example 9.1.1). As usual, we will find that the Basic Approximation Process plays a central role in the analysis.

Section 9.1 Exercises

1. Compile a list of three different kinds of functions that cannot be evaluated using the four basic arithmetic operations.

2. Find $f(1)$, $f(2)$, $f(5)$ and $f(-1)$ for each of the following. Describe the method you used for each.

   (a) $f(x) = 3/\sin(x)$ (b) $f(x) = \tan(x)$ (c) $f(x) = \sqrt{\sin(x^2)}$

   (d) $f(x) = \arcsin(x)$ (e) $f(x) = e^x$ (f) $f(x) = \log(x^2 + 2)$

9.2 Taylor's Polynomial Approximations

We shall use the first three steps of the Basic Approximation Process to develop *polynomial approximations* for any member of the class of infinitely differentiable functions, the class C^∞. Recall that this class includes all the important functions that we have studied: polynomial, rational, trigonometric, logarithmic, and exponential functions. Again, the reason polynomials are used for approximations is the fact that polynomial functions can be *evaluated exactly* using only the arithmetic operations of addition, subtraction, and multiplication.

We begin by introducing some terminology. The nth-degree polynomial $P : \mathbb{R} \to \mathbb{R}$ written in the form

$$c_0 + c_1(x-a) + c_2(x-a)^2 + \cdots + c_n(x-a)^n$$

or in summation notation

$$\sum_{k=0}^{n} c_k(x-a)^k$$

is called a *polynomial in powers of* $(x-a)$.

We use powers of $(x-a)$ since we plan to approximate a function near a by a polynomial.

Before explaining how to expand a polynomial P in powers of $(x-a)$, we illustrate the fact that the coefficients in the expansion depend upon a.

Example 9.2.1 Consider the polynomial $P : \mathbb{R} \to \mathbb{R}$ defined by $P(x) = 2 + 3x + x^2 - 2x^3$.

(a) The expansion of P in powers of $(x-2)$ is:

$$-4 - 17(x-2) - 11(x-2)^2 - 2(x-2)^3$$

(b) The expansion of P in powers of $(x-1)$ is:

$$4 - (x-1) - 5(x-1)^2 - 2(x-1)^3$$

(c) The expansion of P in powers of $(x+1)$ is:

$$2 - 5(x+1) + 7(x+1)^2 - 2(x+1)^3$$

The reader is urged to expand the polynomial expressions in parts (a), (b), and (c) above to check that they all do in fact equal $P(x)$. △

We shall illustrate the process of developing polynomial approximations (that is, determining the coefficients) by considering the question of approximating the square root function

$$f(x) = \sqrt{x}$$

near $x = 1$ with polynomials in powers of $(x-1)$.

The first step in the Basic Approximation Process is to make an approximation. Since the simplest polynomial is a constant, a polynomial of degree 0, we let $P_0 = f(1) = 1$ be the first approximation.

The second step in the Basic Approximation Process is to make a better approximation (one with a smaller error). Since the next simplest polynomial is an affine function, a polynomial of degree 1, we let the second approximation P_1

be the line tangent to the graph of f at the point $x = 1$. This is the approximation given by the Linear Approximation Theorem. That is,

$$P_1(x) = f(1) + f'(1)(x-1) = 1 + \frac{1}{2}(x-1)$$

The third step in the Approximation Process is to develop a convergent sequence of approximations, each having less error than the preceding one. The next simplest polynomial would be one of degree 2, a quadratic. Thus it seems reasonable to find a quadratic that has the same curvature as does the graph of f at the point $x = 1$ as well as the same tangent there. Hence we want a quadratic polynomial in powers of $x-1$ whose zero, first, and second derivatives agree with those of f at $x = 1$. That is, we want to determine the coefficients c_0, c_1, and c_2 of

$$P_2(x) = c_0 + c_1(x-1) + c_2(x-1)^2$$

such that

$$P_2(1) = c_0 = f(1) = 1$$

$$P_2'(1) = c_1 = f'(1) = \tfrac{1}{2}$$

$$P_2''(1) = 2c_2 = f''(1) = -\tfrac{1}{4}$$

Hence $c_0 = 1$, $c_1 = \frac{1}{2}$, and $c_2 = f''(1)/2 = -\frac{1}{8}$, and so our third approximation is

$$P_2(x) = 1 + \frac{(x-1)}{2} - \frac{(x-1)^2}{8}$$

To see how this improves the accuracy, consider evaluating at $x = 1.1$. To five decimal places, $f(1.1) = \sqrt{1.1} = 1.04881$. The first approximation gives $P_1(1.1) = 1.05$, accurate to two decimal digits, and the second approximation gives $P_2(1.1) = 1.04875$, accurate to four decimal digits.

The following table gives the numerical values (accurate to three decimal places) of f and the three polynomial approximations that we have constructed for the values $x = 0, 1, 2$.

x	$f(x)$	$P_0(x)$	$P_1(x)$	$P_2(x)$
0	0	1	0.5	0.375
1	1	1	1	1
2	1.414	1	1.5	1.375

The improvement in the accuracy of the approximation to f shown by the graphs of the polynomials from P_0 to P_2 is geometrically displayed in the following diagram.

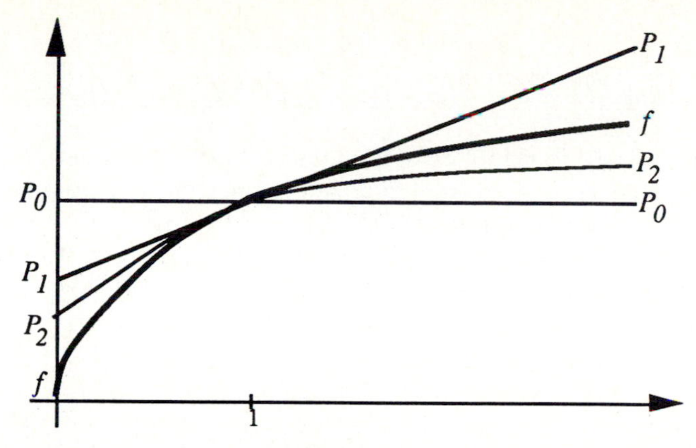

Figure 9.2.2

Based on the above table and diagram, it seems reasonable to continue this pattern of forming the next approximation by constructing a polynomial of the next higher degree whose nonzero derivatives agree with those of f at $x = 1$. We shall do this, even though we have no "nice" geometric interpretation such as tangent slope or concavity for derivatives of any order greater than two. Thus the fourth approximation $P_3(x)$ should be a cubic polynomial in powers of $(x-1)$ whose zero, first, second, and third derivatives agree with those of f at $x = 1$. That is,

$$P_3(x) = c_0 + c_1(x-1) + c_2(x-1)^2 + c_3(x-1)^3$$

such that

$$P_3(1) = c_0 = f(1) = 1$$

$$P_3'(1) = c_1 = f'(1) = \tfrac{1}{2}$$

$$P_3''(1) = 2c_2 = f''(1) = -\tfrac{1}{4}$$

$$P_3'''(1) = 3!c_3 = f'''(1) = \tfrac{3}{8}$$

Hence $c_0 = 1$, $c_1 = \frac{1}{2}$, $c_2 = f''(1)/2! = -\frac{3}{8}$, $c_3 = f'''(1)/3! = \frac{3}{48} = \frac{1}{16}$, and so our fourth approximation is

$$P_3(x) = 1 + \frac{1}{2}(x-1) - \frac{1}{8}(x-1)^2 + \frac{1}{16}(x-1)^3$$

Following the same procedure, we construct the fifth polynomial approximation:

$$\begin{aligned} P_4(x) &= 1 + f'(1)(x-1) + \frac{1}{2!}f''(1)(x-1)^2 + \frac{1}{3!}f'''(1)(x-1)^3 \\ &\quad + \frac{1}{4!}f''''(1)(x-1)^4 \end{aligned}$$

or

$$P_4(x) = 1 + \frac{1}{2}(x-1) - \frac{1}{8}(x-1)^2 + \frac{1}{16}(x-1)^3 - \frac{5}{128}(x-1)^4$$

We are now ready to complete step three of the Approximation Process by generalizing the process of constructing approximating polynomials. We will define a sequence of polynomials $\{P_n(x)\}$ by

$$\begin{aligned} P_n(x) &= f(1) + f'(1)(x-1) + \frac{f''(1)}{2!}(x-1)^2 + \frac{f^{(3)}(1)}{3!}(x-1)^3 \\ &\quad + \cdots + \frac{f^{(n)}(1)}{n!}(x-1)^n \end{aligned}$$

Summation notation will often be used in order to make the expressions more compact. Thus,

$$P_n(x) = \sum_{k=0}^{n} \frac{f^{(k)}(1)}{k!}(x-1)^k$$

The polynomials in the sequence of polynomial approximations $\{P_n(x)\}$ are called *Taylor Polynomials* in powers of $(x-1)$. The approximation $P_n(x)$ is referred to as the *expansion of f about the point $x = 1$ in a Taylor Polynomial of degree n in powers of $(x-1)$*. The formal definition is as follows.

Definition 9.2.3 (Taylor Polynomial) Let f be a function with n derivatives at a. The Taylor Polynomial of degree n in powers of $(x-a)$ is

$$P_n(x) = \sum_{k=0}^{n} \frac{f^k(a)}{k!}(x-a)^k$$

Taylor Polynomials are named after the English mathematician Brook Taylor (1685–1731) who first published (in 1715) a method for expanding a function in a manner similar to the above definition.

In the next section we shall consider the question of error in using the Taylor Polynomials.

Before presenting another example, we shall make some observations.

1. All the Taylor Polynomials give the exact value of the function when evaluated at the point $x = a$ about which f is being expanded (in the previous example the point was $x = 1$).

2. The function f could be expanded about any point in its domain, however for computational reasons the point is chosen so as to minimize the difficulty in evaluating the derivatives.

3. If a function f is to be expanded about the point $x = a$ by Taylor Polynomials in powers of $(x-a)$, then the derivatives of f are all evaluated at $x = a$.

4. All that is needed in order to construct a Taylor Polynomial approximation of a function expanded about a point is to know the values of the derivatives of the function at the point.

Example 9.2.4 We will construct a sequence of Taylor Polynomials that expand $f(x) = 1/(1-x)$ about $x = 0$.

We need to determine the values of the derivatives of f at $x = 0$. Thus we will compute the first few derivatives and look for a pattern.

$$f(x) = (1-x)^{-1} \qquad f(0) = 1$$

$$f'(x) = (-1)(1-x)^{-2}(-1) = (1-x)^{-2} \qquad f'(0) = 1$$

$$f''(x) = (-2)(1-x)^{-3}(-1) = 2(1-x)^{-3} \qquad f''(0) = 2$$

$$f'''(x) = 3!(1-x)^{-4} \qquad f'''(0) = 3!$$

$$f^{(4)}(x) = 4!(1-x)^{-5} \qquad f^{(4)}(0) = 4!$$

The pattern should now be clear: $f^{(n)}(0) = n!$. Thus we define our sequence of Taylor Polynomials $\{P_n(x)\}$ by

$$P_n(x) = \sum_{k=0}^{n} \frac{f^{(k)}(0)}{k!}(x-0)^k = \sum_{k=0}^{n} x^k = 1 + x + x^2 + \cdots + x^n$$

△

Example 9.2.5 Let $f(x) = e^x$. With this function, $f^{(n)}(x) = e^x$ for all n. The natural point to expand f about is $x = 0$ since $f^{(n)}(0) = e^0 = 1$ for all n. Thus the nth-degree Taylor Polynomial about the point 0 for $f(x) = e^x$ is

$$P_n(x) = 1 + \frac{1}{1!}x^1 + \frac{1}{2!}x^2 + \cdots + \frac{1}{n!}x^n = \sum_{k=0}^{n} \frac{x^k}{k!}$$

△

We shall present one more example, this time using the logarithmic function.

Example 9.2.6 We will find a sequence of Taylor Polynomials in powers of $(x-2)$ that expand the logarithmic function $f(x) = \log(x)$ about the point $x = 2$.

We need to determine the derivatives of f. Thus we compute the first few derivatives and look for a pattern.

$$f(x) = \log(x)$$

$$f'(x) = 1/x = x^{-1}$$

$$f''(x) = -x^{-2}$$

$$f'''(x) = 2x^{-3}$$

$$f^{(4)}(x) = -3!x^{-4}$$

$$f^{(5)}(x) = 4!x^{-5}$$

The pattern should now be clear:

$$f^{(k)}(x) = (-1)^{k-1}(k-1)!x^{-k} \quad \text{for } k > 0$$

Therefore,

$$f^{(k)}(2) = \begin{cases} \log(2) & \text{for } k = 0 \\ (-1)^{k-1}(k-1)!2^{-k} & \text{for } k > 0 \end{cases}$$

and so our sequence of Taylor Polynomials $\{P_n(x)\}$ is defined by

$$\begin{aligned} P_n(x) &= \log(2) + \sum_{k=1}^{n} \frac{(-1)^{k+1}(k-1)!2^{-k}}{k!}(x-2)^k \\ &= \log(2) + \sum_{k=1}^{n} (-1)^{k+1} \frac{1}{k2^k}(x-2)^k \end{aligned}$$

△

Section 9.2 Exercises

For Exercises 1 through 6, find the sequence of Taylor Polynomials in powers of $(x - a)$.

1. $f(x) = 1 + x + x^2, \quad a = 2$
2. $f(x) = 2 - x + 2x^2 - x^3, \quad a = -1$
3. $f(x) = \sin(x), \quad a = \pi/2$
4. $f(x) = e^{(x-1)}, \quad a = 1$
5. $f(x) = e^x + x + \sin(x), \quad a = 0$
6. $f(x) = \cos(x), \quad a = \pi/3$

For Exercises 7 through 14, find the Taylor Polynomial of degree n expanded about the point $a = 0$.

8. $f(x) = e^{-x}\cos(x), \quad n = 4$
9. $f(x) = \cos(x)\log(1 + x), \quad n = 5$
10. $f(x) = \sin(x)\sqrt{1 + x}, \quad n = 5$
11. $f(x) = 1/(1 + x + x^2), \quad n = 4$
12. $f(x) = \cos(x) - 1 + x^2/2, \quad n = 4$
13. $f(x) = \tan(x), \quad n = 5$
14. $f(x) = 1/[1 - \sin(x)], \quad n = 4$

15. (This exercise requires using a plotting routine that allows for plotting two graphs on the same set of axes.) An error bound for approximating a function by a Taylor polynomial over a given interval can be obtained experimentally by plotting the graphs of the two functions on the same set of axes and then measuring the largest vertical distance between the two curves.

    (a) Determine experimentally the minimum degree Taylor polynomial for $\sin(x)$ expanded about $a = 0$ that will approximate $f(x) = \sin(x)$ over the interval $[0, \pi]$ with accuracy 0.1.

    (b) Repeat part (a) using $a = \pi/2$.

    (c) Repeat part (a) using $a = \pi$.

9.3 Taylor's Theorem

Throughout this text we have stressed the importance of associating an error bound with an approximation. In this section we will develop an error bound when a function is approximated by Taylor Polynomials. As always, the error is the function minus the approximation.

Definition 9.3.1 *Let f be a function with $n+1$ continuous derivatives at a. The error term $R_n(x)$ in the expansion of f about the point $x = a$ in a Taylor Polynomial of degree n in powers of $(x-a)$ is*

$$R_n(x) = f(x) - P_n(x) = f(x) - \sum_{k=0}^{n} \frac{f^{(k)}(a)}{k!}(x-a)^k$$

The first step is to obtain an expression for $R_n(x)$. This is given in the following important theorem.

Theorem 9.3.2 (Integral Remainder) *If $f : \mathbb{R} \to \mathbb{R}$ has $n+1$ continuous derivatives over a closed interval containing a and x, then*

$$f(x) = \sum_{k=0}^{n} \frac{f^{(k)}(a)}{k!}(x-a)^k + R_n(x)$$

where

$$R_n(x) = \frac{1}{n!}\int_a^x f^{(n+1)}(t)(x-t)^n \, dt$$

Proof: Although there are several proofs of this theorem, none of them is very well motivated. Our proof is by mathematical induction and involves integration by parts. Step I in mathematical induction is to establish the "base case." That is, establish that the desired result is true for some initial value of the index n.

For $n = 0$: substituting $n = 0$ in the statement of the theorem, we see that what needs to be shown is:

$$f(x) = \sum_{k=0}^{0} \frac{f^{(k)}(a)}{k!}(x-a)^k + R_0(x)$$

where $R_0(x) = \int_a^x f'(t)dt$. (Recall that $0! = 1$.)

Since

$$\sum_{k=0}^{0} \frac{f^{(k)}(a)}{k!}(x-a)^k = f(a)$$

the condition that needs to be verified can be expressed as

$$f(x) = f(a) + \int_a^x f'(t)dt$$

This, however, is just the statement of the Fundamental Theorem of Calculus Part II [transpose the $f(a)$ term]. Hence the desired result is true for $n = 0$.

This completes step I; however, in order to better understand the mechanics of the proof, we suggest the reader verify the conclusion directly for $n = 1$, 2, and 3.

> The argument for $n = 1$ begins by applying the technique of integration by parts to R_0. Let
>
> $$\begin{array}{ll} u = f'(t) & dv = dt \\ du = f''(t)dt & v = (t-x) \end{array}$$
>
> [The "unmotivated" part of the proof is in setting $v = (t-x)$ rather than setting $v = t$. The rationale is that choosing the constant of integration to be $-x$ rather than 0 gives the result in the desired form.] The details are now left to the reader.

Now, we return to the proof. Step II in mathematical induction is to show that if the result holds for some index, $n-1$, then the result must hold for the next value of the index n. Thus the result must hold for all values of the index greater than 0, since we have shown that the result holds for 0 (the base case starting point) and that one can proceed from one index value to the next.

Thus we assume that

$$f(x) = \sum_{k=0}^{n-1} \frac{f^{(k)}(a)}{k!}(x-a)^k + R_{n-1}(x)$$

where

$$R_{n-1}(x) = \frac{1}{(n-1)!}\int_a^x f^{(n)}(t)(x-t)^{n-1}dt$$

We now verify that the result holds for the index n. To do this, we apply the technique of integration by parts to $R_{n-1}(x)$. Let

$$\begin{array}{ll} u = f^{(n)}(t) & dv = (x-t)^{n-1}dt \\ du = f^{(n+1)}(t)dt & v = -(x-t)^n/n \end{array}$$

Thus

$$\begin{aligned} f(x) &= \sum_{k=0}^{n-1}\frac{f^{(k)}(a)}{k!}(x-a)^k - \frac{1}{(n-1)!}[\frac{f^{(n)}(t)(x-t)^n}{n}]_a^x \\ &\quad +\frac{1}{n}\int_a^x f^{(n+1)}(t)(x-t)^n dt] \end{aligned}$$

and so

$$f(x) = \sum_{k=0}^{n-1}\frac{f^{(k)}(a)}{k!}(x-a)^k + \frac{f^{(n)}(a)}{n!}(x-a)^n + \frac{1}{n!}\int_a^x f^{(n+1)}(t)(x-t)^n dt$$

or

$$f(x) = \sum_{k=0}^{n}\frac{f^{(k)}(a)}{k!}(x-a)^k + R_n(x)$$

Hence by Mathematical Induction the theorem is true for all nonnegative integers n. □

There is another form of Taylor's Theorem that gives the remainder in terms of a derivative. This is called Lagrange's form for the remainder.

Theorem 9.3.3 (Taylor's Theorem, Derivative Form Remainder)
If $f : \mathbb{R} \to \mathbb{R}$ has $n+1$ continuous derivatives over a closed interval containing a and x, then

$$f(x) = \sum_{k=0}^{n}\frac{f^{(k)}(a)}{k!}(x-a)^k + R_n(x)$$

where

$$R_n(x) = \frac{f^{(n+1)}(c)}{(n+1)!}(x-a)^{n+1}$$

for some point c between x and a.

Proof. Since $f^{n+1}(x)$ is continuous over a closed interval, the Extreme Value Theorem (Theorem 2.6.16) guarantees that $f^{n+1}(x)$ obtains a maximum value, say at $x = c$, in the interval. That is, $f^{n+1}(x) \leq f^{n+1}(c)$, for all x in the interval. Thus,

$$
\begin{aligned}
R_n(x) &= 1/n! \int_a^x f^{n+1}(x-t)dt \\
&\leq \frac{f^{n+1}(c)}{n!} \int_a^x (x-t)^n dt \\
&= \frac{f^{n+1}(c)}{n!} \left[-\frac{(x-t)^{n+1}}{n+1} \right]_a^x \\
&= \frac{f^{n+1}(c)}{(n+1)!}(x-a)^{n+1}
\end{aligned}
$$

□

We repeat, the importance of associating an error bound expression with an approximation has been stressed throughout this text. Taylor Polynomial approximations are no exceptions. The *exact* error is given by $R_n(x)$. Since generally neither the integral nor derivative form of $R_n(x)$ can be evaluated (why?), we need a technique for determining an error bound. The following examples illustrate such a technique.

Example 9.3.4 Suppose we wish to find the *maximum possible error produced when* $f : [-2, 2] \to \mathbb{R}$ defined by $f(x) = e^x$ is approximated using a Taylor Polynomial of degree 4 expanded about the point $a = 0$. The polynomial is (by Example 9.2.5)

$$P_4(x) = 1 + x + \frac{x^2}{2} + \frac{x^3}{6} + \frac{x^4}{24}$$

The remainder is

$$R_4(x) = \frac{e^c}{5!}x^5$$

for some c between x and 0.

The error of the approximation is exactly $e^c x^5/120$, for some number c between 0 and x. Since the theorem does not tell us what c is, our goal is to obtain some *estimates* of the error. *We can live with overestimates of error; however, we do not want to underestimate the error.*

As a conservative estimate, we know that $0 < e < 3$. Therefore, for any c, $-2 < c < 2$, we have $|e^c| < 9$.

For some points c in the interval $[-2, 2]$, the value of e^c will be much smaller; the important point is that *it will never be larger!*

Consequently, we know for all c, $-2 < c < 2$,

$$|R_n(x)| < \frac{9}{120}x^5 = 0.075x^5$$

Well, for $-2 < x < 2$, we have $0 < |x^5| < 32$. Therefore, the worst possible error for any x under consideration is less than $0.075 \cdot 32 = 2.4$.

This example can be extended to higher degree approximates. In general,

$$R_n(x) = \frac{e^c}{(n+1)!} x^{n+1}$$

Therefore, using a polynomial approximation of degree n, the worst error is bounded by

$$|R_n(x)| < \frac{9}{(n+1)!} 2^{n+1}$$

since $e^c < 9$ for $-2 < c < 2$.

For even small values of n the error is modest. For example, for $n = 7$

$$|R_7(x)| < \frac{9}{8!} 2^8 \approx 0.0190476 \quad \text{for} \quad |x| < 2$$

△

Example 9.3.5 This example will (partially) answer the question asked in Example 9.1.3: How can we compute the cosine of a number? We will estimate the maximum possible error when a Taylor Polynomial of degree 5 is used to estimate $f(x) = \cos(x)$. What point should we expand our polynomial about? Since the derivatives of cosine involve either sine or cosine, which are easy to evaluate at 0, the natural choice is to expand $\cos(x)$ about $a = 0$. Taylor Theorem says:

$$f(x) = f(0) + f'(0)x + \frac{f''(0)}{2} x^2 + \frac{f^{(3)}(0)}{6} x^3 + \frac{f^{(4)}(0)}{24} x^4 + \frac{f^{(5)}(0)}{120} x^5 + R_5$$

where

$$R_5(x) = \frac{f^6(c)}{720} x^6$$

for some c between 0 and x.

Now,

$$\begin{array}{ll} \cos(0) = 1 & \cos'(0) = 0 \\ \cos''(0) = -1 & \cos^{(3)}(0) = 0 \\ \cos^{(4)}(0) = 1 & \cos^{(5)}(0) = 0 \end{array}$$

and $\cos^{(6)}(c) = -\cos(c)$, which is between -1 and 1. Since $|\cos^{(6)}(c)| \leq 1$ for any real c,

$$\cos(x) = 1 - \frac{1}{2} x^2 + \frac{1}{24} x^4 + R_5(x)$$

where $|R_5(x)| \leq x^6/720$. If we restrict x to the interval $[-1, 1]$, then the error will be less than $1/720 \approx 0.0014$ for any x in the interval. In particular,

$$\cos(1) \approx 1 - \frac{1}{2} + \frac{1}{24} = \frac{13}{24} = 0.541666\ldots$$

with an error of at most 0.0014. △

In most applied situations, the problem is to find an approximation having an accuracy that satisfies a given error bound. For example, suppose we wish to approximate $\cos(x)$ on the interval $[-2, 2]$ with an error of less than 0.0001 for any x in the interval.

The error term for approximating $\cos(x)$ with the nth order Taylor Polynomial for cosine expanded about $x = 0$ is

$$R_n(x) = \frac{\cos^{(n+1)}(c)}{(n+1)!}x^{n+1}$$

for some c in the interval $[-2, 2]$. Since

$$\left|\frac{\cos^{(n+1)}(c)}{(n+1)!}x^{n+1}\right| \leq \frac{|x|^{n+1}}{(n+1)!} \leq \frac{2^{n+1}}{(n+1)!}$$

for x in $[-2, 2]$, we must find an integer n such that

$$\frac{2^{n+1}}{(n+1)!} \leq 0.0001$$

From the previous example, we know that with $n = 5$, $1/(n+1)! = 1/720 \approx 0.0014$, which is too large. So we can start looking with $n \geq 6$. We find that

$$\frac{2^{9+1}}{(9+1)!} \approx 0.00028 \text{ and } \frac{2^{10+1}}{(10+1)!} \approx 0.00005$$

Thus we need to have 10 terms in the Taylor Polynomial. (Note that half of them will be zero, when $\cos^{(k)}(0) = 0$.) The desired approximation for $\cos(x)$ in $[-2, 2]$ is

$$\cos(x) \approx 1 - \frac{x^2}{2!} + \frac{x^4}{4!} - \frac{x^6}{6!} + \frac{x^8}{8!} - \frac{x^{10}}{10!}$$

Section 9.3 Exercises

In Exercises 1 through 4, estimate the error if a Taylor Polynomial of degree n expanded about $x = a$ is used to estimate f at x.

1. $f(x) = e^x, \quad n = 3, \ a = 0, \ x = 0.25$
2. $f(x) = \sin(x), \quad n = 5, \ a = 0, \ x = 1$
3. $f(x) = \sqrt{x}, \quad n = 4, \ a = 4, \ x = 5$
4. $f(x) = \tan^{-1}(x), \quad n = 5, \ a = 0, \ x = 0.75$
5. Use a Taylor Polynomial expanded about $a = 0$ for the function $f(x) = \exp(x)$ to find an approximation to $\exp(3)$ that is accurate to within 0.0001.

6. Use a Taylor Polynomial expanded about $a = 4$ for the function $f(x) = \sqrt{x}$ to approximate $\sqrt{4.1}$ to within 0.00001.

7. Use a Taylor Polynomial to approximate $\sin(5°)$ to an accuracy of 0.0001.

8. Use a Taylor Polynomial to approximate $\cos(42°)$ to an accuracy of 0.0001.

9. Use a fourth degree Taylor polynomial expanded about $a = 0$ to estimate $\int_{-1}^{1} e^{-x^2} dx$. That is, integrate the Taylor Polynomial to estimate the integral.

10. Use a Taylor Polynomial to estimate $\int_1^4 dx/(1 + x^2)$ with an accuracy of 0.01.

9.4 Taylor's Theorem in Several Variables

Taylor's Theorem, which allows us to approximate a function by a polynomial, also holds for functions of several variables. Let $f : \mathbb{R} \to \mathbb{R}$ be a C^{n+1} function in an open interval containing the point a. Taylor's Theorem says that

$$f(x) = P_n(x) + R_n(x)$$

where P_n is the Taylor Polynomial of degree n for f expanded about a:

$$P_n(x) = f(a) + f'(a)(x - a) + \frac{f^{(2)}(a)}{2!}(x - a)^2 + \cdots + \frac{f^{(n)}(x)}{n!}(x - a)^n$$

and $R_n(x)$ is the remainder:

$$R_n(x) = \frac{f^{(n+1)}(c)}{(n + 1)!}(x - a)^{n+1}$$

where c is some point between a and x.

We wish to extend this result to functions of several variables. Let us see how we can make sense of the terms when f is a function of several variables, $f : \mathbb{R}^m \to \mathbb{R}$. The derivative of f at a is $f'(a) = (D_1 f(a), D_2 f(a), ..., D_m f(a))$, which is a vector in $\mathbb{R}^m$. The points x and a are in $\mathbb{R}^m$, so $x = (x_1, x_2, ..., x_m)$ and $a = (a_1, a_2, ..., a_m)$. The difference $(x - a)$ is the vector difference computed componentwise:

$$x - a = (x_1 - a_1, x_2 - a_2, \ldots, x_m - a_m)$$

We need to interpret a term of the form

$$f^{(k)}(a)(x - a)^k$$

When $k = 1$, we need $f'(a)(x-a) = (x-a)f'(a)$ to be a real number. This is true if the product is the dot product:

$$\begin{aligned}(x-a)\cdot f'(a) &= (x_1 - a_1, \ldots, x_m - a_m)\cdot(D_1 f(a), \ldots, D_m f(a)) \\ &= (x_1 - a_1)D_1 f(a) + \cdots + (x_m - a_m)D_m f(a)\end{aligned}$$

To extend this to higher derivatives, we can take the dot product before taking the power. In order to do this, we need to interpret higher derivatives as higher powers. We will consider differentiation as an operator, transforming functions to functions. Thus D_1, the partial derivative with respect to the first variable, transforms f to the new function $D_1 f$ whose value at x is $D_1 f(x)$. We can do algebraic operations on these transformations:

$$(rD_1 + sD_2)f = rD_1 f + sD_2 f$$

and

$$\begin{aligned}(D_1 + D_2)^2 f &= (D_1 + D_2)(D_1 + D_2)f \\ &= (D_1 D_1 + D_1 D_2 + D_2 D_1 + D_2 D_2)f \\ &= (D_{1,1} + 2D_{1,2} + D_{2,2})f = D_{1,1}f + 2D_{1,2}f + D_{2,2}f\end{aligned}$$

Recall that the gradient of f is the function $\nabla f : \mathbb{R}^m \to \mathbb{R}^m$ given by

$$\nabla f(a) = f'(a) = (D_1 f(a), D_2 f(a), \ldots, D_m f(a))$$

Just as we consider D_1 as a transformation that takes a function $f : \mathbb{R}^m \to \mathbb{R}$ to the function $D_1 f : \mathbb{R}^m \to \mathbb{R}$, we can define

$$\nabla = (D_1, D_2, ..., D_m)$$

as the transformation that takes the function $f : \mathbb{R}^m \to \mathbb{R}$ to the function $\nabla f : \mathbb{R}^m \to \mathbb{R}^m$. Using the definition of the dot product we have

$$x\cdot\nabla = (x_1, \ldots, x_m)\cdot(D_1, \ldots, D_m) = x_1 D_1 + x_2 D_2 + \cdots + x_m D_m$$

Example 9.4.1 Let $f : \mathbb{R}^2 \to \mathbb{R}$ and $a = (a_1, a_2)$ be in $\mathbb{R}^2$. Then

$$\begin{aligned}[(a\cdot\nabla)^2 f](x,y) &= [((a_1, a_2)\cdot(D_1, D_2))^2 f](x,y) = [(a_1 D_1 + a_2 D_2)^2 f](x,y) \\ &= [(a_1^2 D_{1,1} + 2a_1 a_2 D_{1,2} + a_2^2 D_{2,2})f](x,y) \\ &= [a_1^2 D_{1,1}f + 2a_1 a_2 D_{1,2}f + a_2^2 D_{2,2}f](x,y) \\ &= a_1^2 D_{1,1}f(x,y) + 2a_1 a_2 D_{1,2}f(x,y) + a_2^2 D_{2,2}f(x,y) \qquad \triangle\end{aligned}$$

Now we can interpret $f^{(k)}(a)(x-a)^k$ as $[((x-a)\cdot\nabla)^k f](a)$.

Since we always apply differentiation before evaluation, e.g., $Df(a)$ means $[Df](a)$ rather than $D[f(a)]$ (which is zero), we can simplify notation by writing

$$[((x-a)\cdot\nabla)^k f](a) = [(x-a)\cdot\nabla]^k f(a)$$

Definition 9.4.2 (Multivariate Taylor Polynomial) *Let $f : \mathbb{R}^m \to \mathbb{R}$ be a C^n function in an open interval containing a. Then the* n-th degree Taylor Polynomial for f *expanded about a is*

$$P_n(x) = \sum_{k=0}^{n} \frac{1}{k!}[(x-a)\cdot\nabla]^k f(a)$$

Example 9.4.3 If $f : \mathbb{R}^2 \to \mathbb{R}$ is a C^2 function, then the second-degree Taylor Polynomial for f expanded about a is

$$\begin{aligned}
P_2(x) &= f(a) + [(x - a\cdot\nabla)]f(a) + \frac{1}{2!}[(x-a)\cdot\nabla]^2 f(a) \\
&= f(a) + [(x_1 - a_1, x_2 - a_2)\cdot(D_1, D_2)]f(a) \\
&\quad + \frac{1}{2!}[(x_1 - a_1, x_2 - a_2)\cdot(D_1, D_2)]^2 f(a) \\
&= f(a) + [(x_1 - a_1)D_1 + (x_2 - a_2)D_2]f(a) \\
&\quad + \frac{1}{2!}[(x_1 - a_1)D_1 + (x_2 - a_2)D_2]^2 f(a) \\
&= f(a) + (x_1 - a_1)D_1 f(a) + (x_2 - a_2)D_2 f(a) \\
&\quad \frac{1}{2}[(x_1 - a_1)^2 D_{1,1} + 2(x_1 - a_1)(x_2 - a_2)D_{1,2} \\
&\quad + (x_2 - a_2)^2 D_{2,2}]f(a) \\
&= f(a) + (x_1 - a_1)D_1 f(a) + (x_2 - a_2)D_2 f(a) + \frac{1}{2}(x_1 - a_1)^2 D_{1,1} f(a) \\
&\quad + (x_1 - a_1)(x_2 - a_2)D_{1,2} f(a) + \frac{1}{2}(x_2 - a_2)^2 D_{2,2} f(a)
\end{aligned}$$

△

Theorem 9.4.4 (Multivariate Taylor's Theorem) *Let $f : \mathbb{R}^m \to \mathbb{R}$ be defined and have continuous partial derivatives of order $n+1$ in an open rectangle in $\mathbb{R}^m$ containing the points a and x.*

Then $f(x) = P_n(x) + R_n(x)$ where P_n is the nth degree Taylor Polynomial for f about a and

$$R_n(x) = \frac{1}{(n+1)!}[(x-a)\cdot\nabla]^{n+1} f(c)$$

where c is some point on the line segment between a and x.

Proof: We will reduce the problem to functions of one variable by considering the function f restricted to the line segment joining a and x. We define a function $g : \mathbb{R} \to \mathbb{R}^m$ by $g(t) = a + t(x - a)$. This function is a parameterization of the line segment joining a to x: when $t = 0$, $g(0) = a$ and when $t = 1$, $g(1) = x$. The composite function $f \circ g : \mathbb{R} \to \mathbb{R}$ is defined on an interval containing [0,1] and is differentiable. By the Chain Rule,

$$\begin{aligned}(f \circ g)'(t) &= f'(g(t)) \cdot g'(t) = f'(g(t)) \cdot (x-a) \\ &= (D_1 f(g(t)), ..., D_m f(g(t))) \cdot (x-a) \\ &= [(x-a) \cdot (D_1, ..., D_m)] f(g(t)) = [(x-a) \cdot \nabla] f(g(t))\end{aligned}$$

Analogously, for $k = 2, 3, ..., n+1$

$$(f \circ g)^{(k)}(t) = [(x-a) \cdot (D_1, ..., D_m)]^k f(g(t)) = [(x-a) \cdot \nabla]^k f(g(t))$$

By Taylor's Theorem for a function $f \circ g$ of one variable, evaluated at 1 and expanded about 0, we have

$$\begin{aligned}(f \circ g)(1) &= (f \circ g)(0) + \frac{1}{1!}(f \circ g)'(0) + \frac{1}{2!}(f \circ g)''(0) \\ &\quad + \cdots + \frac{1}{n!}(f \circ g)^{(n)}(0) + \frac{1}{(n+1)!}(f \circ g)^{(n+1)}(d)\end{aligned}$$

for some d between 0 and 1. Letting $c = g(d)$ we have the desired result:

$$\begin{aligned}f(x) &= f(g(1)) = (f \circ g)(1) \\ &= \sum_{k=0}^{n} \frac{1}{k!}(f \circ g)^{(k)}(0) + \frac{1}{(n+1)!}(f \circ g)^{(n+1)}(d) \\ &= \sum_{k=0}^{n} \frac{1}{k!}[(x-a) \cdot \nabla]^k f(a) + \frac{1}{(n+1)!}[(x-a) \cdot \nabla]^{n+1} f(c)\end{aligned}$$

□

Example 9.4.5 We will use Taylor's Formula to find the expansion of $f(x, y) = x^2 + xy + y$ in powers of $(x-1)$ and $(y-1)$. Since f is a polynomial of degree 2, the third partial derivatives will be zero and the error term in Taylor's Theorem with $n = 2$ will vanish. Thus with $a = (1, 1)$ we have

$$\begin{aligned}f(x,y) &= f(1,1) + [(x-1, y-1) \cdot \nabla] f(1,1) + \frac{1}{2!}[(x-1, y-1) \cdot \nabla]^2 f(1,1) \\ &= f(1,1) + [(x-1, y-1) \cdot (D_1, D_2)] f(1,1) \\ &\quad + \frac{1}{2!}[(x-1, y-1) \cdot (D_1, D_2)]^2 f(1,1) \\ &= 3 + [(x-1)D_1 + (y-1)D_2] f(1,1) \\ &\quad + \{[(x-1)^2 D_{1,1} + 2(x-1)(y-1)D_{1,2} \\ &\quad + (y-1)^2 D_{2,2}] f(1,1)\}/2 \\ &= 3 + (x-1)D_1 f(1,1) + (y-1)D_2 f(1,1) + [(x-1)^2 D_{1,1} f(1,1) \\ &\quad + 2(x-1)(y-1)D_{1,2} f(1,1) + (y-1)^2 D_{2,2} f(1,1)]/2 \\ &= 3 + (x-1)3 + (y-1)2 + [(x-1)^2 2 + 2(x-1)(y-1)1 \\ &\quad + (y-1)^2 0]/2 \\ &= 3 + 3(x-1) + 2(y-1) + (x-1)^2 + (x-1)(y-1)\end{aligned}$$

(which is equal to the original expression $x^2 + xy + y$). △

Example 9.4.6 The third-degree Taylor Polynomial for $f(x, y) = \exp(x+y)$ at (0,0) is

$$\begin{aligned}
P_3(x,y) &= f(0,0) + [(x,y)\bullet\nabla]f(0,0) + \frac{1}{2}[(x,y)\bullet\nabla]^2 f(0,0) \\
&\quad + \frac{1}{6}[(x,y)\bullet\nabla]^3 f(0,0) \\
&= f(0,0) + [xD_1 + yD_2]f(0,0) + \frac{1}{2}[xD_1 + yD_2]^2 f(0,0) \\
&\quad + \frac{1}{6}[xD_1 + yD_2]^3 f(0,0) \\
&= 1 + xD_1 f(0,0) + yD_2 f(0,0) + \frac{1}{2}[x^2 D_{1,1} f(0,0) \\
&\quad + 2xyD_{1,2} f(0,0) + y^2 D_{2,2} f(0,0)] \\
&\quad + \frac{1}{6}[x^3 D_{1,1,1} f(0,0) + 3x^2 y D_{1,1,2} f(0,0) \\
&\quad + 3xy^2 D_{1,2,2} f(0,0) + y^3 D_{2,2,2} f(0,0)] \\
&= 1 + x + y + \frac{1}{2}(x^2 + 2xy + y^2) + \frac{1}{6}(x^3 + 3x^2 y + 3xy^2 + y^3) \\
&= 1 + (x+y) + \frac{1}{2}(x+y)^2 + \frac{1}{3!}(x+y)^3
\end{aligned}$$

Note that if we find the third-degree Taylor Polynomial of $g(t) = \exp(t)$ at 0 we obtain

$$p(t) = 1 + t + \frac{1}{2}t^2 + \frac{1}{3!}t^3$$

If we substitute $t = (x + y)$ in the above, we obtain the third-degree Taylor Polynomial for $f(x, y) = \exp(x + y)$ at (0,0). Thus in this case the multivariate Taylor Polynomial can also be found from the univariate Taylor Polynomials.

Another way to obtain P_3 is to write

$$f(x,y) = \exp(x+y) = e^{x+y} = e^x e^y$$

The univariate third-degree Taylor polynomials for e^x and e^y are

$$p_1(x) = 1 + x + \frac{1}{2}x^2 + \frac{1}{3!}x^3$$

and

$$p_2(y) = 1 + y + \frac{1}{2}y^2 + \frac{1}{3!}y^3$$

If we multiply these together and throw out the terms of degree greater than 3, we obtain $P_3(x, y)$. △

The above various ways to obtain Taylor Polynomial raises the question of the uniqueness of Taylor Polynomial: can there be *other* polynomial expansions of f about a? Notice that in Taylor Theorem, where

$$f(x) = P_n(x) + R_n(x)$$

$R_n(x)$ has a factor of $(x-a)^{n+1}$. Thus

$$\lim_{x\to a} \frac{R_n(x)}{|x-a|^n} = 0$$

Theorem 9.4.7 (Uniqueness Theorem for Taylor Polynomials) *If f has continuous partial derivatives of order $n+1$ at a and*

$$f(x) = T_n(x) + S(x)$$

where T_n is a polynomial of degree n and $\lim_{x\to a} S(x)/|x-a|^n = 0$, then $T_n = P_n$, the Taylor polynomial of degree n for f at a.

Proof: For simplicity we will assume $a = 0$. Let $T_n(x) - P_n(x) = p(x) + r(x)$, where $p(x)$ is the polynomial of all lowest degree terms [of degree $k \le n$, where k is the least degree of any term in $T_n(x) - P_n(x)$] and all terms in $r(x)$ have degree greater than k. For example, if $T_n(x,y) - P_n(x,y) = x^3y + 2x^2y^2 + xy^4$, then $k = 4$, $p(x,y) = x^3y + 2x^2y^2$ and $r(x,y) = xy^4$.

We will assume $T_n(x)$ is different from $P_n(x)$ and obtain a contradiction. If $T_n(x)$ is different from $P_n(x)$, then $p(x)$ is nonzero for some $x = z$.

$$\lim_{t\to 0} \frac{p(tz) + r(tz)}{|tz|^k} = \lim_{t\to 0} \frac{T_n(tz) - f(tz)}{|tz|^k} + \lim_{t\to 0} \frac{f(tz) - P_n(tz)}{|tz|^k} = 0$$

by the assumptions in the theorem on the error. But

$$\lim_{t\to 0} \frac{p(tz) + r(tz)}{|tz|^k} = \lim_{t\to 0} \frac{p(tz)}{|tz|^k} + \lim_{t\to 0} \frac{r(tz)}{|tz|^k} = \frac{p(z)}{|z|^k} + \lim_{t\to 0} \frac{r(tz)}{|t|^k|z|^k} = \frac{p(z)}{|z|^k}$$

since $p(tz) = t^k p(z)$ and the degree of $r(tz)$ in t is greater than k. Thus $p(z)/|z|^k = 0$, or $p(z) = 0$, which is impossible. Thus $T_n(x) = P_n(x)$ for all x: P_n and T_n are the same polynomial. □

Example 9.4.8 The preceding uniqueness result allows us to use Taylor theorem to compute derivatives rather than using derivatives to compute Taylor polynomial. Suppose we wish to compute all third-order partial derivatives of $f(x,y) = \sin(xy)e^x$ at (0,0). We know the Taylor Polynomials for e^x and $\sin(t)$:

$$e^x = 1 + x + \frac{1}{2!}x^2 + \frac{1}{3!}x^3 + R_3(x)$$

and

$$\sin(t) = t - \frac{1}{3!}t^3 + R_3^*(t) \quad \sin(xy) = xy - \frac{1}{3!}(xy)^3 + R_3^*(xy)$$

We multiply the expansions together, putting into the remainder all terms of degree greater than 3:

$$\begin{aligned} f(x,y) &= \sin(xy)e^x \\ &= [1 + x + \frac{1}{2!}x^2 + \frac{1}{3!}x^3 + R_3(x)][xy - \frac{1}{3!}(xy)^3 + R_3^*(xy)] \\ &= xy + x^2y + r(x,y) = \frac{1}{2!}(2xy) + \frac{1}{3!}(6x^2y) + r(x,y) \end{aligned}$$

The remainder $r(x,y)$ only has terms of degree 4 or more; thus $r(x,y)/|(x,y)|^3$ approaches 0 as (x,y) approaches (0,0). Thus by the uniqueness theorem the polynomial

$$T_3(x,y) = \frac{1}{2!}(2xy) + \frac{1}{3!}(6x^2y)$$

must be the third-degree Taylor Polynomial P_3 for f about (0,0). Equating terms of degree 3 of P_3 and T_3 we have:

$$[(x,y)\cdot(D_1, D_2)]^3 f(0,0) = 6x^2y$$

or,

$$[xD_1 + yD_2]^3 f(0,0) = 6x^2y$$

or,

$$x^3 D_{1,1,1}f(0,0) + 3x^2yD_{1,1,2}f(0,0) + 3xy^2D_{1,2,2}f(0,0) + y^3D_{2,2,2}f(0,0) = 6x^2y$$

Now equating coefficients of terms in x and y,

$$\begin{aligned} D_{1,1,1}f(0,0) = 0, \quad D_{1,1,2}f(0,0) = 2 \\ D_{1,2,2}f(0,0) = 0, \quad D_{2,2,2}f(0,0) = 0 \end{aligned}$$

The reader should try computing these derivatives directly to appreciate the effort saved. △

We can also use Taylor's Theorem to determine the tangent plane to the graph of a function. If $f : \mathbb{R} \to \mathbb{R}$, the first-degree Taylor Polynomial for f about a is

$$P_1(x) = f(a) + f'(a)(x - a)$$

The graph of $y = P_1(x) = f(a) + f'(a)(x-a)$ is the tangent line to the graph of f at $x = a$. (Notice that the slope of the graph of $y = f(a) + f'(a)(x-a)$ at $x = a$ is $\Delta y/\Delta x = y - f(a)/(x-a) = f'(a)$.)

Consider a function $f : \mathbb{R}^2 \to \mathbb{R}$. The first-degree Taylor Polynomial for f about $a = (a_1, a_2)$ is

$$\begin{aligned} P_1(x,y) &= f(a) + [(D_1, D_2)\boldsymbol{\cdot}(x - a_1, y - a_2)]f(a) \\ &= f(a) + (x-a_1)D_1 f(a) + (y - a_2)D_2 f(a) \end{aligned}$$

The graph of the function $z = f(x,y)$ is the surface given by $\{(x,y,z) : F(x,y,z) = z - f(x,y) = 0\}$. Recall that the tangent plane to $F(x,y,z) = 0$ at $(a, f(a)) = (a_1, a_2, f(a))$ is

$$\{(x,y,z) : \nabla F(a, f(a))\boldsymbol{\cdot}(x - a_1, y - a_2, z - f(a)) = 0\}$$

Since

$$\nabla F(a, f(a)) = (-D_1 f(a), -D_2 f(a), 1)$$

the equation of the tangent plane is

$$-(x - a_1)D_1 f(a) - (y - a_2)D_2 f(a) + [z - f(a)] = 0$$

or,

$$z = f(a) + (x - a_1)D_1 f(a) + (y - a_2)D_2 f(a)$$

This is the same as $z = P_1(x,y)$, analogous to the case of a function with one variable.

Example 9.4.9 Let $f(x,y) = (x+y)xy^2$. We want to find the equation of the tangent plane to the graph $z = f(x,y)$ when $(x,y) = (1,2)$. From the above discussion, the equation of the tangent plane is

$$\begin{aligned} z &= f(1,2) + (x-1)D_1 f(1,2) + (y-2)D_2 f(1,2) \\ &= 12 + (x-1)16 + (y-2)16 \end{aligned}$$

△

As a final application of the multivariate Taylor's Theorem we will prove the following theorem characterizing extreme points of a function of two variables. This theorem was stated and used in Chapter 4. In this application a function $f : \mathbb{R}^2 \to \mathbb{R}$ has a critical point at p, i.e., $f'(p) = 0$, and we wish to determine whether f has a minimum or maximum at p.

Theorem 9.4.10 (Second Derivative Test in Two Variables)
Let $f : \mathbb{R}^2 \to \mathbb{R}$ be a C^2 function with critical point p.

(a) *If $D_{1,1}f(p)D_{2,2}f(p) - [D_{1,2}f(p)]^2 > 0$ and $D_{1,1}f(p) < 0$, then f has a local maximum at p.*

(b) *If* $D_{1,1}f(p)D_{2,2}f(p) - [D_{1,2}f(p)]^2 > 0$ *and* $D_{1,1}f(p) > 0$, *then* f *has a local minimum at* p.

(c) *If* $D_{1,1}f(p)D_{2,2}f(p) - [D_{1,2}f(p)]^2 < 0$, *then* f *has a saddle point at* p.

(d) *If* $D_{1,1}f(p)D_{2,2}f(p) - [D_{1,2}f(p)]^2 = 0$, *then no conclusion can be drawn.*

Proof: We will prove part (b), the proofs of the other parts are similar. Applying Taylor Theorem to f at $p = (p_1, p_2)$ with $n = 1$ and $x = p+(r, s) = (p_1+r, p_2+s)$ we obtain

$$\begin{aligned} f(x) &= f(p) + rD_1f(p) + sD_2f(p) \\ &\quad +\{r^2D_{1,1}f(p') + 2rsD_{1,2}f(p') + s^2D_{2,2}f(p')\}/2 \\ &= f(p) + \{r^2D_{1,1}f(p') + 2rsD_{1,2}f(p') + s^2D_{2,2}f(p')\}/2 \end{aligned}$$

where p' is on the line segment between p and $p + (r, s)$. Let $a = D_{1,1}f(p')$, $b = D_{1,2}f(p')$, and $c = D_{2,2}f(p')$, so that

$$f(x) = f(p) + \frac{1}{2}(ar^2 + 2rsb + cs^2)$$

Now, $f(p)$ is a minimum if $f(x) \geq f(p)$ for x near p, i.e., if

$$(ar^2 + 2rsb + cs^2) \geq 0$$

for all r and s small.

The assumption of part (b) of the theorem is that $D_{1,1}f(p) > 0$. Since $D_{1,1}f(x)$ is continuous, $a = D_{1,1}f(p') > 0$ when p' is close to p, i.e., for all r and s sufficiently small. Similarly $ac - b^2 > 0$ for all r and s sufficiently small. Thus for small r and s,

$$\begin{aligned} ar^2 + 2rsb + cs^2 &= \frac{1}{a}(a^2r^2 + 2abrs + b^2s^2 - b^2s^2 + acs^2) \\ &= \frac{1}{a}[(ar + bs)^2 + (ac - b^2)s^2] \geq 0 \end{aligned}$$

Thus for all x near p, $f(x) \geq f(p)$, i.e., f has a local minimum at p. □

Section 9.4 Exercises

1. If $f(x, y) = y\exp(xy)$ and $a = (1, 2)$, find $[a{\bullet}\nabla]f(a)$.

2. If $f(x, y) = 3x^2y + xy + 4xy^2$ and $a = (2, 3)$, find $[a{\bullet}\nabla]f(a)$.

3. If $f(x, y) = \exp(1/2x + 3y)$ and $a = (r, s)$, find $[a{\bullet}\nabla]^2f(x, y)$.

4. If $f(x, y) = xy^3 + x\sin(y)$ and $a = (r, s)$, find $[a{\bullet}\nabla]^2f(x, y)$.

5. Write $f(x, y) = x^2 + 3xy + y^2$ as a polynomial in powers of $(x - 1)$ and $(y - 2)$.

6. Write $f(x, y) = 2x^2 - 3xy + y^2$ as a polynomial in powers of $(x - 1)$ and $(y - 2)$.

7. Find the fourth-degree Taylor Polynomial for $f(x, y) = \sin(x + y)$ at $(0, 0)$.

8. Find the third-degree Taylor Polynomial for $f(x, y) = x \exp(x + y)$ at $(0, 0)$.

9. Find the third-order partial derivatives of $f(x, y) = \sin(x + y)e^x y$ at $(0, 0)$.

10. Find the third-order partial derivatives of $f(x, y) = \cos(x + xy)xy$ at $(0, 0)$.

11. Find the equation of the tangent plane to $f(x, y) = 1 + xy\cos(x)$ at $(x, y) = (0, 1)$.

12. Find the equation of the tangent plane to $f(x, y) = e^x \sin(x + y)$ at $(x, y) = (1, 1)$.

13. Verify the statement in the proof of the Multivariate Taylor's Theorem, Theorem 9.4.4, that

$$(f \circ g)^{(2)}(t) = [(x - a) \cdot \nabla]^2 f(g(t)).$$

9.5 Infinite Series

Introduction

In a previous section, we showed that a function f having n derivatives can be approximated with an nth-degree Taylor's Polynomial,

$$f(x) \approx P_n(x) = \sum_{k=0}^{n} \frac{f^{(k)}(a)}{k!}(x - a)^k$$

Thus for a C^∞ function (i.e., a function that is infinitely differentiable), the above *finite sum* can be expanded to an *infinite sum*:

$$\sum_{k=0}^{\infty} \frac{f^{(k)}(a)}{k!}(x - a)^k$$

This expression containing infinitely many terms is called the *Taylor Series for f expanded about $x = a$*. The sequence $\{P_n\}$ of Taylor Polynomials for f will be considered as a sequence of approximations to the Taylor Series for f. (This is the third step in the Basic Approximation Process.) The sum of a Taylor Series is then the limit of the corresponding sequence of Taylor' Polynomials, provided the sequence converges. If the sequence diverges, then the Taylor Series is said to diverge. The reader should recognize the parallelism in the development of Taylor Series from a sequence of Taylor Polynomials and the development of an improper integral from a sequence of (proper) integrals.

We recall the basic question raised in the introduction of this chapter: given a function $f : A \to B$ and x in A, how can we evaluate $f(x)$? It is clear that an answer is to approximate f with a Taylor Polynomial P_n and then evaluate $P_n(x)$. We also know that this approximation can be made as accurate as we wish (assuming that f is a C^∞ function). A deeper understanding of Taylor Series and convergence is necessary to enable us to pursue our goal of developing techniques to approximate function values *easily.* Our approach will be to start with series in which each term is a constant rather than containing a variable as is the case with Taylor' Series. This approach will enable us to understand and become familiar with the notions of convergence and divergence of series before introducing variables into the discussion.

Definition 9.5.1 (Series) *An* infinite series *or just* series *is an expression of the form* $a_0 + a_1 + a_2 + \cdots + a_n + \cdots$ *or, using the summation notation,*

$$\sum_{k=0}^{\infty} a_k$$

The a_k's are called the *terms* of the series.

The sum of a finite set of numbers is independent of how the numbers are arranged or grouped thanks to the commutative and associative laws of arithmetic, not so with a series (an infinite collection of numbers).

Example 9.5.2 Can a value be assigned to the following series?

$$1 - 1 + 1 - 1 + 1 - 1 + 1 - 1 + 1 \cdots$$

One might say 0 since

$$(1 - 1) + (1 - 1) + (1 - 1) + \cdots = 0 + 0 + 0 + \cdots = 0$$

Another might say 1 since

$$1 + (-1 + 1) + (-1 + 1) + (-1 + 1) + \cdots = 1 + 0 + 0 + 0 + \cdots = 1$$

Still another might say -1 since interchanging the even- and odd-numbered terms and then grouping yields

$$-1 + (1 - 1) + (1 - 1) + (1 - 1) + \cdots = -1 + 0 + 0 + 0 + \cdots = -1$$

$\triangle$

Certainly we should insist that if a set of numbers has a sum, then the sum should be unique. Is the problem with the above series that it has no sum or is the problem in the way in which we grouped or interchanged the terms?

The following example should convince us that at least some series can be summed.

Example 9.5.3 Phred, who is standing two units away from a wall, steps toward the wall in such a manner that the length of each step is one-half of his distance to the wall. If Phred continues stepping indefinitely, the distance he moves should be

$$1 + \frac{1}{2} + \frac{1}{2^2} + \frac{1}{2^3} + \cdots + \frac{1}{2^n} + \cdots$$

Our intuition leads us to say that this series sums to 2. △

It should be no surprise to the reader that we turn to the Basic Approximation Process to develop a definition for the sum of a series. Given the series $\sum_{k=0}^{\infty} a_k$, the idea is to form a sequence of approximations $\{s_n\}$ called *partial sums*, which are defined as follows:

$$\begin{aligned}
s_0 &= a_0 \\
s_1 &= a_0 + a_1 \\
s_2 &= a_0 + a_1 + a_2 \\
s_3 &= a_0 + a_1 + a_2 + a_3 \\
&\vdots \\
s_n &= a_0 + a_1 + a_2 + \cdots + a_n = \textstyle\sum_{k=0}^{n} a_k
\end{aligned}$$

In the previous example,

$$\begin{aligned}
s_0 &= 1 \\
s_1 &= 1 + \tfrac{1}{2} = \tfrac{3}{2} \\
s_2 &= 1 + \tfrac{1}{2} + \tfrac{1}{2^2} = 7/4 \\
s_3 &= 1 + \tfrac{1}{2} + \tfrac{1}{2^2} + \tfrac{1}{2^3} = \tfrac{15}{8} \\
&\vdots \\
s_n &= 1 + \tfrac{1}{2} + \tfrac{1}{2^2} + \tfrac{1}{2^3} + \cdots + \tfrac{1}{2^n} = \tfrac{2^{n+1}-1}{2^n} = 2 - \tfrac{1}{2^n}
\end{aligned}$$

(From physical considerations, we know that the last term in s_n is "half the distance to the wall." Hence this last term also represents the remaining distance. Thus $s_n = 2 - \frac{1}{2^n}$.) Clearly $s_n \to 2$.

Formally, we have:

Definition 9.5.4 (Partial Sum) *Let $\{a_k\}$ be the sequence of terms of the series $\sum_{k=0}^{\infty} a_k$. The* nth partial sum *of the series is $s_n = \sum_{k=0}^{n} a_k$.*

It is essential that the reader distinguish between and understand the roles of the two sequences associated with a series $\sum_{k=0}^{\infty} a_k$:

(1) The sequence $\{a_k\}$ of terms of the series
(2) The sequence $\{s_n\}$ of partial sums of the series, $s_n = \sum_{k=0}^{n} a_k$

In order to help the reader with this distinction, we will usually use k as the subscript for the terms of the series and n as the subscript for the partial sums. Of course, k is really a "dummy" variable (as in the case of integration) since both $\sum_{k=0}^{m} a_k$ and $\sum_{n=0}^{m} a_n$ mean the same thing, $a_0 + a_1 + \cdots + a_m$.

Also note that we start our series with a subscript of zero; thus the nth partial sum s_n is the sum of $n+1$ terms, $a_0, a_1, \ldots, a_n$.

And now for the formal definition of convergence. Since we know about convergence of a *sequence*, we use the concept of partial sums to turn the question of convergence of a *series* into a question about the convergence of the *sequence* of partial sums.

Definition 9.5.5 (Convergence of a Series) *Let $\{s_n\}$ be the sequence of partial sums of the series $\sum_{k=0}^{\infty} a_k$, $s_n = \sum_{k=0}^{n} a_k$. If $s_n \to L$, then the series is said to* converge *to L and L is called the* sum *or* value *of the series.*

If $\{s_n\}$ diverges, then the series is said to diverge. *(A divergent series has no sum.)*

Example 9.5.6 The series $1 - 1 + 1 - 1 + 1 - 1 + \cdots$ diverges since the sequence of partial sums $\{s_n\}$ has the values

$$s_n = \begin{cases} 1 & \text{for } n \text{ odd} \\ 0 & \text{for } n \text{ even} \end{cases}$$

and thus does not converge. △

Example 9.5.7 Applying Taylor's Theorem with the Derivative Form Remainder to $f(x) = e^x$ and evaluating at $a = 0$ and $x = 1$ yields

$$e^1 = \sum_{k=0}^{n} \frac{1}{k!} + R_n(1)$$

where

$$R_n(1) = \frac{e^c}{(n+1)!}$$

for some c between 0 and 1. Since

$$e^c \leq e^1, |R_n(1)| \leq \frac{e}{(n+1)!}$$

and so $\lim_{n\to\infty} R_n(1) = 0$. Thus,

$$\lim_{n\to\infty} \sum_{k=0}^{n} \frac{1}{k!} = e^1$$

or, the series $\sum_{k=0}^{\infty} 1/k!$ converges with sum e:

$$e = \sum_{k=0}^{\infty} \frac{1}{k!}$$

△

Example 9.5.8 This example continues the answer to the question asked in Example 9.1.3 at the beginning of this chapter, and previously discussed in Example 9.3.5: How can we find the cosine of a number?

From Example 9.3.5 we can write

$$\cos(x) = \sum_{k=0}^{n}(-1)^k \frac{x^{2k}}{2k!} + R_n(x)$$

where

$$|R_n(x)| = \left| \frac{\cos^{(2n+1)}(c)}{(2n+1)!} x^{2n+1} \right| \leq \frac{|x|^{2n+1}}{(2n+1)!}$$

for some c between 0 and x. Since

$$\lim_{n \to \infty} |R_n(x)| \leq \lim_{n \to \infty} \frac{|x|^{2n+1}}{(2n+1)!} = 0$$

we have

$$\cos(x) = \sum_{k=0}^{\infty}(-1)^k \frac{x^{2k}}{2k!}$$

In particular,

$$\cos(1) = \sum_{k=0}^{\infty}(-1)^k \frac{1}{2k!} = 1 - \frac{1}{2} + \frac{1}{4!} - \frac{1}{6!} + \cdots$$

Since the error term goes to zero, we can approximate the value of cosine as accurately as we wish with an nth partial sum for a sufficiently large n. △

Example 9.5.9 Consider the series $\sum_{k=0}^{\infty} a_k$ whose terms are those of the unbounded sequence $a_k = k$ for $k = 0, 1, \ldots$. Then the nth partial sum $s_n = \sum_{k=0}^{n} a_k > n$, and so $\lim_{n \to \infty} s_n = \infty$. Thus the series diverges to infinity. We write

$$\sum_{k=0}^{\infty} a_k = \infty$$

Since ∞ is not a number, the series has no sum. △

The sequence of terms of a series need not be unbounded for the series to diverge as shown by the series in Example 9.5.6 [$\sum_{k=0}^{\infty}(-1)^k$].

The major question of interest in the analysis of a series is whether the given series converges or diverges.

Consider a series $\sum_{k=0}^{\infty} a_k$ with positive terms, $a_k \geq 0$.

Series with Positive Terms

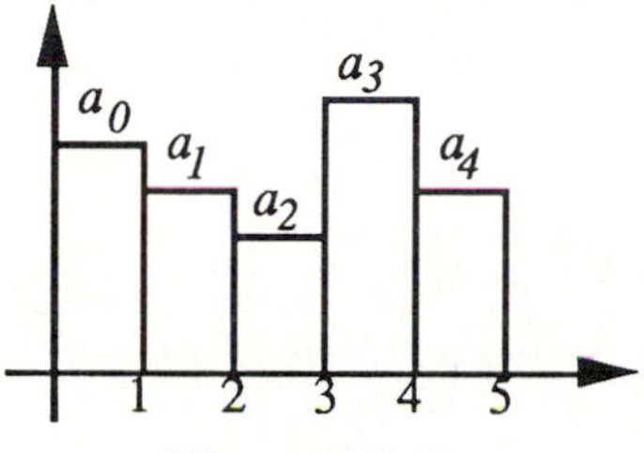

Figure 9.5.10

The area of the rectangle with base from 0 to 1 is a_0, the area of the rectangle with base from 1 to 2 is a_1, etc. The nth partial sum is the sum of the areas of the first $n+1$ rectangles. The sum of the series, if it exists, can be interpreted as the limit of the sum of areas of the rectangles. Clearly, if the series is to converge, the area of the kth rectangle must go to zero as $k \to \infty$; that is, $\lim_{k\to\infty} a_k = 0$.

If some of the terms a_k are negative, then we can represent them with rectangles lying below the x-axis (i.e., "negative area").

Series with Some Negative Terms

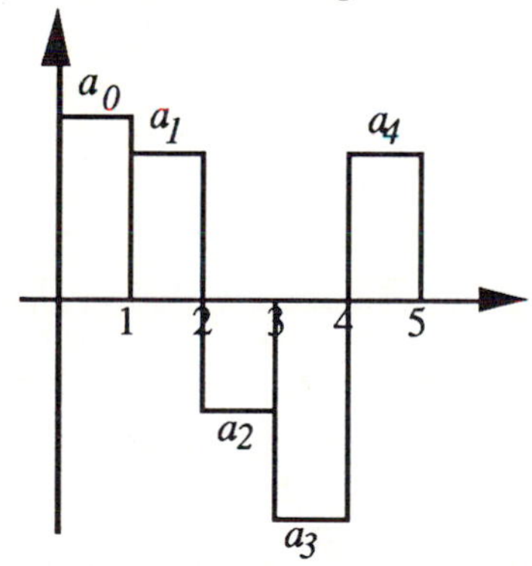

Figure 9.5.11

As the following theorem shows, even with negative terms, if the series $\sum_{k=0}^{\infty} a_k$ converges, then $\lim_{k\to\infty} a_k = 0$.

Theorem 9.5.12 *Let* $\{a_k\}$, $k = 0, 1, 2, \ldots$, *be the infinite sequence of terms associated with the series* $\sum_{k=0}^{\infty} a_k$. *If the series converges, then* $\lim_{k\to\infty} a_k = 0$.

Proof: Let $\{s_n\}$ be the associated sequence of partial sums, $s_n = \sum_{k=0}^{n} a_k$. Suppose that $\sum_{k=0}^{\infty} a_k = A$, i.e., $\lim_{n\to\infty} s_n = A$.

By definition, $s_n = s_{n-1} + a_n$, so $a_n = s_n - s_{n-1}$. Taking limits we get

$$\lim_{k\to\infty} a_k = \lim_{k\to\infty}(s_k - s_{k-1}) = A - A = 0 \qquad \square$$

It is important to note that the converse of this theorem is not necessarily true: if $\lim_{k\to\infty} a_k = 0$, the series need not converge.

Example 9.5.13 Consider the series $\sum_{k=1}^{\infty} 1/k$.

This special series is called the *harmonic series.* Clearly,

$$\lim_{k\to\infty} a_k = \lim_{k\to\infty} \frac{1}{k} = 0$$

However, the harmonic series diverges.

To understand the divergence of this series, consider the terms of the sequence of partial sums of the form s_N, where $N = 2^n$, $n = 1, 2, 3, \ldots$, i.e., $s_2, s_4, s_8, \ldots$. We have

$$s_2 = 1 + \tfrac{1}{2} = \tfrac{3}{2},$$

$$s_4 = s_2 + \tfrac{1}{3} + \tfrac{1}{4} > s_2 + \tfrac{1}{4} + \tfrac{1}{4} = \tfrac{4}{2},$$

$$s_8 = s_4 + \tfrac{1}{5} + \tfrac{1}{6} + \tfrac{1}{7} + \tfrac{1}{8} > s_4 + \tfrac{1}{8} + \tfrac{1}{8} + \tfrac{1}{8} + \tfrac{1}{8} > \tfrac{4}{2} + \tfrac{1}{2} = \tfrac{5}{2}$$

In general,

$$s_N \geq \frac{n+2}{2} \qquad \text{for } N = 2^n$$

Since these partial sums are unbounded, the series must diverge.

The term "harmonic" comes from the field of music and dates back to the Pythagorean School in ancient Greece. The sequence of tones obtained by plucking a taut string, then shortening the string to $\frac{1}{2}$ its length and plucking, then shortening the string to $\frac{1}{3}$ its length and plucking, etc., was named the harmonic (musical) scale by the Pythagoreans. If the string was initially of unit length, then the "sounding lengths" of the harmonic scale is the sequence $\{1, \frac{1}{2}, \frac{1}{3}, \frac{1}{4}, \ldots\}$. The harmonic scale provided a standard for musical instruments until today's "equal temperament" scale was developed in the sixteenth century. $\triangle$

Since convergence of a series depends on the limit of the sequence of partial sums, convergence depends on the behavior of the "tail end" of the series. (Why?) Thus the first finite number of terms can be neglected in determining convergence or divergence. Of course, the sum of a convergent series depends on all of the terms of the series. The following theorem makes these observations explicit.

Theorem 9.5.14 *Let $\{a_k\}$, $k = 0, 1, 2, \ldots$ be an infinite sequence and suppose that $\sum_{k=0}^{\infty} a_k = A$. Then $\sum_{k=N}^{\infty} a_k = A - \sum_{k=0}^{N-1} a_k$.*

The Algebra of Limits Theorem of Chapter 2 can be extended to series, since a series converges if and only if the sequence of partial sums converges.

The Algebra of Series

Theorem 9.5.15 (Algebra Of Series) *Suppose that* $\sum_{k=0}^{\infty} a_k = A$ *and* $\sum_{k=0}^{\infty} b_k = B$, *then*

(a) $\sum_{k=0}^{\infty}(a_k + b_k) = A + B$

(b) $\sum_{k=0}^{\infty} ca_k = c\sum_{k=0}^{\infty} a = cA$ *for all real numbers c.*

Theorem 9.5.15(a) can be paraphrased as: "The limit of the sums is the sum of the limits." We will consider products of series in a later section.

Example 9.5.16 The series $\sum_{k=0}^{\infty} 5/k!$ converges to $5e$. By the above theorem, $\sum_{k=0}^{\infty} 5/k! = 5\sum_{k=0}^{\infty} 1/k! = 5e$ using the result of Example 9.5.7. △

Section 9.5 Exercises

1. Determine which of the following series converge. Remember, you may use Theorem 9.5.12 to tell that a series diverges.

   (a) $\sum_{n=1}^{\infty} 2n/(2n+1)$ (b) $\sum_{k=1}^{\infty} 1/(k^2+5k+6)$ (c) $\sum_{k=3}^{\infty} 1/(2k-1)$

   (d) $\sum_{j=0}^{\infty} \sin(j\pi)$ (e) $\sum_{i=0}^{\infty}(3/2)^i$ (f) $\sum_{d=2}^{\infty}(2/3)^{d-2}$

There is a type of series, called a *telescoping series,* whose partial sums are easy to compute. The following exercises will introduce these series.

2. Consider the series $\sum_{k=1}^{\infty} 1/(k^2+k)$. We note that $1/(k^2+k) = 1/[k(k+1)]$.

   (a) Use the partial fraction transformation to find constants A and B with $1/[k(k+1)] = A/k + B/(k+1)$.

   (b) Using the result of part (a), write the first five partial sums, $s_1, \ldots, s_5$. Simplify each of these expressions (how do they "telescope"?).

   (c) Generalize the results of part (b) to obtain a simple expression for s_n. Determine if the series converges. If it does converge, find the limit.

3. Consider the series $\sum_{k=1}^{\infty} 3/(k^2+4k+3)$. Note that k^2+4k+3 $= (k+1)(k+3)$

   (a) Use the partial fraction transformation to find constants A and B so that

   $$A/(k+1) + B/(k+3) = 3/(k^2+4k+3).$$

   (b) Using the result of part (a), write the first five partial sums, $s_1, \ldots, s_5$. Simplify each of these expressions (how do they "telescope"?).

(c) Generalize the results of part (b) to obtain a simple expression for s_n. Determine if the series converges. If it does converge, find the limit.

4. Use the technique of partial fractions to find a general expression for the sequence of partial sums of each of the following telescoping series.

   (a) $\sum_{k=1}^{\infty} 1/(k^2+7k+12)$ (b) $\sum_{k=3}^{\infty} 1/(k^2+7k+12)$

   (c) $\sum_{k=2}^{\infty} 1/(k^2-1)$ (d) $\sum_{k=5}^{\infty} 1/(k^2-1)$

5. Find the sum of each of the series in the previous exercise.

6. In general, a *telescoping series* is a series $\sum_{k=0}^{\infty} a_k$ for which there is a sequence $\{b_n\}$ such that $a_k = b_{k+1} - b_k$. Show that $s_n = b_{n+1} - b_0$ and that $\sum_{k=0}^{\infty} a_k = \lim_{n\to\infty} -b_0$ if $\{b_n\}$ converges.

7. What is wrong, if anything, with the following argument.
   Let S be the sum of the series $1+2+4+8+\cdots+2^n+\cdots$.
   Since $2S = 2+4+8+\cdots = S-1$, we have $2S = S-1$ and thus $S = -1$.

8. The purpose of this exercise is to develop a feeling for the behavior of the harmonic series, $\sum_{k=1}^{\infty} 1/k$.

   (a) Is the sequence of partial sums, $\{s_n\}$, increasing? Why?

   (b) Compute s_n for $n = 10, 50, 100, 10^5, 10^{10}, 10^{100}, 10^{1,000}, 10^{0,000}$. How does the value of s_n change when n increases by a factor of 10?

   (c) How does the rate of growth of s_n compare with the rate of growth of $\log(n)$? (Refer to Section 3.9 for a discussion on rate of growth.)

   (d) Show that $\log(n) + 1/n < s_n < \log(n) + 1$. Hint: Think of $\log(n)$ as the area under the graph of $f(x) = 1/x$ over the interval $[1, 10]$. Draw pictures for $UF(f,n)$ and $LA(f,n)$ (Chapter 5).

   (e) Using $\log(n)$ to approximate s_n, determine a value of n such that $s_n > 100$. Does using $\log(n)$ to approximate s_n help explain the pattern that you observed in part (b)? Explain.

9. The purpose of this exercise is to develop a feeling for the behavior of the alternating harmonic series, $\sum_{k=1}^{\infty}(-1)^{k+1}1/k$.

   (a) Compute the partial sums, s_n, for $n = 2, 4, 6, 8, 10$. Is $\{s_{2n}\}$ an increasing sequence? Explain.

   (b) Compute s_n for $n = 1, 3, 5, 7, 9$. Is $\{s_{2n+1}\}$ a decreasing sequence? Explain.

   (c) What relationship, if any, exists between $\{s_{2n}\}$ and $\{s_{2n+1}\}$? Explain.

   (d) Assume $\sum_{k=1}^{\infty}(-1)^{k+1}1/k$ converges and has sum L. What relationship exists between $\{s_{2n}\}$, $\{s_{2n+1}\}$, and L? Explain. Use this relationship to determine an error bound for $|\sum_{k=1}^{\infty}(-1)^{k+1}1/k - s_{20}|$? Explain.

(e) Show that the alternating harmonic series converges to log(2). [Hint: Expand $\log(x)$ in Taylor's series about $x = 1$.] Determine n such that s_n for the alternating harmonic series approximates log(2) with accuracy of 0.01.

9.6 Geometric Series

There are two basic questions that can be asked concerning the convergence of any series:

1. Does the series converge?
2. Given that a particular series converges, to what value does the series converge?

We shall develop tests in the next section that will enable us to answer question 1 for certain classes of series. Question 2 is usually much more difficult. In this section we will consider *geometric series,* the most important of those few series for which the answer to the second question is easy. In fact, *the geometric series may be the most important and mostly widely applied of all series.*

A geometric series is a series that is characterized by the fact that every term is a constant multiple of the previous term. That is, a geometric series is a series that can be written as

$$\sum_{k=0}^{\infty} cr^k = c\sum_{k=0}^{\infty} r^k$$

(r is the constant multiple). Thus any geometric series is a constant multiple of the basic geometric series

$$1 + r + r^2 + r^3 + \cdots = \sum_{k=0}^{\infty} r^k$$

which begins with 1.

A few examples of geometric series are:

$$3 - 3 + 3 - 3 + 3 - 3 + \cdots + (-1)^n 3 + \cdots \quad (r = -1)$$

$$1 + 1 + 1 + \cdots + 1^n + \cdots \quad (r = 1)$$

$$\tfrac{7}{10^2} + \tfrac{7}{10^4} + \tfrac{7}{10^6} + \cdots + \tfrac{7}{10^{2n}} + \cdots = \tfrac{7}{10^2}(1 + \tfrac{1}{10^2} + \tfrac{1}{10^4} + \cdots) \quad (r = \tfrac{1}{10^2})$$

$$1 + 2 + 4 + 8 + 16 + 32 + 64 + \cdots + 2^n + \cdots \quad (r = 2)$$

$$1 - 2 + 4 - 8 + 16 - 32 + 64 - \cdots + (-1)^n 2^n + \cdots \quad (r = -2)$$

Example 9.6.1 The infinite decimal 0.333... representing $\frac{1}{3}$ can be expressed as a geometric series since

$$\begin{aligned} 0.333\ldots &= \frac{3}{10} + \frac{3}{10^2} + \frac{3}{10^3} + \cdots = \sum_{k=1}^{\infty} 3\left(\frac{1}{10}\right)^k \\ &= 3\sum_{k=1}^{\infty}\left(\frac{1}{10}\right)^k \end{aligned}$$

In this example, the ratio between successive terms is $\frac{1}{10}$. How could we show that this series sums to $\frac{1}{3}$? (See below.) △

The series $1 + \frac{1}{2} + \frac{1}{2^2} + \frac{1}{2^3} + \cdots$ is a geometric series since each term is one-half the previous term. We were able to sum this series in Example 9.5.3 using a physical interpretation. How could we have determined the sum without relying on the physical interpretation? The following theorem provides the answer.

Theorem 9.6.2 (Geometric Series) *The geometric series* $\sum_{k=0}^{\infty} r^k$ *converges if and only if* $|r| < 1$. *Furthermore, when* $|r| < 1$,

$$\sum_{k=0}^{\infty} r^k = \frac{1}{1-r}$$

and the error associated with the n*th partial sum* s_n *is*

$$\text{Error} = |s_n - \sum_{k=0}^{\infty} r^k| = \frac{|r^{n+1}|}{1-r}$$

Proof: If $r \geq 1$, then $\lim_{n\to\infty} r^n \neq 0$ and thus the series must diverge. (Why?) Thus consider the case where $|r| < 1$. We will develop an expression for s_n and then apply the limit operation. Now

$$s_n = 1 + r + r^2 + r^3 + \cdots + r^n$$

Multiplying s_n by r and then subtracting the result from s_n yields

$$s_n - rs_n = (1 + r + r^2 + \ldots + r^n) - (r + r^2 + r^3 + \ldots + r^{n+1}) = 1 - r^{n+1}$$

Thus (since $r \neq 1$),

$$s_n = \frac{1 - r^{n+1}}{1-r}$$

Since $|r| < 1$, $r^{n+1} \to 0$ as $n \to \infty$ and thus

$$s_n \to \frac{1}{1-r}$$

To obtain the error expression, merely subtract s_n from the limit. □

It is *very important* to note that the summation starts with 1 (i.e., r^0). This is the *standard form.* Thus to evaluate a convergent geometric series, "factor out" the first term and obtain the product of the "first term" and a geometric series in standard form. That is,

$$\sum_{k=4}^{\infty} a_k r^k = a_4 r^4 \sum_{k=0}^{\infty} a_k r^k = a_4 r^4 \frac{1}{1-r} = \frac{a_4 r^4}{1-r} \quad \text{for } |r| < 1$$

Example 9.6.3 We will find the rational fraction represented by the expression 0.78787878... and then, using the error expression from Theorem 9.6.2 on Geometric Series, determine how many terms of the corresponding geometric series are needed to approximate the fraction with $ACC = 0.0004$.

$$\begin{aligned} 0.787878... &= \frac{78}{10^2} + \frac{78}{10^4} + \frac{78}{10^6} + ... = \frac{78}{10^2}\left(1 + \frac{1}{10^2} + \frac{1}{10^4} + ...\right) \\ &= \frac{78}{10^2}\left[\sum_{k=0}^{\infty}\left(\frac{1}{10^2}\right)^k\right] = \frac{78}{100}\left(\frac{1}{1 - 1/100}\right) = \frac{78}{99} \end{aligned}$$

To determine the number of terms necessary to give the desired accuracy, use Theorem 9.6.2 to write down the error

$$\text{Error} = \frac{78}{100}\frac{(1/10^2)^{n+1}}{1 - 1/10^2} < 0.0004$$

and solve for n. This reduces to

$$\frac{78}{99 \cdot 10^{2n+2}} < \frac{4}{10^4} \quad \text{or} \quad \frac{78 \cdot 10^2}{99 \cdot 4} < 10^{2n}$$

which implies that $n \geq 1$. Thus two terms, $n = 0$ and $n = 1$ are necessary to obtain $ACC = 0.0004$. (How many decimal places would this be?) △

The following example illustrates how geometric series can be used to express decimal expressions as rational fractions whenever the decimal expressions are eventually repeating.

Example 9.6.4 We will express 0.135242424... as a rational fraction.

The procedure is to split off the nonrepeating portion, express the repeating portion in terms of a geometric series, and then sum the two parts.

$$0.135242424\ldots = \frac{135}{10^3} + \frac{1}{10^3}(.242424\ldots) = \frac{135}{10^3} + \frac{24}{10^3}\sum_{k=1}^{\infty}\left(\frac{1}{10^2}\right)^k$$

$$= \frac{135}{10^3} + \left(\frac{24}{10^3}\right)\left(\frac{1}{10^2}\right)\sum_{m=0}^{\infty}\left(\frac{1}{10^2}\right)^m = \frac{135}{10^3} + \left(\frac{24}{10^5}\right)\frac{1}{1-1/10^2}$$
$$= \frac{135}{10^3} + \frac{24}{99 \cdot 10^3} = \frac{13389}{99,000}$$

△

Question 9.6.5 Make up five examples of geometric series. Find an example of a series that is not geometric. ◊

Of course, any finite geometric series converges, whatever its ratio might be. A careful examination of the proof of Theorem 9.6.2 on Geometric Series gives

$$\sum_{k=0}^{N} r^k = \frac{1 - r^{N+1}}{1 - r}$$

for $r \neq 1$.

Example 9.6.6 Consider the finite series $\sum_{k=0}^{5}(3)^k$. This finite sum is a finite geometric series with $r = 3$. The sum of the series is

$$\sum_{k=0}^{5}(3)^k = \frac{1-3^6}{1-3} = 364$$

△

Example 9.6.7 If a person enters into a savings program in which she saves one dollar the first month and each succeeding month she saves twice the amount saved the preceding month, how many months will it take to accumulate $1000?

This is a finite geometric series problem with $r = 2$. The question is to determine the smallest N such that

$$\sum_{k=0}^{N} 2^k \geq 1000$$

That is, determine the smallest N such that

$$\frac{1 - 2^{N+1}}{1 - 2} \geq 1000 \quad \text{or} \quad 2^{N+1} \geq 1001$$

Since $2^{10} = 1024$, the answer is $N = 9$. △

In the next example we consider a more sophisticated application of series to a practical problem.

Example 9.6.8 The federal government claims to be able to significantly stimulate the economy by giving each taxpayer a \$1000 rebate. Economists reason that 80% of the \$1000 will be spent by each taxpayer and 80% of this 80% will be spent and so on and so on.

The total amount of economic activity generated by the \$1000 rebate to each taxpayer is, in dollars,

$$1000 + (0.8)1000 + (0.8)^2 1000 + (0.8)^3 1000 + \cdots$$

This sum is 1000 times a geometric series with $r = 0.8$:

$$\text{Total activity } = \sum_{k=0}^{\infty} 1000(0.8)^k = 1000 \sum_{k=0}^{\infty} (0.8)^k = \frac{1000}{1 - 0.8} = 5000$$

An important factor in this computation was the assertion that 80% of the money received by individuals is put back into the economy. This ratio is called the *Marginal Propensity to Consume,* or simply MPC. △

Question 9.6.9 If the MPC in the previous problem is increased to 90%, what would the effect of a \$1000 rebate be? ◇

Example 9.6.10 (A Function Defined by a Series) We have seen that the geometric series (with ratio x) $\sum_{k=0}^{\infty} x^k$ converges when $|x| < 1$. We can consider x a variable and define a function $f : (-1, 1) \to \mathbb{R}$

$$f(x) = \sum_{k=0}^{\infty} x^k \quad \text{for} \quad -1 < x < 1$$

For each value of x in the domain $(-1, 1)$, the series (with ratio x) converges. The function is defined by the use of a series. We know that the series sums to $1/(1 - x)$, so

$$f(x) = 1/(1 - x) \quad \text{for} \quad -1 < x < 1$$

△

Section 9.6 Exercises

1. Write each of the following series in the form $c \sum_{k=0}^{\infty} r^k$.

   (a) $\sum_{k=3}^{\infty} (3)^k (5)^{-k}$ (b) $\sum_{k=0}^{\infty} (\frac{2}{5})^{k-1}$ (c) $\sum_{k=3}^{\infty} 6(\frac{3}{4})^{k-3}$

   (d) $\sum_{k=-2}^{\infty} 3(4)^k$ (e) $\sum_{k=0}^{\infty} \frac{1}{2^k} - \frac{1}{3^k}$ (f) $\sum_{k=24}^{\infty} 5(\frac{2}{3})^{k-4}$.

2. Write out the first three terms and then compute the sum, exactly!

   (a) $\sum_{k=0}^{4} 2^k$ (b) $\sum_{k=0}^{5} 2$ (c) $\sum_{k=3}^{10} 3(\frac{1}{3})^{k+1}$

   (d) $\sum_{j=-2}^{5} 2^{j+2}$ (e) $\sum_{k=4}^{20} 3(0.1)^k$ (f) $\sum_{k=12}^{20} 8(\frac{1}{2})^{k-3}$

3. For each of the following infinite repeating decimals, use power series to convert to a rational number.

   (a) 0.77777 ... (b) 0.16666 ... (c) 0.16232323 ... (d) 0.10123123 ...

4. Determine convergence or divergence for each of the following series. If a series converges, find its sum. If a series diverges, explain why it does. Identify each series as a geometric series, telescoping series, or otherwise.

   (a) $\sum_{k=1}^{\infty} 2^{k-1}/3^k$ (b) $\sum_{k=1}^{\infty} 1/(k^2+9k+20)$ (c) $\sum_{k=1}^{\infty}(k+3)/k^2$

   (d) $\sum_{k=2}^{\infty} k/\log(k)$ (e) $\sum_{k=1}^{\infty} \frac{3}{2^k}$ (f) $\sum_{k=3}^{\infty} (k+1)!/k!$.

5. Determine convergence or divergence for each of the following series. If a series converges, find its sum. If a series diverges, explain why it does. Identify each series as a geometric series, telescoping series, or otherwise.

   (a) $\sum_{k=0}^{\infty}(k^3-(k+1)^3)$ (b) $\sum_{k=7}^{\infty}(-1/7)^k$ (c) $\sum_{k=-2}^{\infty}(3/2)^{k+7}$

   (d) $\sum_{k=1}^{\infty} 1/k^3 - 1/(k+1)^3$ (e) $\sum_{k=0}^{\infty} 3^{k+2}/5^{k-3}$ (f) $\sum_{k=2}^{\infty} 2^{k+3}/3^k$

   (g) $\sum_{k=0}^{\infty}(2^k+5^k)/10^k$ (h) $\sum_{k=3}^{\infty} k^k/k!$ (i) $\sum_{k=0}^{\infty} 1/k!$

6. Using an MPC of 95%, how much economic activity would an injection of $1 billion create.

7. Suppose that you wish to create an additional $50 billion in activity in an economy with an MPC of 90%. How much money needs to be put into the economy?

8. A very rich man plans to give away half of his fortune at the end of each month. Assuming that he starts with $2 billion and earns no extra during the year, how many months will it take to give away $1.9 billion?

9. Two students, Sue and John, flip a coin to see who buys breakfast. The first to flip a head must buy. Sue suggests that John flip first. John protests saying that one-half of the time he will lose on the first flip, but relents when Sue explains to him that she will have the same likelihood of buying if John's flip is a tail. Who gets the better of this deal and by how much? Suggestion: If you have difficulty doing this problem, try it 50 or so times with your roommate, keeping track of how often the person who flips first loses.

10. A square has sides of unit length. Another square is inscribed within the first by placing its corners at the midpoints of the sides of the first square. A third square is constructed within the second in a similar manner, and so on. What is the sum of the areas of all the squares?

11. An equilateral triangle has sides of unit length. Another equilateral triangle is inscribed within the first by placing its vertices at the midpoints of the sides of the first triangle. A third triangle is constructed in a similar manner, and so on. What is the sum of the areas of all the triangles?

12. What is the sum of the perimeters of the squares in Exercise 11?

13. What is the sum of the perimeters of the triangles in Exercise 12?

9.7 Convergence Tests

In this section we will develop some tests for determining convergence or divergence for series of *positive terms.* The tests that we develop will be based primarily on the following observation:

> If $\sum_{k=0}^{\infty} a_k$ is a series of nonnegative terms (i.e., $a_k \geq 0$) , then the sequence of partial sums $\{s_n\}$ is a monotonically increasing sequence and hence converges if and only if it is bounded (see Chapter 2).

Thus we direct our attention to the question of boundedness of the sequence of partial sums. One way to determine boundedness is to compare the given series term by term with a "test series." For example, if we are given the series $\sum_{k=0}^{\infty} a_k$, $a_k \geq 0$ and we know that $\sum_{k=0}^{\infty} b_k$ converges to B and that $a_k \leq b_k$, then

$$s_n = \sum_{k=0}^{n} a_k \leq \sum_{k=0}^{n} b_k \leq B$$

Thus s_n is bounded and hence $\sum_{k=0}^{\infty} a_k$ must converge.

Geometrically, from our interpretation of summing areas, Figure 9.5.10, we see that if $\sum_{k=0}^{\infty} b_k$ converges then $\sum_{k=0}^{\infty} a_k$ converges since the rectangles for a_k are contained within the rectangles for b_k. Furthermore since convergence depends on only the tail end of the series, this containment of rectangles need only be true for k larger than some number N_0. This gives us the following useful theorem.

Theorem 9.7.1 (The Comparison Test) *Let* $\sum_{k=0}^{\infty} a_k$ *and* $\sum_{k=0}^{\infty} b_k$ *be series with* $0 \leq a_k \leq b_k$ *for all* $k > N_0$.

(a) If $\sum_{k=0}^{\infty} b_k$ *converges, then* $\sum_{k=0}^{\infty} a_k$ *converges.*

(b) If $\sum_{k=0}^{\infty} a_k$ *diverges, then* $\sum_{k=0}^{\infty} b_k$ *diverges.*

To use the Comparison Test on a series, we must find a test series whose convergence or divergence is known with which to compare the series to be tested. If we suspect that the series in question converges, we need to find a convergent series, all of whose terms are larger than the series in question. This is part (a) of the theorem. If, on the other hand, we suspect the series in question diverges, we need to find a divergent series, all of whose terms are less than the series in question. This procedure uses part (b) of the theorem. Above all, remember, *all series involved in a Comparison Test must have nonnegative terms.*

Geometric series are often used as test series.

Example 9.7.2 Consider the series $\sum_{k=0}^{\infty} \cos^2(k)/2^k$. Since

$$0 \leq \frac{\cos^2 k}{2^k} \leq \frac{1}{2^k}$$

and $\sum_{k=0}^{\infty} 1/2^k$ converges, $\sum_{k=0}^{\infty} \cos^2 k/2^k$ converges. △

Question 9.7.3 How do we know that $\sum_{k=0}^{\infty} 1/2^k$ converges? ◇

Example 9.7.4 Consider the series $\sum_{k=1}^{\infty} (k+1)/k^2 = \sum_{k=1}^{\infty} 1/k + 1/k^2$.

We recognize $1/k$ as the general term in the harmonic series, so we compare the terms of our series to the terms of the divergent harmonic series:

$$0 \leq \frac{1}{k} \leq \frac{k+1}{k^2}$$

Therefore, the series in question diverges (since the harmonic series diverges).

A slight variation of this last example is to consider the series

$$\sum_{k=0}^{\infty} \frac{2k^3 + 2k + 4}{5k^4 - 5k^2 - 7}$$

Since $(2k^3 + 2k + 4) \ / \ (5k^4 - 5k^2 - 7) \geq (\frac{2}{5})(1/k)$, the series in question diverges. (Why?) △

Generalizing the process used in this last example, we see that given a series whose terms are rational functions of n (i.e., quotient of two polynomials in n), the terms will eventually behave like $1/n^p$ for some number p. Thus another class of test series for the Comparison Test are those of the form $\sum_{k=0}^{\infty} 1/k^p$. These are called *p-series.*

We now pursue the question: *For what values of p does the corresponding p-series converge?*

Extending the idea of comparing areas as was done in developing Theorem 9.5.12, we can compare the rectangles representing a series with the area underneath the graph of a function. Suppose we are given the series

$$\sum_{k=0}^{\infty} a_k \quad a_k \geq 0$$

and that we can find a continuous function $f : [0, \infty) \rightarrow \mathbb{R}$, such that $f(k) = a_k$ for all $k \geq 1$. One way to envision such a function is that its graph will pass through the points (k, a_k) for $k \geq 1$. Furthermore, suppose that *both* $\{a_k\}$ *and* f are nonincreasing. This will cause all of the rectangles representing the sum (after the first, a_0) to fall below the graph of f, as in Figure 9.7.5(a) below.

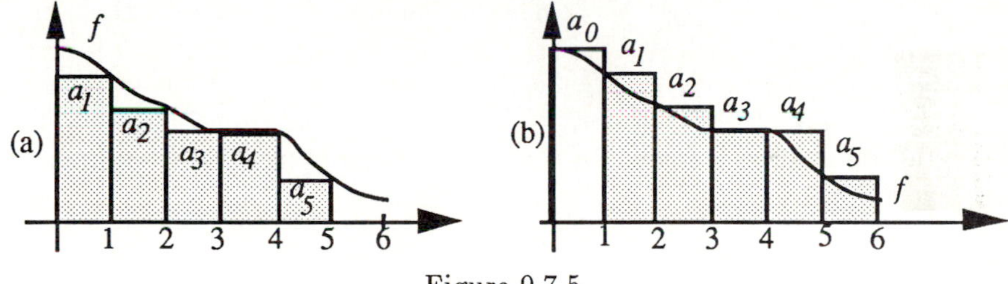

Figure 9.7.5

Therefore, the sum of the series

$$\sum_{k=1}^{\infty} a_k$$

should be less than or equal to the area under the graph of f. Formally, then

$$\sum_{k=1}^{\infty} a_k \leq \int_0^{\infty} f(x)dx$$

By shifting the rectangles to the right we have Figure 9.7.5(b). Thus we see that

$$\int_1^{\infty} f(x)dx \leq \sum_{k=1}^{\infty} a_k \leq \int_0^{\infty} f(x)dx \leq \sum_{k=0}^{\infty} a_k.$$

Since f is continuous on $[0, \infty)$, $\int_0^{\infty} f$ converges if and only if $\sum_{k=1}^{\infty} a_k$ converges. Thus we have that $\int_0^{\infty} f(x)dx$ converges if and only if $\sum_{k=0}^{\infty} a_k$ converges. Note that we could start the index at any integer m, rather than 0.

Theorem 9.7.6 (The Integral Test) *If f is continuous and nonincreasing on the interval $[m, \infty)$ and if $f(k) = a_k \geq 0$ and a nonincreasing sequence for all $k = m, m+1, m+2, \ldots$, then $\sum_{k=m}^{\infty} a_k$ converges if and only if $\int_m^{\infty} f(x)dx$ converges.*

Note carefully that Theorem 9.7.6 applies only to series whose general terms are *nonnegative and nonincreasing.*

Example 9.7.7 We can test the convergence of the *quadratic series* $\sum_{k=1}^{\infty} 1/k^2$ by using the Integral Test. Clearly, $\{1/k^2\}$ is a nonnegative and nonincreasing sequence. Consider the function $f : [1,\infty) \to \mathbb{R}$ defined by $f(x) = 1/x^2$. We have

$$\int_1^{\infty} \frac{1}{x^2} dx = \lim_{A\to\infty} \int_1^{A} \frac{1}{x^2} dx = \lim_{A\to\infty} \left(\frac{-1}{A} - \frac{-1}{1}\right) = \lim_{A\to\infty} \left(1 - \frac{1}{A}\right) = 1$$

Since the integral exists, the quadratic series converges. △

The harmonic series and the quadratic series are both examples of p-series.

The Integral Test provides an easy method for resolving the question concerning convergence of p-series. Consider the function $f : [1,\infty) \to \mathbb{R}$ defined by $f(x) = 1/x^p$.

Clearly, f is nonnegative, continuous, and nonincreasing. We may test the convergence of a p-series by considering $\int_1^{\infty} 1/x^p\, dx$. Recall from Chapter 6 that this integral converges if and only if $p > 1$. The results of the test are summarized in the next theorem.

Theorem 9.7.8 *The series* $\sum_{k=1}^{\infty} 1/k^p$ *converges if and only if* $p > 1$.

Example 9.7.9 (Another Function Defined by a Series) It is possible to define a function by using p-series. Consider the function $f : (1,\infty) \to \mathbb{R}$ defined by $f(x) = \sum_{k=1}^{\infty} 1/k^x$, $x > 1$.

From the previous theorem, the series converges and the sum of the p-series defines the value of f at x. Notice that in this case a function is determined by a series that is not a Taylor's Series. Also note that unlike Example 9.6.10 which also defined a function by a series, no simple expression exists for this f. △

Example 9.7.10 From Example 9.7.9 we know that $f(x) = \sum_{k=1}^{\infty} 1/k^x$ converges and thus is defined for $x > 1$. Suppose we wish to determine $f(3)$ with an accuracy of 0.001. We begin by expressing $f(3)$ as

$$f(3) = \sum_{k=1}^{\infty} \frac{1}{k^3} = \sum_{k=1}^{n} \frac{1}{k^3} + \sum_{k=n+1}^{\infty} \frac{1}{k^3} = \sum_{k=1}^{n} \frac{1}{k^3} + \text{tail}$$

Next we will restrict the tail to be less than 0.001. From the diagram

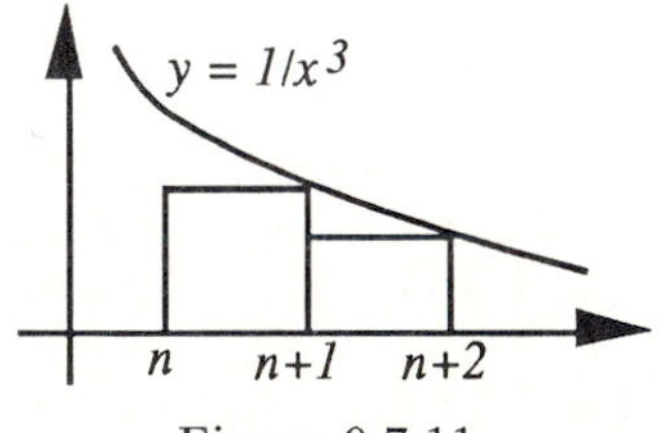

Figure 9.7.11

we see that

$$\text{tail} = \sum_{k=n+1}^{\infty} \frac{1}{k^3} \le \int_n^{\infty} \frac{1}{x^3}dx = \frac{1}{2n^2}.$$

Thus we want $1/2n^2 \le 0.001$ or $2n^2 \ge 1000$ or $n \ge 23$. We thus approximate $f(3)$ as

$$\sum_{k=1}^{23} \frac{1}{k^3} \approx 1.201$$

△

Given a series $\sum_{k=0}^{\infty} a_k$ it is often easier to find a comparison test series $\sum_{k=0}^{\infty} b_k$ where a_k and b_k behave in a similar manner for large values of k rather than where a_k and b_k are related by an inequality. The following variation of the Comparison Test allows for this additional freedom.

Theorem 9.7.12 (The Limit Comparison Test) *Let $\sum_{k=0}^{\infty} a_k$ and $\sum_{k=0}^{\infty} b_k$ be series with positive terms.*

(a) *If $\lim_{k\to\infty} a_k/b_k = L > 0$, then both series converge or both series diverge.*

(b) *If $\lim_{k\to\infty} a_k/b_k = 0$ and if $\sum_{k=0}^{\infty} b_k$ converges, then $\sum_{k=0}^{\infty} a_k$ converges.*

(c) *If $\lim_{k\to\infty} a_k/b_k = \infty$ and if $\sum_{k=0}^{\infty} b_k$ diverges, then $\sum_{k=0}^{\infty} a_k$ diverges.*

Example 9.7.13 Consider the series

$$\sum_{k=0}^{\infty}(4k^7 - 3k^5 + k^4 + 6k^3 - 7) \,/\, (k^{10} - 4k^6 + 7k^2).$$

Since the terms of this series eventually behave like $4/k^3$ (Why?), we may apply the Limit Comparison Test and obtain

$$\lim_{k\to\infty} \frac{(4k^7 - 3k^5 + k^4 + 6k^3 - 7)/(k^{10} - 4k^6 + 7k^2)}{(4/k^3)} = 1$$

Thus the given series converges (since $\sum_{k=1}^{\infty} 4/k^3$ converges). Recall from Chapter 2 that this limit can be evaluated by inspection. In order to appreciate the Limit Comparison Test, the reader should try determining convergence of the given series by the Comparison Test. △

The Ratio Test

One of the most commonly applied tests uses no test series for comparison. We state the test in the form of a theorem.

Theorem 9.7.14 (The Ratio Test) *Let $\sum_{k=0}^{\infty} a_k$ be a series of positive terms, and suppose that $\lim_{k\to\infty} a_{k+1}/a_k = L$.*

(a) *If $L < 1$, the series converges.*

(b) *If $L > 1$, the series diverges.*

(c) *If $L = 1$, the test is inconclusive.*

Proof: Suppose $L < 1$. We wish to show that the series converges. Since L is to the left of 1 on the number line, let $r = (L+1)/2$ be the midpoint between L and 1. Let $ACC = r - L$, the distance from r to L. The reader should show that $ACC > 0$.

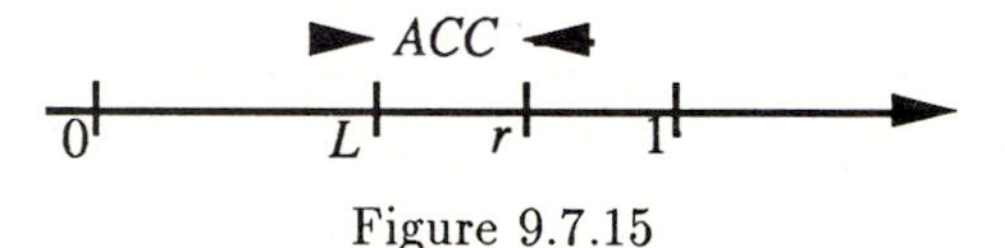

Figure 9.7.15

Now, since the sequence a_{k+1}/a_k converges to L, we can find N such that for all $k \geq N$,

$$|\frac{a_{k+1}}{a_k} - L| \leq ACC$$

In particular, we have $a_{k+1}/a_k < r$ or $a_{k+1} < ra_k$, when $k \geq N$. We have the list of inequalities:

$$\begin{aligned}
&a_{N+1} < ra_N \\
&a_{N+2} < ra_{N+1} < r^2 a_N \\
&a_{N+3} < ra_{N+2} < r^3 a_N \\
&a_{N+4} < ra_{N+3} < r^4 a_N \\
&\vdots
\end{aligned}$$

Now, the series

$$ra_N + r^2 a_N + r^3 a_N + \cdots$$

is geometric with ratio r, and $|r| < 1$. The inequalities allow us to compare the original series to this geometric series. Since the geometric series converges ($|r| < 1$), the series being tested must converge.

To prove part(b) of the theorem suppose that $L > 1$. In much the same way we proved part 1, select $r = (1+L)/2$, the midpoint between 1 and L. As before, let $ACC = L - r > 0$ be the distance from r to L.

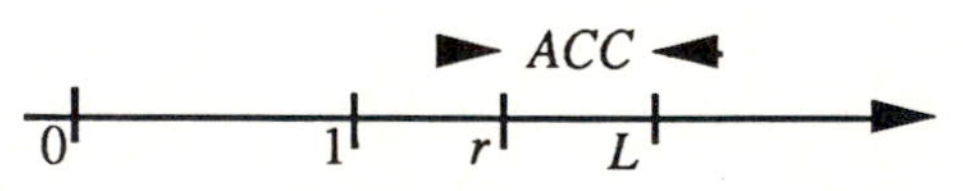

Figure 9.7.16

Again, we can find N such that for all $k \geq N$,

$$|\frac{a_{k+1}}{a_k} - L| \leq ACC$$

In particular, $a_{k+1}/a_k > r$ or $a_{k+1} > ra_k$, for all $k \geq N$. We generate another list of inequalities:

$$\begin{aligned}
&a_{N+1} > ra_N \\
&a_{N+2} > ra_{N+1} > r^2 a_N \\
&a_{N+3} > ra_{N+2} > r^3 a_N \\
&a_{N+4} > ra_{N+3} > r^4 a_N \\
&\vdots
\end{aligned}$$

Again, we form a geometric series (this time with $r > 1$) and compare the series in question to the geometric series. Since the geometric series diverges ($r > 1$), the Comparison Test guarantees the divergence of the series being tested.

To see that part (c) is correct, consider the series

$$\sum_{k=1}^{\infty} \frac{1}{k} \quad \text{and} \quad \sum_{k=1}^{\infty} \frac{1}{k^2}$$

In both cases, $L = 1$. However, the harmonic series diverges and the quadratic series converges. □

Example 9.7.17 Consider the series $\sum_{k=0}^{\infty} 1/k!$. Although the Integral Test would be difficult to apply to this series, the application of both the Comparison and Ratio Tests is very straight forward.

Comparison Test:
For $k \geq 4, 2^k \leq k!$ and thus $1/k! \leq 1/2^k$. Since $\sum_{k=0}^{\infty} 1/2^k$ is a convergent geometric series, $\sum_{k=0}^{\infty} 1/k!$ converges.

Ratio Test:

$$\lim_{k\to\infty} \frac{a_{k+1}}{a_k} = \lim_{k\to\infty} \frac{1/(k+1)!}{1/k!} = \lim_{k\to\infty} \frac{k!}{(k+1)k!} = \lim_{k\to\infty} \frac{1}{k+1} = 0 = L$$

Since $L < 1$, the series $\sum_{k=0}^{infty} 1/k!$ converges (to e as we saw in Example 9.5.7). △

Example 9.7.18 Consider the series $\sum_{k=1}^{\infty} k^k/k!$. Using the Ratio Test we get

$$\lim_{k\to\infty} \frac{(k+1)^{k+1}/(k+1)!}{k^k/k!} = \lim_{k\to\infty} \frac{(k+1)^{k+1}}{(k+1)!} \frac{k!}{k^k} = \lim_{k\to\infty} \left(\frac{k+1}{k}\right)^k = \lim_{k\to\infty} \left(1 + \frac{1}{k}\right)^k$$

(The reader should verify the second equality.)
Also, recall from Chapter 7

$$\lim_{k\to\infty}\left(1+\frac{1}{k}\right)^k = e > 1$$

Therefore, the series diverges. △

Example 9.7.19 Consider the series $\sum_{k=0}^{\infty} 1/(2k+1)$. Using the Ratio Test we get

$$\lim_{k\to\infty}\frac{1/[2(k+1)+1]}{1/(2k+1)} = \lim_{k\to\infty}\frac{2k+1}{2k+3} = 1$$

The test is inconclusive. However, we can use the Limit Comparison test to see that $\sum_{k=0}^{\infty} 1/(2k+1)$ diverges. (How?) △

The last three examples illustrate the fact that some tests work for some series, some for others. In general, Comparison tests are useful when comparing series whose terms consist of polynomial or logarithmic factors. When the terms of the series contain exponential or factorial factors, the Ratio test is usually the best choice.

Section 9.7 Exercises

1. For each of the following series, determine if the series converges or diverges. (Give the reasons for your answers.)

   (a) $\sum_{k=0}^{\infty} k/(k^2+1)$ (b) $\sum_{k=0}^{\infty} 2^{k-1}/[(k+1)(k+2)]$

   (c) $\sum_{n=1}^{\infty} 1/k2^k$ (d) $\sum_{k=1}^{\infty} 1/k^{1/2}$

   (e) $\sum_{k=0}^{\infty} 1/(3k+1)^2$ (f) $\sum_{k=1}^{\infty} 1/[2+3\log(k)]$

   (g) $\sum_{k=1}^{\infty}(1/k)[1/\log(k)]^{3/2}$ (h) $\sum_{k=0}^{\infty}(k!)^2/(2k!)$

   (i) $\sum_{k=0}^{\infty} k!/5^k$ (j) $\sum_{k=1}^{\infty} 3^k k!/k^k$

   (k) $\sum_{k=0}^{\infty} 1/(1+k^{1/2})$ (l) $\sum_{k=1}^{\infty} \log(k)/k^2$

   (m) $\sum_{k=0}^{\infty} 1/(k+25)$ (n) $\sum_{k=0}^{\infty} 3k/(k^3-k+4)$

2. For each of the following series, determine if the series converges or diverges. (Give the reasons for your answers.)

   (a) $\sum_{k=2}^{\infty}(1+1/k)$ (b) $\sum_{k=1}^{\infty}\sin(\pi/2-1/k)$

   (c) $\sum_{k=0}^{\infty}1/(k+6)$ (d) $\sum_{k=2}^{\infty}1/\log(k)$

   (e) $\sum_{k=2}^{\infty}1/\log(k^k)$ (f) $\sum_{k=1}^{\infty}k^2 3^{-k^2}$

   (g) $\sum_{k=0}^{\infty}(3k^4+5k)/[(k^2+1)(k^3+3)]$ (h) $\sum_{k=2}^{\infty}(2k+1)/\sqrt{k^2-2}$

   (i) $\sum_{k=1}^{\infty}ke^{-k}$ (j) $\sum_{k=1}^{\infty}k\sin(1/k)$

   (k) $\sum_{k=0}^{\infty}(4k^3+k)/(k^6-3k+1)$ (l) $\sum_{k=1}^{\infty}\sqrt{2k+4}\,/k^2$

   (m) $\sum_{k=1}^{\infty}(1+1/k)\log(1+1/k)$ (n) $\sum_{k=2}^{\infty}[\log(k)]^{-\log(k)}$

3. We know that the series $\sum_{k=1}^{\infty}1/k^4$ converges. What does it converge to? Use your calculator or computer to approximate the value to within 0.0001.

4. We know that the quadratic series $\sum_{k=1}^{\infty}1/k^2$ converges. What does it converge to? Use your calculator or computer to approximate the value to within 0.01.

9.8 Relative and Absolute Convergence

Alternating Series

So far we have considered series with positive terms. The simplest series containing both positive and negative terms are the *alternating series.*

Definition 9.8.1 *A series is called alternating if the terms alternate in sign.*

The Taylor Series for the natural logarithm function provides an interesting example of an alternating series.

Example 9.8.2 Let $f(x)=\log(x)$. We will expand f in a Taylor Series about the point $x=a$. The first question is: What value should be selected for a? The value for a is chosen in order to simplify the evaluation of the function and its derivatives. Thus for the natural logarithm function, we choose $a=1$. (Note that we cannot choose $a=0$, since the logarithm function is not defined at zero.) Now $f'(x)=1/x$, $f''(x)=-1/x^2$, $f^{(3)}(x)=2!/x^3$, $f^{(4)}(x)=-3!/x^4$, and in general

$$f^{(k)}(x)=\frac{(-1)^{k-1}(k-1)!}{x^k}$$

Taylor's Theorem gives

$$\log(x) = \sum_{k=0}^{n} \frac{f^{(k)}(a)}{k!}(x-a)^k + \frac{f^{(n+1)}(c)}{(n+1)!}(x-a)^{n+1}$$

Setting $a = 1$ and noting that $f^{(0)}(1) = \log(1) = 0$ yields

$$\log(x) = \sum_{k=1}^{n} \frac{(-1)^{k-1}}{k}(x-1)^k + \frac{(-1)^n}{(n+1)c^{n+1}}(x-1)^{n+1}$$

Now substituting 2 for x (and transforming the index by setting $j = k-1$), we obtain a Taylor Series for log(2)

$$\sum_{j=0}^{\infty} \frac{(-1)^j}{j+1} = 1 - \frac{1}{2} + \frac{1}{3} - \frac{1}{4} + \cdots$$

This is called the *alternating harmonic series.* Does this series converge? The remainder term when $x = 2$ is

$$R_n(2) = \frac{(-1)^n}{(n+1)c^{n+1}}$$

where c is between 1 and 2. Thus

$$|R_n(2)| = \frac{1}{(n+1)c^{n+1}} \leq \frac{1}{n+1}$$

(since $1 \leq c \leq 2$).

Thus $\lim_{n\to\infty} R_n(2) = 0$ and

$$\log(2) = \sum_{k=0}^{\infty} \left(\frac{(-1)^k}{k+1}\right) = 1 - \frac{1}{2} + \frac{1}{3} - \frac{1}{4} + \cdots \qquad \triangle$$

Notice that the error term estimate in the above example was very simple, $1/(n+1)$. This is not an accident, as is shown by the following theorem.

Theorem 9.8.3 (Alternating Series) *If the terms of the series* $\sum_{k=0}^{infty}(-1)^k a_k$ *satisfy the conditions*

(a) *All* a_k*'s have the same sign (all positive or all negative)*

(b) $|a_{k+1}| \leq |a_k|$ *for all* k *(terms are nonincreasing in absolute value)*

(c) $\lim_{k\to\infty} a_k = 0$.

then the series $\sum_{k=0}^{\infty}(-1)^k a_k$ *converges and the difference between the limit and the* n*th partial sum is less than* $|a_{n+1}|$. *That is, if the limit is* S, *then*

$$\left| S - \sum_{k=0}^{n}(-1)^k a_k \right| \leq |a_{n+1}|$$

Proof: Assume that $a_k \geq 0$ for all $k \geq 0$. Let s_n be the nth partial sum, $s_n = \sum_{k=0}^{n} a_k$.

Now $s_{2(n+1)} = s_{2n} - (a_{2n+1} - a_{2n+2})$, where the term in parentheses is positive, since $a_{2n+1} > a_{2n+2}$. Thus for every n, $s_{2(n+1)} < s_{2n}$.

Also,

$$s_{2n} = (a_0 - a_1) + (a_2 - a_3) + \cdots + (a_{2n-2} - a_{2n-1}) + a_{2n} > 0$$

since each parenthesized term is positive and a_{2n} is positive.

Thus $\{s_{2n}\}$ is a decreasing sequence of numbers which is bounded below by zero and hence must be convergent, say to s.

Applying a similar analysis to s_{2n+1} we find that $\{s_{2n+1}\}$ is an increasing sequence bounded above, and thus must converge to some number s'. These two limits must be the same since

$$s - s' = \lim_{n\to\infty}(s_{2n}) - \lim_{n\to\infty}(s_{2n+1}) = \lim_{n\to\infty}(s_{2n} - s_{2n+1}) = \lim_{n\to\infty}(-a_{2n+1}) = 0.$$

Given an n, let $t = FLOOR(n/2) = [n/2]$. Then $s_{2t+1} \leq s_n \leq s_{2t}$.

Thus by the Squeeze Theorem on limits, the sequence $\{s_n\}$ must converge to S.

If n is even, the error $S - s_n$ satisfies

$$S - s_n = -a_{n+1} + (a_{n+2} - a_{n+3}) + \cdots \geq -a_{n+1}$$

Since both sides of this inequality are negative (recall that $\{s_n\}$ is a decreasing sequence), $|S - s_n| \leq a_{n+1}$.

Similarly, if n is odd, $S - s_n \leq a_{n+1}$. In either case, $|S - s_n| \leq a_{n+1}$. □

Example 9.8.4 How many terms should we use to approximate log(2) to within 0.0001? Since log(2) is given by the alternating series

$$\log(2) = \sum_{k=0}^{\infty} \frac{(-1)^k}{k+1} = 1 - \frac{1}{2} + \frac{1}{3} - \frac{1}{4} + \cdots$$

the error between the nth partial sum s_n and log(2) is at most $|a_{n+1}|$. For the error to be at most 0.0001, we want

$$|a_{n+1}| \leq 0.0001 \quad \text{or} \quad \frac{1}{n+1} \leq 0.0001 \quad \text{or} \quad n > 10,000$$

Because of the large number of terms necessary to get high accuracy, other methods of approximating log(2) may be used in practice. △

The next example of an alternating series illustrates a method for showing that the sequence $\{a_k\}$ is decreasing.

Example 9.8.5 Consider the series $\sum_{k=1}^{\infty}(-1)^k(k+1)^{1/2}/k$. This is certainly an alternating series. [The term $(-1)^k$ is the clue.] In order to use the alternating series theorem, we must establish that

$$a_k = \frac{(k+1)^{1/2}}{k}$$

is decreasing.

Consider the function $f : [1,\infty) \to \mathbb{R}$ defined by $f(x) = (x+1)^{1/2}/x$. We can establish that $|a_k|$ is decreasing by showing that $f'(x) < 0$. Using the quotient rule for derivatives we get

$$f'(x) = \frac{-x-2}{2x^2(x+1)^{1/2}}$$

Therefore, $f'(x) < 0$ for all $x > 1$ and the sequence $\{a_k\}$ is decreasing.

Clearly, $\lim_{k\to\infty} a_k = 0$, and the alternating series theorem gives the convergence of the series.

If we wish to approximate the sum to within 0.01, i.e., $|s - s_n| \leq 0.01$, where s_n is the nth partial sum, we need

$$\frac{(n+2)^{1/2}}{(n+1)} \leq 0.01$$

Since for large n,

$$\frac{(n+2)^{1/2}}{(n+1)} \approx \frac{(n+2)^{1/2}}{(n+2)} = \frac{1}{(n+2)^{1/2}}$$

we want

$$n + 2 \geq 10,000$$

Again, we need many terms to get an approximation accurate to within 0.01. △

Absolute Convergence

In some sense, it is more difficult for a nonnegative series to converge than for an alternating series to converge. The fact that terms are alternating negative and positive help the partial sums from becoming too large. An example of this observation is the harmonic series which diverges; however, the corresponding harmonic alternating series converges.

For any series $\sum_{k=0}^{\infty} a_k$, alternating or otherwise, we may associate a corresponding series of absolute values, $\sum_{k=0}^{\infty} |a_k|$.

It is important to note the following definition.

Definition 9.8.6 *Let $\sum_{k=0}^{\infty} a_k$ be a series of real numbers.*

(a) *If* $\sum_{k=0}^{\infty} |a_k|$ *converges, then* $\sum_{k=0}^{\infty} a_k$ *is said to be* absolutely convergent.

(b) *If* $\sum_{k=0}^{\infty} |a_k|$ *diverges but* $\sum_{k=0}^{\infty} a_k$ *converges, then* $\sum_{k=0}^{\infty} a_k$ *is said to be* conditionally convergent.

Theorem 9.8.7 (Absolute Convergence) *If a series converges absolutely, then it converges. That is, if* $\sum_{k=0}^{\infty} |a_k|$ *converges, then* $\sum_{k=0}^{\infty} a_k$ *converges.*

Example 9.8.8 The alternating harmonic series is conditionally convergent. △

Example 9.8.9 Consider the series $\sum_{k=1}^{\infty} \cos(k)/k^2$. We begin by testing for the convergence of the corresponding series of absolute values $\sum_{k=1}^{\infty} |\cos(k)|/k^2$. We may compare this series with the quadratic series:

$$\frac{|\cos(k)|}{k^2} \leq \frac{1}{k^2}$$

Since the quadratic series is convergent, our original series is absolutely convergent. △

Now for a few general comments:

1. Any nonnegative series which converges is also absolutely convergent.
2. If a series is absolutely convergent, then the series is convergent. We may think of absolute convergence as a powerful form of convergence.
3. Every convergent series is either conditionally or absolutely convergent.
4. Divergent series are *neither* conditionally *nor* absolutely convergent.

We conclude this section by restating the Comparison and Ratio tests in terms of absolute convergence.

Theorem 9.8.10 (Comparison Test for Absolute Convergence) *Let* $\sum_{k=0}^{\infty} a_k$ *and* $\sum_{k=0}^{\infty} b_k$ *be series with* $|a_k| \leq |b_k|$.

(a) *If* $\sum_{k=0}^{\infty} b_k$ *converges absolutely, then* $\sum_{k=0}^{\infty} a_k$ *converges absolutely.*

(b) *If* $\sum_{k=0}^{\infty} a_k$ *does not converge absolutely, then* $\sum_{k=0}^{\infty} b_k$ *does not converge absolutely.*

Theorem 9.8.11 (Ratio Test for Absolute Convergence) *Let* $\{a_k\}$ *be the sequence of terms of* $\sum_{k=0}^{\infty} a_k$, *and suppose that* $\lim_{k\to\infty} |a_{k+1}|/|a_k| = L$.

(a) *If* $L < 1$, *the series converges absolutely.*

(b) *If* $L > 1$, *the series diverges* (a_n *does not go to 0).*

(c) *If $L = 1$, the test is inconclusive.*

As was the case for techniques of integration, only the most important convergence tests have been considered. No attempt been made to present an exhaustive list of tests.

Having developed a facility for determining convergence or divergence of a series of constants, we are now ready to return (in the next section) to a major emphasis of this chapter, the analysis of Taylor Series.

Section 9.8 Exercises

1. For each of the following series, determine if the series converges conditionally, converges absolutely, or diverges.

   (a) $\sum_{k=1}^{\infty}(-1)^k(k+1)/k$ (b) $\sum_{k=1}^{\infty}(-1)^{k-1}(2k+1)/k$

   (c) $\sum_{j=1}^{\infty}(1/3j - 1/2j)$ (d) $\sum_{k=2}^{\infty}(-1)^k/\log(k)$

   (e) $\sum_{m=0}^{\infty} m(2/3)^{-m}$ (f) $\sum_{k=0}^{\infty}(k+1)/(2k^4-1)$

   (g) $\sum_{k=0}^{\infty}(-1)^{k+1}3^k/(k-2^k)$ (h) $\sum_{n=0}^{\infty}(-1)^n 1/(3n-17)$

   (i) $\sum_{k=1}^{\infty}(-1)^k 3^k/k!$ (j) $\sum_{k=0}^{\infty} 9^k/k!$

   (k) $\sum_{i=1}^{\infty}(-1)^i 4/i^2$ (l) $\sum_{k=1}^{\infty}(-1)^k \log(k)/\sqrt{k}$

2. Give an example for each of the following series or explain why no example exists.

   (a) A conditionally convergent series that is not absolutely convergent.
   (b) An absolutely convergent series that is not conditionally convergent.
   (c) An absolutely convergent series such that the series of absolute values has the same sum (value) as the series itself.
   (d) A nonalternating series that is conditionally convergent, but not absolutely convergent.
   (e) A convergent nonnegative series $\sum_{k=1}^{\infty} a_k$ such that $\sum_{k=1}^{\infty} a_k^2$ diverges.

3. Show that if $\sum_{k=0}^{\infty} a_k$ and $\sum_{k=0}^{\infty} b_k$ are both absolutely convergent, then $\sum_{k=0}^{\infty}(a_k + b_k)$ is absolutely convergent.

4. Show that when $\sum_{k=0}^{\infty} a_k$ converges conditionally (not absolutely) then the series consisting of the positive $a_k's$ diverges to positive infinity and the series consisting of the negative $a_k's$ diverges to negative infinity.

5. Show that if $\sum_{k=0}^{\infty} a_k$ is absolutely convergent, then the sequence of terms $\{a_k\}$ of the series is a bounded sequence.

9.9 Power Series

The previous three sections dealt with series of constants. We shall now consider series whose terms involve variables (e.g., a Taylor Series). Convergence questions for series of constants were of the "yes or no" type. However, for series involving variables, a more interesting type of convergence question is involved. Namely, For what values of the variable does the series converge?

We have already seen two types of series that involve variables, Taylor Series and the geometric series.

Definition 9.9.1 (Power Series) *A* power series *is a series of the form*

$$\sum_{k=0}^{\infty} c_k(x-a)^k = c_0 + c_1(x-a) + c_2(x-a)^2 + \cdots$$

Example 9.9.2 The Taylor Series $\sum_{k=0}^{\infty} f^{(k)}(a)/k!(x-a)^k$ is a power series with coefficients $c_k = f^{(k)}(a)/k!$. △

Example 9.9.3 The geometric series $\sum_{k=0}^{\infty}(x-a)^k$ is a power series with constant coefficient $c_k = 1$. △

Note: Not all series that contain variables are power series. For example, $\sum_{k=0}^{\infty} k/x^k$ is not a power series. (Why?)

The Ratio Test is usually used to determine the values of the variable for which a power series converges unless the series is geometric. In this case, it is often easier to use the geometric series criterion for determining convergence.

The next few examples illustrate the fact that the value(s) of the variable for which a power series converges forms an interval.

Example 9.9.4 We will determine the value(s) of x for which $\sum_{k=0}^{\infty}(x-2)^k$ converges. Since the series is geometric [set $r = (x-2)$], the series converges for $|x-2| < 1$, or $1 < x < 3$, and diverges for all other values of x. Note that the same result can be obtained from the Ratio Test. (Would the result have been any different if the index had begun with $k = 13$ rather than with $k = 0$?) △

Example 9.9.5 We will determine the value(s) of x for which $\sum_{k=2}^{\infty} x^k/k!$ converges. Since this power series is not geometric, we shall apply the Ratio Test.

$$\lim_{k\to\infty} \frac{|x^{k+1}/(k+1)!|}{|x^k/k!|} = \lim_{k\to\infty} \frac{|x|}{k+1} = 0 < 1 \quad \text{for all } x$$

Thus the power series converges for all values of x. From Example 9.2.5 we know that this series is the Taylor Series for $\exp(x) = e^x$. Thus the Taylor Series for the exponential function converges for all real numbers. △

Example 9.9.6 We will determine the value(s) of x for which $\sum_{k=1}^{\infty} k!(x-4)^k$ converges. Since the series is not geometric, we apply the Ratio Test.

$$\lim_{k\to\infty} \frac{|(k+1)!(x-4)^{k+1}|}{|k!(x-4)^k|} = \lim_{k\to\infty} (k+1)|(x-4)| < 1 \quad \text{only for } x = 4$$

Thus the series converges *only* for $x = 4$.
Could this result have been obtained by applying Theorem 9.5.12? △

Example 9.9.7 We will determine the value(s) of x for which $\sum_{k=0}^{\infty} x^k/k$ converges. Since the series is not geometric (Why?), we apply the Ratio Test.

$$\lim_{k\to\infty} \frac{|x^{k+1}/(k+1)|}{|x^k/k|} = \lim_{k\to\infty} \frac{k+1}{k}|x| = |x| < 1$$

for $-1 < x < 1$. Unlike the geometric criterion for convergence, the Ratio Test is inconclusive when the limit is equal to one. Thus it remains to check for convergence at $x = -1$ and $x = 1$. This is done by substituting these values for x into the original power series and obtaining a series of constants which are then analyzed for convergence using the material of the past several sections.

First substitute $x = -1$ into the original power series and obtain $\sum_{k=1}^{\infty}(-1)^k 1/k$, the alternating harmonic series. We know that this series converges (apply the Alternating Series test).

Next substitute $x = 1$ into the original power series and obtain $\sum_{k=1}^{\infty} 1/k$, the harmonic series. We know that this series diverges (apply the p-series test).

Thus the series converges for $-1 \leq x < 1$. △

Observations:

1. A power series $\sum_{k=0}^{\infty} c_k(x-a)^k$ always converges for $x = a$.
2. Both the geometric criterion and the Ratio Test determine an open "interval of convergence" for the power series (except in the "trivial" situation illustrated in the Example 9.9.6). Over this open interval the series converges absolutely.
3. The interval of convergence is centered at $x = a$.
4. The series must be checked for convergence, absolute or conditional, at each endpoint of the interval of convergence determined by the Ratio Test. A test different from the Ratio test must be used to test the endpoints. (Why?)

These observations are formalized in the next two theorems.

Theorem 9.9.8 *If the power series $\sum_{k=0}^{\infty} c_k(x-a)^k$ converges for some $x_1 \neq a$, then the series converges absolutely for all x closer to a, i.e., all x such that $|x-a| < |x_1 - a|$.*

Proof: The idea of this proof is to construct a geometric comparison series whose convergence implies absolute convergence of $\sum_{k=0}^{\infty} c_k(x-a)^k$. Since

$$\sum_{k=0}^{\infty} c_k(x_1-a)^k$$

converges, $\lim_{k\to\infty} c_k(x_1-a)^k = 0$ by Theorem 9.5.12.

Choose an N such that for $k > N$

$$|c_k(x_1-a)^k| < 1$$

Let $M = max\{|c_0|, |c_1(x_1-a)|, |c_2(x_1-a)^2|, \ldots, |c_N(x_1-a)^N|, 1\}$. Thus for all k,

$$|c_k(x_1-a)^k| \leq M$$

Now multiplying and dividing the terms of the series by $c_k(x_1-a)^k$ yields

$$\begin{aligned} |c_k(x-a)^k| &= |c_k(x_1-a)^k \frac{c_k(x-a)^k}{c_k(x_1-a)^k}| \\ &= |c_k(x_1-a)^k| \left| \frac{(x-a)^k}{(x_1-a)^k} \right| \leq M \left| \frac{(x-a)^k}{(x_1-a)^k} \right| \end{aligned}$$

If $|x-a| < |x_1-a|$, then $|(x-a)/(x_1-a)| \leq 1$ and

$$\sum_{k=0}^{\infty} M \left| \frac{(x-a)^k}{(x_1-a)^k} \right| = \sum_{k=0}^{\infty} M \left| \frac{(x-a)}{(x_1-a)} \right|^k$$

is a geometric series with ratio $|(x-a)/(x_1-a)| < 1$, and must converge. Then by the Comparison Test, the series

$$\sum_{k=0}^{\infty} c_k(x-a)^k$$

converges absolutely. □

Let R be the least upper bound of the set of all x such that the power series converges at $a + x$. Thus for any x whose distance from a is less than R (i.e., $|x-a| < R$) the power series converges absolutely. R is called the *radius of convergence* of the power series. We note for emphasis that R may be zero, nonzero finite, or infinite, giving the following theorem.

Theorem 9.9.9 (Power Series Convergence) *Let $\sum_{k=0}^{\infty} c_k(x-a)^k$ be any power series. Then one of the following must hold:*

(a) *The series converges only for $x = a$ ($R = 0$).*

(b) *The series converges absolutely for all x (R is infinite).*

(c) *There is a real number $R > 0$ such that the series converges absolutely for all x with $|x - a| < R$ and diverges for all x with $|x - a| > R$ (R nonzero).*

The interval from $a - R$ to $a + R$ is called the *interval of convergence.* At the endpoints of the interval of convergence the series may either converge or diverge. These points, $a - R$ and $a + R$, must be examined individually.

Example 9.9.10 We will find the power series and radius of convergence of $exp(x) = e^x$. From our previous work (Examples 9.2.5 and 9.9.5) we know that the Taylor Series for e^x is $\sum_{k=0}^{\infty} x^k/k!$ and that this series converges for all real numbers. Thus the radius of convergence is infinity. △

Power Series as Functions

We have shown that every power series has an interval of convergence and for each point in the interval of convergence the power series converges to a unique value. Thus every power series defines a function whose domain is the interval of convergence. For example, the function $g : \mathbb{R} \to \mathbb{R}$ defined by the geometric series $g(x) = \sum_{k=0}^{\infty} x^k = 1/(1 - x)$ has domain $(-1, 1)$. Note that the series $\sum_{k=0}^{\infty} x^k$ does not equal $1/(1-x)$ for x outside of $(-1, 1)$ although the expression makes sense for all $x \neq 1$. For example, if we let $x = 2$, then the series is $1+2+4+\cdots$ which diverges, while the value of the expression is $1/(1-2) = -1$.

The objective of the rest of this section and the next is to show how large numbers of series can be generated from a few known series by applying the operations of addition, scalar multiplication, composition, multiplication, differentiation, and integration. The reader might think back to how the class of polynomials was generated from the constant and identity functions and the operations of addition and multiplication, or how the class of trigonometric functions was generated from the sine and cosine functions and appropriate operations. In addition to the generation of series that are useful and important in themselves, the reader is urged to pay particular attention to the process of generating functions from "base" functions by means of function operations. This process is fundamental in the development of a mathematical theory.

The algebraic operations of series addition and scalar multiplication given in the Algebra of Series Theorem certainly applies to power series. We shall illustrate the composition operation involving a function defined by a power series in the next three examples.

Example 9.9.11 Let g be defined by the geometric series:

$$g(x) = \sum_{k=0}^{\infty} x^k = \frac{1}{1 - x} \quad \text{for } |x| < 1$$

and let h be defined by $h(t) = -2t$. Then the composite function g(h(t)) is defined by $g(h(t)) = \sum_{k=0}^{\infty}(-2t)^k = 1/(1+2t)$.

Note that the domain of the composite function is $(-\frac{1}{2}, \frac{1}{2})$. (Why?) △

Example 9.9.12 Let g be defined by the geometric series, $g(x) = \sum_{k=0}^{k} x^k, |x| < 1$, and let h be defined by $h(t) = t^2$. The composite function $g \circ h$ is then defined by

$$g \circ h(t) = g(h(t)) = \sum_{k=0}^{\infty} t^{2k} = \frac{1}{1-t^2}$$

What is the domain of the composite function? △

Example 9.9.13 Let $g : \mathbb{R} \to \mathbb{R}$ be defined by $g(x) = e^x = \sum_{k=0}^{\infty} x^k/k!$ and let h be defined by $h(t) = -t^2$. The composite function is

$$g \circ h(t) = g(h(t)) = \sum_{k=0}^{\infty}(-1)^k \frac{t^{2k}}{k!}$$

What is the domain of the composite function? △

We have seen that every power series defines a function. We also have seen how, starting with a function that is C^∞ (i.e., a function that is infinitely differentiable), we can obtain its Taylor Series. A more challenging problem is to recognize a function given its power series representation. The process of obtaining a closed form expression for a function defined by a series is similar to integrating by finding an antiderivative. Namely, finding transformations that will convert the given series expression into a known form. We shall illustrate the process with two examples.

Example 9.9.14 Consider the power series

$$f(x) = \sum_{k=2}^{\infty}(-x)^k = x^2 - x^3 + \cdots$$

This is a geometric series with the first term $(-x)^2$ and $r = -x$. To sum the series, we factor out the first term and then sum the resulting geometric series that is in standard form.

$$f(x) = \sum_{k=2}^{\infty}(-x)^k = x^2\sum_{k=0}^{\infty}(-x)^k = \frac{x^2}{1-(-x)} = \frac{x^2}{1+x} \quad \text{for } |x| < 1$$ △

Example 9.9.15 Suppose we wish to find a function that is represented by the series

$$\sum_{k=0}^{\infty} \frac{x^{2k}}{3^k} = 1 + \frac{x^2}{3} + \frac{x^4}{9} + \cdots$$

A simple transformation gives

$$\sum_{k=0}^{\infty} \frac{x^{2k}}{3^k} = \sum_{k=0}^{\infty} \left(\frac{x^2}{3}\right)^k$$

This is a basic geometric series with $r = x^2/3$. Thus

$$\sum_{k=0}^{\infty} \frac{x^{2k}}{3^k} = \frac{3}{3 - x^2} \qquad \text{for } |x| < \sqrt{3}$$

△

Section 9.9 Exercises

For all of the series in Exercises 1 through 20, find the interval of convergence.

1. $\sum_{k=0}^{\infty} x^k/(k+1)$
2. $\sum_{k=1}^{\infty} x^k/\sqrt{k}$
3. $\sum_{k=1}^{\infty} \log(k)x^k/k^2$
4. $\sum_{k=1}^{\infty} x^k/(k^2+k)$
5. $\sum_{m=0}^{\infty} 2x^m/m!$
6. $\sum_{n=0}^{\infty} (1000)^n x^n/(n!)$
7. $\sum_{k=0}^{\infty} k!(x-2)^k$
8. $\sum_{j=0}^{\infty} (j+2)^2(x-2)^j$
9. $\sum_{k=1}^{\infty} x^{k+1}/k^2$
10. $\sum_{k=0}^{\infty} x^{2k}/(k+1)$
11. $\sum_{n=1}^{\infty} 2^n x^n/n$
12. $\sum_{k=2}^{\infty} x^{2k}/\log(k)$
13. $\sum_{n=1}^{\infty} (3x)^n$
14. $\sum_{k=1}^{\infty} k^k x^k/k!$
15. $\sum_{k=0}^{\infty} (x-1)^k$
16. $\sum_{k=0}^{\infty} 5^k x^k/k!$
17. $\sum_{k=1}^{\infty} (k/100)^k x^k$
18. $\sum_{k=4}^{\infty} 1/[\log(k)]^k (x-1)^k$
19. $\sum_{k=1}^{\infty} (-1)^k (2/3)^k (x+1)^k$
20. $\sum_{k=1}^{\infty} \log(k)/2^k (x-2)^k$

For Exercises 21 through 23, find the Taylor series for the function expanded about 0 and the radius of convergence.

21. $f(x) = \sin(x)$
22. $f(x) = e^x \cos(x)$
23. $f(x) = 1/(x+1) - 2$

24. Prove or disprove: If the power series $\sum_{k=0}^{\infty} c_k |x-a|^k$ diverges for $x = x_1$, then the series diverges for all x with $|x-a| > |x_1 - a|$.

9.10 Operations on Series

The development of each major concept in our text has included an "Algebra" theorem which states the rules of applying the concept to algebraic combinations of functions. Thus, if we can apply the concept to basic functions, we can apply it to more complicated functions obtained from these by algebraic operations. This is also true of series, where our basic series are polynomials, geometric series, and the Taylor Series for the sine, cosine, exponential, and logarithm functions. In this section we show how to multiply series and how to apply integration and differentiation to series.

Multiplication of Series

Before stating the definition of the product of two series, let us illustrate polynomial multiplication by multiplying two second degree polynomials together. The reader should verify the following equality in order to gain a feeling for how the coefficients are formed.

$$\begin{aligned}(a_0 + a_1x + a_2x^2) &\times (b_0 + b_1x + b_2x^2)\\ &= a_0b_0 + (a_0b_1 + a_1b_0)x + (a_0b_2 + a_1b_1 + a_2b_0)x^2\\ &\quad + (a_1b_2 + a_2b_1)x^3 + a_2b_2x^4.\end{aligned}$$

The following definition is formed by abstracting the pattern of forming the coefficients in the above product.

Definition 9.10.1 (Product of Series) *Let $\sum_{k=0}^{\infty} a_k$ and $\sum_{k=0}^{\infty} b_k$ be infinite series. The* product *of the series is*

$$\left(\sum_{k=0}^{\infty} a_k\right)\left(\sum_{k=0}^{\infty} b_k\right) = \sum_{k=0}^{\infty} c_k,$$

where $c_k = \sum_{j=0}^{k} a_j b_{k-j}$.

Example 9.10.2 Consider the series $1 + \frac{1}{2} + \frac{1}{3} + \frac{1}{4} + \cdots$ and the series $1 - \frac{1}{2} + \frac{1}{3} - \frac{1}{4} + \cdots$. In the notation of the previous definition, $a_k = 1/(k+1)$ and $b_k = (-1)^k/(k+1)$. To indicate the product we list some terms:

$$\begin{aligned}c_0 &= a_0b_0 = 1\\ c_1 &= a_0b_1 + a_1b_0 = -\tfrac{1}{2} + \tfrac{1}{2} = 0\\ c_2 &= a_0b_2 + a_1b_1 + a_2b_0 = \tfrac{1}{3} - \tfrac{1}{4} + \tfrac{1}{3} = \tfrac{5}{12}\\ c_3 &= a_0b_3 + a_1b_2 + a_2b_1 + a_3b_0 = 0\end{aligned}$$

△

Computing the product of two series is a useful technique for finding the Taylor Series for a function that is the product of two simple functions. The radius of convergence of the product is at least as large as the smaller of the radii of convergence of the two series formed when both functions are expanded about the same point.

Example 9.10.3 Suppose we want to find the Taylor Series for $f(x) = e^x \sin(x)$ expanded about 0. From Example 9.9.10 and the exercises we know that the radii of convergence for e^x and for $\sin(x)$ are infinity and

$$e^x = 1 + x + \frac{x^2}{2!} + \frac{x^3}{3!} + \cdots$$
$$\sin(x) = \sum_{k=0}^{\infty} \frac{(-1)^k}{(2k+1)!} x^{2k+1} = x - \frac{x^3}{3!} + \frac{x^5}{5!} - \cdots$$

The product of the series will be

$$f(x) = e^x \sin(x) = \sum_{k=0}^{\infty} c_k x^k$$

where $c_k = \sum_{j=0}^{k} a_j b_{k-j}$, and where a_k and b_k are the coefficients in the Taylor Series of e^x and $\sin(x)$, respectively.

$$c_0 = a_0 b_0 = 1 \cdot 0 = 0.$$

$$c_1 = a_0 b_1 + a_1 b_0 = 1 \cdot 1 + 0 \cdot 1 = 1$$

$$c_2 = a_0 b_2 + a_1 b_1 + a_2 b_0 = 1 \cdot 0 + 1 \cdot 1 + (\tfrac{1}{2})0 = 1$$

$$c_3 = 1(-\tfrac{1}{3!}) + 1 \cdot 0 + (\tfrac{1}{2!})1 + (\tfrac{1}{3!})0 = \tfrac{1}{3}$$

$$c_4 = 1 \cdot 0 + 1(-\tfrac{1}{3!}) + (\tfrac{1}{2!})0 + (\tfrac{1}{3!})1 + (\tfrac{1}{4!})0 = 0$$

$$c_5 = 1(\tfrac{1}{5!}) + 1 \cdot 0 + (\tfrac{1}{2})(-\tfrac{1}{3!}) + (\tfrac{1}{3!})0 + (\tfrac{1}{4!})1 + (\tfrac{1}{5!})0 = \tfrac{-1}{30}$$

Thus Taylor Series for $f(x) = e^x \sin(x)$ begins

$$f(x) = x + x^2 + \frac{x^3}{3} - \frac{x^5}{30} + \cdots$$

△

Differentiation and Integration

If a function is represented by a power series, is the function differentiable? For example, if

$$f(x) = \sum_{k=0}^{\infty} c_k (x-a)^k$$

for $|x - a| < R$, is f differentiable? Furthermore, is there a series representation of $f'(x)$?

The answers to these questions are highly satisfactory and are given in the next theorem.

Theorem 9.10.4 (Differentiation of Power Series)
If $R > 0$ and $f(x) = \sum_{k=0}^{\infty} c_k(x-a)^k$ converges for $|x-a| < R$, then f is a C^∞ function on $(a-R, a+R)$ and the derivatives are obtained by successive term-by-term differentiation of the series for f. For example,

$$f'(x) = \sum_{k=1}^{\infty} kc_k(x-a)^{k-1} \quad \text{for} \quad |x-a| < R$$

The proof is omitted. There are several points that should be made concerning this theorem.

1. Every power series is differentiable.

2. Power series are differentiable on the *interior* of the interval of convergence. A power series may fail to be differentiable at an endpoint of the interval of convergence.

3. The derivative of a power series is a power series, which is obtained by differentiating the given series term-by-term.

4. Since the derivative of a power series is again a power series, the derivative will be differentiable. Thus functions represented as power series will have derivatives of all orders: they are C^∞ functions.

Example 9.10.5 Consider the function f represented as a geometric series

$$f(x) = \sum_{k=0}^{\infty} x^k \quad \text{for} \quad -1 < x < 1$$

Using the theorem, we can find f' as

$$f'(x) = \sum_{k=0}^{\infty} kx^{k-1} = 1 + 2x + 3x^2 + \cdots$$

Of course, since $f(x) = 1/(1-x)$ for $-1 < x < 1$ and from our knowledge of the derivative we also know that $f'(x) = 1/(1-x)^2$, therefore,

$$\frac{1}{(1-x)^2} = \sum_{k=1}^{\infty} kx^{k-1} \quad \text{for} \quad -1 < x < 1$$

△

Example 9.10.6 We can use the differentiation rule to find functions that certain series represent. Consider the series

$$\sum_{k=2}^{\infty} k(k-1)(x-2)^{k-2}$$

Note that each term is the second derivative of $(x-2)^k$. Since

$$\sum_{k=0}^{\infty}(x-2)^k = \frac{1}{1-(x-2)} = \frac{1}{3-x}$$

we have

$$\sum_{k=0}^{\infty} k(k-1)(x-2)^k = \left(\frac{d^2}{dx^2}\right)\frac{1}{3-x} = \frac{2}{(3-x)^3} \qquad \triangle$$

Question 9.10.7 Over what interval is this function differentiable? $\diamond$

Term-by-term integration of a power series is also valid on the interior of the interval of convergence. We have the following theorem.

Theorem 9.10.8 (Integration of Power Series)
If $R > 0$ and $f(x) = \sum_{k=0}^{\infty} c_k(x-a)^k$ converges for $|x-a| < R$, then the antiderivative of f is given by

$$\int f(x)dx = \sum_{k=0}^{\infty}\left(\frac{c_k}{k+1}\right)(x-a)^{k+1} + C \quad \text{for} \quad |x-a| < R$$

Furthermore, if r and s are in the interior of the integral of convergence, then

$$\int_r^s f(x)dx = \sum_{k=0}^{\infty}\int_r^s c_k(x-a)^k dx$$

Example 9.10.9 By integrating $1/(1+x) = \sum_{k=0}^{\infty}(-1)^k x^k$ for $|x| < 1$, we obtain

$$\log(1+x) = \int\sum_{k=0}^{\infty}(-1)^k x^k dx = \sum_{k=0}^{\infty}\int(-1)^k x^k dx = \sum_{k=0}^{\infty}\frac{(-1)^k x^{k+1}}{k+1} + C$$

The constant of integration C is zero, since when $x = 0$ the equality becomes $0 = 0 + C$. $\triangle$

Example 9.10.10 The Taylor Series for the inverse tangent can be obtained by integration since

$$\begin{aligned}\tan^{-1}(x) &= \int\frac{dx}{1+x^2} = \int\frac{dx}{1-(-x^2)} = \int\sum_{k=0}^{\infty}(-1)^k x^{2k}dx \\ &= \sum_{k=0}^{\infty}\int(-1)^k x^{2k}dx = \sum_{k=0}^{\infty}\frac{(-1)^k x^{2k+1}}{2k+1} + C\end{aligned}$$

Setting $x = 0$ on both sides we have $0 = 0 + C$, thus

$$\tan^{-1}(x) = \sum_{k=0}^{\infty}\frac{(-1)^k x^{2k+1}}{2k+1} \quad \text{for all } x \qquad \triangle$$

As in the discussion of the use of Taylor Polynomial approximations, we can use term-by-term integration to approximate the integral of a function for which no anti-derivative in closed form can be found.

Example 9.10.11 Suppose we want to estimate $\int_0^1 e^{-x^2}dx$.
Since $e^y = \sum_{k=0}^{\infty} y^k/k!$ for all y, we have

$$e^{-x^2} = 1 - x^2 + \frac{x^4}{2!} - \frac{x^6}{3!} + \cdots$$

Term-by-term integration gives us an infinite series that converges to the desired number:

$$\begin{aligned} \int_0^1 e^{-x^2}dx &= \left[x - \frac{x^3}{3} + \frac{x^5}{5\cdot 2!} - \frac{x^7}{7\cdot 3!} + \frac{x^9}{9\cdot 4!} - \cdots\right]_0^1 \\ &= 1 - \frac{1}{3} + \frac{1}{5\cdot 2!} - \frac{1}{7\cdot 3!} + \frac{1}{9\cdot 4!} - \cdots \end{aligned}$$

This is an alternating series, so we know that the error in using four terms is less than the magnitude of the fifth term:

$$\left|\int_0^1 e^{-x^2}dx - \left[1 - \frac{1}{3} + \frac{1}{5\cdot 2!} - \frac{1}{7\cdot 3!}\right]\right| \le \frac{1}{9\cdot 4!}$$

or

$$\left|\int_0^1 e^{-x^2}dx - 0.7429\right| \le 0.005 \qquad \triangle$$

Relation between Taylor Series and Power Series

We have seen how we can start with a power series of radius of convergence R

$$\sum_{k=0}^{\infty} c_k(x-a)^k$$

and obtain a C^∞ function f

$$f(x) = \sum_{k=0}^{\infty} c_k(x-a)^k$$

Given the C^∞ function f, we can obtain its Taylor Series

$$\sum_{k=0}^{\infty} \frac{f^{(k)}(a)}{k!}(x-a)^k$$

which is a power series with radius of convergence R. *On the interval of convergence, these two series are the same.*

To see this, we must show that if

$$f(x) = \sum_{k=0}^{\infty} c_k(x-a)^k \quad \text{for } |x-a| < R$$

then

$$c_k = \frac{f^{(k)}(a)}{k!}$$

We differentiate the series

$$f(x) = c_0 + c_1(x-a) + c_2(x-a)^2 + c_3(x-a)^3 + \cdots$$

term by term and evaluate at $x = a$, obtaining:

$$\begin{aligned}
&f(a) = c_0 \\
&f'(a) = c_1 \\
&f''(a) = 2c_2 \\
&f^{(3)}(a) = 2 \cdot 3c^3 \\
&\vdots \\
&f^{(k)}(a) = k!c_k
\end{aligned}$$

Thus

$$c_k = \frac{f^{(k)}(a)}{k!}$$

Since a power series expansion for a function f over its interval of convergence is the Taylor series expansion for f, the coefficients of the power series may be used to determine the derivatives of the function. That is if $\sum_{k=0}^{\infty} a_k(x-a)^k$ defines the function f, then over the interval of convergence

$$\sum_{k=0}^{\infty} a_k(x-a)^k = f(x) = \sum_{k=0}^{\infty} \frac{f^k(a)}{k!}(x-a)^k.$$

Thus,

$$a_k = \frac{f^k(a)}{k!} \quad \text{or} \quad f^k(a) = k!a_k.$$

Example 9.10.12 We will find the thirtieth derivative of $f(x) = e^{x^2/3}$ evaluated at $x = 0$.

Since $e^u = \sum_{k=0}^{\infty} u^k/k!$ for all u, we have, substituting $u = x^2/3$,

$$f(x) = e^{x^2/3} = \sum_{k=0}^{\infty} \frac{x^{2k}}{3^k k!}$$

Thus $f^{(30)}(0) = 30!/[3^{15}(15!)]$. (Why 15! ?) △

Example 9.10.13 (Generating Functions) In Chapter 1 we introduced the Fibonacci Sequence $\{r_k\}$, where $r_0 = 0$, $r_1 = 1$, and $r_k = r_{k-1} + r_{k-2}$ for $k \geq 2$. (The next term in the sequence is obtained by adding together the two previous terms.) In this example we will use the algebra of series to obtain a formula for r_k.

We will use the technique of *generating functions.* A sequence $\{a_k\}$ is an *infinite* set of numbers, so it is often difficult to deal with. The generating function of a sequence $\{a_k\}$ is the function defined by the associated power series,

$$f(x) = \sum_{k=0}^{\infty} a_k x^k$$

This series will at least converge at $x = 0$, but in this application we are not interested in the interval of convergence: we will operate on the series without regard to the radius of convergence. If the generating function f of $\{a_k\}$ can be expressed as a simple formula, then we have transformed the infinite set $\{a_k\}$ into the simple formula for f. We can recover the individual term a_k from f by differentiating the series $a_k = f^{(k)}(0)/k!$.

We will find the generating function f of the Fibonacci Sequence $\{r_k\}$ and then use $r_k = f^{(k)}(0)/k!$ to find a formula for r_k. The generating function is

$$f(x) = r_0 + r_1 x + r_2 x^2 + \cdots = \sum_{k=0}^{\infty} r_k x^k$$

We will substitute the initial values of r_0, r_1 and the recurrence formula $r_k = r_{k-1} + r_{k-2}$ into f and then simplify:

$$\begin{aligned} f(x) &= 0 + x + \sum_{k=2}^{\infty} r_k x^k \\ &= x + \sum_{k=2}^{\infty} [r_{k-1} + r_{k-2}] x^k \\ &= x + \sum_{k=2}^{\infty} r_{k-1} x^k + \sum_{k=2}^{\infty} r_{k-2} x^2 \end{aligned}$$

We will now transform the two series on the right-hand side so that the summation subscript begins at $k = 0$. To do this, we need to adjust each term so that it is of the form $r_k x^k$ by factoring out an x from the first term and an x^2 from the second term:

$$\begin{aligned} f(x) &= x + [r_1 x^2 + r_2 x^3 + \cdots] + [r_0 x^2 + r_1 x^3 + \cdots] \\ &= x + x[r_1 x^1 + r_2 x^2 + \cdots] + x^2[r_0 + r_1 x^1 + \cdots] \\ &= x + x f(x) + x^2 f(x) \end{aligned}$$

(since $r_0 = 0$). Now we have an equation which we can solve for f:

$$f(x) = x + xf(x) + x^2 f(x)$$

This gives

$$f(x) = \frac{x}{1 - x - x^2}$$

the generating function for the Fibonacci Sequence.

We now use $r_k = f^{(k)}(0)/k!$ to obtain a formula for r_k. To make the differentiation easier, we will use the partial fraction transformation on $x/(1 - x - x^2)$. First we use the quadratic formula to find the roots of $x^2 + x - 1$:

$$x = \frac{-1 \pm \sqrt{5}}{2}$$

So

$$\begin{aligned} x^2 + x - 1 &= \left[x - \frac{1}{2}(-1 + \sqrt{5})\right]\left[x - \frac{1}{2}(-1 - \sqrt{5})\right] \\ &= \left[x + \frac{1}{2}(1 - \sqrt{5})\right]\left[x + \frac{1}{2}(1 + \sqrt{5})\right] \end{aligned}$$

To simplify the expressions, let $g_1 = (1 + \sqrt{5})/2$ and $g_2 = (1 - \sqrt{5})/2$. Thus the partial fraction decomposition is

$$\begin{aligned} \frac{x}{1 - x - x^2} &= \frac{-x}{x^2 + x - 1} = \frac{-x}{(x + g_1)(x + g_2)} \\ &= \frac{A}{x + g_1} + \frac{B}{x + g_2} = \frac{-g_1}{\sqrt{5}} \frac{1}{x + g_1} + \frac{g_2}{\sqrt{5}} \frac{1}{x + g_2} \\ &= \frac{1}{\sqrt{5}} \left(\frac{-g_1}{x + g_1} + \frac{g_2}{x + g_2} \right) \end{aligned}$$

Computing the first few derivatives we have:

$$f'(x) = \frac{1}{\sqrt{5}} \left(\frac{g_1}{(x + g_1)^2} - \frac{g_2}{(x + g_2)^2} \right)$$

$$f''(x) = \frac{1}{\sqrt{5}} \left(\frac{-2g_1}{(x + g_1)^3} + \frac{2g_2}{(x + g_2)^3} \right)$$

$$f'''(x) = \frac{1}{\sqrt{5}} \left(\frac{3!g_1}{(x + g_1)^4} - \frac{3!g_2}{(x + g_2)^4} \right)$$

Setting $x = 0$ and using $g_1 g_2 = -1$ we obtain:

$$r_1 = \frac{f'(0)}{1!} = \frac{1}{\sqrt{5}}(g_1 - g_2)$$

$$r_2 = \frac{f''(0)}{2!} = \frac{1}{\sqrt{5}}(g_1^2 - g_2^2)$$

$$r_3 = \frac{f'''(0)}{3!} = \frac{1}{\sqrt{5}}(g_1^3 - g_2^3)$$

We can see that the formula is:

$$r_k = \frac{1}{\sqrt{5}}(g_1^k - g_2^k)$$

(The reader may wish to verify this formula for the first few k.)

Let us review the steps in finding a formula for the terms a_k in a sequence:

1. Form the generating function f of the sequence:

$$f(x) = \sum_{k=0}^{\infty} a_k x^k$$

2. Manipulate the equation for f so that it involves f alone (no series).
3. Solve the equation for f.
4. Obtain a_k from $a_k = f^{(k)}(0)/k!$. △

Section 9.10 Exercises

For Exercises 1 through 6, find the first four nonzero terms in the Taylor Series expansions of the function expanded about 0, and the radius of convergence.

1. $f(x) = \sin(x)\cos(x)$
2. $f(x) = e^x \cos(x)$
3. $f(x) = e^x/(1-x)$
4. $f(x) = \sin(x)/(1-x)$
5. $f(x) = e^x \log(1+x)$
6. $f(x) = \sin(x)\log(1+x)$

In Exercises 7 through 10, use the basic geometric series to find a function that is represented by the series. Specify the domain of the each function.

7. $\sum_{k=0}^{\infty} x^{k+1}$
8. $\sum_{k=1}^{\infty} x^{k-1}/3^k$
9. $\sum_{k=1}^{\infty}(-1)^{k-1}x^{2k-1}$
10. $\sum_{k=0}^{\infty}(3-x)^k$

In Exercises 11 through 14, use term-by-term differentiation to find a function that is represented by the series. Specify the domain of each function.

11. $\sum_{k=1}^{\infty}(-k)(-x)^{k-1}$

12. $\sum_{k=0}^{\infty}(k+1)x^k$

13. $\sum_{k=1}^{\infty}(-1)^k 2kx^{2k-1}$

14. $\sum_{k=1}^{\infty} 2k/3^k x^{2k-1}$

In Exercises 15 through 18, use term-by-term integration to find a function that is represented by the given series. Specify the domain of each function.

15. $\sum_{k=0}^{\infty} x^{k+1}/(k+1)$

16. $\sum_{k=0}^{\infty} (-1)^k/(k+2)x^{k+2}$

17. $\sum_{k=0}^{\infty} (-1)^k/(2k+1)x^{2k+1}$

18. $\sum_{k=0}^{\infty} -1/(k+1)(1-x)^{k+1}$

In Exercises 19 through 26, find a function that is represented by the given series. Specify the domain of each function.

19. $\sum_{k=1}^{\infty} (-1)^{k-1}/2^k(x-2)^{k-1}$

20. $\sum_{k=1}^{\infty}(x^{k-1}+x^k)$

21. $\sum_{k=1}^{\infty} x^{k-1}/5^k$

22. $\sum_{k=3}^{\infty}(k-1)(k-2)x^{k-3}$

23. $\sum_{k=3}^{\infty} (k-1)(k-2)/5^k x^{k-3}$

24. $\sum_{k=1}^{\infty} x^{3k-1}$

25. $\sum_{k=1}^{\infty} 4^{k+1}x^{k-1}$

26. $\sum_{k=1}^{\infty} 4^k/kx^k$

In Exercises 27 through 30 approximate the integral using series to within 0.01.

27. $\int_0^1 e^{-x^3}dx$

28. $\int_0^1 \sin(x^2)dx$

29. $\int_0^1 \sin(\sqrt{x})dx$

30. $\int_0^1 \tan^{-1}(x^2)dx$

31. $\int_0^1 \sqrt{x}\sin(x)dx$

32. $\int_0^1 \sin(x)\cos(x)dx$

In Exercises 33 through 36, find the function that is represented by the series. Specify the domain of each function.

33. $\sum_{k=3}^{\infty} kx^k$

34. $\sum_{k=1}^{\infty}(-1)^{k+1}x^{2k}/k$

35. $\sum_{k=1}^{\infty} x^{4k}/(4k)$

36. $\sum_{k=0}^{\infty}(-1)^k 9^k x^{2k+1}/(2k)!$

37. Find the thirty-third derivative of $f(x) = \sum_{k=0}^{\infty} 2^k x^k$ evaluated at $x = 0$.

38. Find the hundred and first derivative of $f(x) = \sin(x)\cos(x)$ evaluated at $x = 0$. Hint: Expand f in a Taylor series about $a = 0$. Recall that $\sin(x)cos(x) = 1/2sin(2x)$.

39. Use the technique of generating functions to solve the recurrence relation $a_k = ka_{k-1}$ where $a_0 = 1$. (You should be able to guess the answer!)

40. [Project] Prepare a class presentation on the number e. Include some historical facts, describe several uses of e [e.g., growth-decay (biology), interest computations (economics)], and complete the proof outlined below showing that e is an irrational number.

Outline of a proof by contradiction showing that e is an irrational number.

(a) Assume that e is a rational number.

(b) Show that $1/e = \sum_{k=0}^{\infty}(-1)^k/k!$.

(c) Use the error bound determination from alternating series to show that

$$0 < \left| \frac{1}{e} - \sum_{k=0}^{n-1} \frac{(-1)^k}{k!} \right| < \frac{1}{n!}.$$

Therefore for $n \geq 2$,

$$0 < (n-1)! \left| \frac{1}{e} - \sum_{k=0}^{n-1} \frac{(-1)^k}{k!} \right| < \frac{1}{n} \leq \frac{1}{2}.$$

(d) Note that $(n-1)! \sum_{k=0}^{n-1}(-1)^k/k!$ is an integer. (Why?)

(e) Since e is assumed to be a rational number, there is integer n such that $(n-1)!/e$ is an integer. (Why?)

(f) Thus for a sufficiently large integer n, $|(n-1)!/e - (n-1)! \sum_{k=0}^{n-1}(-1)^k/k!|$ is an integer strictly between 0 and $\frac{1}{2}$, a contradiction.

Appendix A

TRIGONOMETRY

Angles may be measured in either degrees or radians, and it is often necessary to convert between the two. There are $360°$ or 2π radians in a circle. This relationship is obtained by measuring the length of an arc on the unit circle. The circumference of a circle is $2\pi r$, where r is the radius. Thus for the unit circle, $r = 1$, the circumference is 2π. The distance along an arc is proportional to the angle the arc subtends. Thus

$$\frac{\text{Distance along arc}}{\text{Circumference of circle}} = \frac{\text{angle}}{360}.$$

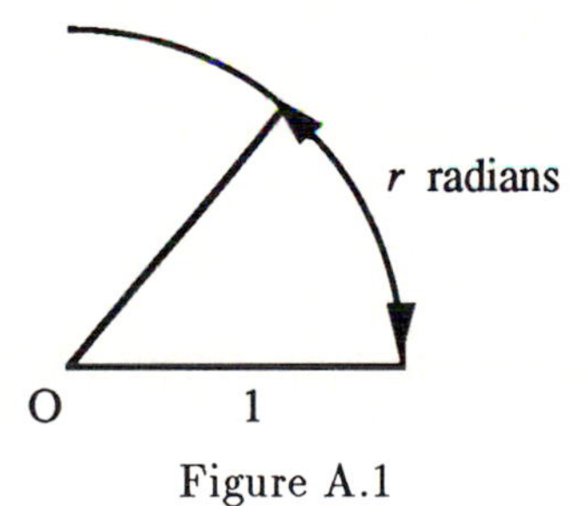

Figure A.1

The conversion factors are π radians $= 180°$, or 1 radian $= 180/\pi$ degrees $\approx 57.2958°$.

Let the plane have an xy-coordinate system with a unit circle centered at the origin. Consider a point P on the unit circle with coordinates (x, y). We wish to define the trigonometric functions of a real number, θ, considered as an angle measured counterclockwise from the x-axis along the unit circle to P. To do this, we construct a right triangle by drawing a line from the point P to the x-axis. Then we have a right triangle of sides x and y, and hypotenuse 1. We define the sine of $\theta, \sin(\theta)$, as the ratio of the opposite side and the hypotenuse of the right triangle, $\sin(\theta) = y/1 = y$. The cosine of θ is defined as the ratio of the adjacent side and the hypotenuse, $\cos(\theta) = x/1 = x$.

Definition of Trigonometric Functions

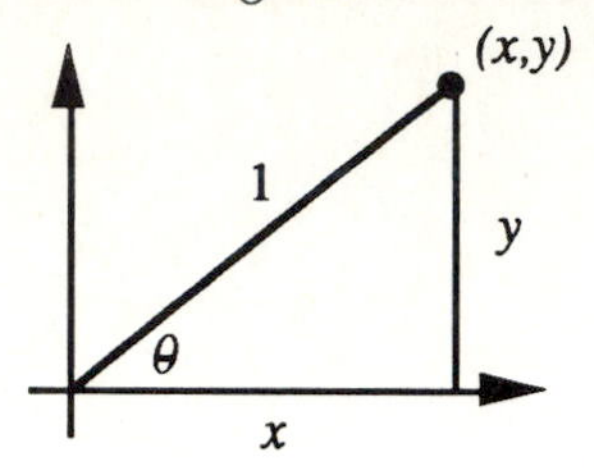

Figure A.2

The other trigonometric functions are defined in terms of the sine and cosine:

The secant $\sec(\theta) = 1/\cos(\theta)$
The cosecant $\csc(\theta) = 1/\sin(\theta)$
The tangent $\tan(\theta) = \sin(\theta)/\cos(\theta)$
The cotangent $\cot(\theta) = \cos(\theta)/\sin(\theta)$

Since sine and cosine can be zero, these definitions for tan and cot only apply when the denominator is nonzero.

Some useful values of sine for various angles in radians are:

$$\begin{array}{ll} \sin(0) = 0 & \sin(\pi/6) = 1/2 \\ \sin(\pi/4) = 1/\sqrt{2} & \sin(\pi/3) = \sqrt{3}/2 \\ \sin(\pi/2) = 1 & \end{array}$$

The sum of the angles in a triangle is π radians; the two angles which are not $\pi/2$ must sum to $\pi/2$. In Figure A.0.13 defining the trigonometric functions, $\phi = \pi/2 - \theta$. Thus $\cos(\theta) = x = \sin(\theta)$. This gives the relation $\cos(\theta) = \sin(\pi/2 - \theta)$. With this, we can find values for cosine, for example

$$\cos\left(\frac{\pi}{3}\right) = \sin\left(\frac{\pi}{2} - \frac{\pi}{3}\right) = \sin\left(\frac{\pi}{6}\right) = \frac{1}{2}$$

Similarly we can find

$$\tan\left(\frac{\pi}{3}\right) = \frac{\sin(\pi/3)}{\cos(\pi/3)} = \frac{\sqrt{3}/2}{1/2} = \sqrt{3}$$

We will consider the graphs of the trigonometric functions. From the definition we see that $\sin : \mathbb{R} \to [-1, +1]$. Since a complete trip around the circle is done in 2π radians, the trigonometric functions will repeat every 2π radians, i.e., $\sin(x) = \sin(x + 2\pi)$. The graph of sin is a periodic wave. Since $\cos(x) = \sin(\pi/2 - x)$, the graph of cosine is the same as the graph of sin, offset by $\pi/2$.

Graphs of sin and cos

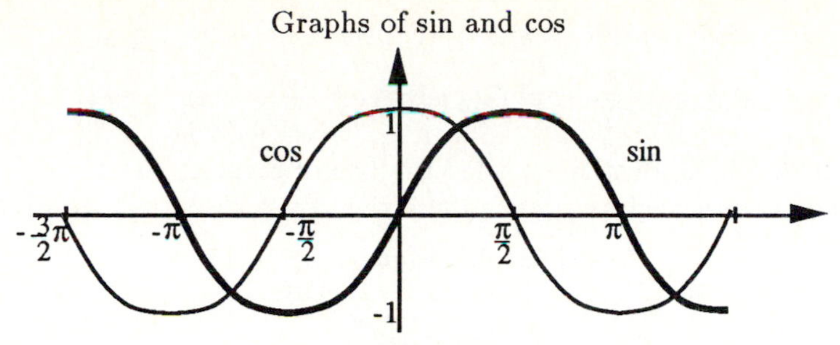

Figure A.3

Another feature of the graph of sin is that $\sin(-x) = -\sin(x)$. A function f for which $f(-x) = -f(x)$ is called an *odd* function. Similarly, $\cos(-x) = \cos(x)$. A function f for which $f(-x) = f(x)$ is called an *even* function.

There are many relationships between the trigonometric functions. We will only discuss the most important. The most useful relationship is the *Pythagorean Theorem.* The Pythagorean Theorem says that in a right triangle, the sum of the squares of the sides is equal to the square of the hypotenuse. Applying this result to the right triangle in Figure A.0.13 we have $x^2 + y^2 = 1$. In terms of the definition of sine and cosine, this gives

1. $\sin(\theta)^2 + \cos(\theta)^2 = 1$

The next trigonometric identities are for sums:

2. $\sin(x + y) = \sin(x)\cos(y) + \sin(y)\cos(x)$
3. $\sin(x - y) = \sin(x)\cos(y) - \sin(y)\cos(x)$
4. $\cos(x + y) = \cos(x)\cos(y) - \sin(x)\sin(y)$
5. $\cos(x - y) = \cos(x)\cos(y) + \sin(x)\sin(y)$

Applying 2 to the case when $x = y$ we have

6. $\sin(2x) = 2\sin(x)\cos(x)$

Or, as a half-angle formula, replacing x by $x/2$

7. $\sin(\mathrm{x}) = 2\sin(x/2)\cos(x/2)$

Applying identity 4 to the case when $x = y$ we have

8. $\cos(2\mathrm{x}) = \cos^2(x) - \sin^2(x)$

If in identities 2 and 3 we let $a = x + y$ and $b = x - y$ so that $x = (a + b)/2$ and $y = (a - b)/2$, then subtracting identity 3 from identity 2 gives:

9. $\sin(a) - \sin(b) = 2\sin\left(\frac{a-b}{2}\right)\cos\left(\frac{a+b}{2}\right)$

Doing the same with identities 3 and 4 we have

10. $\cos(a) - \cos(b) = -2\sin\left(\frac{a-b}{2}\right)\sin\left(\frac{a+b}{2}\right)$

The next and final identity is called the law of cosines. Let a triangle have sides a, b, and c with angles opposite these sides of A, B, and C, respectively.

Law of Cosines

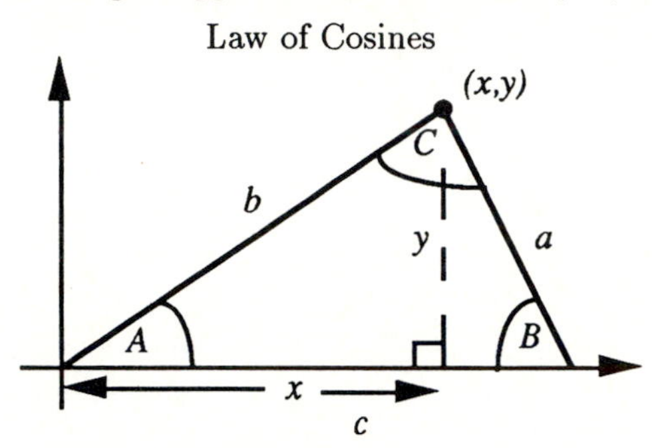

Figure A.4

By the Pythagorean Theorem we have:

$$x^2 + y^2 = b^2$$

and

$$y^2 + (c - x)^2 = a^2$$

Expanding the second of these,

$$a^2 = c^2 - 2cx + x^2 + y^2 = c^2 - 2cx + b^2$$

Since $\cos(A) = x/b$, we have $x = b\cos(A)$. Substituting into the equation for a^2 gives the *law of cosines*:

11. $a^2 = b^2 + c^2 - 2bc \cdot \cos(A)$

Note that when $A = 90°$ we obtain the Pythagorean Theorem.

Question A.5 (a) Find values x and y such that $\sin(x + y) \neq \sin(x) + \cos(y)$.
(b) Find a value x such that $\sin(2x) \neq 2\sin(x)$. ◇

Appendix B

COMPUTER ALGEBRA SYSTEMS

B.1 Introduction

The purpose of this appendix is to introduce you to Computer Algebra Systems. It is best if you can follow along on your own system, but it can also give you an idea of the capabilities of a Computer Algebra System even if none is available.

You are probably familiar with the use of computers for numerical work and for graphics. A Computer Algebra System (CAS) is a program that can perform symbolic tasks such as algebra, symbolic differentiation and symbolic integration, as well as numerical computation and graphics. Before the 1980s these systems were primarily used by researchers on large computer systems. Since then Computer Algebra Systems have become available on MS-DOS and Macintosh microcomputers and on minicomputers. Since they can perform most of the computational tasks traditionally taught in calculus, a modern calculus course must emphasize concepts and applications, not computations.

B.2 CAS Notation

The notation used in a CAS is similar to the standard mathematical notation and there are built-in functions which perform many of the common mathematical operations, such as the trigonometric functions, logarithms, etc. In this appendix we will primarily use the notation of the CAS *Maple*. For information on your particular system, see the user manual or the on-line help.

There are basically three ways in which a user interacts with a computer system or program: *command line* (such as UNIX, VMS, MS-DOS), *icons* (such as Macintosh), or *menu*. The menu systems will list some choices, and the user selects the desired choice. In this manual we will assume a command line CAS;

they are the most common and are easiest to describe in a text. You will have to refer to the locally available information on how to interact with your system.

B.3 Interaction with a CAS

We will give an example of using a CAS, with all input and output shown. (After this section we will only show the interesting output.) You may wish to follow along on your CAS as you read. We will use the syntax of Maple; see your user's manual for the corresponding commands in other CAS notations. In the next section we will discuss some of the operations in more detail.

This example will be a simple analysis of a polynomial:

$$x^4 - 6x^3 + 3x^2 + 26x - 24$$

After starting the system and getting a prompt, >, we enter the polynomial and assign it a name, p.

```
> p := x^4 - 6*x^3 + 3*x^2 + 26*x - 24;
```

$$p := x^4 - 6x^3 + 3x^2 + 26x - 24$$

The first line shows (in `typewriter` font) what the user entered. The second line shows the response by the CAS. The user enters the expressions in one-dimensional format, where exponentiation is indicated by a "^". The CAS responds in a two-dimensional format. Now we *factor* p:

```
> factor(p);
```

$$(x-4)(x+2)(x-1)(x-3)$$

and *expand* the result as a check:

```
> expand(");
```

$$x^4 - 6x^3 + 3x^2 + 26x - 24$$

The """ refers to the last expression displayed by the CAS. Another way to find the roots of p is to solve the equation $p = 0$ for the variable x:

```
solve(p=0,x);
```

$$4, -2, 1, 3$$

We can check this by substituting 4 in place of x in the expression p:

```
> subs(x=4,p);
```

$$0$$

Finally, we will graph the polynomial on an interval containing the roots:

```
> plot(p,x=-3..5);
```

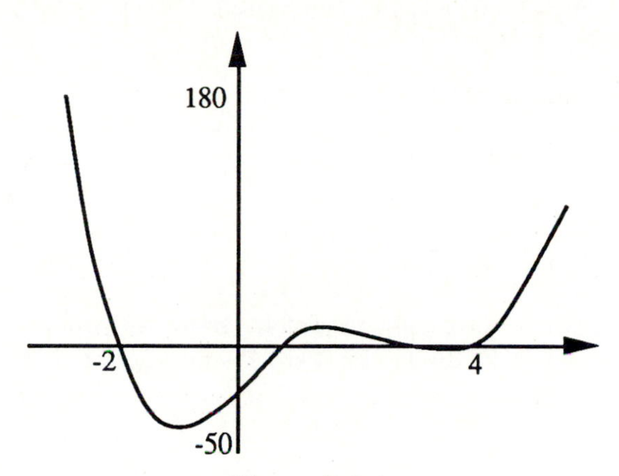

Figure B.3.1

Graphics vary substantially among systems, so you should do some experimentation here: try graphing the polynomial with different domains (say `x=-10..10`) and with the various graphics options.

B.4 Operations

We will discuss some of the basic operations of Computer Algebra Systems in the rest of this Appendix. There are several other sources for information on commands: on-line help and the user's manual.

Arithmetic

Every computer algebra system offers two types of numbers: exact integer and decimal. An integer can have arbitrarily many digits, limited by the storage capacity of the system. Computing the factorial of 5 is easy for human or computer:

```
> 5!;
```

$$120$$

When first using a CAS it is important to experiment with problems with a known solution. This allows you to be sure that you are entering the data correctly and to check system behavior. Once we have confidence in our ability to enter the values in the correct format, and in the ability of the CAS to compute the correct answer, we can try a problem we would not attempt to do by hand. For example, a CAS can compute the factorial of 100:

```
> 100!;
```

93326215443944152681699238856266700490715968264381621\
46859296389521759999322991560894146397615651828625369\
792082722375825118521091686400000000000000000000000000\

The output is too long for one line, so it is continued on several lines, in this case with "\" indicating continuation.

The usual arithmetic operations are available: $+, -, *, /, \wedge, !$. The exponentiation operation is the circumflex"$\wedge$" or double asterisks"$**$": $2 \wedge 3 = 2 ** 3 = 2^3 = 8$. The division operation / results in a rational number rather than a decimal. Rational arithmetic is also done exactly.

Decimal numbers can have high precision. Suppose you want to know the decimal expansion of π to 20 digits. First set the number of decimal digits to 20 by entering:

```
> Digits := 20;
```

then evaluating π:

```
> evalf(Pi);
```

$$3.1415926535897932385.$$

Exercises

1. What is 50!?

2. What is 60!?

3. How many digits are there in 40!, 50!?

Suppose we wish to continue investigating the number of digits in $n!$. Counting the number of digits is tedious and error-prone. Perhaps there is a CAS command to help? Most CAS have on-line help; or you can use the manual. The user may have to guess the name of a command or topic.

4. Does your CAS have a command to determine the number of digits in an integer? What might be its name? [(Some possibilities are "digit(s)", "size", "length", ...]

5. How does the number of digits in $n!$ grow? For example, if n is doubled, does the number of digits approximately double, triple, square, cube, or what?

6. What is π accurate to fifty digits?

Graphing

Most computer algebra systems have graphic as well as numerical and symbolic capabilities. It is often useful to graph a function to obtain information about it. Usually it is possible to graph several functions on the same set of axes by using a list:

```
> plot({x∧2,x},x = -1..1);
```

which returns the familiar graph of the parabola and straight line segment superimposed.

Now for an application of graphing. Suppose we wish to approximate the maximum of the function defined by $f(x) = x^2 \sin(x) - x$ on the interval $[0, 3]$ with an error bound of 0.03. We can define an algebraic expression, and give it a name for later reference:

```
> ex := x ∧ 2 * sin(x) - x;
```

One way to approximate the maximum over $[0, 3]$ is to graph it:

```
> plot(ex,x = 0..3);
```

We obtain the graph:

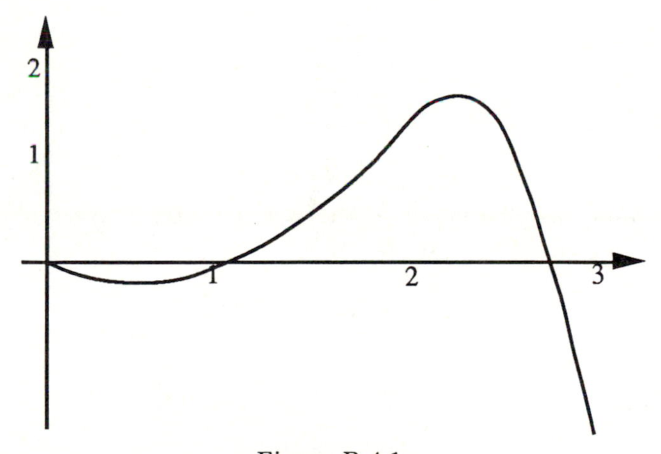

Figure B.4.1

The maximum appears to occur between 2.0 and 2.5 with a value between 1.5 and 2.0. We can replot with a smaller range to get a more accurate estimate:

```
> plot(ex,x = 2.0..2.5);
```

giving

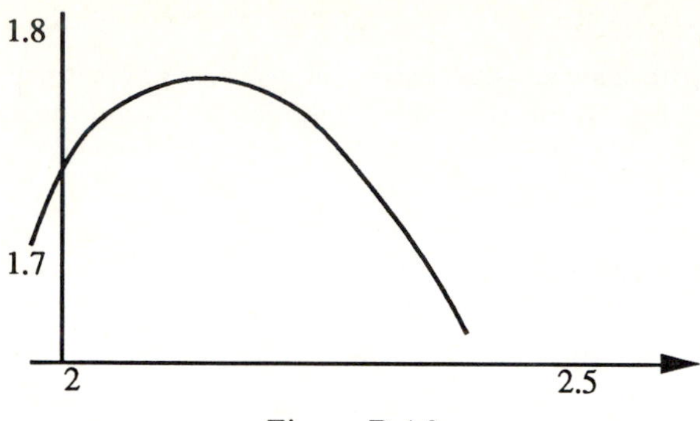

Figure B.4.2

The maximum appears to occur between 2.1 and 2.2 with a value between 1.70 and 1.75. Continuing this process we can get as accurate an estimate as we want for the maximum.

Exercises:

(7) Find the smallest positive root of the function $f(x) = x^2 \sin(x) - x$. [Clearly $f(0) = 0$. We want the next value greater than zero where f is zero.]

Factoring

If we are given a polynomial of degree n, i.e., $a_n x^n + a_{n-1}x^{n-1} + \cdots + a_1 x + a_0$, then we can find the roots if we can factor the polynomial. We are all familiar with the quadratic equation which gives the roots of a second-degree polynomial. There are analogous formulas for the roots of third- and fourth-degree polynomials. However, it is known that no general formula exists for roots of polynomials of degree greater than four. Most computer algebra systems have built-in procedures to attempt factoring, but they do not always succeed. Consider the polynomial $x^5 - 17x^3 + 12x^2 + 52x - 48$. We can give the polynomial a name and try factoring it as before:

```
> poly := x∧5 -17*x∧3 +12*x∧2 + 52*x -48;
> factor(poly,x);
```

$$(x-3)(x-2)(x-1)(x+2)(x+4)$$

If the polynomial cannot be factored, then the graph can be sketched to approximate the roots, as in the preceding section.

Exercises

8. Factor $2x^3 + 5x^2 + 4x + 1$.

9. Factor $6x^6 + 5x^5 + 5x^3 - 6x^2$.

10. Factor $64x^5 + 432/5x^5 - 1956/5x^3 + 1027x^2 - 840x + 225$.

11. Factor $6x^4 - 7/2x^3 + 5/2x^2 - 7/6x + 1/6$.

Solving Equations

We will show how to solve equations with the aid of a CAS, using an example which can easily be done by hand. Again, when first using a CAS it is important to experiment with problems with a known solution.

Suppose we wish to solve the following equation for x in terms of y:

$$\frac{4x^2 - y^2}{2xy + y^2} = 0$$

We can use the CAS to give the expression on the left-hand side a name:

```
> z := (4*x∧ 2 - y∧2)/(2*x*y + y∧2);
```

We will apply the CAS **solve** command to the equation:

```
> solve(z=0,x);
```

$$\frac{y}{2}$$

Is this solution valid for all values of y? No, we see that if we substitute $x = 0$, $y = 0$ into the original expression we get an indeterminate form:

> subs($\{x = 0, y = 0\}, z$);

returns $\frac{0}{0}$. Thus even though the CAS returned the correct symbolic answer, the set of values for the variable y was not specified. *The user must interpret the answer to see when it makes sense.*

One way to eliminate extraneous solutions when the function is a rational function (a ratio of polynomials) is to factor the numerator and denominator first:

```
> factor(z);
```

$$\frac{2x-y}{y}$$

From this we can see that the solution is $y = 2x$ except when $x = y = 0$.

Exercises

Solve $f(x) = 4$ in Exercises 12 through 14 by doing the following:

(a) Remove common factors by factoring the rational function f

(b) Approximate, by superimposing the graphs of f and $y = 4$, where the graph of f crosses the line $y = 4$

(c) Check your answer in (b) by using the `solve` command.

12. $f(x) = (x^3 - 3/2x^2 - x)/(x^2 + x - 6)$
13. $f(x) = (x^4 + 3x^3 + 2x^2)/(x^2 + 3x + 2)$
14. $f(x) = (x^3 + 1/3x^2 + 5/9x + 1/9)/(x^5 - x)$

Limits

A CAS can take limits. For example, to find the limit of $f(x) = x^2 + 1$ as x approaches 1, we enter

```
> limit(x∧2+1,x=1);
```

which returns 2.

```
> limit(sin(x)/x,x=0);
```

returns 1. The CAS probably uses L'Hospital's Rule (see Section 3.9) for this type of problem.

Exercises

Find the limits of the following functions.

15. $f(x) = [x^2 \sin(x) - 7x]/[x \cos(x)]$ at $x = 0$
16. $f(x) = x \sin(1/x)$ at $x = 0$

Differentiation

A CAS can easily compute derivatives, since only the derivatives of a few basic functions (sin, cos, log, exp) and rules (for sums, products, ratios and compositions) are needed. For example, we can enter

```
> diff(x∧2*sin(exp(x)),x);
```

and obtain

$$2x \sin(\exp(x)) + x^2 \cos(\exp(x)) \exp(x)$$

We can differentiate with respect to several variables:

```
>diff(cos(y∧2+x*y),x,y);
```

returns

$$-\cos(y^2 + xy)(2y + x)y - \sin(y^2 + xy)$$

Lagrange multiplier problems are usually difficult and time consuming because they involve setting up the problem; performing differentiations to obtain a system of equations; solving the system of equations for potential extrema;

and examining points for extrema. The major difficulties occur in the third step. Since, in general, the system of equations is nonlinear, even solving a linear system is time-consuming. This means that only a few examples can be given, carefully selected so that the resulting system of equations is easy to solve. With a CAS, students can focus on the essential steps in the process.

Lagrange Multipliers Example

Suppose we want to find the extrema of the function

$$f(x,y,z,t) = x^2 + 2y^2 + z^2 + t^2$$

subject to the conditions

$$x + 3y - z + t = 2 \quad \text{and} \quad 2x - y + z + 2t = 4$$

From geometric considerations, we expect a single minimum since the conditions define a plane in $\mathbb{R}^4$ and the level sets of f are ellipses in the plane. Using a CAS, we would define the expressions:

```
> f := x∧2 + 2*y∧2 + z∧2 + t∧2;
> g1 := x + 3*y - z + t - 2;
> g2 := 2*x - y + z + 2*t - 4;
```

The steps outlined above are carried out in the following manner.

1. Set up the problem:

```
> h := f + lambda1*g1 + lambda2*g2;
```

2. For each variable, compute the partial derivative:

```
> dhx := diff(h,x);
> dhy := diff(h,y);
> dhz := diff(h,z);
> dht := diff(h,t);
> dhl1 := diff(h,lambda1);
> dhl2 := diff(h,lambda2);
```

3. Solve the six linear equations in six unknowns:

```
> solve({dhx=0,dhy=0,dhz=0,dht=0,dhl1=0,
> dhl2=0},{x,y,t,lambda1,lambda2});
```

The CAS returns the exact solution(s) in a list:

$$\{x = \frac{67}{69}, y = \frac{2}{23}, z = \frac{14}{69}, t = \frac{67}{69},$$

$$lambda1 = -\frac{26}{69}, lambda2 = \frac{18}{23}\}.$$

4. Evaluate f at the point obtained in step 3:

```
> subs({x=67/69,y=2/23,z=14/69,t=67/69},f);
```

which returns 134/69. Thus, we have the minimum value of f subject to the given constraints.

Exercises

17. Find the minimum of the function

$$f(x,y,z,t) = x^2 + y^2 + z^2 + t^2$$

subject to the conditions

$$x + y - z + 2t = 2 \text{ and } 2x - y + z + 3t = 3$$

18. Find the minimum of the function

$$f(x,y,z,t) = 2x^2 + y^2 + z^2 + 2t^2$$

subject to the conditions

$$x + y + z - t = 1, \ 2x + y - z + 2t = 2, \text{ and } x - y + z - t = 4$$

Integration

Integration is harder for Computer Algebra Systems as well as for people since there are no simple rules which can produce an antiderivative even if it exists. However, there capabilities are comparable to those of experienced people, although they sometimes fail when people would succeed. For example, suppose we want to find an antiderivative of $f(x) = 2x\cos(x^2+x)+\cos(x^2+x)$. Many Computer Algebra Systems will try to integrate each term separately, and probably fail. However an observant person would factor before attempting to integrate:

```
>int(factor(2*x*cos(x∧2+x)+cos(x∧2+x)),x);
```

returns $\sin(x^2+x)$. (Note that the functions of a CAS can be composed, just as functions from $\mathbb{R}$ to $\mathbb{R}$.) Most nice functions do not have a nice antiderivative, so numerical integration is often necessary (see Section 5.4). A CAS will often have a command to perform numerical integration. If none is available, there may be a **sum** function. For example, to compute a Riemann Sum for $f(x) = e^{\sin(x)}$ on the interval $[0,1]$ with $n = 10$ subdivisions and evaluating at the right-hand endpoints of the subinterval, we need to compute $\sum_{k=1}^{10} f(k/10) \cdot 1/10$. Using the **sum** function we would enter:

```
>sum(exp(sin(k/10))/10, k=1..10);
```

Exercises

19. Using a built-in numerical integration function or the `sum` function, approximate $\int_0^1 e^{\sin(x)}dx$ to within 0.01.

20. Using a built-in numerical integration function or the `sum` function, approximate $\int_0^1 \sin(\cos(x))dx$ to within 0.01.

Appendix C

ANSWERS TO QUESTIONS

Chapter 1

(1.2.9) We need two conditions to hold for r to be the lub for a set:
(1) $r \geq s$ for every number s in the set S,
(2) If t is any upper bound for S (i.e., $t \geq s$ for every number s in the set S) then $t \geq r$.

(1.2.13) The Cartesian product $\{x, y\} \times \{z\} = \{(x, z), (y, z)\}$.

(1.2.34) The distance between (1,2,3,4) and (0,0,0,0) is $[1^2 + 2^2 + 3^2 + 4^2]^{1/2} = [1 + 4 + 9 + 16]^{1/2} = \sqrt{30} \approx 5.48$.

(1.4.11) (a) This correspondence is a one-to-one and onto function. The inverse function has graph $\{(c, 1), (d, 2), (b, 3), (a, 4)\}$.

(b) This correspondence is a many-to-one function. It is not one-to-one since both 1 and 2 in the domain correspond to a in the range. It is not onto since no element in the domain corresponds to b in the target.

(c) This correspondence is not a function since 1 in the domain corresponds to both a and c in the range.

(1.4.22) Let the function have a finite set as domain, or consider the inverse tangent function.

(1.5.17) (a) We can obtain a new theorem by replacing "increasing" by "decreasing" where *indicated,* giving:

Theorem C.1 *If* $f : \mathbb{R} \to \mathbb{R}$ *is strictly* decreasing, *then* f *is one-to-one.*

(b) Again we can change the proof by replacing "increasing" by "decreasing" and "<" by ">" where *indicated,* giving:

Proof: What we need to show is that if $f(x) = f(y)$ for some x and y in Domain(f), then $x = y$. That is, if two points, x and y, are mapped to the same point, then x and y must actually be identical. Since x and y are real numbers one of the relations $x = y$, $x < y$, or $x > y$ must hold. (This fact is called the *Law of Trichotomy.*) If $x < y$, then by the definition of "strictly *decreasing,*" $f(x) > f(y)$. This is impossible, since we have assumed that x and y map to the same point, $f(x) = f(y)$. Similarly, if $y < x$ then $f(y) > f(x)$ which is also impossible. Thus the only possibility is $x = y$, which is what we wanted to show. □

A second approach to the proof would be to utilize the existing Theorem 1.5.16 and note that f is strictly decreasing if and only if $-f$, defined by $(-f)(x) = -f(x)$, is strictly increasing.

Proof: We want to show that if f is strictly decreasing then f is one-to-one. Suppose $f(x) = f(y)$. We need to show that $x = y$. Since $f(x) = f(y)$, multiplying by -1 we have $-f(x) = -f(y)$. Since $-f$ is one-to-one, this implies that $x = y$. □

(1.5.27) (a) Since cosine is strictly decreasing between $\theta = 0$ and $\theta = \pi$, we can obtain an inverse function $\arccos : [-1, 1] \to [0, \pi]$.

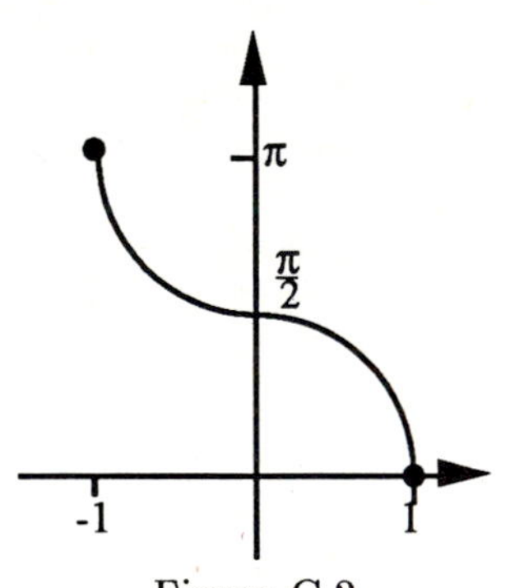

Figure C.2

(b) $\cos(0) = 1$ so $\arccos(1) = 0$; $\cos(\pi/4) = 1/\sqrt{2}$ so $\arccos(1/\sqrt{2}) = \pi/4$; $\cos(0) = 1$ so $\arccos(1) = 0$; $\cos(\pi) = -1$, so $\arccos(-1) = \pi$.

(1.6.11) The sketch of a curve passing through the given data points, that reflects the periodic nature of the change in hours of sunlight, suggests using the sine function as the basic model. Thus we transform the relation $f(t) = \sin(t)$ through shifting and stretching to "fit" the given data. Since the vertical distance is 6 (from 9 to 15), we stretch vertically by a factor of

3 replacing $\sin(t)$ by $3\sin(t)$. Since we want a period of one year, 365 days, we stretch horizontally by multiplying t by $2\pi/365$. We now shift the origin to $(80, 12)$ by replacing t by $t-80$ and adding 12. This yields the transformed relation $g(t) = 3\sin(2\pi(t-80)/360)+12$. The desired function is then $g : [0, 365] \to [9, 12]$ defined by $g(t) = 3\sin(2\pi(t-80)/365) + 12$.

Chapter 2

(2.1.3) Prove that $\sqrt{3}$ is irrational.

The proof of the irrationality of $\sqrt{2}$ is mimicked.

Suppose $\sqrt{3}$ is written in the reduced form $\sqrt{3} = a/b$. Squaring, we obtain $a^2 = 3b^2$. Since a^2 is 3 times b^2, a^2, and hence a, is a multiple of 3. Letting $a^2 = 3p$, we get

$$9p^2 = 3b^2 \quad \text{or} \quad 3p^2 = b^2$$

We must conclude that b^2, and hence b, is a multiple of 3. However, a and b are to have no common factors! We must conclude that $\sqrt{3}$ is irrational.

Chapter 3

(3.4.6) (a) Yes, fg is differentiable. The differentiability at $x = 0$ is established by computing the limit as in the definition of the derivative. Since g is not differentiable at 0, the hypothesis of theorem 3.4.1 is not satisfied, so the Theorem does not apply. Thus there is no contradiction.

(b) Yes, $f \circ g(x) = (|x|)^2 = x^2$ is differentiable. Since g is not differentiable, the hypothesis of the Chain Rule is not satisfied and thus the Chain Rule does not apply. Therefore no contradiction exists.

(3.5.3) $f(x) = \sin(x)$, $f'(x) = \cos(x)$, $f''(x) = -\sin(x)$, $f^{(3)}(x) = -\cos(x)$, and $f^{(4)}(x) = \sin(x)$

(3.8.20) The derivative f' vanishes at a, c, and e, so these are the critical points. Since f' is increasing near c, f has a local minimum at c. (See Figure 3.8.19.) Similarly, f' is decreasing near a, so f has a local maximum at a. Since $f' \geq 0$ near e, the graph of f does not turn around, so f has neither a minimum nor a maximum at e.

(3.8.28) $f(x) = 13(x-3)^4 + 117x - 32$. Since $f''(x) = 156(x-3)^2 > 0$ for x not equal to 3, $x = 3$ is an inflection point but the concavity of the graph does not change at this point.

Chapter 4

(4.1.4) Translate the vector with tail at $P = (-1,3)$ and head at $Q = (2,0)$ to the origin by reducing the x-components by -1 and the y-components by 3. The position of the tail becomes $(-1-(-1), 3-3) = (0,0)$ and the position of the head becomes $(2-(-1), 0-3) = (3,-3)$. The coordinate representation of $\vec{PQ}$ is $(3,-3)$.

(4.1.7) Translate the vector $\vec{PQ}$, with $P = (1,2)$ and $Q = (4,9)$, to the origin obtaining $\vec{V} = (3,7)$. The slope is

$$\frac{\Delta y}{\Delta x} = \frac{7}{3}$$

[$= (9-2)/(4-1)$ using the original vector $\vec{PQ}$].

(4.1.11) Translate $\vec{PQ}$, $P = (-1,1)$ and $Q = (-3,6)$, to the origin obtaining $\vec{V} = (-3-(-1), 6-1) = (-2,5)$. $-4\vec{V} = -4(-2,5) = (8,-20)$.

(4.1.14) (a) $\vec{A}$ and $-\vec{A}$ have the same length and the opposite direction.

$$|-\vec{A}| = |-1||\vec{A}| = |\vec{A}|$$

If $\vec{A} = (a_1, a_2)$ then $-\vec{A} = (-a_1, -a_2)$.

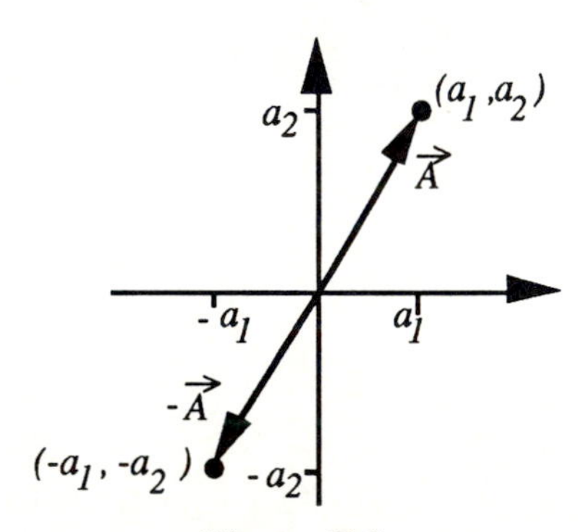

Figure C.3

(b) Let $\vec{A} = (a_1, a_2)$ and $\vec{B} = (b_1, b_2)$. $\vec{B} - \vec{A} = (b_1 - a_1, b_2 - a_2)$ is the vector with tail at $A = (a_1, a_2)$ and head at $\vec{B} = (b_1, b_2)$.

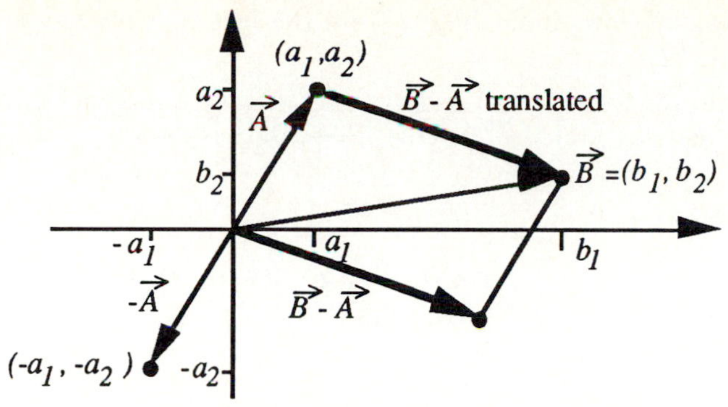

Figure C.4

(4.1.18) Let $\vec{V} = (v_1, v_2, \ldots v_n)$ be any vector with n components.
(a) Then

$$\begin{aligned} r(s\vec{V}) &= r(sv_1, sv_2, \ldots, sv_n) = (rsv_1, rsv_2, \ldots, rsv_n) \\ &= (rs)(v_1, v_2, \ldots, v_n) = (rs)\vec{V} \end{aligned}$$

(b) If $|\vec{V}| = 0$, then

$$0 = |\vec{V}|^2 = [v_1^2 + v_2^2 + \cdots + v_n^2]$$

If the sum of squares is zero, then each term must be zero. Thus each $v_i = 0$ for $i = 1, \ldots, n$. Thus

$$\vec{V} = (v_1, v_2, \ldots, v_n) = (0, 0, \ldots, 0) = \vec{O}$$

(4.1.21) Let $\vec{V}$ be any nonzero vector. Then $|\vec{V}| \neq 0$ [by Question 4.1.18 (b)]. We can multiply $\vec{V}$ by $1/|\vec{V}|$ to obtain $\vec{V}/|\vec{V}|$. The length of $\vec{V}/|\vec{V}|$ is $|\vec{V}|/|\vec{V}| = 1$.

(4.1.26) Let $\vec{A} = (a_1, a_2, \ldots, a_n)$ and $\vec{B} = (b_1, b_2, \ldots, b_n)$ be two vectors in $\mathbb{R}^n$. Then

$$\begin{aligned} \vec{A}\bullet\vec{B} &= (a_1, a_2, \ldots, a_n)\bullet(b_1, b_2, \ldots, b_n) \\ &= \sum_{i=1}^{n} a_i b_i = \sum_{i=1}^{n} b_i a_i \\ &= \vec{B}\bullet\vec{A} \end{aligned}$$

(4.1.27) This result is proved by applying the symmetry property and distributive property of Theorem 4.1.25 twice.

$$\begin{aligned} (\vec{A}+\vec{B})\bullet(\vec{A}+\vec{B}) &= \vec{A}\bullet(\vec{A}+\vec{B}) + \vec{B}\bullet(\vec{A}+\vec{B}) \\ &= (\vec{A}+\vec{B})\bullet\vec{A} + (\vec{A}+\vec{B})\bullet\vec{B} \\ &= \vec{A}\bullet\vec{A} + \vec{B}\bullet\vec{A} + \vec{A}\bullet\vec{B} + \vec{B}\bullet\vec{B} \\ &= \vec{A}\bullet\vec{A} + 2\vec{A}\bullet\vec{B} + \vec{B}\bullet\vec{B} \end{aligned}$$

(4.1.30) This result follows immediately from the fact that for any θ, $|\cos(\theta)| \leq 1$.

$$|\vec{A}\cdot\vec{B}| = |\vec{A}||\vec{B}||\cos(\theta)| \leq |\vec{A}||\vec{B}|$$

(4.2.4)

$$\begin{aligned} h(t) &= f(t)\cdot g(t) = (t, t^2, 3t)\cdot(\sin(t), 2t, 4) \\ &= t\sin(t) + 2t^3 + 12t = t(\sin(t) + 2t^2 + 12) \end{aligned}$$

(4.2.14) We can compare the graph to the standard parabola parallel to the x-axis, $x = y^2$, by completing the square:

$$x = 6 - 2y^2 - y \Longrightarrow x - \frac{49}{8} = -2\left(y + \frac{1}{4}\right)^2$$

Thus the vertex has been moved from $(0,0)$ to $(\frac{49}{8}, -\frac{1}{4})$. The factor -2 reverses the direction and stretches it:

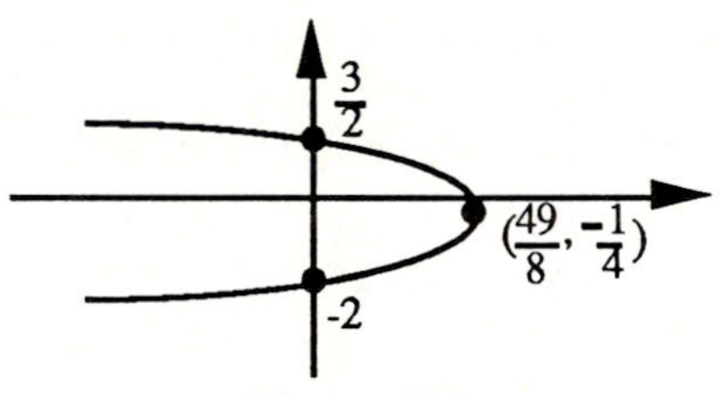

Figure C.5

(4.2.28) The velocity of the particle at $f(t) = (t^2, 3t, \sin(t))$ is $f'(t) = (2t, 3, \cos(t))$. The acceleration of the particle at $f(t)$ is $f''(t) = (2, 0, -\sin(t))$. The velocity and acceleration at $t = 1$ are $f'(1) = (2, 3, \cos(1))$ and $f''(1) = (2, 0, -\sin(1))$.

(4.2.31) Let $h : \mathbb{R} \to \mathbb{R}$ and $f : \mathbb{R} \to \mathbb{R}^n$ with $f(t) = (f_1(t), f_2(t), \ldots, f_n(t))$, h differentiable at t and f differentiable at $h(t)$. Then

$$(f \circ h)(t) = (f_1 \circ h(t), \ldots, f_n \circ h(t))$$

So

$$\begin{aligned} (f \circ h)'(t) &= ((f_1 \circ h)'(t), \ldots, (f_n \circ h)'(t)) \\ &= (f_1'(h(t))h'(t), \ldots, f_n'(h(t))h'(t)) \\ &= (f_1'(h(t)), \ldots, f_n'(h(t)))h'(t) \\ &= f'(h(t))h'(t) \end{aligned}$$

(4.2.38) A point (x, y, z) is on the desired line if and only if the vector from $(1,2,1)$ to (x, y, z) is some multiple t of the vector $(3,2,-1)$. This gives the relation: $(x, y, z) - (1,2,1) = t(3,2,-1)$ for some t. Thus the function $f(t) = (1,2,1) + t(3,2,-1)$ has the correct range.

(4.2.39) $f(t) = (3t, t^2, t+1)$ has velocity $f'(t) = (3, 2t, 1)$. $|f'(t)| = [3^2+4t^2+1^2]^{1/2} = [10+4t^2]^{1/2}$. The speed when $t = 1$ is $|f'(1)| = [10+4]^{1/2} = \sqrt{14} \approx 3.74$. The unit tangent at $t = 1$ is $f'(1)/|f'(1)| \approx (0.80, 0.53, 0.27)$.

(4.4.20)

$$\begin{aligned} D_2h(1,2,3) &= g'(4,5)\bullet(D_2u(1,2,3), D_2v(1,2,3)) = (1,-4)\bullet(-2,-3) \\ &= 1(-2) + (-4)(-3) = 10 \end{aligned}$$

(4.5.23) The graph of f is given by the zero set of $g(x, y, z) = x^2 + 2xy + y - z$. A vector perpendicular to this surface is $g'(x, y, z) = \nabla g(x, y, z) = (2x + 2y, 2x + 1, -1)$. If $x = 1$ and $y = 0$ then $z = 1$ and the normal vector is $g'(1,0,1) = (2,3,-1)$. A point (x, y, z) is on the tangent plane if and only if the vector from $(1,0,1)$ to (x, y, z) is perpendicular to the normal vector $g(1,0,1) = (2,3,-1)$. Thus (x, y, z) is on the tangent plane if and only if

$$[(x, y, z) - (1,0,1)]\bullet(2,3,-1) = 0$$

This gives the equation of the tangent plane at (1,0,1) as

$$2(x-1) + 3y - 1(z-1) = 0$$

or

$$2x + 3y - z = 1$$

(4.6.1) A local minimum occurs at p if for all x near p, $f(x) \geq f(p)$.

(4.7.6) We want to minimize

$$f(x, y, z) = x^2 + y^2 + z^2$$

subject to two conditions:

$$g_1(x, y, z) = x + y + z - 1 = 0 \quad \text{and} \quad g_2(x, y, z) = x^2 + y^2 - z^2 = 0$$

Lagrange Multipliers give: $\nabla f(x, y, z) = r\nabla g_1(x, y, z) + s\nabla g_2(x, y, z)$. The gradients are: $\nabla f(x, y, z) = (2x, 2y, 2z)$, $\nabla g_1(x, y, z) = (1,1,1)$, and $\nabla g_2(x, y, z) = (2x, 2y, -2z)$.
Equating components gives five unknowns x, y, z, r, s and five equations:

1. $2x = r + s2x$
2. $2y = r + s2y$
3. $2z = r - s2z$
4. $x + y + z = 1$
5. $x^2 + y^2 - z^2 = 0$

Solving the first three equations for x, y, and z, respectively, we have:

1′. $x = r/(2-2s)$
2′. $y = r/(2-2s)$
3′. $z = r/(2+2s)$

The first two only make sense if $s \neq 1$. If $s = 1$, then by equation 2, $r = 0$. From equation 3, $z = 0$. If $z = 0$, then from equation 5, $x = y = 0$ which contradicts equation 4.

Thus $s \neq 1$. The right-hand sides of eequations 1′ and 2′ are equal, so $x = y$. Substituting $x = y$ into equations 4 and 5 gives $2x + z = 1$ and $2x^2 = z^2$; or $2x^2 = z^2 = (1-2x)^2 = 1 - 4x + 4x^2$. Thus $2x^2 - 4x + 1 = 0$ and $x_1 = [2+\sqrt{2}]/2$ or $x_2 = [2-\sqrt{2}]/2$. If $x = x_1$, then $y = x_1$ and $z = 1 - x - y = -1 - \sqrt{2}$. If $x = x_2$, then $y = x_2$ and $z = 1 - x - y = -1 + \sqrt{2}$. We thus have two solutions:

$$\begin{aligned} x = y &= [2+\sqrt{2}]/2, \quad z = -1-\sqrt{2} \\ x = y &= [2-\sqrt{2}]/2, \quad z = -1+\sqrt{2} \end{aligned}$$

With the first solution, $x^2 + y^2 + z^2 = 6 + 4\sqrt{2}$; with the second solution, $x^2 + y^2 + z^2 = 6 - 4\sqrt{2}$. The second solution gives the minimum distance.

Chapter 5

(5.1.2) $\{2, 2.8284, 3.0614, 3.1214, 3.1365, 3.1403, 3.1413\}$
$\{4, 3.3137, 3.1825, 3.3152, 3.1441, 3.1422, 3.1417\}$

(5.1.3) Upper Approx. 5,375,000; Lower Approx. 4,825,000.
Error < Upper Approx. − Lower Approx. = 550,000.

(5.4.22) If $f(x) = \sin(x^3)$, then

$$f^{(2)}(x) = -9x^4 \sin(x^3) + 6x \cos(x^3)$$

and

$$f^{(4)}(x) = 81x^8 \sin(x^3) - 324x^5 \cos(x^3) - 180x^2 \sin(x^3)$$

Thus $|f^{(2)}(x)| \leq 9 + 6 = 15$ and $|f^{(4)}(x)| \leq 81 + 324 + 180 = 585$ on [0,1]. Using the error estimate for the Trapezoidal Rule, we want $15 \cdot 1/12n^2 \leq 0.01$ or $n \geq 12$. Using the error estimate for Simpson's Rule, we want $585 \cdot 1/180n^4 < 0.01$ or $n \geq 4.2$ (or $n \geq 6$ since n must be even).

(5.5.8) We will divide $[-1,4]$ into subintervals on which FLOOR is constant and apply Theorem 5.5.7, giving

$$\begin{aligned}
\int_{-1}^{4} FLOOR(t)dt &= \int_{-1}^{0} FLOOR(t)dt + \int_{0}^{1} FLOOR(t)dt \\
&\quad + \int_{1}^{2} FLOOR(t)dt + \int_{2}^{3} FLOOR(t)dt \\
&\quad + \int_{3}^{4} FLOOR(t)dt \\
&= \int_{-1}^{0} -1dt + \int_{0}^{1} 0dt + \int_{1}^{2} 1dt + \int_{2}^{3} 2dt \\
&\quad + \int_{3}^{4} 3dt \\
&= -1+0+1+2+3=5
\end{aligned}$$

(5.5.12) No. If the average value of f is greater than or equal to the average value of g on $[a,b]$, then $1/(b-a)\int_a^b f(x)dx \geq 1/(b-a)\int_a^b g(x)dx$ which implies that $\int_a^b [f(x)-g(x)]dx \geq 0$. But this does not imply that $f(x)-g(x) \geq 0$ for all x. Consider the example when $f,g:[0,1] \rightarrow \mathbb{R}$, $f(x)=1-x$ and $g(x)=x$.

(5.6.13) We can parameterize the graph of f by $g(t)=(t,f(t))$ for t in $[a,b]$. Then the length of the graph is

$$\int_a^b [(t')^2+(f'(t))^2]^{1/2}dt = \int_a^b [1+(f'(t))^2]^{1/2}dt.$$

(5.6.29) We will use disks with holes, integrating with respect to x. The volume ΔV_k of a disk with a hole is the volume of the outer disk of radius r_2 minus the volume of the inner disk of radius r_1.

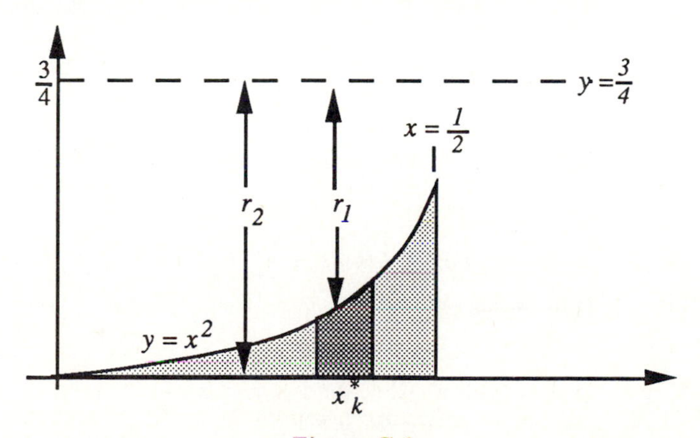

Figure C.6

Thus,

$$\Delta V_k \approx \pi r_2^2 \Delta x_k - \pi r_1^2 \Delta x_k \approx \left[\pi\left(\frac{3}{4}\right)^2 - \pi\left(\frac{3}{4} - y\right)^2\right] \Delta x_k.$$

Writing all expressions in terms of x we have

$$\Delta V_k \approx [\pi(\frac{3}{4})^2 - \pi(\frac{3}{4} - x^2)^2]\Delta x_k$$

Summing and taking the limit we have

$$V = \pi \int_0^{\frac{1}{2}} \left[\left(\frac{3}{4}\right)^2 - \left(\frac{3}{4} - x^2\right)^2\right] dx$$

Chapter 6

(6.1.10) From the definition of the FLOOR function (see Example 1.5.10) we have the following facts:

$$FLOOR(t) = \begin{cases} -1 & \text{for } -1 \le t < 0 \\ 0 & \text{for } 0 \le t < 1 \\ 1 & \text{for } 1 \le t < 2 \\ 2 & \text{for } 2 \le t < 3 \\ 3 & \text{for } 3 \le t < 4 \end{cases}$$

We shall compute $F(x)$ over each of the unit intervals separately and then construct the graph.

For (0,1): $F(x) = \int_{-1}^{x} FLOOR(t)dt = \int_{-1}^{0}(-1)dt + \int_0^x (0)dt = -1.$

For [1,2): $F(x) = \int_{-1}^{x} FLOOR(t)dt = \int_{-1}^{0}(-1)dt + \int_0^1 (0)dt + \int_1^x (1)dt = -1 + 0 + x - 1 = x - 2.$

For [2,3): $F(x) = \int_{-1}^{x} FLOOR(t)dt = \int_{-1}^{0}(-1)dt + \int_0^1 (0)dt + \int_1^2 (1)dt + \int_2^x (2)dt = 2x - 4.$

For [3,4): $F(x) = \int_{-1}^{x} FLOOR(t)dt = \int_{-1}^{0}(-1)dt + \int_0^1 (0)dt + \int_1^2 (1)dt + \int_2^3 (2)dt + \int_3^x (3)dt = 3x - 7.$

Therefore, $$F(x) = \begin{cases} -1 & \text{for } 0 < x < 1 \\ x - 2 & \text{for } 1 \le x < 2 \\ 2x - 4 & \text{for } 2 \le x < 3 \\ 3x - 7 & \text{for } 3 \le x < 4 \end{cases}$$

Graph of $F(x)$

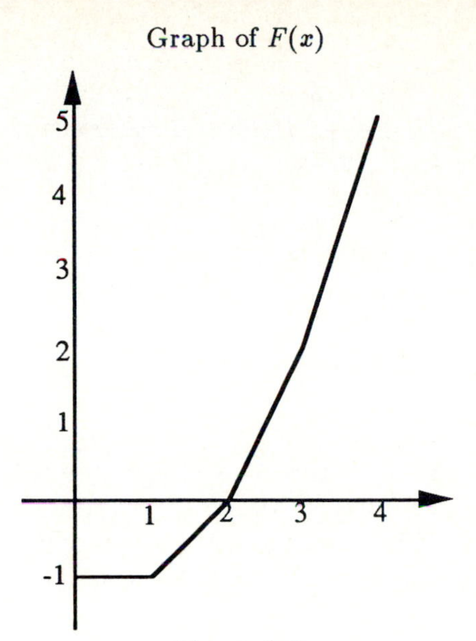

Figure C.7

(6.1.12) We need to compute $\frac{1}{6}\int_{-2}^{4}|x-2|dx$. We remove the absolute value sign by decomposing the integral into two parts, where $|x-2|=-(x-2)$ (on $[-2,2]$) and where $|x-2|=(x-2)$ (on $[2,4]$):

$$\int_{-2}^{4}|x-2|dx=\int_{-2}^{2}-(x-2)dx+\int_{2}^{4}(x-2)dx$$

Now we can apply the Fundamental Theorem of Calculus, Part II, since $x^2/2-2x$ is an antiderivative of $x-2$:

$$\int_{-2}^{4}|x-2|dx=-(x^2/2-2x)]_{-2}^{2}+\left(\frac{x^2}{2}-2x\right)]_{-2}^{2}=8+2=10$$

Thus the average value is $\frac{10}{6}$.

(6.1.15) The constant of integration plays no role. If a constant is added to a function, the graph is shifted vertically. Thus the interval where a function is increasing does not change if a constant is added to a function.

(6.1.18) *Proof.* Let c be a fixed point in the domain of f and let x be any point in the domain of f. Define the function F by $F(x)=\int_c^x f(t)dt$. In order to show that F is continuous, we let a be an arbitrary point in the domain of f and show that F is continuous at $x=a$. That is, we want $\lim_{x\to a}F(x)=F(a)$, or, $\lim_{x\to a}[F(x)-F(a)]=0$.

Now,

$$F(x) - F(a) = \int_c^x f(t)dt - \int_c^a f(t)dt = \int_a^x f(t)dt$$

Thus, we want to show that $\lim_{x \to a} \int_a^x f(t)dt = 0$. Note, that since f is integrable, f is bounded. Thus, there exists numbers m and M such that $n \leq f(t) \leq M$ for all t in the domain of f. Hence,

$$m(x-a) = \int_a^x mdt < \int_a^x f(t)dt < \int_a^x Mdt = M(x-a)$$

Now, applying the Squeeze Theorem as $x \to a$, we have $\lim_{x \to a} \int_a^x f(t)dt = 0$. □

[Note that when f is continuous, Part I of the Fundamental Theorem of Calculus says the integral $\int_a^x f(t)dt$ is differentiable and hence, continuous.]

(6.3.10) There are two special cases in which there are general algorithms for evaluating integrals of the form $\int \sin^m(x)\cos^n(x)dx$.

Case 1: m or n is an odd positive integer. Let $n = 2p + 1$ and let m be an arbitrary rational number, $m \neq -1$. (This restriction will be removed in Chapter 7 to allow m to be any real number.) Using the transformation $\cos^2(x) = 1 - \sin^2(x)$, the integrand may be written as:

$$\begin{aligned} \sin^m(x)\cos^{2p+1}(x) &= \sin^m(x)[\cos^2(x)]^p\cos(x) \\ &= \sin^m(x)[1 - \sin^2(x)]^p\cos(x) \end{aligned}$$

The right-hand side can be expanded into a linear combination of terms of the form $c_k \sin^k(x)\cos(x)$ and each term can then be transformed into a standard form by applying the transformation $u = \sin(x)$. Similar results are obtained if m is an odd integer.

Case 2: m and n are both even positive integers. Using the transformations

$$\sin^2(x) = \frac{1 - \cos(2x)}{2} \quad \text{and} \quad \cos^2(x) = \frac{1 + \cos(2x)}{2}$$

the integrand, $\sin^m(x)\cos^n(x)$, can be expressed as a linear combination of cosine terms raised to smaller powers. Each term can then be treated by the algorithm in case 1, or further reduced by applications of this case.

(6.3.11) Since all six trigonometric functions can be expressed in terms of sines or cosines, every product of trigonometric functions can be expressed in the form $\sin^m(x)\cos^n(x)$. (Note that m and n are not necessarily positive integers.)

(6.4.8) If $\int_{-\infty}^{c} f(x)dx$ and $\int_{c}^{\infty} f(x)dx$ *both* converge for any real number c, then $\int_{-\infty}^{\infty} f(x)dx$ converges and is defined to be

$$\int_{-\infty}^{\infty} f(x)dx = \int_{-\infty}^{c} f(x)dx + \int_{c}^{\infty} f(x)dx.$$

If either integral on the right-hand side diverges, so does $\int_{-\infty}^{\infty} f(x)dx$.

(6.4.12) For positive values of p, the expression $1/x^p$ is unbounded as $x \to 0$. Thus, $\int_0^1 1/x^p dx$ is an improper integral, and, therefore,

$$\begin{aligned}\int_0^1 \frac{1}{x^p}dx &= \lim_{a\to 0+}\int_a^1 \frac{1}{x^p}dx = \lim_{a\to 0+} \left.\frac{1}{(1-p)x^{p-1}}\right]_a^1 \\ &= \lim_{a\to 0+}\left[\frac{1}{1-p} - \frac{1}{(1-p)a^{p-1}}\right]\end{aligned}$$

This limit is $1/(1-p)$ for $p < 1$ and nonexistent for $p > 1$.

Chapter 7

(7.2.1) Let f satisfy the logarithmic transformation (Eq. 7.1). The reader should keep in mind that f satisfies the property that $f(x^n) = nf(x)$ for any integer n (using properties (5) and (6)). We write $x^{p/q}$ as $(x^{1/q})^p$ and then apply the function f. That is,

$$(**) f(x^{p/q}) = f((x^{1/q})^p) = pf(x^{1/q}).$$

We shall now compute $f(x^{1/q})$ separately and then substitute into (**). Writing $x = x^{q/q} = (x^{1/q})^q$ and then applying f yields

$$f(x) = f((x^{1/q})^q) = qf(x^{1/q})$$

and so $f(x^{1/q}) = 1/qf(x)$.

Now, substituting into (**) yields

$$f(x^{p/q}) = p(f(x^{1/q})) = p(\frac{1}{q})f(x) = \frac{p}{q}f(x).$$

(7.2.5) Trapezoidal Approximation to Log

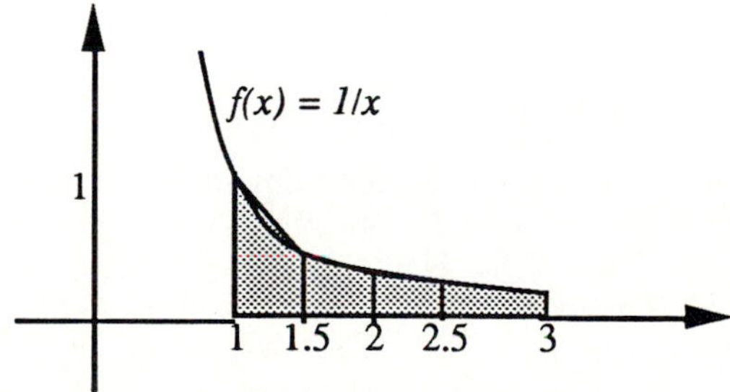

Figure C.8

$$\begin{aligned}\log(3) &= \int_1^3 \frac{1}{t}dt \approx (\frac{1}{2})(\frac{1}{2})[f(1) + 2f(\frac{3}{2}) + 2f(2) + 2f(\frac{5}{2}) + f(3)] \\ &= \frac{1}{4}[1 + 2(\frac{2}{3}) + 2(\frac{1}{2}) + 2(\frac{2}{5}) + \frac{1}{3}] \\ &= \frac{67}{60}.\end{aligned}$$

(7.2.8) Statements (a), (b), and (c) result from making the following substitutions in the appropriate expressions: log for f, $f(x)$ for a, and $g(x)$ for b.
Part a: Apply the above substitutions in (Eq. 7.1).
Part b: Apply the above substitutions in property 5. ($f(a/b) = f(a)-f(b)$)
Part c: Apply the above substitutions in Question ??. ($f(x^{p/q})) = p/qf(x)$)

(7.3.3) The idea is to first apply the logarithmic function to $\exp(x)/\exp(y)$ in order to express the quotient as a difference and then to apply the exp function to the result. That is,

$$\log(\frac{\exp(x)}{\exp(y)}) = \log(\exp(x)) - \log(\exp(y)) = x - y.$$

Thus,

$$\frac{\exp(x)}{\exp(y)} = \exp(\log(\frac{\exp(x)}{\exp(y)})) = \exp(x - y).$$

(7.5.2) Population growth is not normally monotonic. The limiting factors on the growth of some populations annihilate the population entirely, in other populations a wave pattern is produced.

Other equations satisfying assumptions (a) and (b) can be obtained by modifying the differential equation to be $p'(t) = kp^r(t)(C - p^r(t))$, for $r > 0$. The effect of increasing r is to "flatten out" the initial and final stages.

(7.5.4) The exponential function grows faster than the polynomial function since

$$\lim_{x\to\infty} e^x/x^n = \lim_{x\to\infty} e^x/n! = \infty$$

after n applications of L'Hospital's Rule.

A polynomial function grows faster than the logarithmic function since

$$\lim_{x\to\infty} \frac{\log_k(x)}{x} = \lim_{x\to\infty} \frac{1}{x\log(k)} = 0$$

by applying L'Hospital's rule and using the fact that $\log_k(x) = \frac{\log(x)}{\log(k)}$.

(7.5.7) For interest compounded quarterly:

$$\begin{aligned} S &= \$10,000\frac{1.015^{80}-1}{0.06} = \$381,777.13. \\ P &= \$381,777.13(1.015)^{-80} = \$116,018.13. \end{aligned}$$

For interest compounded continuously:

$$\begin{aligned} S &= \$10,000\frac{e^{1.2}-1}{0.06} = \$386,686.15. \\ P &= \$386,686.15e^{-1.2} = \$116,467.63. \end{aligned}$$

Chapter 8

(8.2.12) Three-dimensional analogue of Theorem 8.2.6. If the differentiable function $F = (f_1, f_2, f_3)$ has a potential function, then

$$D_2f_1 = D_1f_2, D_3f_1 = D_1f_3 \text{ and } D_3f_2 = D_2f_3$$

Three-dimensional analogue of Theorem 8.2.8. If f_1, f_2, and f_3 are C^1 functions defined over an open rectangle in $\mathbb{R}^3$ and if

$$D_2f_1 = D_1f_2 \quad D_3f_1 = D_1f_3 \quad \text{and } D_3f_2 = D_2f_3$$

then $F = (f_1, f_2, f_3)$ has a potential function.

(8.3.11) Area $= \int_{-1}^{1}\int_{l(x)}^{u(x)}(1)dydx$, where $l(x) = -\sqrt{1-x^2}$ and $u(x) = -\sqrt{1-x^2}$. So, Area $= 2\int_{-1}^{1}(1-x^2)^{1/2}dx$. Apply the transformation $x = \sin(u)$ and evaluate the indefinite integral.

$$\begin{aligned} Area &= 2\int \cos^2(u)du = 2\int \frac{1+\cos(2u)}{2}du = u + \frac{1}{2}\sin(2u) + C \\ &= u + \sin(u)\cos(u) + C \end{aligned}$$

[The last step used the trigonometric identity $\sin(2u) = 2\sin(u)\cos(u)$.] Next (re)transform to the x variable and evaluate at the limits of integration:

$$\begin{aligned} Area &= \sin^{-1}(x) + x(1-x^2)^{1/2}]_{-1}^{1} = \sin^{-1}(1) - \sin^{-1}(-1) \\ &= \frac{1}{2}\pi + \frac{1}{2}\pi = \pi \end{aligned}$$

Chapter 9

(9.6.5) $\sum_{k=0}^{\infty} 2(0.3)^k$, $\sum_{k=0}^{\infty} 5(2/3)^k$, $\sum_{k=0}^{\infty} 2^k$, $\sum_{k=0}^{\infty} 3(1.1)^k$, $\sum_{k=0}^{\infty}(-1)^k$. $\sum_{k=0}^{\infty} 2k(2/3)^k$ is not geometric.

(9.6.9) $10,000 in activity

(9.7.3) It is geometric with $r = 1/2$.

(9.10.7) (2,4)

Trigonometry Appendix

(A.5) (a) Let $x = y = \pi/2$. Then $\sin(x+y) = \sin(\pi) = 0$ but $\sin(x) + \sin(y) = 1 + 1 = 2$.

(b) With $x = \pi/2$, $\sin(2x) = \sin(\pi) = 0$ but $2\sin(x) = 2$.

Appendix D

ANSWERS TO EXERCISES

Chapter 1

Section 1.2

1.

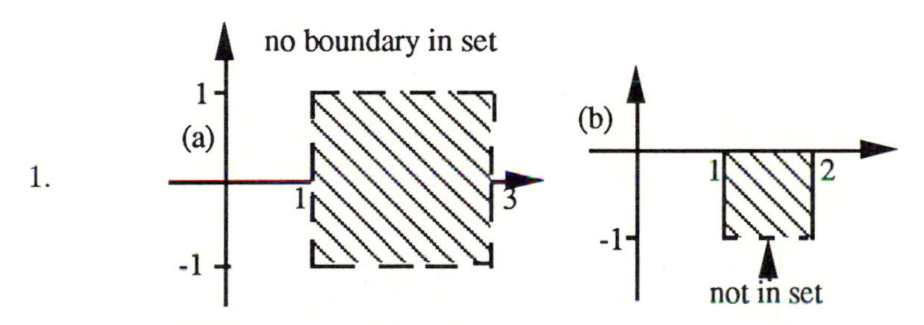

Figure D.1

3. **(a)** $[1,3] \times [1,2]$ **(c)** $[1,3] \times [-1,1]$

4. (a) $\{(x,y) : 0 \leq x \leq 2 \text{ and } 0 \leq y \leq x\}$
   (b) $\{(x,y) : -1 \leq x \leq 1 \text{ and } 0 \leq y \leq x^2\}$

5. (a) T (c) T

6. (a) lub $= 0$, glb $=-3$ (c) glb $= -12$, lub $= 10$ (g) lub $= 5$, glb $= -1$

7. (a) {(1,3), (1,6), (2,3), (2,6)} (c) {(a,a), (a,b), (b,a), (b,b)}
   (e) {(2,1), (2,3), (2,7), (6,1), (6,3), (6,7)}
   (g) $\{(2,1,a), (2,1,b), (2,3,a), (2,3,b), (4,1,a), (4,1,b)\ (4,3,a), (4,3,b)\}$
   (i) $\{(a,b,c,1), (a,b,c,2), (a,b,d,1), (a,b,d,2), (a,b,e,1), (a,b,e,2), (a,c,c,1), (a,c,c,2), (a,c,d,1), (a,c,d,2), (a,c,e,1), (a,c,e,2)\}$

9. (a) $[-1,2] \times [-1,3]$ (b) $(-1,2) \times (-1,3)$

11. (a) 3, 6, 4, 8 (b) kmn, $kmnp$

12. (a) 6 (c) 5 (e) 2

13. (a) {1, 2, 3,...} (c) {1} (e) $\{x : x > 1\}$ (g) $(6, 2, 5, -1)$

Section 1.3

1.

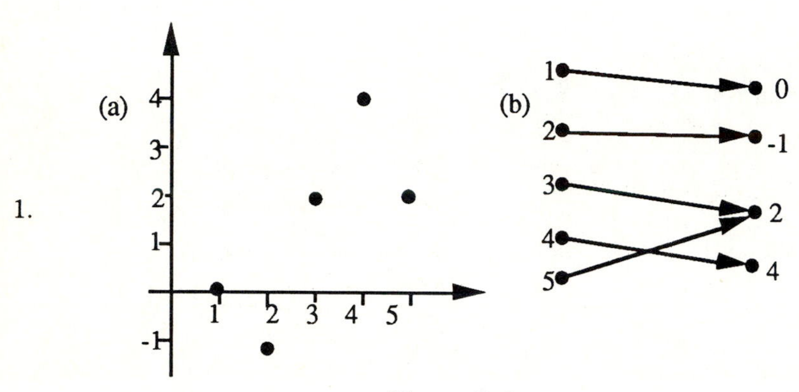

Figure D.2

(c) The domain is {1, 2, 3, 4, 5}. (d) The range is $\{0, -1, 2, 4\}$.

3. (a) Yes (c) No, since 1 corresponds to both d and y.

4. (a) $A = C^2/4\pi, C > 0$ (c) $r = (3V/4\pi)^{1/3}, V > 0$
(e) $S = 4\pi(3V/4\pi)^{2/3}, V > 0$ (h) $A = h(P - 2h)/2, 0 < h < P$

5. (a) $S = 2A/\pi, A \geq 0$ (c) $V = \frac{7}{4}\pi r^2(4 - r), 0 \leq r \leq 4$
(e) $A = hw/2, h > 0$ and $w > 0$

6. (a) 5 (c) -1 (e) $8 - 3x$ (g) $5 - 3y$ (i) $5 - 3y^2$ (k) $5 - 3x - 3y$

7. (a) 1 (c) $8x^3$ (e) $8y^3$ (g) $1 - 3x + 3x^2 - x^3$ (i) y^{12} (k) x^9

9. (a) (1,0) (c) (9,6x) (f) $((x+1)^2, 2(x+1)(y-2))$
(g) $(x^2, 2x^2y)$ (i) (0,0) (k) $(t^4, 4t^3)$

10. (a) 3 (c) $3x^2 + 3hx + h^2$ (e) 0

11. (a) $f : \mathbb{R} \to \mathbb{R}$, Domain$(f) = \mathbb{R} - \{1\}$ (c) $f : \mathbb{R} \to \mathbb{R}^2$, Domain$(f) = \mathbb{R}$.
(e) $f : \mathbb{R}^2 \to \mathbb{R}$, Domain$(f) = \{(x, y) : x \neq 0, y \neq 0\}$

12. (a) $a(n) = \begin{cases} 0 & \text{if } n < 200 \\ 550n & \text{if } n = 200 \\ [550 - 2(n-200)]n & \text{if } 200 < n \leq 400 \end{cases}$

(b) Cost(length,width) = 17·width + 14·length. Domain(Cost) is $(0, \infty) \times (0, \infty)$

Section 1.4

1. (a)

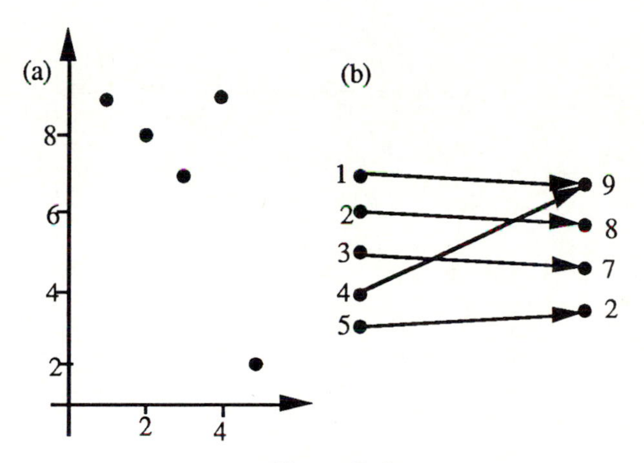

Figure D.3

(c) The domain is {1, 2, 3, 4, 5} (d) The range is {2, 7, 8, 9}

(e) The function is not one-to-one since $f(4) = f(1) = 9$

3. (a) function, one-to-one, onto, {(a,1), (b,2), (c,3), (d,4)}
(c) function, one-to-one, onto, (a,4), (b,3), (c,2), (d,1)

4. Source = $\mathbb{R}^2$, target = $\mathbb{R}$, natural domain = $\mathbb{R}^2$, range = $\mathbb{R}$, not one-to-one, but is onto.

6. Source = $\mathbb{R}$, target = $\mathbb{R}^2$ natural domain = $\mathbb{R}$, range is a parabola, one-to-one, not onto.

7. (a) No. The altitude is $(t-1)^2 + 1$ which is a minimum at $t = 1$. Thus the cow will always descend. (c) $(x_0 - 1, \sqrt[3]{x_0})$

Section 1.5

1. 1, 1, 2, 3, 5, 8, 13, 21, 34, 55 (3) $CEIL(10.1) = 11$.

8. (a) Period is $2\pi/3$ (c) $\sin(x)\cos(x) = \frac{1}{2}\sin(2x)$, so the period is π.
(e) h is identically 1, so it is not periodic.

9. (a) $f(x) = -x/3 + \frac{4}{3}$.

10. (a)

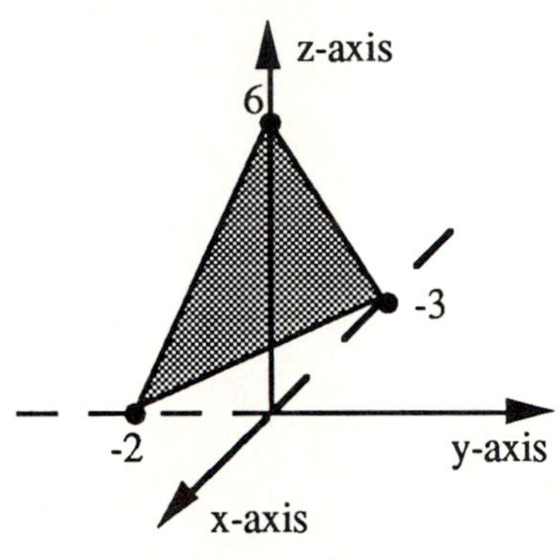

Figure D.4

11. (a) At most twice. (Give an example where it changes direction only once.)

12. (a) $\mathbb{R} - \{-1, -3\}$
(c) The denominator factors to $(x+1)(y-2)$. Thus the function is undefined whenever $x = -1$ or y = 2, which is a pair of intersecting lines in the plane. The natural domain is $\mathbb{R} - \{(x, y) : x = -1 \text{ or } y = 2\}$.

15. (a) e (b) 1

18. No, since f is decreasing for $x < 0$ and increasing for $x > 0$.

19. (a) False. (c) False. (e) True.

20. (a) $f(x) = \begin{cases} -3x/\sqrt{91} & \text{for } -\sqrt{91} \leq x \leq 0 \\ \sin(x) & \text{for } x > 0 \end{cases}$

21. (c) $f(x) = x$ iff $x^2 - 2 = x$ iff $x^2 - x - 2 = 0$ iff $x = -1$ or $x = 2$.

23. The graph of $g(x) = e^x$ will lie between the graphs of $f(x) = 2^x$ and $h(x) = 3^x$.

Section 1.6

1. (a) $f(x) = \sin(2\pi x/5)$.
(c) It is the graph of sine shifted left by 1 unit. The range is $[-1, 1]$.
(e) It is the graph of sine scaled vertically by a factor of 4 and shifted up by 2 units. The range is $[4(-1) + 2, 4(1) + 2] = [-2, 6]$.

2. (a) $\sqrt{3x - 1}$ (c) $\sqrt{2 - x^2}$ (e) $9x + 4$

3. (a) 5 (c) Undefined. (e) Undefined. (g) 5 (i) 4 (k) 2

4. (a) 1 (c) 1 (e) 1

6. (a)

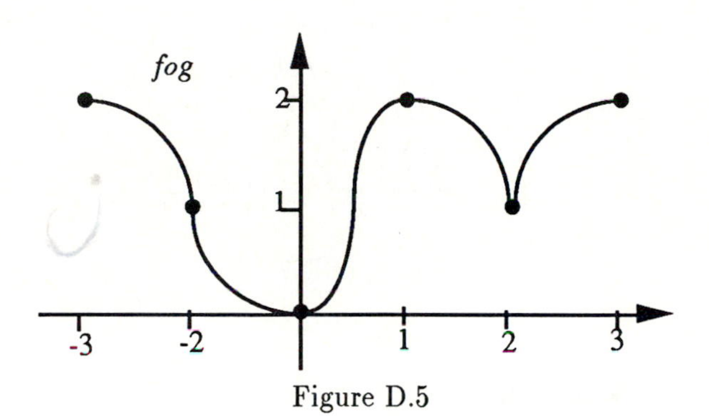

Figure D.5

7. (a)

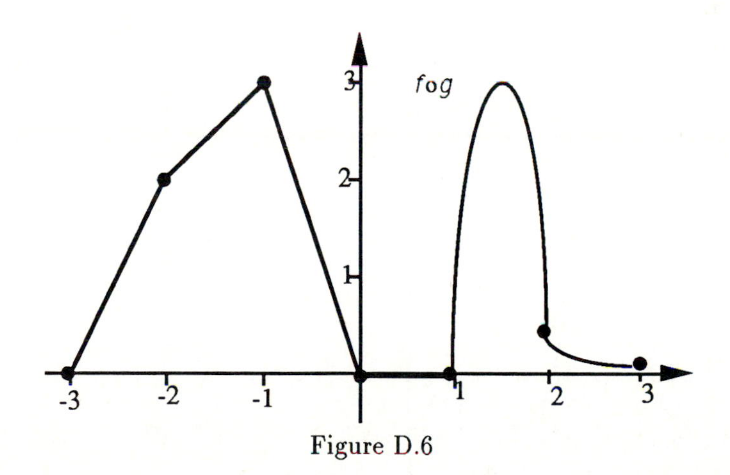

Figure D.6

8. (a) Let the two affine functions be constants.

13.

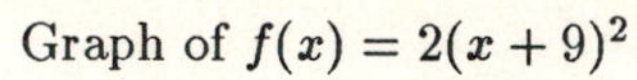

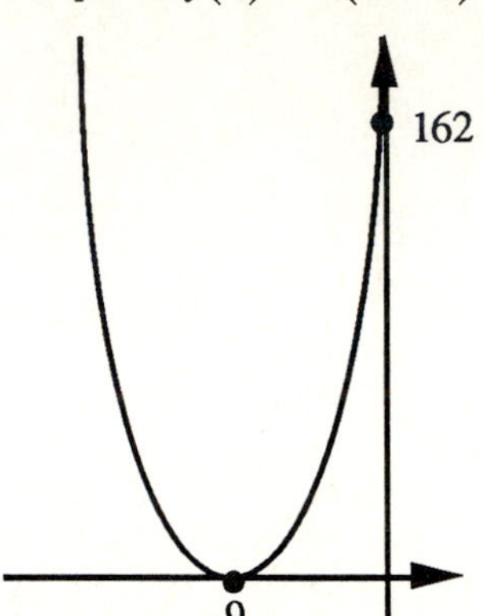

Figure D.7

15.

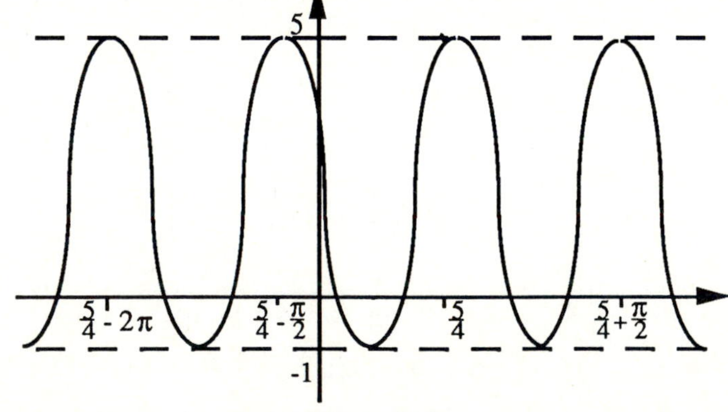

Figure D.8

17.

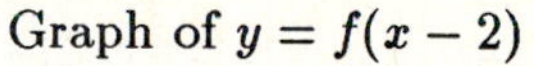

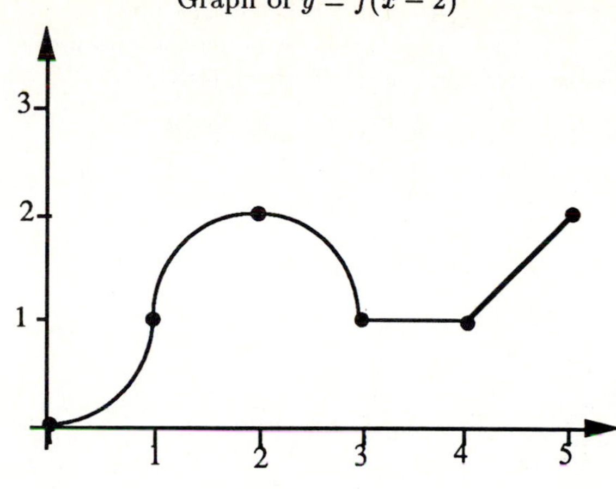

Figure D.9

18. $f(x) = (x-1)^2 + 2$

(21) $f(x) = \sin(\pi x) + 3$

19. $f(x) = (x-1)^2 + 2$

21. $f(x) = \sin(\pi x) + 3$

23. (a) $f(x) = \sin(2\pi x/9) + 1$ (c) Let the long side be along the y-axis, then the depth is $d(x,y) = (1+y/6) - (x-1)^2(1+y/6)$ for (x,y) in $[0,2] \times [0,6]$.

27. Yes, $\{-x+1, x, -x+1, x, \ldots\}$ since $f(x) = -x+1$, $f(f(x)) = x$, $f(f(f(x))) = -x+1$, etc. Thus f composed with itself an even number of times will yield x and f composed with itself an odd number of times will yield $-x+1$. Thus the sequence is periodic with period 2.

Section 2.2

1. (a) $-1, 0, \frac{1}{3}, \frac{1}{2}, \frac{3}{5}; \frac{4}{5}$
   (b) $\frac{5}{4}, \frac{20}{7}, \frac{45}{10}, \frac{80}{13}, \frac{125}{16}; \frac{500}{31}$
   (c) $-\frac{1}{10}, \frac{1}{20}, -\frac{1}{30}, \frac{1}{40}, -\frac{1}{50}; \frac{1}{100}$
   (d) $(-1,2), (1,\frac{3}{2}), (-1,\frac{4}{3}), (1,\frac{5}{4}), (-1,\frac{6}{5}); (1,\frac{11}{10})$
   (e) $(-1,2,0), (1,1,0), (-1,\frac{2}{3},0), (1,\frac{2}{4},0), (-1,\frac{2}{5},0); (1,\frac{2}{10},0)$

2. (a) Converges to 1, $N \geq 2/ACC$ (b) Diverges (to positive infinity).
   (c) Converges to 0, $N \geq ACC/10$.
   (d) Diverges (first component diverges).
   (e) Diverges (first component diverges).

3. (a) Converges to 0. (b) Diverges [$x_n = 2(-1)^n$].
   (c) Diverges. (d) Diverges (first component is unbounded).
   (e) Diverges (second component is unbounded).

5. (a) Diverges, second component is unbounded.
   (c) Converges to 1, $n \geq 2/ACC$.
   (e) Diverges, $e_n \approx (4/3)^n$ for n large.

6. 10th term;

$$|\frac{2^n - 1}{2^n} - 1| = |1 - \frac{1}{2^n} - 1| = \frac{1}{2^n} < \frac{1}{1024} < \frac{1}{1000}$$

   for $n \geq 10$.

7. $s_n = 0.777...7$ (n 7's)

9. (s_n, t_n) where $s_n = 0.333...3$ (n 3's) and $t_n = 0.625$ for $n > 3$.
   (0.33333, 0.625).

14. Convergent; $\{s_n\}$ is an increasing sequence that is bounded above by 1. Note that $s_n = s_{n-1}$ plus one-half the distance between 1 and s_{n-1}. Thus $s_n < 1$.

15. (a) If $\{s_n\}$ is an unbounded sequence in $\mathbb{R}^n$, then $\{s_n\}$ is a divergent sequence.
    (b) At least one of the component sequences is unbounded.

17. (a) If $\{s_n\}$ is a convergent sequence in $\mathbb{R}^3$, then $\{s_n\}$ is a bounded sequence.
    (b) It must be shown that the "hypothesis true–conclusion false" case cannot exist. that is, the case "$\{s_n\}$ is convergent and $\{s_n\}$ is unbounded" cannot exist.

18. (a) Center coordinate system at the center of the pond with the x-axis aligned east-west. Then $s_n = ((1+8/n)\cos(n\pi/2), (1+8/n)\sin(n\pi/2))$ for $n \geq 1$ gives a sequence of positions, starting one unit north of the pond.

19. Each component of the sequence is a bounded, monotone sequence, and thus converges.

Section 2.3

1. $a, b, c, d, f, i, j, k, m, o$

3. sequences (e), (g), and (l) all diverge to $+\infty$, (h) diverges to $-\infty$, (n) diverges to neither.

5. (a) Diverges to $-\infty$. (b) Converges to 0. (c) Converges to 0.
   (d) Converges to -2. (e) Converges to 3/4. (f) Converges to (0,0).
   (g) Converges to (1,0).

9. (a) Given any $ACC > 0$, there is an N such that $||s_n| - |L|| < ACC$, for all $n > N$.

11. (a) If $r(n) \to 0$, then degree $p(n) <$ degree $q(n)$.
If degree $p(n) <$ degree $q(n)$, then $r(n) \to 0$.

12. Yes; s_n/t_n approaches 2, which is the ratio of their weights.

Section 2.4

1. (a) $f(s_n) = \frac{2}{n} - 1$; 1, 0, $-\frac{1}{3}$, $-\frac{1}{2}$, $-\frac{3}{5}$
(b) $g(a_n) = n + 2 - \frac{1}{n+2}$; $\frac{8}{3}$, $\frac{15}{4}$, $\frac{24}{5}$, $\frac{35}{6}$, $\frac{48}{7}$
(c) $f(b_n) = 1 - (-1)^n n$; 2, -1, 4, -3, 6
(d) $f(c_n) = 1 - n^2$; 0, -3, -8, -15, -24
(e) $g(d_n) = 1 + 1/n$; 2, $\frac{3}{2}$, $\frac{4}{3}$, $\frac{5}{4}$, $\frac{6}{5}$
(f) $g(d_n) = \begin{cases} 1/n^2 & \text{for odd } n > 0 \\ 1 + 1/n & \text{for even } n > 0 \end{cases}$
1, $\frac{3}{2}$, $\frac{1}{9}$, $\frac{5}{4}$, $\frac{1}{25}$

2. (b) Diverges to infinity. (d) 0 (o) 0 (q) $(\frac{5}{2}, 4, 2)$ (t) 0

3. (a) -1 (b) 0 (c) 1 (d) 0 (e) $(-1, 1)$

5. (a) 1 (b) $\frac{1}{3}$ (c) 0

7. Let $f : \mathbb{R} \to \mathbb{R}$. Then the limit of f as x approaches positive infinity is L, written $\lim_{x\to\infty} f(x) = L$ if for *any* sequence $\{t_n\}$ diverging to positive infinity, $\lim_{n\to\infty} f(t_n) = L$.
(a) 1

9. -1

11. Center the coordinate system below the flagpole. Then a suitable height function is

$$h(x, y) = \begin{cases} 15 - 5x^2/16 & \text{for } |x| \le 4, |y| \le 4, (x, y) \ne (0, 0) \\ 25 & \text{for } (x, y) = (0, 0) \end{cases}$$

The limit is different from the value only for $(a, b) = (0, 0)$.

Section 2.5

1. (a) $\mathbb{R}$ (b) $\mathbb{R}$ (c) $x \ne 1$ (d) $\mathbb{R}$ (e) $x \ne -y, x \ne 0$ (f) $xy \ne 0$ (g) $x \ne 0$ ((h) $y \ne 0, x \ne (n - 1/2)\pi$ (i) $x \ne z$ (j) $y \ne (n + 1/2)\pi$, $x \ne -1$

3. $A = -1$, $\frac{1}{2}$

5. $A = 25$, $B = \frac{39}{4}$

7. (a) $\lim_{x\to p} f(x) = f(p) = \cos(p)$

8. (a) $\lim_{x\to p} f(x) = f(p)$

9. Center the coordinates below the flagpole with the roof sloping in the x-direction. Let the height (accurate to 0.01) be

$$h(x,y) = \begin{cases} 15.25 - 5|x|/3.8 & \text{for } 0.2 \leq |x| \leq 4 \text{ and } 0 \leq |y| \leq 4 \\ 15 & \text{for } 0 \leq x \leq 0.2 \text{ and } |y| \geq 0.2 \\ 25 - 50\max\{|x|,|y|\} & \text{for } |x| < 0.2 \text{ and } |y| < 0.2 \end{cases}$$

Section 2.6

1. $f(0) < 0$ and $f(2) > 0$.

3. $f(0) < 0$ and $f(1) > 0$.

7. For "large" values of the variable, the polynomial has the numerical sign of the highest powered term, and for "small" values of the variable, the polynomial has the numerical sign of the lowest powered term.

9. The interval [0,4] is not contained in domain(f) since f is not defined at $x = 3$.

11. (a) $[-3,-1], [-2,-1], [-2,-3/2], [-2,-7/4], [-2,-15/8]$; $-31/16 \approx -1.9$
(c) $[1,2], [3/2,2], [3/2,7/4], [3/2,13/8], [3/2,25/16]$, $[49/32,25/16], [49/32,99/64], [49/32,99/64], [49/32,197/128]$, $[49/32,393/256]$; $393/256 \approx 1.54$

13. Nine questions. Apply the Bisection Algorithm to the length (five times) and then to the width (four times).

Section 2.7

4. (a) hor. asy. $y = 3$, vert. asy. $x = 0$, intercept $x = -1/3$.
(b) hor. asy. $y = -1$, vert. asy. $x = 0$
(c) hor. asy. $y = 0$, vert. asy. $x = -3, 2$
(d) hor. asy. $y = 0$, vert. asy. $x = 1$
(e) hor. asy. $y = 2$, vert. asy. $x = 1, -1/2$
(f) hor. asy. $y = 1/2$, vert. asy. $x = 0$
(g) hor. asy, $y = 0$, vert. asy. $x = -1, 2, 3$
(h) hor. asy. $y = 1$, vert. asy. $x = -3, -1$
(i) no hor. asy., vert. asy. $x = 1$
(j) hor. asy. $y = 0$, vert. asy. $x = 2$

Section 3.2

1. (a) 30 ft/sec (b) 20 ft/sec 3. (a) 13 ft/sec (b) 13 ft/sec

5. 98 7. $y = -x + 8$ 9. $y = -3$ 11. $y = 2x - 2$

13. $a = 2, b = -1, c = 0$ 15. $(-3, 0)$

17. (a) F (b) T (c) F (d) T (e) F 19. (a) 16.08 π (b) 16 π

21. (a) 192 ft/sec (b) 128 ft/sec (c) 8 sec (d) 1024 ft

Section 3.3

1. (a) 0 (c) $4x^3$ (e) $\cos(x)$ (g) $2x + 2$

3.

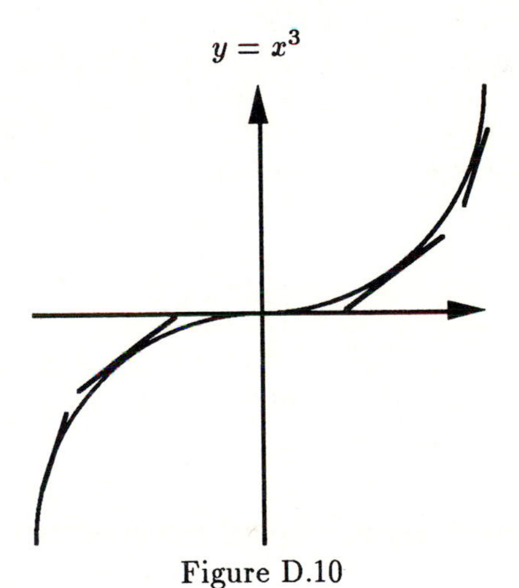

Figure D.10

5. These are just different interpretations of the same mathematical concept.

6. (a) $f(x) = x$ (c) the function whose value is 1 at the rationals and 0 at the irrationals (e) $f(x) = x^2$ (g) $f(x) = x^2$

7. T, F, F, F, F, T

10. (a) $\mathbb{R}$ (b) $\mathbb{R}$ (c) $\{x : x \geq -2, x \neq 2\}$ (d) No (e) Yes (h) No (i) Yes (11) c

13. The graph of f' looks roughly like:

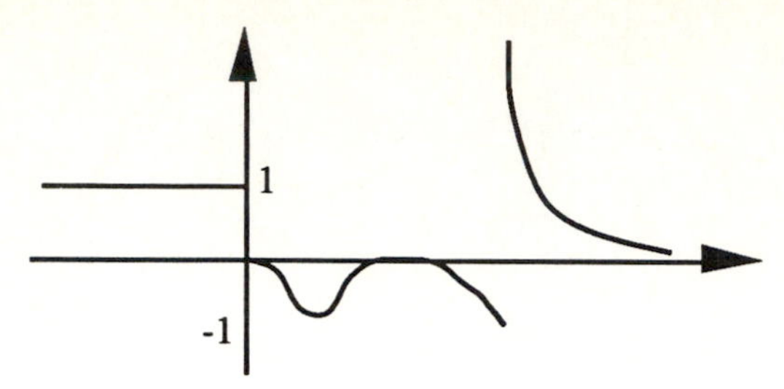

Figure D.11

15. (a) $y = f(x) = x^2$ so the equation of the tangent line is $y = f'(1)(x-1) + f(1) = 2(x-1) + 1$.
(c) $y = f(x) = \sin(x)$ so the equation of the tangent line is $y = f'(\pi/4)(x - \pi/4) + f(\pi/4) = (1/\sqrt{2})(x - \pi/4) + 1/\sqrt{2}$.

16. The ramp should be tangent to the shore at $x = 2$, so the equation is $y = 3(2)^2(x-2) + 2^3$.

Section 3.4

1. $dy/dx = 21x^6 - 1/\sqrt{x}$ 3. $dy/dx = x^2\cos(x) + 2x\sin(x) - \sec^2(x)$

5. $dy/dx = ((4x^5+5)(6x-6) - (3x^2-6x)20x^4)/(4x^5+5)^2$ 7. $dy/dx = (x\cos(x) - \sin(x))/x^2$

9. $dy/dx = (1+e^x)/(x+e^x)$ 11. $dy/dx = (\cos(x) - x\sin(x))\cos(x\cos(x))$

13. $dy/dx = 3x^2\sin(x)\sec^2(x^3) + \cos(x)\tan(x^3)$

15. $-e^{3x^2}\sin(x) + 6xe^{3x^2}\cos(x)$ 17. $6e^{2x}\sin^2(e^{2x})\cos(e^{2x})$

19. $4(e^{3x}\log(x))^3(e^{3x}/x + 3e^{3x}\log(x))$ 21. $f'(1) = 7$

23. $f'(2) = 1/4$ 25. $D_t s = 32t + 30$ 27. $D_x y = -\sin^2(x) + \cos^2(x)$

29. $b = 2$ and $d = -7$ 31. F, T, F, T, F 33. $f'(0) = 0$

35. (b) Applying the definition of differentiability yields:

$$\begin{aligned}
(f-g)'(x) &= \lim_{h\to 0} \frac{(f-g)(x+h) - (f-g)(x)}{h} \\
&= \lim_{h\to 0} \frac{f(x+h) - g(x+h) - f(x) + g(x)}{h} \\
&= \lim_{h\to 0} [\frac{f(x+h) - f(x)}{h} - \frac{g(x+h) - g(x)}{h}] \\
&= \lim_{h\to 0} \frac{f(x+h) - f(x)}{h} - \lim_{h\to 0} \frac{g(x+h) - g(x)}{h} \\
&= f'(x) - g'(x)
\end{aligned}$$

38. **(a)** $\arctan'(x) = 1/(1+x^2)$ on all of $\mathbb{R}$. 39. $-\frac{1}{4}$

41. The problem can be done by calculus or geometry. Use both methods as a check on your answer.

42. (a) $y = 1/4(x-3)+1$ (c) $y = -2/\sqrt{3}(x-1/2)+\pi/3$

Section 3.5

1. $f'(x) = 10x + 3/x^2$; $f''(x) = 10 - 6/x^3$

3. $g'(x) = 1/2x^{-1/2} + 12x^2\sec^2(4x^3)$;
$g''(x) = -1/4x^{-3/2} + 288x^4\sec^2(4x^3)\tan(4x^3) + 24x\sec^2(4x^3)$

5. $u'(x) = 3(1-1/x^2)[x+(1/x)]^2$; $u''(x) = 6(x+1/x)(1-1/x^2)^2 - 6/x^3(x+1/x)^2$

7. $dy/dx = -\cot(x)\csc(x)$

9. $dy/dx = (10y+3)/(1-10x)$ 11. $dy/dx = -y/x$

13. 21 15. 0 17. The best is $d'(t) > 0$ and $d''(t) > 0$.

19. (a) The number of infected students is growing at a constant rate
(d) The number of infected students is growing, but at a decelerating rate

21. (a) $f'(t) > 0$ and $f''(t) < 0$ (c) $f'(t) < 0$ and $f''(t) < 0$

21. 21 23. 4.5 sec

25. 10π cm^2/min 27. decreasing at a rate of 4π

29. Since $\sin(y) = x$, $\cos(y)dy/dx = 1$. Thus....

Section 3.6

1. $c = 1$ 3. $c = (12/5)^{1/2}$ 5. $c = 1$

7. $c = \arccos(-2/\pi)$ 9. $c = 9/4$

11. Increasing for $x < -\sqrt{2}$ or $x > \sqrt{2}$, decreasing for $-\sqrt{2} < x < \sqrt{2}$.

13. Increasing for $-2 < x < 0$, decreasing $x < -2, x > 0$.

17. (a) $(x-1)/(x+1) = (-1+2x) + ERROR(x)$ (b) $ERROR(0.03) = -0.0017$

19. (a) T; (c) F (e)T. 21. T, F, T, T, F, F, T

Section 3.7

1. $x^4 + \sin(x) - \cos(x) + c$ 3. $-1/(2x^2) - 1/(4x^4) + c$

5. $(t+1)^3/3 + c$ 7. $\tan(y) + y + c$ 9. $-1/2\cos(2x) + c$

11. $h(x) = x^5/5 - 3x^4/4 + 4x^3/3 - x^2 + 4x + 1$

13. $r(x) = 5x^2 + \sin(x) - 20x - 200.54$ 17. $g(x) = x^4 + x^3/6 - 13x/6 + 1$

19. Differentiate the suggested antiderivative, $1/2[f(x)]^2 + c$, to obtain the function $f(x)f'(x)$.

21. Substitute $\sin'(x)$ for $\cos(x)$ to obtain $\int \sin'(x)\sin(x)dx$ and now apply the result of Exercise 19.

23. Expressing time in seconds, $v(t) = at$, 60 m/sec equals $v(60)$, and so $a = 1$ m/sec/sec. Thus $x(60) = (1/2)at^2 = (1/2)3600 = 1800$ m. The car has gone 1800 m in the minute.

24. (a) $\int 2x\cos(x^2)dx = \int d/dx \sin(x^2)dx = \sin(x^2) + C$

Section 3.8

1. Global minimum at $(-3/2, -9/4)$. 3. No extrema values.

5. Increasing $x < 1, x > 3$; decreasing $1 < x < 3$;
concave up $x > 2$; concave down $x < 2$;
local maximum (1,5); local minimum (3,1);
inflection point (2,3).

7. Increasing $x < 3$; decreasing $x > 3$;
concave up $x < 3, x > 3$; global maximum (3,2).

9. Increasing $-\sqrt{3} < x < 0, x > \sqrt{3}$; decreasing $x < -\sqrt{3}$, $0 < x < \sqrt{3}$;
concave up $x < -1, x > 1$; concave down $-1 < x < 1$;
local maximum $(0, -1)$; global minima $(-\sqrt{3}, -13/4)$, $(\sqrt{3}, -13/4)$;
inflection points $(-1, -9/4), (1, -9/4)$.

11. Increasing $-1 < x < 0, x > 1$; decreasing $x < -1, 0 < x < 1$;
concave up $x < -1/\sqrt{3}$, $x > 1/\sqrt{3}$; concave down $-1/\sqrt{3} < x < 1/\sqrt{3}$;
local maximum (0,0); global minimum $(-1, -1)$, $(1, -1)$;
inflection points $(-1/\sqrt{3}, -1/3)$, $(1/\sqrt{3}, -1/3)$.

13. Decreasing $x < -3$, $x > -3$;
concave up $x > -3$; concave down $x < -3$;
no extrema; no inflection points.

15. There is a local maximum at d and a local minimum at 0.

19. We want $f(2) = 4$ and $f'(2) = 3 \cdot 2^2 + 2a2 = 0$ so $a = -3$ and $b = 8$.

21. Volume $= \pi r^2 h = 100$, material $= 2\pi r^2 + 2\pi rh = 2\pi r^2 + 200/r$; ans: $r = [50/\pi]^{1/3}$, $h = 2[50/\pi]^{1/3}$

23. 6 by 2.5 in

25. Volume $= x$ by $(10 - 2x)$ by $(16 - 2x)$ where $x = 2$ in.

27. Sides $= \frac{40}{3}$; base $= \frac{40}{3}$. 28. F, F, T, T, F, T, T, F

Section 3.9

1. 1 3. 7/6 5. -1/2 7. 0 9. 0

11. 2 13. 0 15. 0 17. $2^{-3/2}$ 19. (a) F

21. (a) After n applications of L'Hospital's Rule,

$$\lim_{x\to\infty} e^x/x^n = \lim_{x\to\infty} e^x/n! = \infty$$

and so e^x grows faster than x^n.
(b) One application of L'Hospital's Rule yields

$$\lim_{x\to\infty} \log(x)/x^n = \lim_{x\to\infty} 1/nx^n = 0$$

Thus x^n grows faster than $\log(x)$. This result is also clear from the concavity of the two functions, x^n is concave upward (for $n > 1$) and $\log(x)$ is concave downward.

Section 3.10

1. $D_x f(x,y) = 6x + y$; $D_y f(x,y) = x$; $D_{x,x} f(x,y) = 6$
$D_{x,y} f(x,y) = 1$; $D_{y,y} f(x,y) = 0$; $D_{y,x} f(x,y) = 1$

3. $D_x f(x,y) = -y^2/(x-y)^2$; $D_y f(x,y) = x^2/(x-y)^2$;
$D_{x,x} f(x,y) = 2y^2/(x-y)^3$; $D_{y,y} f(x,y) = 2x^2/(x-y)^3$;
$D_{x,y} f(x,y) = D_{y,x} f(x,y) = -2xy/(x-y)^3$

5. $D_x f(x,y,z) = 3y^2x^2$; $D_y f(x,y) = 2yx^3$;
$D_{x,x} f(x,y,z) = 6y^2x$; $D_{y,y} f(x,y) = 2x^3$;
$D_{x,y} f(x,y,z) = D_{y,x} f(x,y) = 6yx^2$
$D_z f(x,y,z) = 1$;
$D_{z,z} f(x,y,z) = D_{x,z} f(x,y,z) = D_{y,z} f(x,y,z) = 0$

7. $D_x f(x,y,z) = -\sin(z)\sin(xy)y$; $D_y f(x,y,z) = -\sin(z)\sin(xy)x$;
$D_z f(x,y,z) = \cos(z)\cos(xy)$; $D_{x,x} f(x,y,z) = -\sin(z)\cos(xy)y^2$;
$D_{y,y} f(x,y,z) = -\sin(z)\cos(xy)x^2$; $D_{z,z} f(x,y,z) = -\sin(z)\cos(xy)$;
$D_{x,y} f(x,y,z) = D_{y,x} f(x,y,z) = -\sin(z)\sin(xy) - \sin(z)\cos(xy)xy$;
$D_{x,z} f(x,y,z) = D_{z,x} f(x,y,z) = -\cos(z)\sin(xy)y$;
$D_{y,z} f(x,y,z) = D_{z,y} f(x,y,z) = -\cos(z)\sin(xy)x$;

9. $D_x f(x,y,z) = 2(xy+z)y$; $D_y f(x,y,z) = 2(xy+z)x$;
$D_z f(x,y,z) = 2(xy+z)$; $D_{x,x} f(x,y,z) = 2y^2$; $D_{y,y} f(x,y,z) = 2x^2$;
$D_{z,z} f(x,y,z) = 2$; $D_{x,z} f(x,y,z) = D_{z,x} f(x,y,z) = 2y$;
$D_{x,y} f(x,y,z) = D_{y,x} f(x,y,z) = 2(xy+z) + 2xy$;
$D_{y,z} f(x,y,z) = D_{z,y} f(x,y,z) = 2x$

11. (a) $f(x,y) = x|y|$

13. If $(x,y) \neq (0,0)$, then $D_x f(x,y) = y(x^4 + 4x^2y^2 - y^4)/(x^2+y^2)^2$ and $D_x f(0,0) = 0$. So, $D_x f(0,y) = -y$ for all y, $D_{y,x} f(0,y) = -1$, and $D_{y,x} f(0,0) = -1$. On the other hand, if $(x,y) \neq (0,0)$ then $D_y f(x,y) = x(x^4 - 4x^2y^2 - y^4)/(x^2+y^2)^2$ and $D_y f(x,y) = 0$. So, $D_y f(x,0) = x$ for all x, $D_{x,y} f(x,0) = 1$, and $D_{x,y} f(0,0) = 1$.

15. $f(x,y) = x^2y^2/4 + xy^3/3 + 3y + 3$ is a solution.

Chapter 4

Section 4.1

1. $(2,-3)$ 3. $\sqrt{11}$ 5. $(-1,6,13)$ 7. $-3\vec{\imath} + \vec{\jmath}$

9. $(2,3,4)/\sqrt{29}$ 13. $\arccos(5/\sqrt{84}) \approx 0.99$ radians

15. (0,0,1) 25. (a) 2 27. $\vec{Z}$ = (0,0,1,0) is not unique.

29. $\vec{Z}$ = (0,2,1) is not unique.

31. (a) The (maximum) height is $10\cos(\theta°)$, and the usable volume is $1000(2 - \sin(\theta°))\cos(\theta°)$.

Section 4.2

1. (a) $f(t) = (t, 1/t)$ for $t > 0$.
(c) An ellipse can be parameterized in a manner similar to a circle. Let $f(t) = (\cos(t)/2, \sin(t)/3)$ for all t in $[0, 2\pi)$.

2. (a) The parabola $x = y^2$. 5. $f'(t) = (1, 2t, -\sin(t))$

7. $g(t) = (1,3,1) + (3,6,1)(t-1) = (3t-2, 6t-3, t)$

9. 0.0100001 **(11)** [0.95,1.05]

13. Velocity = $f'(t) = (0, 6t, \cos(t))$, acceleration = $f''(t) = (0, 6, -\sin(t))$.

15. $(f \circ h)'(0) = f'(h(0))h'(0) = (3, 4h(0))h'(0) = (3,8)3 = (9,24)$

19. $f(t) = (1,2,3) + t(1,-1,1)$

21. $f(1) = (1,4,1)$, $f'(t) = (1,3,2t)$, $f'(1) = (1,3,2)$
$LINE(t) = (1,4,1) + t(1,3,2)$

23. $[(1,2,4) + (-2,1,-3)]/2 = (-1/2, 3/2, 1/2)$

25. All points on the line segment are of the form $(2,1,-1) + t[(0,2,2) - (2,1,-1)]$ for some t in [0,1]. But there is no solution to $(1,1,1) = (2,1,-1) + t[(0,2,2) - (2,1,-1)]$ with t in [0,1].

27. (a) It lands approximately at coordinates (7.5, 12.5, 0).

Section 4.3

1. A unit vector in the direction of (1,1) is $u = (1/\sqrt{2}, 1/\sqrt{2})$, and $D_u f(x,y) = (2x-1)/\sqrt{2}$.

3. A unit vector in the direction of (0,2,0) is (0,1,0), and $D_u f(x,y,z) = 3$.

5. Use the methods of Chapter 1 to translate the example given in this section.

6. (a) The slope is about -5.7 in direction v and -6 in direction u; so the cow should go in direction v.

Section 4.4

1. $\nabla(f) = (2xy\cos(xy) + 2\sin(xy), 2x^2\cos(xy))$

3.

$$\begin{aligned} f(x,y) &= f(1,1) + f'(1,1)\bullet[(x,y)-(1,1)] + ERROR(x,y) \\ &\approx f(1,1) + f'(1,1)\bullet[(x,y)-(1,1)] \\ &= 4 + (2,6)\bullet(x-1, y-1) = 4 + 2(x-1) + 6(y-1) \\ &= 2x + 6y - 4 \end{aligned}$$

5.

$$\begin{aligned} \frac{df}{dt} &= \frac{\partial f}{\partial x}\frac{dx}{dt} + \frac{\partial f}{\partial y}\frac{dy}{dt} \\ &= 2y^2 9t^2 + 4xy\cos(t) = 18t^2\sin^2(t) + 4(3t^3+1)\sin(t)\cos(t) \end{aligned}$$

7.

$$\begin{aligned} \frac{\partial f}{\partial x} &= \frac{\partial f}{\partial u}\frac{\partial u}{\partial x} + \frac{\partial f}{\partial v}\frac{\partial v}{\partial x} \\ &= (1+v)2x\cos(y) + u\cdot 3 = (1+3x+y)2x\cos(y) + 3x^2\cos(y) \end{aligned}$$

9.

$$\frac{dV}{dt} = \frac{\partial V}{\partial r}\frac{dr}{dt} + \frac{\partial V}{\partial h}\frac{dh}{dt} = \pi[\frac{2rh}{3}1 + \frac{r^2}{3}2] = 350\pi$$

11.

$$\begin{aligned} DIST(t) &= (x(t)^2 + y(t)^2)^{1/2}, \\ \frac{d}{dt}DIST(t) &= (D_x DIST(t), D_y DIST(t)) \cdot (x'(t), y'(t)) \\ &= \frac{x}{(x^2+y^2)^{1/2}}x'(t) + \frac{y}{(x^2+y^2)^{1/2}}y'(t) \\ &\approx -9.84 \text{ m/sec} \end{aligned}$$

13. $\text{curl}(f) = (x^2 - x^2, xy - 2xy, 2xz - xz) = (0, -xy, xz)$

17. $\text{div}(f) = y^2 - x\sin(y)$

Section 4.5

1. $\nabla(f) = (2xy + 3, x^2)$. Evaluating at (1,2) gives (7,1). A unit vector is $(7,1)/\sqrt{50}$.

3. (a) Parabola in the xz-plane; (c) parabola in xy-plane; (e) no solutions; (g) parabola in the plane $x = -2$; (i) no solutions; (k) parabola in the plane $z = 9$; (m) parabolas; (o) parabolas.

4. (a) parabola in the y-z plane; (c) two straight lines in the xy-plane; (e) parabola in the plane $x = 5$; (g) parabola in the plane $y = 2$; (i) parabola in the plane $y = -8$; (k) hyperbola in the plane $z = -5$; (m) parabolas; (o) hyperbolas.

5. (a) A sphere of radius 1 centered at the origin.

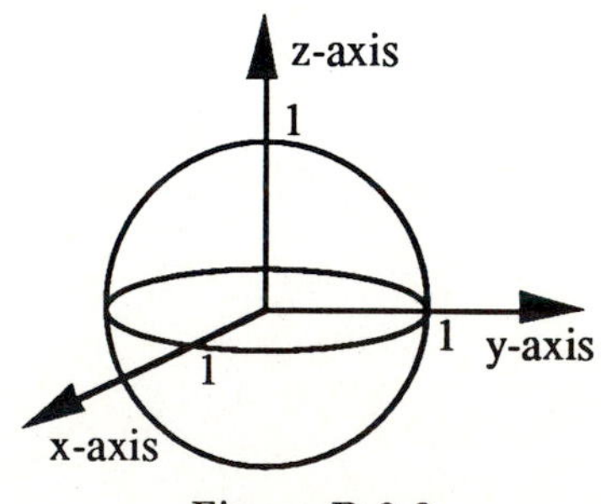

Figure D.0.3

(b) An ellipsoid with x-radius 2, y-radius 3, and z-radius 1 centered at the origin.

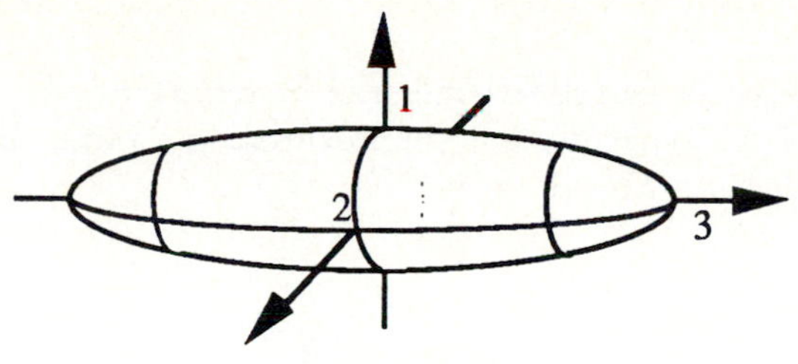

Figure D.0.4

6. (a) An hyperboloid of two sheets centered at the origin.

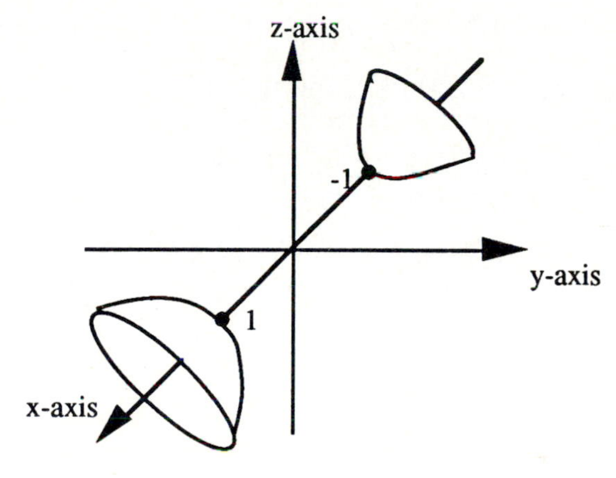

Figure D.0.5

(b) An hyperboloid of one sheet centered at the origin.

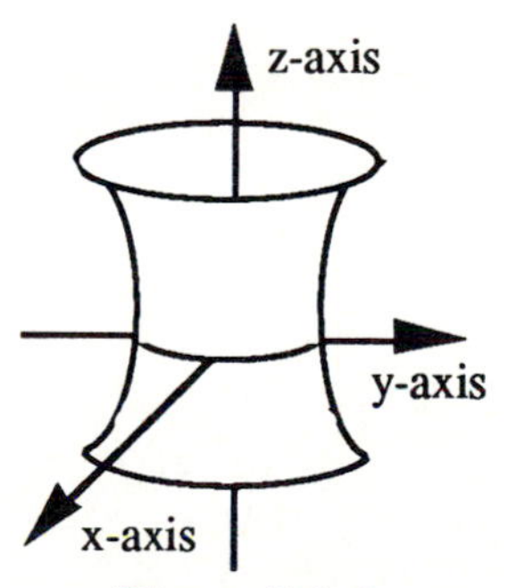

Figure D.0.6

7. $1 - x = (y - 2)/2 = z - 3$

9. $\text{grad}(f) = (2x, 2y, -2z)$, so $(4, 2, -2)$ is perpendicular to the surface at (2,1,1). The equation of the plane is $[(x, y, z) - (2, 1, 1)]\cdot(4, 2, -2) = 0$, or $2x + y - z = 4$.

11. $(x-1)/2 = (y-2)/4 = 5-z$.

13. The slope as a level set is $\{(x,y,z) : F(x,y,z) = z - f(x,y) = 0\}$. The equation of the tangent plane (approximating a ramp) is $10(x-1) + 4(y-1) + (z-1) = 0$.

Section 4.6

1. $(2,1)$, $(-2,1)$ are saddle points and $(0,0)$ is a local minimum

3. Minimize the distance squared:

$$(x+1)^2 + (y-4)^2 + (z-2)^2 = (x+1)^2 + (y-4)^2 + (3y-2x+5)^2$$

Ans: $(x,y) = (12/7, -1/14)$; dist $= 19/\sqrt{14}$.

5. (a) The trough should be a 4 by 4 square, 2 feet deep. (c) The lowest point is along the east-west road, 9.5 miles east of the barn.

7. Minimize $[1-(m+b)]^2 + [3-(2m+b)]^2 + [3-(4m+b)]^2$.
Ans: $m = 4/7$, $b = 1$.

Section 4.7

1. Maximize $x^2 + xy + y^2 + yz + z^2$ subject to $x^2 + y^2 + z^2 = 1$.
ans: $1 + 1/\sqrt{2}$.

3. Minimize $xy + 2yz + 2xz$ subject to $xyz = 32$.
Ans: $x = 4$, $y = 4$, $z = 2$. (Does the use of Lagrange Multipliers make this problem easier?)

5. Length = width = 2; height = 4

7. (a) $h \approx 13.28$, $r \approx 4.43$. (c) $-\sqrt{1/14}(100, 200, 300)$

Chapter 5

Section 5.1

3. Approx. 3,000,000 sold; the error is less than 175,000 homes.

Section 5.2

1. 165 3. 63 5. 570

7.

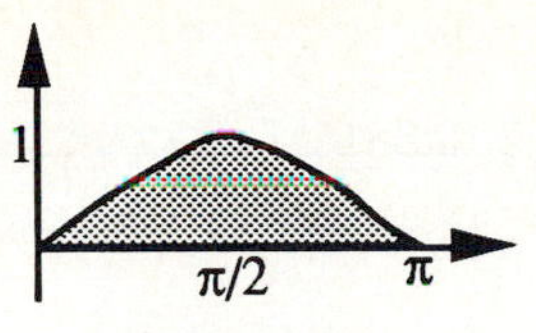

Figure D.12

9. $UA(f,6) = 27.5625$ 11. 15/8, 35/8 13. 10.8

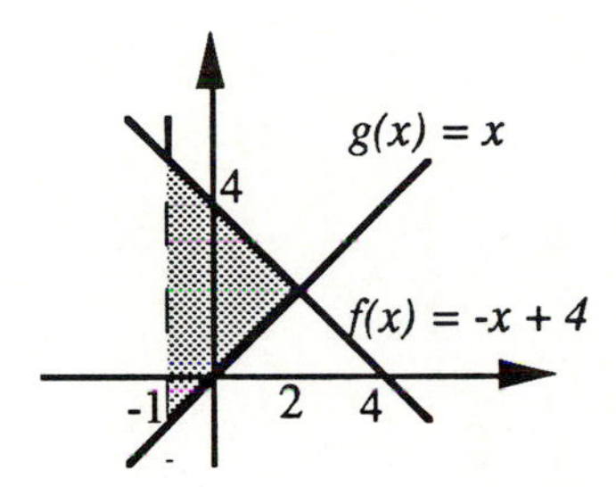

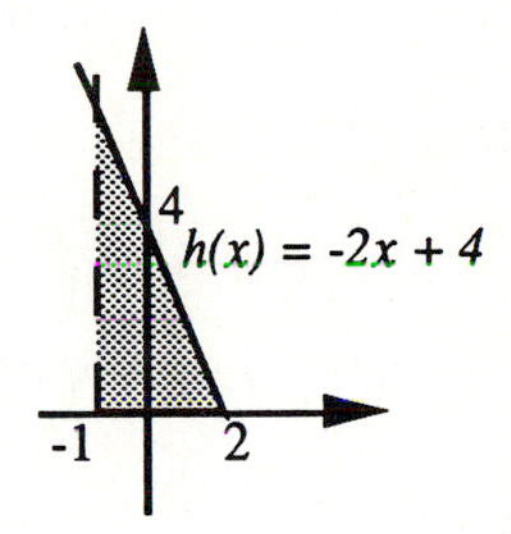

Figure D.13

15. $\lim_{n\to\infty} \sum_{k=1}^{n} h(-1 + k3/n)3/n = 9$ 17. (a) 4/5   (b) 1

19. 3/2

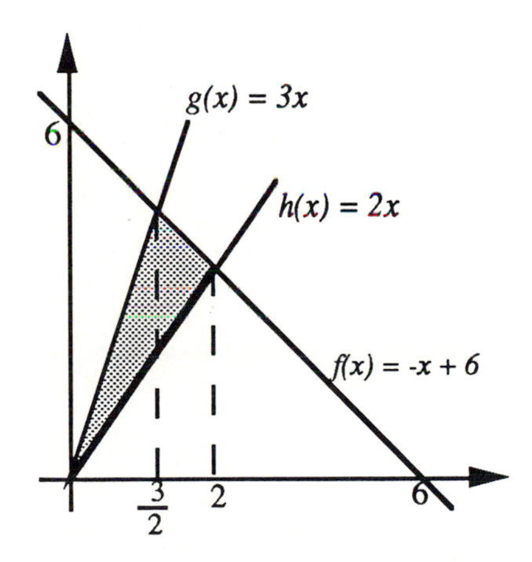

Figure D.14

21. 1/2   23. 15/4

Section 5.3

1. $-5/32$

3. $|\cos(-2)| + 2|\cos(-1)| + 1/2(1) + 9/2|\cos(2)| + 2|\cos(6)|$

4. 0.505   6. -0.67 9. 15   11. 24

Section 5.4

1. 1.151479, 1.147782 3. 2.681517, 2.834873

5. 5/12, 5/72 7. 90 9. 6

Section 5.5

3. 8 5. $-5/3$ 7. 16/3

Section 5.6

1. $f(t) = (t, t)$ 3. $f(t) = (\cos(t), \sin(t))$

5. $f(t) = (\sin(t) - 4, t)$, t in [2,5]

7. $\int_{-1}^{3} 2\sqrt{4+t^2}dt$ 9. $\int_0^2 \sqrt{2}dt$
11. $\int_{-1}^{3} \sqrt{1+\cos^2(t)}dt$ 15. $2\sqrt{\pi^2+4}$
17. $2 + 2\sqrt{2-\sqrt{2}} + \sqrt{2}$ 19. $\sqrt{8+\pi^2} + \sqrt{4+\pi^2}$

25. $S_1 = 2\pi \int_{-1}^{1} t^2\sqrt{1+4t^2}dt > 2\pi \int_{-1}^{1}(t^2-1)\sqrt{1+4t^2}dt$

27. $\int_0^{\pi/4} \tan^2(t)dt$ 29. $\int_0^2 \pi t^4 dt$
31. $\int_{-1}^{2} \pi t^2 dt$ 33. $\int_0^2 2\pi x^{3/2} dx$

35. $\int_0^5 2\pi(8-x)(1+\sin(x))dx$

Section 6.1

1. $F(x) = 3x - 3$ 3. $F(x) = x^2$
5. $F(x) = 1.5x^2 - 2x + 0.5$ 7. $F(x) = -0.5x^2 + x + 7.5$
9. 12 11. 4
13. 1/2 15. 32
17. 16 19. 0
21. 14 23. 14/3
25. 2 27. 9/4
29. $-9/2$ 31. 15/4
33. $2/x$ 35. $3x^2/\sqrt{1+x^3}$
37. $1/x$ 39. $x - 1$
42. False; $f(x) = x$, $[-1, 1]$ 44. True; $|0| = 0$
46. False; $f(x) = x$, $[-1, 1]$ 48. True; by linearity of the integral
50.(b) True 50.(d) True

Section 6.2

1. $-\sin(x)\sin(y)dy + \cos(x)\cos(y)dx$ 3. $2x\sin(y)dx + (x^2\cos(y) + 3y^{1/2})dy$

5. $\frac{yz(y+z)}{(x+y+z)^2}dx + \frac{xz(x+z)}{(x+y+z)^2}dy + \frac{xy(x+y)}{(x+y+z)^2}dz$ 7. $-\frac{x}{3}\cos(3x) + \frac{1}{9}\sin(3x) + C$

9. $-\frac{x}{5}\cos(5x) + \frac{1}{25}\sin(5x) + C$ 11. $\frac{1}{2}(x - \sin(x)\cos(x)) + C$

13. $(2 - x^2)\cos(x) + 2x\sin(x) + C$ 15. $\frac{13}{42}$

17. $\frac{1}{5}(-3\sin(2x)\cos(3x) + 2\cos(2x)\sin(3x)) + C$

19. $x^2e^x - 2xe^x + 2e^x + C.$ 21. 1 23. 0.0104 25. 0.000241

Section 6.3

1. $(x-1)^6/6 + C$ 3. $2\sin(\sqrt{x}) + C$ 5. $[\arctan(x^2)]/2 + C$

7. $\frac{(2x+1)^{3/2}}{3} + C$ 9. $(\sqrt{x^2-4})(x^4 - 8x^2 - 64)/5 + C$ 11. $64/3$

13. $(16\sqrt{2} - 14)/3$ 15. $x - \arctan(x) + C$ 17. $\frac{x^4}{2} + x^2 + C$

19. $\arctan(x-1) + C$ 21. $[x(x\arctan(x) - 1) + \arctan(x)]/2 + C$

23. $2\sqrt{ax+b}((ax+2b)/(a^2(ax+b)) + C \quad (a > 0)$ 25. $\sec(x) + C$

27. $\frac{\sin(2x)}{4} + \frac{x}{2} + C$ 29. $\frac{-\cos^5(x)}{5} + \frac{2\cos^3(x)}{3} - \cos(x) + C$

31. $\frac{-\cos(3x)}{3} + \frac{\cos^3(3x)}{9} + C$ 33. $\frac{-\cos^5(x+1)}{5} + \frac{2\cos^3(x+1)}{3} - \cos(x+1) + C$

35. $\frac{\cos(x)}{2} - \frac{\cos(3x)}{6} + C$ 37. $\frac{\cos(x)}{2} - \frac{\cos(5x)}{10} + C$

39. $\frac{9\arcsin(x/3)}{2} + \frac{x(9-x^2)^{1/2}}{2} + C$ 41. $[\arctan((x+1)/2)]/2 + C$

43. $\frac{3\tan^5(x) - 5\tan^3(x) + 15\tan(x)}{15} - x + C$ 45. $-\arcsin(x/2) - \frac{(4-x^2)^{1/2}}{x} + C$

46. (e)

$$B_n = \begin{cases} \frac{(2^k k!)^2}{(2k+1)!} & \text{if } n = 2k+1,\ k \text{ a positive integer} \\ \frac{(2k)!}{(2^k k!)^2}\,\frac{\pi}{2} & \text{if } n = 2k,\ k \text{ a positive integer} \end{cases}$$

Section 6.4

1. Diverges	3. π	5. Diverges	7. 1/2
9. Diverges	11. Converges	13. Converges	15. Diverges
17. 3a/2	19. (a) $\int_0^\infty \sin(x)dx$	20.(a) true	20.(b) false

Chapter 7

Section 7.2

1. $6x/(3x^2+4)$

3. $3(3x^2+3)/(x^3+3x)$

5. $[(2x\log(\sin(x)+4)+x^2\cos(x))/(\sin(x)+4)]\ /\ (x^2\log(\sin(x)+4))$

7. $(6x^4+60x^3+37x^2+380x-6)\ /\ ((x+10)(x^2+6))$

9. $-x[x^2\sin(x)+2x\sin(x)-x\cos(x)-4\cos(x)]/(x+2)^2$

11. $y/(xy-x)$

13. $2\cos(x)y\log(y)/(y-2\sin(x))$

15. $\log^3(x)+3/x$

21. (a) improper at $x=0,\infty$: diverges
(b) improper at $x=1$; diverges
(c) improper at $x=0$; -4

Section 7.3

1. $e^{x^2}(2x^2+1)$

3. $e^x\tan(e^x)+e^{2x}sec^2(e^x)$

5. $\log(4)4^x+2x$

7. $[y-\log(2)\sec(x)2^{\sec(x)}\tan(x)]/(-x)$

9. $2x+3/(x+2)$

11. $e^0=1$

13. 2

15. e

19. $2e^2$

21. 6.521610, 6.391210

23. 9+ years at simple interest (or approximately 8 years and 8 months compounded continuously.

25. Yes

29. (a) Diverges, (b) converges, (c) converges, (d) converges

Section 7.4

1. $e^{3x+2}/3 + C$
3. $3^{x^2}/2\log(3) + C$
5. $-1/2\log(4)4^{2x} + C$
7. $x - 3e^{-x} + C$
9. $\log^2(x)/2 + C$
11. $x^2/2 + 7\log(|x|) + C$
13. $(2x+1)e^{-2x}/4 + C$
15. $x\tan(x) + \log(|\cos(x)|) + C$
17. $x\log(x) - x + C$
19. $[\log(|x-4|) - \log(|x+4|)]/8 + C$
21. $3\log(|x-2|) - \log(|x|)/2 + C$
23. $2\log(|x-2|) + 3\log(|x+1|) + C$
25. $[-4\log(|x+2|) + 9\log(|x-3|)]/5 + x + C$
27. $[-\log(x^2+4) - \arctan(x/2) + 2\log(|x-1|)]/10 + C$
29. $-3\log(x^2+1)/2 + 3\log(|x|) + \arctan(x) + C$
31. $3/8\log(|x + \sqrt{1+x^2}|) + x/4(x^2+1)^{3/2} + 3x/8\sqrt{1+x^2} + C$
33. $[(6x^2-3)\sin(2x) + 6x\cos(2x) - 4x^3]/(-24) + C$
35. $(3x-1)e^{3x}/9 + C$
37. $x/2\sqrt{4+x^2} - 2\log(|x/2 + 1/2\sqrt{4+x^2}|) + C$
39. $\log(9/8)$ 41. $y = x$ 43.(b) $\sinh'(x) = \cosh(x)$.
45. **(a)** $\tanh(x) = \sinh(x)/\cosh(x)$ for all real x.
46. (a) $\tanh'(x) = \text{sech}^2(x)$. 47.(a) $\cosh(x) + C$

Section 7.5

3. $(t^*, c/2)$ where $t^* = p^{-1}(c/2)$
7. n^n grows faster than $n!$ since $\lim_{n\to\infty} n!/n^n \approx \sqrt{2\pi}\lim_{n\to\infty} n^{1/2}/e^n = 0$ Also note:

$$\frac{n!}{n^n} = \left(\frac{1}{n}\right)\left(\frac{2}{n}\right)\left(\frac{3}{n}\right)\cdots\left(\frac{n}{n}\right) < \frac{1}{n} \to 0.$$

9. $\log_2(x)$ grows faster than $\log_2(\log_2(x))$. Apply L'Hospital's Rule to $\lim_{x\to\infty}\log_2[\log_2(x)]/\log_2(x)$.
11. \$106,151.49 13. \$503.02 15. \$334.27, \$332.46 17. \$4,118.12
19. $y(x) = -1/5e^{4x} + 11/5e^{-x}$ 21. $y(x) = 2e^x - 1/2e^{-x} + \frac{1}{2}e^{3x}$

Section 8.1

1. 0   3. π 5. 9   7. 31   9. $4\pi^2$

Section 8.2

1. (a) $g(x,y) = (x^3 + y^3)/3 + xy + 2.$ (b) No potential function can exist.

2. (a) 0   (c) -1 (e) 14   (g) An open rectangle in the domain of F contains the curve, e.g., $(0.5, 2.5) \times (-0.5, 2.5)$, so we need only apply the Corollary, 8.2.13, obtaining 0.

5. 72 3/5

7. F, T, F, T, T, T, F [for part a, note that the existence of partial derivatives does not imply continuity–see Section 4.4]

Section 8.3

1. 7   3. 2   5. $[1 - \cos(2)]/2$ 7. 110
9. 20   11. 8   13. 0   15. 128/15
17. 24   19. 19   21. 8   23. $\int_0^{\sqrt{2}}(\int_{y^2}^2 f(x,y)dx)dy$

25. $\int_{-3}^{3}\int_{-\sqrt{9-x^2}}^{\sqrt{9-x^2}}\int_1^{5-x} dzdydx = 24\pi.$

Section 8.4

1.(a) $-dx \wedge dy$ 1.(c) $4dx \wedge dy \wedge dz$
2.(a) $d\alpha = -xdx \wedge dy$ 2.(c) $d\alpha = x\cos(x+z)dx \wedge dy \wedge dz$
7. 0   9. 0
11. $4\pi/\sqrt{3}$ 13. $3\pi/8$

17. Hint: Apply Green's Theorem to the rightmost term.

19. Hint: Apply Green's First Identity.

Section 8.5

1. 4π 3. $(\pi + 2)/8$ 5. 9

7. 6π 9. $\pi/12$ 11. 6π

13. $a^5\pi/5$ 15. $(\pi/4)\sin(4)$ 17. 8π (19) 32/9

21. Upper $= 5\pi$, Lower $= \pi$.   23. $64\pi/3$ 25. πa^3 27. $4\pi\log(b/a)$

29. (c) $dx \wedge dy = udu \wedge dv.$

Chapter 9

Section 9.2

1. $7 + 5(x-2) + (x-2)^2$

3. $a_0 = 1,\ a_1 = 0,\ a_2 = -1/2,\ a_3 = 0,\ a_4 = 1/24$

5. $P_4(x) = 1 + 3x + x^2/2 + x^4/24$ 7. $P_5(x) = x + x^2 + x^3/3 - x^5/30$

9. $P_5(x) = x - x^2/2 - x^3/6 + 3x^5/40$ 11. $P_4(x) = 1 - x + x^3 - x^4$

13. $a_0 = 0,\ a_1 = 1,\ a_2 = 0,\ a_3 = 1/3,\ a_4 = 0,\ a_5 = 2/15$

Section 9.3

1. 0.00025 (many possible answers) 3. 0.000053 5. 14 9. 23/15

Section 9.4

1. $10e^2$

3. $(1/4r^2 + 3rs + 9s^2)\exp(1/2x + 3y)$

5. $f(x,y) = x^2 + 3xy + y^2 = 11 + 8(x-1) + 7(y-2) + (x-1)^2$
   $+ 3(x-1)(y-2) + (y-2)^2$

7. $P_4(x,y) = x + y - 1/6(x+y)^3$

9. $D_{1,1,1}f(0,0) = 0 \quad D_{1,1,2}f(0,0) = 2$
   $D_{1,2,2}f(0,0) = 2 \quad D_{2,2,2}f(0,0) = 0$

11. $z = 1 + x$

Section 9.5

1. (a) Diverges (c) diverges (e) diverges

2. (a) $A = 1,\ B = -1.$
   (b) $s_1 = 1 - 1/2,\ s_3 = 1 - 1/,\ s_5 = 1 - 1/6$
   (c) $s_n = 1 - 1/n + 1$, so the limit is 1.

4. (a) $s_n = 1/4 - 1/n + 4$ (c) $s_n = 1/2(3/2 - 1/n - 1/n + 1)$ 5.(a) 1/4
   (b) 3/4

Section 9.6

(1) **(a)** $\sum_{k=0}^{\infty}(3/5)^3(3/5)^k$ **(c)** $\sum_{k=0}^{\infty}6(3/)^n$ **(e)** Can't be done!

(2) **(a)** 31 **(c)** 3280/59049 **(e)** $3(0.1)^4(1-(0.1)^{17})/0.9$

(3) **(a)** 7/9 **(b)** 1607/9900

(4) **(a)** convergent geometric **(c)** divergent otherwise **(e)** convergent geometric

(5) **(a)** divergent, other, $\lim_{k\to\infty} \neq 0$ **(c)** divergent, geometric, $r > 0$
(e) $(3^5)(5^4)/2$, geometric **(g)** 13/4, sum of two geometric series
(i) e

(6) \$20 billion **(8)** 5 months **(10)** 2

(12) $4+4/\sqrt{2}+4/2+4(1/\sqrt{2})^3+\cdots = 4\sum_{k=0}^{\infty}(1/\sqrt{2})^k = 4\sqrt{2}/(\sqrt{2}-1)$

Section 9.7

1. (a) Diverges, compare with $\sum_{k=1}^{\infty} 1/k$ (c) Converges, ratio test
(e) Converges, compare with $\sum_{k=1}^{\infty} 1/k^2$ (g) Converges, integral test
(i) Diverges, ratio test (k) Diverges, comparison test (with $\sum_{k=1}^{\infty} 1/\sqrt{k}$)
(m) Diverges, limit comparison test

2. (a) Diverges (c) Converges (e) Diverges, by integral test (g) diverges
(i) Converges (k) Converges, by limit comparison test (m) Diverges

Section 9.8

1. (a) Diverges (c) Diverges (e) Diverges (g) Diverges (i) Converges absolutely (k) Converges absolutely

Section 9.9

1. $[-1,1)$ 3. $[-1,1]$ 5. All reals 7. 2 9. $[-1,1]$

11. $[-1/2,1/2)$

13. $(-1/3,1/3)$ 15. (0,2) 17. 0

19. $(-5/2,1/2)$ 21. Infinity 23. 2

Section 9.10

1. $\sin(x)\cos(x) = x - 2x^3/3 + 2x^5/15 - 4x^7/315 + \cdots$; radius = infinity.

3. $e^x/1 - x = 1 + 2x + 5/2x^2 + 8/3x^3 + \cdots$; radius = 1.

5. $e^x \log(1+x) = x + x^{2/2} + x^{3/3} + 3x^5/40 + \cdots$; radius = 1.

7. $f : (-1,1) \to \mathbb{R}$, $f(x) = x/(1-x)$

9. $f : (-1,1) \to \mathbb{R}$, $f(x) = x/(1+x^2)$

11. $f : (-1,1) \to \mathbb{R}$, $f(x) = -1/(1+x)^2$

13. $f : (-1,1) \to \mathbb{R}$, $f(x) = -2x/(1+x^2)^2$

15. $f : (-1,1) \to \mathbb{R}$, $f(x) = -\log(1-x)$

17. $f : (-1,1) \to \mathbb{R}$, $f(x) = \arctan(x)$

19. $f : (0,4) \to \mathbb{R}$, $f(x) = 1/x$

21. $f : (-5,5) \to \mathbb{R}$, $f(x) = 1/5 - x$

23. $f : (-5,5) \to \mathbb{R}$, $f(x) = 2/(5-x)^3$

25. $f : (-1/4, 1/4) \to \mathbb{R}$, $f(x) = 16/1 - 4x$

27. 0.81 29. 0.60 31. 0.3

33. $f(x) = (3x^3 - 2x^4)/(1-x)^2$ 35. $f(x) = -1/4log(1-x^4), |x| < 1$

37. $33!2^{33}$

INDEX